AF697305

WITHDRAWN

Water – From Interfaces to the Bulk

Heriot-Watt University, UK

27–29 August 2008

FARADAY DISCUSSIONS

Volume 141, 2009

RSC Publishing

The Faraday Division of the Royal Society of Chemistry, previously the Faraday Society, founded in 1903 to promote the study of sciences lying between Chemistry, Physics and Biology.

EDITORIAL STAFF

Editor
Philip Earis

Assistant editor
Madelaine Chapman

Publishing assistant
Claire Springett

Team leader, serials production
Joanna Pugh

Technical editor
Christina Hodkinson

Publisher
Janet Dean

Faraday Discussions (Print ISSN 1359-6640, Electronic ISSN 1364-5498) is published 3 times a year by the Royal Society of Chemistry, Thomas Graham House, Science Park, Milton Road, Cambridge, UK CB4 0WF.
Volume 141 ISBN-13: 978 1 84755 8367

2009 annual subscription price: print+electronic £592, US $1,160; electronic only £533, US $1045. Customers in Canada will be subject to a surcharge to cover GST. Customers in the EU subscribing to the electronic version only will be charged VAT. All orders, with cheques made payable to the Royal Society of Chemistry, should be sent to RSC Distribution Services, c/o Portland Customer Services, Commerce Way, Colchester, Essex, UK CO2 8HP.
Tel +44 (0) 1206 226050;
E-mail sales@rscdistribution.org

If you take an institutional subscription to any RSC journal you are entitled to free, site-wide web access to that journal. You can arrange access *via* Internet Protocol (IP) address at www.rsc.org/ip. Customers should make payments by cheque in sterling payable on a UK clearing bank or in US dollars payable on a US clearing bank. Periodicals postage is paid at Rahway, NJ and at additional mailing offices. Airfreight and mailing in the USA by Mercury Airfreight International Ltd., 365 Blair Road, Avenel, NJ 07001, USA.

US Postmaster: send address changes to *Faraday Discussions*, c/o Mercury Airfreight International Ltd., 365 Blair Road, Avenel, NJ 07001. All despatches outside the UK by Consolidated Airfreight.

PRINTED IN THE UK

Faraday Discussions documents a long-established series of *Faraday Discussion* meetings which provide a unique international forum for the exchange of views and newly acquired results in developing areas of physical chemistry, biophysical chemistry and chemical physics.

ORGANISING COMMITTEE, Volume 141

Chair
Martin McCoustra (Edinburgh, UK)

Editor
Colin Bain (Durham, UK)
Victoria Buch (Jerusalem, Israel)
John Finney (London, UK)
Jean-Pierre Hansen (Cambridge, UK)
Georg Held (Reading, UK)
Andrea Russell (Southampton, UK)
Richard Wheatley (Nottingham, UK)

FARADAY STANDING COMMITTEE ON CONFERENCES

Chair
C D Bain (Durham, UK)

K J Edler (Bath, UK)
A J Orr-Ewing (Bristol, UK)
G Jackson (Imperial, UK)
A Rodger (Warwick, UK)

© The Royal Society of Chemistry 2009. Apart from fair dealing for the purposes of research or private study, or criticism or review, as permitted under the Copyright, Designs and Patents Act 1988 and Related Rights Regulations 2003, this publication may only be reproduced, stored or transmitted, in any form or by any means, with the prior permission in writing of the Publishers or in the case of reprographic reproduction in accordance with the terms of licences issued by the Copyright Licensing Agency in the UK. US copyright law applicable to users in the USA. The Royal Society of Chemistry takes reasonable care in the preparation of this publication but does not accept liability for the consequences of any errors or omissions.

Royal Society of Chemistry: Registered Charity No. 207890.

∞The paper used in this publication meets the requirements of ANSI/NISO Z39.48-1992 (Permanence of Paper).

Water - From Interfaces to the Bulk

Faraday Discussions

www.rsc.org/faraday_d

A General Discussion on Water – From Interfaces to the Bulk was held at Heriot-Watt University, UK on 27th, 28th and 29th August 2008.

RSC Publishing is a not-for-profit publisher and a division of the Royal Society of Chemistry. Any surplus made is used to support charitable activities aimed at advancing the chemical sciences. Full details are available from www.rsc.org

CONTENTS

ISSN 1359-6640; ISBN 978-1-84755-836-7

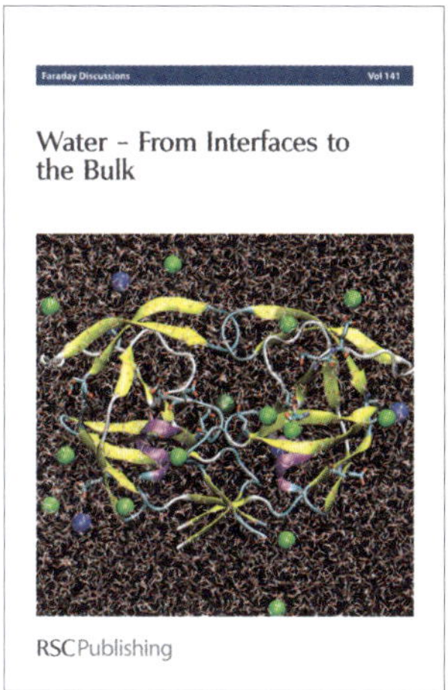

Cover

See Pavel Jungwirth, *Faraday Discuss.*, 2009, **141**, 9–30. Snapshot from a molecular dynamics simulation showing the distribution of sodium (green) and potassium (blue) ions at the water/protein (HIV protease) interface.

Image reproduced by permission of Professor Pavel Jungwirth, from *Faraday Discuss.*, 2009, **141**, 9.

INTRODUCTORY LECTURE

PAPERS AND DISCUSSIONS

 This journal is © The Royal Society of Chemistry 2009

CONCLUDING REMARKS

ADDITIONAL INFORMATION

Spiers Memorial Lecture
Ions at aqueous interfaces

Pavel Jungwirth*

Received 23rd September 2008, Accepted 25th September 2008
First published as an Advance Article on the web 20th October 2008
DOI: 10.1039/b816684f

Studies of aqueous interfaces and of the behavior of ions therein have been profiting from a recent remarkable progress in surface selective spectroscopies, as well as from developments in molecular simulations. Here, we summarize and place in context our investigations of ions at aqueous interfaces employing molecular dynamics simulations and electronic structure methods, performed in close contact with experiment. For the simplest of these interfaces, *i.e.* the open water surface, we demonstrate that the traditional picture of an ion-free surface is not valid for large, soft (polarizable) ions such as the heavier halides. Both simulations and spectroscopic measurements indicate that these ions can be present and even enhanced at surface of water. In addition we show that the ionic product of water exhibits a peculiar surface behavior with hydronium but not hydroxide accumulating at the air/water and alkane/water interfaces. This result is supported by surface-selective spectroscopic experiments and surface tension measurements. However, it contradicts the interpretation of electrophoretic and titration experiments in terms of strong surface adsorption of hydroxide; an issue which is further discussed here. The applicability of the observed behavior of ions at the water surface to investigations of their affinity for the interface between proteins and aqueous solutions is explored. Simulations show that for alkali cations the dominant mechanism of specific interactions with the surface of hydrated proteins is *via* ion pairing with negatively charged amino acid residues and with the backbone amide groups. As far as halide anions are concerned, the lighter ones tend to pair with positively charged amino acid residues, while heavier halides exhibit affinity to the amide group and to non-polar protein patches, the latter resembling their behavior at the air/water interface. These findings, together with results for more complex molecular ions, allow us to formulate a local model of interactions of ions with proteins with the aim to rationalize at the molecular level ion-specific Hofmeister effects, *e.g.* the salting out of proteins.

1. Introduction

The structure, dynamics and chemistry of aqueous interfaces have attracted considerable attention in recent years, as also reflected in the theme of the 141st Faraday Discussion. The present Discussion focuses on a range of important areas concerning interfacial behavior of aqueous systems; from the surface charge of liquid water, over ice interfaces, to confined (nanoscale) systems, and to electrochemical and biological interfaces. Overviews of latest scientific problems concerning interfacial

Institute of Organic Chemistry and Biochemistry, Academy of Sciences of the Czech Republic and Center for Complex Molecular Systems and Biomolecules, Flemingovo nám. 2, 16610 Prague 6, Czech Republic. E-mail: pavel.jungwirth@uochb.cas.cz

This journal is © The Royal Society of Chemistry 2009

water including many of these topics can also be found in the April 2006 special issue of *Chemical Reviews* and in dedicated monographs.[1,2] Within the present contribution we narrow the focus to issues connected with our studies of ions at aqueous interfaces.

Aqueous electrolytes including their interfaces emerged as subjects of intensive scientific scrutiny already in the nineteen-hundreds with the basic theory thereof being built up by the first half of the 20th century.[3,4] More recently, studies of aqueous electrolytes and specific ion effects have been undergoing a remarkable renaissance with a particular emphasis on interfacial systems.[5] The reason for the renewed interest in aqueous interfaces is at least twofold. First, processes at aqueous interfaces have been shown to play a prominent role in widely studied problems in a variety of fields ranging from chemistry of atmospheric aerosols and heterogeneous catalysis[6–8] to biophysics and biochemistry.[9] Second, new surface-selective spectroscopic techniques and computational methods have allowed for investigations of aqueous interfaces with atomistic detail revealing how incompletely they have been understood so far.[10]

The present paper focuses on our recent attempts to characterize ions at aqueous interfaces at a molecular level. Since it is not supposed to be a review but rather an account of our limited contributions, it does not aim at providing a complete overview of the present state of matters within this ever broadening field. Solely due to our particular choice of systems under investigation several exciting areas concerning *e.g.* ions at water/non-polar liquid or water/solid (insulator or metal) interfaces will be omitted or only briefly touched upon here. This paper thus provides a limited and necessarily personally-biased account on the development of our understanding of interfacial behavior of aqueous ions based on molecular simulations, which we performed in close contact with surface-selective spectroscopic experiments.

Other researchers in the field might tell the story of ions at aqueous interfaces differently (and better); however, we hope that useful general patterns will emerge from this somewhat personal narrative. Chronologically, we started about a decade ago with investigations of inorganic ions at the air/water interface, being motivated by unresolved issues concerning heterogeneous chemistry of atmospheric aqueous sea-salt aerosols.[11,12] Having collected a "database" of computational results which allowed us to unravel certain general patterns of ionic surface affinities, our interest has shifted lately toward the behavior of ions at the interface between a biomolecule and the surrounding aqueous solution, focusing on specific ion effects on proteins.[13–15] This path has lead us (somewhat accidentally) from the simplest aqueous interface to arguably the most complex ones. Since the reader may find it pedagogically helpful to move from simple to complex interfaces, we stick to this route also in this paper. In the following chapters we thus discuss first the behavior of ions at the air/water interface, proceeding later to issues concerning ions at hydrated proteins. Doing this, we present also the latest (mostly unpublished) computational results, putting special emphasis on controversial and not completely resolved issues.

2. Inorganic ions at the air/water interface

2.1 Simulations of surfaces of aqueous salt solutions

At the simplest aqueous interface, *i.e.* that between water and air or vapor, the behavior of salt ions should be according to the textbook knowledge rather straightforward.[4,16–18] Namely, viewing this interface as a plane between a high and low dielectric constant media, ions should be repelled from it into the water phase by the electrostatic image force (see Fig. 1). This situation is pertinent to inorganic ions. Large organic ions, *e.g.* containing long alkyl chains, are also subject to the image force; however, they are attracted to the surface due their hydrophobicity.[19] But even for inorganic ions the picture turns out to be more complicated than Fig. 1 suggests. These ions are indeed repelled from the surface by the image force;

 This journal is © The Royal Society of Chemistry 2009

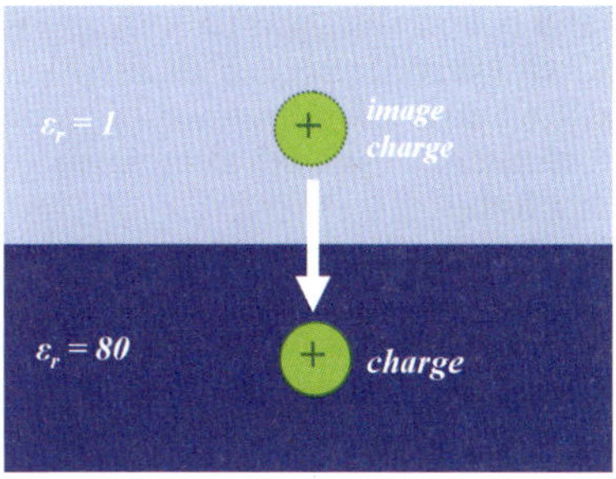

Fig. 1 A schematic picture of the water surface as a flat, sharp interface between two dielectric continua with $\varepsilon_r = 80$ (water) and $\varepsilon_r = 1$ (air or vapor).[21] A model spherical ion is repelled from this surface by the electrostatic image force.

however, the free energy penalty for bringing a monovalent ion from the aqueous phase right to the interface is only of the order of several kT (the rest of the free energy penalty being paid only when the ion is fully desolvated in the air).[20] Such a relatively weak electrostatic repulsion from the water surface can in principle be overwhelmed by other forces connected with ion-specific interaction with water, *e.g.* due to ion cavitation and polarization.[21] Ion-specific dispersion forces can also come into play,[22] but are likely of secondary importance at the surface of strongly polar solvents.

Consequently, large and polarizable (soft) inorganic ions, such as heavier halides, can be found in the topmost water layer and can even exhibit surface enhancement, followed by subsurface depletion.[12,23–26] This is demonstrated in Fig. 2, which shows density profiles of surface-attracted bromide and surface-repelled sodium in a slab of roughly molar aqueous solution of NaBr.[12] As a matter of fact, the surface affinity of iodide, bromide, and to a lesser extent also of chloride, was first observed in pioneering polarizable molecular dynamics (MD) simulations of water clusters,[27–30] computationally only being found later for extended aqueous slabs.[12,23–26] We should stress here, that even for the largest and softest investigated halide, *i.e.* iodide, the interfacial enhancement is rather modest, the peak surface concentration exceeding that of the bulk by no more than a factor of 2–3.[12] Inorganic ions thus do not behave as typical surfactants with many orders of magnitude of surface enhancement; nevertheless, some of them can act as surface active species. This concerns particularly

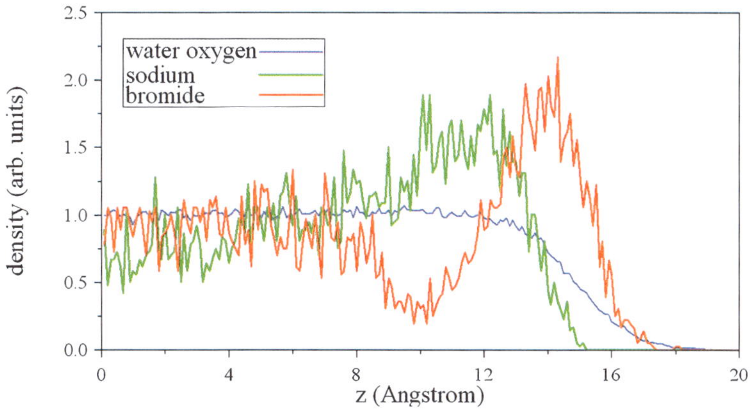

Fig. 2 Density profiles of bromide, sodium, and water oxygens in a slab of 1.2 M aqueous NaBr.[12] Distributions of ions and water are plotted in layers parallel to the surface; from the center of the slab across the interface into the vapor (only half of the slab is depicted).

anions, which are large and polarizable (with few exceptions such as fluoride). In contrast, small hard positively charged ions, *e.g.* alkali cations or fluoride are always repelled from the surface.[21,31] Surface depletion is also pertinent to polyvalent ions, where much stronger (compared to monovalent ions) image charge repulsion overwhelms surface driving forces.[32]

Affinity for the air/water interface was also observed in MD simulations of more complex inorganic salts containing molecular ions. There, the situation is complicated by difficulties in deriving accurate polarizable force fields for non-spherical ions possessing internal structure. Therefore, only a limited set of such ions has been investigated so far. Surface affinity has been observed, *e.g.* for azide[33] and thiocyanate.[34] For the latter case, the concentration dependence of the surface enhancement has been investigated in the molar range showing, similarly to heavier halides, a gradual saturation of the surface effect. A particularly difficult case has been that of nitrate anion, where first calculations indicated some surface enhancement.[35] Subsequent more converged simulations with a refined force field revealed that nitrate is repelled, albeit weakly from the top layer and rather prefers to reside in the immediate subsurface.[36,37] This demonstrates the sensitivity of the results on simulation conditions, in particular on the force field. Rough estimates of ionic surface affinities can be already obtained from nanosecond simulations with relatively simple (but polarizable) potentials with several hundreds of water molecules in the unit cell of the slab. However, to be more quantitative, force fields must be refined and longer simulations with larger unit cells are needed.

2.2 Experimental considerations for surfaces of aqueous salt solutions

The above computational findings of surface affinities of large and soft inorganic ions were first met with skepticism, since they were in conflict with the textbook notion of an essentially ion-free electrolyte surface due to image charge repulsion.[16–18] Moreover, inorganic salts in general increase the surface tension of water. It follows from the Gibbs adsorption equation[38]

$$\partial\gamma/\partial\mu = 1/kT \times \partial\gamma/\partial(\ln fx) = -\Gamma \qquad (1)$$

that a positive change in surface tension γ with salt ion chemical potential μ means a negative surface excess Γ of the ions. (The chemical potential of the ion species is related to the logarithm of its mole fraction x times activity coefficient f, the latter being close to unity at small concentrations.) Note, however, that the surface excess is an integral quantity over the whole interfacial region. As such, it can still be negative if surface accumulation of ions is overcompensated by subsurface depletion. Qualitatively, such a situation is indeed observed, *e.g.* for heavier halides (Fig. 2), although it is difficult to obtain a converged value of the subsurface ion depletion computationally due to its slow convergence into the bulk and the necessarily finite width of the simulated slab.[21,25]

The calculated surface enhancement (*e.g.* for bromide and iodide) or at least presence at the interface (*e.g.* for chloride) helped to rationalize atmospheric observations and laboratory experiments concerning heterogeneous chemistry on aqueous sea salt aerosols.[6,11] Subsequently, surface selective spectroscopies, such as vibrational sum frequency generation (VSFG),[39–42] second harmonic generation (SHG),[43–45] photoelectron spectroscopy (PES),[46–48] and atomic scattering techniques[49,50] allowed for an indirect or direct observation of ions within the interfacial layer between an aqueous electrolyte solution and its vapor. All these techniques showed that inorganic salt ions can influence the interfacial layer. SHG and PES methods were, moreover, capable of directly detecting the interfacial ionic signal as a function of bulk concentration, revealing surface adsorption of anions like bromide, iodide or thiocyanate, but not of alkali cations.[44,48,51] Although more work is needed to obtain fully quantitative results for each individual system, we

 This journal is © The Royal Society of Chemistry 2009

can conclude that the affinity of large, soft inorganic anions for the air/water interface is reasonably well established by now, both computationally and experimentally.

2.3 Simulations of hydronium and hydroxide at aqueous interfaces

Aqueous electrolytes encompass not only salts but also solutions of acids and basis. Strong acids and bases practically completely dissociate in water releasing into the solution two important ions which have been not discussed yet—hydronium and hydroxide. Even pure water contains these ions, albeit at a very low amounts of 10^{-7} M. The issue of surface affinity of H_3O^+ and OH^- has been subject of vigorous computational and experimental investigations recently.[52–61] Researchers generally agree that these two ions exhibit a non-trivial surface behavior. However, as reflected also in this Faraday Discussion, there is disagreement not only on the quantitative amount of ionic surface adsorption but also on a more qualitative issue, namely which of these two ions exhibits larger surface enhancement.[62–64]

Similarly as for salt ions, the first computational indication about an interesting surface behavior of hydronium and hydroxide came from medium-sized to large cluster studies.[65–69] Note that one needs at least about 20 water molecules in order to start distinguishing between surface and internal ion solvation. For hydronium, calculations revealed a strong preference for cluster surfaces.[65–69] This may came as a surprise since naively one could assume that proton as a small ion with zero polarizability should prefer internal solvation. The trick is, of course, that hydrated proton chemically reacts with water, the two limiting resulting structures being the Eigen (H_3O^+) and Zundel ($H_5O_2^+$) cations,[70,71] which dramatically alters its properties and interfacial behavior (*vide infra*). For hydroxide, computational studies show a systematic shift from an affinity for the surface to delocalization over the whole cluster upon increasing the cluster size.[69,72,73]

While cluster studies can provide useful hints, there is need for investigating extended aqueous surfaces in order to directly connect with experimentally scrutinized surfaces of bulk aqueous systems. MD simulations of hydronium in aqueous slabs employing a polarizable classical force field[52,54] or a more sophisticated effective valence bond (EVB) scheme[53,60] allowing for proton transfer, revealed that hydronium is attracted to or at least not repelled from the water surface. In contrast, hydroxide was found to be weakly repelled from the surface although it did occasionally penetrate all the way to the topmost water layer.[54]

Most recently, we have performed a series of simulations[58,62,74] to check the above early predictions. The principal aim of these calculations has been to establish the free energy difference connected with transferring hydronium or hydroxide from the aqueous bulk to the surface. Our approach to this problem has been a pragmatic combination of three computational methods—classical MD with polarizable potentials, *ab initio* MD based on density functional theory (DFT) methods, and accurate *ab initio* (MP2 or coupled cluster) calculations. A similar methodology has been already applied in our studies of salt ions at the air/water interface[21] with that the combination of empirical and electronic structure techniques becomes particularly topical for the "difficult", chemically non-trivial hydronium and hydroxide ions. In the following paragraphs we, therefore, briefly summarize the employed computational strategy.

Each of the above three computational approaches has its advantages and disadvantages. Classical MD simulations can be easily performed for relatively large systems containing hundreds to thousands of water molecules in periodic unit cells mimicking extended aqueous interfaces. We have been employing unit cells of the form of a rectangular prism where the system acquires the shape of a periodic slab with a bulk region between two air/water interfaces (Fig. 3). Depending on the actual system size, the area of each of these interfaces in the unit cell has varied between 3 and 9 nm^2. This computational approach is fast enough to allow collecting data for varying amounts of dissolved species from just a single ion to several moles

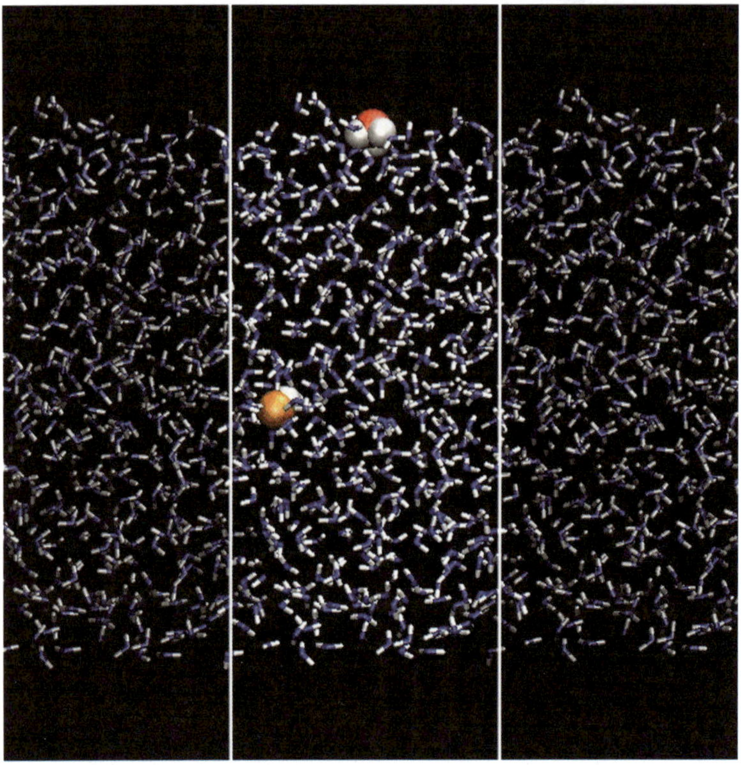

Fig. 3 Water slab containing 432 water molecules and a single hydronium and hydroxide in a unit cell of dimensions of approximately 19 × 19 × 237 Å.[58] White lines define the boundary between the the central and neighboring cells.

per litre. Also, long enough (nanosecond to microsecond) runs can be performed so that thermodynamic quantities such as free energy differences can be converged either directly or using advanced simulation techniques. The basic drawback of classical MD is that the empirical potential does not allow for proton hopping unless special approaches such as the EVB are employed.[60,75,76] Neither does it properly describe the partial charge transfer character of strong ion–water interactions, although a polarizable force field to a certain extent accounts for this effect.[64]

Ab initio MD allows one to get rid of an empirical potential and to describe explicitly the electronic structure of the whole system accounting naturally for proton hopping and charge transfer effects. The price is a dramatic increase in computational load. Consequently only smaller slab systems with up to about a couple of hundred water molecules in the unit cell can be studied.[77] Also, the accessible picosecond rather than nanosecond timescales allow for studying dynamics but typically not thermodynamics. Nevertheless, a most recent "tour de force" *ab initio* MD simulation showed that free energy differences are becoming within reach of this method.[78] Another issue is that *ab initio* MD techniques are based on DFT methods, typically employing non-hybrid functionals such as BLYP or PBE.[79] These functionals are necessarily of an approximate character (*e.g.* in many cases they tend to overdelocalize the electron density) which also translates to the accuracy with which they describe aqueous systems. Currently, hybrid functionals, which provide generally a better description (*e.g.* by reducing the self-interaction error), are being introduced to *ab initio* MD. Finally, accurate *ab initio* calculations at the MP2 or even coupled cluster (CC) level can be employed for small cluster systems containing

 This journal is © The Royal Society of Chemistry 2009

a hydronium or hydroxide ion with several water molecules. Such calculations provide valuable information *per se* and also serve as benchmarks for developing and testing empirical potentials for the ionic product of water.[62]

In our studies of hydronium and/or hydroxide in aqueous slabs, we have most heavily relied on classical MD simulations employing polarizable force fields.[54,58,62,64,74] Using a thermodynamic integration method we established for the hydronium cation either in the Eigen or Zundel form a free energy surface minimum of -2 to -3 kcal mol^{-1}. This surface enhancement is in accord with results of EVB calculations, which allow for proton hopping.[53,60] In contrast, we found that hydroxide has a very weak preference for the bulk of about 1 kcal mol^{-1} (with the statistical error of all of our free energy estimates being ± 1 kcal mol^{-1} with 85% confidence). In other words, simulations show that hydronium accumulates at the water surface, while hydroxide has at best the same concentration there as in the bulk. Therefore, in the topmost layer of water there shall be an excess of hydronium over hydroxide. We denote this situation as an "acidic" surface of water.[58] We use the quotation marks since pH is strictly speaking a bulk property related to proton activity, nevertheless, we operationally define as "acidic" an aqueous surface with a number excess of hydronium over hydroxide.

A simple qualitative rationalization of this effect is as follows. While H_3O^+ makes three strong donating hydrogen bonds, its oxygen is a dysfunctional hydrogen bond acceptor due to its very low negative charge compared, *e.g.* to the O atom of a water molecule (Fig. 4). From this perspective, hydronium has been labeled as the smallest "amphiphilic" species[53] and as such prefers to reside at the surface with the O atom exposed into the vapor. Qualitatively, the situation for OH^- is the opposite with its oxygen being an excellent hydrogen-bond acceptor and hydrogen a poor donor. However, quantitatively this "amphiphilic" effect is weaker since the H atom is small and so is the reduction of its partial charge compared to hydrogen in a water molecule. Hydroxide is, therefore, able to develop also a donating hydrogen bond, albeit a weak one (Fig. 4). As a result, it does not exhibit surface enhancement although it can occasionally penetrate all the way to the surface layer.

Classical MD allows us, provided long (beyond nanosecond) simulations are performed, to obtain the free energy difference between an ion solvated in the bulk and at the surface. However, these simulations are only as good as the empirical interaction potential and without sufficient care a "garbage in, garbage out" scenario may become applicable. A dynamical approach which bypasses the use of empirical potentials is the *ab initio* MD where energies and forces are calculated using DFT-based methods.[80] We employed the BLYP functional to investigate the behavior of hydronium or hydroxide in clusters with 48 water molecules and in small aqueous slabs with 72 waters.[58,62] Picosecond timescale trajectories revealed that hydronium

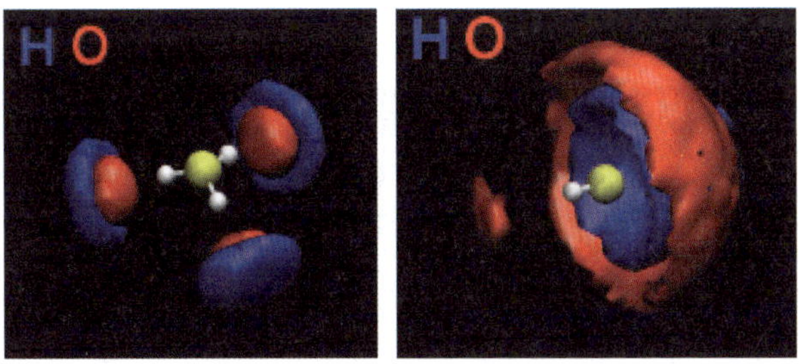

Fig. 4 Solvation shells around bulk-located aqueous hydronium (left) and hydroxide (right) from polarizable MD simulations.[62] Color coding: ion oxygens—yellow, ion hydrogens—white, water oxygens—red, and water hydrogens—blue. While H_3O^+ forms 3 strong and very directional donating and no accepting hydrogen bonds, OH^- exhibits 4–5 accepting hydrogen bonds and a weak donating one.

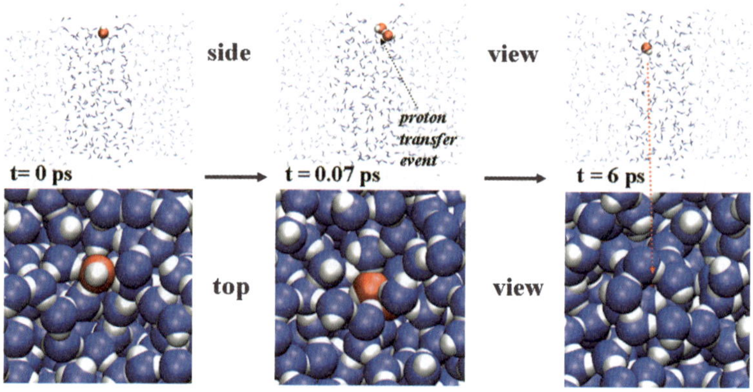

Fig. 5 Three snapshots from a *ab initio* MD trajectory of a single hydronium initially placed on the surface of an aqueous slab with 216 water molecules in a $15 \times 15 \times 70$ Å unit cell.[81] OH^- moves swiftly from the surface to the subsurface, with proton transfer accelerating its mobility. It becomes fully hydrated with no direct exposure to the vapor phase and remains so for the rest of the 10 ps trajectory.

prefers to reside at the surface, while hydroxide tends to move from the surface to the subsurface layers. For a set of optimized cluster structures we estimated the surface preference of hydronium to be at least as strong (if not stronger) using DFT that that obtained using polarizable empirical potentials. Most recently, we have launched a larger scale *ab initio* simulation of a single OH^- ion in an aqueous slab containing 216 water molecules in a unit cell.[81] Such size has been shown previously to be sufficient for an aqueous system to develop a well-defined bulk region between the two water/vapor interfaces.[77] In a simulation lasting ten picoseconds we see a hydroxide ion initially placed at the surface starting to "dive" into the subsurface within 50 fs and then exploring this region of the slab (Fig. 5). This indicates that there are no special "binding sites" for hydroxide at the water surface and that the free energy profile connected with the motion of OH^- toward the bulk lacks an appreciable surface minimum. This result is also supported by most recent massive (~500 ps) *ab initio* MD simulations of a very similar system which allowed to extract the free energy difference between a surface and bulk position of hydroxide.[78] The obtained value is roughly zero, this result falling within the statistical error of 1 kcal mol^{-1} of the value obtained from MD with an empirical polarizable potential.[58,62,78] Nevertheless, the aqueous hydroxide described within DFT has a more delocalized charge (particularly at the interface) than its counterpart modelled with a polarizable force field, which is a factor which tends to reduce its repulsion from the surface.[78]

Finally, we performed accurate benchmark *ab initio* calculations at the MP2 and CCSD(T) levels employing the aug-cc-pvtz basis set for small clusters containing

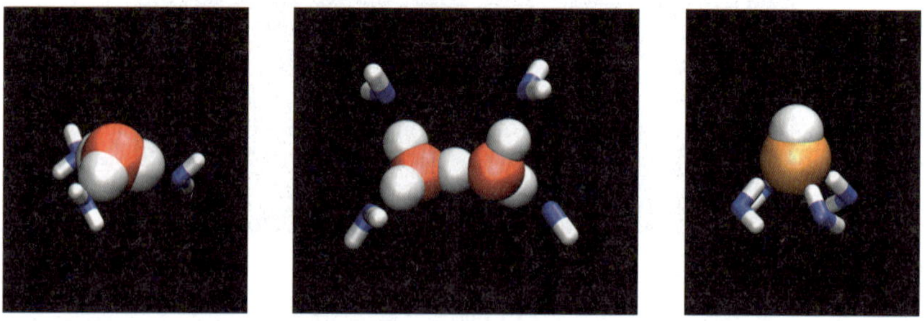

Fig. 6 *Ab initio* MP2/aug-cc-pvdz structures of H_3O^+ with three, $H_5O_2^+$ with four, and OH^- with four water molecules, used for benchmarking our polarizable force field.[58]

 This journal is © The Royal Society of Chemistry 2009

a single proton or hydroxide ion and up to six water molecules,[58] similarly to previous computational studies on these systems.[82,83] Optimized geometries of H_3O^+ with three, $H_5O_2^+$ with four, and OH^- with four tightly bound water molecules are depicted in Fig. 6. Note that the latter structure is not the global minimum but rather a hypercoordinated geometry close to those being prevalent in the liquid.[84] Such calculations serve as benchmarks allowing to refine polarizable force fields for hydronium (both in the Eigen and Zundel forms) and for hydroxide.[58]

2.4 Experiments concerning hydronium and hydroxide at aqueous interfaces

The present computationally based picture of an "acidic" surface of water with an excess of hydronium over hydroxide in the topmost layer is in accord with a broad range of surface-selective spectroscopic measurements. Indirectly, the effect of hydronium or hydroxide on water molecules at the interface between air and acid or base solution was investigated using the vibrational sum frequency generation (VSFG) spectroscopy.[54,57,59] These measurements reveal a strong effect of hydronium but only a weak influence of hydroxide on the water molecules in the interfacial layer. Similarly, isotope-exchange IR measurements of cold aqueous nanoparticles show enhanced isotopic scrambling at the surface traced to an increased presence there of protons but not of hydroxide.[58,62] The second harmonic generation (SHG) spectroscopy allows one to probe interfacial OH^- directly *via* excitation to charge-transfer-to-solvent states.[61] A concentration dependence study in the millimolar to molar range provides data that are best fitted by a Langmuir adsorption isotherm with zero or positive value of the surface free energy (compared to the bulk) and clearly excludes strong surface adsorption beyond ΔG of -1 kcal mol^{-1}.[61] This technique also allows one to indirectly probe interfacial concentration of hydronium by comparing a salt and acid with a common anion.[55] Such a study of aqueous HI *vs.* NaI or KI has revealed a strong interfacial enhancement of hydronium with respect to alkali cations.[55] Most recently, the surface behavior of OH^-, which has a distinct signature both in the valence and core regions, has been probed directly by photoelectron spectroscopy (PES) in aqueous microjets.[81] For a proper choice of photon wavelengths PES probes only the top ~5–10 Å of the liquid, *i.e.* about the two to three highest water layers.[51] A concentration study shows a linear dependence of the surface signal on the amount of hydroxide in the bulk from which no surface adsorption of OH^- is inferred.[81] Additionally, no PES signature of hydroxide at the surface of neat water has been observed.[85]

A direct way of relating the above molecular level calculations and experiments to macroscopic observations is *via* surface tension. Measurements show that adding alkali hydroxides to water leads, as it does for inorganic salts, to an increase in surface tension.[86,87] In this context it is interesting to note that for some salts a tiny dip in surface tension by about 0.02% has been observed in the millimolar region (the so-called Jones–Ray effect) in the 1930s.[88] This effect was later interpreted as an artifact of the measurements;[89] however, it has been invoked recently in interpretation of SHG experiments.[90,91] Whether the Jones–Ray effect is real or artificial, for the sake of present discussion it suffices to say that it has only been observed in salt and not in hydroxide solutions. In contrast to hydroxides, acids—including strong, fully dissociating ones such as HCl, HBr or HI—decrease the surface tension of water.[86] A straightforward application of the Gibbs adsorption equation [eqn (1)] to these results leads to a conclusion that hydroxide is repelled from the water surface while hydronium is attracted to it. Surface tension is, however, an integral property of the whole interfacial layer and we have shown before that for large, soft inorganic ions a surface tension increase does not necessarily exclude a surface peak.[21] Nevertheless, in order to maintain a net negative surface excess such a peak has to be relatively small (say maximally a several fold increase compared to the bulk) and should be overcompensated by subsurface depletion. Therefore, it is safe to conclude at this point that based on surface tension measurements hydronium exhibits surface activity, while hydroxide is repelled from or at most shows a small peak at water surface.

While calculations and surface selective spectroscopies are in agreement with surface tension experiments, comparison with other macroscopic measurements is more controversial. Measurements of small water droplets sprayed in the laboratory or in nature (the so called waterfall effect), which have been performed in the late 19th and early 20th centuries and have been revisited recently, showed that these water particles mostly carry a negative charge.[92,93] In addition, electrophoretic mobility measurements of gas bubbles in water yielded negative ζ-potentials of about −35 mV,[56,94] while even lower negative values were obtained for oil droplets in water at pH = 7.[95,96] An isoelectric point (IP) around pH = 4 was observed for gas bubbles in water.[56,94] The titration curve tended to level off on both sides of the IP, indicating surface accumulation of negative charge above and positive charge below this point. Similar behavior was also observed for oil droplets in water, where it was deduced from pH-stat measurements that the surface concentration of the negative charge at pH = 7 amounts to −5 to −7 μC cm^{-2}.[63,95,96] This corresponds to about one negative charge per 3 nm^2 or, assuming (rather generously) a 1 nm width of the interfacial layer of neat water, to a roughly 0.5 M of surface charge. This negative surface charge, which has been assumed as originating from OH^- ions,[56,94–97] has also been invoked in interpreting disjoining pressure measurements determining stabilities of thin aqueous films.[98,99]

It is instructive to relate the above suggested surface adsorption of hydroxide to surface tension effects. Using the Gibbs adsorption equation [eqn (1)] one can directly deduce the surface tension change with pH, which would be caused by a net surface excess of hydroxide of 3×10^{13} cm^{-2} (corresponding to the negative surface charge of −5 μC cm^{-2} observed for pH neutral water[63,95,96]). At pH = 7 one would thus obtain a change in surface tension of ~3 mN m^{-1} per pH unit. The accuracy of surface tension measurement is typically in the 0.01–0.1 mN m^{-1} range; therefore, such a large change should be easily detectable. Instead, there is no observable difference in surface tension within a broad range of pH values around pH 7 and, as for inorganic salts, one needs orders of magnitude higher (at least millimolar or larger) concentrations of hydroxide or hydronium to start detecting measurable changes in surface tension.[86,87] The only way to make a surface excess of OH^- of 3×10^{13} cm^{-2} consistent with a vanishing change of surface tension with pH around pH = 7 is to assume an unrealistically (several millimeter) thick subsurface layer completely depleted of hydroxide.

The suggested interpretation of the above macroscopic experiments in terms of an extremely strong (*i.e.* by about six orders of magnitude) surface enhancement of hydroxide[56,94–97] is thus at odds with results of calculations, surface-selective spectroscopies, and surface tension measurements which span a broad range of concentrations from neat water to several moles per litre. Full resolution of the existing controversy may be difficult since electrophoretic mobilities and titration measurements are experiments which neither directly reveal the chemical nature of the observed interfacial charge, nor tell us exactly where and within how thick a layer the charge resides.

What can be done on the computational side to enhance our understanding of the behavior of hydronium and hydroxide at aqueous interfaces? Many of the macroscopic measurements are performed for the oil/water rather than the air/water interface. Simulations are, therefore, being performed for different aqueous hydrophobic interfaces, in particular those between water and oil (alkane) and water and a rigid hydrophobic wall (graphene).[64,74,100,101] Classical MD simulations employing polarizable potentials show that hydronium and hydroxide exhibit at the oil/water interface a behavior very similar to that at the air/water interface, *i.e.* surface attraction for the former but not for the latter ion[64] (see Fig. 7). The situation is somewhat different at the interface between water and a rigid hydrophobic wall, which causes density oscillations and strong orientational structuring within the interfacial water

 This journal is © The Royal Society of Chemistry 2009

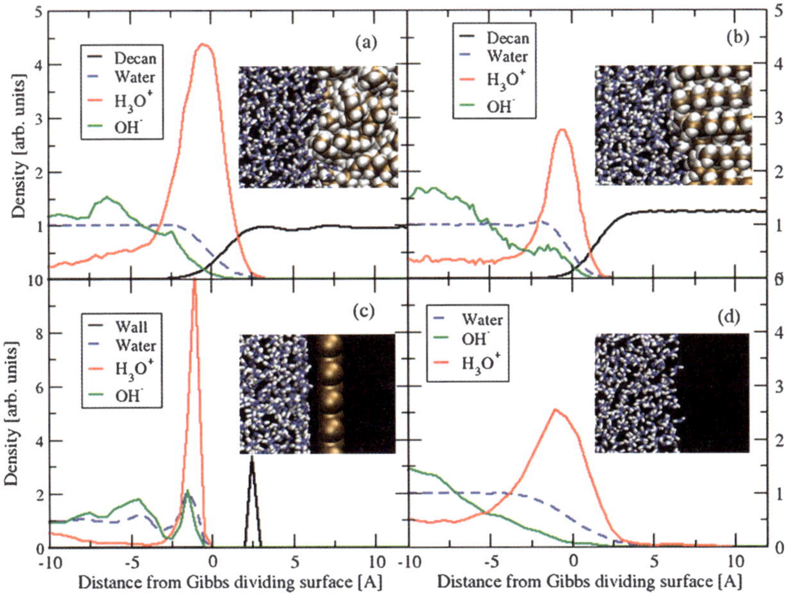

Fig. 7 Density profiles of hydronium, hydroxide, water oxygens, and alkane carbons across the (a) water/disordered decane, (b) water/ordered decane, (c) water/carbon wall, and (d) water/vapor interfaces.[64] Insets show snapshots of these four interfaces.

layers.[74] As a result, hydronium exhibits a strong and sharp surface peak, while hydroxide is also slightly enhanced in the subsurface layer (Fig. 7). A more general picture of the behavior of H_3O^+ and OH^- is thus emerging from these calculations which, admittedly, does not bring us much closer to the resolution of the existing controversy with some macroscopic experiments.

In the near future, progress in hardware and software will allow simulating larger systems (*i.e.* bigger unit cells in periodic systems) for longer timescales and with improving description of the intermolecular interactions. *Ab initio* MD simulations going beyond the limitations of empirical force fields will become an invaluable tool for obtaining accurate thermodynamic quantitities, particularly if the accuracy of the employed density functionals is further improved. This will lead to a refinement of the present view on hydronium, hydroxide and other ions at aqueous interfaces; however, it is unlikely that the computational picture will change qualitatively.

3. Beyond the air/water interface

Do the lessons we have learnt (or have not learnt yet) from ion affinities to the free surface of water help us in understanding their behavior at more complex aqueous interfaces? These are the interfaces between water and non-polar liquids, metals, insulating solids including ice, or large (bio)molecules.[102–104] The prominent question concerning ions at these interfaces is as follows: how transferable is the finding from the air/water interface that the surface affinity is connected with ionic size, polarizability, and other specific ion issues such as ion–water hydrogen-bonding patterns? As an example, we have seen in the previous section that, *e.g.* the behavior of hydronium and hydroxide at the soft oil/water interface is similar to that at the open water surface, while a hard hydrophobic wall behaves somewhat differently. Several other cases of ions at these aqueous interfaces have been studied;[105,106] nevertheless, at present our understanding falls short of being able to answer the above question fully. Rather, we report here on our limited explorations in this direction focusing solely on the last of the above-mentioned interfaces. To this end, we have started recently a systematic investigation of interfaces between proteins and their aqueous environment aimed at understanding surface affinities of biologically relevant ions.

3.1 Ions at the protein/water interface

The influence of salt ions on collective properties of hydrated proteins (*i.e.* Hofmeister effects[107,108] is viewed traditionally as coming from specific ion–water interactions in the aqueous bulk.[109] Indeed, ions are classified in textbooks as either cosmotropes ("structure makers") or chaotropes ("structure breakers") according to their ability to structure water molecules around themselves.[110] In simple terms, cosmotropes should organize layers of water molecules around themselves, "stealing" thus effectively the solvent from proteins, which leads to salting-out, while chaotropes are not able to do that. However, there is mounting experimental evidence that monovalent ions (even cosmotropic ones) are not able to affect more than their immediate solvent shell.[111–113] Therefore, other possible explanations of salt action on protein are being looked for. If it is not the bulk solvent, then the prime suspect for Hofmeister effects becomes the interface between the protein and the surrounding salt solution.[9]

As a first case study[14] we have investigated the behavior of sodium and potassium at surfaces of hydrated proteins. Na^+ and K^+ are the biologically most relevant monovalent salt cations. Moreover, they represent simple hard atomic ions the behavior of which at the air/water interface is straightforward—they are both almost equally repelled from it.[21,31] Surface tension arguments invoking the air/water interface have been used recently to rationalize interactions of ions (both cations and anions) with surfaces of hydrated proteins.[114,115] In a similar spirit, an old model representing globular proteins as spheres with a low dielectric constant and with a dipole or a smeared charge distribution[116] has been recently revived and dressed with dispersion interactions to account for specific ion effects.[117] We found it, therefore, instructive to explore ion-specific interactions of Na^+ and K^+ with proteins using MD simulations and to check the applicability of the above models. To this end we selected a set of five very different proteins—actin as a structural protein, bovine pancreatic trypsin inhibitor (BPTI) as an enzyme-inhibiting protein, ubiquitin as the "kiss of death" protein marker, rubredoxin as an electron-transfer system, and ribonuclease A as an enzyme.[14] Each of these systems was simulated in a mixed solution of varying but equal amounts (typically 0.25 M) of NaCl and KCl for several nanoseconds to ensure statistically meaningful results

Fig. 8 summarizes the affinities of sodium and potassium for surfaces of aqueous proteins.[14] They are expressed in terms of distributions of individual ions (in layers with a given distance from the protein surface) and integrals thereof. The latter are the cumulative sums providing the number of ions of each type which can be found within a certain distance from the surface of the protein. From the first column of Fig. 8 two non-trivial results emerge. First, both cations exhibit affinity for protein surface. Second, this surface enhancement is about a factor of two stronger for sodium than for potassium and this remarkable ion specificity is preserved for all the aqueous proteins under study. Analyzing the MD trajectories it is easy to trace the origins of this effect. Affinity and selectivity comes for alkali cations primarily from interactions with the carboxylic side chain groups of the negatively charged glutamate and aspartate residues (second column of Fig. 8) and from the electronegative oxygens of the backbone amide group (third column), the rest of the protein surface being relatively unimportant (last column).

Similar results are obtained from MD simulations of short polypeptides and individual glutamates and aspartates and from *ab initio* calculations of cationic pairing with acetate and formate in water.[14] The computational predictions are also supported by conductivity measurements and X-ray absorption spectroscopy (XAS).[14,118,119] Finally, the preferential affinity of sodium over potassium for the COO^- group is in accord with the empirical law of matching water affinities.[120]

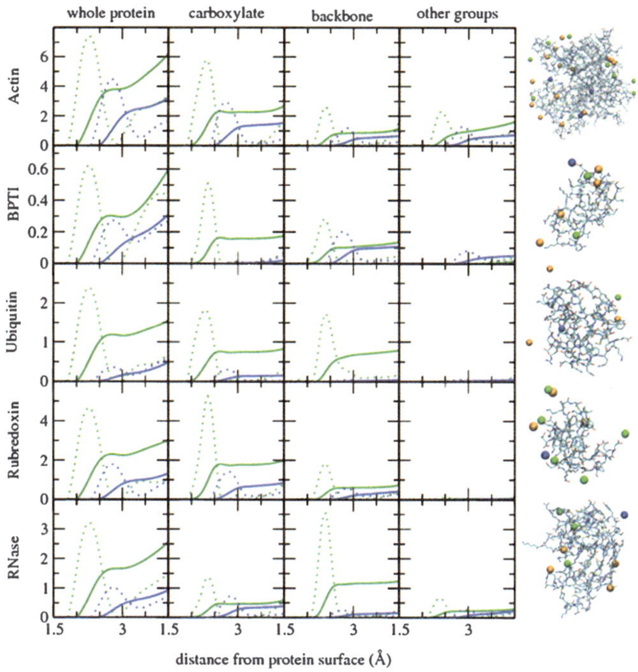

Fig. 8 Distribution functions (dotted lines) and the corresponding integrated sums (full lines) for sodium (green) and potassium in the vicinity of aqueous proteins (adapted from ref. 14 © 2006 by the National Academy of Sciences of the USA). Note that both cations exhibit affinity for the protein surface, which is stronger for Na^+ than for K^+, and that the hotspots for cation binding are the negatively charged carboxylates of glutamates and aspartates, as well as the amide oxygens at the backbone.

According to this rule of thumb, cations preferentially interact in water with anions or anionic groups with similar hydration enthalpies. In other word, a small cation (sodium) exhibits a stronger affinity for a small anionic group (COO^-) than a larger cation (potassium). We can thus conclude that, unlike the air/water interface, the protein/water interface exhibits affinity and selectivity for alkali cations. This effect is primarily due to the presence COO^- and C=O groups at the protein surface. It is, therefore, an example of ion-pairing rather than segregation of ions at an interface between low and high dielectric constant media. Due to this, and due to the fact that binding between alkali cations and negatively charged groups at the protein surface is not very strong, a local ion-pairing model for these interactions can be utilized at least as a first approximation. Therefore, a lot can be learned already from studying interactions of alkali cations with aqueous negatively charged polypeptides, terminated amino acids, or even carboxylate anions.[14]

Let us now switch the electric polarity and ask ourselves which *anions* preferentially interact with *positively* charged amino acid residues. As a first choice for addressing this question, we employed the series of halide anions—fluoride, chloride, bromide, and iodide.[121] Among amino acids there are three residues which bear a positive charge on their side chain: arginine with guanidinium, lysine with ammonium, and protonated histidine with the imidazolium group. Using extensive MD simulations lasting tens to hundreds of nanoseconds and employing both polarizable and non-polarizable force fields, we have thus investigated interactions of these (terminated) amino acids with halides in aqueous solutions. Results of these simulations are presented in Fig. 9 as 3D density plots, *i.e.* distributions of each of the halide anions around the three amino acids. These were obtained

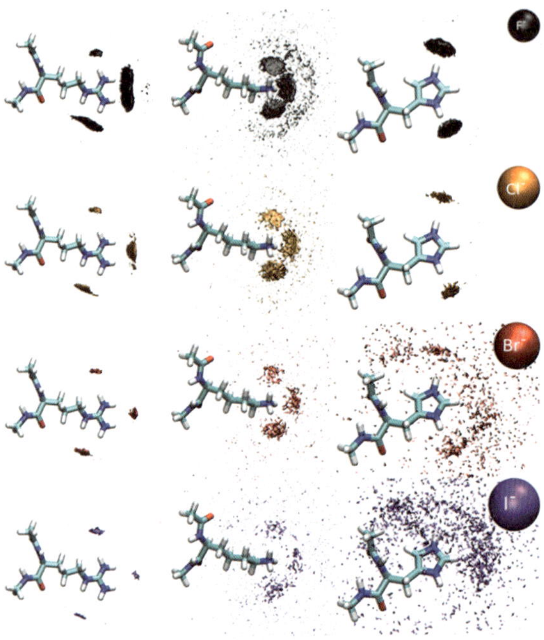

Fig. 9 Density distributions of fluoride (black), chloride (yellow), bromide (red), and iodide (blue) around aqueous arginine, lysine, and protonated histidine.[121]

with a non-polarizable force field, the results with polarizable potentials being similar (*vide infra*).

The following conclusions can be drawn from these simulations. Halides exhibit a tendency to pair with the cationic side chain groups. This tendency decreases as we move down the periodic table, fluoride showing the strongest affinity for the positively charged side chain groups. While proteins have only one anionic side chain group (COO^-), there are three positively charged amino acids. Among these, guanidinium interacts most strongly with the lighter halides, followed by imidazolium and ammonium. Qualitatively, the results are the same for both polarizable and non-polarizable force fields. Quantitatively, inclusion of polarizability reduces the strength of the ion pairing. However, it underlines another effect present for heavier halides, namely the tendency of Br^- and I^- to accumulate in the vicinity of the non-polar regions of the (terminated) amino acids (Fig. 9). This is a behavior similar to their affinity to the air/water interface which we observed previously. Additionally, halides (particularly larger ones) tend to interact with the acidic hydrogen of the amide group. For soft ions, the combined effect of ion pairing and segregation at the interface between water and non-polar protein surface region thus hints to a need to extend the local model which we invoked earlier for the description of the interaction of alkali cations with aqueous proteins. Further refinement of this model may also be necessary for more complex molecular ions, particularly polyvalent ones (*e.g.* alkali earth dications or sulfate) which have a tendency towards very strong ion pairing interactions, both with charged amino acid residues and with counter-ions in the bulk solvent.[13,122–124]

3.2 Ion-mediated protein interactions

While the affinity of ions for aqueous proteins is an interesting issue *per se*, from the biophysical and biochemical point of view a more important question is how specific ion interactions influence processes involving proteins such as association

 This journal is © The Royal Society of Chemistry 2009

(salting-out), crystallization, denaturation, and enzymatic activity. As a matter of fact, it was *via* studying the salting-out of proteins that specific ion effects were discovered 120 years ago. In 1888, Franz Hofmeister with his students at the Pharmacological Institute in Prague ordered separately common anions and cations according to their ability to salt-out egg white from aqueous solutions.[107,108] The result is the lyotropic or Hofmeister series, which separately for the most common anions and cations reads as $SO_4^{2-} > Cl^- > NO_3^- > Br^- > I^- > ClO_4^- > SCN^-$ and $NH_4^+ > K^+ > Na^+ > Mg^{2+}$. It follows from it that strongly hydrated anions and weakly hydrated cations are usually efficient salting-out agents, with the effect of anions being much more pronounced than that of cations. It should be stressed that the Hofmeister series is an empirical rule rather than a "law of nature" and as such has many exceptions. Nevertheless, it represents a very useful concept for rationalizing specific ion effects, which has been invoked also in other phenomena than salting-out, such as enzymatic activities, crystallization of proteins, swelling of tissues, ion exchange, surface tension, *etc.* Note, however, that the concept of the Hofmeister series is hardly applicable to situations where a very particular geometric arrangement of the protein is built in order to accommodate a given ion such as is the situation for ion channels or for metal-ion binding positions in active sites of enzymes.

With the protein surface rather than the aqueous bulk being established as the primary suspect for Hofmeister effects[9] it is thus tempting to ask ourselves how the ion–protein interactions characterized in the previous section translate to protein–protein interactions. To this end, we have recently started investigating association of model proteins: lysozymes, in aqueous solutions using atomistic MD together with a coarse-grained Monte Carlo approach.[125] Lysozyme is a rather exceptional protein in that it carries a large positive charge at neutral and slightly

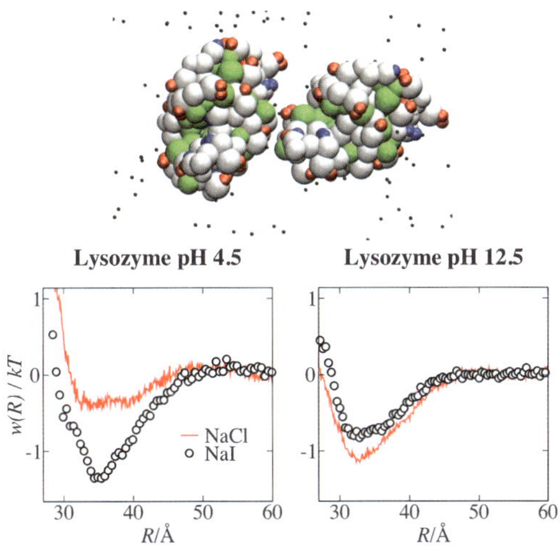

Fig. 10 Association of two lysozyme molecules within a coarse-grained model in aqueous salt solutions.[125,127] At pH 4.5 iodide is more effective salting-out agent causing a deeper free energy minimum than chloride. At a (unrealistically high) pH of 12.5 the order of the anions reverses. Reprinted Fig. 2 with permission from M. Lund, P. Jungwirth and C. E. Woodward, *Phys. Rev. Lett.*, **100**, 258105 (2008). Copyright 2008 by the American Physical Society.†

† Readers may view, browse, and/or download material for temporary copying purposes only, provided these uses are for noncommercial personal purposes. Except as provided by law, this material may not be further reproduced, distributed, transmitted, modified, adapted, performed, displayed, published, or sold in whole or part, without prior written permission from the American Physical Society.

acidic pH (+9 at pH = 4.5). This charge can be effectively screened (neutralized) by salt anions which leads to protein association in aqueous solutions. Measurements of the second virial coefficient show that in this case iodide is a more efficient salting-out agent than chloride in a reversed Hofmeister ordering.[126] Our coarse-grained Monte Carlo calculations using effective ion–amino acid interactions extracted from atomistic MD reproduce this effect showing a deeper free energy minimum for an associated lysozyme pair in NaI than in NaCl.[125] This is despite the fact that our MD simulations show a stronger binding of Cl^- over I^- to positively charged amino acid residues.[121] The key here is the stronger binding of iodide to non-polar residues, which overwhelms the ion-pairing effect. At a (somewhat unrealistic) pH of 12.5 leading to a net charge on lysozyme of −5, the subtle balance between ion-pairing and segregation at non-polar parts of the protein surface reverses and chloride becomes more efficient than iodide in salting-out (Fig. 10),[127] which is the normal Hofmeister ordering.

We have also recently started addressing questions concerning how specific ions affect enzymatic activities and protein denaturation processes. To this end we have performed several case studies, including investigations of the effect of complex molecular ions such as choline and sulfate on the enzymatic activity of the horse-radish peroxidase[13] and most recently also the influence of sodium *vs.* potassium on the activity of the HIV protease and the denaturation abilities of guanidinium and tetrapropylammonium halides and sulfates for model peptides such as mellitin and alanine-based α-helix. Classical MD simulations do not allow us to follow the enzymatic reactions (for which quantum chemical or mixed approaches are needed) directly, nevertheless they provide information about binding patterns of these ions in the vicinity of the reaction center. Concerning denaturation, MD simulations can indicate binding sites where the process can start. In the course of these ongoing studies we are learning that the behavior of molecular ions is more complex than that of their monovalent atomic counterparts. For example, guanidinium behaves as a rather "promiscuous" ion, which can associate both with negatively charged and non-polar residues and can even engage in homo-ion pairing. In addition, divalent ions such as sulfate can create "binding pockets" from positively charged surface residues, distorting thus the protein structure. They also tend to strongly associate with counter-cations in the solutions, which negatively influences the ability of both ions to interact with the protein. While we are only starting to categorize these effects, it is clear already now that they will limit the applicability of the simple local model for ion–protein interactions and of the applicability of separate cationic and anionic Hofmeister series.

4. Conclusions

The field of ions at aqueous interfaces is undergoing a remarkable renaissance thanks to modern computational and spectroscopic techniques, which allow scrutinizing the interfacial region with molecular resolutions. Concerning the air/water interface, the traditional concept of an inactive, ion-free surface of aqueous electrolytes is yielding to a more complex picture, in which inorganic ions with large size and polarizability or specific ("amphiphilic") interactions with water can appear or even be enhanced at the surface. While the new picture is reasonably well established for inorganic salt ions, the situation is much more controversial for the inherent water ions, hydronium and hydroxide. Researchers agree in general that these ions exhibit an interesting surface behavior; however, there is both quantitative and qualitative disagreement between molecular level calculations, surface-selective spectroscopies and surface tension measurements on one side, and electrophoretic and titration experiments, as well as disjoining pressure measurements, on the other side. The former methods predict a surface enhancement of one to two orders of magnitude of hydronium and virtually no surface enhancement or even weak repulsion of hydroxide, while the latter interpret the measured negative surface charge in

 This journal is © The Royal Society of Chemistry 2009

terms of roughly six orders of magnitude surface enhancement of hydroxide. Here, we discussed this controversy in detail and while we fell short of fully resolving it, we found the former picture being consistently supported by surface- and molecularly-resolved computational and experimental evidence.

The free water surface is but the simplest interface and it is interesting to investigate how much it can teach us about ionic behavior at more complex aqueous interfacial systems. The closest to the open aqueous surface is the interface between water and soft hydrophobic material such as oil or liquid alkanes. These interfaces are indeed structurally not too far from each other and, as a result, ionic interfacial affinities are also similar. A hard (rigid) hydrophobic interface, such that between water and graphene is already slightly different due to enhanced structuring and density oscillations of water, which provide additional (sub)surface driving force for ions.

Leaving aside other very interesting aqueous interfaces, such as that between water and a solid insulator or metal, we focused on the interface between a protein and its aqueous environment. This interface is much more complex than the air/water or oil/water interfaces; nevertheless, lessons learned about them do come useful. Namely, to the first approximation, the surface of a hydrated protein can be viewed as composed of separate non-polar and polar or charged patches. At the former, ions segregate similarly as at the above hydrophobic interfaces. At the latter, a more appropriate picture is that of ion-pairing with charged or polar amino acid side-chain groups and the polar amide group at the protein backbone. Such an approach is clearly not aimed at describing very selective ion bindings, *e.g.* at the active sites of enzymes or in ion channels, which are due to particular structural and electronic arrangements of whole segments of the biomolecule. It should be rather applicable for rationalizing the more general but still ion-specific salt effects observed for salting-out, denaturation, crystallization, enzymatic activity, *etc.* as described phenomenologically *via* the Hofmeister series. In answering the question of why various ions of the same valency affect processes involving proteins differently, we are only now starting to see the light at the end of the tunnel, namely the molecular mechanisms responsible for the ion-specific behavior.

Acknowledgements

Our calculations presented here are results of an extensive collective effort. I have been blessed by a fantastic generation of students and postdocs (several of them being independent researchers now), including Robert Vácha, Jan Heyda, Mikael Lund, Martina Roeselová, Martin Mucha, Luboš Vrbka, Babak Minofar, Lukasz Cwiklik and Barbara Jagoda-Cwiklik. I am also indebted to my collaborators, in particular to Doug Tobias, Victoria Buch, Anne Milet, Liem Dang, Barbara Finlayson-Pitts, Werner Kunz, Lai-Sheng Wang, Bernd Winter, Phil Mason, and Jiří Vondrášck. Support from the Czech Ministry of Education (grant LC512), from the Czech Science Foundation (grant 203/08/0114), and from the US-NSF (grant CHE 0431312) is gratefully acknowledged. Work accomplished in the Institute of Organic Chemistry and Biochemistry was also supported *via* Project Z40550506.

References

1 *Water in confining geometries* (Springer series in cluster physics), ed. V. D. Buch and J. P. Devlin, Springer, Berlin, 2003.
2 I. O. Brovchenko, A. Oleinikova, *Interfacial and confined water*, Elsevier, Amsterdam, 2008.
3 P. H. Debye and E. Hückel, The theory of electrolytes. I. Lowering of freezing point and related phenomena, *Phys. Z.*, 1923, **24**, 185–206.
4 L. Onsager and N. N. T. Samaras, The surface tension of Debye–Hückel electrolytes, *J. Chem. Phys.*, 1934, **2**, 528–536, p. 528–536.
5 D. J. Tobias and J. C. Hemminger, Chemistry—getting specific about specific ion effects, *Science*, 2008, **319**(5867), 1197–1198.

6 J. H. Hu, Q. Shi, P. Davidovits, D. R. Worsnop, M. S. Zahniser and C. E. Kolb, Reactive uptake of Cl_2(g) and Br_2(g) by aqueous surfaces as a function of Br^- and I^- ion concentration—the effect of chemical reaction at the interface, *J. Phys. Chem.*, 1995, **99**(21), 8768–8776.
7 B. J. Finlayson-Pitts, The tropospheric chemistry of sea salt: a molecular-level view of the chemistry of NaCl and NaBr, *Chem. Rev.*, 2003, **103**(12), 4801–4822.
8 B. J. Finlayson-Pitts and J. C. Hemminger, Physical chemistry of airborne sea salt particles and their components, *J. Phys. Chem. A*, 2000, **104**(49), 11463–11477.
9 Y. J. Zhang and P. S. Cremer, Interactions between macromolecules and ions: the Hofmeister series, *Curr. Opin. Chem. Biol.*, 2006, **10**(6), 658–663.
10 P. Jungwirth, B. J. Finlayson-Pitts and D. J. Tobias, Introduction: structure and chemistry at aqueous interfaces, *Chem. Rev.*, 2006, **106**(4), 1137–1139.
11 E. M. Knipping, M. J. Lakin, K. L. Foster, P. Jungwirth, D. J. Tobias, R. B. Gerber, D. Dabdub and B. J. Finlayson-Pitts, Experiments and simulations of ion-enhanced interfacial chemistry on aqueous NaCl aerosols, *Science*, 2000, **288**(5464), 301–306.
12 P. Jungwirth and D. J. Tobias, Molecular structure of salt solutions: a new view of the interface with implications for heterogeneous atmospheric chemistry, *J. Phys. Chem. B*, 2001, **105**(43), 10468–10472.
13 L. Vrbka, P. Jungwirth, P. Bauduin, D. Touraud and W. Kunz, Specific ion effects at protein surfaces: a molecular dynamics study of bovine pancreatic trypsin inhibitor and horseradish peroxidase in selected salt solutions, *J. Phys. Chem. B*, 2006, **110**(13), 7036–7043.
14 L. Vrbka, J. Vondrasek, B. Jagoda-Cwiklik, R. Vacha and P. Jungwirth, Quantification and rationalization of the higher affinity of sodium over potassium to protein surfaces, *Proc. Natl. Acad. Sci. U. S. A.*, 2006, **103**(42), 15440–15444.
15 P. Jungwirth and B. Winter, Ions at aqueous interfaces: from water surface to hydrated proteins, *Annu. Rev. Phys. Chem.*, 2008, **59**, 343–366.
16 J. E. B. Randles, Ionic hydration and the surface potential of aqueous electrolytes, *Discuss. Faraday Soc.*, 1957, (24), 194–199.
17 J. E. B. Randles, Structure at the free surface of water and aqueous electrolyte solutions, *Phys. Chem. Liq.*, 1977, **7**, 107.
18 N. K. Adam, *The Physics and Chemistry of Surfaces*, Oxford University Press, London, 1941.
19 T. Hrobarik, L. Vrbka and P. Jungwirth, Selected biologically relevant ions at the air/water interface: a comparative molecular dynamics study, *Biophys. Chem.*, 2006, **124**(3), 238–242.
20 V. S. Markin and A. G. Volkov, Quantitative theory of surface tension and surface potential of aqueous solutions of electrolytes, *J. Phys. Chem. B*, 2002, **106**(45), 11810–11817.
21 P. Jungwirth and D. J. Tobias, Specific ion effects at the air/water interface, *Chem. Rev.*, 2006, **106**(4), 1259–1281.
22 M. Bostrom, D. R. M. Williams and B. W. Ninham, Surface tension of electrolytes: specific ion effectsexplained by dispersion forces, *Langmuir*, 2001, **17**(15), 4475–4478.
23 L. X. Dang and T. M. Chang, Molecular mechanism of ion binding to the liquid/vapor interface of water, *J. Phys. Chem. B*, 2002, **106**(2), 235–238.
24 G. Archontis, E. Leontidis and G. Andreou, Attraction of iodide ions by the free water surface, revealed by simulations with a polarizable force field based on Drude oscillators, *J. Phys. Chem. B*, 2005, **109**(38), 17957–17966.
25 T. Ishiyama and A. Morita, Molecular dynamics study of gas-liquid aqueous sodium halide interfaces. I. Flexible and polarizable molecular modeling and interfacial properties, *J. Phys. Chem. C*, 2007, **111**(2), 721–737.
26 G. L. Warren and S. Patel, Comparison of the solvation structure of polarizable and nonpolarizable ions in bulk water and near the aqueous liquid/vapor interface, *J. Phys. Chem. C*, 2008, **112**(19), 7455–7467.
27 L. Perera and M. L. Berkowitz, Structure and dynamics of $Cl^-(H_2O)_{20}$ clusters—the effect of the polarizability and the charge of the ion, *J. Chem. Phys.*, 1992, **96**(11), 8288–8294.
28 L. Perera and M. L. Berkowitz, Many-body effects in molecular-dynamics simulations of $Na^+(H_2O)_n$ and $Cl^-(H_2O)_n$ clusters, *J. Chem. Phys.*, 1991, **95**, 1954; L. Perera and M. L. Berkowitz, Many-body effects in molecular-dynamics simulations of $Na^+(H_2O)_n$ and $Cl^-(H_2O)_n$ clusters, *J. Chem. Phys.*, 1993, **99**(5), 4236–4237, (erratum).
29 L. X. Dang and D. E. Smith, Molecular dynamics simulations of aqueous ionic clusters using polarizable water, *J. Chem. Phys.*, 1993, **99**(9), 6950–6956.
30 S. J. Stuart and B. J. Berne, Surface curvature effects in the aqueous ionic solvation of the chloride ion, *J. Phys. Chem. A*, 1999, **103**(49), 10300–10307.

 This journal is © The Royal Society of Chemistry 2009

31 T. M. Chang and L. X. Dang, Recent advances in molecular simulations of ion solvation at liquid interfaces, *Chem. Rev.*, 2006, **106**(4), 1305–1322.
32 S. Gopalakrishnan, P. Jungwirth, D. J. Tobias and H. C. Allen, Air/liquid interfaces of aqueous solutions containing ammonium and sulfate: spectroscopic and molecular dynamics studies, *J. Phys. Chem. B*, 2005, **109**(18), 8861–8872.
33 X. Yang, B. Kiran, X. B. Wang, L. S. Wang, M. Mucha and P. Jungwirth, Solvation of the azide anion (N_3^-) in water clusters and aqueous interfaces: a combined investigation by photoelectron spectroscopy, density functional calculations, and molecular dynamics simulations, *J. Phys. Chem. A*, 2004, **108**(39), 7820–7826.
34 P. B. Petersen, R. J. Saykally, M. Mucha and P. Jungwirth, Enhanced concentration of polarizable anions at the liquid water surface: SHG spectroscopy and MD simulations of sodium thiocyanide, *J. Phys. Chem. B*, 2005, **109**(21), 10915–10921.
35 P. Salvador, J. E. Curtis, D. J. Tobias and P. Jungwirth, Polarizability of the nitrate anion and its solvation at the air/water interface, *Phys. Chem. Chem. Phys.*, 2003, **5**(17), 3752–3757.
36 B. Minofar, R. Vacha, A. Wahab, S. Mahiuddin, W. Kunz and P. Jungwirth, Propensity for the air/water interface and ion pairing in magnesium acetate *vs.* magnesium nitrate solutions: molecular dynamics simulations and surface tension measurements, *J. Phys. Chem. B*, 2006, **110**(32), 15939–15944.
37 J. L. Thomas, M. Roeselova, L. X. Dang and D. J. Tobias, Molecular dynamics simulations of the solution/air interface of aqueous sodium nitrate, *J. Phys. Chem. A*, 2007, **111**(16), 3091–3098, p. 3091–3098.
38 J. W. Gibbs, *The Collected Works of J. Willard Gibbs*, Longmans, New York, 1928.
39 P. B. Miranda and Y. R. Shen, Liquid interfaces: a study by sum-frequency vibrational spectroscopy, *J. Phys. Chem. B*, 1999, **103**(17), 3292–3307.
40 M. J. Shultz, C. Schnitzer, D. Simonelli and S. Baldelli, Sum frequency generation spectroscopy of the aqueous interface: ionic and soluble molecular solutions, *Int. Rev. Phys. Chem.*, 2000, **19**(1), 123–153.
41 D. F. Liu, G. Ma, L. M. Levering and H. C. Allen, Vibrational spectroscopy of aqueous sodium halide solutions and air/liquid interfaces: observation of increased interfacial depth, *J. Phys. Chem. B*, 2004, **108**(7), 2252–2260.
42 E. A. Raymond and G. L. Richmond, Probing the molecular structure and bonding of the surface of aqueous salt solutions, *J. Phys. Chem. B*, 2004, **108**(16), 5051–5059.
43 P. B. Petersen and R. J. Saykally, Confirmation of enhanced anion concentration at the liquid water surface, *Chem. Phys. Lett.*, 2004, **397**(1–3), 51–55.
44 P. B. Petersen and R. J. Saykally, On the nature of ions at the liquid water surface, *Annu. Rev. Phys. Chem.*, 2006, **57**, 333–364.
45 H. T. Bian, R. R. Feng, Y. Y. Xu, Y. Guo and H. F. Wang, Increased interfacial thickness of the NaF, NaCl and NaBr salt aqueous solutions probed with non-resonant surface second harmonic generation (SHG), *Phys. Chem. Chem. Phys.*, 2008, **10**(32), 4920–4931.
46 S. Ghosal, A. Shbeeb and J. C. Hemminger, Surface segregation of bromine in bromide doped NaCl: Implications for the seasonal variations in Arctic ozone, *Geophys. Res. Lett.*, 2000, **27**(13), 1879–1882.
47 R. Weber, B. Winter, P. M. Schmidt, W. Widdra, I. V. Hertel, M. Dittmar and M. Faubel, Photoemission from aqueous alkali metal–iodide salt solutions using EUV synchrotron radiation, *J. Phys. Chem. B*, 2004, **108**(15), 4729–4736.
48 S. Ghosal, J. C. Hemminger, H. Bluhm, B. S. Mun, E. L. D. Hebenstreit, G. Ketteler, D. F. Ogletree, F. G. Requejo and M. Salmeron, Electron spectroscopy of aqueous solution interfaces reveals surface enhancement of halides, *Science*, 2005, **307**(5709), 563–566.
49 O. Hofft, A. Borodin, U. Kahnert, V. Kempter, L. X. Dang and P. Jungwirth, Surface segregation of dissolved salt ions, *J. Phys. Chem. B*, 2006, **110**(24), 11971–11976.
50 J. H. Kim, T. Shin, K. H. Jung and H. Kang, Direct observation of segregation of sodium and chloride ions at an ice surface, *ChemPhysChem*, 2005, **6**(3), 440–444.
51 B. Winter and M. Faubel, Photoemission from liquid aqueous solutions, *Chem. Rev.*, 2006, **106**(4), 1176–1211.
52 L. X. Dang, Solvation of the hydronium ion at the water liquid/vapor interface, *J. Chem. Phys.*, 2003, **119**(12), 6351–6353.
53 M. K. Petersen, S. S. Iyengar, T. J. F. Day and G. A. Voth, The hydrated proton at the water liquid/vapor interface, *J. Phys. Chem. B*, 2004, **108**(39), 14804–14806.
54 M. Mucha, T. Frigato, L. M. Levering, H. C. Allen, D. J. Tobias, L. X. Dang and P. Jungwirth, Unified molecular picture of the surfaces of aqueous acid, base, and salt solutions, *J. Phys. Chem. B*, 2005, **109**(16), 7617–7623.
55 P. B. Petersen and R. J. Saykally, Evidence for an enhanced hydronium concentration at the liquid water surface, *J. Phys. Chem. B*, 2005, **109**(16), 7976–7980.

56 M. Takahashi, ζ-potential of microbubbles in aqueous solutions: electrical properties of the gas/water interface, *J. Phys. Chem. B*, 2005, **109**(46), 21858–21864.
57 T. L. Tarbuck, S. T. Ota and G. L. Richmond, Spectroscopic studies of solvated hydrogen and hydroxide ions at aqueous surfaces, *J. Am. Chem. Soc.*, 2006, **128**(45), 14519–14527.
58 V. Buch, A. Milet, R. Vacha, P. Jungwirth and J. P. Devlin, Water surface is acidic, *Proc. Natl. Acad. Sci. U. S. A.*, 2007, **104**(18), 7342–7347.
59 C. J. Tian, N. Ji, G. A. Waychunas and Y. R. Shen, Interfacial structures of acidic and basic aqueous solutions, *J. Am. Chem. Soc.*, 2008, **130**, 13033–13039.
60 J. M. J. Swanson, C. M. Maupin, H. N. Chen, M. K. Petersen, J. C. Xu, Y. J. Wu and G. A. Voth, Proton solvation and transport in aqueous and biomolecular systems: Insights from computer simulations, *J. Phys. Chem. B*, 2007, **111**(17), 4300–4314.
61 P. B. Petersen and R. J. Saykally, Is the liquid water surface basic or acidic? Macroscopic *vs.* molecular-scale investigations, *Chem. Phys. Lett.*, 2008, **458**(4–6), 255–261.
62 R. Vacha, V. Buch, A. Milet, P. Devlin and P. Jungwirth, Autoionization at the surface of neat water: is the top layer pH neutral, basic, or acidic?, *Phys. Chem. Chem. Phys.*, 2007, **9**, 4736–4747.
63 J. K. Beattie, Comment on Autoionization at the surface of neat water: is the top layer pH neutral, basic, or acidic? by R. Vacha, V. Buch, A. Milet, J.P. Devlin and P. Jungwirth, *Phys. Chem. Chem. Phys.*, 2007, **9**, 4736, *Phys. Chem. Chem. Phys.*, 2008, **10**(2), 330–331.
64 R. Vacha, D. Horinek, M. L. Berkowitz and P. Jungwirth, Hydronium and hydroxide at the interface between water and hydrophobic media, *Phys. Chem. Chem. Phys.*, 2008, **10**, 4975–4980.
65 I. Kusaka and D. W. Oxtoby, Evaluating free energy, enthalpy, and entropy of protonated water clusters by a grand canonical Monte Carlo simulation, *J. Chem. Phys.*, 2000, **113**(22), 10100–10104.
66 J. W. Shin, N. I. Hammer, E. G. Diken, M. A. Johnson, R. S. Walters, T. D. Jaeger, M. A. Duncan, R. A. Christie and K. D. Jordan, Infrared signature of structures associated with the $H^+(H_2O)_n$ (n = 6 to 27) clusters, *Science*, 2004, **304**(5674), 1137–1140.
67 S. S. Iyengar, T. J. F. Day and G. A. Voth, On the amphiphilic behavior of the hydrated proton: an *ab initio* molecular dynamics study, *Int. J. Mass Spectrom.*, 2005, **241**(2–3), 197–204.
68 C. J. Burnham, M. K. Petersen, T. J. F. Day, S. S. Iyengar and G. A. Voth, The properties of ion–water clusters. II. Solvation structures of Na^+, Cl^-, and H^+ clusters as a function of temperature, *J. Chem. Phys.*, 2006, **124**(2), 024327.
69 E. Brodskaya, A. P. Lyubartsev and A. Laaksonen, Investigation of water clusters containing OH^- and H_3O^+ ions in atmospheric conditions. A molecular dynamics simulation study, *J. Phys. Chem. B*, 2002, **106**(25), 6479–6487.
70 M. Eigen, *Angew. Chem., Int. Ed. Engl.*, 1964, **3**, 1.
71 G. Zundel, Hydrogen bonds with large proton polarizability and proton transfer processes in electrochemistry and biology, *Adv. Chem. Phys.*, 2000, **111**, 1.
72 H. M. Lee, P. Tarkeshwar and K. S. Kim, Structures, energetics, and spectra of hydrated hydroxide anion clusters, *J. Chem. Phys.*, 2004, **121**(10), 4657–4664.
73 A. Vegiri and S. V. Shevkunov, Hydration shell structure of the $OH^-(H_2O)_n$ (n = 1–15) clusters from a model potential energy function, *J. Chem. Phys.*, 2000, **113**(19), 8521–8530.
74 R. Vacha, R. Zangi, J. B. F. N. Engberts and P. Jungwirth, Water structuring and hydroxide ion binding at the interface between water and hydrophobic walls of varying rigidity and van der Waals interactions, *J. Phys. Chem. C*, 2008, **112**(20), 7689–7692.
75 A. Warshel and R. M. Weiss, An empirical valence bond approach for comparing reactions in solutions and in enzymes, *J. Am. Chem. Soc.*, 1980, **102**(20), 6218–6226.
76 M. A. Lill and V. Helms, Molecular dynamics simulation of proton transport with quantum mechanically derived proton hopping rates (Q-HOP MD), *J. Chem. Phys.*, 2001, **115**(17), 7993–8005.
77 I. F. W. Kuo and C. J. Mundy, An *ab initio* molecular dynamics study of the aqueous liquid/vapor interface, *Science*, 2004, **303**(5658), 658–660.
78 D. J. Tobias, C. J. Mundy, M. E. Tuckerman and L. X. Dang, private communication, 2008.
79 D. Marx and J. Hutter, *Ab initio* molecular dynamics: theory and implementation, in *Modern Methods and Algorithms of Quantum Chemistry*, ed. J. Grotendorst, NIC, FZ Juelich, 2000, p. 301–449.
80 J. VandeVondele, M. Krack, F. Mohamed, M. Parrinello, T. Chassaing and J. Hutter, QUICKSTEP: fast and accurate density functional calculations using a mixed Gaussian and plane waves approach, *Comput. Phys. Commun.*, 2005, **167**(2), 103–128.
81 R. Vacha, B. Winter and P. Jungwirth, *Chem. Phys. Lett.*, 2008, to be submitted.
82 W. H. Robertson, E. G. Diken, E. A. Price, J. W. Shin and M. A. Johnson, Spectroscopic determination of the OH^- solvation shell in the $OH^-\cdot(H_2O)_n$ clusters, *Science*, 2003, **299**(5611), 1367–1372.

 This journal is © The Royal Society of Chemistry 2009

83 J. M. Headrick, E. G. Diken, R. S. Walters, N. I. Hammer, R. A. Christie, J. Cui, E. M. Myshakin, M. A. Duncan, M. A. Johnson and K. D. Jordan, Spectral signatures of hydrated proton vibrations in water clusters, *Science*, 2005, **308**(5729), 1765–1769.
84 M. E. Tuckerman, D. Marx and M. Parrinello, The nature and transport mechanism of hydrated hydroxide ions in aqueous solution, *Nature*, 2002, **417**(6892), 925–929.
85 B. Winter, R. Weber, W. Widdra, M. Dittmar, M. Faubel and I. V. Hertel, Full valence band photoemission from liquid water using EUV synchrotron radiation, *J. Phys. Chem. A*, 2004, **108**(14), 2625–2632.
86 P. K. Weissenborn and R. J. Pugh, Surface tension of aqueous solutions of electrolytes: Relationship with ion hydration, oxygen solubility, and bubble coalescence, *J. Colloid Interface Sci.*, 1996, **184**(2), 550–563.
87 *International Critical Tables*, ed. E. W. Washburn, McGraw-Hill, New York, 1928, vol. 4.
88 G. Jones and W. A. Ray, The surface tension of solutions of electrolytes as a function of the concentration. I. A differential method for measuring relative surface tension, *J. Am. Chem. Soc.*, 1937, **59**(1), 187–198.
89 I. Langmuir, Repulsive forces between charged surfaces in water, and the cause of the Jones–Ray effect, *Science*, 1938, **88**(2288), 430–432.
90 P. B. Petersen, J. C. Johnson, K. P. Knutsen and R. J. Saykally, Direct experimental validation of the Jones–Ray effect, *Chem. Phys. Lett.*, 2004, **397**(1–3), 46–50.
91 P. B. Petersen and R. J. Saykally, Adsorption of ions to the surface of dilute electroyte solutions: The Jones–Ray effect revisited, *J. Am. Chem. Soc.*, 2005, **127**(44), 15446–15452.
92 H. W. Gilbert, Electrical charges arising at a liquid/gas interface, *Proc. Phys. Soc., London*, 1924, **37**, 195–214.
93 L. Zilch, J. S. Maze, J., G. Ewing and M. F. Jarrold, *J. Phys. Chem. C*, 2008, submitted.
94 A. S. Najafi, J. Drelich, A. Yeung, Z. H. Xu and J. Masliyah, A novel method of measuring electrophoretic mobility of gas bubbles, *J. Colloid Interface Sci.*, 2007, **308**(2), 344–350.
95 J. K. Beattie, The intrinsic charge on hydrophobic microfluidic substrates, *Lab Chip*, 2006, **6**(11), 1409–1411.
96 J. K. Beattie and A. M. Djerdjev, The pristine oil/water interface: surfactant-free hydroxide-charged emulsions, *Angew. Chem., Int. Ed.*, 2004, **43**(27), 3568–3571.
97 K. G. Marinova, R. G. Alargova, N. D. Denkov, O. D. Velev, D. N. Petsev, I. B. Ivanov and R. P. Borwankar, Charging of oil/water interfaces due to spontaneous adsorption of hydroxyl ions, *Langmuir*, 1996, **12**(8), 2045–2051.
98 C. Stubenrauch and R. von Klitzing, Disjoining pressure in thin liquid foam and emulsion films—new concepts and perspectives, *J. Phys.: Condens. Matter*, 2003, **15**(27), R1197–R1232.
99 K. A. Karraker and C. J. Radke, Disjoining pressures zeta potentials and surface tensions of aqueous non-ionic surfactant/electrolyte solutions: theory and comparison to experiment, *Adv. Colloid Interface Sci.*, 2002, **96**(1–3), 231–264.
100 K. N. Kudin and R. Car, Why are water-hydrophobic interfaces charged?, *J. Am. Chem. Soc.*, 2008, **130**(12), 3915–3919.
101 R. Zangi and J. Engberts, Physisorption of hydroxide ions from aqueous solution to a hydrophobic surface, *J. Am. Chem. Soc.*, 2005, **127**(7), 2272–2276.
102 S. S. Durand-Vidal, J.-P. Simonin and P. Turq, *Electrolytes at interfaces*, Springer, Berlin, 2000.
103 W. Stumm, *Chemistry of the solid–water interface*, Wiley, New York, 1992.
104 A. H. Wachter, *Liquid Interfaces in Chemical, Biological, and Pharmaceutical Applications*, Taylor & Francis, New York, 2001.
105 G. M. Luo, S. Malkova, J. Yoon, D. G. Schultz, B. H. Lin, M. Meron, I. Benjamin, P. Vanysek and M. L. Schlossman, Ion distributions near a liquid/liquid interface, *Science*, 2006, **311**(5758), 216–218.
106 C. D. Wick and L. X. Dang, Recent advances in understanding transfer ions across aqueous interfaces, *Chem. Phys. Lett.*, 2008, **458**(1–3), 1–5.
107 W. Kunz, J. Henle and B. W. Ninham, 'Zur Lehre von der Wirkung der Salze' (about the science of the effect of salts): Franz Hofmeister's historical papers, *Curr. Opin. Colloid Interface Sci.*, 2004, **9**(1–2), 19–37.
108 F. Hofmeister, Zur Lehre von der Wirkung der Salze, *Arch. Exp. Pathol. Pharmakol.*, 1888, **24**, 247.
109 R. L. Baldwin, How Hofmeister ion interactions affect protein stability, *Biophys. J.*, 1996, **71**(4), 2056–2063.
110 K. A. B. Dill and S. Bromberg, *Molecular Driving Forces: Statistical Thermodynamics in Chemistry & Biology*, Taylor & Francis, London, 2002.
111 A. W. Omta, M. F. Kropman, S. Woutersen and H. J. Bakker, Negligible effect of ions on the hydrogen-bond structure in liquid water, *Science*, 2003, **301**(5631), 347–349.

This journal is © The Royal Society of Chemistry 2009

112 J. D. Smith, R. J. Saykally and P. L. Geissler, The effects of dissolved halide anions on hydrogen bonding in liquid water, *J. Am. Chem. Soc.*, 2007, **129**(45), 13847–13856.
113 R. Mancinelli, A. Botti, F. Bruni, M. A. Ricci and A. K. Soper, Hydration of sodium, potassium, and chloride ions in solution and the concept of structure maker/breaker, *J. Phys. Chem. B*, 2007, **111**(48), 13570–13577.
114 L. M. Pegram and M. T. Record, Hofmeister salt effects on surface tension arise from partitioning of anions and cations between bulk water and the air/water interface, *J. Phys. Chem. B*, 2007, **111**(19), 5411–5417.
115 L. M. Pegram and M. T. Record, Partitioning of atmospherically relevant ions between bulk water and the water/vapor interface, *Proc. Natl. Acad. Sci. U. S. A.*, 2006, **103**(39), 14278–14281.
116 J. G. Kirkwood and J. B. Shumaker, Forces between protein molecules in solution arising from fluctuations in proton charge and configuration, *Proc. Natl. Acad. Sci. U. S. A.*, 1952, **38**(10), 863–871.
117 M. Bostrom, D. R. M. Williams and B. W. Ninham, Special ion effects: why the properties of lysozyme in salt solutions follow a Hofmeister series, *Biophys. J.*, 2003, **85**(2), 686–694.
118 J. S. Uejio, C. P. Schwartz, A. M. Duffin, W. S. Drisdell, R. C. Cohen and R. J. Saykally, Characterization of selective binding of alkali cations with carboxylate by X-ray absorption spectroscopyof liquid microjets, *Proc. Natl. Acad. Sci. U. S. A.*, 2008, **105**(19), 6809–6812.
119 E. F. Aziz, N. Ottosson, S. Eisebitt, W. Eberhard, B. Jagoda-Cwiklik, R. Vacha, P. Jungwirth and B. Winter, Cation-specific interactions with carboxylate in amino acid and acetate aqueous solutions: X-ray absorption and *ab initio* calculations, *J. Phys. Chem. B*, 2008, **112**, 12567–12570.
120 K. D. Collins, G. W. Neilson and J. E. Enderby, Ions in water: Characterizing the forces that control chemical processes and biological structure, *Biophys. Chem.*, 2007, **128**(2–3), 95–104.
121 J. Heyda, T. Hrobarik and P. Jungwirth, *J. Phys. Chem. A*, 2008, submitted.
122 P. E. Mason, J. W. Brady, G. W. Neilson and C. E. Dempsey, The interaction of guanidinium ions with a model peptide, *Biophys. J.*, 2007, **93**, L4–L6.
123 C. E. Dempsey, P. E. Mason, J. W. Bracly and G. W. Neilson, The reversal by sulfate of the denaturant activity of guanidinium, *J. Am. Chem. Soc.*, 2007, **129**(51), 15895–15902.
124 P. E. Mason, C. E. Dempsey, G. W. Neilson and J. W. Brady, Nanometer-scale ion aggregates in aqueous electrolyte solutions: Guanidinium sulfate and guanidinium thiocyanate, *J. Phys. Chem. B*, 2005, **109**(50), 24185–24196.
125 M. Lund, P. Jungwirth and C. E. Woodward, Ion specific protein assembly and hydrophobic surface forces, *Phys. Rev. Lett.*, 2008, **100**(25), 258105.
126 R. Piazza and M. Pierno, Protein interactions near crystallization: a microscopic approach to the Hofmeister series, *J. Phys.: Condens. Matter*, 2000, **12**(8A), A443–A449.
127 M. Lund and P. Jungwirth, Patchy proteins, anions and Hofmeister series, *J. Phys.: Condens. Matter*, 2008, submitted.

 This journal is © The Royal Society of Chemistry 2009

The surface of neat water is basic

James K. Beattie,* Alex M. Djerdjev and Gregory G. Warr

Received 28th March 2008, Accepted 28th May 2008
First published as an Advance Article on the web 18th September 2008
DOI: 10.1039/b805266b

Theoretical studies which conclude that the surface of neat water is acidic (with a pH $\leq$ 4.8), due to the preferential adsorption of hydronium ions, are contrary to the available experimental evidence. Air bubbles in water have a negative charge, as do hydrophobic oil drops in water, and streaming potential measurements on inert surfaces such as Teflon in water show a similar negative surface charge. In each case the pH dependence of the zeta potential has an isoelectric point between pH 2–4. An isoelectric point of pH 4 implies a preference for hydroxide over protons of 10^6, the opposite of what was inferred from the theoretical simulations. Water behaves similarly at all inert hydrophobic interfaces with the preferential adsorption of hydroxide ions to give a negatively charged surface at neutral pH. The surface-charge density at the oil/water interface in mM salt solutions is -5 to -7 μC cm^{-2}, which corresponds to one hydroxide ion on every 3 nm^2 of the surface. The homogenisation of an inert oil such as hexadecane in water in the absence of any salt or base still leads to formation of an emulsion. The hydroxide adsorbed on the large surface area of the emulsion greatly exceeds that present at 10^{-7} M in neutral water; it is created by the increased autolysis of water, driven by the strong adsorption of hydroxide ions at the oil/water interface. These surfactant-free, salt-free emulsions are stable for some hours, with protons as the only counterions to the negative hydroxide surface.

Jungwirth's simulations

Pavel Jungwirth and his co-authors posed a question as the title of a *Physical Chemistry Chemical Physics* article: 'Autoionization at the surface of neat water: is the top layer pH neutral, basic, or acidic?'[1] These same authors had already provided their answer to this question in an article in the *Proceedings of the National Academy of Science* entitled 'Water surface is acidic'.[2] Their conclusion was based on molecular dynamics simulations which showed no preference for hydroxide ions at the air/water interface, but a preference for hydronium ions to be at the interface rather than in the bulk—by a factor of at least 150. This corresponds to a surface pH at least as low as 4.8, with a lower bound of pH 1.9.

The authors acknowledge that their conclusion contradicts 'macroscopic measurements' on oil droplets and air bubbles that indicate a negative charge on these particles. Indeed, their conclusion is contrary to many pieces of experimental evidence that indicate conclusively that the surface of neat water is basic, charged negatively by the preferential adsorption of hydroxide ions.[3]

School of Chemistry, University of Sydney, NSW, 2006, Australia. E-mail: j.beattie@chem.usyd.edu.au; Fax: +61 (0)2 9351 3329; Tel: +61 (0)2 9351 3797

This journal is © The Royal Society of Chemistry 2009

pH dependence of zeta potentials

Gas bubbles

It has been known since 1861 that air bubbles in water migrate toward the positive electrode in an electrophoresis cell, and hence must be negatively charged.[4] In 1914 McTaggart used a cell rotating on its horizontal axis to counteract the buoyancy of the bubble and thus make more accurate measurements.[5] Graciaa *et al.* further improved this technique by eliminating the effects of electro-osmotic flow to claim the first reliable determination of the zeta potential of an air bubble: −65 mV.[6]

Others have used the horizontal deflection on the path of a rising bubble caused by an electric field to make similar measurements. In a recent study Takahashi measured a zeta potential of −100 mV at pH 10, which decreased with decreasing pH to −35 mV at pH 5.8 and reached an isoelectric point at about pH 4.[7] Other measurements have indicated that the isoelectric point could be as low as pH 2–3.[8] What is not in doubt, however, is that at near-neutral pH the gas bubbles are negatively charged. The negative charge is independent of the identity of the gas (for air, hydrogen, oxygen or nitrogen) as would be expected.[9] The zeta potential decreases with increasing electrolyte concentration, in the manner typical of double-layer compression.

Thin films

Another way to investigate the surface charge at the air/water interface is to measure the disjoining pressure of a thin film. At high pHs the charges on the interfaces keep the surfaces apart, but as the pH is lowered these repulsive forces decrease and the film ultimately collapses. Exerowa and co-workers showed that this occurs below pH 5.[10] Other workers have subsequently confirmed these observations.[11]

Oil drops

Drops of oil such as hexadecane with a low solubility in water behave similarly to air bubbles, although with some subtle differences. This means that general conclusions about the oil/water interface can be applied to the air/water interface. Thus it has long been known that oil drops in the absence of surfactants spontaneously acquire a negative charge and migrate towards the positive electrode in an electrophoresis cell. In 1938 Carruthers showed that the electrophoretic mobilities of solid octadecane particles and liquid octadecane droplets were almost identical, being negatively charged above an isoelectric point of about pH 2.5 in 0.01 M NaCl.[12] The pH dependence of the electrophoretic mobility, which increases from the isoelectric point to a limiting value above pH 8–10, suggested that hydroxide ion is responsible for the negative charge.[13] More recently, Marinova and colleagues showed, by measurements in 10^{-3} M Na_2CO_3 solution at pH 9.8, that hydrogen carbonate and carbonate ions are not responsible for this pH dependence. They found no difference in the zeta potential from an emulsion of xylene droplets at the same pH and ionic strength but in the absence of carbonate.[14] They showed further that the zeta potential decreased (became less negative) as the concentration of aqueous NaCl electrolyte was increased from 0.1 to 10 mM, consistent with typical double-layer compression, but not with adsorption of chloride ions. They also took care to exclude ionic and surface-active impurities. The consistent results obtained by different authors with different methods of preparing emulsion samples, with different oils, and the general agreement with previous observations made with different materials and different techniques indicate that adventitious contamination is not the cause of this effect.

A quantitative measurement of the adsorption of hydroxide ions at the oil/water interface was possible with experiments on surfactant-free emulsions of a few vol% oil-in-water.[15] As these emulsions are being formed by the homogenisation of oil in surfactant-free water containing $\sim 10^{-4}$ M NaCl, the adsorption of hydroxide ions at

 This journal is © The Royal Society of Chemistry 2009

the newly created surface is large enough to create a measurable change in the pH of the dispersion. The protons released by the formation of the hydroxide ions cause the pH to drop. (The opposite would occur if protons were preferentially adsorbed at the interface.) By adding base to keep the pH constant in a pH-stat experiment, the total surface charge of the emulsion can be obtained. If, at the same time, the size of the spherical drops is measured, the surface area can be calculated and hence the surface charge density obtained. This turns out to be −5 to −7 $\mu C\ cm^{-2}$ (−0.05 to −0.07 $C\ m^{-2}$) for emulsions of hexadecane, perfluoromethyldecalin or squalene formed at pH 7–9.

The large negative zeta potential of the hexadecane emulsion becomes less negative as acid is added, with an isoelectric point obtained by extrapolation of less than pH 4. The isoelectric point cannot be measured directly in this case because the emulsion would coalesce; its stability depends entirely on the surface charge created by the adsorbed hydroxide ions. The pH dependence of the zeta potential is essentially independent of the counter-anion in the system, whether it is chloride, iodide or perchlorate. This indicates that these anions do not compete with hydroxide at the interface in creating the surface charge.

Solid surfaces

Inert hydrophobic solids, with no ionisable groups, also develop a negative charge in contact with water. Indeed, this was explicit in the early work of Carruthers described above, who found that the zeta potentials of solid particles and liquid drops of octadecane were almost identical.[12] The pH dependence of this surface charge, measured by streaming potentials or streaming currents on a solid plane surface, displays the same characteristics as those shown by air bubbles and oil drops—a large negative value in alkaline solutions, decreasing to an isoelectric point at about pH 4.[16] The isoelectric point measured on a Teflon surface was independent of the KCl concentration, indicating that chloride ion was not responsible for the negative charge. This spontaneous creation of a surface charge, presumably by the adsorption of hydroxide ion, accounts for the otherwise puzzling observation of electro-osmotic flow in microfluidic channels formed by uncharged substrates such as Teflon, polyethylene and poly(vinyl chloride).[17]

Self-assembled monolayers on solid supports can also be used to create hydrophobic surfaces. Both a C_{18}-trimethoxysilane on silica[18] and a C_{18}-thiol on gold[19] show negative surface charges at neutral pH with an isoelectric point at about pH 4.

Hydroxide adsorption

The pH dependence of the zeta potentials of air bubbles, oil drops and inert solid surfaces led naturally to the explanation that hydroxide ion is responsible. Thus, among others, McTaggart wrote in 1914 about gas bubbles: "since in water a surface is always negative, one may suppose the existence of a selective adsorption of OH^- ions".[5] In 1941 Dickinson wrote: "It has long been known that droplets of oil, dispersed in water, carry an electrical charge which is usually negative. This charge is thought to be due to the adsorption of hydroxyl ions, and gives rise to an electrical double layer across the interface."[13] In 1969 Exerowa claimed that the potential of the diffuse double layer affecting thin films "is formed by the adsorption of OH^- ions".[10] The title of the paper in 1996 by Marinova *et al.* reads "Charging of oil/water interfaces due to spontaneous adsorption of hydroxyl ions."[14] This idea has been repeated more frequently recently.

In response to criticism of their claim that the air/water interface is acidic, rather than basic in accord with the macroscopic experimental evidence, Jungwirth and co-authors advanced the idea that the structure of the double layer contains protons at the surface 'in the topmost layer', in accordance with their calculations, and hydroxide ions towards the bulk in a subsurface layer.[20] They then suggest that an

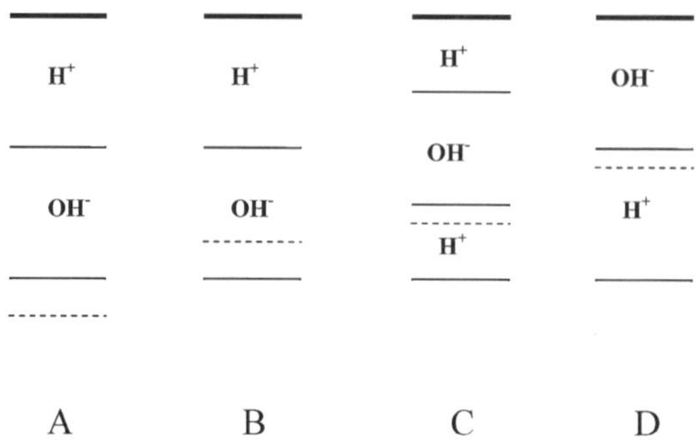

Fig. 1 Alternative arrangements of protons and hydroxide ions in an electrically neutral, salt-free double layer at the air/water interface with air at the top. The dashed line denotes the shear plane.

ambiguity in the interpretation of the calculations and the experimental results arises, since "due to the neutrality of the whole interface, techniques scrutinising it from the vapour side can see it as acidic, while approaching it from the aqueous bulk side it will appear as basic". This explanation regrettably contains a fundamental error in colloid science, as can be illustrated in Fig 1.

Because of electroneutrality, the number of protons must equal the number of hydroxide ions in the double layer. If the shear plane, from which the zeta potential is measured, is outside the double layer, as indicted by the dashed line in Fig 1A, then the zeta potential would be zero; the system would always be at the isoelectric point, contrary to the experimental observations. But if the shear plane were inside the hydroxide layer, as in Fig 1B, then the net charge inside the shear plane would be positive and the zeta potential in neutral solution would also be positive, again contrary to experiment. The only way to have protons in the 'topmost layer' and a negative zeta potential is to postulate a triple-layer structure (Fig 1C)—some protons on top, then hydroxide ions to dominate the net charge to give a negative zeta potential, and the remaining protons in the diffuse layer to satisfy electroneutrality. But this elaboration of the original explanation is an *ad hoc* artifice. A simpler explanation is to accept that the molecular dynamics calculations are somehow flawed and that the double layer exists as illustrated in Fig 1D—with hydroxide ions adsorbed on the surface and protons in the diffuse layer—consistent with the available experimental evidence.

Surface spectroscopy

One criticism of the surface hydroxide hypothesis is that there is no spectroscopic evidence for an enhanced concentration of hydroxide ions at the interface. Surface spectroscopic techniques can selectively interrogate a surface for particular species, and do detect the presence of hydronium ions at the air/water interface in acidic solutions. One of the errors in the arguments of Jungwirth and co-authors was to extrapolate this evidence from acidic solutions to the properties of the interface of near-neutral water.[21] In fact, the evidence for the presence of hydronium ions at the interface appears only below pH 2,[22] consistent with the evidence of an isoelectric point between pH 2–4, with a positive charge in more acidic conditions and a negative charge in near-neutral and basic conditions.

There is some spectroscopic evidence for the presence of surface hydroxide above pH 13,[22] but none between pH 2 and 13. This does not refute the hydroxide adsorption hypothesis, however, but may only indicate that any spectroscopic signal for surface hydroxide is weak. It may be difficult to distinguish it from water, as the surface contains adsorbed hydroxide over most of the pH range.

 This journal is © The Royal Society of Chemistry 2009

There is also surface spectroscopic evidence for the adsorption of some anions at the interface.[23] Their adsorption is weaker than that of hydroxide, however, as can be inferred from the following comparison. The adsorption of iodide at low concentrations at the air/water interface follows a Langmuir adsorption isotherm with a half-adsorption concentration of about 10^{-3} M.[24] In contrast, the adsorption of hydroxide ion on oil, from the pH dependence of the zeta potential, reaches half-saturation at about pH 5, or at 10^{-9} M hydroxide ion concentration, six orders of magnitude lower. The iodide is adsorbing at an interface already charged by hydroxide ions over most of the pH range. This explains why the zeta potential of the oil drops is independent of the identity of the anion,[15] since, as described above, the surface charge is determined largely by the hydroxide ion and only slightly by other anions.

Emulsions stabilised by the autolysis of water

The pH-stat experiments mentioned above were conducted in 0.2 mM NaCl solutions, so the sodium cation served as the counterion to the hydroxide ion in the double layer. The amount of base added to maintain the pH at 9 or 7 was of the order of 10^{-4} M, far greater than that available at these pHs. This suggests that when the new oil/water surface is formed during the homogenisation, and adsorbs a hydroxide ion, that hydroxide ion is formed by the autolysis of water. The addition of base then neutralises the protons formed by this autolysis. This further suggests that it might be possible to form an emulsion with hydroxide ions producing the surface charge and hydronium ions serving as the counterions, that is, an emulsion stabilised solely by the autolysis of water.

When 2 vol% of hexadecane is passed through a high-pressure homogeniser with water under a N_2 atmosphere at pH 7, the resulting dispersion becomes white and remains so for at least several hours (Fig 2). Examination of this dispersion with back-scattered light indicates an emulsion with an average droplet size of about 1 μm. An electroacoustic signal is generated by the dispersion; the dynamic mobility spectrum indicates that the droplets are negatively charged (Fig. 3). The Acousto-Sizer-II software gives a good fit to the spectrum for the 2 vol% hexadecane sample

Fig. 2 An emulsion prepared by homogenisation of 2 vol% hexadecane in water in the absence of any surfactant or salt, stabilised by the autolysis of water.

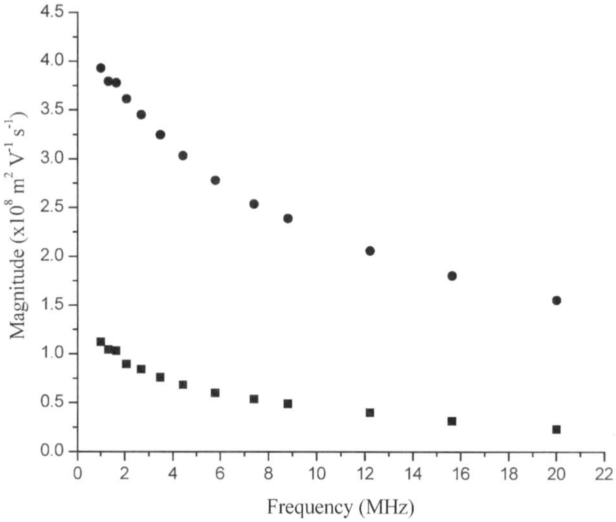

Fig. 3 The magnitudes of the dynamic mobility spectrum for a salt-free, 2 vol% hexadecane-in-water emulsion as formed (lower curve) and after addition of NaOH(aq) to pH 9 (upper curve).

with a log-normal median size of 1.1 μm and a zeta potential of −20 mV, although these parameters must be viewed with caution due to the very low ionic strength of the dispersion, with the consequent thick double layer.

Unlike the pH-stat experiments in mM salt solutions, pH is not a useful parameter in these salt-free dispersions. Measurement of pH in water with very low conductivity is problematic, and the effect of the double layer on the activity of the hydronium ions adds an additional complication. Since charge separation is occurring as the emulsion is formed, with hydroxide ions on the surface and protons in the diffuse layer, the conductivity of the dispersion should increase. This was indeed observed with a conductivity at the beginning of the homogenisation of 7.6×10^{-6} S m^{-1} (only slightly higher than that of pure water) increasing to 7.4×10^{-5} S m^{-1} at the end of the homogenisation. Quantitative interpretation of this result is also problematic, again because of the uncertain effects of the double layer.

The hydronium counterions can be replaced with Na^+ by addition of NaOH. As NaOH solution is added to the dispersion the zeta potential becomes more negative and the pH becomes measurable. To reach pH 9 requires 6×10^{-5} M NaOH in excess of the 1×10^{-5} M NaOH necessary to adjust the pH from 7 to 9 (Fig 4). Examination of the emulsion at pH 9 by light back-scattering indicated that the droplet size had not altered. The dynamic mobility spectrum displayed an increased amplitude to give a zeta potential of −63 mV, again with a droplet size of 1.1 μm (Fig 3).

The surface area of this emulsion calculated from the diameter of 1.1 μm is about 1.1×10^6 cm^2 L^{-1}. (It is this large surface area of the relatively concentrated emulsion that allows quantitative titration of the surface charge.) The NaOH required to neutralise these hydronium ions and replace them in the double layer by Na^+ indicates a surface charge density in the original emulsion of 5 μC cm^{-2}. The same surface charge density was measured previously when the emulsion was prepared directly in 0.2 mM NaCl at pH 9.[15] Once the pH is increased to pH 9 the emulsion is stable for days. The comparison with the emulsions prepared directly at pH 9 is only semiquantitative, however, for those made with only protons initially in the double layer are more polydisperse and with larger drops, a consequence of their lower zeta potential.

 This journal is © The Royal Society of Chemistry 2009

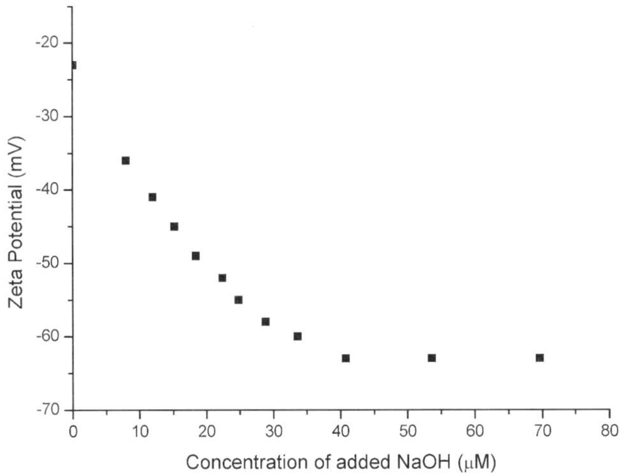

Fig. 4 The increase in the negative zeta potential of the salt-free, hexadecane-in-water emulsion as the concentration of NaOH is increased to pH 9.

These are remarkable emulsions. The input of energy from the homogeniser creates new surface area to which hydroxide ions are strongly adsorbed. The hydroxide ions are formed by the autolysis of water, leaving protons as the only counterions in the diffuse double layer. The homogenisation is less efficient than when it is conducted in dilute salt solutions and at higher pH, because the emulsions stabilised solely by the autolysis of water are more susceptible to coalescence. Surprisingly, however, charge separation occurs and protons and hydroxide ions exist in close proximity without recombination. This phenomenon has already been suggested to interpret the surface conductivity of water at Teflon surfaces, described below, with both protons and hydroxide ions claimed to reside in the stagnant layer.[16]

The observation of salt-free, surfactant-free emulsions corroborates that of Pashley[25] that such can be formed, but contradicts his claim that degassing of the water is required. We have not investigated whether degassing would lead to increased stability of these emulsions. Our results support the interpretation of Eastoe and co-workers who showed that degassing was not responsible for the observation of turbidity in the oil/water mixtures, but that a freeze–thaw cycle was necessary. Presumably the freeze–thaw process provides the necessary energy input to cause a small amount of homogenisation.[26]

Conclusions

All of the above evidence leads inexorably to the unambiguous conclusion that the surface of neat water in air is basic, not acidic. Hydroxide ions are very strongly adsorbed; the negative surface charge that produced is neutralised only by protons at pH 2–4. They are so strongly adsorbed that it enhances the autolysis of water: surfactant-free, oil-in-water emulsions can be formed in neat, salt-free water where the negative surface charge of the hydroxide is balanced by an equal charge of protons—water is acting as the surfactant to stabilise the oil drops against coalescence.

Surface charge

The surface charge density at the interface of water with insoluble oil drops is −5 to −7 μC cm^{-2}. This has not yet been measured for gas bubbles, but the similarity of the pH dependence of the zeta potentials of oil drops and gas bubbles suggests that it will be about the same. For the oil drops much of the charge is compensated inside of the shear plane by counterion condensation. For hexadecane emulsions in 0.2 mM NaCl the diffuse layer charge calculated from the zeta potential is only $-0.6\ \mu C\ cm^{-2}$. For thin air/water films in 0.3 mM electrolyte in the presence of non-ionic surfactants a charge density calculated from the disjoining pressure is $-0.25\ \mu C\ cm^{-2}$,[11] *i.e.* of a similar magnitude.

Zimmerman and co-workers have used a clever microslit electrokinetic apparatus to estimate the surface charge at the water interface with Teflon.[16] By measuring the streaming currents and potentials in channels of varying thicknesses from 1 to 60 μm, the contribution of the surface conductivity can be separated from the total channel conductivity. When the surface conductivity is combined with zeta potential measurements it is then possible to evaluate both the excess charge and the compensated charge in the stagnant layer at the surface. In neutral solution in 0.1 mM KCl the total surface charge is $-0.39\ \mu C\ cm^{-2}$, with about one-third compensated, while in KOH at pH 9 the total surface charge is slightly larger ($-0.46\ \mu C\ cm^{-2}$), with about one-quarter compensated. These surface charges are lower limits, as they are obtained by assuming that the ion mobilities in the stagnant layer are the same as in the bulk, but there is evidence that the mobilities are somewhat smaller. More evidence is needed on this question. There may be a difference between a solid and a fluid interface; this could be related to the observation that the isoelectric points for the solids are clearly about pH 4, while those for oils and gases appear to be <3. Nevertheless, the general conclusion from the work on Teflon surfaces is that hydroxide ion is preferentially adsorbed by a factor of 10^6 over protons, and that potassium and chloride ions do not compete. Furthermore, these authors interpret data on the surface conductivity of 10^{-5} M HCl solutions as indicating the presence of both hydroxide ions and protons in the stagnant layer on Teflon.

Theories

The results described above indicate why attempts to model the air/water interface with molecular dynamics simulations have been futile. The simulation boxes are just too small to capture the electrostatic effects created by hydroxide ions on every 3 nm^2 (300 $Å^2$) of surface. A simulation described in 2004 as 'exceedingly large-scale' and 'a significant leap forward in size' used 216 waters in a box $15 \times 15 \times 71$ Å.[27] The work of Jungwirth and co-authors used 72 waters in a box $13 \times 16 \times 11$ Å.[2] Neither of these is quite large enough to capture the area occupied by even one hydroxide ion.

There have been numerous speculations about how and why hydroxide ions are attracted to the interface, but none seem to have provided a convincing molecular description. Many depend on models of water at the interface that ignore the presence of the strong electrostatic effects of the hydroxide ions, and hence are initially flawed. A hydroxide ion every 3 nm^2 is separated from the next by less than 20 Å or 6–8 water molecules. The hydrogen-bonded structure of every water at the interface will be affected by this charged species. A successful model of water at the air/water interface must begin with a charged surface, not a neutral one.

Experimental methods

Hexadecane was purified by five passages through a column of basic alumina. Freshly boiled Milli-Q water was protected from CO_2 absorption with a guard tube containing soda lime until it was used in the homogenisation. Emulsions were prepared by passage of the oil/water mixture through a four-stage piston homogeniser as described previously.[15] Dynamic mobility spectra were measured on a prototype of the AcoustoSizer-II (Colloidal-Dynamics) and light back-scattering measurements were made with a Malvern HPPS instrument.

 This journal is © The Royal Society of Chemistry 2009

Acknowledgements

This work was supported by the Australian Research Council. We thank Colloidal-Dynamics Pty Ltd for use of the AcoustoSizer-II.

References

1 R. Vácha, V. Buch, A. Milet, J. P. Devlin and P. Jungwirth, *Phys. Chem. Chem. Phys.*, 2007, **9**, 4736.
2 V. Buch, A. Milet, R. Vacha, P. Jungwirth and J. P. Devlin, *Proc. Natl. Acad. Sci. U. S. A.*, 2007, **104**, 7342.
3 J. K. Beattie, The Intrinsic Charge at the Hydrophobe/Water Interface, in *Colloid Stability—The Role of Surface Forces—Part II*, ed. T. Tadros, Wiley-VCH, Weinheim, 2007, pp. 153–164.
4 G. Quincke, *Ann. Phys. Chem.*, 1861, **113**, 513.
5 H. A. McTaggart, *Philos. Mag.*, 1914, **27**, 297.
6 A. Graciaa, G. Morel, P. Saulner, J. Lachaise and R. S. Schechter, *J. Colloid Interface Sci.*, 1995, **172**, 131.
7 M. Takahashi, *J. Phys. Chem. B*, 2005, **109**, 21858.
8 C. Li and P. Somasundaran, *J. Colloid Interface Sci.*, 1991, **146**; A. Graciaa, P. Creux, and J. Lachaise, Electrokinetics of Bubbles, in *Encyclopedia of Surface and Colloid Science*, ed. A. T. Hubbard, Marcel Dekker, New York, 2002, vol. 2, pp. 1876–86.
9 P. Creux, J. Lachaise, A. Graciaa and J. K. Beattie, *J. Phys. Chem. C*, 2007, **111**, 3753.
10 D. Exerowa, *Kolloid Z. Z. Polym.*, 1969, **232**, 703.
11 K. A. Karraker and C. J. Radke, *Adv. Colloid Interface Sci.*, 2002, **96**, 231.
12 J. C. Carruthers, *Trans. Faraday Soc.*, 1938, **34**, 300.
13 W. Dickinson, *Trans. Faraday Soc.*, 1941, **37**, 140.
14 K. G. Marinova, R. G. Alargova, N. D. Denkov, O. D. Velev, D. N. Petsev, I. B. Ivanov and R. P. Borwankar, *Langmuir*, 1996, **12**, 2045.
15 J. K. Beattie and A. M. Djerdjev, *Angew. Chem., Int. Ed.*, 2004, **43**, 3568.
16 R. Zimmermann, S. Dukhin and C. Werner, *J. Phys. Chem. B*, 2001, **105**, 8544.
17 J. K. Beattie, *Lab Chip*, 2006, **6**, 1409.
18 A. Hozumi, H. Sugimura, Y. Yokoygawa, T. Kameyama and O. Takai, *Colloids Surf., A*, 2001, **182**, 257.
19 R. Schweiss, P. B. Welzel, C. Werner and W. Knoll, *Langmuir*, 2001, **17**, 4304.
20 R. Vacha, V. Buch, A. Milet, J. P. Devlin and P. Jungwirth, *Phys. Chem. Chem. Phys.*, 2008, **10**, 332.
21 J. K. Beattie, *Phys. Chem. Chem. Phys.*, 2008, **10**, 330.
22 T. L. Tarbuck, S. T. Ota and G. L. Richmond, *J. Am. Chem. Soc.*, 2006, **128**, 14519.
23 P. B. Petersen and R. J. Saykally, *J. Am. Chem. Soc.*, 2005, **127**, 15446.
24 P. B. Petersen, J. C. Johnson, K. P. Knutsen and R. J. Saykally, *Chem. Phys. Lett.*, 2004, **397**, 46.
25 R. M. Pashley, *J. Phys. Chem. B*, 2003, **107**, 1714.
26 G. R. Burnett, R. Atkin, S. Hicks and J. Eastoe, *Langmuir*, 2004, **20**, 5673; J. Eastoe and C. Ellis, *Adv. Colloid Interface Sci.*, 2007, **134–135**, 89.
27 I. F. W. Kuo and C. J. Mundy, *Science*, 2004, **303**, 658.

Negative charges at the air/water interface and their consequences for aqueous wetting films containing surfactants

Katarzyna Hänni-Ciunel, Natascha Schelero and Regine von Klitzing*

Received 29th May 2008, Accepted 30th June 2008
First published as an Advance Article on the web 9th October 2008
DOI: 10.1039/b809149h

Negative charges at the air/water interface have been proven previously by the authors *via* disjoining pressure measurements of wetting films. In the present study the surface charges of wetting film are modified in order to control the film stability. Studies at different pH values support the assumption that the negative charges are caused by an excess of hydroxide ions. Addition of different types of salt anions seems to lead to an increase in surface charges with decreasing hydration shell. Both water-soluble and water-insoluble amphiphiles replaces the charges. The air/water surface remains negatively charged in case of non-ionic surfactant (dodecyl-β maltoside, $C_{12}G_2$) and anionic ampihiphiles (stearic acid and SDS). The charge will be reversed in case of cationic amphiphiles. Under all of those conditions, films against likely charged solid surfaces are stable and unstable against oppositely charged solid substrates. For this purpose, the charge of silicon wafers was precoated with polyelectrolytes before. At concentrations well below the CMC cationic surfactants are not able to reverse the charges at the air/water interface. Respective wetting films are less stable and show a partial dewetting even on likely charged surfaces. These results lead to the conclusion that the films are stabilised by electrostatic repulsion. Potential adsorption of water-soluble surfactants on oppositely charged surfaces is also taken into account and is studied by contact-angle measurements.

1 Introduction

Wetting phenomena play an important role in many technical applications like flotation and cleaning processes of surfaces. Thereby the stability of wetting films between the vapor phase and the solid phase plays a decisive role. It depends on the forces acting between its two interfaces and is characterised by the disjoining pressure (Π) and its dependence on film thickness. Generally, the disjoining pressure in the liquid film is dominated by long-range repulsive electrostatic interactions and a repulsive van der Waals contribution, both contributing to the classical Derjaguin, Landau, Verwey and Overbeek (DLVO) theory.[1,2] In experimental studies of film stability, different techniques have been developed, depending on the type of studied system. For the investigation of interfacial forces in thin liquid films the *thin-film pressure balance* (TFPB) in various modifications[3,4] has proved to be a suitable technique. With this method the disjoining pressure isotherm (Π *vs.* film thickness h) of a mechanically stable film can be recorded. The TFPB is used for free-standing foam

Stranski-Laboratorium für Physikalische und Theoretische Chemie, Institut für Chemie, Technische Universität Berlin, Strasse des 17. Juni 124, D-10623 Berlin, Germany. E-mail: klitzing@chem.tu-berlin.de; Fax: +49 (0)30 31426602; Tel: +49 (0)30 31423476

This journal is © The Royal Society of Chemistry 2009

films (air/liquid/air). Since foam films are symmetric films, the surface charge cannot be determined with this kind of measurement.

In the present work we have used a modified TFPB technique for wetting films (air/liquid/solid) to probe the existence of surface charge at air/water interfaces and to control it. The strategy is to measure aqueous wetting films against silicon surfaces, and to modify the surface charge of both surfaces. The solid silicon surface will be coated with different polyelectrolytes. In order to change the charge at the air/water interface, surfactants of different charge will be added to the solution from which the wetting film is formed. This gives one the opportunity to determine the sign of the charge at the air/water interface but also to control the stability of the films by independent modification of surface charges.

Former studies of aqueous wetting films showed that they are stable on negatively charged surfaces and unstable on positively charged surfaces.[5,6] Coatings by polyelectrolytes on solid substrates show a tendency for a higher contact angles for polycation-terminated coatings (25–50°) in comparison to polyanion coated ones (0–30°).[7] Hence, the surfaces are hydrophilic in both cases and the difference cannot explain the strong stability of water films on negatively charged surfaces and the instability on positively charged surfaces. Also the orientation of water molecules measured by near-edge X-ray absorption fine-structure (NEXAFS)[8] or sum-frequency generation (SFG) spectroscopy[9] and the resulting dipole cannot decide on stabilisation or destabilisation over several 10s of nm. Recent molecular dynamics (MD) simulations of the free surface of water suggest that the molecules in the top-most layer are oriented mainly with the hydrogen atoms upwards, *i.e.* pointing away from the aqueous half-space. This orientation causes a surface dipole moment and a surface potential of the order of +500 mV for the free liquid surface.[10]

The results of former studies on aqueous wetting films on coated silicon surfaces indicate that the interactions within the film are dominated by electrostatic forces. That is also confirmed by the screening effect with increasing ionic strength.[5] Electrostatic repulsion within the films can only be explained by "free" charges at the air/water interface, but so far their origin is still under debate. The oil/water interface is a hydrophobic interface like the air/water interface and is also assumed to be negatively charged. The negative zeta-potential value of oil droplets depends on the pH and is attributed to the specific location of hydroxide ions at the oil/water interface.[11,12]

In the present paper the charge of the solid and the liquid interface are modified in order to control the stability of wetting films. The silicon surface is modified *via* adsorption of polyelectrolyte multilayers with different sign of charge of the outermost layer. The charge of the liquid surface is varied by changing pH and the type of anion and by adding amphiphiles of different sign of charge and hydrophobicity.

2 Experimental

2.1 Modification of the solid surface

For reasons of better contrast, silicon wafers instead of glass wafers were used as solid substrates.[13] Either they were used without any further coating or the surface was modified by the adsorption of polyelectrolyte layers according to the layer-by-layer method invented by Decher and others in the early 90s.[14–17] Polyethylene imine (PEI), polyallylamine hydrochloride (PAH) and poly(styrene sulfonate) sodium salt (PSS) were obtained from Aldrich (Steinheim, Germany). The molecular weight of PEI is 750.000 g mol^{-1} and 70.000 g mol^{-1} in the case of PSS. NaCl was purchased from Merck. The silicon wafers were provided by Wacker Siltronic AG, Burghausen (Germany) and cleaned for 30 min in a 1 : 1 H_2O_2–H_2SO_4 mixture. The polyelectrolyte films were deposited on the silicon wafers by immersion for 20 min into aqueous solutions containing 10^{-2} monomol l^{-1} (concentration of monomer units) of the respective polyelectrolyte and by rinsing with Milli-Q-water after each deposition

 This journal is © The Royal Society of Chemistry 2009

step. Either a single PEI or PAH layer or a double layer of PEI and PSS was deposited. The films were dried in an air stream after completion of the deposition.

2.2 Thin-film pressure balance (TFPB)

The interaction within the wetting and foam films were studied by a modified thin-film pressure balance (TFPB) method, using the porous-plate technique.[13,6] The film is formed in the hole drilled into a porous glass plate. The plate is enclosed in a hermetically sealed stainless steel cell and the film holder is connected to the atmospheric pressure by a capillary tube. Before the film is formed the film holder was immersed into the respective aqueous solution for 2 h. After pulling the film holder out of the solution at atmospheric pressure a so-called droplet of the solution is formed on the hole of the fritted glass disc. After closing the cell and before starting the measurement, the system was equilibrated for 30 min. Then, the disjoining pressure Π as a function of film thickness h is determined (disjoining pressure isotherm).[18–20] Usually the TFPB method is used for the investigation of free-standing films (foam films). For studies of wetting films the TFPB apparatus is adapted by attaching a solid substrate (the bare or modified silicon wafer) to the bottom side of the porous glass plate by adhesion.[5] A thin liquid film is formed on the top side of the wafer by adjusting the gas pressure within the cell by a piston pump. The film thickness is measured interferometrically[3,20] by white light reflected from the top and bottom surface of the film. The light beam passes through an objective and an interference filter (550 nm) and is detected by a photomultiplier. Simultaneously, the thinning of the film is recorded by video microscopy.

2.3 Preparation of wetting films

Water-soluble surfactants studied in the present paper are dodecyl-β maltoside ($C_{12}G_2$) from Glycon (Luckenwalde), dodecyltrimethylammonium bromide ($C_{12}TAB$) and dodecylsulfonate (SDS) from Sigma-Aldrich. The respective concentrations in water were below the critical micellar concentration (cmc) and were 5×10^{-5} M in the case of $C_{12}G_2$, 10^{-4} M for $C_{12}TAB$ and 3×10^{-4} M in the case of SDS. 0.1 M NaCl was added to each solution in order to obtain a fixed ionic strength.

In case of the water-insoluble amphiphiles the preparation of the wetting films was more complicated. In analogy to Langmuir films, dioctadecyldiammonium bromide (DODAB) or stearic acid (both from Sigma-Aldrich) were dissolved in chloroform (Merck, Darmstadt, Germany) and spread at the air/water interface of the water droplet formed on the hole of the porous glass disc. After evaporation of the chloroform a densely packed amphiphile layer was built. In order to determine the amphiphile concentration for getting a densely packed amphiphile layer, the respective Langmuir isotherms were taken into account. They were measured by Symnitz in the group of Brezesinski (Max Planck Institute of Colloid and Interfaces, Potsdam, Germany). The Langmuir isotherms (lateral pressure *vs.* area) for both amphiphiles have been measured with the Langmuir balance (Riegler & Kirstein, Potsdam, Germany) at constant temperature of 20 °C. According to the isotherm data and X-ray diffraction measurements, above the pressure of 15 mN m^{-1} DODAB forms a two-dimensional crystalline phase on the liquid surface with the alkyl chains tilted by approx. 40° with respect to the surface normal[21,22] and forms a stable monolayer up to 40 mN m^{-1}. From the isotherms, the pressure of 25 mN m^{-1}, which creates the densely packed DODAB monolayer, and 40 mN m^{-1}, for the stearic acid in the condensed state, are chosen. At this pressure the area per molecule is 55.2 Å^2 for DODAB and 22.5 Å^2 for stearic acid.

2.4 Contact-angle measurements

The contact angle of water on coated and bare substrates was measured by the sessile drop method using an OCA drop 15 plus instrument from Dataphysics (Filderstadt,

Table 1 Contact angle (Θ) of H_2O droplet on differently coated silicon wafers before and after the exposure to different surfactants. The net charge of the respective outermost layer of the coating and the respective surfactant at pH 5.5 are given in brackets

	Surfactant treatment			
Outermost layer	Untreated surface	$C_{12}G_2$	$C_{12}TAB$ (+)	SDS (−)
PEI (+), Θ/°	27 ± 1	44 ± 5	28 ± 3	37 ± 4
PSS (−), Θ/°	28 ± 3	32 ± 2	34 ± 2	30 ± 3
PDADMAC (+), Θ/°	46 ± 2	48 ± 2	46 ± 2	53 ± 6
Bare Si (−), Θ/°	0	24 ± 3	32 ± 2	24 ± 5

Germany). Measurements were made at room temperature in a closed quartz glass cell, which contained a reservoir of water to ensure a water-saturated atmosphere. Generally, the substrates were incubated in the chamber for 15 min before deposition of the water droplets. Water droplets (Millipore-Q) of *ca.* 10 μl were then placed on the substrate by means of a syringe through a small hole in the Teflon lid of the cell tightly fitting the needle of the syringe. The contact angle of sessile drops was determined using an image-analysis program. Independent measurements were made for 3 droplets on each of 3 newly prepared surfaces, and the final mean value was evaluated.

The contact-angle measurements were done on bare or coated silicon wafers. To check if some surfactant molecules could adsorb at the silicon/liquid interface during the TFPB measurements, the contact angle of a water droplet was measured either directly after wafer preparation (referred to as "untreated surface" in Table 1) or after the additional immersion into the respective surfactant solution for 30 min (referred to as "surfactant treatment" in Table 1). The time period of 30 min corresponds to the contact time of the wafer with the respective solution in a TFPB experiment, before the TFPB measurement is started. Then, the wafer is rinsed with water and dried with a gentle active-carbon-filtered air stream.

3 Results

3.1 Water wetting films

In contrast to foam films, wetting films of pure water can be stable. Fig. 1 shows the isotherm of a pure water film (open diamonds) against a silicon substrate. The water has a pH of about 5.5. At this pH, the silicon surface is negatively charged. For the following measurements at different pH the ionic strength was fixed at 10^{-3} M by addition of an excess of NaCl. At pH 8.9 the film has the largest thickness, and the film becomes thinner with decreasing pH. At pH 4.3 the film breaks already at a disjoining pressure of 400 Pa. It should be noted that the pure water film (pH 5.5) has almost the same thickness as the film at pH 8.9 and an ionic strength of 10^{-3} M. Obviously, the effect of increasing pH compensates the effect of increasing ionic strength.

Fig. 2 shows disjoining pressure isotherms at a fixed ionic strength of 10^{-4} M, but different salt anions. The cation was sodium in all three cases, but the anion was changed from iodide *via* chloride to fluoride. The thickness of the NaI-containing film is the thickest and with NaF the thinnest. The Debye length varies between 25 and 30 nm. The theoretical Debye length of the corresponding bulk solution is about 30 nm.

3.2 Comparison of a $C_{12}G_2$ foam film and a $C_{12}G_2$ wetting film

In Fig. 3 the disjoining pressure isotherms of an aqueous free-standing foam film and a wetting film on silicon are shown. Both films are stabilised by 5×10^{-5} M $C_{12}G_2$

 This journal is © The Royal Society of Chemistry 2009

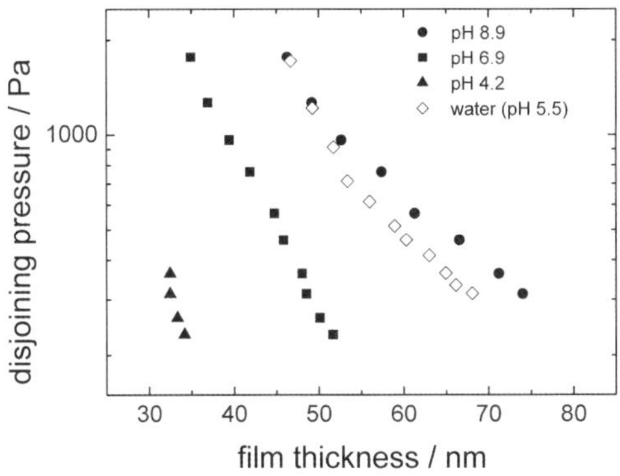

Fig. 1 Disjoining pressure isotherms of aqueous wetting films at different pH. The ionic strength was fixed at 10^{-3} M. For comparison the pure water film (open diamonds) is shown.

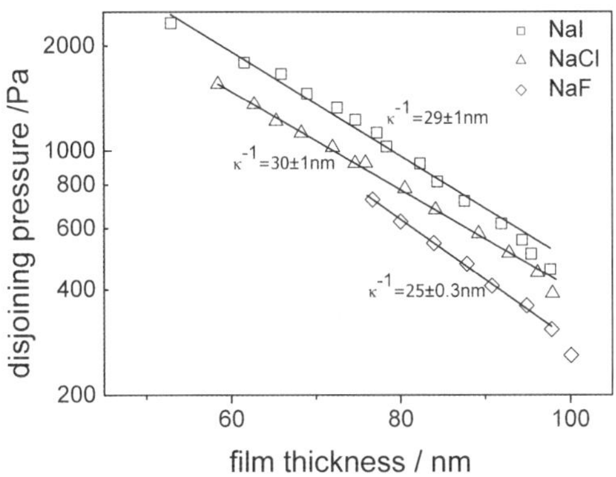

Fig. 2 Disjoining pressure isotherms of aqueous wetting films after addition of different types of salt at a fixed ionic strength of 10^{-4} M.

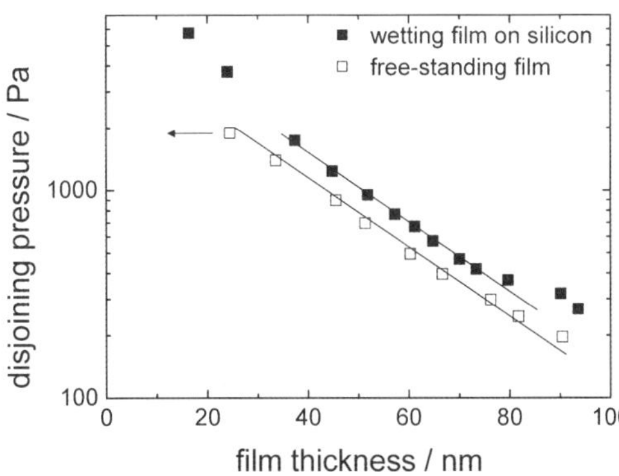

Fig. 3 Disjoining pressure isotherms of free-standing film (open symbols) and wetting film on silicon surface (filled symbols) formed from aqueous solutions containing 5×10^{-5} $C_{12}G_2$ and 10^{-4} M NaCl. The arrow indicates the rupture of the free-standing film. The solid lines are fits by a simple exponential decay leading to a Debye length of about 26 nm for the foam film and about 27 nm for the wetting film.

and 10^{-4} M NaCl. The data points are averages of three separate experiments. The wetting film is stable at pressures above 6000 Pa. It couldn't be destroyed in the experiment, while the foam film ruptured at about 2000 Pa, marked with an arrow in Fig. 3. The wetting film is about 10 nm thicker than the foam film. The thickness of both types of film is in the typical range for common black films (CBF). The logarithmic plot shows a linear decay, which is a an indication of stabilisation by electrostatic double layer forces. The electrolyte NaCl used at a concentration 10^{-4} M gives one the opportunity to have a known salt concentration for further calculations. The foam film isotherm can be fitted until its rupture at a thickness of about 25 nm by a simple exponential decay. The wetting film can be fitted by the same function down to a thickness of 40 nm, but at smaller thicknesses the disjoining pressure isotherm shows a larger slope.

The fits lead to a Debye length of 27 nm for the wetting film and of 26 nm for the foam film, which are both in quite good agreement with the theoretical value of about 30 nm calculated for an ionic strength of about 10^{-4} M. Since the experiments were carried out at a pH of about 5.5 (water), one can assume that the bare silicon surface is negatively charged. That would mean that the opposing liquid/air surface of the wetting film should be also negatively charged. Hence the same film should rupture on a positively charged surface, which was studied in the following.

3.3 Wetting surfactant films on modified silicon surfaces

3.3.1 Non-ionic surfactant. The charge modification of the silicon surface was achieved by the layer-by-layer technique.[14] A positive surface charge was obtained by adsorption of PEI. Branched PEI was always used for generating the first adsorbed layer on a cleaned silicon surface, since it is well-known for having a pronounced adhesion of all following polyelectrolyte layers onto the solid surface. An explanation is that the branched PEI brings a lot of charges onto the surface.[17] In order to compare the wetting behavior on other polycations, wetting films against PEI/PSS/PAH or against PEI/PSS/PDADMAC were investigated. PSS-terminated coatings lead to a negative charge of the solid surface. Exemplarily for wetting films against polycation-terminated coatings, the rupture of an aqueous $C_{12}G_2$ film against a PEI-coated silicon surface is shown in Fig. 4a. Also, on the other polycation-terminated coatings the $C_{12}G_2$ films were unstable. On a PSS-terminated coating like PEI/PSS the $C_{12}G_2$ film is stable (Fig. 4b). Aqueous $C_{12}G_2$ wetting films on negatively charged surfaces (silicon or PSS) drain regularly. The disjoining pressure isotherm of a stable $C_{12}G_2$ wetting film on silicon with PSS as outermost layer is shown in Fig. 5 and it is compared with the isotherm for surfactant wetting film on a bare silicon surface. The disjoining pressure isotherms of both wetting films become again steeper for thin films at thicknesses below 40 nm. Therefore not all of the curve could be fitted by a simple exponential decay. The isotherm of the

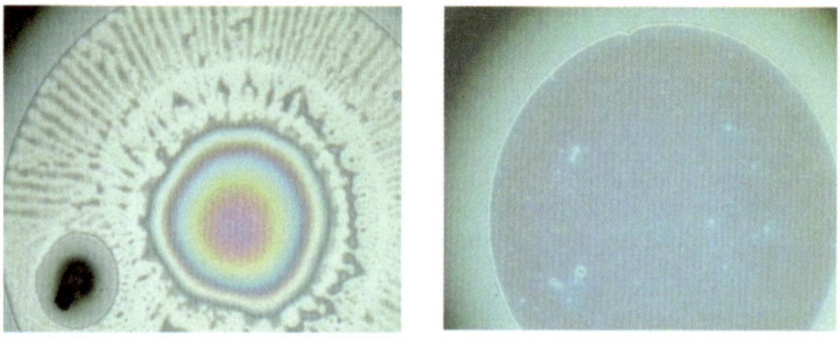

a b

Fig. 4 Photos of aqueous $C_{12}G_2$ wetting films against precoated silicon surfaces: (a) already-ruptured wetting film on PEI-coated silicon surface; (b) stable wetting liquid film on PSS-terminated coating. The film solution contained 5×10^{-5} M $C_{12}G_2$ and 10^{-4} M NaCl.

 This journal is © The Royal Society of Chemistry 2009

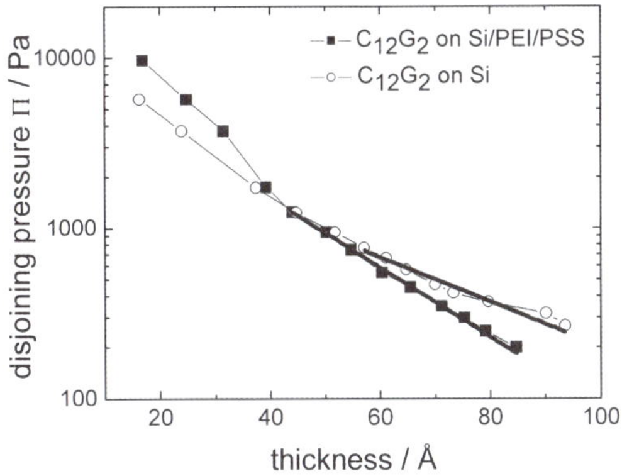

Fig. 5 Disjoining pressure isotherms of an aqueous wetting containing 5×10^{-5} M $C_{12}G_2$ and 10^{-4} M NaCl on pure silicon (open symbols) and on silicon coated with PEI/PSS (PSS was the outermost layer, filled symbols). The solid lines are fits by a simple exponential decay.

surfactant wetting film on the PSS-terminated coating shows a larger slope than the one determined on bare silicon. The Debye length κ^{-1} obtained from the experimental data is 21 nm in case of the wetting film against PSS-terminated coating and about 27 nm in case of the bare silicon surface, as mentioned above. In the following the effect of charge modification of (solid and liquid) interfaces is shown. The theoretical value is 30 nm as already mentioned above.

3.3.2 Ionic surfactants. The tendency to be able to stabilise the wetting films by tuning the interfacial charges has been confirmed in the experiment with the anionic surfactant SDS. The films were formed from aqueous solutions containing 3×10^{-4} M SDS and 10^{-4} M NaCl. In this case the air/liquid interface is negatively charged due to the adsorption of SDS. The SDS wetting film remains stable on PSS-terminated coatings due to electrostatic repulsion of both negatively charged film interfaces and a dewetted surface is observed on PEI-coated silicon.

Following the presumption that the stability of wetting films depends on the sign of surface charges, the sign of charge at the air/liquid interface was modified as well by a positively charged surfactant. Fig. 6 shows the behavior of a wetting film which was formed from aqueous solution containing 10^{-4} M positively charged $C_{12}TAB$ in 10^{-4} M NaCl on precoated silicon. The wetting film at low pressure shows the opposite behavior compared to the ones stabilised by SDS. Against a PSS-terminated coating the $C_{12}TAB$ film ruptured immediately as the pressure was applied (Fig. 6a). This result was expected on basis of the results described in the previous

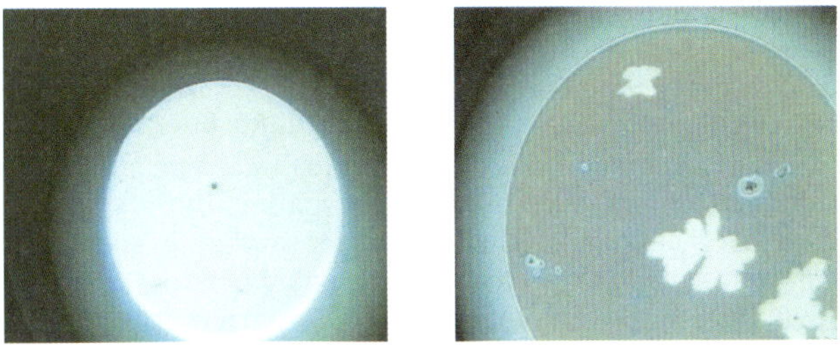

a b

Fig. 6 Wetting films containing 10^{-4} M $C_{12}TAB$ and 10^{-4} M NaCl on precoated silicon: (a) ruptured wetting film on PSS-terminated coating; (b) wetting film against a PEI-coated silicon wafer.

chapters, and can be explained by electrostatic attraction between the oppositely charged surfaces. A liquid film of the same composition on a positively charged PEI surface is stable at low pressure. However, further pressure increase to about 300 Pa gives rise to the formation of instability sites, appearing as small dewetted islands (Fig. 6b). They are stable for more than half an hour at constant pressure, but grow as the pressure is increased. Of course, the instability of an aqueous film containing cationic surfactant on a positively charged surface was unexpected and is discussed below.

3.3.3 Contact-angle measurements. Beside electrostatic interactions between the opposing interfaces the film stability can be influenced by hydrophobic interactions due to the adsorption of surfactant molecules at the solid surface. In particular, in case of oppositely charged surfactants and coatings the surfactant adsorption and hydrophobisation of the silicon surface could have an pronounced effect. The different contact angles of water droplets on bare and modified silicon surfaces are shown in Table 1 before and after treatment with different surfactant molecules.

The most pronounced effect occurs for the bare silicon surface. After immersion in a surfactant the contact angle increases from about 0° (complete wetting) to a partial wetting and a contact angle of at least 24°. The largest contact angle (34°) occurs after exposure to the cationic surfactant $C_{12}TAB$. In case of PEI-coated silicon wafers the contact angle increases from 27 to 44° and 37° for the neutral $C_{12}G_2$ and the anionic SDS respectively. Only for the positively charged $C_{12}TAB$ does the contact angle remain more or less the same. For a PSS-terminated coating almost no effect is observed for SDS, while the largest increase in contact angle is observed in case of the positively charged $C_{12}TAB$. In case of the PDADMAC-terminated coating only an obvious increase in contact angle is observed after the exposure to SDS. To summarise, there is a tendency towards the adsorption of water-soluble surfactants at oppositely charged surfaces. This leads to the conclusion that an additional destabilisation due to hydrophobisation of the solid interface in the presence of an oppositely charged surfactant cannot be excluded. However, the change in contact angle after contact with surfactant solution is rather minor.

3.4 Aqueous wetting films of insoluble amphiphiles

The study of various surfactant wetting films indicates that their stability is mainly governed by the electrostatic interactions between the air/liquid and liquid/solid interface. However, the adsorption of water-soluble surfactant molecules at the solid interface within the film can involve additional hydrophobic effects and influence the film stability. Thus, further study focuses on the interactions between two charged film interfaces concerning a pure water film, where additional adsorption effects should be absent. The charge of the solid surface was again adjusted by the deposition of a PEI or a PEI/PSS layer. The application of a monolayer of insoluble amphiphiles allows specific tuning of the charge density at the air/water interface without changing the properties of the opposing solid interface. The successful formation of densely packed amphiphile monolayers on the water surface was obviously not possible for the calculated amount of the water-insoluble substance. It was necessary to use an amount of amphiphile one order of magnitude higher in order to observe an effect for different amphiphiles and different coatings of the silicon surface. One reason could be that there is no mechanical barrier for the spread amount of amphiphile against flowing out of the small area of the liquid film toward the porous material of the film holder. Spread amphiphiles may flow through the pores of the fritted glass. Fig. 7 shows the behavior of water films with an insoluble amphiphile, either DODAB or stearic acid, spread on it. The wetting film with a DODAB monolayer is stable against a PEI precoated silicon surface (Fig. 7a). In case of a PSS-terminated coating the wetting film with DODAB is unstable (Fig. 7b). If the DODAB is replaced by the oppositely charged stearic acid the stability behaviour is the

 This journal is © The Royal Society of Chemistry 2009

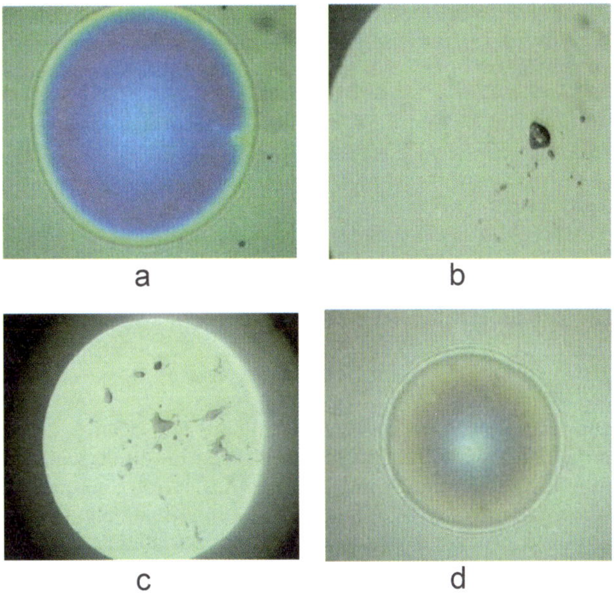

Fig. 7 Pure water wetting films with a spread monolayer of water-insoluble amphiphile: (a) DODAB/water film against a PEI-precoated Si surface. (b) DODAB/water film against a PEI/PSS-precoated Si surface. (c) Stearic acid/water film against a PEI-precoated Si surface (d) Stearic acid/water film against a PEI/PSS-precoated Si surface.

opposite: the film is unstable on PEI-precoated Si surfaces (Fig. 7c) while it is stable on PSS-terminated coatings (Fig. 7d). To summarise, aqueous wetting films with insoluble amphiphiles at the air/liquid interface are stable, if the outermost layer of the silicon coating and the amphiphiles have the same charge, and they are unstable in case of oppositely charged amphiphiles and surface coatings. The results clearly demonstrate the correlation between the wetting films' stability and the electrostatic interactions between charged film interfaces. The quantitative measurements on stable water films covered with amphiphile monolayer were not successful due to the missing reproducibility of the determined film thickness values. This can be attributed to several factors. One could be the missing mechanical barrier from the amphiphilic molecules, already mentioned above. Too low concentrations result in turn in some instability sites due to disturbances in the charge density, as was observed for the originally calculated amount of the amphiphile, although the amphiphile and opposing solid surface had the same charge (*e.g.* DODAB/water film against a PEI-terminated coating).

4 Discussion

4.1 Pure water wetting films

The results support earlier findings of the authors that an aqueous wetting film on a negatively charged surface is electrostatically stabilised[5] due to negative charges at the air/water interface. The dominant role of electrostatics was proven by the addition of salt. With increasing pH the film becomes thicker and more stable. This indicates that the negative charge at the air/water interface is caused by an excess of hydroxide ions at the surface similar to that at the oil/water interface.[11,12] In the acid regime, in addition to the decrease in charge at the liquid interface, the silicon surface charge is also close to its isoelectric point which leads to destabilisation of the film.

The addition of different monovalent salts has a pronounced effect on the film thickness. With increasing polarisability of the anion and decreasing hydration shell the film becomes thicker. This leads to the conclusion that the excess of anions and therefore the surface charge increases with decreasing size of ion hydration shell. One explanation might be that the ions with a larger hydration shell (cosmotropic ions) prefer to stay in water, while more "hydrophobic ions" with a smaller hydration shell (chaotropic ions) rather adsorb at the air/water interface. Positive adsorption of charged species of high polarisability, such as iodide and other salt ions, in the interface region was also concluded from molecular simulation studies by Jungwirth and Tobias.[23,24] The Debye length is not strongly affected by the choice of salt as expected. For iodide and chloride it is the same as in the respective bulk solution (30 nm). For fluoride it is slightly lower with 25 nm.

4.2 Water–surfactant wetting films

Pure water films do not form stable free-standing films, but a small amount of non-ionic surfactant [below the critical micellar concentration (cmc)] is enough to stabilise foam films. Even though the surfactant is neutral, long-range electrostatic repulsion between surfaces of free-standing films containing alkyl polyglycol ethers (abbreviated as C_nE_m)[25,26] and alkyl glucosides (abbreviated as C_nG_m)[27,28] has been observed. As the amount of non-ionic surfactants on the surface increases, the electrostatic repulsion between the interfaces and the film stability decline. The surface potential (between 30 and 50 mV) of symmetric films containing non-ionic surfactant decreases as the surface coverage by the surfactant increases.[27,4] The reason for the charge effects still remains argumentative.[29]

From experiments on *symmetric* films like foam films, one cannot obtain the sign of the surface potential or the surface charges. In contrast to that, in an *asymmetric* wetting film the sign of the surface charge of the liquid/air interface can be easily determined, if the sign of charge of the opposing solid surface is known. In addition, the latter one can be controlled and modified independently from the liquid interface. The fact that a $C_{12}G_2$ film is stable on a negatively charged surface, but unstable on a positively charged one leads to the conclusion that the liquid/air interface of an aqueous $C_{12}G_2$ film at those low surfactant concentrations is negatively charged. This implies that the bare air/water interface is also negatively charged and that the $C_{12}G_2$ partially replace them at low surfactant concentrations.

Films containing the negatively charged surfactant SDS or the water-insoluble amphiphile stearic acid are stable against negative surfaces but unstable against positively charged surfaces. Films of positively charged surfactant like C_{12}TAB or DODAB show the opposite behaviour. The fact that C_{12}TAB films show a partial dewetting on positively charged surfaces, where one would expect complete wetting (*i.e.* a stable film) can also be explained by electrostatic interactions. The former negative charge of the water/air interface is reversed due to the adsorption of the positively charged surfactant, but at those low concentrations the charge is not high enough to stabilise the film completely. This then leads to islands of instability. Similar observations were made for lower concentrations of DODAB. These results fit the behavior of foam films containing positively charged surfactants. Studies on alkyl trimethylammonium bromide foam films report a charge reversal from negative to positive at the air/solution interface. Films stabilised by cationic surfactants exhibit a minimum in the surface potential at a certain surfactant concentration, which indicates a reversal of the surface potential from negative to positive sign with increasing surfactant concentration.[30–32] An analogous relation was found for the quartz/solution interface[33] as the concentration of the surfactant increases. Respective studies on the effect of surfactant concentration in wetting films are underway.

Fits of the disjoining pressure isotherms of the $C_{12}G_2$ foam film and $C_{12}G_2$ wetting film lead to similar Debye lengths of 26 and 27 nm respectively. That means that the average ionic strength is about the same in both films. The values for the

 This journal is © The Royal Society of Chemistry 2009

Debye length are close to the theoretical value for an ionic strength of 10^{-4} M of the surfactant solutions. However for films against a PEI/PSS-precoated silicon surface the Debye length is lower (21.2 nm) as against the silicon surface. A reason for that deviation could be the larger amount of counter ions in the polymer coating in comparison to the bare silicon surface. In water they could contribute to the diffuse double layer in front of the surface and to additional screening of the surface potential. The $C_{12}G_2$ wetting film is thicker than the $C_{12}G_2$ foam film, which is due to the higher surface potential of the silicon surface (about −80 mV) than the one of the water/air interface (−30 to −50 mV).

The fit in Fig. 3 is done by a simple exponential decay. That assumes that the linearisation of the Poisson–Boltzmann equation is allowed. This simple approach is sufficient for fitting the disjoining pressure isotherms of a foam film, but it does not fully describe the disjoining pressure isotherms of the wetting films. Disjoining pressure isotherms of wetting films differ systematically from an exponential decay at film thicknesses below 40 nm. They show a steeper slope in this region, which would mean an additional repulsion force.

Could other forces play a decisive role for the film stability?

Steric repulsion becomes only important at film thicknesses of twice the adsorbed molecules (several nm in the present case) and can be neglected in films of a thickness of several 10s of nm as presented here.

In *symmetric* systems, the Hamaker constant has a *positive* value,[34,1,35] *i.e.* the resulting negative van der Waals forces always have an attractive contribution to the disjoining pressure. In aqueous wetting films on silicon the van der Waals forces are repulsive with a negative Hamaker constant.[34,1]

Another reason for the deviation could be undulation forces.[36] The principle is that due to the introduction of a solid surface this surface cannot react to transverse fluctuations of the opposing fluid interface like a second fluid interface might do. This leads to repulsive interactions and therefore to thicker films and also to more stable films. Foam films, which would not have such pronounced undulation forces, already ruptures at 2000 Pa, while the wetting film is stable up to a pressure above 6000 Pa. But usually these undulation forces become important at even thinner films.[37] Thus, the effect requires more detailed clarification.

Beside electrostatic interactions, hydrophobic interactions also have to be taken into account, especially for water-soluble surfactants. Under the assumption that an increase in contact angle means an adsorption of surfactant at the silicon interface, one can conclude from these results that the adsorption takes place in the case of oppositely charged surface coatings and surfactants. This would mean that in the case of oppositely charged surfactants and surfaces the fast rupture of the wetting film in the TFPB experiments could be also explained by the increased hydrophobisation of the solid surface. The adsorption of surfactant molecules is much lower if the surfactant and the terminated layer of the coating have the same charge, but in the present case the effect of hydrophobisation is minor.

Especially in the case of a polymer-coated silicon surface another effect has to be taken into account. An adsorption of surfactant with the same charge is possible, since the coating by the outer polyelectrolyte layer, which determines the net charge, is not complete. Either uncomplete coating of PEI on silicon or the strong interdigitation between the oppositely charged polyelectrolytes (PEI and PSS) in the next step leads locally to the occurrence of positive and negative charges. This could be the reason why the wetting film is less stable in the case of surfactants and terminated layers of the silicon coating with the same charge than in case of pure water or neutral surfactant ($C_{12}G_2$).

5 Conclusions

Results on interactions and stability of aqueous wetting films are presented. pH adjustment of water wetting films indicates that the negative charges at the air/water interface are caused by an excess of hydroxide ions. The addition of different types of monovalent salts gives an ion-specific effect. With increasing ion size, *i.e.* decreasing hydration shell, the anions show a stronger tendency to adsorb at the surface as predicted by theoretical simulations. The negatively charged air/water interface has certain consequences for the wetting and foam films containing surfactants. Addition of non-ionic or anionic surfactant leads to a stable wetting film against a negatively charged solid surface, and the film is unstable against positively charged surfaces. Adding positively charged surfactant leads to the opposite behavior. At low concentrations of cationic surfactant the charge reversal seems to be incomplete, which leads to less stable films. The dominating effect of electrostatic interactions is proven by water-insoluble amphiphiles, where the adsorption of amphiphiles at the opposing solid interfaces and additional hydrophobisation can be neglected.

Wetting films are thicker by about 10 nm than foam films of the same surfactant solution, which is due to the higher surface potential than of the air/liquid interface. Another difference between both types of films seems to be the influence of undulation forces which are assumed to become pronounced in wetting at film thicknesses below 40 nm. They lead to thicker films than predicted by a simple exponential decay in disjoining pressure, as it is fitting for foam films. The ion concentration is the same in both films and very similar to the bulk. It is increased if the solid surface is coated with a polyelectrolyte layer.

The qualitative observation of the pure water wetting films with specifically charged interfaces let us conclude that the stability is dominated by long-range electrostatic interactions.[5,6,38] It is possible to foresee and to control the behavior simply by adjusting the surface charge of each interface using charged species adsorbed or associated with the surfaces. This might be of great interest not only for technical applications like the cleaning of solid surfaces (dishes, tiles), but also for basic research.

Acknowledgements

Christian Symietz (Max Planck Institute of Colloid and Interfaces, Potsdam, Germany) is acknowledged for Langmuir isotherms of DODAB and stearic acid on water. We thank W. Jaeger for the polyelectrolyte samples PDADMAC and Wacker Siltronic AG for providing the silicon wafers. Financial support from the Deutsche Forschungsgemeinschaft (Graduiertenkolleg 827 Transport phenomena with moving boundaries), the Sonderforschungsbereich 448 (Project B10), and the Fonds der Chemischen Industrie is gratefully acknowledged.

References

1 J. Israelachvili, *Intermolecular and Surface Forces*, Academic Press, New York, 1992.
2 V. Bergeron, *J. Phys.: Condens. Matter*, 1999, **11**, R215.
3 A. Scheludko, *Adv. Colloid Interface Sci.*, 1967, **1**, 391.
4 C. Stubenrauch, J. Schlarmann and R. Strey, *Phys. Chem. Chem. Phys.*, 2002, **4**, 4504.
5 K. Ciunel, M. Armlin, G. H. Findenegg and R. von Klitzing, *Langmuir*, 2005, **21**, 4790.
6 R. von Klitzing, Effect of interface modification on the forces in foam films and wetting films, *Adv. Colloid Interface Sci.*, 2005, **114/115**, 253.
7 K. Hänni-Ciunel, G. H. Findenegg and R. v. Klitzing, *Soft Mater.*, 2007, **5**, 61.
8 K. R. Wilson, M. Cavalleri, B. S. Rude, R. D. Schnaller, A. Nillson, L. G. M. Petterson, T. Catalano, J. D. Bozek and R. J. Saykally, *J. Phys.: Condens. Matter*, 2002, **14**, L221.
9 E. A. Raymond, T. L. Tarbuck, M. G. Brown and G. L. Richmond, *J. Phys. Chem. B*, 2003, **107**, 546.
10 S. I. Mamatkulov, P. K. Khabibullaev and R. Netz, *Langmuir*, 2004, **20**, 4756.
11 K. G. Marinova, R. G. Alagrova, N. D. Denkov, O. D. Velev, D. N. Petsev, I. B. Ivanov and R. P. Borwankar, *Langmuir*, 1996, **12**, 2045.
12 J. K. Beattie and A. M. Djerdjev, *Angew. Chem., Int. Ed.*, 2004, **43**, 3568.
13 D. Rapoport, K. Hänni-Ciunel and R. von Klitzing, unpublished.

 This journal is © The Royal Society of Chemistry 2009

14 G. Decher, *Science*, 1997, **210/211**, 831.
15 R. v. Klitzing, B. Tieke, Polyelectrolyte Membranes, in *Polyelectrolytes with Defined Molecules Architecture I*, ed. M. Schmidt, Advances in Polymer Science (vol. 165), Springer, Heidelberg, 2004, pp. 177–210.
16 P. A. Neff, A. Naji, C. Ecker, R. v. Klitzing and A. R. Bausch, Electrical Detection of Self-Assembled Polyelectrolyte Multilayers by a Thin Film Resistor, *Macromolecules*, 2006, **39**, 463–466.
17 R. von Klitzing, Internal Structure of polyelectrolyte multilayer assemblies, *Phys. Chem. Chem. Phys.*, 2006, **8**, 5012.
18 K. Mysels and M. N. Jones, *Discuss. Faraday Soc.*, 1966, **42**, 42.
19 D. Exerova and A. Scheludko, *Chim. Phys.*, 1971, **24**, 47.
20 A. Scheludko and D. Platikanov, *Kolloidn. Zh.*, 1961, **175**, 150.
21 C. Symietz, M. Schneider, G. Brezesinski and H. Möhwald, *Macromolecules*, 2004, **37**, 3865.
22 K. Vorberg, PhD Thesis, Polyelektrolytadsorptionan geladenen Lipid-Monoschichten, Potsdam University, 1998.
23 P. Jungwirth and D. J. Tobias, *J. Phys. Chem. B*, 2001, **105**, 10468.
24 P. Jungwirth and D. J. Tobias, *J. Phys. Chem. B*, 2002, **106**, 6361.
25 T. Kolarov, R. Cohen and D. Exerowa, *Colloids Surf.*, 1989, **42**, 49.
26 K. A. Karraker and C. J. Radke, *Adv. Colloid Interface Sci.*, 2002, **96**, 231.
27 V. Bergeron, A. Waltermo and P. M. Claesson, *Langmuir*, 1996, **12**, 1336.
28 O. J. Rojas, C. Stubenrauch, J. Schulze-Schlarmann and P. M. Claesson, *Langmuir*, 2005, **21**, 11836.
29 C. Stubenrauch and R. Klitzing, *J. Phys.: Condens. Matter*, 2003, **15**, R1197.
30 D. Balzer, *Langmuir*, 1993, **9**, 3375.
31 T. Kolarov, R. Yankov, N. E. Espionova, D. Exerova and Z. M. Zorin, *Colloid Polym. Sci.*, 1993, **271**, 519.
32 R. Aveyard, B. P. Binks, J. Esquena, P. D. I. Fletcher, P. Bault and P. Villa, *Langmuir*, 2002, **18**, 3487.
33 I. P. Sergeeva, V. D. Sobolev and N. V. Churaev, *Kolloidn. Zh.*, 1990, **52**, 1129.
34 J. Lyklema, *Fundamentals of Interface and Colloid Science, Vol. I*, Academic Press, London, 1991.
35 L. Bergström, *Adv. Colloid Interface Sci.*, 1997, **70**, 125.
36 I. B. Ivanov and D. S. Dimitrov, *Thin Liquid Films*, ed. I. B. Ivanov, Marcel Dekker, New York, 1988, p. 379.
37 K. Mecke, personal communication, 2006.
38 M. Krasowska, M. Kolasinska, P. Warszynski and K. Malysa, *J. Phys. Chem. C*, 2007, **111**, 5743.

Water-mediated ordering of nanoparticles in an electric field

Dusan Bratko,[*ab] **Christopher D. Daub**[a] **and Alenka Luzar**[*a]

Received 29th May 2008, Accepted 19th June 2008
First published as an Advance Article on the web 30th September 2008
DOI: 10.1039/b809135h

Interfacial polar molecules feature a strongly anisotropic response to applied electric field, favoring dipole orientations parallel to the interface. In water, in particular, this effect combines with generic orientational preferences induced by spatial asymmetry of water hydrogen bonding under confined geometry, which may give rise to a Janus interface. The two effects manifest themselves in considerable dependence of water polarization on both the field direction relative to the interface *and* the polarity (sign) of the field. Using molecular simulations, we demonstrate strong field-induced orientational forces acting on apolar surfaces through water mediation. At a field strength comparable to electric fields around a DNA polyion, the torques we predict to act on an adjacent nanoparticle are sufficient to overcome thermal fluctuations. These torques can align a particle with surface as small as 1 nm^2. The mechanism can support electrically controlled ordering of suspended nanoparticles as a means of tuning their properties and can find application in electro-nanomechanical devices.

1 Introduction

Electrowetting on macroscopic surfaces is usually characterized in terms of contact angle, $0 \leq \theta_c \leq 180°$, defined by Young's equation $\cos\theta_c = (\gamma_{sv} - \gamma_{sl})/\gamma_{lv}$, where γ_{ab} is the surface free energy between the solid (s), liquid (l) and vapor (v) phases.[1,2] A change from negative to positive $\cos\theta_c$ implies a transition from (partially) drying to wetting behavior. In the present context, when characterizing the interactions of solid surfaces with water, we refer to surfaces with $\theta_c > 90°$ as hydrophobic and those with $\theta_c < 90°$ as hydrophilic. Discussion of alternative conventions can be found in ref. 3 and 4. Surface wettability can be enhanced by application of electric voltage driving dipolar water molecules to the field-exposed region.

The macroscopic relation due to Young and Lippmann[5] can describe electrocapillarity in a planar confinement:

$$\cos\theta_c = \frac{\gamma_{sv} - \gamma_{sl}}{\gamma_{lv}} - \frac{W_{el}(V)}{2\gamma_{lv}} = \cos\theta_c^o - \frac{W_{el}(V)}{2\gamma_{lv}} \tag{1}$$

Here $W_{el}(V)$ is the change in electrostatic free energy per unit area, associated with surface spreading of the liquid, wetting both walls (hence the factor 1/2), V is the voltage across the interface, and θ_c^o the contact angle in the absence of electric field. The form of W_{el} depends on system geometry and material properties but is

[a]*Department of Chemistry, Virginia Commonwealth University, Richmond, VA, 23284-2006, USA. E-mail: aluzar@vcu.edu*
[b]*Department of Chemical Engineering, University of California, Berkeley, CA, 94720-1462, USA. E-mail: dnb@berkeley.edu*

This journal is © The Royal Society of Chemistry 2009

generally presumed to be proportional to the areal electric capacitance of the interface, c, and the potential drop across the interface squared, $W_{el} \sim -\frac{c}{2}V^2$.[1,2]

Molecular level studies of these phenomena,[6–9] however, can bring in an entirely new perspective. In our recent work we demonstrated the importance of molecular scale events in understanding and making quantitative predictions of electrowetting phenomena in a nanoscale system.[8,9] In particular, we highlighted the coupling between interfacial hydrogen bonds and molecular alignment in electric field. This coupling modulates the dependence on field direction and polarity in contrast with conventional picture in a macroscopic system discussed above; system size plays a crucial role.

These findings are based on our molecular simulations performed in two very different scenarios: a sessile nanodrop on an atomic graphene surface[8] and water in a nanopore.[9] The droplet scenario[8] is a miniature of the experimental setup of a drop inside a capacitor described by Bateni *et al.*[10,11] Except for using a uniform capacitor field, it resembles the case of a dielectric drop atop an electrode.[1] Here, surface tensions change since the field tends to be stronger at droplet interfaces. In the second scenario,[9] water from a field-free bulk phase is driven into a planar hydrophobic confinement exposed to electric field. In both scenarios, we demonstrated a strong dependence of wetting on the *direction* of the applied electric field, distinct from the macroscopic picture where the effect depends only on the field squared [eqn (1)]. In a nanoscale confinement, water wets a surface better when the applied electric field is parallel than when it is perpendicular to the interface.

In a perpendicular field, polarity can also make a difference. Even in the absence of directional forces between the walls and liquid molecules, wetting by water is asymmetric, enhanced at the wall at which the field points into the aqueous phase and almost unchanged at the opposite side[9] (Janus interface).[12] We explain the sign-sensitive response to the field[9] in terms of orientation preferences of surface molecules.[13–16] We predicted such preferences, associated with water hydrogen bonding, long ago from mean-field arguments[13,15] and they were confirmed in simulation[14] and experiment.[16] The observed anisotropy in electrowetting is a new nanoscale phenomenon that has so far been elusive as, in current experimental setups, surface molecules represent a very low fraction of all molecules[10] affected by the field. This is different from spectroscopic techniques where molecules will sense the confinement effect without necessarily residing in the solid/liquid contact layer.[17,18]

Field direction and polarity are therefore important determinants of electrowetting effects in a nanoporous material. Conversely, for fixed field direction and freely rotating nanotubes (pores), or nanoparticles, surface energetics will favor the alignment of the particle surface with the electric field suggesting a novel mechanism of addressing structural order in a nanomaterial through controlled molecular or supramolecular charge distribution. We explore such a scenario in this work.

We consider situations where the external field bears no direct influence on nonpolar nanopore walls; hence the presence of water is essential. The torque τ on a wall of area A is $\tau(E,\phi) = -A\partial\gamma_{sl}(E,\phi)/\partial\phi$, where ϕ is the angle of the field direction measured clockwise relative to the wall. Free-energy differences between aligned and perpendicular nanopores we showed in our earlier work[9] indicate that the effect may suffice to align a water-filled nanotube in a field E of strength comparable to that in the double layer next to a DNA polyion,[19] synthetic polyelectrolyte,[20] or ionic-surfactant assembly,[21,22] $E\varepsilon_r \equiv E_0 \approx 1$ V Å^{-1} (ε_r is the relative permittivity).

We use molecular simulations to assess the magnitude of this solvent-mediated effect. Specifically, we estimate the strength of orientational forces water can mediate between external field and nanopores. For a range of imposed field directions and strengths up to those next to ionized polymers,[19] ion channels[23,24] and electrochemical electrodes,[25] we determine forces and torques on water-filled nanopores with parallel walls for different wall–water affinities (measured in terms of varying water contact angle) and at several distances apart. For the above field strengths, we show that the work necessary to rotate the nanopore by an angle $\delta\phi$ away

 This journal is © The Royal Society of Chemistry 2009

from full alignment, $w(E,\phi) = -\int_0^{\delta\phi} \tau(E,\phi)\mathrm{d}\phi$, exceeds thermal energy, k_BT, for wall areas as small as $\sim$1 nm^2 and angular displacement $\sim\delta\phi$ $\mathcal{O}(10°)$.

The planar pore models we employ are defined in the next section. We then present results illustrating the role of field direction on asymmetric water structure, hydrogen-bond populations, and water polarization in a field-exposed nanopore. We show how these effects relate to free-energy dependence on the angle between the applied field and pore walls, leading to a torque that drives the pore into alignment with the field. Our principal results demonstrate a notable water-mediated force toward interface alignment with the field. We show that only contact-layer water is engaged in this mediation; however, the attraction between water molecules and confinement walls bears only a minor influence. While we exemplify the aligning mechanism in the case of nanosized planar confinements, the effect is clearly more general and will operate at all smooth aqueous interfaces. Here, we do not consider systems with specific directional forces between wall molecules and water but it is clear that the introduction of such forces would result in additional orientational trends superimposed on the more general effect addressed in the present study.

2 Dependence of free energy on field direction

The driving force toward interface alignment with the applied field E_0 can be quantified in terms of the change in wall/water surface tension as a function of the angle ϕ between field direction and the interface. When water in a planar, field-exposed confinement maintains equilibrium with a field-free bulk phase, the pertinent work function is the wetting surface free energy, σ, here defined as

$$\sigma = \left(\frac{\partial \Omega}{\partial A}\right)_{\mu,D,T} \tag{2}$$

where Ω is the grand potential and A the water/wall contact area (the wetted area of confinement walls). Furthermore, μ is chemical potential, T temperature and D the width of the confinement. For constant D, area A is related to volume V of confined water, $A = 2V/D$, leading to an alternative expression for σ:

$$\sigma = \frac{D}{2}\left(\frac{\partial \Omega}{\partial V}\right)_{\mu,D,T} = -\frac{D}{2}P_{\|} \tag{3}$$

where $P_{\|} = (P_{xx} + P_{yy})/2$ represents the average of the lateral components of the pressure in the confinement. In the absence of the field ($P_{\|} = P_{xx} = P_{yy}$), the quantity σ can be identified with the surface tension difference $\Delta\gamma = \gamma_{sl} - \gamma_{sv}$. As we have shown in the preceding work,[9] when confined water is polarized by applied electric field, the wetting surface free energy, σ, depends on the angle ϕ measuring the direction of the field relative to confinement surfaces. In general, surfaces aligned with the field ($\phi = 0$) appear more hydrophilic than the ones perpendicular ($\phi = \pi/2$) to the applied field. In the present study, we determined the dependence of σ on ϕ for a set of surfaces with different zero-field contact angles $\theta_c \approx 135$, 93 and 69°. In view of the strong dependence of pressure and σ on the liquid density in the pore, itself a function of ϕ, these calculations require open ensemble (μVT) simulations. Knowing σ as a function of angle ϕ in an open (μVT) system also provides information about the torque τ acting on non-aligned confinement surfaces at specified field strength E_0 and the angle between the surface and the field, ϕ:

$$\frac{\tau}{A} = -\left(\frac{\partial \sigma}{\partial \phi}\right)_{\mu,V,T} = -\frac{1}{A}\left(\frac{\partial \Omega}{\partial \phi}\right)_{\mu,V,T} \tag{4}$$

If we are interested only in the torque exerted by the field, and not in the absolute values of σ, direct calculation of τ is also possible. Beginning with the relation

$$\Omega(\mu,V,T) = -k_{\rm B}T\ln\Xi \tag{5}$$

where $\Xi = \sum_N \sum_{states\ i} e^{-\frac{U_i}{k_{\rm B}T}+\frac{\mu N}{k_{\rm B}T}}$ is the grand-canonical partition function.

We can express the torque (per unit area A) as

$$\frac{\tau}{A} = -\frac{1}{A}\frac{\partial\Omega}{\partial\phi} = \frac{k_{\rm B}T}{A}\frac{\partial}{\partial\phi}\ln\Xi = \frac{k_{\rm B}T}{A}\frac{\sum_N \sum_{states\ i} -\frac{1}{k_{\rm B}T}\frac{\partial U_i}{\partial\phi}e^{-\frac{U_i}{k_{\rm B}T}+\frac{\mu N}{k_{\rm B}T}}}{\Xi} = -\frac{1}{A}\left\langle\frac{\partial U}{\partial\phi}\right\rangle_{\mu,V,T} \tag{6}$$

Note the difference between the grand potential derivative $\langle\partial U/\partial\phi\rangle_{\mu VT}$ in eqn (6) and the energy derivative $(\partial\langle U\rangle/\partial\phi)_{\mu VT}$. An analogous relation for fixed NVT obtains the torque in the canonical ensemble. Our simulation results (Section 3) confirm equality of torques determined as derivatives of σ [eqn (4)] with those from direct calculation [see eqn (6)].

3 Models and methods

3.1 Model systems

The model system consists of a confined water slab between parallel insulating walls whose uniformly distributed sites interact with water molecules through Lennard-Jones potential but carry no charges or dipoles. The thickness of the walls exceeds the range of water–water and water–wall interacting-site pair potentials. To describe the interactions between water molecules we apply the SPC/E [26] potential, which has been commonly used in related studies.[6–9,25,27] The presence of external field may suggest the use of polarizable models; however, even for the nonscreened fields E_0, the strengths we consider are weak compared to molecular and ionic fields that lead to significant polarization of a water molecule. According to Fig. 6 of ref. 28, for a fluctuating-charge (TIP4P-FQ)[29] polarizable force field of water, the strongest field considered in the present study ($E_0 = 0.4$ V Å^{-1}) will produce about 1% change in the average dipole moment of a water molecule. Further, recent comparisons between different interaction-site simulation models of water show that calculated surface tensions of water are relatively robust with respect to details of the model potentials.[30]

Interaction between water molecules and smooth confinement walls is described by the integrated (9-3) Lennard-Jones potential[9,14,31–33]

$$u_{\rm w}(z) = A\left(\frac{\sigma_{\rm wO}}{D'\pm z}\right)^9 - B\left(\frac{\sigma_{\rm wO}}{D'\pm z}\right)^3 \tag{7}$$

Here, $A = 4\pi\rho_{\rm w}\varepsilon_{\rm wO}\sigma_{\rm wO}^3/45$ and $B = 15A/2$,[33] $D' = D/2$, z is the distance of the water oxygen from the slit midplane with walls at $z = \pm D'$, $\rho_{\rm w}$ is the presumed uniform number density of interacting sites of wall material, $\varepsilon_{\rm wO}$ and $\sigma_{\rm wO}$ are Lennard-Jones potential depth and size parameters for water oxygen–wall site pair, and the sum or difference is used in the denominators to describe interactions with a wall at $-D'$ or D' respectively. $\varepsilon_{\rm wO}$ and $\sigma_{\rm wO}$ are Berthelot's means for water-oxygen and wall-site Lennard-Jones parameters given in Table 1. According to the above definitions, the slit width actually occupied by water molecules will be close to $D - \sigma_{\rm wO}$.

To describe hydrophobic, paraffin-like walls, w(1), we used Lennard-Jones parameters from Lee *et al.*[14] Walls with increased hydrophilicities, w(2) and w(3), were

Table 1 Lennard-Jones parameters of different atomic species, s (oxygen and hydrogen atoms of water and interacting sites of the walls, w(i))

s	ε_s/kJ mol^{-1}	q_s/e_o	σ_s/Å
O	0.6502	−0.8476	3.1656
H	—	0.4238	—
w(1)	0.6483	—	3.754
w(2)	3.45	—	3.754
w(3)	5.00	—	3.754

modeled following the approach of Jaffe and coworkers,[34] who calibrated the strength of water–wall Lennard-Jones interaction to match the desired water contact angle on the specified surface. Using wall site density ρ_W, size σ_W and well depth $\varepsilon_{w(1)O}$ from the hydrocarbon wall model,[33] and values for $\varepsilon_{w(2)O}$ and $\varepsilon_{w(3)O}$, interpolated from Table IV of ref. 34 we obtain the following wall/water contact angles: $\theta_c(1) = 135°$, $\theta_c(2) = 93°$, and $\theta_c(3) = 69°$, all with error bars of $\pm 3°$. These contact angles were estimated by Young's equation and respective interfacial tension differences, $(\gamma_{sv} - \gamma_{sl})$, along with the simulated surface tension for SPC/E water calculated at identical conditions in our earlier work, described in the supplement to ref. 9.

We consider two gap widths $D = 16.4$ and 27 Å. In the hydrophobic confinement, w(1), the lower of the two widths is just above the threshold wall separation of mechanical instability (spinodal) for capillary evaporation.[4,33,35] We explore the effect of the field on the water-phase behavior in a separate study.[36] Finite size effects were mitigated by the use of periodic boundary conditions along lateral directions; a 1 nm spherical truncation of intermolecular potentials was implemented following previous studies.[9,31] Temperature was held at 298 K in all calculations.

We maintain equilibrium with ambient, field-free aqueous environment by using open ensemble Monte Carlo simulations (GCMC) following the procedures we described earlier.[9,32,33,37,38] Excess chemical potential for water $\mu_{ex} = -12.1k_BT$ was selected to secure vanishing pressure in bulk water in the absence of applied field. Wetting free energies, σ, are calculated from average lateral components of the pressure tensor, $P_{\|}$, eqn (3).

For the analysis of water hydrogen bonding we use our usual definition based on geometric criteria[39] with the cutoff value $R_{OH}{}^c = 2.45$ Å, the OH intermolecular distance dictated by the first minimum in the corresponding radial distribution function for SPC/E water, and a maximum angle between O–O vector and the covalent O–H bond of $\psi^c = 30°$. This is the angle at which the average number of hydrogen bonds per water molecule is within 10% of the asymptotic value for large ψ^c. Interestingly, this threshold angle does not change up to an applied field E_0 of 1.5 V Å^{-1}.[36]

3.2 Pressure tensor calculation

Average pressure tensor components $P_\perp = P_{zz}$ and $P_{\|} = (P_{xx} + P_{yy})/2$ are calculated from energy differences ΔU_α associated with uniform scaling of molecular coordinates α ($\alpha = z$ or x,y) and volume change ΔV_α. Here, ΔU_α comprises changes in intermolecular and water–wall interactions. Scaling of molecular center-of-mass positions has no effect on interactions with the applied electric field.

$$P_{\alpha\alpha} = \rho kT + \lim_{\Delta V_\alpha \to 0} \frac{kT \ln\left\langle \exp\left(-\frac{\Delta U_\alpha}{kT}\right)\right\rangle}{\Delta V_\alpha} \tag{8}$$

Forward and backward scaling was employed for improved accuracy. A coordinate scaling factor $f = 1 \pm \delta$ with $\delta = 10^{-5}$ was chosen empirically for optimal compromise between round-off errors (decreasing with increasing δ) and the numerical accuracy of the finite difference approximation employed in eqn (8) (improved upon decreasing δ). Variation of δ within the interval $10^{-6} \leq \delta \leq 10^{-4}$ revealed no appreciable effect on calculated pressure components The calculated normal component of the pressure tensor, $P_{\perp}$, agreed within numerical uncertainty with wall pressure calculated directly from wall/water forces as described in our earlier work.[33]

4 Results and discussion

We first present our results for water density profiles in a field-exposed slit for different angles ϕ of incoming field of strength $E_0 = 0.2$ V Å^{-1}. The slit width is 1.64 nm and the walls are composed of hydrophobic material with water contact angle 135°, model w(1) in Table I. As shown in Fig. 1a, water density in the interior of the slit oscillates around that of the field-free bulk phase equilibrated with the

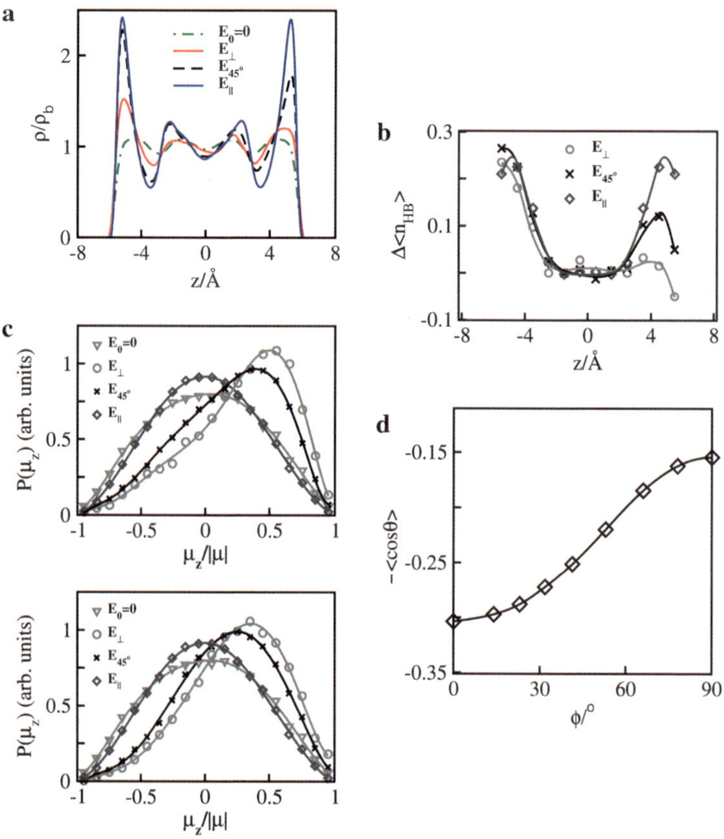

Fig. 1 (a) Water density profiles in a 16.4 Å wide slit subject to electric field $E_0 = 0.2$ V Å^{-1} for different angles ϕ between the walls and incoming field. The green curve shows the density profile at $E_0 = 0$. (b) The change in the average number of hydrogen bonds per molecule due to the applied field $E_0 = 0.2$ V Å^{-1} at different field angles ϕ relative to the walls. (c) Average projections of molecular dipole normal to the wall in the absence of applied field and in field of strength $E_0 = 0.2$ V Å^{-1} for different field/wall angles ϕ. Comparison between l.h.s. (upper graph) and r.h.s. (lower graph) walls shows a stronger alignment with normal field on the left wall where the field orients water hydrogens into the liquid. (d) Extent of dipole alignment with the field, measured in terms of $\langle \cos\theta \rangle = \langle \boldsymbol{E}_0\boldsymbol{\mu} \rangle / |E_0\mu|$, as a function of the angle between the direction of incoming field and confinement walls, ϕ. Confinement width $D = 16.4$ Å and the water–wall contact angle is 135°.

 This journal is © The Royal Society of Chemistry 2009

confinement but is elevated in the immediate vicinity of the walls. The density increase depends on the angle ϕ, being most pronounced when the field is parallel to the walls ($\phi = 0$). In perpendicular field, electrostriction is weaker overall and strong asymmetry in the density profile is observed.[9] Similar asymmetry has been observed by other groups.[6,7,25] We show that density remains almost unaffected by the field on the r.h.s. wall where molecular alignment with the field is least favorable for hydrogen bonding. The curve corresponding to $\phi = \pi/4$ interpolates between the two extreme cases corresponding to $\phi = 0$ and $\pi/2$.

Field-induced changes in hydrogen-bond populations are shown in Fig. 1b. We demonstrate stronger wetting to coincide with higher population of hydrogen bonds. Interestingly, the number of hydrogen bonds per molecule $\langle n_{\mathrm{HB}}(z)\rangle$ appears to increase in the field in all cases shown. A slight reduction is observed only when the perpendicular field tends to orient both water hydrogens toward the r.h.s. wall, an orientation allowing at most two hydrogen bonds per molecule. In examples shown in Fig. 1, the field is either parallel to the confinement walls ($E_z = 0$) or contains a positive component perpendicular to walls ($E_z > 0$), *i.e.* pointing from left to right. Nearly unperturbed hydrogen bonding at the negative z wall with the field at an angle $\phi = \pi/4$ reflects the flexibility of the hydrogen-bond network. Water molecules can keep hydrogen bonds while accommodating small perturbations from the optimal molecular alignment with the walls, but resist following the field when the deviation approaches $\pi/2$. Because of angle preferences imposed by hydrogen bonds, in a perpendicular field, polarization of water is stronger at the left wall where the fields directs water hydrogens away from the wall (Fig. 1c). When the field is parallel to the walls, on the other hand, the tendency to optimize hydrogen bonding cooperates with already existing trend toward near-parallel alignment.[14,16,31]

Consistent with easier water polarization in the direction along the surfaces, Fig. 1d, electrostriction is strongest when the field is parallel to confinement walls. An interesting behavior is revealed when monitoring the average numbers of water molecules, $\langle N\rangle$, accommodated in slits with varied wall contact angles, Fig. 2. A comparison between wall types w(1), w(2), and w(3) (see Table 1) with contact angles $\theta(1) = 135°$, $\theta(2) = 93°$, and $\theta(3) = 69°$ shows expected differences in $\langle N\rangle$ in the absence of the field. Application of a perpendicular field results in considerable electrostriction in the hydrophobic pore characterized by high compressibility.[33,40] This compressibility is associated with the existence of depleted water layers next to strongly hydrophobic plates ($\theta = 135°$).[41,42] The difference of ~3.4 confined water

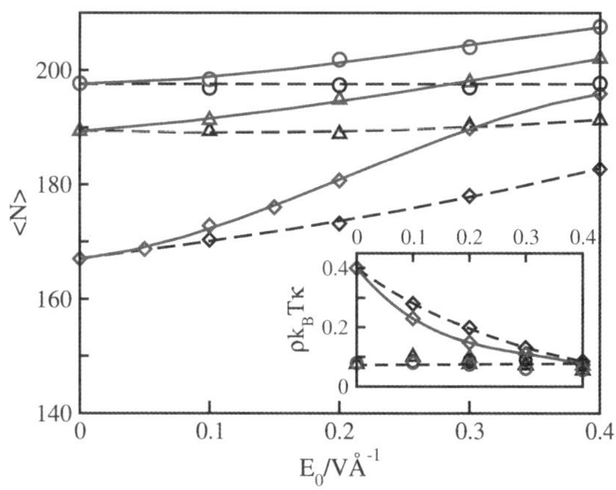

Fig. 2 Average number of water molecules $\langle N\rangle$, and reduced compressibilities $\rho k_{\mathrm{B}}T\kappa = (\langle N^2\rangle - \langle N\rangle^2)/N$ in an open slit with periodic box dimensions 21 × 21 × 16.4 Å as a function of field strength E_0 for wall–water contact angles $\theta_c = 135°$ (diamonds), 93° (triangles) or 69° (circles) under perpendicular (dashed) or parallel fields (solid lines).

molecules per nm^2 surface area is associated with different densities of water in the first solvation layers at hydrophilic ($\theta_c = 69°$) and hydrophobic ($\theta_c = 135°$) confinement surfaces. In the field, available space is gradually used up to accommodate additional molecules attracted into the system. Pores with contact angles $\theta = 93$ or 69°, on the other hand, are filled more tightly even in the absence of the field. Low compressibility of the liquid next to hydrophilic walls reflects weak density fluctuations comparable to those in the bulk phase. A similar regime is approached in hydrophobic pores after we suppress fluctuations by applying an external electric field to the system.

In view of the low initial compressibilities in pores with lower contact angles, only a slight density increase is observed in the perpendicular field. In a parallel field, electrostriction is noticeable even in the hydrophilic pore, $\theta = 69°$, again reflecting stronger affinity for water when the field is applied laterally along the walls.

Enhanced solvation of walls aligned with the field suggests considerable work is needed to rotate the plates away from a field-aligned orientation. Wetting surface free energies, σ, therefore depend on the angle ϕ between the field and the walls. This dependence is illustrated in Fig. 3a and 3b for all three wall types at two pore widths $D = 16.4$ and 27 Å. For narrow hydrophobic pores, it has been shown that electric field reduces the threshold wall separation for capillary evaporation.[43] Our results demonstrate that the effect will also depend on field direction, being maximal when the field is parallel to the walls, as is the case in ion channels.

Interestingly, the incentive toward wall alignment with the field depends only weakly on the pore width and water–wall contact angle. The net effect therefore originates almost exclusively in the first solvation layers, regardless of the amount of intervening liquid between them. Apparent insensitivity to the nature of wall material, on the other hand, confirms that the observed changes in σ derive primarily from the properties of the liquid, namely its polarity (favoring parallel dipole/wall alignment),[44] as well as orientation dependence of hydrogen bonding.[13–15] Inclusion of specific, angle-dependent wall–water interactions,[45] present in most hydrophilic materials, would clearly alter and possibly strengthen the dependence of wetting surface free energies on the field angle ϕ.

According to eqn (4), the angle dependence of the wetting free energy leads to a torque acting on confinement walls. In Fig. 4 and 5, we present the torque dependence on angle ϕ measuring the deviation from surface alignment with the field. The curves shown here were obtained by direct calculation of torques on confined water molecules [eqn (6)]. They are consistent with results of the alternative calculation, eqn (4), based on the slopes of wetting free energies presented in Fig. 3. The torque vanishes in free energy extrema corresponding to full alignment ($\phi = 0°$) or

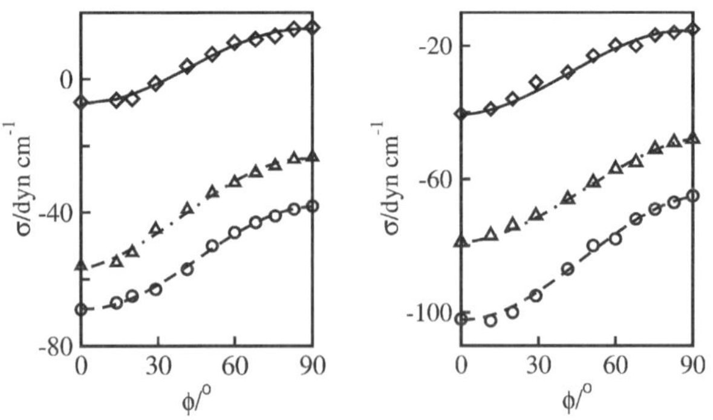

Fig. 3 Wetting free energies of slits of width $D = 16.4$ (left) or 27 Å (right) in electric field of strength $E_0 = 0.2$ V Å^{-1} as functions of field–wall angle ϕ. Wall–water contact angles are 135° (diamonds), 93° (triangles) or 69° (circles).

 This journal is © The Royal Society of Chemistry 2009

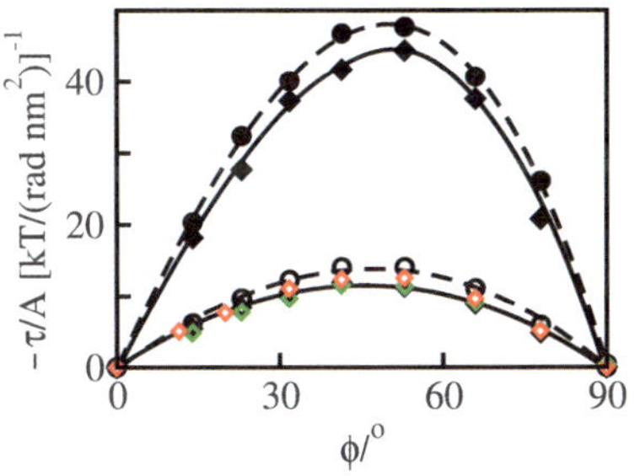

Fig. 4 Calculated torques on confined water–slab systems subject to applied fields $E_0 = 0.2$ V Å^{-1} (open symbols) or 0.4 V Å^{-1} (solid symbols). Diamonds correspond to walls with contact angle $\theta_c = 135°$ and circles are for $\theta_c = 69°$, all for open (μVT) systems with wall separation D = 16.4 Å. The green diamonds describe a closed (NVT) system with number of molecules the same as in the field-free systems, while the red diamonds describe a wider pore (D = 27 Å), both with hydrophobic walls (θ_c = 135°).

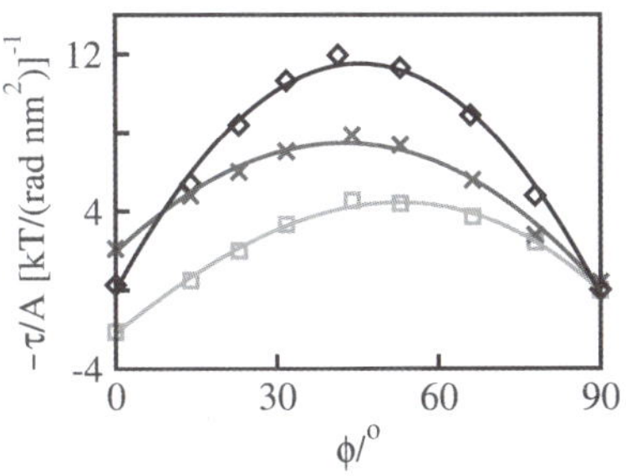

Fig. 5 Total torques on a confined, 16.4 Å wide water–slab system subject to applied field E_0 = 0.2 V Å^{-1} (diamonds), equal to the sum of torques acting on the left-hand-side (crosses) and right-hand-side (squares) surfaces. For nonzero angles ϕ, the perpendicular component of the field is pointing towards the right side. The wall–water contact angle is 135°.

perpendicular orientation ($\phi = 90°$). In strong fields, the maximal torque [steepest slope of $\sigma(\phi)$] is observed around $\phi \approx 53°$, the limiting angle for a water dipole to accommodate up to three hydrogen bonds.[13,15] This is different from a symmetric dipolar fluid such as Stockmayer fluid where we find (results not shown) the maximal torque at the expected angle $\phi \approx 45°$.

In agreement with observed variations of wetting free energies, Fig. 3, the torques are virtually independent of confinement width and wall contact angle. Physical reasons for this independence have been discussed in the context of wetting free energies (Fig. 3). In addition, we observe very small effect of ensemble conditions (NVT *vs.* μVT) on calculated torques.

Unlike calculations of surface wetting free energies, amenable only for the confinement as a whole, direct torque calculations allow us to separate forces acting on individual walls. Since any significant torque contributions come from water mediation of field forces in surface layers next to either of the walls, net forces on either of the plates can be obtained as sums of molecular contributions allocated to the two halves of the slab divided by the pore mid-plane. Results of such a calculation are presented in Fig. 5. The net torque comprises two unequal contributions from separate plates. In the parallel alignment, the vanishing total torque consists of two finite and oppositely equal torques on the two walls. They are explained by the fact that the spontaneous water alignment differs slightly from exact parallel one,[14,31] causing weak, oppositely equal torques on the two walls. The observed asymmetry in curves $\tau(\phi)$ is directly related to the asymmetry of water molecules and their angle preferences imposed by hydrogen bonding.[13–15]

Fig. 4 shows a strong dependence of the torque on the field strength. Doubling the field strength brings the value of calculated torques per area from $\sim$10 k_BT rad^{-1} nm^{-2} at E_0 = 0.2 V Å^{-1} to $\sim$40 k_BT rad^{-1} nm^{-2}, confirming our prediction that τ/A should be proportional to E_0^2. This result, verified in numerous additional calculations (not shown), is consistent with weak-field predictions of polarization energy of water from the Langevin equation, and this is the energy that competes with hydrogen bonding. Deviations from these simple predictions could be expected for more realistic interface models including orientation dependent wall–water interactions.[45] Increasing isotropic wall attractiveness alone gives slightly higher values of torques at hydrophilic walls compared to hydrophobic ones. This is because stronger attraction to the surfaces enhances the impact of orientational preferences of water molecules in the wall–water contact layer compared to more loose contacts next to hydrophobic plates.

So far, our work has only addressed effects unrelated to specific properties of the surfaces. The two general mechanisms contributing to observed torques are based on (i) hydrogen-bond-related orientational preferences of interfacial water molecules that already exist in the absence of the field, and (ii) the universal advantage of aligning dipoles of any kind *along*, rather than normal to, the interface. The general propensity to parallel alignment, supporting favorable head-to-tail stacking of dipoles, has been analyzed thoroughly in contexts of interfacial structural ordering[44] and anisotropic dielectric response at interfaces.[46] Our calculations using Stockmayer fluid in a field-exposed confinement reveal the existence of similar torques as observed in water; however, these results (not shown) were always insensitive to the sign (polarity) of the field, a phenomenon attributed to asymmetric interactions of surface water molecules.

Lastly we mention the roles of particular water models and boundary conditions we used. We have tested the field response of TIP4P[47] water, and found it to be only marginally more affected by the field as compared to SPC/E, suggesting that our results are not overly sensitive to the water model used. Further, our studies avoid field strengths capable of any significant polarization of water molecules that would warrant the use of polarizable models of water.[29,48] To assess the impact of truncation of molecular interactions, we have also carried out a number of simulations using the adaptation of 3D-Ewald summation for a slab finite in one of the three spatial dimensions.[49] These additional calculations confirmed (results not shown) the same qualitative behavior, with *amplified* dependence on field direction observed upon removing the pair-potential cutoff we used in most calculations. We also note that in our studies of confinement effects we employ lateral periodicity primarily as a mathematical simplification rather than to describe a truly infinite system. As such, the use of truncated potentials has been argued to produce representative results for nano-sized aqueous confinements encountered in biosystems and in nanomaterials.[50]

5 Concluding remarks

In this work we discuss a new mechanism to orient nanoparticles by an applied electric field when the particles carry no charges or dipoles of their own. Coupling to the field can be accomplished through solvent-mediated interaction between the applied electric field and a nanoparticle. The key to this phenomenon is the orientational preference of solvent molecules in the surface layer. This preference manifests in angle-dependent water–particle potentials that enable interfacial solvent to mediate field-induced orientational forces between dipolar molecules and the nanoparticle. In the case of water, orientation-dependent molecule–wall interactions combine the general propensity of dipolar fluids to spontaneously polarize in a lateral direction[44] with specific effects of water asymmetry and hydrogen bonding.[13,15] The former effect alone can induce the preference for dipolar alignment with confinement

 This journal is © The Royal Society of Chemistry 2009

walls and the latter is responsible for polarity dependence (Janus interface).[9] These mechanisms are not limited to confinements, but can equally operate on isolated surfaces.

To maintain the alignment of the nanopore within a tolerance $\delta\phi$, work against the field, $-\int_0^{\delta\phi} \tau(E,\phi)\mathrm{d}\phi$, must appreciably exceed thermal energy, $k_\mathrm{B}T$. For field strength E_0 of 0.2 V Å^{-1} and for a nanopore area of 1 nm^2, the simulated torques suggest a typical angle deviation $\delta\phi \approx \pm 20°$. For a field $E\varepsilon_\mathrm{r}$ next to a DNA polyion of $\mathcal{O}(1)$ V Å^{-1}, for example, this tolerance diminishes by an order of magnitude, $\delta\phi \approx \pm 2°$, and the nanoparticles would be fully aligned. In practice, the accessible field strengths, E_0, will ultimately determine whether the suggested mechanism can one day find use in nanomechanical devices.

We can envisage increasing the sensitivity to the field by using solvents with greater molecular asymmetry and/or dipole moments than water (*e.g.* DMSO, DMF). Orientational forces are likely to discriminate not only between filled and empty nanopores but also between pores containing solvents of different polarities. While the bigger solvent asymmetry only controls the magnitude of the Janus-interface effect,[9] the solvent polarity combined with molecular asymmetry controls the magnitude of the torque. We plan to explore the use of alternative solvents in our future studies.

Acknowledgements

This work was supported by the National Science Foundation through awards CHE-0718724 (to A. L.) and CBET-0432625 (D. B.).

References

1 B. Shapiro, H. Moon, R. L. Garrell and C. J. Kim, *J. Appl. Phys.*, 2003, **93**, 5794–5811.
2 F. Mugele and J. C. Baret, *J. Phys.: Condens. Matter*, 2005, **17**, R705–R774.
3 T. Hayashi, A. J. Pertsin and M. Grunze, *J. Chem. Phys.*, 2002, **117**, 6271–6280.
4 A. Luzar, *J. Phys. Chem. B*, 2004, **108**, 19859–19866.
5 For a tutorial review on electrowetting, including the Young–Lippmann equation, see: R. Shamai, D. Andelman, B. Berge and R. Hayes, *Soft Matter*, 2008, **4**, 38–45.
6 J. Dzubiella and J. P. Hansen, *J. Chem. Phys.*, 2005, **122**, 234706.
7 S. Vaitheeswaran, H. Yin and J. C. Rasaiah, *J. Phys. Chem. B*, 2005, **109**, 6629–6635.
8 C. D. Daub, D. Bratko, K. Leung and A. Luzar, *J. Phys. Chem. C*, 2007, **111**, 505–509.
9 D. Bratko, C. D. Daub, K. Leung and A. Luzar, *J. Am. Chem. Soc.*, 2007, **129**, 2504–2510.
10 A. Bateni, S. Laughton, H. Tavana, S. S. Susnar, A. Amirfazli and A. W. Neumann, *J. Colloid Interface Sci.*, 2005, **283**, 215–222.
11 A. Bateni, S. S. Susnar, A. Amirfazli and A. W. Neumann, *Langmuir*, 2004, **20**, 7589–7597.
12 X. Y. Zhang, Y. X. Zhu and S. Granick, *Science*, 2002, **295**, 663–666.
13 A. Luzar, S. Svetina and B. Zeks, *Chem. Phys. Lett.*, 1983, **96**, 485–490.
14 C. Y. Lee, J. A. McCammon and P. J. Rossky, *J. Chem. Phys.*, 1984, **80**, 4448–4455.
15 A. Luzar, S. Svetina and B. Zeks, *J. Chem. Phys.*, 1985, **82**, 5146–5154.
16 Q. Du, E. Freysz and Y. R. Shen, *Science*, 1994, **264**, 826–828.
17 W. H. Thompson, *J. Chem. Phys.*, 2004, **120**, 8125–8133.
18 W. H. Thompson, *J. Chem. Phys.*, 2002, **117**, 6618–6628.
19 K. Wagner, E. Keyes, T. W. Kephart and G. Edwards, *Biophys. J.*, 1997, **73**, 21–30.
20 D. Bratko and D. Dolar, *J. Chem. Phys.*, 1984, **80**, 5782–5789.
21 A. Luzar and D. Bratko, *J. Chem. Phys.*, 1990, **92**, 642–648.
22 D. Bratko, C. E. Woodward and A. Luzar, *J. Chem. Phys.*, 1991, **95**, 5318–5326.
23 D. Bostick and M. L. Berkowitz, *Biophys. J.*, 2003, **85**, 97–107.
24 A. Philippsen, W. P. Im, A. Engel, T. Schirmer, B. Roux and D. J. Muller, *Biophys. J.*, 2002, **82**, 1667–1676.
25 K. J. Schweighofer, X. F. Xia and M. L. Berkowitz, *Langmuir*, 1996, **12**, 3747–3752.
26 H. J. C. Berendsen, J. R. Grigera and T. P. Straatsma, *J. Phys. Chem.*, 1987, **91**, 6269–6271.
27 M. Kiselev and K. Heinzinger, *J. Chem. Phys.*, 1996, **105**, 650–657.
28 K. L. Yang, S. Yiacoumi and C. Tsouris, *J. Chem. Phys.*, 2002, **117**, 337–345.
29 S. W. Rick, S. J. Stuart and B. J. Berne, *J. Chem. Phys.*, 1994, **101**, 6141–6156.

30 F. Chen and P. E. Smith, *J. Chem. Phys.*, 2007, **126**, 221101.
31 J. C. Shelley and G. N. Patey, *Mol. Phys.*, 1996, **88**, 385–398.
32 K. Leung, A. Luzar and D. Bratko, *Phys. Rev. Lett.*, 2003, **90**, 065502.
33 D. Bratko, R. A. Curtis, H. W. Blanch and J. M. Prausnitz, *J. Chem. Phys.*, 2001, **115**, 3873–3877.
34 R. L. Jaffe, P. Gonnet, T. Werder, J. H. Walther and P. Koumoutsakos, *Mol. Simul.*, 2004, **30**, 205–216.
35 T. M. Truskett, P. G. Debenedetti and S. Torquato, *J. Chem. Phys.*, 2001, **114**, 2401–2418.
36 D. Bratko, C. D. Daub and A. Luzar, *Phys. Chem. Chem. Phys.*, 2008, DOI: 10.1039/b809072f.
37 A. Luzar and D. Bratko, *J. Phys. Chem. B*, 2005, **109**, 22545–22552.
38 D. Bratko and A. Luzar, *Langmuir*, 2008, **24**, 1247–1253.
39 A. Luzar, *J. Chem. Phys.*, 2000, **113**, 10663–10675.
40 N. Giovambattista, P. J. Rossky and P. G. Debenedetti, *Phys. Rev. E*, 2006, **73**, 041604.
41 K. Lum, D. Chandler and J. D. Weeks, *J. Phys. Chem. B*, 1999, **103**, 4570–4577.
42 K. Lum and A. Luzar, *Phys. Rev. E*, 1997, **56**, R6283–R6286.
43 J. Dzubiella, R. J. Allen and J. P. Hansen, *J. Chem. Phys.*, 2004, **120**, 5001–5004.
44 S. H. L. Klapp and M. Schoen, *J. Chem. Phys.*, 2002, **117**, 8050–8062; V. A. Froltsov and S. H. L. Klapp, *J. Chem. Phys.*, 2007, **126**, 114703.
45 A. Pertsin and M. Grunze, *J. Phys. Chem. B*, 2004, **108**, 16533–16539.
46 V. Ballenegger and J. P. Hansen, *J. Chem. Phys.*, 2005, **122**, 114711.
47 W. L. Jorgensen, J. Chandrasekhar, J. D. Madura, R. W. Impey and M. L. Klein, *J. Chem. Phys.*, 1983, **79**, 926–935.
48 L. X. Dang and T. M. Chang, *J. Chem. Phys.*, 1997, **106**, 8149–8159.
49 I. C. Yeh and M. L. Berkowitz, *J. Chem. Phys.*, 1999, **111**, 3155–3162.
50 J. L. England, S. Park and V. S. Pande, *J. Chem. Phys.*, 2008, **128**, 044503.

 This journal is © The Royal Society of Chemistry 2009

Ultrafast phase transitions in metastable water near liquid interfaces

Oliver Link,[a] Esteban Vöhringer-Martinez,[a] Eugen Lugovoj,[a] Yaxing Liu,[a] Katrin Siefermann,[a] Manfred Faubel,[b] Helmut Grubmüller,[c] R. Benny Gerber,[d] Yifat Miller[d] and Bernd Abel*[a]

Received 8th July 2008, Accepted 9th July 2008
First published as an Advance Article on the web 20th October 2008
DOI: 10.1039/b811659h

Electron spectroscopy for chemical analysis (ESCA) is a powerful tool for the quantitative analysis of the composition and the chemical environment of molecular systems. Due to the lack of compatibility of liquids and vacuum, liquid-phase ESCA is much less well established. The chemical shift in the static ESCA approach is a particularly powerful observable quantity for probing electron orbital energies in molecules in different molecular environments. Employing high harmonics of 800 nm (40 eV), near-infrared femtosecond pulses, and liquid-water microbeams in vacuum we were able to add the dimension of time to the liquid interface ESCA technique. Tracing time-dependent chemical shifts and energies of valence electrons in liquid interfacial water in time, we have investigated the timescale and molecular signatures of laser-induced liquid–gas phase transitions on a picosecond timescale.

1 Introduction

Liquid–gas phase transitions in water are an everyday occurrence. Beyond equilibrium thermodynamics extreme states of matter, *i.e.* the dynamics of superheated metastable water are of fundamental as well of technical interest but the molecular picture in general is not well resolved.[1,2]

Liquid water has many unusual properties when compared with simple organic liquids and bulk water as well as single water molecules play a decisive role in many chemical[3] and biological systems.[4] The dynamical structure of water and the molecular picture of its important role in chemical and biological systems, often associated with a distinctive local structure of hydrogen bonds,[5,6] has been studied with powerful spectroscopic techniques in the past;[7–10] however, this yielded a degree of understanding which is often insufficient. Water at extreme conditions (*e.g.* temperature and pressure) is also a field of continuous strong interest.[11–13] In particular, water in systems at very high temperatures and pressures was investigated without providing general conclusive molecular pictures. Its hydrogen-bond structure has been investigated theoretically and experimentally.[14] If water is heated

[a]*Institut für Physikalische Chemie der Universität Göttingen, Tammannstrasse 6, 37077 Göttingen, Germany. E-mail: babel@gwdg.de; Fax: +49 (0)551 393150; Tel: +49 (0)551 393106*
[b]*MPI für Dynamik und Selbstorganisation Göttingen, Bunsenstrasse 10, 37073 Göttingen, Germany*
[c]*MPI für biophysikalische Chemie Göttingen, Am Fassberg 11, 37077 Göttingen, Germany*
[d]*Department of Chemistry, Hebrew University of Jerusalem, Israel*

This journal is © The Royal Society of Chemistry 2009

above its critical point it forms a supercritical (sc) phase. Sc water plays an important role in many areas such as high-temperature chemistry,[15] hazardous-waste treatment and possibly even the origin of life.[16] The dynamics of water and solutes have been studied mostly in static cells with spectroscopic techniques in the time[17] and frequency domains.[15] Although the liquid–gas phase transition in water appears to be a trivial process in terms of thermodynamics, it is anything but trivial if one asks about the timescale of such a process, especially for superheated water, *i.e.* water heated significantly above its boiling point.

If short infrared pulses resonant on the water absorption of the OH-stretch vibration at around 3 μm are employed, it is possible to heat the water at a rate faster than the thermal expansion rate and to prepare extreme states of water, well above the boiling point and even significantly above the critical point.[7] We will refer to these states as *metastable*, *superheated* and in the case of $T > T_{\text{critical}} = 373.9\ ^\circ\text{C}$ as *supercritical*.[2] If water is prepared at these extreme states, the question is how the phase and the hydrogen-bonding network evolve in time. For water heated above its boiling point and below the critical point the system is expected to display a fast first-order phase transition (correlated with an enthalpy of transformation).[2] The question to ask here is: how fast? Is it possible to observe a spectral signature in an ultrafast experiment that provides the timescale and the spectral signatures of such fast phase transitions, as well as information about the evolving hydrogen-bonding network? In addition, what is the dynamics of a system heated above the critical point? At the critical point the system should display a continuous phase transition, also called a second-order phase transition. Such a transition is not correlated with latent heat.[1]

In this contribution we report investigations of the molecular photoemission signature of the evolution of the phase and the hydrogen-bonding network of superheated water below and above the critical point, as well as its timescales. The

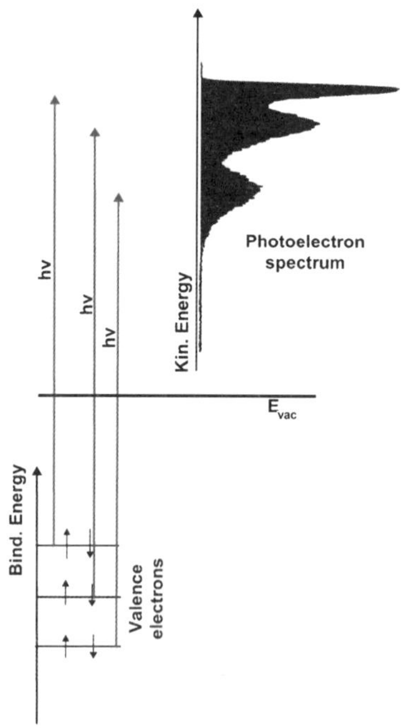

Fig. 1 Scheme illustrating the principle of photoelectron spectroscopy (high-energy photon produces an electron with a kinetic energy determined by the difference of the photon energy and the binding energy of the electron).

 This journal is © The Royal Society of Chemistry 2009

combination of several powerful technologies, *i.e.* photoelectron spectroscopy near volatile liquid interfaces in vacuum,[18] ultrafast pump–probe spectroscopy, and table-top high-harmonics generation of soft X-ray radiation[19] enabled us to add the dimension of time to the liquid interface ESCA (electronic spectroscopy for chemical analysis)[20] or liquid-phase photoelectron emission spectroscopy (*cf.* Fig. 1).

2 Experimental approach

In the present experiments a commercial Ti : sapphire laser system (Hurricane, Spectra Physics) was employed, delivering short pulses with a pulse length of 100 fs and a pulse energy of 1.3 mJ at a repetition rate of 1 kHz. The laser system employs the chirped pulse amplification (CPA) technique which incorporates typically four components, namely a Ti : sapphire oscillator, a grating stretcher, a regenerative amplifier and a grating compressor (Fig. 2). The internal laser setup was slightly modified by exchanging a mirror with a beamsplitter in front of the internal compressor. Three thirds of the internally amplified radiation were reflected by the beamsplitter and temporally compressed by the internal compressor to give 780 mW

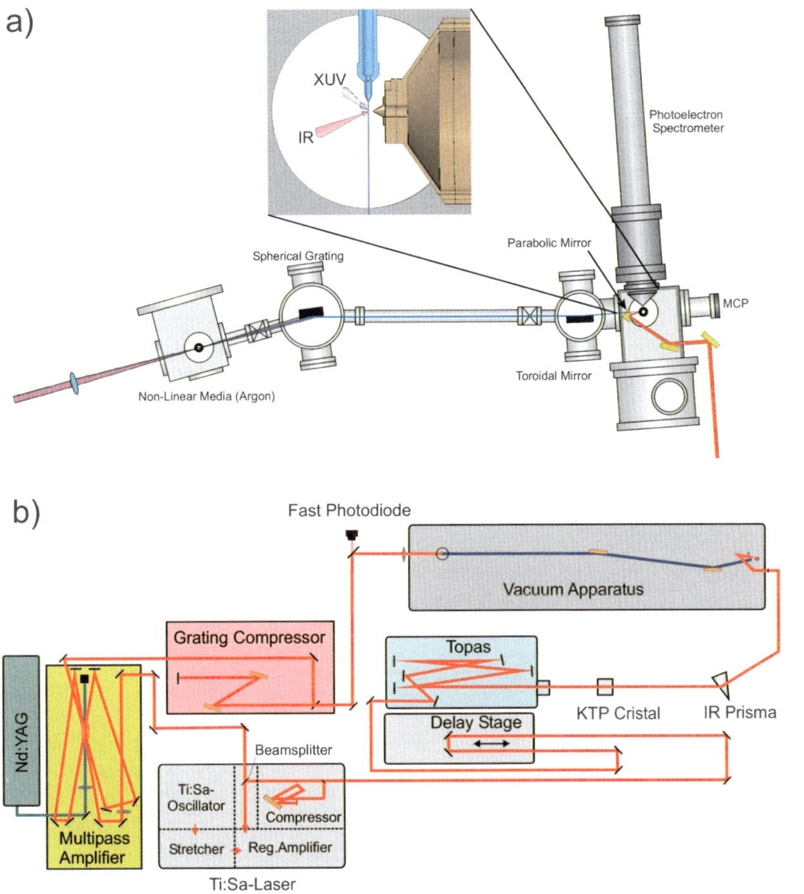

Fig. 2 (a) Schematic drawing of the vacuum apparatus consisting of four chambers and the TOF photoelectron spectrometer. In the inset the overlap of the IR-pump and the extreme ultraviolet or soft X-ray (EUV or XUV) probe pulse on the water filament is depicted. (b) Scheme of the optical setup, providing the IR-pump pulse (2.6–3.0 μm) and the EUV-probe pulse (39 eV).

of the standard 100 fs Hurricane output. The remaining 380 mW of the amplified but uncompressed pulses were transmitted and coupled out towards an external multipass amplifier. There an amplification up to 1.8 W takes place during three passes through an additional Ti : sapphire crystal (pumped by 10 W of an ORC 1000, Clark). Subsequent compression in an external grating compressor led to pulses of 1.2 mJ energy at 107 fs pulse duration. The pulses are slightly longer than the standard Hurricane pulses due to gain-narrowing taking place in the multipass amplifier.

The 780 mW of the regular Hurricane output were directed to an optical parametric amplifier (TOPAS, Light Conversion) for photon conversion into the near infrared (Fig 2). As the standard model of the TOPAS only supports IR generation until 2.5 μm due to the materials used, for extension of the wavelength region the TOPAS output was employed as the input for a second nonlinear process. In a subsequent KTP-crystal placed 15 cm behind the TOPAS exit an additional wave mixing process of the generated near IR around 1.1 μm and the remaining fundamental radiation at 800 nm took place.[21] Difference frequency generation and parametric amplification lead to radiation in the desired wavelength region at 2.6–3.0 μm. By optimizing the TOPAS settings pulse energies larger than 10 μJ could be obtained (up to 30 μJ at 2.7 μm). Finally a quartz prism (Infrasil) was used to separate the IR from the fundamental radiation and from other wavelengths resulting from the different wave-mixing processes.

The externally amplified 1.5 mJ pulses coming from the multipass amplifier are used for the generation of high harmonics.[22,23] The essential peak power of $>10^{14}$ W cm^{-2} is obtained by focusing the short pulses with a 150 mm lens into an argon-filled capillary with a 2.3 mm inner diameter resulting in peak intensities of 8×10^{14} W cm^{-2}. The argon pressure inside the capillary is about 50–200 mbar. The radiation enters and leaves the capillary through laser-drilled holes which have a diameter of 100 μm. In order to optimize the high harmonic output the phase mismatch between the fundamental and the harmonic radiation in the interaction region must be minimized.[24] This can be achieved by adapting the argon pressure and by tuning the parameters of the incident light in front of the focusing lens.

The vacuum chamber is pumped by a roots pump (350 m^3 h^{-1}) to remove the argon gas load from the chamber and keeping the background pressure at a level of 10^{-2} mbar in order to minimize the reabsorption of the XUV-radiation, which is less than 10% under the present conditions.

In the high-harmonics generation process several harmonics of the fundamental radiation are generated such that a dispersive element is necessary to separate a particular harmonic frequency. Therefore a spherical diffraction grating (700 lines per mm) is illuminated at a deviation angle of 161.8° exploiting the total external reflection. The grating has a radius of 17 100 mm and is therefore not focusing but collimating the high-energy radiation. To protect the grating from thermal damage by the fundamental radiation a 150 nm aluminum filter can be introduced in the optical path in front of the grating. In this experiment the 25th harmonic of the 800 nm radiation was used which corresponds to a wavelength of 32.1 nm and to a photon energy of 38.6 eV.

In the experimental chamber the IR-pump and the XUV-probe beam are overlapped in space and time on the liquid target (Fig. 2). The target is a liquid microjet which was introduced by Faubel and coworkers[25–27] (Fig. 3). A flux of 0.3–0.4 ml min^{-1} of the solvent (*i.e.* water) is pressed through a micronozzle with a high-pressure liquid chromatography (HPLC)-pump (Economy 2/ED, Techlab Instruments). The nozzle diameter is in the range of 10 μm giving a liquid filament in the experimental chamber of more or less the same diameter (Fig. 3). At velocities of 20–50 m s^{-1} the water filament remains continuous for about 2–3 mm until it breaks up into a stream of droplets (Fig. 3b) which is frozen out in a cooling trap with liquid nitrogen. A second cooling trap inside the chamber acts as a strong pump for gaseous water and allows in combination with a turbopump (1600 l s^{-1}) to keep the background pressure in the working apparatus at a level of 10^{-4} mbar.

 This journal is © The Royal Society of Chemistry 2009

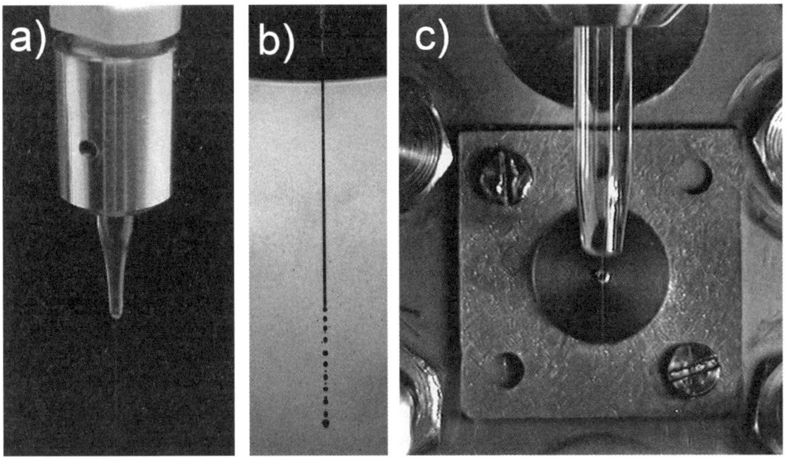

Fig. 3 (a) The quartz micronozzle producing a small water filament of ≈10 μm diameter. (b) The stable filament and its decomposition into a stream of droplets after a certain length. (c) The micro water beam in front of the spectrometer entrance where a skimmer of 100 μm diameter is mounted.

The IR-pump pulse is coupled into the experimental chamber *via* a CaF_2-window and focused onto the liquid filament with a 60°-off-axis parabolic mirror. The mirror has an aluminum surface and a focal length of 34 mm focusing the IR beam down to 70 μm (somewhat larger than the filament's diameter). The nearly collimated XUV-probe beam is directed in vacuum, under grazing incidence, onto a toroidal mirror which has an imaging length of $2f = 360$ mm (f being the focal length).[28] The mirror is covered with gold, has a size of 20 × 100 mm, and radii of curvature of 24.1 and 5375.3 mm respectively. The focus dimensions and quality depend strongly on the incident angle and thus on precise pre-alignment. The focal point is slightly asymmetric with a focal diameter in the spectrometer plane of 100–120 μm. The XUV-radiation can be monitored with a XUV-sensitive CCD camera (OD2562Z, Proxitronic) which is installed on the optical axis on the backside of the liquid beam chamber (Fig. 2). Thus it is possible to see the filament's shadow in the XUV-image which is useful for alignment purposes.

For detection of the emitted photoelectrons a time-of-flight electron spectrometer (Kaesdorf) is mounted to the side (perpendicular to the beam and the optical axis) of the experimental chamber (Fig. 2). A skimmer of 100 μm diameter separates the experimental chamber from the spectrometer and is part of a differential pumping stage towards a second inner aperture of 2 mm diameter. A vacuum better than 10^{-7} mbar can be maintained in the spectrometer under operation. The photoelectrons are ejected from the irradiated side of the liquid filament pointing towards the micro skimmer. Field-free conditions are maintained by a 3D Helmholtz coil construction to compensate for the earth magnetic field and other disturbing magnetic fields.

Under these conditions photoelectrons are sampled by the micro skimmer and propagate through the entrance of the spectrometer over a 60 cm long free-flight distance until they are finally accelerated (6.9 cm, 280 V) for higher detection efficiency towards the final MCP-detector. The 40 mm diameter MCP accepts a solid angle of 1.74 msr of the initial electron emission. The resolution of the spectrometer in this liquid-beam photoemission configuration has been determined to be 0.2 eV.

If the pump and probe pulses are temporally overlapped, the spatial overlap on the beam can be detected (in the photoelectron spectrum) by an appearance of electrons with high kinetic energy generated and accelerated by a mixed

This journal is © The Royal Society of Chemistry 2009

IR–XUV-multiphoton absorption process. This effect (signal) can also be employed for cross-correlation measurements ($\approx$500 fs for the present case) to estimate the pulse length of the XUV-probe pulse and the overall time resolution.[29]

3 Computational methods

Molecular dynamics simulation

All the simulations were carried out with the GROMACS software package.[30,31] Thereby Newton's equations of motion were propagated with a time-step of 1 fs using the SETTLE algorithm[30] to constrain bond lengths and angles. The water model used in the present MD calculations was the SPC model.[32] Test simulations employing other models (*e.g.* TIP4P)[33] gave comparable results. The water box system with dimensions of 30 $\times$ 4 $\times$ 4 nm consisted of 16 184 water molecules. It was equilibrated with a Berendsen thermostat and barostat at 300 K and 1 bar for 1 ns. Afterwards the system was heated up to the required temperature and post-equilibrated for 200 ps at constant volume. To allow an expansion of the dense phase into vacuum, the water box was translated by 300 nm in the direction of the long edge and the total length of the box was elongated to 500 nm. The phase transitions were simulated in a micro-canonical ensemble with the same time step for 100 ps. The short-range Coulomb interactions were calculated explicitly up to a distance of 1.2 nm, and for the long-range interactions the particle mesh Ewald (PME) method was used.[30]

Lifetimes of hot water aggregates

The starting structure of the different water aggregates were taken from *ab initio* geometry optimization[34] and minimized in the classical force field with the steepest descent algorithm to a convergence of 0.1 kJ mol^{-1}. After minimization, each atom was assigned a random velocity from a Maxwell–Boltzmann distribution at 775K. Therefore, each atom was given a random velocity from a one-dimensional Maxwell distribution in each direction (x,y,z) and the total kinetic energy of all atoms was calculated. This total kinetic energy with the number of degrees of freedom in the system was used to calculate the achieved temperature. The ratio of the achieved and the desired temperature T served as scale factor for all velocities.

For the simulations a time step of 0.1 fs for integration of Newton's equation was used and 101 trajectories with a total length of 100 ps each were simulated. The position of the atoms were recorded every 2 fs. The lifetime of an aggregate was defined as the time after which one of the O–O distances was longer than 4 Å during at least 200 fs.

4 Results & discussion

In the present experiments we used an ultrafast femtosecond near IR laser pulse (2.5–3 μm) to excite water resonantly *via* its OH-stretch vibration inducing subsequent ultrafast heating to internal energies between about 10 and 45 kJ mol^{-1} corresponding to hot transient phases between 450 and 1100 K. The evolution of the superheated metastable water phase is probed with pulsed ultrafast photoelectron emission spectroscopy, employing EUV radiation produced *via* high-harmonics generation (HHG) and a liquid water jet in vacuum.[25,35] The temperature as well as thermodynamic properties of the beam in vacuum can be characterized and controlled very well, as has been shown by us and others.[25,27] In the present experiment only liquid water has been investigated. For the ultrafast liquid interface ESCA technique to work the infrared pump (near 3 μm) and the EUV (high harmonics) "laser-like" probe beam (near 40 eV photon energy) have to be overlapped nearly collinearly in space and time on the microjet. The EUV probe beam alone in the present case ionizes water and produces photoelectrons from the $^1b_1$, $^3a_1$, $^1b_2$,

 This journal is © The Royal Society of Chemistry 2009

and $^2a_1$ (*cf.* Fig. 4, at negative delays) valence electrons of liquid water and some gas-phase water around the liquid microbeam.[25,36] Photoelectrons near the liquid interface (about 3 monolayers of water) can escape into vacuum nearly unperturbed[37] and can be sampled with a 100 µm (micro) skimmer at a distance close to the microbeam (100–500 µm) and their kinetic energy can be measured in a time-of-flight electron spectrometer. The electron spectrometer, differentially pumped between skimmer and spectrometer, enables us to record the entire photoelectron spectra (multiplex advantage) and a reference spectrum at negative delay times in single shots at 1 kHz repetition rate that are averaged for predefined time-steps of the delay of the two femtosecond (fs) pulses. The time-of-flight data are converted into kinetic energies of the photoelectrons and in turn into binding energies of the valence electrons depending upon local intramolecular electron densities and the molecular environment (*e.g.* hydrogen bonding). With this technology, time-dependent photoelectron spectra as a function of delay time between the IR-pump and the EUV probe pulses have been recorded. A time-series of spectra (with *perpendicular* polarization of pump and probe beams) is displayed in Fig. 4.

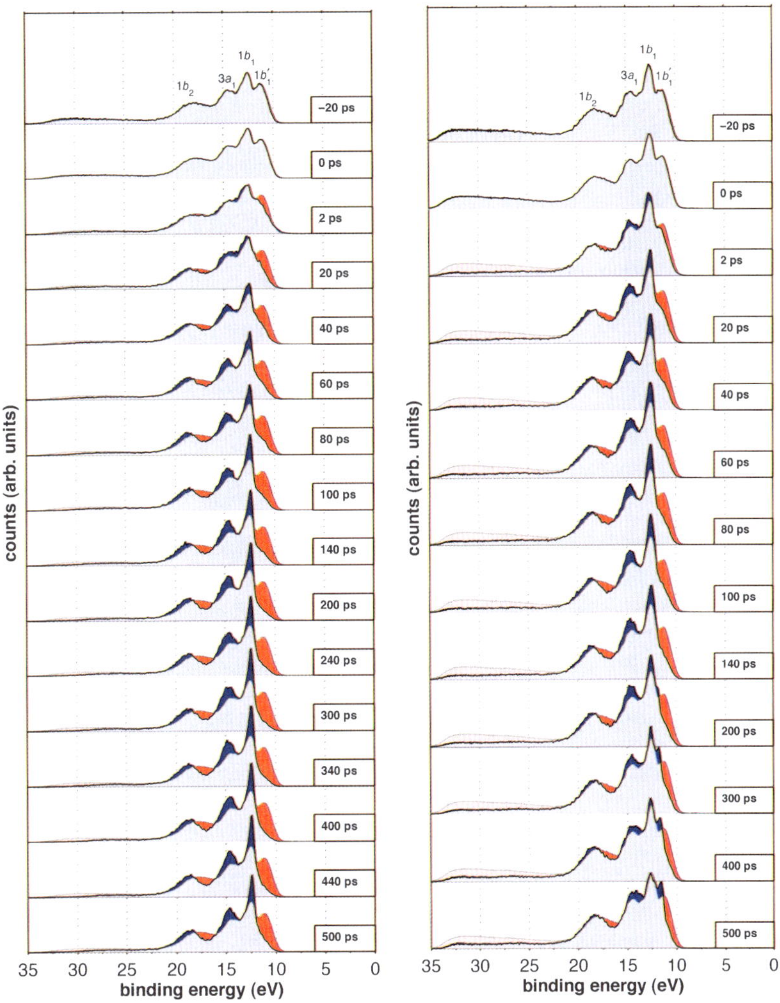

Fig. 4 Time-resolved photoelectron spectra of liquid and metastable water (2700 nm: left and 2900 nm: right).

The broad features of the $^1b_1$, $^3a_1$, $^1b_2$ and $^1b_{1,liq.}$ photoelectrons can be resolved. Prominent features of a photoelectron spectrum of water for the valence electrons are known and easily assignable to the $^1b_1$, $^3a_1$ and $^1b_2$ orbitals of isolated water in the gas phase.[25,36] In the present case a superposition of spectra from the liquid and the gas phase of water is observed. While somewhat narrower photoemission lines of the $^1b_1$, $^3a_1$ and $^1b_2$ electrons in the gas phase are overlapped with shifted (towards lower binding energies) and broader $^1b_{1,liq.}$, $^3a_{1,liq.}$ and $^1b_{2,liq}$ lines of liquid water, the $^1b_{1,liq.}$ photoemission line is resolved best because it is the line with the largest liquid–gas shift of all valence orbitals and observable in a region of least overlap.[25,38] The upper limit of the energy resolution in the photoelectron spectra has been measured for the narrowest lines to be 0.2 eV at present.

For clarity, the time-resolved spectra in Fig. 4 (solid lines, colored in blue) have been plotted with a color code displaying the differences to the reference spectrum at negative delays. Red displays a *decrease* and blue an *increase* in intensity relative to the reference spectrum. All spectra are normalized on a shot-to-shot basis to the corresponding reference spectra at negative delays and averaged until a satisfactorily signal-to-noise ratio is obtained. Usually, 100 000 shots are averaged for one time point. This procedure enabled us to eliminate shot-to-shot variations and energy drifts of the high harmonic radiation and the pump pulses. Tracing time-dependent chemical shifts and binding energies of valence electrons in superheated water in time, we have investigated the timescale and molecular signatures of the evolution of superheated water (with energy contents varied *via* the known absorption coefficient of water and the wavelength of the tunable IR-laser, see ref. 13 and references cited there) on a picosecond timescale with a time-resolution of about 500 fs (to be improved soon *via* multilayer mirrors[39] or a second grating[40]). If intensities at a particular chemical shift (*i.e.* binding energy) are monitored, time-profiles can also obtained instead of time-dependent spectra.

The positions (binding energies) of particular interest in the spectrum belong to: the low-energy wing of the (static) liquid $^1b_{1,liq.}$ photoemission line (10.5 eV), its maximum (11.16 eV),[38] a position between the $^1b_1$ and the $^1b_{1,liq.}$ emission maximum (11.6 eV) and the $^1b_1$ gas-phase emission maximum (12.6 eV). While the energy windows at 10.5 eV and at 11.16 eV both monitor the decay of the liquid phase, the 11.6 eV energy monitors the appearance of the supercritical phase and the presence of clusters and water aggregates at later times. The chemical shift at 12.6 eV monitors the increase of gas-phase water. As we watch the spectra as a function of time (*e.g.* in Fig. 4, left panel), the initial shift towards higher binding energy and a decay of the liquid $^1b_{1,liq.}$ peak at 11.16 eV is obvious (red). At the same time an increase of the $^1b_1$ gas-phase peak at 12.6 eV occurs (blue). Beyond these changes also intensity at 14.5 and 18 eV builds up which belongs to the gas-phase photoemission peak of the $^3a_1$ and $^1b_2$ electrons of water. At 17 eV intensity of the $^1b_{2,liq.}$ orbital vanishes as a function of time. These overall features (initial shift and decay of the liquid-phase photoemission) is a clear signature for a fast evaporation of the liquid after excitation with an IR-femtosecond pulse resonant with the OH-stretch of water in the liquid phase. The internal energies of the excited water phase have been estimated from the known absorption coefficients of water[13] and the penetration depth and have been determined to be ≈ 10 and ≈ 45 kJ mol^{-1} corresponding to ≈ 450 and ≈ 750 K respectively. While the first traces of Fig. 4 (left panel) correspond to temperatures of 450 K (superheated but below the critical temperature $T_{crit.}$) the right panel displays experiments for 750 K corresponding to a supercritical fluid. Beyond the overall features discussed above a new feature is transient intensity in a range of the chemical shift (*i.e.* at 11.6 eV) between the liquid-phase $^1b_{1,liq.}$ peak at 11.16 eV and the gas-phase peak at 12.6 eV. It is known that in this energy range photoemission of clusters occurs ($^1b_1$).[41,42] These features and trends are easily observed in the time-resolved traces of the chemical shift as a function of time (Fig. 4). For excitation at 2700 nm the intensity at 10.5, 11.16 and 11.6 eV is decreasing with a fast component within 5–10 ps and a slower component on a 500 ps timescale. At the same time

 This journal is © The Royal Society of Chemistry 2009

the gas-phase photoemission of the $^1b_1$ at 12.6 eV corresponding to gas-phase water increases. For excitation at 2900 nm near the maximum of the liquid-water absorption the intensity at 10.5 and 11.16 eV decreases on a faster (5–10 ps) and then apparently slower (but overall still fast) timescale of 100–150 ps. Contrary to the previous case, intensity increases again in the intermediate range of 11.6 eV for longer times while the gas-phase intensity at 12.6 eV ($^1b_1$) first increases and then decreases again (*cf.* Fig. 4, right panel). We attribute this feature tentatively to the appearance of new species, namely hot water aggregates or hot clusters, between the liquid and gas phases from a dispersion of the liquid and/or an ultrafast re-condensation of hot gas-phase molecules.

The preliminary conclusion from the experiment is that a superhot sub-critical or supercritical phase is formed with a characteristic spectral signature, *i.e.* an early shift towards higher binding energies. The origin of this chemical shift may be attributed to a weakening of the hydrogen-bonding network and the interaction being correlated with fewer hydrogen bonds per water molecule. Depending upon the energy content of the hot water phase, (i) the liquid disappears in a process that may be called fast evaporation or (ii) the liquid is dispersed. In the former case the loss of the liquid phase and the gain of the gas phase is irreversible. In the supercritical case the liquid-phase photoemission vanishes and the gas phase builds up temporarily. On a timescale of a few hundred picoseconds the gas phase vanishes again and photoemission in the 11.6 eV energy range monitors more or less large clusters[41] formed either from the disintegrating liquid or through re-condensation of the gas phase or both, when aggregates from the disintegrating liquid serve as nucleation seeds. The disappearance of the gas phase may suggest the latter scenario at later times. The experiment shows unambiguously that a superheated sub-critical and supercritical water phase have a finite lifetime and that the spectral features and its timescales can be observed with time-resolved ESCA. While the evolution of the sub-critical phase reresembles fast (normal) evaporation, the evolution of the supercritical phase results in a very fast dispersion of the fluid into a distribution of monomers and hot aggregates. This effect is gradually observed when changing the energy content of the water through changing the excitation wavelength of the IR laser from 2650 to 2900 nm (maximum of absorption). It must be kept in mind that in the present experiment mixed spectra of gas phase and liquid phase are always observed. It is also important to note that we observe approximately the same near-interface molecules (initially in the first three monolayers[37] of the liquid and then in the dense, evolving gas phase) throughout the evaporation process, *i.e.* the escape depth of electrons observed in our experiment depends upon the number density perpendicular to the water surface.

In order to obtain more insight into the processes the time-resolved data have been compared with molecular dynamics simulations of the process (evolution of a hot near-interface water phase for different internal energies).

The simulations are carried out with the GROMACS software package and the details are given in the computational methods section. For different times the evolution of the (extremely) heated water phase is displayed in Fig. 5. The water molecules are color-coded in such a way that the color corresponds to the thermodynamic state in the schematic phase diagram of water (shown in the inset) *if thermodynamic equilibrium is assumed.* Under this assumption the local density and the kinetic energy of each molecule was calculated and averaged over 2 ps. With the temperature calculated from the kinetic energy of the molecules (rigid SCP model for water), the pressure was calculated from the local density employing an extended equation of state for liquid water.[43] According to the pressure and temperature the assignment according to the phase diagram was made. From the simulation we see that the evolution of the superheated phase is mainly an expansion of a supercritical phase for quite some time. However, the density is quickly changing and the phase is obviously heterogeneous, which manifests itself in the presence of various transient aggregates, which form and partially dissociate again on a picosecond timescale.

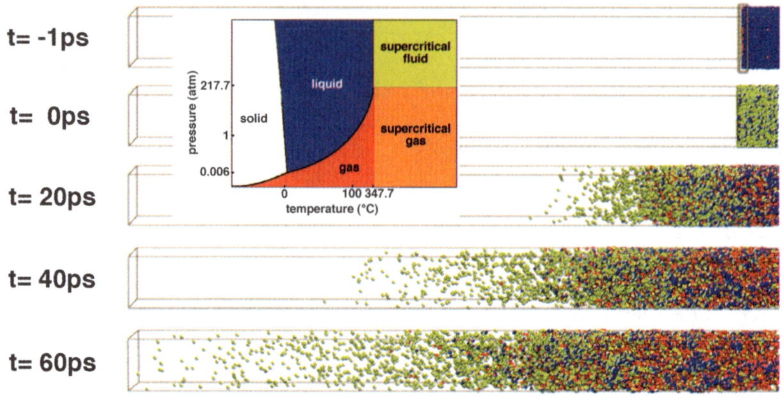

Fig. 5 Snaphots of the left side of the simulation box (total dimension 12 × 12 × 500 nm: shown 12 × 12 × 130 nm) with approx. 140 000 water molecules for different times. The snapshot at $t = -1$ ps represents the equilibrated box at 300 K and the one at $t = 0$ ps after the instantaneous heating-up process. The other snapshots are taken out of the simulation at constant energy and volume (NVE).

In order to shed light on the lifetimes of aggregates in the expanding overheated phase and the question whether such aggregates live long enough to be detected *via* ultrafast ESCA, we have calculated lifetimes of highly excited (high temperature) dimers, trimers and tetramers. Fig. 6 shows the distribution of lifetimes of dimers, trimers and tetramers for trajectories (775 K) for which dissociation could be observed (non- reactive trajectories were omitted). For the dimer (Fig. 6, upper panel) half of the dissociated aggregates have a lifetime smaller than 2 ps. The trimer and tetramer (Fig. 6, middle and lower panels) show lifetime distributions which are shifted to longer lifetimes. Only 20% of the dissociated trimer has a lifetime smaller than 2 ps and only 10% in the case of the tetramer. The remaining trajectories of the trimer and the tetramer are distributed over a wide range compared to the dimer and in the case of the tetramer even 30% of the trajectories exhibit an aggregate lifetime larger than 24 ps. In addition, the fraction of nondissociating trajectories increases for increasing aggregate size.

These results show that aggregates at supercritical conditions (starting initially in their minimum-energy configurations) have predicted lifetimes on the order of a few to tens of picoseconds, depending critically on the size of the aggregates. Furthermore, it is expected that larger aggregates consisting of 5–20 molecules exhibit even larger lifetimes and there should be no problems in detecting them in the present type of experiment. Because the initial conditions of the simulation may not necessarily reflect the real conditions in the experiment (in terms of aggregate formation) the lifetimes obtained may be regarded as rough estimates rather than precise predictions.

The current results strongly suggest that the phase evolution of superheated water is not necessarily a trivial process starting from the hot water phase and leading rapidly to isolated molecules in the gas phase. Depending upon the initial temperature a fast "normal evaporation" or a dispersion-like evolution of the phase may be observed in real-time with partial large heterogeneities in the aggregate size. In the future we aim to shed more light onto this phenomenon of transient heterogeneity in an expanding supercritical (hydrogen-bonded) phase which may be special for water and which may change for other systems depending upon the hydrogen-bonding interactions.

5 Conclusions and outlook

In summary, we have demonstrated that ultrafast electronic spectroscopy for chemical analysis (ESCA) near liquid interfaces (being conceptually an ultrafast extreme

 This journal is © The Royal Society of Chemistry 2009

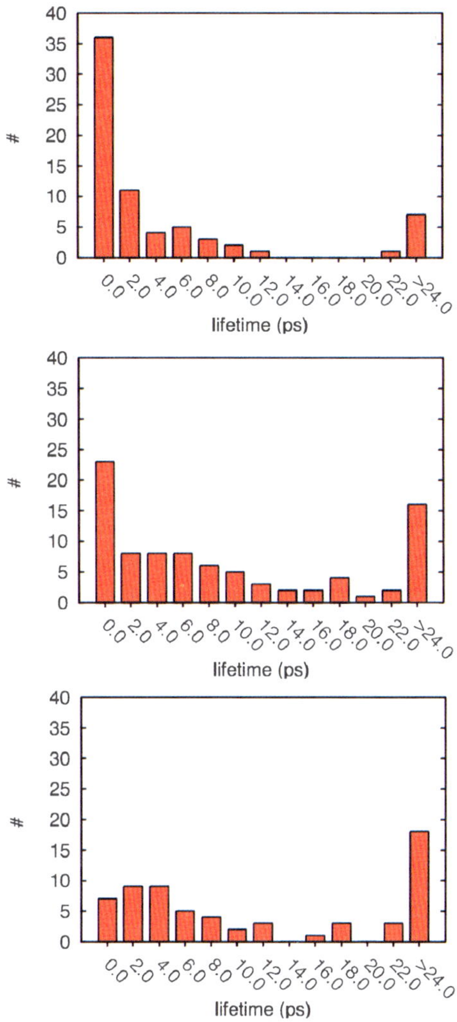

Fig. 6 Distribution of lifetimes of the dissociated aggregates (starting at minimum energy configuration). Upper panel: dimer (number (#) of dissociated aggregates: 70 out of 101). Middle panel: trimer (number of dissociated aggregates: 88 out of 101). Lower panel: tetramer (number of dissociated aggregates: 64 out of 101).

ultraviolet photoelectron spectroscopy) is a powerful new two-dimensional spectroscopic tool that can monitor electron orbital energies and chemical shifts in time in the condensed phase near liquid interfaces, and that it enables a new avenue for monitoring ultrafast processes in chemistry and biologic model systems near water and other liquid interfaces. In a first application of the technique, we have shed light onto the dynamics and evolution of superheated and supercritical water phases and related phase transitions which are interesting both from a fundamental point of view and for certain analytical applications.[44] With regard to the evolution of superheated phases we found that the ultrafast process is not necessarily a trivial one starting from the metastable phase and leading directly to isolated molecules. The features of the evolution appear to be governed by hydrogen-bond dynamics and energetics.

Together with theory, the main future goal of our group will be to develop the current technique into an ultrafast photoemission spectroscopy (in the spirit of liquid interface ESCA, but going much beyond), namely to trace molecular orbitals (and their autocorrelation functions) during molecular transformations near liquid interfaces. One of the main challenges will be the clear distinction of liquid surface phenomena and processes in the bulk (below the first mono layer), which could in principle be addressed in EUV wavelength-dependent experiments (due to different probing and electron escape depths).

Acknowledgements

This work was supported by the Deutsche Forschungsgemeinschaft (SPP1134, "Investigation of Transient Structures of Photo-Induced Chemical Reactions in Solution with Ultrafast XUV-Photoelectron Spectroscopy", Project AB63/7-1,2, SFB 755 "Nanoscale Photonic Imaging", Project B2, GK782 "Dynamics and spectroscopy of molecular aggregates, coils and networks"), the Fonds der Chemischen Industrie and the Volkswagenstiftung. The joint research project was also financially supported by the State of Lower Saxony and the Volkswagen Foundation, Hannover, Germany.

References

1 H. E. Stanley, *Introduction to Phase Transitions and Critical Phenomena*, Oxford University Press, Oxford, 1971.
2 P. G. Debenedetti, *Metastable Liquids: Concepts and Principles*, Princeton Academic Press, Princeton, New Jersey, 1996.
3 E. Vöhringer-Martinez, B. Hansmann, H. Hernandez-Soto, H. Hernandez, J. S. Francisco, J. Troe and B. Abel, *Science*, 2007, **315**, 497–501.
4 B. L. de Groot and H. Grubmüller, *Science*, 2001, **294**, 2353–2357.
5 B. Winter, E. F. Aziz, U. Hergenhahn, M. Faubel and I. V. Hertel, *J. Chem. Phys.*, 2007, **126**, 124504.
6 P. Wernet, D. Nordlund, U. Bergmann, M. Cavalleri, M. Odelius, H. Ogasawara, L. A. Nyslund, T. K. Hirsch, L. Ojambe, P. Glatzel, L. G. Pettersson and A. Nilsson, *Science*, 2004, **304**, 995–999.
7 Z. Wang, Y. Pang and D. D. Dlott, *J. Phys. Chem. A*, 2007, **111**, 3196–3208.
8 S. Ashihara, N. Huse, A. Espagne, E. T. Nibbering and T. Elsaesser, *J. Phys. Chem. A*, 2007, **111**, 743–746.
9 M. L. Cowan, B. D. Bruner, N. Huse, J. R. Dwyer, B. Chugh, E. T. Nibbering, T. Elsaesser and R. J. Miller, *Nature*, 2005, **434**, 199–202.
10 S. Woutersen and H. J. Bakker, *Phys. Rev. Lett.*, 2006, **96**, 138305.
11 A. F. Goncharov, N. Goldmann, L. E. Fried, C. Crowhurst, I.-F. Kuo, W. C. J. Mundy and J. M. Zaug, *Phys. Rev. Lett.*, 2005, **94**, 125508.
12 F. Bencivenga, A. Cunsolo, M. Krisch, G. Monaco, G. Ruocco and F. Sette, *Phys. Rev. E*, 2007, **75**, 051202.
13 A. Charvat, B. Stasicki and B. Abel, *J. Phys. Chem. A*, 2006, **110**, 3297–3306.
14 N. Goldmann and L. E. Fried, *J. Chem. Phys.*, 2007, **126**, 134505.
15 N. Akiya and P. E. Savage, *Chem. Rev.*, 2002, **102**, 2725.
16 L. J. Rothschild and R. L. Mancinelli, *Nature*, 2001, **409**, 1092–1101.
17 T. B. Brill, *J. Phys. Chem. A*, 2000, **104**, 4343.
18 B. Winter and M. Faubel, *Chem. Rev.*, 2006, **106**, 1176–1211.
19 P. Agostini and L. F. DiMauro, *Rep. Prog. Phys.*, 2004, **67**, 813.
20 M. Drescher, *Z. Phys. Chem.*, 2004, **218**, 1147–1168.
21 U. Emmerichs, S. Woutersen and H. J. Bakker, *J. Opt. Soc. Am.*, 1997, **14**, 1480.
22 M. Ferray, A. L. L' Huillier, X. F. Li, L. A. Lompre and G. Mainfray, *Phys. Rev. A*, 1989, **39**, 5751.
23 E. Seres, J. Seres, F. Krausz and C. Spielmann, *Phys. Rev. Lett.*, 2004, **92**, 163002.
24 *Coherent Sources of XUV Radiation*, ed. P. Jaegle, Springer, New York, 2006.
25 M. Faubel, in *Adv. Series in Physical Chemistry*, ed. C. Y. Ng, World Scientific, New York, 2000.
26 M. Faubel and T. Kisters, *Nature*, 1989, **339**, 527–529.
27 M. Faubel, S. Schlemmer and J. P. Toennies, *Z. Phys. D: At., Mol. Clusters*, 1988, **10**, 269.
28 J. Jasny, U. Teubner, W. Theobald, C. Wäulker, J. Bergmann and F. P. Schäfer, *Rev. Sci. Instrum.*, 1994, **65**, 1631.

 This journal is © The Royal Society of Chemistry 2009

29 M. Hentschel, R. Kienberger, C. Spielmann, G. A. Reider, N. Milosevic, T. Brabec, P. Corkum, U. Heinzmann, M. Drescher and F. Krausz, *Nature*, 2001, **414**, 509–513.
30 H. J. C. Berendsen, D. van der Spoel and R. van Drunen, *Comput. Phys. Commun.*, 1995, **91**, 43–56.
31 E. Lindal, B. Hess and D. van der Spool, *J. Mol. Model.*, 2001, **7**, 306–317.
32 B. Guilliot, *J. Mol. Liq.*, 2002, **101**, 219–260.
33 W. L. Jorgensen, J. Chandrasekar, J. D. Madura, R. W. Impey and M. Klein, *J. Chem. Phys.*, 1983, **79**, 926–935.
34 M. W. Schmidt, K. K. Baldridge, J. A. Boatz, S. T. Elbert, M. S. Gordon, J. H. Jensen, S. Koseki, N. MAtsunaga, K. A. Nguyen, S. Su, T. L. Windus, M. Dupois and J. A. Montgomery, *J. Comput. Chem.*, 1993, **14**, 1347–1363.
35 A. Charvat, E. Lugovoj, M. Faubel and B. Abel, *Rev. Sci. Instrum.*, 2004, **75**, 1209.
36 M. Faubel, B. Steiner and J. P. Toennies, *J. Chem. Phys.*, 1997, **106**, 9013–9031.
37 Y. Itikawa and N. Mason, *J. Phys. Chem. Ref. Data*, 2005, **34**, 1–22.
38 H. Agren and V. Carravetta, *Mol. Phys.*, 1985, **55**, 901.
39 Y. Lim, T. Westerwalbesloh, A. Aschentrup, A. Wehmeyer, G. Haindl, U. Kleineberg and U. Heizmann, *Appl. Phys. A*, 2001, **72**, 121.
40 L. Nugent-Glandorf, M. Scheer, D. A. Samuels, A. M. Mulhisen, E. R. Grant, X. Yang, V. M. Bierbaum and S. R. Leone, *Phys. Rev. Lett.*, 2001, **87**, 193002.
41 L. Belau, K. R. Wilson, S. R. Leone and M. Ahmed, *J. Phys. Chem. A*, 2007, **111**, 10075–10083.
42 O. Björnholm, F. Federmann, S. Kakar and T. Moller, *J. Chem. Phys.*, 1999, **111**, 546–550.
43 C. A. Jeffery and P. H. Austin, *J. Chem. Phys.*, 1999, **110**, 484–496.
44 A. Charvat and B. Abel, *Phys. Chem. Chem. Phys.*, 2007, **9**, 3335–3360.

This journal is © The Royal Society of Chemistry 2009

General discussion

Professor Finney opened the discussion of the introductory lecture by Professor Jungwirth: The ΔG values given for both H_3O^+ (−3 kcal mol^{-1} and OH^- (+1 kcal mol^{-1}) each came with an "error" of ±1 kcal mol^{-1}. Assuming this "error bar" is a standard error, this means that the 95% probability ranges of both these values overlap. Bearing in mind also the fact that these ΔG values are of the order of a few kT, this implies that the physical implications of the outcome of the simulations will be very sensitive to the simulations themselves (force fields, long range corrections, sample size *etc.*). I worry that too strong conclusions might be being drawn from the simulations.

Also, calculations were reported for (a) the singly hydrated proton H_3O^+ and (b) the unhydrated OH^- group. Are these necessarily the entities that should be considered? Have calculations been done for *e.g.* $H_5O_2^+$, $H_7O_3^+$, $H_3O_2^-$ *etc.*? If so, what were the results?

Professor Jungwirth replied: The confidence level of our ±1 kcal mol^{-1} statistical error in the free energy calculation is better than 85% (see also analogous calculations by Kubar *et al.*[1] or Yang *et al.*[2]). Therefore, our free energy values for hydronium and hydroxide with their statistical uncertainties do not overlap with each other, not even in the 95% probability range mentioned by Professor Finney. Concerning the hydrated proton, we performed classical MD calculations for both limiting cases, *i.e.* H_3O^+ (Eigen) and $H_5O_2^+$ (Zundel). The surface free energy minima for the two species are within 1 kcal mol^{-1} from each other. We did not simulate the delocalized form of hydroxide, $H_3O_2^-$. It would be worthwhile doing so, particularly in the light of *ab initio* DFT-based MD simulations which suggest an abundance of a rather delocalized form of hydroxide at the interface.

1 T. Kubar, M. Hanus, F. Ryjacek and P. Hobza, *Chem.–Eur. J.*, 2006, **12**, 280.
2 W. Yang, R. Bitetti-Putzer and M. Karplus, *J. Chem. Phys.*, 2004, **120**, 2618.

Dr Bakó opened the discussion of the paper by Dr Beattie: What is the driving force for increasing autolysis of water on the large surface area? We know that similar process can found on the metal(Pd, Rh)/water interface and in this case the driving force is the strong metal–H interaction. I do not know any specific interactions in this case.

Dr Beattie responded: My colleague Dr Angus Gray-Weale has developed a model that accounts quantitatively for the pH dependence of the ζ-potential of the hexadecane oil-in-water emulsions; it is presented in our poster. The force that attracts hydroxide ions in aqueous solution to regions of low dielectric constant arises because hydroxide ions strongly depress the dielectric constant of the solution. The ion constrains its neighbouring water molecules and these then cannot respond so easily to an external field. In water the fluctuating water dipoles are correlated and so are attracted to each other, similar to the fluctuating electronic dipoles that produce the dispersion interaction between two atoms. Suppression of the molecular dipole-moment fluctuations around an ion costs free energy, and this cost is least when the ion is close to a low dielectric surface, because a low dielectric is a region of low dipole-moment fluctuation.

Professor Debenedetti continued the discussion of the paper by Professor Jungwirth: Are your results sensitive to the way you treated electrostatic interactions (Ewald sum, reaction field) and to different models of water?

Professor Jungwirth answered: Qualitatively, results for slabs where Ewald summation is employed and for isolated large clusters (with no Ewald, of course) are similar, although quantitatively ion surface propensities tend to be stronger in finite-size clusters due to curvature effects (see, *e.g.* Stuart and Berne[1]).

As for water models, there is consistency among results with different polarizable potentials (POL3, Dang-Chang, *etc.*). Non-polarizable potentials (*e.g.*, SPCE) typically lead to smaller surface propensities of ions.

1 S. J. Stuart and B. J. Berne, *J. Phys. Chem. A*, 1999, **103**, 10300.

Professor Bain continued the discussion of the paper by Dr Beattie: Professor Jungwirth raised the question of why the surface tension, γ, of water was insensitive to pH if there was a surface excess charge due to hydroxide adsorption of $-5\ \mu C\ cm^{-2}$. You argued in response that the surface was already saturated with OH^- around neutral pH. I don't believe that this explains the independence of γ with respect to pH. According to the Gibbs equation for a 1 : 1 electrolyte, $d\gamma/d\ln c = -2RT\Gamma$, where the solution is ideally dilute and Γ is the surface excess (~0.5 $\mu mol\ m^{-2}$ according to your estimation). This equation gives a surface tension variation of around $-5\ mN\ m^{-1}$ per pH unit, which far exceeds that observed experimentally. The surface tension variation is determined not by the change in the hydroxide adsorption with pH, but by the surface excess of OH^-. The physical reason is that the surface tension involves the free energy of transfer of ions from the bulk solution to the surface. Changing the pH changes the activity of the ions in solution, even if it has little effect on the surface excess. Analogously, surfactant solutions show the greatest slopes in a γ *vs.* ln c plot when the surface excess is approaching the plateau in the adsorption isotherm (just below the critical micelle concentration).

Dr Beattie responded: Thank you for that clarification. The determination of the surface charge density of the hexadecane emulsion drops was made with a background electrolyte concentration of 0.2 mM NaCl, so the activity of the Na^+ remained constant. This removes the factor of 2 from the Gibbs equation (see Hunter[1]), but still leads to an interfacial tension variation of 2–3 $mN\ m^{-1}$ per pH unit from the surface concentration of the hydroxide. But the interfacial tension is a measure of the surface excess of ions integrated across the entire interfacial width, and so must take into account any compensating depletion of ions as well. There does not appear to be any report of the pH dependence of the interfacial tension of the alkane/water interface, either in the presence or absence of salt.

There also appears to be no determination of the surface charge of the air/water interface due to adsorbed hydroxide ions. We infer that it is similar to that of the alkane/water interface from the similarity of the pH dependence of the ζ-potentials of the two systems. There is scant information on the pH dependence of the surface tension. One report[2] indicates that there is no difference between the surface tension of water and air or nitrogen, without specifying any pH values. This is consistent with the assertion by Professor Jungwirth that the surface tension is independent of pH, again based on a comparison between air and nitrogen atmospheres.[3] Even this simple comparison is complicated, however, by the realisation that under nitrogen the only ions present are hydroxide and hydronium and the Debye length is very large, while under air there are also bicarbonate co-ions and the Debye length is somewhat shorter.

Manciu and Ruckenstein[4] were able to model the Jones–Ray effect (the decrease in the surface tension with electrolyte concentration around 1 mM salt) by postulating the adsorption of hydroxide ions and screening by the added electrolytes. But the surface charge density required was an order of magnitude lower than that inferred from the ζ-potentials. Karraker and Radke[5] also modelled the Jones–Ray effect by postulating an adsorbed hydroxide layer, again with a larger adsorption equilibrium constant and a smaller surface charge density than is inferred

 This journal is © The Royal Society of Chemistry 2009

from the oil emulsion results. Clearly a thorough investigation of the pH dependence of the surface tension and of the Jones–Ray effect is required.

1 R. J. Hunter, *Foundations of Colloid Science*, Oxford University Press, Oxford, 2nd edn, 2001, p. 69.
2 N. R. Pallas and B. A. Pethica, *Colloids Surf.*, 1983, **6**, 221.
3 P. Jungwirth, *Faraday Discuss.*, 2009, DOI: 10.1039/b816684f.
4 M. Manciu and E. Ruckenstein, *Adv. Colloid Interface Sci.*, 2003, **105**, 63.
5 K. A. Karraker and C. J. Radke, *Adv. Colloid Interface Sci.*, 2002, **96**, 231.

Dr Christenson continued the discussion of the paper by Professor Jungwirth: Why did you not show the surface tension *vs.* concentration curves for, say LiCl and HCl instead of LiOH and HCl. With the same anion the suggestions that Li^+ and H^+ are oppositely adsorbed is much more convincing. Li^+ is negatively adsorbed whereas H^+ is positively adsorbed. This offers better support for your contention that the surface of water is "acidic".

Professor Jungwirth answered: Indeed, comparing surface tension trends for aqueous LiCl (increasing) and HCl (decreasing) suggests that Li^+ is depleted from the interface while H^+ is adsorbed. In the same spirit, comparing KOH with KCl which both exhibit a comparable increase in surface tension with concentration (see *e.g.* Weissenborn and Pugh[1]) suggests that OH^- and Cl^- have a similar negative net surface excess.

1 P. K. Weissenborn and R. J. Pugh, *J. Colloid Interface Sci.*, 1996, **184**, 550.

Professor Tobias commented on the paper by Dr Beattie: I have a comment concerning the molecular structure of aqueous interfaces and its manifestation in the surface potential, which may be relevant to the interpretation of electrokinetic quantities such as the ζ-potential. Although the ζ-potential is a measure of the movement of charge, the organization of (electrically neutral) water molecules presumably affects how the charges sensed by measurement of the ζ-potential are screened. Second harmonic generation spectroscopic measurements reported by Eisenthal and co-workers in 1988[1] have demonstrated that water molecules are oriented at the air/water interface such that their dipole moments are pointed, on average, into the bulk. This implies that the air side is negative relative to the bulk water side (*i.e.* the surface potential is positive). Mundy and co-workers[2] have recently suggested, using first principles molecular dynamics simulations, that the surface potential is slightly negative (*ca.* −18 mV), and that the shape of the charge density is important (*i.e.* the dipole orientation is not the whole story). In our first principles MD simulation of the air/water interface, carried out using the same protocols as Kathmann *et al.*,[2] we have observed that the water dipole have a broad orientational distribution, but the average orientation is consistent with the SHG data of Eisenthal *et al.*[1] Shen and co-workers have very recently reported vibrational sum frequency generation spectra that show that the presence of ions changes the net orientation of water molecules at the air/water interface.[3] In particular, the presence of hydroxide anions at the interface appears to orient the water molecules such that their dipoles are pointed more directly into the bulk solution.

1 M. C. Goh, J. M. Hicks, K. Hemnitz, G. R. Pinto, K. Bhattacharyya, K. B. Eisenthal and T. F. Heinz, *J. Phys. Chem.*, 1988, **92**, 5074.
2 S. M. Kathmann, I.-F. W. Kuo and C. J. Mundy, *J. Am. Chem. Soc.*, 2008, DOI: 10.1021/ja802851w.
3 C. Tian, N. Ji, G. A. Waychunas and Y. R. Shen, *J. Am. Chem. Soc.*, 2008, **30**, 13033.

Professor Kim continued the discussion of the paper by Professor Jungwirth: In X-ray/neutron reflectivity and ellipsometry experiments for interfacial water at the hydrophobic surface, it has been observed that there is a water layer with a density lower than the bulk water. However, I am not sure if your data are showing it or not. If so, how does the segregation of ions correlate with the density-deficient layer? Can the ion segregation and density deficiency in the interfacial water layer at hydrophobic surfaces be explained with the same thermodynamic origin?

Professor Jungwirth answered: In our *PCCP* article[1] we have investigated several aqueous hydrophobic interfaces. A soft hydrophobic interface, such as that between water and oil or long chain alkane behaves similarly to the air/water interface, *i.e.* there is a ~5 Å interfacial region where the water density smoothly drops to zero while the alkane density smoothly increases to its bulk value. Water density oscillations are weak (if any); however, they may be somewhat enhanced if one looks at the intrinsic rather than geometric interface. The interface between water and a hard, rigid hydrophobic environment (*e.g.* graphene) is different. Water is strongly depleted next to the rigid hydrophobic surface and its density then exhibits strong oscillations. Water is simply more structured next to a hard *vs.* soft hydrophobic environment. This also influences surface segregation of ions, including hydronium and hydroxide. At soft hydrophobic interfaces there is a broad peak of hydronium and no peak of hydroxide. Next to the hard wall the hydronium peak narrows and a smaller peak of hydroxide appears. These peaks coincide with the first density maximum of water after the surface depletion layer next to the hard wall.

1 R. Vácha, D. Horinek, M. L. Berkowitz and P. Jungwirth, *Phys. Chem. Chem. Phys.*, 2008, **10**, 4975.

Dr Zaccai said: I have a comment regarding Professor Jungwirth's protein calculation. Experimental work on proteins from halophilic organisms (which live in saturated NaCl and accumulate saturated KCl in their cytoplasms) has shown how they are stabilised by binding hydrated salt ions. They are more stable in the presence of Na^+ than K^+ but dynamically softer in the presence of K^+ and so more suitable for enzyme activity.[1–4]

1 C. Ebel, L. Costenaro, M. Pascu, P. Faou, B. Kernel, F. Proust-De Martin and G. Zaccai, *Biochemistry*, 2002, **41**, 13234.
2 D. Madern, C. Ebel and G. Zaccai, *Extremophiles*, 2000, **4**, 91.
3 C. Ebel, P. Faou, B. Kernel, G. Zaccai, *Biochemistry*, 1999, **38**, 9039.
4 S. B. Richard, D. Madern, E. Garcin and G. Zaccai, *Biochemistry*, 2000, **39**, 992.

Professor Jungwirth replied: Thank you and it would be very interesting to try to relate the dynamical properties of these proteins to the Na^+ *vs.* K^+ binding patterns.

Dr Feibelman asked: What charge would go to the surface in D_2O? What is the magnitude of the zero point effects?

Professor Jungwirth responded: The effect of zero point vibrations is implicitly included in most empirical force fields (due to fitting to experimental measurables) but not in DFT; nevertheless, the results are semi-quantitatively consistent. Therefore, I assume that the effect of deuteration on ionic surface affinities will be small. But it would be certainly worthwhile (albeit computationally costly) to check it, *e.g.* by path integral simulations.

Professor Klein said: Professor Jungwirth, your simulations suggest that the surface of the water/air interface is positively charged (though the bulk of the experimental indications show the opposite). What would be the case (according to simulations) at the interface between pure water and a metal interface? We have recently

 This journal is © The Royal Society of Chemistry 2009

carried out direct measurements for such a situation at a gold surface, which show unequivocally that it is negatively charged, which we attribute to OH^- adsorption.[1]

1 L. Chai and J. Klein, *Langmuir*, 2007, **23**, 7777.

Professor Jungwirth replied: We have not studied the water/metal interface which I assume will have properties very different from those of the water/vapor interface due to the presence of free electrons and other chemical and steric issues concerning the former system.

Professor Klein also asked: I realize that simulations are limited in terms of real times that can be examined; but have simulations of water given any insight into the huge variation in relaxation/exchange rates for hydration water about different ions (which can vary from *ca.* 10^9 s^{-1} for the alkali metal ions to *ca.* 10^{-5} s^{-1} for some trivalent ions—a dynamic range of *ca.* 14 orders of magnitude)?

Professor Jungwirth responded: Molecular dynamics simulations reproduce reasonably well the exchange and relaxation times of water molecules around monovalent ions such as alkali cations or halide anions. They also correctly predict a dramatic decrease in water exchange and a strong stiffening of the solvent shell upon moving to polyvalent ions (such as magnesium).

Professor Lyashchenko commented in addition: Of course the question about the ions on the water/protein surface is very interesting, but it is very complicated case. Can you say something about the ions on water/ice interfaces? It is a typical hydrophilic surface and maybe it is more easy to analyse. On the other hand there is a specification for different ions: for example in the case of ammonium fluoride we have the solid solution with a high solubility in ice (7.44 mol%). In other cases we have a very small solubility of salts.[1]

1 A. Lyashchenko and G. Malenkov, *J. Struct. Chem. (Russian)*, 1969, **10**, 724.

Professor Jungwirth responded: Unlike water, ice is an extremely poor solvent for ions. Solubilities of ions in ice are (with few exceptions) below the micromolar range (see *e.g. Physics of Ice*[1]). Ions are, therefore, expelled from the ice phase in a process called brine rejection, which we have simulated recently.[2] We did not, however, investigate in detail the issue of ion segregation at a stable ice/water interface, which I think will be a more subtle issue than their segregation at the air/water interface.

1 V. F. Petrenko and R. W. Whitworth, *Physics of Ice*, Oxford University Press, Oxford, 1999.
2 L. Vrbka and P. Jungwirth, *Phys. Rev. Lett.*, 2005, **95**, 148501.

Professor Buch continued the discussion of the paper by Dr Beattie by commenting: Empirical, *ab initio*, EVB (empirical valence bond), on-the-fly dynamics consistently indicate an H_3O^+ preference for the surface. The extent of the surface affinity of OH^- or lack thereof is still uncertain, but ~10 kcal mol^{-1} is out of the question based both on computational and SFG results.

Professor Buch then asked: Is it possible that negative charging of bubbles originates from oxiolized organic impurities which become surfactants with COO^- groups pointing into solution?

Dr Beattie replied: Contamination is potentially a problem for low surface area measurements, but it is much less of a problem with the very large surface areas of the emulsions. We have purified hexadecane by multiple passages through alumina, but see no difference from measurements with the unpurified oil. The most compelling evidence against contamination as the source of the surface charge, however, is the striking agreement between ζ-potential measurements on emulsions containing on the order of 10^{14} droplets per litre and on single millimetre drops in a rotating electrophoresis cell.[1] The extension of this inference to bubbles is based on the similarities between the pH dependence of the ζ-potentials of oil drops and gas bubbles.

1 P. Creux, J. Lachaise, A. Graciaa, J. K. Beattie and A. Djerdjev, in preparation.

Dr Salmeron continued the discussion of the paper by Professor Jungwirth: The presence of foreign molecules and ions on the surface of water is a very real situation that is not easily acknowledged or measured quantitatively. In my group we have used X-ray photoelectron spectroscopy "*in situ*" on water surfaces. Invariably "cleanly" prepared surfaces became contaminated with hydrocarbon species, which then influence many properties of water: for example, the pre-melting of ice (*via* dangling H-bonds) and the segregation of ions.

Dr Cacace remarked: I wish to comment on one aspect of Professor Jungwirth's presentation concerning the description of local effect *vs.* "generalized" effects which could be correct descriptors of the Hofmeister series.

One of the interesting properties of the Hofmeister series resides in its ample range of applicability: its effects have been monitored on a wide number of biopolymers (nucleic acids, polysaccharides and of course proteins) but also seem to extend their reach to seemingly unrelated phenomena such as cloud formation and cement hardening. A list of the various examples of the Hofmeister series applicability may be found in a review article.[1] To get an insight into the effect of Hofmeister salts on macromolecules, it could be advantageous to investigate the behaviour of their components (aminoacids, nucleotides, *etc.*) in a direction similar to that already pointed out by Dr Zaccai, concerning the stability of proteins from halophiles in the present discussion.

1 M. G. Cacace, E. M. Landau and J. J. Ramsden, *Q. Rev. Biophys.*, 1997, **30**(3), 241.

Professor Jungwirth responded: I fully agree with this comment and suggestions therein. Indeed, Hofmeister series have been invoked for explaining a plethora of ion-induced effects in biochemistry and physical chemistry. The problem is that we are still lacking a complete molecular understanding which is likely due to the fact that several physical effects (ion-pairing, ion segregation at hydrophobic surfaces, *etc.*) are involved. This also leads to the fact that we sometimes observe a "reverse" or "bell-shaped" Hofmeister series. Quantifying ion affinities also to components of the studied macromolecules is exactly in the spirit of our investigations.[1,2]

1 L. Vrbka, J. Vondrasek, B. Jagoda-Cwiklik, R. Vácha and P. Jungwirth, *Proc. Natl. Acad. Sci. U. S. A.*, 2006, **103**, 15440.
2 J. Heyda, T. Hrobarik and P. Jungwirth, *J. Phys. Chem. B.*, submitted.

Mr Chen asked: It seems like quantum effects like zero-point energy (as mentioned earlier) and electron delocalisation effects leading to solvated electrons being formed in water may play important roles in the simulations. How much delocalisation has been seen in *ab initio* molecular dynamics simulations, and how much of it is a genuine physical effect and how much is due to computational artifice, given that functionals like BLYP (which was used in Professor Jungwirth's simulations) tend to overly delocalise electrons in general?

Professor Jungwirth replied: Electron delocalization is an important issue in simulations of ions like hydronium and hydroxide, which can be further complemented by zero-point nuclear delocalization of protons. GGA DFT functionals such as

 This journal is © The Royal Society of Chemistry 2009

BLYP or PBE indeed tend to overestimate the electron delocalization. In contrast, classical force fields artificially overlocalize these ions (albeit part of the electronic delocalization is recovered in polarizable calculations). Remarkably, the results concerning surface preferences of hydronium and hydroxide are semi-quantitatively consistent within these two approaches.

Dr Paulaitis enquired: In your talk, you compared ion partitioning at four water/hydrophobic material interfaces, and concluded that the one interface exhibiting behavior significantly different from the other three was the interface consisting of liquid water and a purely repulsive, hydrophobic wall. Unlike the water vapor/liquid and the liquid water/(liquid or solid) alkane interfaces, ions did not partition at this interface. Your explanation for the different behavior is the low water density—presumably, approaching that of water vapor—adjacent to the purely repulsive, hydrophobic wall. However, this cannot be the entire explanation since you also show that ions do partition at the water vapor/liquid interface. Indeed, for depletion layer of significant size, one might expect ion partitioning at an interface consisting of liquid water and a purely repulsive, hydrophobic wall to exhibit behavior intermediate to that between the water vapor/liquid interface and the liquid water/solid alkane interface, depending on the magnitude of the attractive water–wall interactions. Can you provide a further explanation why your simulation results for ion partitioning in the vicinity of this interface is so different from the other three interfaces?

Professor Jungwirth replied: Let me clarify two issues:

Firstly, the hydrophobic wall was rigid but not purely repulsive, having a standard carbon-like Lennard-Jones potential.

Secondly, hydronium and hydroxide ions do partition at this rigid wall; actually they do so more strongly than at the water/vapor or water/alkane interfaces. However, the character of this partitioning and the driving force is somewhat different. Next to the rigid wall, there is a layer which is depleted of water followed by an oscillatory density profile of a relatively structured water. Both hydronium and hydroxide exhibit a narrow peak (significantly larger for hydronium) which overlaps with the first peak in water density. Therefore, we conclude that for the rigid wall an additional effect comes into play which is connected with the structuring and density oscillations of the interfacial water. For further details please refer to our paper mentioned earlier.[1]

1 R. Vácha, D. Horinek, M. L. Berkowitz and P. Jungwirth, *Phys. Chem. Chem. Phys.*, 2008, **10**, 4975.

Dr Patel communicated to Professor Jungwirth: In neat water, autolysis leads to the formation of hydronium and hydroxyl ions. Experiments seem to suggest that the hydroxyl ions go to the interface and along with the hydronium ions form a double layer in order to satisfy charge neutrality. Does the small concentration of hydronium and hydroxyl ions in neat water preclude studying this double layer by simulations? If so, is it possible that the discrepancy between simulations and experiments is due to the fact that simulations study a solitary hydronium/hydroxyl ion at an interface, whereas experiments probe a double layer?

Professor Jungwirth communicated in reply: It is my understanding that the surface potential of pure water is primarily due to water dipoles (and possibly higher multipoles) and that the effect of hydronium and hydroxide at pH 7 is minor. Simulations with both the hydronium and hydroxide eventually lead to recombination (*ab initio* MD) or formation of a contact ion pair (classical MD). However, if the ions are kept apart a similar interfacial behavior is observed as for "solitary" hydronium and hydroxide.

Dr Cabrera-Sanfelix communicated: Could it be possible that the negative surface change is due to the orientation of the surface hydroniums? That is, the emergent H_3O^+ forms extended H-bonds with the H_2O molecules underneath, compensating its positive charge, keeping the O atom exposed on the water/air interface.

Professor Jungwirth communicated in response: H_3O^+ oriented with oxygen toward the vapor could indeed have a negative contribution to the surface potential from the (slightly) negatively charged O atom. However, this will only contribute to surface potential and not to the surface charge (or ζ-potential) since the O-atom does not physically detach from the rest of the ion.

Professor Chaplin remarked: The conclusion that the surface of pH-neutral water is positive cannot be fairly concluded from current modelling studies. The pHs of the modeled systems appear to be well below the isoelectric point as determined experimentally. For example, if there were just one hydronium ion to 71 water molecules (as in ref. 1), the pH would be 0.1 and even if there were 1000 water molecules the pH (1.3) is still far below an isoelectric point of pH 3. Therefore, such modelling does not predict the effects under neutral conditions. Note that the presence of non-combining hydroxide ions does not change this argument. In addition, the effects of the gas phase are not included in the modelled surface. It is well known that gas adsorbs preferentially at the water/gas interface[2] and such adsorption must affect the surface. The counter-argument that all the experimental evidence by many researchers over many years in many places (and which is self-consistent) is faulty due to adventitious negatively charged adsorbents (all coincidently apparently possessing the same properties and concentrations) has no foundation apart from wishfulness.

1 V. Buch, A. Milet, R. Vácha, P. Jungwirth and J. P. Devlin, *Proc. Natl. Acad. Sci. U. S. A.*, 2007, **104**, 7342.
2 R. Vácha, P. Slavek, M. Mucha, B. J. Finlayson-Pitts and P. Jungwirth, *J. Phys. Chem. A*, 2004, **108**, 11573.

Professor Jungwirth replied: It is true that simulating submicromolar amounts of hydronium and hydroxide corresponding to pH 7 is not directly computationally feasible. However, the surface adsorption of hydronium and lack of strong adsorption of hydroxide is remarkably revealed in simulations of slabs with varying sizes and ion concentrations as well as in simulations of large clusters, using both classical and *ab initio* molecular dynamics. It is also true that atmospheric gases (weakly) adsorb to water surface; however, their effect on surface ion segregation is negligible at ambient conditions. Finally, it is not true that all experimental evidence points to strong surface adsorption of hydroxide. On the contrary, spectroscopic experiments (*e.g.* second harmonic generation[1] and photoelectron spectroscopy[2]) show no or at most very weak surface adsorption of hydroxide. Moreover, adding alkali hydroxides to water results in an increase in surface tension which implies *via* the Gibbs adsorption equation a net surface depletion.[3] This does not in principle rule out a small surface peak compensated by subsurface depletion (as shown *e.g.* for iodide). However, the surface excess of hydroxide of -5 to -7 μC cm^{-2} deduced from titration measurements by Beattie *et al.*[4] is so large that it would lead in pure water either to a centimetre-thick subsurface layer completely depleted of hydroxide or to an unrealistic decrease of surface tension with increase of pH (units of mN m^{-1} per units of pH).

1 P. B. Petersen and R. J. Saykally, *Chem. Phys. Lett.*, 2008, **458**, 255.
2 R. Weber, B. Winter, P. M. Schmidt, W. Widdra, I. V. Hertel, M. Dittmar and M. Faubel, *J. Phys. Chem. A*, 2004, **108**, 2625; and *Chem. Phys. Lett.*, to be submitted.
3 P. K. Weissenborn and R. J. Pugh, *J. Colloid Interface Sci.*, 1996, **184**, 550.

 This journal is © The Royal Society of Chemistry 2009

4 J. K. Beattie, A. M. Djerdjev and G. G. Warr, *Faraday Discuss.*, 2009, DOI: 10.1039/b805858j.

Dr Fraxedas asked: Concerning the calculated halogen–biomolecule surface interactions, the molecular structures used probably correspond to their stable configurations, *i.e.* in liquid media (water containing ions). Biomolecules in "vacuum" would collapse. Is the liquid "jellium" taken into account in the simulations?

Professor Jungwirth replied: For simulations of ion–protein interactions in water we started with crystal structures from the Protein Data Base (PDB). These structures were stable in our aqueous simulations at the ~10 ns timescale.

Ms Mohana Sundaram asked: In your simulation study of specific binding of halide ions to the amino acids—arginine, lysine and histidine—you observed more fluoride ions closer to these charged amino acids than iodide ions. Can this observation be explained by the fact that fluoride is a smaller ion than iodide; hence, it can approach the charged species more closely? If not, what measure did you employ to correct for this size difference to show that the selective partitioning of fluoride ions is due to the specific interaction? Note the recent discussion[1] of the excluded volume correction needed for a meaningful interpretation of partitioning of glycerol and water on a protein surface.

1 I. L. Shulgin and E. Ruckenstein, *J. Phys. Chem. B*, 2008, **112**, 5876.

Professor Jungwirth responded: The issue of different sizes is an important one when comparing two ions in terms of their strength of ion pairing with a third ion or charged group. Indeed, fluoride penetrates closer to the positively charged side chain groups than iodide; however, the observed preference for F^- cannot be explained only by this effect. This is clear from integrated amounts of anions in the vicinity of the charged side chains, where integration is carried out up to the first minimum after the first maximum at the corresponding distribution functions.[1] In other words, we integrate over a larger region for larger anions; nevertheless, there is still more paired F^- than heavier halides. Note that this agrees with measurements of activity and osmotic coefficients, which also indicate that smaller halides pair more strongly with the ammonium cation than larger halides.

1 J. Heyda, T. Hrobarik and P. Jungwirth, *J. Phys. Chem. A*, submitted.
2 S. Lindenbaum and G. E. Boyd, *J. Phys. Chem.*, 1964, **68**, 911.

Dr Christenson communicated: In her paper Professor Dr von Klitzing mentions the decrease in surface potential of symmetric, free-standing films of alkyl polyglycol ethers and alkyl glucosides. We have found a similar decrease in surface potential and surface charge of mica surfaces on adsorption of several different types of nonionic compounds, *i.e.* pentaoxyethylene dodecyl ether[1] as well as sucrose and cyclodextrin.[2] We also suggested that results with dextran[3] could be interpreted as being due to a lowering of surface potential. We assumed at the time that the reason for the surface potential decrease was a reduction in the effective dielectric constant near the surface. This would make the presence of charges at the surface less favourable.

1 M. W. Rutland and H. K. Christenson, *Langmuir*, 1990, **6**, 1083.
2 P. M. Claesson, H. K. Christenson, J. M. Berg and R. D. Neuman, *J. Colloid Interface Sci.*, 1995, **172**, 415.
3 E. Perez and J. E. Proust, *J. Phys. Lett.*, 1985, **46**, 79.

Professor Dr von Klitzing communicated in reply: That is an interesting point. We have no experimental evidence for a decrease in dielectric constant, but we cannot exclude it. Perhaps an experiment with pyrene or another sensor for micropolarity could be helpful. On the other hand the hydrophobic pyrene could be wrapped by surfactant which could pretend an additional decrease in micropolarity.

Professor Dr von Klitzing then opened the discussion of the paper by Professor Luzar: What is the structure of water close to the molecule/water interface? Is there depletion? If not, why is the orientation relative to the molecule so important?

Professor Luzar responded: Fig. 1a–d in my paper illustrates the water structure next to the hydrophobic particles we study. The hydration water layer, albeit somewhat depleted in comparison with a hydrophilic surface, generally adheres to the wall. Note that the density of water in the first hydration layer at surfaces with contact angle 135° is still about 75% of the density at the most attractive/hydrophilic case we study (contact angle of 69°). The orientation effects would clearly weaken with increasing depletion. An orientation preference of water molecules does, however, exist at liquid/vapor interfaces as well.[1] This is due to strong hydrogen bond forces in water, favoring an interface with single broken hydrogen bond both for the liquid/vapor interface[1] and near a large hydrophobic region.[2,3]

1 A. Luzar, S. Svetina and B. Zeks, *Chem. Phys. Lett.*, 1983, **96**, 485.
2 A. Luzar, S. Svetina and B. Zeks, *J. Chem. Phys.*, 1985, **82**, 5146.
3 C. Y. Lee, J. A. McCammon, and P. J. Rossky, *J. Chem. Phys.*, 1984, **80**, 4448.

Professor Rovere asked: What could be the effect of changing the hydration inside the pore?

Professor Luzar answered: In our calculations hydration is modulated through contact angle variation. Within the range of contact angle between 69 and 135°, only small changes in calculated torques were observed (see our Fig. 4).

Dr Patel remarked: It was commented that the mean water density profiles in the slit between hydrophobic walls shown in your Fig. 1a do not show depletion (absence of layering), in contrast with previous results near extended hydrophobic solutes. I would like to point out that previous studies have indeed observed depletion near an extended "ideal" hydrophobic surface, which has no attractions. However, in the presence of typical dispersive attractions between the wall and the water molecules, as in the present study, one does not see depletion, but a layering of the mean water density into solvation shells.

Professor Luzar replied: Yes. Let me rephrase your comment so there is no ambiguity left due to semantics. For purely repulsive extended surfaces (hard walls) in water at ambient conditions (*i.e.* close to liquid–vapor coexistence) the liquid is prone to recede from the surface with a density profile resembling a free vapor/liquid interface, and the contact density is given by the contact theorem.[1,2] This results in long wavelength fluctuations at the emerging liquid/water interface, analogous to capillary waves, in addition to small length scale fluctuations near the large hydrophobic solute.[3] Capillary waves quench the oscillations in the water density profile.

Note that Fig. 1a in our paper presents results for a strongly hydrophobic surface characterized by a contact angle of 135°. The application of a perpendicular electric field results in considerable electrostriction in such a hydrophobic pore characterized by high compressibility (Fig. 2 in our work and Fig. 3a in ref. 4). As we have already pointed out in this paper, this compressibility is associated with the existence of *depleted water layers* next to such strongly hydrophobic walls.

 This journal is © The Royal Society of Chemistry 2009

Specifically, Fig. 2 shows approximately 1 molecule nm^{-2} surface excess at hydrophilic walls (contact angle = 69°) and ~2.5 molecules nm^{-2} depletion at weakly attractive hydrophobic walls (contact angle = 135°) relative to walls characterized by varying $\cos\theta_c(\Delta\gamma = 0)$.

1 R. Evans, in *Liquids at Interfaces : Proceedings of the Les Houches Summer School, Course XLVIII, May–June, 1988*, ed. J. Charvolin, J. F. Joanny, and J. Zinn-Justin, North Holland, Amsterdam, 1990, and references therein.
2 F. H. Stillinger, *J. Solution Chem.*, 1973, **2**, 141.
3 See, for example, K. Lum and A. Luzar, *Phys. Rev. E*, 1997, **56**, R6283.
4 D. Bratko, C. D. Daub, K. Leung and A. Luzar, *J. Am. Chem. Soc.*, 2007, **129**, 2504.

Professor Debenedetti said: Firstly, do you have an intuition as to how non-smooth surfaces (especially real hydrophilic surfaces, with anisotropic interactions) would enhance or suppress the effect you report?

Secondly, torque is a vector not a scalar. In your paper, the formalism you present yields a scalar (the magnitude of the torque). Have you considered a more general formalism where torque appears as a vector?

Professor Luzar replied: As to your first question, in the molecular dynamics simulations (which we are currently performing) using molecularly rough surfaces, we observe similar behavior as that observed with the smooth ones presented in this paper. Inclusion of specific, angle-dependent wall–water interactions, present in most hydrophilic materials, would clearly alter and possibly strengthen the dependence of wetting surface free energies on the field angle.

With respect to the torque, we have not considered a more general formalism. We only give the magnitude since there is no ambiguity about the direction of the aligning effect.

Professor Yamaguchi asked: How does the interparticle distance affect the torque? When the distance is large enough, does the torque not work?

Professor Luzar answered: As we point out in our paper, the new mechanisms we predict are not limited to confinements, and can equally operate on isolated surfaces. The incentive toward wall alignment with the field depends only weakly on the pore width (see *e.g.* our Fig. 3). The net effect therefore originates almost exclusively in the first solvation layer, regardless of the amount of the intervening liquid between them.

Professor Klein enquired: Did you notice any difference in the dynamic behaviour of molecules when subjected to high fields? I ask because at values of the field of $E \approx 10^9$ V m^{-1} (similar to some of your simulations) the energy of the water dipole, E^D, becomes comparable with its thermal energy k_BT, suggesting that the dynamic properties of water might be strongly perturbed.

Professor Luzar replied: In this paper we do not address dynamic issues, but are currently studying them.

We notice slower molecular dynamics only in electric fields 2 to 3 times stronger than reported here. Note that polarization field reduces the average field on individual molecules by approximate factor of $1/\varepsilon_r$, where ε_r is the relative permittivity. At applied fields $|\boldsymbol{E}_0| = \mathcal{O}(10^9 \text{ V m}^{-1})$, the dipole energy $\boldsymbol{E}_0\boldsymbol{D}/\varepsilon_r$ ($\boldsymbol{D}$ stands for the dipole moment) is of the order of $0.1k_BT$, which is small compared to intermolecular interactions, including hydrogen bonds.

Dr Feibelman asked: Should the orientational forces reduce the temperature at which ice nucleates at the surface?

Professor Luzar answered: We studied field strengths too weak to result in electro-freezing conditions. For comparison, in bulk water at 298 K we observe electrofreezing at very high field strengths $E_0 = E\varepsilon_r = \mathcal{O}(1\ \text{V Å}^{-1})$ and above,[1] in agreement with others.[2,3]

1 D. Bratko, C. D. Daub and A. Luzar, *Phys. Chem. Chem. Phys.*, 2008, DOI: 10.1039/b809072f.
2 G. Sutmann, *J. Electroanal. Chem.*, 1998, **450**, 289.
3 I. C. Yeh and M. L. Berkowitz, *J. Chem. Phys.*, 1999, **110**, 7935.

Professor Jungwirth asked: Are the relevant electric fields realizable by macroscopic devices or rather by charged nanostructures?

Professor Luzar replied: Both. Our actual, dielectrically screened fields,[1] $E = E_0/\varepsilon_r$, are comparable to those present next to an electrified AFM tip.[2] In nature on the other hand there are even stronger local fields in electric double layer of a DNA molecule,[3] or a synthetic polyelectrolyte,[4] and in ion channels.[5]

1 See also C. D. Daub, D. Bratko, K. Leung and A. Luzar, *J. Phys. Chem. C*, 2007, **111**, 505.
2 A. Phillippsen, W. Im, A. Engel, T. Schirmer, B. Roux and D. J. Muller, *Biophys. J.*, 2002, **82**, 1667.
3 K. Wagner, E. Keyes, T. W. Kephart and G. Edwards, *Biophys. J.*, 1997, **73**, 21.
4 D. Bratko and D. Dolar, *J. Chem. Phys.*, 1990, **80**, 5782.
5 J. Dzubiella and J. P. Hansen, *J. Chem. Phys.*, 2005, **122**, 234700.

(C5:60) **Dr Cabrera-Sanfelix** communicated: How much does this reorientation of the nanoparticle depend on the volume of water surrounding it? Could it be possible to have the same effect only with 2 or 3 ML of water adsorbed on the nanoparticle?

Also, in your abstract you talk about a "Janus interface". Could you give an explanation about it?

Professor Luzar communicated in reply: The effect originates primarily in the first hydration layer, and the amount of surrounding water is not essential. As we discuss at several places in our Faraday paper,[1] when an electric field is applied in the direction perpendicular to the confining hydrophobic surfaces, the competition between field-induced alignment and orientational preference of interfacial water molecules relative to the surfaces results in asymmetric wettability of opposing surfaces.[2] Granick was the first to call such an asymmetric hydrophobic/hydrophilic confinement a "Janus interface".[3] The name comes from Roman mythology where Janus, the god of gates and doors, was usually depicted with two heads looking in opposite directions. Note that Granick starts in his experiment (addressing completely different physical phenomenon) with one hydrophobic and one hydrophilic surface, while we start with two hydrophobic surfaces and under the influence of an electric field end up with water favorably attracted to the surface on one side (hydrophilic) but abhorred on the other (hydrophobic).

1 D. Bratko, C. D. Daub and A. Luzar, *Faraday Discuss.*, 2009, DOI: 10.1039/b809135h
2 D. Bratko, C. D. Daub, K. Leung and A. Luzar, *J. Am. Chem. Soc.*, 2007, **129**, 2504.
3 X. Y. Zhang, Y. X. Zhu and S. Granick, *Science*, 2002, **295**, 663.

Mr Hooton addressed Professor Luzar: I don't know much about this work or area, but a comment was made about two gaps of 16.4 and 27 Å and it was suggested that these were very small.

This is just a comment. I understand that the work is investigating how an electric field effects the orientation of water molecules, and in turn how the water molecules affect nanoparticles, such as DNA. This work uses a water slab in a periodic boundary condition unit cells of 16.4 and 27 Å. These dimensions seem very small, if this is a study of bulk water systems and general behaviour in what appears to be

 This journal is © The Royal Society of Chemistry 2009

a classical simulation using molecular mechanics and Lennard-Jones potential.[1,2] These references do simulations of protein folding, and molecular dynamics, and they use bigger unit cells.

I don't know a lot about it, but I've worked with period boundary conditions and we used 67 Å unit cells. 16.4 Å is approximately ⅓ of the width of a histone complex, which is 55 by 110 Å. It's also only about 16 times the diameter of a hydrogen atom, which has a Bohr radius of 0.5 Å, so it's probably only about 5 times the diameter of a water molecule, but maybe about 3 times to 6 times the actual length of a water molecule as the crow flies! Are these sizes not too small for general bulk water studies of how electric fields affect water, and how water affects macromolecules if this is what they are trying to do? The water slab is not much wider than a water molecule.

1 I. Daidone, A. Amadei, D. Roccatano and A. Di Nola, *Biophys. J.*, 2003, **85**, 2865.
2 A. Caflisch and M. Karplus, *Proc. Natl. Acad. Sci. U. S. A.*, 1994, **91**, 1746.

Professor Luzar answered: The water-mediated mechanism we report is of interest when acting on particles with no charges of their own. DNA is clearly not in this class.

Selected gap widths are shown to be sufficient to capture what are essentially surface effects on confinement walls. Our trial calculations for wider systems increasing the width of the intervening aqueous film bear a minimal effect on orientational forces we determined and do not justify the associated computational cost. I also addressed this in my earlier answer to Professor Yamaguchi.

Dr Christenson opened the discussion of the paper by Professor Abel: Fig. 5 of your paper uses the phase diagram of bulk water for a comparison with your results. Yet you are probing perhaps two layers of water molecules at the surface of water. Isn't your system therefore likely to be rather different in behaviour? Or perhaps hardly relevant at all?

Professor Abel replied: In the experiment we probe approximately 3 monolayers of water in the vicinity of the interface but with significant contributions of bulk and interface water (which shows up in the chemical shifts). The MD calculations are made to shed light into the time evolution of the hot water phase and in order to explain the inhomogeneity in case of supercritical heating. We do not simulate any observable parameters. The color-coding is not an exact assignment to the equilibrium phases in the inset of Fig. 5 but is instead a color code for the density and the kinetic energy of the molecules. Then we correlate the pressure and temperature from an extended equation of state with conditions in an equilibrium phase. So, the point here is not to pretend that we have near-equilibrium conditions but to understand the dynamics of the system. We measure the interface molecules and significant contributions from the bulk and the MD calculations provide the whole overall picture. Therefore, I think that the calculations are important and that we do not compare different things.

Dr Bovensiepen remarked: My first question is: how is the excited state characterised after a resonant vibrational excitation? What is the time-dependent evolution of the temperature or energy content?

Secondly, since an extreme non-equilibrium state is analysed in the experiment what can we learn about the phase transitions which occur in the thermodynamic equilibrium?

Professor Abel answered: In water the evolution of intramolecular and intermolecular vibrational energy redistribution is on the order of 50–500 fs. This is known for example from experiments of the Elsässer group in Berlin. Our experiment sets in or starts its observation when this is over and energy equilibration is accomplished. In the future we plan to do the same experiments on a much shorter time scale in order to address this question (and shorter timescales) further. At the moment the observable we measure does not depend upon the history of IVR and VET among molecules. I think that we can hardly learn anything about phase transitions under equilibrium conditions at relatively moderate conditions (temperatures and pressures). We indeed learn something about the lifetime of superheated water (which is controlled by hydrogen bonding at these temperatures) which is not well understood at all. We also learn some things about the supercritical phase which is a medium for many chemical reactions but which is not necessarily well understood. If you look in the literature you find that people are arguing that (for several reasons) it should simply not be possible to heat a liquid above its critical point. This was shown here together with the timescale of the phase evolution.

In summary, it is the more extreme states of matter that we would like to investigate rather than "normal" phase transitions. In this context it should be noted that this is also true for strongly cooled liquid water phases, which are accessible with our technology. I believe that much less is known about extreme non-equilibrium states than for equilibrium states and that they are the more interesting things to study.

Dr Feibelman asked: Can the presence of OH at surfaces not be seen? Or can it be discounted given the 1.6 eV splitting of the O(1s) X-ray absorption lines from OH and H_2O? We know the O absorption cross section and thus, given the supposed concentration of OH, we know how intense its signal should be.

Professor Abel responded: In principle this may be seen in an ESCA experiment if the concentration is high enough (its concentration in 55 mol per litre of water in water). In the past the latter was often a problem even when people worked with quite high concentrations of precursors.

Dr Feibelman commented: Hot water doesn't explain the chirality of life. Astrophysical seeding seems a more likely explanation. See, *e.g.* papers by Cronin and Pizzarello[1] and by Engel and Macko.[2]

1 J. R. Cronin and S. Pizzarello, *Science*, 1997, **275**, 951.
2 M. H. Engel and S. A. Macko, *Nature*, 1997, **389**, 265.

Professor Abel replied: I think that the question is very interesting but beyond the scope of our contribution. Our (small) side point in the article was that supercritical water in deep-sea vents (maybe near special minerals) may have played a role in the origin of life. This is of course not a strong or well-founded point of the paper. We or anyone else can—of course—not prove this easily, but any discovery of non-racemic amino acids in meteorites also cannot prove any astrophysical seeding.

Dr Coudert said: You put a strong emphasis on the time evolution of the system after deposition and the timescales associated with the evolution. It seems to me that this evolution is bound to depend on the volume into which the energy is deposited into the water jet. To what extent do you control this energy deposition, either by focusing the laser pulse or by hitting only the side of the jet? Have you observed or do you plan to look into the dependence of the observed timescales on energy deposition?

Professor Abel answered: Yes, that is exactly true. In comparison with the experiment at 2700 nm a wavelength of 2900 nm penetrates less deeply into the water phase and consequently the pulse energy is released in a smaller volume. With this we control our energy deposit as well as the enthalpy in the irradiated volume and in turn the temperature. Focusing the laser gives us more degrees of

 This journal is © The Royal Society of Chemistry 2009

freedom for the variation of internal energy in the water phase, although this approach is only possible if plasma formation can be effectively be suppressed and avoided.

In summary, the experiments shown here are an investigation of the dependence of the phase evolution of superheated water for different internal (or deposited) energies. For most of the experiments (in particular for the smaller penetration depths) we irradiated the beam from one side only: the side facing towards the skimmer. This is obligatory because we have to collect photoelectrons from the irradiated side.

Mr Hooton commented: Professor Abel talked about H_2O molecules hydrogen-bonding in metastable supercritical water, to each other. He said the interactions are "not trivial". Rowe published a theoretical paper[1] in 2001 showing that the "dwell time" of H_2O molecules on the surface of protein, in solution, stabilises the protein in H_2O solution. This may be related to the problems in Professor Abel's work. I've seen other papers published since 2001 on protein hydration and stability which also do not reference Rowe's paper. I wonder whether the interaction theory and the "dwell-time" in Rowe's paper may have a connection to Professor Abel's work.

1 A. J. Rowe, *Biophys. Chem.*, 2001, **93**(2–3), 93.

Professor Abel responded: I think that the paper you cite may be interesting but I do not think that it could explain hydrogen bonding in very hot phases of water or provide more insight than our MD calculations. In principle we can look at all details there. A comparison with these is just difficult because of the very different system conditions (with the temperature being just one of them).

Professor Havenith remarked: It would be nice to compare the microdroplet experiment with experiments in bulk. T-jump experiments have been reported before based upon ultrafast heating in a cell by short infrared laser pulses. From these experiments it is known that it is crucial to illuminate the cell with IR radiation from both sides in order to have a uniform temperature distribution, otherwise the absorption of IR radiation will lead to a non-uniform temperature distribution within the droplet. Have you considered this point and how does a non-uniform energy deposition and the energy transport in the droplet influence your final results?

Professor Abel replied: The classical T-jump experiments and new highly sophisticated laser-driven T-jump experiments may be considered as complementary with respect to the conditions (T,p) and the time scales (nanoseconds *vs.* femtoseconds).

Inhomogeneous temperature distributions (due to hot spots in the laser profile *etc.*) are certainly a complication in standard T-jump experiments. In normal T-jump experiments the cells are pumped from both sides because of the small penetration depth of IR radiation into normal water.

Here the pump area and the pump depth (for the water beam) is very much larger than the probe area and particularly the probe depth. Along the short probe depth (a few mono layers) near the interface the conditions may safely be considered to be "homogeneous" in terms of heating the beam. In fact, we do heat the beam from one side only that is exposed to the micro skimmer (for the highest temperatures). The penetration depth is in any case so large and the probe depth so small that along the probe depth the conditions are constant. The energy transport is in general slow in comparison with the time scale of the experiment.

Dr Cacace said: I have three short questions concerning your interesting experimental setup.

Firstly, do you observe cavitation effects?

Secondly, which is the time scale of the process or processes involved during the passage of the jet through the observation "window"?

Thirdly, is the system amenable to fast kinetics investigations?

Professor Abel answered: We do not exactly observe cavitation but if cavitation in general is accompanied by more or less large fluctuations especially in the supercritical phase the "dispersion effect" we maybe observe (at least to some extent) a manifestation of cavitation.

The motion of the water beam (beam velocity 20 m s^{-1}) relative to the time scale of the experiment is slow. It may be regarded as "not moving" during the experiment.

The system is indeed designed for fast kinetic investigations related to chemistry at interfaces. One nice example shown was the dynamics of the solvated electron in liquid water.

Professor McCoustra commented: The TR-PES technique gives excellent information on the time evolution of the electronic structure of the system. Vibrational spectroscopy is a good probe of hydrogen-bonding in liquids. Have you made, or are you aware of, any TR-vibrational spectroscopic measurements to correlate with your electronic information?

Professor Abel answered: Time-resolved vibrational spectroscopy is indeed a very powerful tool to shed light into the dynamics that happens before the evolution of the hot phase. Such experiments have been performed by a number of groups who wanted to understand the inter- and intramolecular dynamics of water. Such experiments have also been performed with static cells in which people investigated supercritical water or reactions in supercritical water. However, I am not aware of experiments that probe such extreme conditions as we do. Such experiments would not be possible in static cells but would require free (very flat) jets owing to the small penetration depth of IR radiation in water.

Professor Dr von Klitzing commented: Is it possible with your set-up to get information about critical phenomena, *i.e.* about hydrogen bonding during critical fluctuations?

Professor Abel responded: The idea of the experiment was indeed the observation of fluctuations and hydrogen bonding in the supercritical phase of a free jet (such an experiment is not possible in a static cell). The dispersion effect we observed in the "supercritical" experiment in comparison with the lower temperature experiment is due to the appearance of a large number of clusters which emerge from the disintegrating liquid (likely driven by fluctuations) and from re-clustering in the hot evolving (supercritical) water phase.

Professor Chaplin remarked: Reference to conditions "above the critical point" are dubious as we only know the temperature and not the density or pressure within the surface layer of the liquid water microbeam.

Professor Abel answered: From the experiment we estimate the enthalpy content of the irradiated phase and in turn its temperature. From the theory (MD simulations) and an extended equation of state for water we obtain the "unknown" parameters.

Professor Lyashchenko said: I wish to ask about the timescale in the case of metastable water in your experiments. Is it a microscopic "memory" of water (approximately the same as in other structural processes in water) or is it macroscopic memory? Also, what was the real temperature of the spectral experiments?

 This journal is © The Royal Society of Chemistry 2009

Professor Abel answered: You have seen the time scales of the evolution of the high enthalpy water phases. With regard to your questions I can only comment on the time evolution of the hot phase of water and to some extent on its hydrogen bonding. It is difficult to judge if this (what we see) is related to any memory effect. In any case we observe microscopic features of the hot evolving phase. The temperature of the water leaving the nozzle before the laser irradiation is around 273 K, or slightly below due to evaporative cooling (the water is, however, always in its liquid state in our experiments). The conditions after the laser irradiation are given in the paper.

Mr Willard continued the discussion of the paper by Professor Luzar: By applying an electric field across your simulations you change the relative free energies of the states where the nanoparticles are aligned and anti-aligned with the field. This free energy difference arises because water molecules at the solute surface prefer to form hydrogen bonds along the solute/solvent interface. This essentially biases the dipoles of interfacial water molecules to align with the solute surface. In the presence of an electric field water molecules prefer to align their dipoles with the field. In liquid water a competition will arise between the formation of hydrogen bonds and the alignment of the dipoles with the electric field. Hydrogen bonds are relatively strong and typically dominate this competition, but if the field is large enough this competition can lead to a qualitative change in the hydrogen bonding structure of water. Since the hydrogen bonding structure plays a crucial role in the alignment of these nanoparticles, might there be a turnover in the field-induced response of these nanoparticles as the strength of the field is increased and network of hydrogen bonds becomes more disrupted?

Professor Luzar replied: This is a good question. We analyzed this competition more systematically in a parallel study.[1] We confirm beyond doubt that the effects you anticipate require much stronger electric fields. No significant change in the average number of hydrogen bonds or their strength is found within the range of field strengths considered in the present work. The plasticity of hydrogen bond network and fluctuations about the apparent tetrahedral coordination measured by the distributions of the O–O–O angles in the triplets of nearest neighbor molecules do not change at field strength $E_0 = E\varepsilon_r < 1\ \text{V Å}^{-1}$.

1 D. Bratko, C. D. Daub and A. Luzar, *Phys. Chem. Chem. Phys.*, 2008, **10**, 6807.

Dr Rao communicated, regarding the paper by Professor Abel: The authors mentioned using resonant photons for excitation using the O–H stretch frequency. Resonant photons (even if very weak in their energy) can cause dramatic effects on the structure of water to the extent that they can lead to dissociation. Please refer to the dissociation to salt water using radio frequency by Roy *et al.*[1]. Did you consider acid dissociation in your technique?

1 R. Roy, M. L. Rao and J. Kanzius, *Mater. Res. Innov.*, 2008, **12**, 3.

Professor Abel communicated in reply: We actually do measure the structural changes in the water as a result of the intense excitation *via* IR radiation and as a result of the making and breaking of hydrogen bonds (not dissociation of covalent bonds in the water molecules). The evolution of the hot water phase is observable in the chemical shift, which we observe for selected time points. It is known that the ionic product of water changes significantly when water is heated up to the critical point. This we cannot see due to the low concentration of ions in the water beam (as opposed to the 55 mol l^{-1} concentration of water). Any other real homolytic dissociation of water—if possible at all under the conditions present in our experiments—produces small amounts of products, which are also difficult to observe with the present technique. For high temperature products in superheated water we refer to Charvat *et al.*[1]

1 A. Charvat, B. Stasicki and B. Abel, *J. Phys. Chem. A*, 2006, **110**, 3297.

 This journal is © The Royal Society of Chemistry 2009

Hydration dynamics of purple membranes

Douglas J. Tobias,[*a] Neelanjana Sengupta[a] and Mounir Tarek[b]

Received 3rd June 2008, Accepted 19th June 2008
First published as an Advance Article on the web 26th September 2008
DOI: 10.1039/b809371g

Protein motions occur over many decades of time, from femtoseconds to seconds and longer. Fast (ps–ns) protein motion enabled by water dynamics on the surfaces of protein molecules is generally believed to be required for biological function. The coupling between water and protein dynamics for soluble proteins in aqueous solution is relatively well understood, while the couplings between protein, lipid, and water dynamics in membranes are only beginning to be unravelled. We report a molecular dynamics simulation study of the dynamics of water hydrating purple membranes (PM). We validate our simulations by comparing the temperature and hydration dependence of crystal lattice parameters and water and protein/lipid dynamics with neutron diffraction and spectroscopic data. We proceed to examine the temperature dependence of several time-dependent quantities describing different aspects of water dynamics, and identify and characterize the correlations between water dynamics and the motion of the protein and lipid components of PM. Our results point to a correlation between the solvation dynamics of lipid molecules and the dynamics of both the protein and lipid components.

Introduction

In order to function, proteins must undergo conformational transitions that move them between substates on their free-energy landscape.[1–3] The conformational space that may be visited corresponds to energy minima arranged in a hierarchical fashion,[4] in which the small number of substates that interconvert slowly *via* large-scale conformational changes are differentiated from the very large number of substates that interconvert by motions on the ps-to-ns timescales.[3,5,6] The large conformational changes are naturally often associated with protein function.[3,7] The precise connection between biological function and fast protein conformational fluctuations is only beginning to be understood, and continues to be a subject of on-going research and debate.[8–13] The fast fluctuations are generally considered to provide the lubrication that promotes larger-scale conformational changes and are, as such, a prerequisite to function.[3] A concrete example of the interplay of fast (ps–ns) dynamics and slow (μs–ms), large-scale conformational changes in an enzyme, demonstrated by a combination of NMR experiments and molecular dynamics (MD) simulations, was recently reported.[14]

Protein transitions between substates were explicitly demonstrated by the variation of the mean-square displacements (MSDs) of individual atoms of metmyoglobin with temperature in X-ray scattering experiments.[1] Subsequent pioneering investigations of myoglobin using Mössbauer spectroscopy[15] and inelastic neutron

[a]*Department of Chemistry, Institute of Surface and Interface Science, Center for Biomembrane Systems, University of California, Irvine, California, 92697-2025, USA*
[b]*Equipe de Dynamique des Assemblages Membranaires, Unité Mixte de Recherche CNRS/UHP 7565, Nancy Université, BP 239, 54506 Vandoeuvre-lès-Nancy cedex, France*

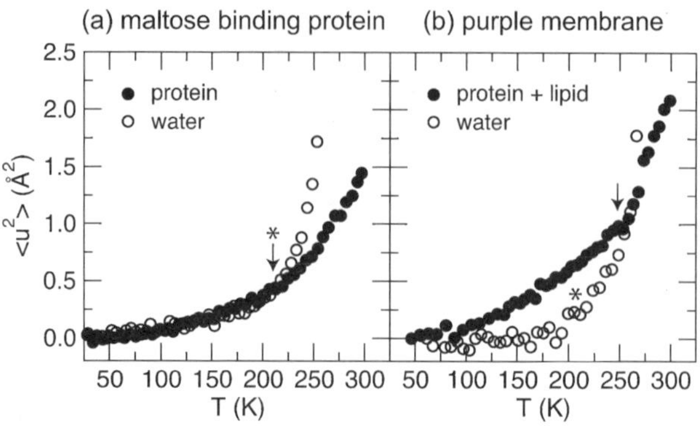

Fig. 1 Mean-squared fluctuations of nonexchangeable H atoms in protein/lipid (filled circles) and water molecules (open circles) in (a) the soluble protein, maltose binding protein (MBP),[46] and (b) the protein bacteriorhodopsin and its surrounding lipids in purple membrane.[44] The fluctuations were measured by neutron scattering on the backscattering spectrometer IN16 at the Institut Laue-Langevin, which probes motions on up to nanosecond time scales. In each system, protein/lipid motion and water motion were measured separately by employing selective deuteration. Inflections in the protein/lipid data are indicated by arrows, and in the water data by asterisks.

scattering experiments[16] allowed estimates of the temperature dependence of MSDs corresponding to motions taking place on the 1 to 100 ns time scale and the 10 ps time scale respectively. These experiments clearly identified a transition, referred to as the dynamical transition, as a hydrated protein molecule is heated from low temperature. This is exemplified by the neutron scattering data from a recent investigation of the soluble maltose binding protein that are shown in Fig. 1a.[17] Below the dynamical transition temperature, a protein is considered to be in a glass-like state, in which its motion is predominantly vibrational, and above it is in a liquid-like state, in which it undergoes anharmonic and diffusive motions.[18]

In light of the fact that the function of biological macromolecules depends critically on the level of their hydration (see review by Rupley and Careri[19]), the dynamical transition has been used extensively to investigate the coupling of the dynamics of a protein molecule and its environment.[16,20–26] The transition has been shown to be diagnostic of the dramatic effect that changing the solvent environment has on the dynamics of soluble proteins: the transition is absent in dehydrated proteins and the extent of motion above the transition temperature, as well as the transition temperature itself, depend on solvent viscosity.[16,20–29]

Water has been called the "lubricant of life" due to its role in promoting fast protein conformational fluctuations that may be important in protein-folding processes and biological function,[30] and the concept of "slaving" has been invoked to discuss the control of protein motion by solvent motion.[31,32] The correlations between solvent mobility and the protein dynamical transition has been recently discussed in terms of concepts from glass physics.[33–37] At the molecular level, time-resolved mean-squared displacements of protein hydration water derived from neutron scattering experiments on a soluble protein have revealed the onset of water translational diffusion at the protein dynamical transition.[12] Direct measurements of protein and water motion by neutron scattering on selectively deuterated samples, shown in Fig. 1b, have recently directly demonstrated the coincidence of transitions in protein and water.[17] MD simulations have elucidated the microscopic mechanism by which the solvent controls fast dynamics in soluble proteins. As first suggested in a phenomenological model proposed to describe the dynamical transition,[21] the onset of water translational diffusion triggers relaxation of the protein–water hydrogen-bond network and enables the dynamical transition

 This journal is © The Royal Society of Chemistry 2009

to occur.[24,25] An analogous mechanism appears to apply in the case of other hydrogen-bonding, small molecule solvents, such as glycerol.[26,38]

The role of hydration on protein function is not nearly as well characterized in complex systems, such as in the cell. *In vivo*, there are many other macromolecular components, *e.g.* cofactors, carbohydrates, nucleic acids and membranes that are equally important for stability and function. A recent study has characterized global water dynamics in bacterial cells by magnetic resonance relaxation measurements, and has shown that a large fraction (~85%) of the water displays bulk-like behavior, and the remainder is dynamically retarded by interactions with biomolecules, even more so (by about a factor of three) than is well established by *in vitro* experiments on small, monomeric proteins in solution.[39] A similar conclusion was reached based on a neutron scattering study, which demonstrated that water dynamics in *E. coli* is dominated by a bulk-like component.[40] A microscopic picture of the behavior of water near biomolecules in a cellular environment remains to be developed. Key questions to be answered are: (i) what is the interplay between water dynamics and protein dynamics in the presence of a variety of molecular components in supramolecular assemblies, and (ii) to what extent are the same mechanisms described above for a two-component protein–water system still at play in a cellular environment? We can begin to answer these questions by considering the case of a protein in its native, hydrated membrane environment.

Considerable progress has been made through experimental studies of dynamical transitions of the light-driven proton pump, bacteriorhodopsin (bR) in purple membrane (PM). A small (26 kDa) protein containing seven transmembrane helices, bR is a homologue of the sensory rhodopsins, and a structural prototype of the large family of G protein-coupled receptors.[41] bR is the only protein in the purple membrane (PM) of *H. salinarium*, and is organized with the membrane lipids in a highly-ordered two-dimensional lattice.[42] Given the relative ease with which intact PM can be isolated or reconstituted, and in light of the roles that bR plays as the simplest biological energy-conversion machine and as a prototype for signal transduction proteins, bR has served as a paradigm for studying the relation between protein dynamics and function in membranes.

The single-particle dynamics of PM have been studied extensively by incoherent neutron scattering under varying conditions of temperature, hydration, and lipidation (see Gabel *et al.*[43] and Fitter *et al.*[44] for reviews). Data from the most recent study of the dynamical transition is shown in Fig. 1b.[45] Selective deuteration was employed to separately measure the PM and water dynamics in essentially identical samples. The mean-squared fluctuations of PM (protein and lipid) on the ns time-scale show an inflection at around 120 K attributed to methyl group rotations,[46] and a pronounced dynamical transition at about 260 K. A surprising result is that an apparent transition in the water dynamics, which occurs at 200 K, does not coincide with the PM transition (Fig. 1b).[47] The temperature dependence of the lamellar spacing of PM stacks suggests that the water transition at 200 K is associated with the onset of water translational mobility.[48]

The ability of bR to complete its photocycle is strongly correlated with the onset of the additional motion that accompanies the transition in the PM at ~260 K.[49] The solvent dependence of proton-pumping activity suggests that biological function is correlated with the transition, which is suppressed by dehydration.[44,50] In addition to a threshold level of hydration, experiments suggest that the lipid matrix is crucial in promoting the full range of mobility required for bR membrane: delipidation of purple membrane leads to a significant reduction in internal dynamics, with a concomitant attenuation of activity.[44,50]

The results summarized in the last two paragraphs show that the dynamical transition of PM that occurs at around 260 K is required for proper bR function, and that the transition is hydration and lipid dependent. They also suggest that, in contrast to the case of soluble proteins, the PM transition is not directly coupled to the onset of water translational mobility. Thus, it appears that the protein, lipid and water dynamics are coupled in a manner that has yet to be worked out. The work reported here is a step toward gaining a better understanding of the coupling between protein/lipid dynamics and water dynamics.

Specifically, we report an extensive molecular dynamics simulation study of PM over a wide range of temperature and multiple levels of hydration. We validate our simulations by comparing the temperature and hydration dependence of crystal lattice parameters and water and protein/lipid dynamics with neutron diffraction and spectroscopic data. We then proceed to examine the temperature dependence of several time-dependent quantities describing different aspects of water dynamics, and identify and characterize the correlations between water dynamics and the motion of the protein and lipid components of PM. Our results point to a connection between the solvation dynamics of lipid molecules and the dynamics of both the protein and lipid components.

Methods

The systems simulated were derived from the PM model reported by Baudry *et al.*[51] A detailed description of the construction of the PM model was given by Baudry *et al.*,[51] so only a brief summary is given here. The simulations reported here consisted of a single bacteriorhodopsin trimer, 28 lipid molecules, 41 sodium ions (for electroneutrality), and either 1315, 1924 or 2408 water molecules, placed in a hexagonal unit cell, which was reproduced in three-dimensions using periodic boundary conditions. The three hydration levels were chosen to correspond to experimental samples with interlamellar spacings of roughly 54, 58 and 62 Å respectively.[48] The lipids included 21 phosphatidyl glycerol phosphate (PGP) molecules and three squalene (SQU) molecules arranged on the periphery of the bR trimer, one PGP molecule inside the bR trimer on the intracellular side of the membrane, and three sulfated triglycoside (GLC) lipids inside the trimer on the extracellular side. All of the lipids contained the same branched, saturated acyl chains. Molecular graphics images of individual lipid molecules and the entire PM unit cell used in the simulation are shown in Fig. 2. Snapshots of the central simulation cells and periodic images are shown in Fig. 3 to depict the molecular packing and the disposition of the water molecules in the interlamellar space. All of the molecular graphics images reported in this paper were prepared using the program VMD.[52]

MD simulations of the system with 1924 water molecules were carried out at 19 temperatures (100, 110, 120, 130, 140, 150, 160, 170, 175, 200, 210, 220, 230, 240, 250, 260, 270, 280 and 296 K) spanning the experimentally observed inflections in plots of the hydrogen atom mean-squared fluctuations *vs.* temperature (Fig. 1b). For further validation against experimental data, and to investigate the hydration dependence of PM and water dynamics, simulations of the system with 1315 water molecules were performed at 7 temperatures (150, 175, 200, 225, 250, 275 and 300 K), and a single simulation of the system with 2408 water molecules was performed at 296 K.

The CHARMM22[53] force fields for proteins and nucleic acids were used to model bR and the sugar moieties in the glycolipids, and the lipids and water molecules were modelled using the CHARMM 27[54] and TIP3P[55] force fields respectively. The MD simulations, which were performed using the NAMD program,[56] were maintained at constant temperature using Langevin dynamics, and a constant pressure of 1 atm using the Nosé–Hoover Langevin-piston algorithm.[57,58] The central simulation cell was allowed to fluctuate anisotropically during the constant pressure simulations. The equations of motion were integrated using the Verlet-I/r-RESPA multiple-time-step algorithm,[59,60] with timesteps of 4 fs for the long-range nonbonded forces, 2 fs for the short-range nonbonded forces and 1 fs for the bonded forces. The SHAKE algorithm[61] was used to constrain the lengths of all bonds to H atoms. Electrostatic interactions were computed using the smooth particle-mesh Ewald sum,[62] and the van der Waals interactions and real-space part of the Ewald sum were smoothly

 This journal is © The Royal Society of Chemistry 2009

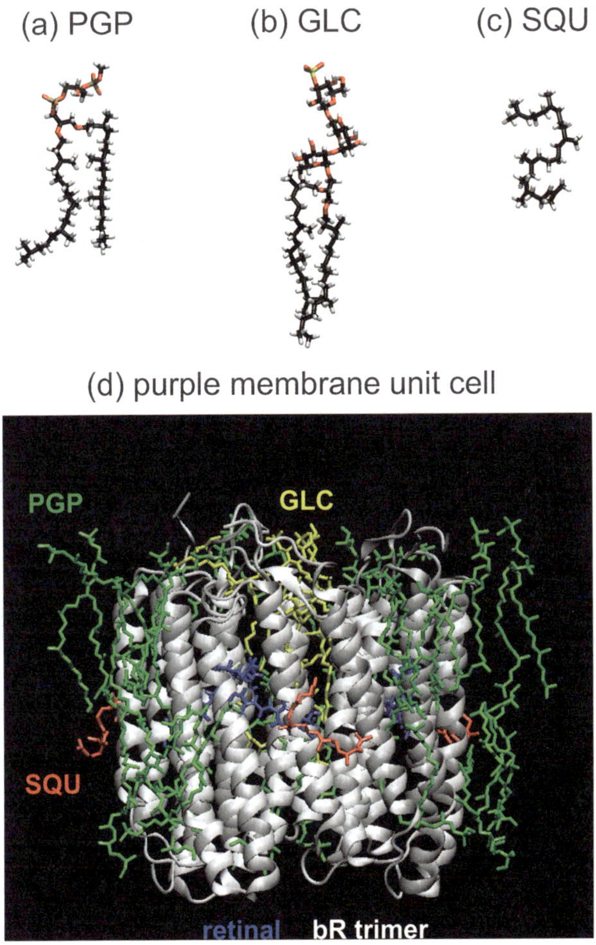

Fig. 2 Molecular graphics images of lipid components of the purple membrane model used in the MD simulations: (a) phosphatidyl glycerol phosphate lipids (PGP); (b) squalene (SQU); (c) sulfated triglycoside lipids (GLC). In (a)-(c), carbon atoms are colored black, hydrogen white, oxygen red, and hydrogen white. (d) Purple membrane unit cell used in the MD simulations. The backbone of the bR trimer is drawn as a cartoon representation, with the α-helices depicted as ribbons and colored white. The lipid molecules and retinal chromophores embedded in the bR molecules are drawn in licorice representation. Retinal is colored blue, PGP green, SQU red, and GLC yellow.

switched to zero over the region 10–11 Å. The simulations were carried out for 5 ns following at least 5 ns of equilibration (sufficient to converge the lattice parameters and protein root-mean-squared deviation from the initial structure) at each temperature and level of hydration.

Results and discussion

Temperature dependence of the lattice parameters

The lengths of the in-plane lattice vectors (a and b), the interlamellar d-spacing (which is the length of the c lattice vector), and the angle (γ) between the a and b lattice vectors are shown for each system simulated as a function of temperature in Fig. 4. The near-equality of the in-plane cell lengths for each simulated system at each temperature (Fig. 4a), and the persistence of the γ angle near a value of

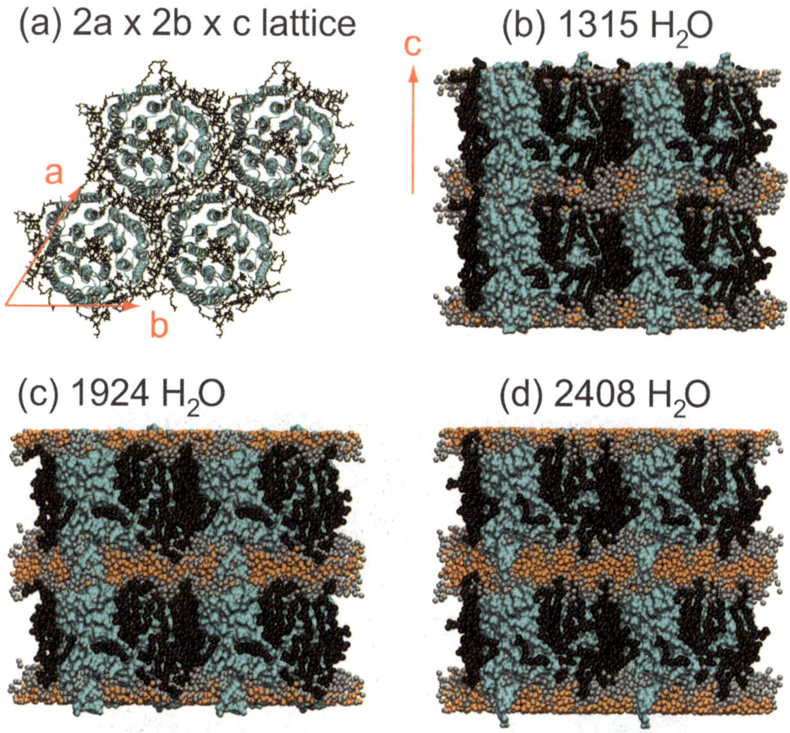

Fig. 3 (a) Top view (down the c axis) of four unit cells ($2a \times 2b \times c$) of the hexagonal PM lattice, with the water molecules removed for viewing clarity. The a and b axes (lattice vectors) are indicated by red arrows. The lipids are drawn in licorice representation and colored black, and the bR molecules are drawn in ribbon representation and colored cyan. (b)–(d) Side view (normal to the c axis) of four unit cells ($a \times 2b \times 2c$) of the PM lattice in systems with 1315, 1924 and 2408 water molecules respectively. The bR molecules are drawn in surface representation and colored cyan and the lipids are drawn with van der Waals spheres and colored black. Water molecules that are "near" (water O within 4 Å) protein and/or lipid molecules are depicted as gray spheres, and the remainder of the water molecules as orange spheres.

60° (Fig. 4c), indicate that the hexagonal symmetry of the unit cell was maintained very well during constant-pressure simulations in which all the lattice parameters were permitted to fluctuate independently. Between 100 and 300 K, the in-plane lattice parameter measured by neutron diffraction increases monotonically from 61.3 to 62.6 Å.[48] The simulation data shown in Fig. 4a reproduce this trend very well.

The average d-spacings of the three simulated systems (Fig. 4b) remain very close to the target values, which are experimentally determined values from different sample preparations. Our system with 2408 water molecules is meant to correspond to PM samples prepared at 100% relative humidity at 300 K. The average d-spacing from our simulation of this "fully hydrated" sample at 296 K, 62.7 Å, is very close to the values of 62–62.5 Å measured by neutron diffraction.[47,48] When a fully hydrated sample is cooled slowly from room temperature, the d-spacing remains constant at around 62 Å until around 250 K, when it abruptly drops to below 55 Å, and decreases to 54 Å at 120 K as the water that is not strongly associated with the membranes crystallizes out as ice.[48] It is possible to prepare samples that have a d-spacing of ~54 Å over a broad temperature range (50–300 K).[47,63] The water present in a sample with a d-spacing of ~54 Å, assumed to be roughly two hydration layers, is strongly associated with the membranes. Our system with 1315 water molecules is meant to model this situation. The average d-spacing of 54.2 Å at 150 K, and the slow increase with temperature, are in excellent agreement with neutron

 This journal is © The Royal Society of Chemistry 2009

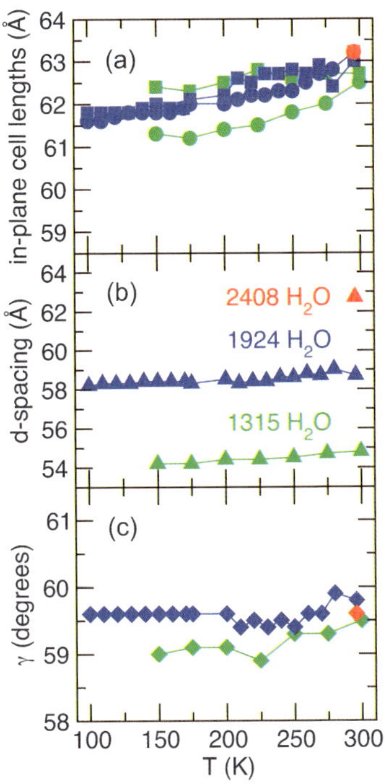

Fig. 4 Temperature dependence of PM lattice parameters from constant-temperature and constant-pressure MD simulations: (a) in-plane lattice parameters, *a* (filled squares) and *b* (filled circles); (b) interlamellar or *d*-spacing (equivalent to the length of the *c* axis); (c) angle (γ) between the *a* and *b* axes.

diffraction data.[48] When a fully hydrated system at 300 K is flash cooled, the *d*-spacing determined by neutron diffraction is 57.5 Å at 100 K, and changes very little as the system is warmed up to 200 K.[48] Our system with 1924 water molecules is meant to model this experimental sample, and the *d*-spacing of 58.2 Å is in good agreement with the neutron data, as is its very weak dependence on temperature. Between 200–270 K the simulated system does not correspond with the flash-cooled experimental sample because above 200 K in the latter the *d*-spacing drops between 200–250 K as water leaves the PM stacks and crystallizes as bulk ice, which melts and rehydrates the PM stacks to the *d*-spacing of 62–63 Å characteristic of the fully hydrated system at ~270 K.

In the experimental samples, the *d*-spacing, and hence the water content in the PM stacks, varies considerably between 200–250 K, depending on the sample preparation and history.[48] However, as will be shown below, the mobility on ps–ns time scales of both the water and PM (protein and lipid) is remarkably insensitive to water content over this temperature range. We will therefore refer primarily to simulations of both our systems with 1315 water and 1924 water molecules to expose details of the water dynamics in this temperature range.

Hydration dependence of the dynamical transition of purple membrane: MD *vs.* neutron scattering

The temperature dependence of the mean-squared fluctuations computed from MD trajectories of the nonexchangeable H atoms in PM is compared to related quantities measured by elastic neutron scattering in Fig. 5. The MD results (Fig. 5a), which are averaged over 5 ns, cannot be quantitatively compared to the neutron results

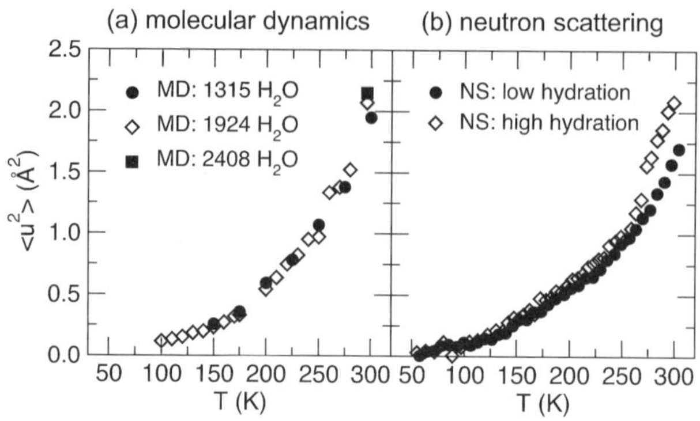

Fig. 5 Mean-squared fluctuations of nonexchangeable H atoms of protein and lipid molecules in PM: (a) computed from MD trajectories, averaged over 1 ns blocks; (b) extracted from elastic neutron scattering data on the IN16 spectrometer at the Institut Laue-Langevin.

(Fig. 5b) because the latter are determined from the *Q*-dependence of the counts contained within the energy broadened elastic peak. The width of the resolution function can be associated with a time scale, which is on the order of ns for the data shown in Fig. 5b, but the neutron spectrum contains contributions from a range of time scales weighted by a decaying function that is the Fourier transform of the energy resolution function.[64] Nonetheless, it is possible to undertake a qualitative comparison of the temperature and hydration dependence of H atom fluctuations measured by neutron scattering and computed from our simulations.

At low hydration (corresponding to samples with a *d*-spacing of ~54 Å), the neutron data reveal that the H atom fluctuations increase with temperature, and there are two weak inflections, one at ~120 K and the other at ~260 K, corresponding to methyl group rotations and a dynamical transition in the PM protein and lipid molecules respectively.[47] At higher hydration, the evolution of the fluctuations with temperature is essentially the same up to the dynamical transition temperature (~260 K). This is expected because the *d*-spacing, and hence the water content, of the sample prepared at high hydration decreased rapidly with temperature from ~62 to ~54 Å below ~270 K. Above the dynamical transition temperature, the amplitudes of H atom motion on the ns time scale are somewhat larger in the higher hydration system.

The MD simulations reproduce both the temperature and hydration dependence of the H atom fluctuations remarkably well, aside from the fact that the inflection attributed to methyl group rotations occurs at too high a temperature (~160 K in the simulations *vs.* 120 K in the neutron data), and the temperature at which the dynamical transition appears to occur, ~250 K, is a little low. The similarity of the fluctuations in the MD simulations up to ~250 K shows that the protein and lipid dynamics are not sensitive to addition of water beyond the hydration shells below the dynamical transition temperature. In contrast, the simulations show that the fluctuations increase steadily with water content above the dynamical transition temperature (*e.g.* ~300 K), in agreement with the neutron data.

Water translational motion in PM stacks *vs.* in bulk water

One of the goals of the work reported in this paper is to gain insight into the role of hydration in the dynamical transition of PM. To this end, we will examine the temperature dependence of several time-dependent quantities describing different aspects of water dynamics, and identify correlations between changes in water dynamics and the motion of the protein and lipid components of PM. We begin with the mean-squared displacements of water O atoms, which describe the

 This journal is © The Royal Society of Chemistry 2009

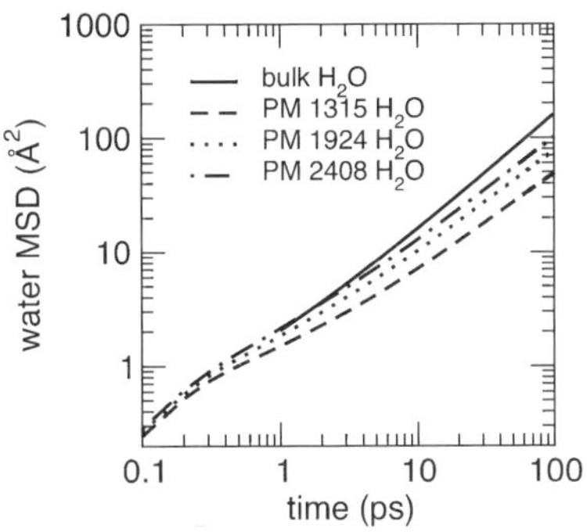

Fig. 6 Time dependence of water O atom mean-squared displacements computed from MD trajectories of bulk TIP3P water and TIP3P water in PM stacks.

translational motion of water molecules. Fig. 6 compares the MSDs at 300 K computed from MD simulations of water in PM stacks at three levels of hydration to bulk TIP3P water. All of the MSDs display a rapid (sub-ps) increase due to ballistic motion of each water molecule within the cage formed by its neighbors. After a few ps, the MSD of bulk water increases linearly with time [MSD(t) $\sim$ t], as expected for Brownian motion. In the PM stacks, the time-dependence of the MSDs is sub-linear, and obeys a power law [MSD(t) $\sim t^{\alpha}$, $\alpha < 1$], which is a signature of the "anomalous" diffusion that is characteristic of water in confined space and water near biomolecular surfaces.[65–70] The largest deviation from linear behavior is observed in the system with 1315 water molecules, in which most of the water is in contact with lipid and/or protein molecules (Fig. 3b). In this system, it takes a water molecule roughly three times as long to achieve a translational displacement equal to its diameter (MSD $\sim$10 Å^2) than it does in bulk water. The water translational diffusion becomes more "normal" (approaches Brownian motion) as the water content between the PM stacks increases. As we shall see, even the water molecules that are not in contact with the protein and lipid molecules display anomalous diffusion due to confinement within the PM stacks.

The temperature dependence of water MSDs on a time scale of tens of ps in PM stacks is shown in Fig. 7. The MSDs averaged over all water molecules in the systems at low and intermediate hydration are plotted in Figs. 7a and 7b, respectively. Although after the first few ps the value of the MSD at a particular time and temperature is slightly lower in the system with less water molecules, the temperature dependence of the MSD(t) is qualitatively very similar in the two systems. After the first few ps, the MSD(t) are relatively flat below 200 K. At 200 K and above, the MSD display a pronounced upturn that signals the entrance into the diffusive regime on a time scale of $\sim$10 ps.

Fig. 7c and 7d separately display the temperature dependence of the MSDs of water molecules that are in contact with protein and/or lipid molecules in the membranes, and the rest of the water molecules between the PM stacks. Here, the "near" water molecules are defined as those whose water O atoms are within 4 Å of any protein or lipid heavy atom, and "far" water molecules are those that are further than 4 Å away. Although more sophisticated definitions of the extent of the water hydration shell have been employed,[67,70] we have found that the value of 4 Å is a sensible choice that leads to good agreement with magnetic relaxation dispersion data on the retardation of water molecules near protein surfaces.[68,71] While the time and temperature dependence is qualitatively similar for both near and far water molecules, the latter undergo substantially larger displacements at a given time and temperature. Thus, water molecules that are not directly in contact with protein and lipids display enhanced translational mobility compared to water molecules in contact with protein and lipids. Although the "far" water molecules

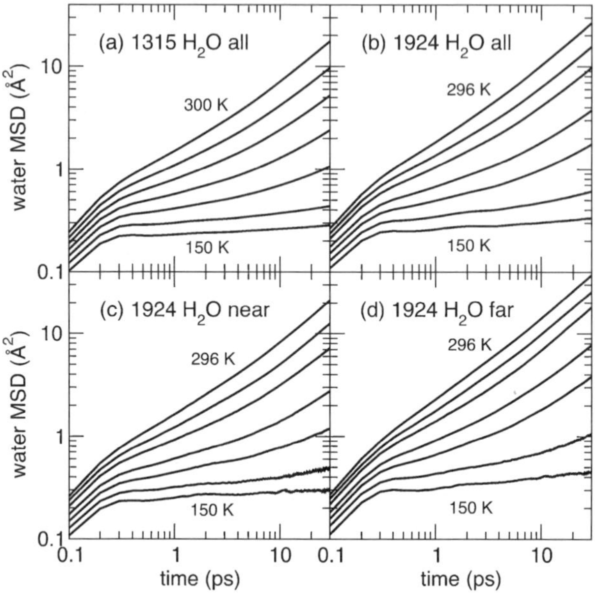

Fig. 7 Time dependence of O atom mean-squared displacements computed from MD trajectories of water in PM stacks: (a) all water in the system with 1315 water molecules; (b) all water in the system with 1924 water molecules; (c) water near (within 4 Å) of protein and/or lipid molecules and (d) the remaining water (beyond 4 Å or protein and/or lipid) in the system with 1924 water molecules. In (a), the curves from top to bottom are at T = 300, 275, 250, 225, 200, 175 and 150 K. In (b)–(d), the curves from top to bottom are at T = 296, 270, 250, 220, 175 and 150 K.

are more translationally mobile than water molecules in contact with the membranes, they are still less mobile than bulk water due to confinement between the membranes in the stacks. This can be seen by comparing the MSD(t = 30 ps) ~40 Å^2 for the "far" water molecules in the PM stack with 1924 water to the corresponding value of 45 Å^2 for bulk water at 30 ps.

We now turn our attention to the temperature dependence of the water mobility, which is conveniently depicted by the values of the MSDs at a particular instant in time. The MSDs at t = 30 ps are plotted in Fig. 8a as a function of temperature for

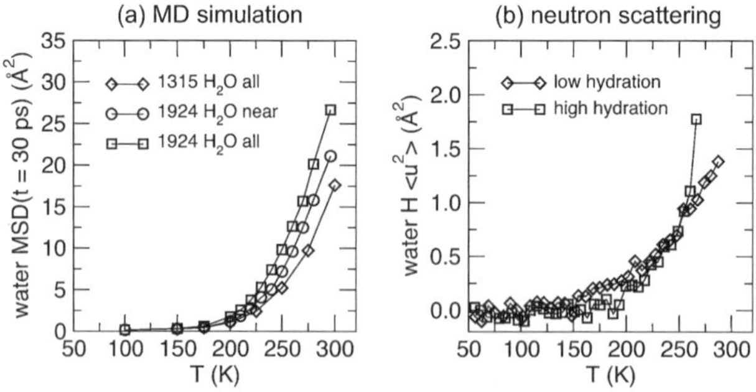

Fig. 8 (a) Temperature dependence of water O mean-squared displacements at t = 30 ps from MD simulations averaged over all water in the system with 1315 water molecules (diamonds), and all (squares) and near (circles) water in the system with 1924 water molecules. (b) Water mean-squared fluctuations measured by neutron scattering experiments on PM samples at low and high hydration.[45,47]

 This journal is © The Royal Society of Chemistry 2009

all the water molecules in the systems with 1315 and 1924 water molecules, and for the water molecules near protein and lipid molecules in the 1924 water system. Essentially the same curves would be obtained if another time was chosen, except the scale on the MSD(t) axis would be different. All three curves plotted in Fig. 8a exhibit a rapid increase in the rate of change of the MSDs as a function of temperature around 200 K. This change of dynamics at around 200 K appears to be similar to the behavior of the protein atomic fluctuations at the dynamical transition (Fig. 1 and 5), and signals the onset of water translational diffusion. The data plotted in Fig. 8a also show that this apparent change of dynamics occurs at roughly the same temperature, regardless of the hydration level, and that the water mobility increases with the amount of water between the PM stacks, regardless of whether only water molecules near protein and lipid or all of the water molecules in the system are considered.

In Fig. 8b the mean-squared fluctuations of water H atoms, measured by elastic neutron scattering at an energy resolution corresponding to roughly 1 ns, are plotted as a function of temperature for PM stacks prepared at room temperature at two levels of hydration.[45,47] The temperature dependence of the neutron and simulation data is qualitatively similar, in the sense that both sets of data display hydration independent inflections indicating the onset of increased mobility at ~200 K, and the water mobility increases with hydration at high temperatures, but there are two notable differences. First, according to the simulation data, water mobility increases with hydration at 200 K, whereas the neutron data show that the water mobility in the higher hydration sample exceeds that of the low hydration sample only above 250 K. To resolve this apparent discrepancy, we recall that the d-spacing, and hence water content, of the two experimental samples is the same below ~250 K.[45,47] Over the temperature range (>250 K) where the water content is different, the neutron data show that water mobility at a given temperature is higher in the system that contains more water, in agreement with the simulations. Second, the simulations predict that the extent of water motion on a time scale of tens of ps is roughly an order of magnitude greater than the fluctuations measured by neutron scattering on the ns timescale above 200 K. This discrepancy is largely due to the differences in the way the quantities are measured. The simulation data are computed from trajectories in real space, while the neutron data are obtained by fitting the Q-dependence of the intensity of the resolution-broadened elastic scattering, I_{el}, to the Debye–Waller form, $I_{el}(Q) = A\exp(-Q^2\langle u^2\rangle/6)$, over a restricted range of Q (*e.g.* $Q^2 < 0.9$ Å^2).[45,47] To check for consistency with the neutron data, we have computed the resolution broadened dynamic structure factors, $S(Q,E)$, for water H atoms in the system with 1924 water molecules at 296 K as described elsewhere,[64,72] using the energy resolution of the neutron spectrometer on which the measurements reported in Fig. 8b were made. Fitting the elastic intensities, $I_{el} = S(Q,0)$, obtained at three values of Q^2 between 0.30 and 0.90 Å^2 to the Debye–Waller form, we obtain a value of $\langle u^2\rangle = 2.4$ Å^2, which is in reasonable agreement with the neutron data, and much lower than the ~20 Å^2 computed directly from the trajectory.

Both neutron scattering experiments and MD simulations agree that in purple membranes there is a hydration independent change in dynamics of lipid and protein dynamics at ~260 K (Fig. 5) and a change in water dynamics at ~200 K (Fig. 8).[45] This is in contrast to the situation for soluble proteins (*e.g.* MBP), where transitions in protein and water dynamics both occur at ~200 K (Fig. 1a).[17] This suggests that there is a difference in the coupling of water and protein/lipid dynamics in purple membranes and soluble proteins.[45] To gain more insight into the nature of the coupling of water and protein/lipid dynamics in purple membranes, we now proceed to a more detailed examination of the changes of water dynamics with temperature.

The first question we seek to answer is: what is the nature of the water displacements that are apparently increasing rapidly with temperature above 200 K (Fig. 8)? A useful quantity for addressing this question is the van Hove correlation

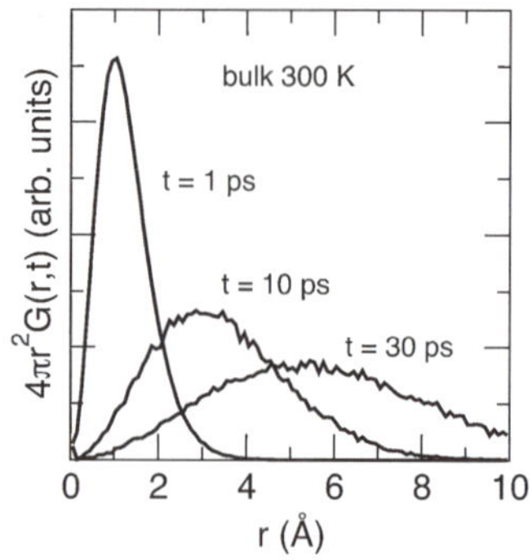

Fig. 9 Water displacement distribution functions for bulk TIP3P water at 300 K.

function, $G(r,t)$. The time and space Fourier transform of $G(r,t)$ is the dynamic structure factor probed by neutron spectroscopy. The quantity $4\pi r^2G(r,t)$, the water "displacement distribution function", is the probability that a tagged particle at the origin at $t = 0$ has been displaced by a distance r at time t. The time evolution of $4\pi r^2G(r,t)$ for bulk water at 300 K is plotted in Fig. 9. On the 1 ps time scale, water molecules are confined to and vibrating within the cage formed by their neighbors, which is manifested in $4\pi r^2G(r,t = 1\ \text{ps})$ by a narrow peak at $r = 1$ Å. After 10 ps, the peak has moved out to ~3 Å, indicating that most of the water molecules have been displaced by a distance roughly equal to the diameter of a water molecule, and after 30 ps the peak has been displaced by roughly two water diameters. The $G(r,t)$ plotted in Fig. 9 for bulk water at 300 K roughly display behavior expected for a Brownian particle, namely a unimodal Gaussian distribution of displacements that moves to larger distances and broadens with increasing time.

The functions $4\pi r^2G(r,t)$ for water molecules in the PM system with 1315 waters at several temperatures are plotted in Fig. 10 for $t = 1$, 10, 30 and 100 ps. Recall that in

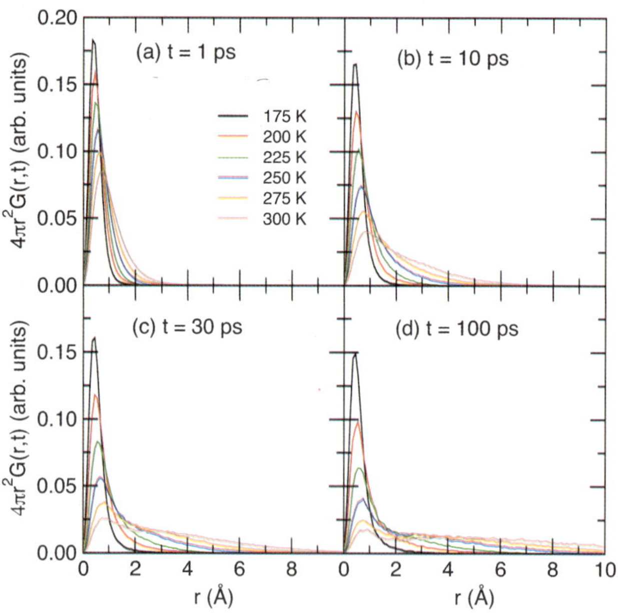

Fig. 10 Water displacement distribution functions for water in the PM system with 1315 water molecules.

this system most of the water molecules are in contact with protein and/or lipid molecules (Fig. 3b). On the time scale of 1 ps, hydration water behaves qualitatively similar to bulk water, in the sense that the distributions are characteristic of vibrational motion within solvent cages, formed by other water molecules and/or protein/lipid atoms, with widths that increase with temperature. At longer times, it is evident from a quick inspection of the data in Fig. 10 that the time dependence of the displacement distributions of hydration water in the PM system is very different from that of bulk water at all temperatures considered, including 300 K. At 175 K, the single narrow peak in $4\pi r^2 G(r,t)$ at all time scales considered indicates that water molecules remained confined in their solvent cage at $T < 200$ K. At 200 K and at times longer than 10 ps we start to see that a small fraction of water molecules has escaped from their solvent cages. Above 200 K, this fraction grows steadily with temperature. The somewhat abrupt increase of the MSDs of water molecules on time scales of >10 ps with temperature at $T \sim 200$ K and above (Fig. 7 and 8) can be attributed to this mobile fraction of water molecules. Thus, the increase in the MSDs with temperature is not due to an increase in mobility of all the water molecules, but rather is due to an increase in the population of mobile water molecules. Indeed, the data in Fig. 10 shows that a significant population of water molecules remains confined on time scales of up to 100 ps even at 300 K.

The analyses presented in this section have demonstrated that MD simulations are capable of reproducing the temperature dependence of water mobility measured by neutron scattering in a complex assembly of biological molecules, namely purple membrane stacks. In addition, the analyses shed light onto the nature of the increase in mobility with temperature that is observed at $T > 200$ K (Fig. 8). It does not appear from the analyses presented thus far that the change of protein dynamics (*i.e.* dynamical transition) that is observed at $T \sim 250$ K is accompanied by a concomitant change in water dynamics. In the next section we examine protein and lipid solvation dynamics and identify changes that do indeed occur at $T \sim 250$ K, which might be relevant to the protein/lipid dynamical transition in PM.

Protein and lipid solvation dynamics

The dynamics of water solvation shells can be quantified in various ways. In this section, we consider two. First, to quantify the decay of the population of a particular shell of water molecules solvating protein and lipid atoms in PM, we have calculated the following "survival" time correlation function:

$$R(t) = \frac{1}{N(t)} \sum_{t_0} \sum_{j=1}^{N_w} p_j(t_0, t_0 + t)$$

Here, N_w is the total number of water molecules in the system. The function p_j assumes a value 1 if the water molecule j resides within the solvation layer (within 4 Å of a protein or lipid heavy atom) continuously for a time t, starting from an arbitrary time origin t_0, and 0 otherwise. The arbitrary time origin is moved over the trajectory, and $N(t)$ is the number of segments corresponding to each time span t.

The normalized survival correlation function, $C_S(t) = R(t)/R(0)$, is plotted for water around protein and lipid groups at several temperatures in Fig. 11. The decay of $C_S(t)$ is caused by the departure of water molecules present in the solvation shell at $t = 0$. As expected, the rate of decay of the correlation function increases with temperature. However, the temperature dependence is different for protein and lipid solvation shells. On one hand, in the case of the protein solvation shell, the rate of decay of $C_S(t)$ appears to increase uniformly with temperature. On the other hand, the rate of decay of the lipid solvation shell appears to increase abruptly between 220 and 250 K. This suggests that a change in lipid solvation dynamics could be associated with the protein/lipid dynamical transition at $\sim$250 K.

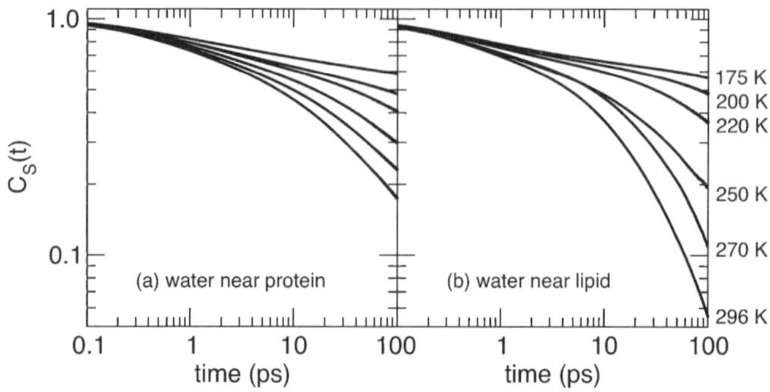

Fig. 11 Water solvation-shell survival functions for water molecules (a) near protein and (b) near lipid molecules in PM, computed from MD trajectories of the system with 1924 water molecules at (from top to bottom) 175, 200, 220, 250, 270 and 296 K.

In previous investigations, we have proposed a role for protein–solvent hydrogen-bond dynamics in the dynamical transition of soluble proteins.[17,24,26] The idea is that the structural relaxation (onset of anharmonic and diffusive motion) of the protein that occurs at the dynamical transition is associated with relaxation of the protein–solvent hydrogen-bond network. A quantity that is useful for characterizing the relaxation of the hydrogen bond network is the following correlation function:[73,74]

$$C_{\mathrm{HB}}(t) = \langle h(0)h(t)\rangle / \langle h \rangle.$$

Here $h(t)$ is a hydrogen-bond population operator, which is equal to 1 if a given donor–acceptor (D–A) pair is hydrogen bonded at time t, and 0 otherwise. The geometric criterion we use states that a hydrogen bond is formed if the D–A distance is less than a cutoff value that depends on the identity of D and A, and the D–H–A angle is greater than 150°. The function $C_{\mathrm{HB}}(t)$ is the probability that a random D–A pair which is hydrogen bonded at time 0 is still bonded at time t, regardless of whether or not the bond was broken at intermediate times. Thus, beyond an initial transient period, the decay of $C_{\mathrm{HB}}(t)$ is not determined by fast hydrogen-bond breaking by simple water rotation/libration (without significant translational displacement), but rather by rearrangement of the hydrogen-bond network.

The function $C_{\mathrm{HB}}(t)$ is plotted separately for protein–water and lipid–water hydrogen bonds at several temperatures in Fig. 12. Beyond the initial, transient (sub-ps) decay, the hydrogen-bond correlations display a secondary decay whose

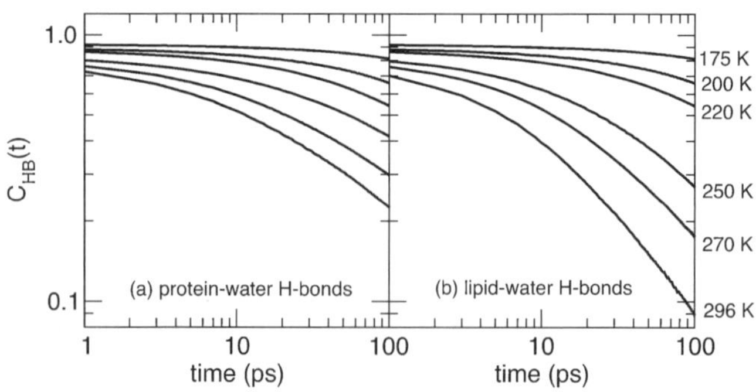

Fig. 12 (a) Protein–water and (b) lipid–water hydrogen-bond correlation functions computed from MD trajectories of the system with 1924 water molecules at (from top to bottom) 175, 200, 220, 250, 270 and 296 K.

 This journal is © The Royal Society of Chemistry 2009

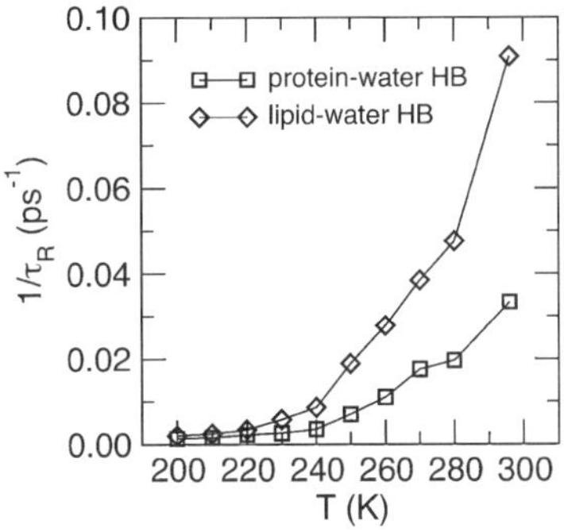

Fig. 13 Temperature dependence of the inverse of the correlation times of the protein–water and lipid–water hydrogen-bond correlation functions.

rate is temperature-dependent. As in the case of the solvation-shell survival correlation functions shown in Fig. 11b, there is a jump in the rate of decay of the lipid–water hydrogen-bond correlation function below $T = 250$ K that is not observed in the corresponding protein–water data. To put this statement on a more quantitative footing, we define a relaxation time, τ_R, as the time at which $C_{HB}(t)$ decays to 1/e, *i.e.* $C_{HB}(\tau_R) = e^{-1}$. The temperature dependence of the inverse of this relaxation time, $1/\tau_R$, which is akin to a relaxation rate, is shown in Fig. 13. Both the protein–water and lipid–water hydrogen-bond data display inflections at ~240 K, which is just below the temperature of the lipid/protein dynamical transition (~250 K in the simulations). The change in slope at the inflection is much greater for lipid–water hydrogen bonds than for protein–water hydrogen bonds. Thus, these data, along with the hydration shell survival functions, suggest a connection between the dynamics of water molecules solvating lipid molecules and the dynamical transition of protein and lipid molecules in purple membranes.

Conclusions

We have performed MD simulations of purple membranes, which are two-dimensional crystals of lipids and trimers of the protein bacteriorhodopsin, in order to investigate the dynamics of water molecules in the interlamellar space of purple membrane stacks. We have validated the simulations of this complex assembly of biomolecules by comparison with neutron diffraction and neutron spectroscopic data. The latter are particularly useful because they probe dynamics on time scales that are readily accessible by MD simulations. The temperature and hydration dependence of the lattice parameters in constant-temperature and constant-pressure simulations with fully flexible simulation cells were in excellent agreement with the diffraction data. The temperature dependence of the mean-squared fluctuations of H atoms in the membrane components (protein and lipid) and water molecules was in good agreement with corresponding results obtained from the Q-dependence of elastic neutron scattering. In particular, both the neutron data and MD simulations revealed inflections in the membrane and water fluctuations at ~250 and ~200 K respectively. The simulations also reproduced the increased amplitudes of protein/lipid and water motion with increasing hydration above ~250 K.

The goal of the study was to gain insight into the apparent difference in the role of water in the dynamical transitions of soluble and membrane proteins. In the case of soluble proteins, it has been well established that the protein dynamical transition is correlated with a transition in solvent dynamics that occurs at the same temperature.[12] The transition in the solvent dynamics has been identified as the onset of translational diffusion.[17,24–26] In contrast, in PM, the dynamical transition in the membrane (protein and lipid molecules) occurs at ~250 K, while the change in water dynamics evident in neutron scattering data occurs at ~200 K.

Our simulations have provided insight into the nature of the additional motion of water between PM stacks that appears to set in above 200 K. Displacement distribution functions revealed that water molecules begin to escape from the cages formed by other water molecules and protein/lipid atoms on time scales greater than a few ps at ~200 K. The enhanced translational mobility of water above 200 K, seen in both the mean-squared amplitudes extracted from neutron data and the mean-squared displacements computed from the MD trajectories, can be attributed to an increase in the population of water molecules that escape their cages with increasing temperature.

Analysis of water dynamics solvating protein and lipid molecules has revealed some changes that are correlated with the dynamical transition at ~250 K. Solvation dynamics was quantified using water survival (in the solvation shell) and protein–water and lipid–water hydrogen-bond correlation functions. Not surprisingly, all of the correlation functions considered decayed more rapidly with increasing temperature. The rate of decay of the survival correlation function for water solvating protein and the protein–water hydrogen-bond correlation function exhibited a relatively smooth increase with temperature. On the other hand, the rate of decay of the survival correlation function for water solvating lipid and the lipid–water hydrogen-bond correlation function began to increase more rapidly with increasing temperature just below 250 K. This suggests that dynamical transition in PM at 250 K is correlated with changes in the dynamics of water molecules solvating the lipids in the membrane. A hypothesis that we are testing in on-going work is that the dynamical transition in the protein is enabled by the melting of the acyl chains of the lipid molecules that is, in turn, accompanied by a breakup of the interactions between the lipid polar head groups by enhanced mobility in their solvation shells. Whether the chain melting triggers a change in solvation dynamics or *vice versa* remains to be determined.

Acknowledgements

This work was supported by the National Science Foundation (grants CHE-0417158 and CHE-0750175 to D. J. T.). We thank Martin Weik, Kathleen Wood, Joe Zaccaï, and Frank Gabel for fruitful collaborations and discussions, Emad Tajkhorshid for providing the coordinates for the PM model, and Jessica Morgan for help setting up and performing the MD simulations. D. J. T. thanks the Université Joseph Fourier for a visiting professorship during which some of the work reported here was carried out.

References

1 H. Frauenfelder, S. G. Sligar and P. G. Wolynes, *Science*, 1991, **254**, 1598.
2 H. Frauenfelder, *Proc. Natl. Acad. Sci. U. S. A.*, 2002, **99**, 2479.
3 K. A. Henzler-Wildman and D. Kern, *Nature*, 2007, **450**, 964.
4 A. Ansari, J. Berendzen, S. F. Bowne, H. Frauenfelder, I. E. Iben, T. B. Sauke, E. Shyamsunder and R. D. Young, *Proc. Natl. Acad. Sci. U. S. A.*, 1985, **82**, 5000.
5 P. W. Fenimore, H. Frauenfelder, B. McMahon and R. D. Young, *Proc. Natl. Acad. Sci. U. S. A.*, 2004, **101**, 14408.
6 P. W. Fenimore, H. Frauenfelder, B. H. McMahon and R. D. Young, *Physica A*, 2004, **351**, 1.
7 A. K. Dioumaev and J. K. Lanyi, *Proc. Natl. Acad. Sci. U. S. A.*, 2007, **104**, 9621.
8 B. F. Rasmussen, A. M. Stock, D. Ringe and G. A. Petsko, *Nature*, 1992, **357**, 423.
9 J. Heberle, J. Fitter, H. J. Sass and G. Büldt, *Biophys. Chem.*, 2000, **85**, 229.
10 R. M. Daniel, R. V. Dunn, J. L. Finney and J. C. Smith, *Annu. Rev. Biophys. Biomol. Struct.*, 2003, **32**, 69.
11 V. Kurkal-Siebert, R. M. Daniel, J. L. Finney, M. Tehei, R. V. Dunn and J. C. Smith, *J. Non-Cryst. Solids*, 2006, **352**, 4387.

 This journal is © The Royal Society of Chemistry 2009

12 W. Doster and M. Settles, *Biochim. Biophys. Acta*, 2005, **1749**, 173.
13 R. E. Lechner, J. Fitter, N. A. Dencher and T. Hauss, *Phys. B*, 2006, **385–386**, 835.
14 K. A. Henzler-Wildman, M. Lei, V. Thai, S. J. Kerns, M. Karplus and D. Kern, *Nature*, 2007, **450**, 838.
15 H. Keller and P. G. Debrunner, *Phys. Rev. Lett.*, 1980, **45**, 68.
16 W. Doster, S. Cusack and W. Petry, *Nature*, 1989, **337**, 754.
17 K. Wood, A. Frölich, A. Paciaroni, M. Moulin, M. Härtlein, G. Zaccaï, D. J. Tobias and M. Weik, *J. Am. Chem. Soc.*, 2008, **130**, 4586.
18 W. Doster, S. Cusack and W. Petry, *Phys. Rev. Lett.*, 1990, **65**, 1080.
19 J. A. Rupley and G. Careri, *Adv. Protein Chem.*, 1991, **41**, 37.
20 J. Smith, K. Kuczera and M. Karplus, *Proc. Natl. Acad. Sci. U. S. A.*, 1990, **90**, 9135.
21 W. Doster and M. Settles, in *Hydration Processes in Biology: Experimental and Theoretical Approaches*, ed. M.-C. Bellissent-Funel, IOS Press, Amsterdam, 1999.
22 A. M. Tsai, D. A. Neumann and L. N. Bell, *Biophys. J.*, 2000, **79**, 2728.
23 A. L. Lee and A. J. Wand, *Nature*, 2001, **411**, 501.
24 M. Tarek and D. J. Tobias, *Phys. Rev. Lett.*, 2002, **88**, 138101.
25 A. L. Tournier, J. Xu and J. C. Smith, *Biophys. J.*, 2003, **85**, 1871.
26 M. Tarek and D. J. Tobias, *Eur. Biophys. J.*, 2008, **37**, 701.
27 A. Paciaroni, S. Cinelli and G. Onori, *Biophys. J.*, 2002, **83**, 1157.
28 G. Caliskan, R. M. Briber, D. Thirumalai, V. Garcia-Sakai, S. A. Woodson and A. P. Sokolov, *J. Am. Chem. Soc.*, 2006, **128**, 32.
29 I. J. Finkelstein, A. M. Massari and M. D. Fayer, *Biophys. J.*, 2007, **92**, 3652.
30 L. D. Barron, L. Hecht and G. Wilson, *Biochemistry*, 1997, **36**, 13143.
31 P. W. Fenimore, H. Frauenfelder, B. McMahon and F. G. Parak, *Proc. Natl. Acad. Sci. U. S. A.*, 2002, **99**, 16047.
32 H. Frauenfelder, P. W. Fenimore and B. H. McMahon, *Biophys. Chem.*, 2002, **98**, 35.
33 J.-M. Zanotti, M.-C. Bellissent-Funel and S. H. Chen, *Europhys. Lett.*, 2005, **71**, 91.
34 S.-H. Chen, L. Liu, E. Fratini, P. Baglioni, A. Faraone and E. Mamontov, *Proc. Natl. Acad. Sci. U. S. A.*, 2006, **103**, 9012.
35 P. Kumar, Z. Yan, L. Xu, M. G. Mazza, S. V. Buldryev, S.-H. Chen, S. Sastry and H. E. Stanley, *Phys. Rev. Lett.*, 2006, 177802.
36 J. Swenson, H. Jansson and R. Bergman, *Phys. Rev. Lett.*, 2006, **96**, 247802.
37 M. Lagi, X. Chu, C. Kim, F. Mallamace, P. Baglioni and S.-H. Chen, *J. Phys. Chem. B*, 2008, **112**, 1571.
38 T. E. Dirama, G. A. Carri and A. P. Sokolov, *J. Chem. Phys.*, 2005, 122.
39 E. Persson and B. Halle, *Proc. Natl. Acad. Sci. U. S. A.*, 2008, **105**, 6266.
40 M. Jasnin, M. Moulin, M. Haertlein, G. Zaccaï and M. Tehei, *EMBO Rep.*, 2008, **9**, 543.
41 D. Oesterhelt, *Curr. Opin. Struct. Biol.*, 1998, **8**, 489.
42 J. K. Lanyi and H. Luecke, *Curr. Opin. Struct. Biol.*, 2001, **11**, 415.
43 F. Gabel, D. Bicout, U. Lehnert, M. Tehei, M. Weik and G. Zaccai, *Q. Rev. Biophys.*, 2002, **35**, 327.
44 J. Fitter, R. E. Lechner and N. A. Dencher, *J. Phys. Chem. B*, 1999, **103**, 8036.
45 K. Wood, M. Plazanet, F. K. Gabel, B. D. Oesterhelt, D. J. Tobias, G. Zaccai and M. Weik, *Proc. Natl. Acad. Sci. U. S. A.*, 2007, **104**, 18049.
46 J. H. Roh, V. N. Novikov, R. B. Gregory, J. E. Curtis, Z. Chowdhuri and A. P. Sokolov, *Phys. Rev. Lett.*, 2005, **95**, 038101.
47 K. Wood, M. Plazanet, F. Gabel, B. Kessler, D. Oesterhelt, G. Zaccai and M. Weik, *Eur. Biophys. J.*, 2008, **37**, 619.
48 M. Weik, U. Lehnert and G. Zaccai, *Biophys. J.*, 2005, **89**, 3639.
49 M. Ferrand, A. J. Dianoux, W. Petry and G. G. Zaccai, *Proc. Natl. Acad. Sci. U. S. A.*, 1993, **90**, 9668.
50 J. Fitter, S. A. W. Verclas, R. E. Lechner, H. Seelert and N. A. Dencher, *FEBS Lett.*, 1998, **433**, 321.
51 J. Baudry, E. Tajkhorshid, F. Molnar, J. Phillips and K. Schulten, *J. Phys. Chem. B*, 2001, **105**, 905.
52 W. Humphrey, W. Dalke and K. Schulten, *J. Mol. Graphics*, 1996, **14**, 33.
53 A. D. MacKerell Jr, D. Bashford, M. Bellott, R. L. Dunbrack Jr, J. Evanseck, M. J. Field, S. Fischer, J. Gao, H. Guo, S. Ha, D. Joseph-McCarthy, L. Kuchnir, K. Kuczera, F. T. K. Lau, C. Mattos, S. Michnick, T. Ngo, D. T. Nguyen, B. Prodhom, W. E. Reiher III, B. Roux, M. Schlenkrich, J. C. Smith, R. Stote, J. Straub, M. Watanabe, J. Wiorkiewicz-Kuczera, D. Yin and M. Karplus, *J. Phys. Chem. B*, 1998, **102**, 3586.
54 S. E. Feller, *Curr. Opin. Colloid Interface Sci.*, 2000, **5**, 217.
55 W. L. Jorgensen, J. Chandrasekhar, J. D. Madura, R. W. Impey and M. L. Klein, *J. Chem. Phys.*, 1983, **79**, 926.

56 L. Kalé, R. Skeel, M. Bhandarkar, R. Brunner, A. Gursoy, N. Krawetz, J. Phillips, A. Shinozaki, K. Varadarajan and K. Schulten, *J. Comput. Phys.*, 1999, **151**, 283.
57 G. J. Martyna, D. J. Tobias and M. L. Klein, *J. Chem. Phys.*, 1994, **101**, 4177.
58 S. E. Feller, Y. Zhang, R. W. Pastor and B. R. Brooks, *J. Chem. Phys.*, 1995, 4613.
59 M. E. Tuckerman and B. J. Berne, *J. Chem. Phys.*, 1992, **97**, 1990.
60 H. Grubmüller, H. Heller, A. Windemuth and K. Schulten, *Mol. Simul.*, 1991, **6**, 121.
61 J.-P. Ryckaert, G. Ciccotti and H. J. C. Berendsen, *J. Comput. Phys.*, 1977, **23**, 327.
62 U. Essmann, L. Perera, M. L. Berkowitz, T. Darden and L. G. Pedersen, *J. Chem. Phys.*, 1995, **103**, 8577.
63 R. E. Lechner, J. Fitter, N. A. Dencher and T. Hauss, *J. Mol. Biol.*, 1998, **277**, 593.
64 M. Tarek and D. J. Tobias, in *Neutron Scattering in Biology, Techniques and Applications*, ed. T. Gutberlet, J. Fitter and J. Katsaras, Springer, Heidelberg, 2006.
65 M.-C. Bellissent-Funel, J. M. Zanotti and S. H. Chen, *Faraday Discuss.*, 1996, **103**, 281.
66 M. Settles and W. Doster, *Faraday Discuss.*, 1996, **103**, 269.
67 R. Abseher, H. Schreiber and O. Steinhauser, *Proteins: Struct., Funct., Genet.*, 1996, **25**, 366.
68 D. J. Tobias, I. W. Kuo, A. Razmara, and M. Tarek, in *Water in Confining Geometries*, ed. J. P. Devlin and V. Buch, Springer, Heidelberg, 2003.
69 A. R. Bizzarri, C. Rocchi and S. Cannistraro, *Chem. Phys. Lett.*, 1996, **263**, 559.
70 F. Pizzitutti, M. Marchi, F. Sterpone and P. J. Rossky, *J. Phys. Chem. B*, 2007, **111**, 7584.
71 N. Sengupta, S. Jaud and D. J. Tobias, *Biophys. J.*, 2008, in press.
72 M. Tarek, G. J. Martyna and D. J. Tobias, *J. Am. Chem. Soc.*, 2000, **102**, 10450.
73 A. Luzar and D. Chandler, *Nature*, 1996, **379**, 55.
74 A. Luzar and D. Chandler, *Phys. Rev. Lett.*, 1996, **76**, 928.

 This journal is © The Royal Society of Chemistry 2009

From shell to cell: neutron scattering studies of biological water dynamics and coupling to activity†

A. Frölich,‡[a] F. Gabel,[a] M. Jasnin,[b] U. Lehnert,§[a] D. Oesterhelt,[c] A. M. Stadler,[b] M. Tehei,¶[b] M. Weik,[a] K. Wood||[ab] and G. Zaccai*[ab]

Received 1st April 2008, Accepted 19th June 2008
First published as an Advance Article on the web 31st October 2008
DOI: 10.1039/b805506h

An integrated picture of hydration shell dynamics and of its coupling to functional macromolecular motions is proposed from studies on a soluble protein, on a membrane protein in its natural lipid environment, and on the intracellular environment in bacteria and red blood cells. Water dynamics in multimolar salt solutions was also examined, in the context of the very slow water component previously discovered in the cytoplasm of extreme halophilic archaea. The data were obtained from neutron scattering by using deuterium labelling to focus on the dynamics of different parts of the complex systems examined.

Introduction

Understanding the dynamic state of water in biological systems and its influence on macromolecular activity is a major scientific challenge. The huge diversity displayed by living organisms is based on a few unifying principles. All known living organisms are constituted of cells or have been through a cellular stage. They require the full set of a variety of macromolecules of specific chemical type (DNA, RNA, proteins...), ions, specific small molecules and water, in order to thrive and reproduce. Life on Earth has colonized every possible niche in conditions that span extremes in solvent salinity and pH values, temperature and pressure. A scoop of soil, a sample of water, of ice, of air or even of rock from just about anywhere on Earth—including the poles, the tops of the highest mountain ranges, the deepest ocean trenches, neutral or alkaline salt lakes, or very hot acidic volcanic springs—are all likely to reveal biological activity. Ice usually has liquid inclusions, and organisms that live at temperatures well above 100 °C are under high pressure, in which the water is still liquid. As was debated at a Royal Society discussion published in 2004,[1] the availability of liquid water may well be the only environmental requirement for Life. The extreme halophilic archaea live in salt lakes and saline ponds in essentially saturated salt water; they counterbalance the osmotic pressure due to the external multimolar

[a] *Institut de Biologie Structurale, UMR 5075, CEA–CNRS–UJF, 41 rue Jules Horowitz, 38027 Grenoble, France*
[b] *Institut Laue Langevin, 6 rue Jules Horowitz, BP 156 Grenoble Cedex 9, France. E-mail: zaccai@ill.fr; Tel: +33 (0)4 76 20 76 79*
[c] *Max Planck Institut für Biochemie, Am Klopferspitz 18, D-82152 Martinsried, Germany*

† All the authors contributed equally to the work.
‡ Present address: Universität Karlsruhe, Germany
§ Present address: McKinsey & Co., London, UK
¶ Present address: AINSE, School of Chemistry, University of Wollongong, Australia
|| Present address: University of Groningen, The Netherlands

This journal is © The Royal Society of Chemistry 2009

NaCl concentration by accumulating multimolar KCl in their cytoplasm. Considering the inside of a cell is already very crowded with macromolecules, which occupy close to 30% of its volume, what is the state of the water compared to that in the bulk liquid? There are about 55 moles in a kilogramme of water. In a six molal solution of a fully ionised monovalent salt, therefore, there are 4.5 water molecules per ion. Ions may be coordinated by six water molecules, so that each and every water molecule has at least one ion as one of its nearest neighbours, leading to significantly reduced water activity. The existence of extreme halophiles proves that Life has adapted to reduced water activity, but there is no evidence that water can be done away with altogether in biological activity.

In this paper, we present and discuss results on water dynamics in different biological systems, and on the coupling between biological activity and the environment around macromolecules: water in the case of soluble proteins, water and lipids in the case of membrane proteins.

Neutron scattering is a particularly suited experimental technique to probe water dynamics under different conditions. Slow neutrons scatter off bound protons exchanging energy in the thermal range corresponding to the picosecond–nanosecond (ps–ns) time-scale, and momentum corresponding to the 0.1 nm length-scale. Energy-resolved incoherent neutron scattering, especially in the elastic (EINS) and quasielastic (QENS) scattering modes, have been informing us on water dynamics under different conditions for decades. Because the method relies on incoherent scattering, samples need not be crystalline or even monodisperse and measurements can be performed on highly complex systems such as living cells[2] or stacks of natural membranes.[3] The incoherent neutron scattering cross section of hydrogen is more than an order of magnitude larger than that of other atomic nuclei usually found in biological material and their isotopes, including deuterium, and the development of *in vivo* specific deuterium-labelling methods permitted the focus on water dynamics in complex biological samples, and its coupling with biological function and activity.

While a consensus has emerged that water molecules close to protein surfaces are dynamically slowed down with respect to bulk water molecules, quantitative aspects and the mechanisms involved remain the objectives of active research (reviewed by Ball[4]). Reports on the dynamics of water molecules close to protein surfaces measured by neutron spectroscopy were presented at a previous Faraday Discussion.[5,6] The factors found for the slowing down of the diffusion coefficient of hydration water compared to bulk water varied between 2 and 100, depending on the protein, temperature and hydration conditions. Bon *et al.*[7] studied hydration water in lysozyme crystals by QENS. They distinguished two water populations, denoted as "first" and "second" hydration shells, with diffusion coefficients slowed down by a factor of 10 and 50 compared to bulk water respectively. By combining EINS data from two spectrometers, Gabel and Bellissent-Funel[8] reported diffusion coefficients of deuterated C-phycocyanin hydration water, in the presence of trehalose molecules, which are slowed down by a factor of 10–15 with respect to bulk water. As has been pointed out, however, in Gabel *et al.*[9] and by Gabel and Bellissent-Funel,[8] results obtained at different energy resolutions, different wave-vector transfers and using different fit models are not directly comparable.

On the scale of the biological cell, water constitutes the malleable matrix that envelops and sustains intracellular components and biological activity. Considering that macromolecular concentration in the cytoplasm can reach 400 mg ml^{-1}, to what extent is water dynamics perturbed by steric confinement between macromolecular surfaces? Does the cell 'tame' water behaviour as some believe (reviewed by Ball[4])? While widely differing opinions concerning the dynamic state of water in the cell are represented in the biology community, there has been very little work on atomic-scale measurements of intracellular water dynamics in close to physiological conditions. Below, we present and discuss recent neutron scattering experiments performed on different cell systems, which shed light on *in vivo* intracellular water dynamics.

 This journal is © The Royal Society of Chemistry 2009

Hydration water is an integral and essential part of a protein structure. Without it the macromolecule would not be folded correctly. It is essential for the protein to fulfil its biological function. Devoid of hydration water, proteins are inactive[10] and lack essential motions, thus implicitly suggesting a close relationship between protein and hydration-water dynamics. The term 'slaving' has been used to express that water can impose its dynamical fingerprint on a protein.[11,12] Cryo-temperature-dependent neutron scattering experiments revealed the solvent dependence of dynamical transitions in soluble proteins[13–21] and in RNA[22] and provided insights into the coupling between them. Molecular dynamics simulations suggested the onset of water translational diffusion to be at the origin of the dynamical transition.[23,24] In contrast, however, the onset of hydration-water translational diffusion at 200 K, in a membrane, was not observed to trigger a dynamical transition in the membrane protein,[25,26] suggesting a more complex interaction between protein, water and the membrane lipid environment. Recent neutron scattering studies, using specific deuterium labelling, confirmed that a dynamical transition coincides with the onset of hydration water translational diffusion, in the case of a soluble protein,[27] but not in the case of a membrane protein.[3,28]

By using specific deuteration and samples with controlled hydration, water dynamics was measured directly by neutron scattering in the hydration shell around a soluble protein and a natural membrane and correlated with macromolecular dynamics, as well as *in situ* in prokaryotic and red blood cells.

Maltose binding protein (MBP) is a soluble monomeric protein of 41 kDa. It plays an essential role in the metabolism of *E. coli*.[29] *In vivo* deuteration of MBP allowed the direct extraction of hydration-water dynamics in neutron scattering experiments on the deuterated protein hydrated in H_2O. Water dynamics thus obtained were compared to protein dynamics obtained by measuring the natural abundance of MBP in D_2O.[27]

The purple membrane (PM) of *H. salinarum* is formed of a highly ordered two-dimensional lattice of archaeal specific lipids and the retinal binding protein, bacteriorhodopsin (BR).[30] BR functions as a light-driven proton pump. Its activity can be followed quite sensitively by spectroscopic methods because it is associated with a millisecond photocycle of protein conformational changes and retinal colour changes. Integral membrane proteins are active in a complex environment constituted of lipid hydrocarbon chains, lipid head groups and hydration water in contact with different parts of the protein. By using fully deuterated membranes hydrated in H_2O and natural abundance membranes hydrated in D_2O, the coupling between hydration water and membrane dynamics has been explored.[3,28] In another set of experiments, the dynamic coupling between membrane protein and lipids was observed by applying *in vivo* deuterium labelling and *in vitro* reconstitution of PM. The samples allowed the observation of the lipid and protein dynamics, separately.

In vivo neutron scattering measurements in the cytoplasm of deuterated prokaryotic organisms revealed very different water-dynamic behaviour in bacteria and extreme halophilic archaea. In order to examine the effect of the high salt environment within halophilic archaea, water dynamics was measured in saturated KCl and NaCl solutions. The water dynamics inside red blood cells (RBC), which contain mainly the equivalent of a 300 mg ml^{-1} solution of hemoglobin, was also measured.

The joint analysis of the results allowed us to propose a coherent view of water dynamics in the hydration shell, the lipid environment of a membrane protein and the intracellular milieu in various cell types, as well as of the dynamic coupling between biological activity and environment.

Materials and methods

Samples

The expression and purification of deuterated and natural abundance MBP has been described before.[27] For neutron scattering experiments, powders of deuterated and natural abundance MBP were hydrated to a level of 0.4 g water per g MBP in H_2O and D_2O respectively (for details see Wood *et al.*[27]).

PM was extracted from natural abundance or fully deuterated cultures of *H. salinarum* as described before.[31,32] The experiments were performed on membrane-water stacks in H_2O or D_2O with lamellar spacings of 54 Å, corresponding to an inter-membrane water layer of 5 Å, and a hydration level of approximately 0.1 g water per g membrane.[28] The specifically labelled lipid and BR samples were prepared by delipidation and reconstitution as described in detail by Lehnert.[33] Three reconstitutions were prepared: (i) a control sample of hydrogenated BR and hydrogenated lipids (H-BR–H-lip or H–H); (ii) a sample with hydrogenated BR and deuterated lipids (H-BR–D-lip or H–D); (iii) a sample with deuterated BR and hydrogenated lipids (D-BR–H-lip or D–H). Several parameters were tested in order to find optimum reconstitution conditions. When the buoyant density (from a sucrose gradient) of the fully hydrogenated reconstitution (H–H) corresponded to native PM, the conditions were considered as correct and applied to the hybrid deuterated PM samples. The experimental values agreed well with the expected ones indicating that the lipid-to-BR ratio corresponded to that in natural PM. X-ray and neutron diffraction indicated poor lattice formation in the reconstituted samples. Light-induced cross-linking indicated that the probability of trimer formation in the reconstituted sample was smaller compared to dimer formation, in contrast to native PM. The UV absorption spectrum indicated a shift of the bound retinal peak to 567 nm, compared to 570 nm in native PM, as observed before in reconstituted samples. The lipid composition of the reconstituted samples was investigated in the negative-ion mode of electrospray ionisation (ESI) mass spectrometry. Traces of hydrogenated phospho- and glyco-lipids were observed indicating that they were not removed during the delipidation process. The hydrogenated contribution accounted for about 5% of the total. It arises from lipid strongly associated to the protein and that was not removed in the delipidation process. The special interaction between the glycolipids and the protein has been discussed by Weik *et al.*[34]

Deuterated *E. coli* (BLE21(DE3) strain) were cultivated as described previously.[35] Cell pellets containing H_2O or D_2O buffer were prepared in such a way as to reduce strongly the quantity of extracellular water in the sample, which was found to represent less than 7% of the total water (see Jasnin *et al.*[35] for details).

Deuterated *H. marismortui* cells were grown at 37°C to an optical density of 0.8–1 (late logarithmic phase) in the medium described before,[36] in which yeast extract was replaced by deuterated algal extract produced at the Max Planck Institut für Biochemie (Martinsried, Germany). For neutron scattering experiments, cells were pelleted in D_2O or H_2O buffer (see Tehei *et al.*[36] for details).

For the RBC experiments, samples of human venous blood were taken from healthy adults. The cells were washed with phosphate buffer solution (PBS) and were gassed with CO to increase the stability of hemoglobin. The glycocalyx matrix was removed as described elsewhere.[37] Half of the cells were then washed with D_2O PBS buffer until the level of H_2O was below 0.1%. No cell lysis was detected during the preparation and the shape of the cells was checked with optical microscopy.

Neutron scattering

Elastic incoherent neutron scattering (EINS) experiments provide information on atomic mean-square displacements (MSD) in a sample as a function of temperature.[38] These are obtained from a Gaussian approximation plot of the scattered elastic intensity dependence on scattering vector amplitude,[39] Q: $\ln I(Q) = 1/6 \langle u^2\rangle Q^2$. In an incoherent scattering situation, waves scattered by a single nucleus at different positions as it moves in time interfere to give a Q-dependent scattering pattern. The pattern then contains information on the shape of the volume traced by the moving particle in the same way as a small-angle scattering curve contains information on the shape of a particle in solution. The Gaussian approximation

 This journal is © The Royal Society of Chemistry 2009

given above is therefore a direct analogy to the Guinier approximation, with the radius of gyration squared defined as half the MSD value; the domain of validity of the approximation is for $\langle u^2\rangle Q^2$ smaller than or about 2.[39] The time-scale examined of 1 ns on IN16 (http://www.ill.eu/in16/home/) at the Institut Laue Langevin (ILL) depends on the energy resolution of the spectrometer, while the range of the scattering vector, Q, in which the scattered intensity is measured, defines the length-scale which is of the order of 1 Å for IN16. Plots of MSD as a function of absolute temperature are called elastic scans. MSD are in absolute $Å^2$ units. An effective dynamic resilience expressed as an elastic force constant (in Newtons per metre) can be calculated from the inverse of the slope of $\langle u^2\rangle$ *versus* T.[38,40] Elastic scans for proteins and membranes usually show a break in slope at about 200 K, which has been interpreted as a dynamical transition from a harmonic low-temperature regime to an anharmonic regime that has been modelled, for example, by a double-well free-energy potential.[15] In a quasi-harmonic approximation, the dynamical transition is associated with a significant decrease in resilience.

An exhaustive description of quasielastic incoherent neutron scattering (QENS) can be found in Bée.[41] QENS is measured as a function of both energy transfer and scattering vector, Q. On the energy scale, the elastic scattering appears as a peak centred on zero energy transfer whose width corresponds to the spectrometer resolution. The QENS appears as a significantly broader peak, also centred on zero energy transfer. The shape of the QENS peak and its Q variation contain information on the time and geometry associated with the motion of the scattering particle. In the case of simple exponential diffusion processes, the QENS spectra can be fitted mathematically by Lorentzian functions. By applying various models, such as the jump diffusion, Singwi–Sjölander and Sears models (see Bée[41] for details), to fit the Lorentzian width as a function of Q, it is possible to calculate rotational relaxation and correlation times, translational residence time and translational and rotational diffusion coefficients for the scattering particle. Instruments used for QENS experiments were IN16 at ILL (energy resolution 0.9 μeV, corresponding to a time-scale of 1 ns, see above), BSS backscattering spectrometer (energy resolution also of 0.9 μeV, at the Jülich Neutron Centre, see http://www.jcns.info/), IN6 at ILL (energy resolution of 90 μeV, corresponding to a time scale of ~10 ps, see http://www.ill.eu/in6/home/), TOFTOF at the Munich FRM2 reactor (energy resolution of 100 μeV, similar to IN6; see ref. 42 for instrumental details), IRIS at the ISIS spallation neutron source at the Rutherford Laboratory (energy resolution of 17 μeV, corresponding to a time scale of ~50 ps, see http://www.isis.rl.ac.uk/molecularSpectroscopy/iris/), and IN5 at the ILL (http://www.ill.eu/in5/), which can be tuned for different energy resolution and Q-range values.

Data from the following neutron scattering experiments are discussed in the paper. In order to avoid laborious repetition and because different experimental conditions applied to the different studies, more detailed descriptions of experimental conditions are given in the relevant Figure caption or results section.

- Elastic scans on deuterated or natural abundance maltose binding protein in H_2O or D_2O collected on the IN16 spectrometer.
- Elastic scans on native and fully deuterated purple membrane in H_2O or D_2O collected on the IN16 spectrometer.
- Elastic scans on specifically deuterated (in the lipid or protein parts), reconstituted purple membrane collected on the IN16 spectrometer.
- QENS scans on IN6 and IRIS to measure water diffusion inside deuterated *E. coli*.
- QENS scans on TOFTOF to measure water diffusion inside red blood cells.
- QENS scans on IN16 and IN6 to measure water diffusion in deuterated *H. marismortui*, with a control experiment using BSS on deuterated *E. coli*.
- QENS scans on IN5 to measure water diffusion in concentrated solutions of NaCl and KCl.

Results

Hydration-water dynamics and coupling to the soluble maltose binding protein from *E. coli*

Water structure[43,44] and dynamics[5,8,9] at the vicinity of a soluble protein surface are different from in the bulk liquid state. MBP powders hydrated at 0.4 g water per g protein contain one hydration layer per protein molecule. The deuterated MBP in H_2O gave access to the dynamics of hydration water, which contributes 73% to the total incoherent scattering cross section but only 2% in natural abundance MBP in D_2O. The MSD of water and protein motions, plotted in Fig. 1A, are similar up to 220 K, at which temperature they both exhibit a transition.[27] Above 220 K, the water MSD are above those of the protein.

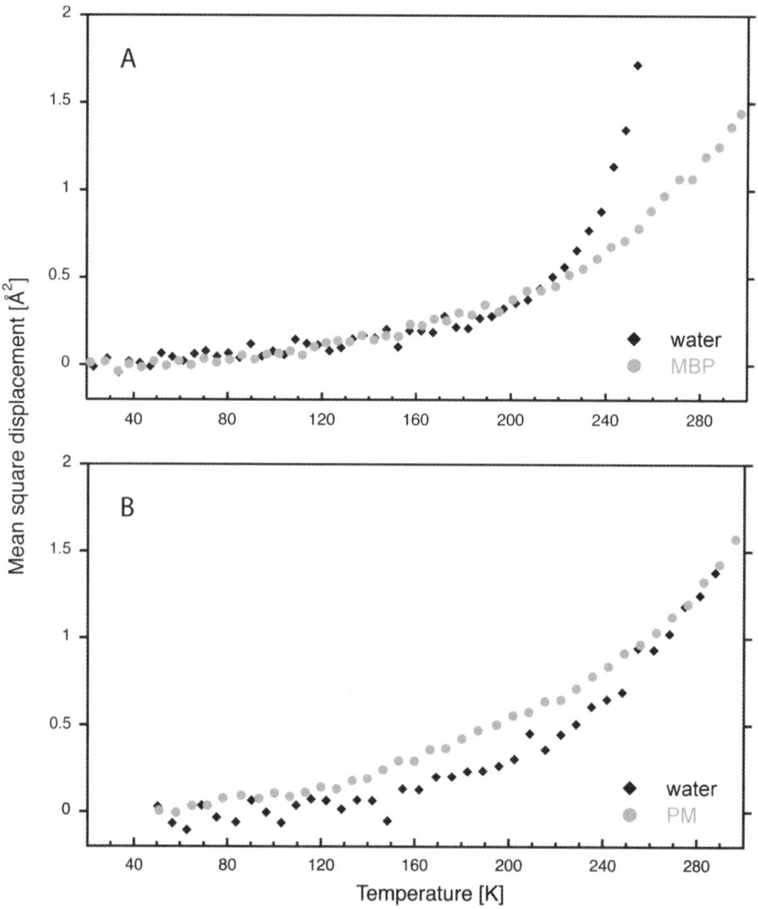

Fig. 1 (A) Mean square displacements (MSD) of maltose binding protein (MBP) and its hydration water.[27] Dynamic transitions in the hydration water and in MBP take place at similar temperatures (~220 K). (B) MSD of purple membrane (PM) and its hydration water.[28] Dynamical transitions take place at 200 K in the hydration water and at 120 and 260 K in PM. The elastic scans were carried out on the IN16 spectrometer with an energy resolution of 0.9 μeV (full-width at half-maximum of the elastic peak and a wavelength of 6.275 Å[63]). MSD were extracted from data in the following Q [Q with units of Å^{-1}] ranges: $0.2 < Q^2 < 1.8$ for hydrogenated MBP in D_2O (panel A, grey circles), $0.2 < Q^2 < 1.1$ for deuterated MBP in H_2O (panel A, black diamonds), $0.2 < Q^2 < 1.5$ for hydrogenated PM in D_2O (panel B, grey circles), $0.2 < Q^2 < 0.9$ for deuterated PM in H_2O (panel B, black diamonds).

 This journal is © The Royal Society of Chemistry 2009

Hydration water dynamics in purple membrane and coupling to bacteriorhodopsin and membrane lipids

PM stacks were hydrated to a level that corresponds to one hydration layer per membrane surface (*i.e.* 5 Å of inter-membrane water). EINS experiments on deuterated PM hydrated in H_2O permitted us to access the dynamics of the first hydration layer directly. In the sample, 60% of the total incoherent scattering cross section was calculated to be due to the contribution of inter-membrane water. In contrast, as with the MBP samples discussed above, in natural abundance PM hydrated in D_2O the contribution from hydration water is negligible and the total incoherent scattering cross section is strongly dominated by the PM (77% BR protein, 23% lipids), thus reflecting membrane dynamics. The elastic temperature scans are presented in Fig. 1B.[28] Hydration-water MSD increase linearly up to 200 K, at which temperature there is a break in slope suggesting a dynamical transition in water dynamics. The PM MSD show breaks in slope at 120 and 260 K, but not at 200 K.

The *in vitro* reconstituted PM samples were characterised by UV spectrophotometry, cross-linking experiments for BR trimer formation, mass spectrometry for lipid composition and deuterium labelling, and X-ray diffraction for lattice formation and order (see Materials and methods). Studying the reconstituted hydrogenated BR and hydrogenated lipids sample (H-BR–H-lip) probed the global dynamics of the whole membrane (68% of the incoherent scattering cross section is from BR, 31% from lipids), whilst in the hydrogenated BR, deuterated lipid sample (H-BR–D-lip) BR clearly dominated the signal (98%). Using deuterated BR with hydrogenated lipids (D-BR–H-lip) provided information on the lipid dynamics, which represent 92% of the incoherent scattering cross section. Fig. 2 presents the MSD measured for the H-BR–H-lip and D-BR–H-lip reconstituted samples at 93% relative humidity which represent, respectively, dynamics of global PM and of the native archaeal lipids in the membrane on the ns time-scale. The H-BR–D-lip sample MSD are not shown; they were very close and slightly below the H-BR–H-lip values, as expected from the BR-to-lipid ratio in the membrane composition. The MSD of the H-BR–H-lip sample are very similar to those of natural PM. The two-dimensional lattice in the reconstituted samples, however, was not as well ordered as in the natural membrane, indicating that membrane

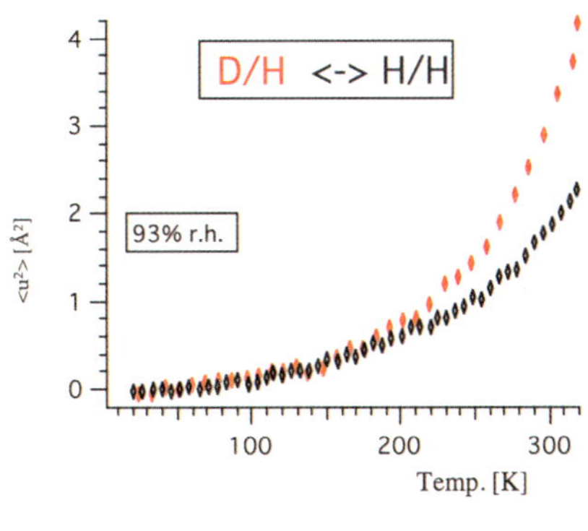

Fig. 2 Mean-square displacements (MSD) extracted from the analysis of the elastically scattered neutron intensity, measured on the IN16 spectrometer at the ILL (energy resolution, full-width at half-maximum, of 0.9 µeV). The MSD, $\langle u^2 \rangle$, were calculated using the Gaussian approximation: $I_{el,T}(Q,\omega = 0) = A\exp\{-\langle u^2 \rangle Q^2/6\}$, in which $I_{el,T}(Q,\omega = 0)$ is the elastically scattered intensity at temperature T; Q and ω are the momentum and energy transfer, respectively, and A is a constant. Data are shown for the H-BR–H-lip (H–H) and D-BR–H-lip (D–H) reconstituted samples (see Materials and methods). Samples were hydrated to 93% relative humidity in D_2O.

internal dynamics in the ns time-scale is not influenced by lattice order. As seen in Fig. 2, between 20 and about 200 K, the MSD of both samples are the same, with a deviation from harmonic behaviour at about 150 K: PM and its lipid components have similar vibrational dynamics at very low temperatures and display the anharmonicity associated with methyl rotations from about 150 K.[14,45] We recall that PM lipids do not have fatty acid chains but phytol chains that are rich in methyl groups. Above 200 K, the MSD of the two samples become clearly different, with the archaeal phytol chains displaying greatly increased dynamics compared to the average membrane. In both samples shown in Fig. 2, further onsets of large-amplitude motions occur at 260 K, suggesting coupled increases in dynamics in the lipid and protein components of PM.

Intracellular water dynamics in *E. coli* and red blood cells

E. coli water dynamics was studied in fully deuterated cell pellets resuspended in H_2O and D_2O buffers respectively. A subtraction of the cell spectra measured in D_2O from the cell spectra measured in H_2O, scaled by the sample mass, provided a good approximation to the scattering signal from the water present in the samples. A sample of H_2O buffer alone was measured as well, and used as a reference for interpreting the data. The measurements were carried out on two neutron spectrometers, IN6 and IRIS (see Materials and methods), to cover diffusive motions from those of bulk to interfacial water. IRIS and IN6 QENS spectra were well fitted using a single and two Lorentzian functions, respectively (see Jasnin *et al.*[35] for details). The Lorentzian extracted on IRIS was attributed to translational motions of cell water. Its half-width at half-maximum (HWHM), Γ_T, was extracted and best fitted using a jump diffusion model,[41] which describes diffusion between sites for the water protons with a mean residence time, τ_0, at each site. We found the following values for the translational diffusion coefficients, D_T, and associated τ_0: $D_T = 1.53 \times 10^{-5}$ cm^2 s^{-1}, $\tau_0 = 2.63$ ps at 281 K and $D_T = 2.39 \times 10^{-5}$ cm^2 s^{-1}, $\tau_0 = 2.16$ ps at 301 K. D_T values are very close to those of bulk water at corresponding temperatures, with residence times about twice as long. The rotational water contribution emerged from the broad Lorentzian extracted from IN6 QENS spectra. We found the following rotational correlation times, $\tau_{cor,R}$: $\tau_{cor,R} = 1.96$ ps at 281 K, and 1.54 ps at 301 K. The values were close to the values extracted for the buffer under the same conditions and of the same order as the values measured for bulk water by QENS[46] and NMR.[47] From the two sets of data, we concluded that *E. coli* water dynamics is dominated by a bulk-like water component at physiological temperature. A similar conclusion was reached for water in RBC, indicating that the result was not restricted to bacteria (see below).

The diffusion of cytoplasmic water in whole human RBC was measured by QENS. RBC grow in the bone marrow of mammals, and deuterated material is not available so far. Natural abundance cells were measured in H_2O and D_2O buffers with high precision and the scattering data were subtracted from each other, to yield intensities representing the dynamics of water only. As a reference, H_2O buffer solution was measured. The experiment was performed on the time-of-flight neutron spectrometer TOFTOF[42] at the Munich FRM2 reactor, with an energy resolution of 100 μeV, in the temperature range 290 to 320 K. The measured spectra in the Q-range from 0.5 to 1.5 Å^{-1} were well fitted from −1.5 to +1.5 meV with a narrow and a broad Lorentzian function plus linear background, corresponding to translational and rotational motions of the water molecules respectively. The HWHM profile as a function of Q^2 of the narrow Lorentzian are in agreement with a jump diffusion model.[41] The translational diffusion coefficient, D_T, and the residence time, τ_0, of water were obtained from the fits. The diffusion coefficients of cytoplasmic water are only slightly reduced compared to those obtained for the buffer solution, but interestingly the residence times of cytoplasmic water are on average five times higher than in buffer solution (details are given in Stadler *et al.*[48]).

 This journal is © The Royal Society of Chemistry 2009

Extreme halophiles: a special case?

Intracellular water dynamics in *H. marismortui*, an extreme halophile originally isolated from the Dead Sea, was studied by QENS. Water motions in centrifuged cell pellets were measured by means of two spectrometers, IN6 and IN16, sensitive to motions with time-scales of 10 ps and 1 ns respectively (see Materials and methods). From IN6 time-of-flight data, using the model of Singwi and Sjölander,[49] a translational diffusion constant of 1.3×10^{-5} cm^2 s^{-1} was determined at 285 K for *H. marismortui* cells. The value is close to that found previously for other cells and close to that for bulk water, as well as that of the water in the 3.5 M NaCl solution bathing the cells.[50] A very slow water component was discovered from the IN16 data (energy resolution of 0.9 μeV). At 285 K, the values of HWHM are independent on Q^2. The water protons of this component displayed a residence time of 411 ps (compared with a few ps in bulk water). At 300 K, the residence time dropped to 243 ps and was associated with a translational diffusion of 9.3×10^{-8} cm^2 s^{-1}, or 250 times lower than that of bulk water. The slow water accounts for about 76% of cell water in *H. marismortui*. No such slow water was found in *E. coli*, measured on the BSS backscattering spectrometer (energy resolution of 0.9 μeV, at the Jülich Neutron Centre (see http://www.jcns.info/)). It was hypothesized that the slow mobility of a large part of *H. marismortui* cell water indicates a specific water structure responsible for the large amount of K^+ bound within these extreme halophilic cells.

In order to estimate the influence of the solvent ions on the intracellular water dynamics in the extreme halophiles, we studied the diffusion coefficients of water in molar solutions of KCl and NaCl by QENS. To this end, a 3M KCl and a 4M NaCl (each at 50 mM Tris, pH = 7.6, $T = 298$ K) were measured on the instrument IN5 at the ILL (see Materials and methods for web details), at an energy resolution of 63 μeV ($\lambda = 6$ Å) in the wave-vector range $0.26 < Q < 1.87$. The quasielastic spectra were fitted in the energy transfer range −0.5 to +0.5 meV with a scattering law constituted of a linear background, an elastic intensity and a single quasielastic contribution, interpreted within the Singwi–Sjölander model.[49] The model assumes that individual water molecules diffuse during a time τ_1 and vibrate during a time τ_0 around an equilibrium position. In the approximation $\tau_1 \gg \tau_0$, we extracted the apparent diffusion coefficient, D_{app}, from the HWHM of the QENS spectra as a function of Q (Fig. 3).

The following apparent diffusion coefficients (combining translational and rotational motions) were measured (the bulk water value is from the literature): D_{app} (4M NaCl in H_2O, $T = 298$ K) = 1.5×10^{-5} cm^2 s^{-1}; D_{app} (3M KCl in H_2O, $T = 298$ K) = 2.0×10^{-5} cm^2 s^{-1} and D_{app} (bulk water,[5] $T = 298$ K) = 2.3×10^{-5} cm^2 s^{-1}. The results showed clearly that the presence of molar salt concentration of NaCl or KCl hardly affects the diffusive properties of water. The findings supplement QENS data published earlier at different salt concentrations and temperatures.[51] Only a weak dependence of the water diffusion coefficient on NaCl concentration (up to 6 M NaCl) was found. In the case of KCl (up to 3.2 M), the diffusion coefficient was similar to that of bulk water. Furthermore, characteristic water frequencies (in the range from 100 to 1000 cm^{-1}) persisted up to 4.6 M KCl and 0.5 M NaCl respectively. It is only in the case of very small or highly charged ions (Li^+, Mg^{2+}, La^{3+} *etc.*) that the authors could find a decrease of the water diffusion coefficient at higher concentrations (>1 M).

It can be concluded that the presence of high salt concentrations in the cytoplasm of extreme halophilic archaea, multimolar NaCl and KCl, is not responsible by itself for the very slow water component observed by Tehei *et al.*[36]

Discussion

The hydration shell is vital to a macromolecule's biological activity. Without hydration water, proteins would lack not only their correctly folded structure but also the

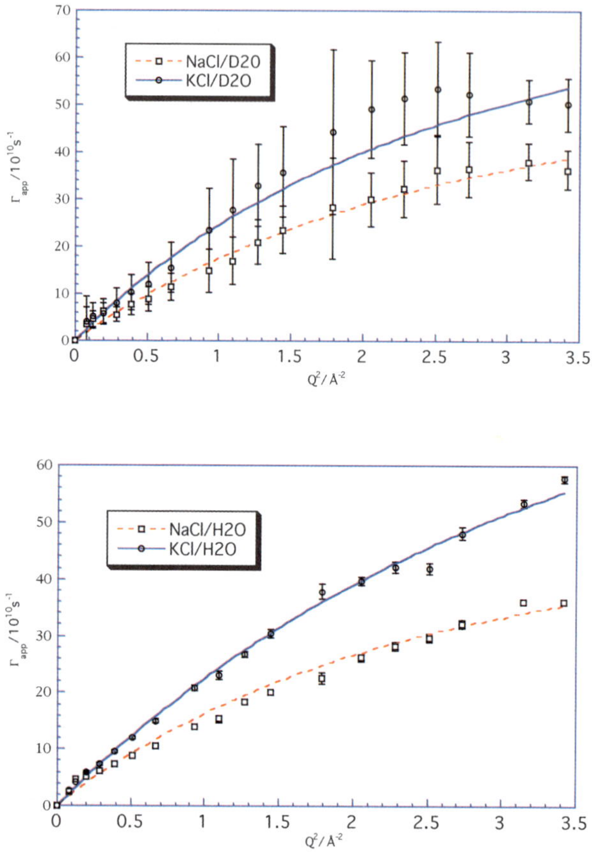

Fig. 3 Q^2-Dependence of the Lorentzian half-widths at half-maximum (HWHM) extracted from the quasielastic spectra of the multimolar NaCl and KCl solutions measured on IN5 (see Materials and methods) with an energy resolution of 63 μeV. 4M NaCl and 3M KCl solutions were investigated both in D_2O (top) and H_2O (bottom). The continuous lines represent the fits obtained with the Singwi–Sjölander model,[49] which were used to extract the apparent diffusion coefficients.

conformational flexibility that *brings them to life* and allows their biological activity. Consequently, protein and hydration water dynamics are supposed to be intimately coupled. One way of exploring the coupling has exploited the so-called protein dynamical transitions, characteristic changes in MSD that appear in at temperatures between about 180 and 250 K. Do hydration-water MSD show similar behaviour, and if so, does the 'transition' temperature coincide with the protein dynamical transition temperature? The questions were addressed in the soluble MBP[27] and in PM,[3,28] by using neutron scattering combined with deuterium labelling to examine the dynamics of each component separately. The first layer of hydration water in both PM and MBP showed a characteristic change in MSD at temperatures between 200 and 220 K (Fig. 1) that was attributed to the onset of translational diffusion[3,27] as proposed earlier.[23] Water and protein transitions in MBP appeared at the same temperature (Fig. 1A), as expected from the intimate coupling between hydration water and soluble protein dynamics, which has been reported frequently in the literature (see *e.g.* Doster *et al.*;[15] Cordone *et al.*;[52] Fitter;[16] Réat *et al.*;[19] Tsai *et al.*;[21] Vitkup *et al.*;[53] Zaccai;[38] Paciaroni *et al.*;[18] Fenimore *et al.*;[11] Chen *et al.*;[13] Roh

 This journal is © The Royal Society of Chemistry 2009

et al.;[20] Swenson *et al.*;[54] Joti *et al.*;[17] and Doster[14]). In the case of PM, however, the hydration water and membrane transitions were found to be separated by as much as 60 K (Fig. 1B). The onset of water translational motion in the first[3,28] and second[25,26] hydration layers at 200 K did not trigger a dynamical transition in PM. The data from specifically labelled reconstituted PM permitted the separated observation of lipid and BR dynamics and showed that the membrane protein dynamics is coupled to the dynamics of its lipid environment, as has been suggested,[55,56] rather than to that of inter-membrane hydration water. The lipid component of PM constitutes only 25% of the membrane mass, and most of the lipid molecules are in contact with protein. The neutron data nevertheless established that the lipids in PM, above 200 K, display higher MSD and softer resilience than BR (Fig. 2). As discussed above, 200 K is the temperature of onset of hydration water translational diffusion. It appears, therefore, that lipid dynamics responds to the water. At 260 K there is another break in the MSD of the lipids towards a domain of even smaller resilience (larger slope with temperature), which coincides with a similar dynamic transition in BR. Early neutron diffraction experiments on PM, using H_2O–D_2O exchange to highlight water location, and their comparison with data from lipid bilayer models, had shown the extent of hydration penetrating into the membrane around lipid head groups, and its dependence on the relative humidity of the sample environment.[57,58] Following these structural observations and the observation that PM activity was inhibited in the dry membrane, it has been speculated that a flexible lipid environment induced by high head-group hydration was essential for BR functional dynamics.[55] The parallel transitions in the lipid and BR MSD data at 260 K, observed in the reconstituted PM samples (Fig. 2), supports the hypothesis. At 200 K, the onset of translational diffusion in inter-membrane hydration water apparently induced some flexibility in the membrane lipids but not sufficiently to free high-amplitude protein motions. At 260 K, perhaps related to a 'melting' of head-group hydration water, both the lipid and BR components of the membrane display a dynamical transition to lower-resilience, high-amplitude MSD motions.

At the intracellular level, the studies on *E. coli* and RBC revealed that water dynamics is similar to the dynamics of water in the bulk state (Fig. 4D and zoomed in insets A, B, C). Our studies contributed to dismantle the concept that the cell somehow 'tames' water by modifying its dynamics compared to bulk water.[59] They confirmed the importance of hydration degree for water dynamics in confined geometries; in deuterated C-phycocyanin, for example, increasing water mobility had already been observed when hydration coverage reached one water layer.[60] The first macromolecular hydration layer accounts for about 0.4 to 0.5 g of water per g of macromolecule. Macromolecular concentrations in *E. coli* and RBC correspond to about 300 to 400 mg ml^{-1}, *i.e.* to four to six water layers around macromolecular surfaces. The analysis established that because of the relatively high hydration level and resulting low water confinement, intracellular water forms a network of communication that is as fluid as bulk water. In both *E. coli* and RBC, however, residence times were found to be increased by a factor of 2 and 5, respectively, suggesting that water molecules spend longer times in the first hydration shell of macromolecular structures than in the bulk phase.

The extreme halophilic archaea appeared to present an exceptional case with respect to cytoplasmic water mobility. The presence of multimolar NaCl and KCl salt concentrations in their cytoplasm is clearly not responsible by itself for the very slow water component observed by Tehei *et al.*[36] (Fig. 3). It was hypothesised that the slow mobility of a large part of *H. marismortui* cell water indicated a specific water structure, which would also be responsible for the large amount of K^+ bound within the extreme halophile cells. The absolute requirement of *H. marismortui* for a high salt environment and its ability to bind potassium ions specifically appear to be closely related to the low mobility of water in the cells. Halophilic proteins have been shown to have special hydration and ion-binding properties associated

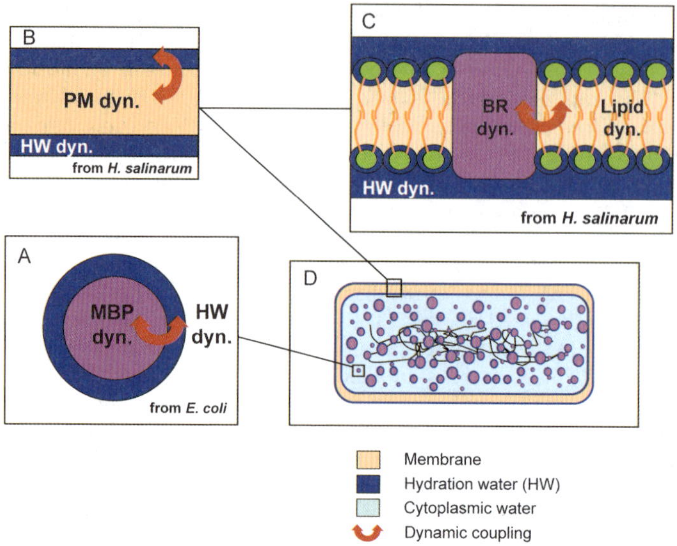

Fig. 4 Summary Figure, which shows hydration and coupling 'from shell to cell'. The Figure illustrates the main points presented and discussed in the paper, drawn in a large schematic cell. (A) Hydration around a soluble protein and coupling with protein dynamics. (B) Hydration around a membrane not coupled to membrane protein dynamics. (C) Dynamic coupling between protein and lipid in the membrane. (D) Channels of free water flowing around the hydrated macromolecules in the cell and the special case of the extreme halophile.

with an excess of negative charge in carboxylic groups on their surface.[61] The interactions may be similar to those of structured water around potassium ions and protein carboxylic groups observed by MacKinnon[62] in the potassium channel protein. It would be of significant interest to measure water mobility in these systems, in order to assess if similar mechanisms are responsible for the slow water component in the extreme halophile.

Acknowledgements

We thank Heloisa Nunes Bordallo, Bruno Demé, Bernhard Frick, Marek Koza, Mark F. Telling and Michaela Zamponi for their help during data collection at the Institut Laue-Langevin, at ISIS and at the Jülich Neutron Centre. This work was supported by the Commissariat à l'Energie Atomique, the Centre National de la Recherche Scientifique, the Université Joseph Fourier, an Agence Nationale de la Recherche Grant (project number JC05_45685), the European Union under the DLAB contracts HPRI-CT-2001-50035 and RII3-CT-2003-505925, and the Integrated Infrastructure Initiative for Neuron Scattering and Muon Spectroscopy (NMI3).

References

1 *Philos. Trans. R. Soc. London, Ser. B*, 2004, **359**, 1–395.
2 M. Tehei, B. Franzetti, D. Madern, M. Ginzburg, B. Z. Ginzburg, M. T. Giudici-Orticoni, M. Bruschi and G. Zaccai, *EMBO Rep.*, 2004, **5**, 66–70.
3 K. Wood, M. Plazanet, F. Gabel, B. Kessler, D. Oesterhelt, D. J. Tobias, G. Zaccai and M. Weik, *Proc. Natl. Acad. Sci. U. S. A.*, 2007, **104**, 18049–18054.
4 P. Ball, *Chem. Rev.*, 2008, **108**, 74–108.
5 M. C. Bellissent-Funel, J. M. Zanotti and S. H. Chen, *Faraday Discuss.*, 1996, **103**, 281–294.
6 M. Settles and W. Doster, *Faraday Discuss.*, 1996, **103**, 269–280.
7 C. Bon, A. J. Dianoux, M. Ferrand and M. S. Lehmann, *Biophys. J.*, 2002, **83**, 1578–1588.

 This journal is © The Royal Society of Chemistry 2009

8 F. Gabel and M. C. Bellissent-Funel, *Biophys. J.*, 2007, **92**, 4054–4063.
9 F. Gabel, D. Bicout, U. Lehnert, M. Tehei, M. Weik and G. Zaccai, *Q. Rev. Biophys.*, 2002, **35**, 327–367.
10 J. A. Rupley and G. Careri, *Adv. Protein Chem.*, 1991, **41**, 37–172.
11 P. W. Fenimore, H. Frauenfelder, B. H. McMahon and R. D. Young, *Proc. Natl. Acad. Sci. U. S. A.*, 2004, **101**, 14408–14413.
12 I. E. Iben, D. Braunstein, W. Doster, H. Frauenfelder, M. K. Hong, J. B. Johnson, S. Luck, P. Ormos, A. Schulte, P. J. Steinbach, A. H. Xie and R. D. Young, *Phys. Rev. Lett.*, 1989, **62**, 1916–1919.
13 S. H. Chen, L. Liu, E. Fratini, P. Baglioni, A. Faraone and E. Mamontov, *Proc. Natl. Acad. Sci. U. S. A.*, 2006, **103**, 9012–9016.
14 W. Doster, *Eur. Biophys. J.*, 2008, **37**, 591–602.
15 W. Doster, S. Cusack and W. Petry, *Nature*, 1989, **337**, 754–756.
16 J. Fitter, *Biophys. J.*, 1999, **76**, 1034–1042.
17 Y. Joti, H. Nakagawa, M. Kataoka and A. Kitao, *Biophys. J.*, 2008.
18 A. Paciaroni, S. Cinelli and G. Onori, *Biophys. J.*, 2002, **83**, 1157–1164.
19 V. Réat, R. Dunn, M. Ferrand, J. L. Finney, R. M. Daniel and J. C. Smith, *Proc. Natl. Acad. Sci. U. S. A.*, 2000, **97**, 9961–9966.
20 J. H. Roh, J. E. Curtis, S. Azzam, V. N. Novikov, I. Peral, Z. Chowdhuri, R. B. Gregory and A. P. Sokolov, *Biophys. J.*, 2006, **91**, 2573–2588.
21 A. M. Tsai, D. A. Neumann and L. N. Bell, *Biophys. J.*, 2000, **79**, 2728–2732.
22 G. Caliskan, R. M. Briber, D. Thirumalai, V. Garcia-Sakai, S. A. Woodson and A. P. Sokolov, *J. Am. Chem. Soc.*, 2006, **128**, 32–33.
23 M. Tarek and D. J. Tobias, *Phys. Rev. Lett.*, 2002, **88**, 138101.
24 A. L. Tournier, J. Xu and J. C. Smith, *Biophys. J.*, 2003, **85**, 1871–1875.
25 M. Weik, *Eur. Phys. J. E: Soft Matter*, 2003, **12**, 153–158.
26 M. Weik, U. Lehnert and G. Zaccai, *Biophys. J.*, 2005, **89**, 3639–3646.
27 K. Wood, A. Frölich, A. Paciaroni, M. Moulin, M. Hartlein, G. Zaccai, D. J. Tobias and M. Weik, *J. Am. Chem. Soc.*, 2008, **130**, 4586–4587.
28 K. Wood, M. Plazanet, F. Gabel, B. Kessler, D. Oesterhelt, G. Zaccai and M. Weik, *Eur. Biophys. J.*, 2008, **37**, 619–626.
29 W. Boos and H. Shuman, *Microbiol. Mol. Biol. Rev.*, 1998, **62**, 204–229.
30 D. Oesterhelt, *Curr. Opin. Struct. Biol.*, 1998, **8**, 489–500.
31 D. Oesterhelt and W. Stoeckenius, *Methods Enzymol.*, 1974, **31**, 667–678.
32 H. Patzelt, A. S. Ulrich, H. Egbringhoff, P. Düx, J. Ashurst, B. Simon, H. Oschkinat and D. Oesterhelt, *J. Biomol. NMR*, 1997, **10**, 95–106.
33 U. Lehnert, Thesis in Biophysics, Université Joseph Fourier, Grenoble, 2002.
34 M. Weik, H. Patzelt, G. Zaccai and D. Oesterhelt, *Mol. Cell*, 1998, **1**, 411–419.
35 M. Jasnin, M. Moulin, M. Haertlein, G. Zaccai and M. Tehei, *EMBO Rep.*, 2008, **9**, 543–547.
36 M. Tehei, B. Franzetti, K. Wood, F. Gabel, E. Fabiani, M. Jasnin, M. Zamponi, D. Oesterhelt, G. Zaccai, M. Ginzburg and B. Z. Ginzburg, *Proc. Natl. Acad. Sci. U. S. A.*, 2007, **104**, 766–771.
37 A. Elgsaeter and D. Branton, *J. Cell Biol.*, 1974, **63**, 1018–1036.
38 G. Zaccai, *Science*, 2000, **288**, 1604–1607.
39 V. Réat, G. Zaccai, M. Ferrand and C. Pfister, in *Biological Macromolecular Dynamics: Workshop on Inelastic and Quasielatic Neutron Scattering in Biology*, ed. S. Cusack, H. Büttner, M. Ferrand, P. Langan and P. Timmins, Adenine Press, New York, 1996, pp. 117–122.
40 D. J. Bicout and G. Zaccai, *Biophys. J.*, 2001, **80**, 1115–1123.
41 M. Bée, *Quasielastic Neutron Scattering, Principles and Applications in Solid State Chemistry, Biology and Materials Science*, Adam Hilger, Bristol and Philadelphia, 1988.
42 T. Unruh, J. Neuhaus and W. Petry, *Nucl. Instrum. Methods Phys. Res., Sect. A*, 2007, **580**, 1414–1422, and erratum 585:1201.
43 F. Merzel and J. C. Smith, *Proc. Natl. Acad. Sci. U. S. A.*, 2002, **99**, 5378–5383.
44 D. I. Svergun, S. Richard, M. H. Koch, Z. Sayers, S. Kuprin and G. Zaccai, *Proc. Natl. Acad. Sci. U. S. A.*, 1998, **95**, 2267–2272.
45 J. H. Roh, V. N. Novikov, R. B. Gregory, J. E. Curtis, Z. Chowdhuri and A. P. Sokolov, *Phys. Rev. Lett.*, 2005, **95**, 038101.
46 J. Teixeira, M. C. Bellissent-Funel, S. H. Chen and A. J. Dianoux, *Phys. Rev. A*, 1985, **31**, 1913–1917.
47 G. Sposito, *J. Chem. Phys.*, 1981, **74**, 6943–6949.
48 A. M. Stadler, J. P. Embs, I. Digel, G. Artmann, G. Zaccai and G. Bueldt, *J. Am. Chem. Soc.*, 2008, submitted.
49 K. S. Singwi and A. Sjölander, *Phys. Rev.*, 1960, **119**, 863–871.

50 B. N. Brockhouse, *Nuovo Cimento, Suppl.*, 1958, **9**, 45.
51 P. Leung and G. Safford, *J. Phys. Chem.*, 1970, **74**, 3696–3709.
52 L. Cordone, M. Ferrand, E. Vitrano and G. Zaccai, *Biophys. J.*, 1999, **76**, 1043–1047.
53 D. Vitkup, D. Ringe, G. A. Petsko and M. Karplus, *Nat. Struct. Biol.*, 2000, **7**, 34–38.
54 J. Swenson, H. Jansson and R. Bergman, *Phys. Rev. Lett.*, 2006, **96**, 247802.
55 G. Zaccai, *J. Mol. Biol.*, 1987, **194**, 569–572.
56 M. Kamihira and A. Watts, *Biochemistry*, 2006, **45**, 4304–4313.
57 G. Zaccai and D. J. Gilmore, *J. Mol. Biol.*, 1979, **132**, 181–191.
58 P. K. Rogan and G. Zaccai, *J. Mol. Biol.*, 1981, **145**, 281–284.
59 P. Mentré, *Cell. Mol. Biol.*, 2001, **47**, 709–715.
60 M. C. Bellissent-Funel, J. M. Zanotti and S. H. Chen, *Faraday Discuss.*, 1996, **103**, 281–294.
61 D. Madern, C. Ebel and G. Zaccai, *Extremophiles*, 2000, **4**, 91–98.
62 R. MacKinnon, *FEBS Lett.*, 2003, **555**, 62–65.
63 B. Frick and M. Gonzalez, *Phys. B*, 2001, **301**, 8–19.

 This journal is © The Royal Society of Chemistry 2009

Time scales of water dynamics at biological interfaces: peptides, proteins and cells

Johan Qvist, Erik Persson, Carlos Mattea and Bertil Halle*

Received 15th April 2008, Accepted 16th May 2008
First published as an Advance Article on the web 29th September 2008
DOI: 10.1039/b806194g

Water ^2H and ^{17}O spin relaxation is used to study water dynamics in the hydration layers of two small peptides, two globular proteins and in living cells of two microorganisms. The dynamical heterogeneity of hydration water is characterized by performing relaxation measurements over a wide temperature range, extending deeply into the supercooled regime, or by covering a wide frequency range. Protein hydration layers can be described by a power-law distribution of rotational correlation times with an exponent close to 2. This distribution comprises a small fraction of protein-specific hydration sites, where water rotation is strongly retarded, and a dominant fraction of generic hydration sites, where water rotation is as fast as in the hydration shells of small peptides. The generic dynamic perturbation factor is less than 2 at room temperature and exhibits a maximum near 260 K. The dynamic perturbation is induced by H-bond constraints that interfere with the cooperative mechanism that facilitates rotation in bulk water. Because these constraints are temperature-independent, hydration water does not follow the super-Arrhenius temperature dependence of bulk water. Water in living cells behaves as expected from studies of simpler model systems, the only difference being a larger fraction of secluded (strongly perturbed) hydration sites associated with the supramolecular organization in the cell. Intracellular water that is not in direct contact with biopolymers has essentially the same dynamics as bulk water. There is no significant difference in cell water dynamics between mesophilic and halophilic organisms, despite the high K^+ and Na^+ concentrations in the latter.

Introduction

At the molecular level, most of biology happens at interfaces where water makes contact with macromolecules or molecular aggregates. We therefore need to understand in detail how water behaves at such interfaces. Taking bulk water as the reference state, we want to know to what extent and in what way the physical properties of water are modified at the interface. Liquid water has an exceptionally high cohesive energy density which tends to minimize the effect of solutes on its H-bonded network structure. The perturbation induced by a biological interface is therefore of short range, essentially limited to the first coordination shell.[1,2] Even within this hydration layer, only subtle structural changes are usually seen.[3]

The dense H-bond network makes water structurally robust without compromising fluidity. Cold water has nearly the same short-range structure as ice I_h, but molecular rotation is six orders of magnitude faster in the liquid. Molecular mobility remains high in liquid water because the H-bond network is restructured by a cooperative

Center for Molecular Protein Science, Department of Biophysical Chemistry, Lund University, SE-22100 Lund, Sweden. E-mail: bertil.halle@bpc.lu.se

mechanism where H-bond partners are interchanged in a concerted way, thereby circumventing the high energy barriers that would have to be surmounted if several H-bonds were broken simultaneously.[4–6] An interface interferes with this cooperative mechanism, leading to a dynamic perturbation that is usually much larger and more amenable to experimental characterization than the structural perturbation.

Rotational diffusion is a more localized motion than translational diffusion and it therefore reports more accurately on the local mobility of the hydration layer. By ^{17}O NMR, water molecules can be monitored selectively in systems of arbitrary complexity and the single-molecule rotational correlation time τ can be accurately determined from spin relaxation experiments.[7] Provided that solute–water hydrogen exchange is not an issue, ^{2}H NMR can be used in the same way. Water exchange between the hydration layer and bulk water is invariably fast on the NMR time scale so the measured spin relaxation rate is a population-weighted average over these environments. The consequent loss of "spatial resolution" is not a serious handicap, because the dynamic perturbation is essentially confined to the (first) hydration layer. This has been established by experiments on simple model interfaces[1,2] and by molecular dynamics simulations.[8–10] Given an estimate of the number (ν_{H}) of water molecules in the hydration layer, *e.g.* from a simulation, we can obtain, in an essentially model-free way, the rotational correlation time $\langle\tau_{\mathrm{H}}\rangle$ averaged over all sites in the hydration layer. The rotational correlation time τ_0 in bulk water is determined from a pure-water reference sample that is routinely measured in parallel with the sample. As a convenient measure of water dynamics in the hydration layer, we use the dynamic perturbation factor (DPF), defined as

$$\xi_{\mathrm{H}} \equiv \frac{\langle\tau_{\mathrm{H}}\rangle}{\tau_0} \tag{1}$$

The most important biological interface is that between water and proteins. Protein surfaces have a complex topography, leading to a strong dynamical heterogeneity in the hydration layer[10–12] with rotational correlation times ranging from picoseconds to nanoseconds (at room temperature). Very little experimental information is available about this dynamical heterogeneity. NMR cannot frequency-resolve sub-nanosecond correlation times (as can be done for less mobile internal water molecules[7,13]), so a different strategy is needed to dissect the hydration-layer average $\langle\tau_{\mathrm{H}}\rangle$. The strategy adopted here for the proteins ubiquitin and β-lactoglobulin is to measure the ^{17}O spin relaxation rate at a fixed high frequency over a wide temperature range extending into the deeply supercooled regime.

Protein surfaces differ in topographical details but their average properties, such as polarity and curvature, are similar. We therefore expect both protein-specific and generic contributions to the hydration-layer-averaged DPF. One aim of this study is to isolate the generic component and to compare it with small solutes. For this purpose, we studied, over the same wide temperature range, the hydration dynamics of two small peptides: *N*-acetylglycine-*N′*-methylamide (NAGMA) and *N*-acetylleucine-*N′*-methylamide (NALMA). The hydration shells of these peptides contain only 5–10% of the number of water molecules in the hydration layers of the two proteins (Fig. 1). Furthermore, the peptides have a relatively simple surface topography, free from concave depressions (grooves and pockets). For these reasons, we can neglect dynamical heterogeneity in the peptide hydration shell.

Our results show that, excluding a small fraction of protein-specific hydration sites, the protein hydration layer differs little from the hydration shell of small peptides. In other words: the vast majority of the water molecules at a protein surface "don't know" if they are hydrating a protein or a small peptide. New principles need not be invoked, except to account for a small fraction of protein-specific hydration sites. Of course, these few hydration sites may sometimes be important for protein function.

Both peptides and proteins were studied here in dilute aqueous solution. We have thus characterized water dynamics in "free" hydration layers sandwiched between

 This journal is © The Royal Society of Chemistry 2009

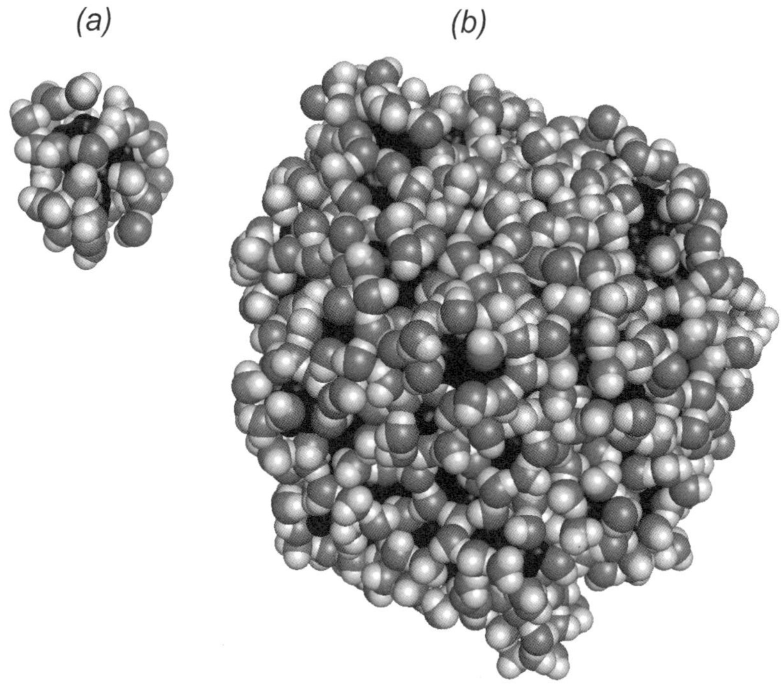

Fig. 1 Hydration shells of (a) NALMA and (b) BLG taken from MD simulations at 300 K. The solute, represented by its molecular surface, is covered by 43 (NALMA) or 740 (BLG) water molecules, the O atoms of which are within 3.3 (O), 3.5 (N) or 5.0 Å (C) of a non-H solute atom. The Figure was rendered with PyMOL (www.pymol.org).

the solute's surface and a bulk-water phase. These model systems are well-defined, but one could question their relevance for what goes on in the real, *in vivo* world. In a typical biological cell, ~30% of the volume is occupied by proteins and other biomolecules that self-assemble into supramolecular structures of stupendous complexity. The remaining 70% is water, some or all of which might differ substantially from bulk water. The view that most cell water differs radically from bulk water has a long history[14–16] and has recently received support from a quasielastic neutron scattering (QENS) study of the extreme halophilic archaeon *Haloarcula marismortui*, which indicated that the mobility of most of the cell water is two orders of magnitude lower than in bulk water.[17] If correct, this finding would force us to rethink much of what we have learnt from studies of model systems such as dilute protein solutions.

To clarify the contentious cell water issue, we have used ^{2}H NMR to selectively study water dynamics in living cells of *Escherichia coli* and *H. marismortui* cultivated in D_2O. To resolve water dynamics on different time scales, we recorded the ^{2}H spin relaxation rate over five orders of magnitude in frequency. Such magnetic relaxation dispersion (MRD) data have not been reported previously for any microorganism. The essentially model-independent picture of cell water dynamics that emerges from these NMR data differs drastically from the view supported by QENS measurements.[17]

Materials and methods

Sample preparation

The peptides NAGMA and NALMA (>99% purity, Bachem) were dissolved in D_2O (99.9% ^{2}H, CIL) at pH* 4.5 (pH* is the pH meter reading, uncorrected for H/D isotope effects). At this pH, the amide deuterons exchange too slowly to contribute to the measured water ^{2}H relaxation rate. The peptide concentrations, determined gravimetrically and by ^{1}H NMR spectroscopy, were kept below 220 mM to avoid self-association and hydration shell overlap. In this dilute regime, the relaxation rate R_1 increases linearly with solute concentration.

Highly purified ubiquitin (expressed in *E. coli*)[13] and bovine β-lactoglobulin (BLG) isoform A (Sigma) were dissolved in H_2O (19% ^{17}O, Isotec) at pH 5.0 and 2.7 respectively. Protein concentrations were determined by amino acid analysis. We verified that neither protein undergoes cold denaturation in the investigated temperature range.

Relaxation measurements at temperatures below the equilibrium freezing point of water were made on emulsified aqueous solutions with heptane as the carrier phase and sorbitan tristearate as the emulsifier.[18] Control experiments gave identical results for solution and emulsion samples at the same temperature.

E. coli (strain K-12 RV308) and *H. marismortui* from DSMZ were cultured aerobically at 310 K and pH* 7.6 in rich media prepared with D_2O (99.9% ^{2}H, Spectra Stable Isotopes). After incubation, the cell suspensions were centrifuged and the cell pellets washed twice with D_2O–saline buffers. Water-content determination, elemental analysis and amino acid analysis were performed on a portion of the cell mass. To minimize cell death, all NMR measurements were completed within 6 h. Control experiments showed that 70 ± 7% of the *E. coli* cells were viable after this period. Furthermore, repeated measurements of R_1 at a fixed high frequency showed no significant variation during the 6 h period.

NMR experiments

For the peptide and protein solutions, the relaxation rate R_1 of the water ^{2}H or ^{17}O longitudinal magnetization was measured at 55.5 or 81.3 MHz respectively. R_1 was determined with 0.5–1.0% accuracy from three-parameter fits to single-exponential inversion–recovery curves with 30 delay times in nonmonotonic order. At each temperature, measurements of R_1^0 for a pure-water reference sample were alternated with solution R_1 measurements. The sample temperature was regulated to ±0.1 K by a precooled stream of dry air and was determined before and after R_1 measurements with a thermocouple in an NMR tube containing a water–ethanol mixture.

For the cell samples, the water ^{2}H relaxation rate R_1 was measured from 1.5 kHz to 76.8 MHz using a field-cycling spectrometer and five fixed-field spectrometers. As a control, the water ^{17}O relaxation rate R_1 was measured at 67.8 MHz on an *E. coli* sample, showing that labile hydrogens make a negligible contribution to the water ^{2}H rate at the highest frequency. All measurements were performed at 300.0 ± 0.1 K with an estimated uncertainty in R_1 of less than 1%.

Molecular dynamics (MD) simulations

The rectangular simulation box contained 1631 (NAGMA) or 1858 (NALMA) SPCE water molecules and one peptide molecule with geometrical and Lennard-Jones parameters from the general Amber force field and partial charges determined with the AM1-BCC method. The MD simulations were performed at 300 K and 1 atm with periodic boundary conditions, particle-mesh Ewald summation for long-range electrostatics, 12 Å cutoff for nonbonded interactions, constrained X–H bonds and 2 fs time step. Atomic coordinates were saved every 1 ps of the stable 1 ns production trajectory.

The number (ν_H) of water molecules in the hydration shell of the peptides was determined to be the mean number of water molecules satisfying at least one of the following geometric criteria: $R(O_W–O) < 3.3$ Å, $R(O_W–N) < 3.5$ Å and $R(O_W–C) < 5.0$ Å. These cutoff distances are close to the first minimum in the O_W–X radial distribution functions.

For the proteins, the number of water molecules in the hydration layer was calculated as $\nu_H = A_S/a_W$, where A_S is the solvent-accessible surface area (SASA) of the protein and a_W is the amount of SASA occupied by one water molecule on average. Applying this procedure to the peptides and comparing with the MD-derived ν_H values, we find $a_W = 10.75$ Å^2. We thus obtained $\nu_H = 443$ for ubiquitin and $\nu_H = 735$ for BLG. For BLG, we obtained $\nu_H = 745$ from a 4 ns MD trajectory[19] (using the same cutoffs as for the peptides). The close agreement with the SASA-derived ν_H value indicates that the a_W value established for peptides also applies to proteins.

Results and discussion

Peptide hydration dynamics

The hydration-layer DPF ξ_H for NAGMA and NALMA was determined from the linear variation of the water ^{2}H relaxation rate R_1 with solute concentration according to:

$$\frac{R_1 - R_1^0}{R_1^0} = \frac{\nu_H}{N_W}\,(\xi_H - 1) \tag{2}$$

where N_W is the water/peptide mole ratio in the sample and ν_H is the number of water molecules in the hydration shell. The hydration number was determined from MD simulations of ~30 mM peptide solutions (see Materials and methods), yielding $\nu_H = 33.3$ (NAGMA) and 42.6 (NALMA). This corresponds to a single layer of water molecules covering the solute (Fig. 1a).

The temperature dependence of the DPFs for NAGMA and NALMA is shown in Fig. 2. The physical significance of the DPF maximum at ~256 K is clear from eqn (1), which yields

$$\frac{d\xi_H(T)}{d\,T} = \frac{\xi_H(T)}{k_B\,T^2}\,[E_0(T) - E_H(T)] \tag{3}$$

where E_H is the apparent Arrhenius activation energy for the hydration shell, defined as

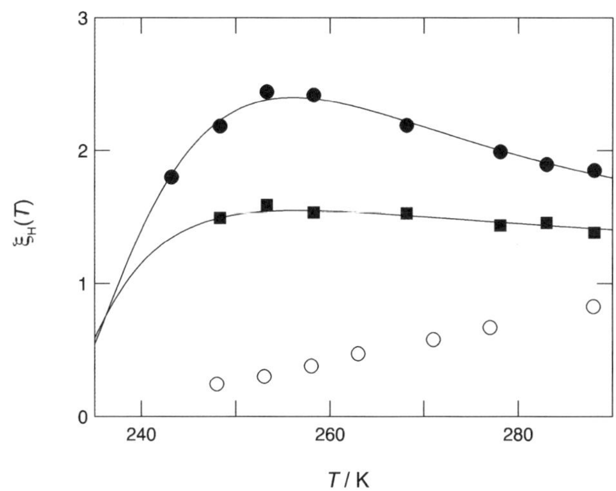

Fig. 2 Temperature dependence of the hydration-shell DPF for NAGMA (■) and NALMA (●), derived from water ^{2}H spin relaxation measurements at 55.5 MHz on dilute (<0.22 M) aqueous solutions. The curves were obtained from separate fits to the monotonic temperature dependence of $\langle\tau_H\rangle$ and τ_0. The open circles are results derived from QENS data for 1.0 M NALMA[28] as described in the text.

$$E_{\mathrm{H}}(T) = -k_{\mathrm{B}} T^2 \frac{\mathrm{d}\ln\langle\tau_{\mathrm{H}}\rangle}{\mathrm{d}T} \tag{4}$$

and similarly for the bulk-water activation energy E_0. The DPF maximum thus occurs at the crossover temperature T_X where these two activation energies are equal.

At room temperature, water rotation in the peptide hydration shell is slowed down by less than a factor of 2 as compared to bulk water (Table 1), as previously found for a wide range of organic solutes.[20–23] As is generally the case, we see that the most hydrophobic solute (NALMA) has the largest effect on water dynamics. But even at the maximum, the largest DPF is only 2.4. Furthermore, our data suggest that water rotation is *faster* in the hydration shell than in bulk water (that is, $\xi_H < 1$) at temperatures below 237–238 K (Fig. 2). This can be inferred by extrapolating the monotonic curves obtained from separate fits to the temperature dependence of $\langle\tau_H\rangle$ and τ_0. The curves shown in Fig. 2 were calculated from these fitted curves by means of eqn (1).

When bulk water is cooled, the apparent activation energy E_0 increases because the liquid structure is gradually transformed toward more open configurations with higher tetrahedral order. This subtle and gradual structural change interferes with the cooperative rotation mechanism, causing τ_0 to increase dramatically.[24] In the hydration shell, the slowing down of water rotation can also be attributed to interference with the cooperative rotation mechanism, partly because of the reduced number of nearby water molecules with which to swap H-bonds and partly because H-bond partners in the solute are either absent (at apolar sites) or else are geometrically constrained (at polar sites). Because these constraints are essentially temperature-independent, the activation energy E_H does not depend as strongly on temperature as E_0.

The hydration dynamics of NAGMA and NALMA have previously been studied by QENS and MD simulations[25–29] at or above the concentrations (1.0 M for NAGMA and 0.5 M for NALMA) where we find that R_1 no longer depends linearly on solute concentration. Whereas our results pertain to the hydration shell of the "isolated" solute, the QENS and MD results may thus be influenced by peptide self-association and hydration shell overlap. For ~1.0 M NAGMA, the simulation gave $\langle\tau\rangle = 2.27$ ps for the average over all $N_W = 55$ water molecules.[29] Attributing the deviation from bulk-water dynamics, with $\tau_0 = 1.90$ ps, to the $\nu_H = 33.3$ (Table 1) water molecules in the hydration shell, we find that the MD result corresponds to $\xi_H = 1.32$, not far from our (dilute-solution) result $\xi_H = 1.37$ (Table 1). Similarly, for ~0.5 M NALMA ($N_W = 92.4$), the simulation[29] yields $\xi_H = 1.46$ while we find $\xi_H = 1.70$ (Table 1).

For a 1.0 M NALMA solution (where $N_W = 45.9 \approx \nu_H$ so we can set $\langle\tau\rangle = \langle\tau_H\rangle$), QENS data[28] acquired in the temperature range 248–288 K gave rotational correlation times substantially *shorter* than the rotational correlation time τ_0 for bulk H_2O.[24] The resulting DPF has a qualitatively different temperature dependence and differs by an order of magnitude at low temperatures from our NMR results

Table 1 Generic hydration dynamics of proteins and peptides

			Ubiquitin		BLG	
Property	NAGMA	NALMA	$p = 0.5$	$p = 0.9$	$p = 0.5$	$p = 0.9$
ν_H	33.3	42.6	222	399	368	662
ξ_H(298 K)	1.37	1.70	1.21	1.86	1.26	2.09
$\xi_H(T_X)$	1.55	2.40	1.58	2.43	1.62	2.70
T_X/K	256.6	256.0	262.5	262.5	263.1	263.1

 This journal is © The Royal Society of Chemistry 2009

(Fig. 2). This discrepancy may result from the high peptide concentration used in the QENS study, but more likely is a consequence of the strong model-dependence in the interpretation of the QENS data.[29,30]

Protein hydration dynamics

The water ^{17}O spin relaxation rate R_1 in a protein solution exceeds the bulk-water value R_1^0, not only because water molecules in the hydration layer rotate more slowly than in bulk water, but also because a small number of water molecules are buried inside the protein. Because they interact strongly with the protein, the orientational time correlation function of these internal water molecules has a long-time tail that decays with the protein's rotational correlation time τ_R.[7] Because τ_R is more than three orders of magnitude longer than τ_0, even a single internal water molecule can have a large effect on the observed magnetic relaxation dispersion (MRD) profile $R_1(\omega_0)$ (Fig. 3). Here, we are interested in water dynamics in the external hydration layer and we therefore measure R_1 at a fixed high frequency (81.3 MHz) where the frequency-dependent internal-water contribution to R_1 is negligibly small (Fig. 3).

Even though the internal-water contribution is eliminated by measuring at a high frequency, R_1 still depends on the resonance frequency ω_0 because a small fraction of the several hundred water molecules in the hydration layer have correlation times of order $1/\omega_0 = 2$ ns. We are thus led to introduce an *apparent* DPF:

$$\begin{aligned} \xi_H(\omega_0, T) &\equiv 1 + \frac{N_W}{\nu_H}\left[\frac{R_1(\omega_0, T) - R_1^0(T)}{R_1^0(T)}\right] \\ &= \frac{1}{5\,\tau_0(T)}\left\langle \frac{\tau_H(T)}{1 + [\omega_0\,\tau_H(T)]^2} + \frac{4\,\tau_H(T)}{1 + 4\,[\omega_0\,\tau_H(T)]^2}\right\rangle \end{aligned} \tag{5}$$

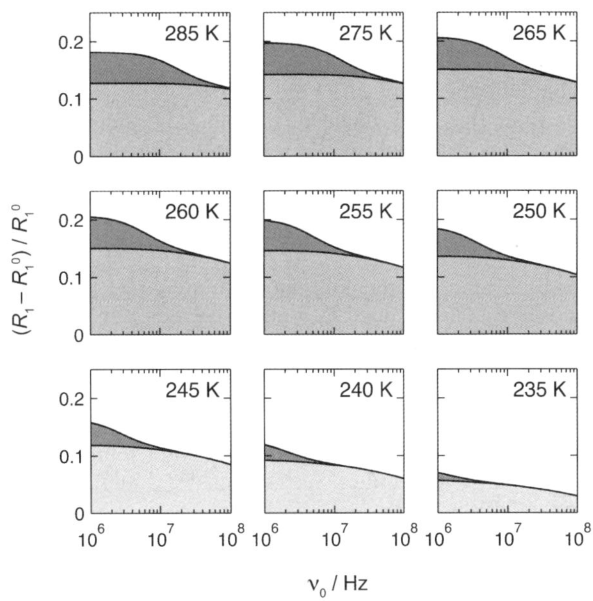

Fig. 3 Normalized water ^{17}O MRD profiles at different temperatures for a dilute (5 mM) aqueous solution of ubiquitin, showing the contributions from the external hydration layer (light gray) and from the single internal water molecule (dark gray). The curves were calculated with parameter values determined here and elsewhere.[13]

The number ν_H of water molecules in the protein's hydration layer was calculated from the solvent-accessible surface area (SASA) of the protein and, for BLG, also from MD simulations. The two methods agreed to within 1% (see Materials and methods).

As seen from eqn (5), the apparent DPF reduces to the true DPF in eqn (1) in the limit $\omega_0 = 0$, or when $(\omega_0\tau_H)^2 \ll 1$ for all sites in the hydration layer. However, at such low frequencies, the relaxation enhancement $R_1 - R_1^0$ is usually dominated by internal water molecules (Fig. 3). For proteins, the apparent DPF in eqn (5) is therefore the only way to quantitatively characterize water dynamics in the external hydration layer. Although the apparent DPF is perfectly well defined, its dependence on the resonance frequency makes it less useful for comparing results obtained on different NMR spectrometers or with other techniques. Moreover, it is not straightforward to interpret the temperature dependence of the apparent DPF. For these reasons, we adopt a model that allows us to convert the apparent DPF to a true DPF (and to determine the parameters in the model). Two model assumptions are needed.

First, we need to specify the mathematical form of the distribution $f(\tau_H,T)$ of correlation times in the hydration layer. Guided by MD simulations that indicate a power-law distribution in the picosecond–nanosecond range,[11,12] we assume that $f(\tau_H,T) \propto \tau_H^{-\nu}$ for $\tau_- < \tau_H < \tau_+$. Second, we must specify how the limits $\tau_\pm$ vary with temperature. For simplicity, we assume that they obey the Arrhenius law: $\tau_\pm(T) \propto \exp[E_H^\pm/(k_BT)]$. Consistent with experimental observations, we also postulate that $\tau_-(T^*) = \tau_0(T^*)$ and $\tau_+(T^*) = \tau_R(T^*)$ at the reference temperature $T^* =$ 293.2 K. The model then contains only two free parameters, which we choose as the power-law exponent ν and the activation energy E_- at the short-τ end of the distribution. Our calculations indicate that the results of the model-dependent analysis are robust. Specifically, they are not significantly altered by allowing for a linear temperature dependence in the activation energies $E_\pm$ or by reasonable variations in the values of the limits $\tau_\pm$ at the reference temperature T^*.

Fig. 4a shows the apparent DPF obtained *via* eqn (5) from water ^{17}O relaxation measurements on dilute solutions of β-lactoglobulin (BLG) and ubiquitin in a 50 K temperature range extending down to 238 K. As for the peptides (Fig. 2), we observe a maximum. The model fits shown in Fig. 4a yield $\nu = 2.15$ for BLG and 2.32 for ubiquitin and $E_- = 27$ kJ mol^{-1} for both proteins. These exponents are close to the value ($\nu = 2.3$) obtained from an MD simulation of cytochrome *c*.[11] The slightly smaller exponent for BLG, corresponding to a wider distribution, is consistent with the known[19] presence in BLG of an unusually large fraction of hydration sites with correlation times of order 1 ns.

From the model parameters, we can calculate the temperature variation of the true (rather than apparent) DPF $\xi_H(T)$. As seen from Fig. 4b, the DPF is substantially larger for BLG than for ubiquitin and both are larger than for the peptides (Fig. 2). These differences are caused by a minor fraction of protein-specific hydration sites with correlation times of order $1/\omega_0$ or longer, as seen directly in low-temperature MRD profiles.[31] Since the power-law exponent is close to 2, equal intervals of $\log(\tau_H)$ make equal contributions to $\langle\tau_H\rangle$ so the long-τ_H tail of the distribution contributes substantially to the DPF. In MD simulations,[12,32,33] the water molecules with the longest correlation times are generally found in concave depressions or pockets on the protein surface. Such secluded hydration sites do not occur in the peptides so the DPF is smaller.

The majority of the several hundred water molecules in the protein's hydration layer are associated with convex protrusions on the surface, where the local environment differs little from the peptide's hydration shell (Fig. 1). To compare this more exposed part of the protein hydration layer with the peptides, we calculate the *partial* DPF $\xi_H(T,p)$ for the most mobile fraction p of the ν_H water molecules in the hydration layer. This quantity is shown in Fig. 4b for $p = 0.5$ and in Fig. 5 as a function of p at two temperatures. We see that the DPFs for the two proteins are nearly the same

 This journal is © The Royal Society of Chemistry 2009

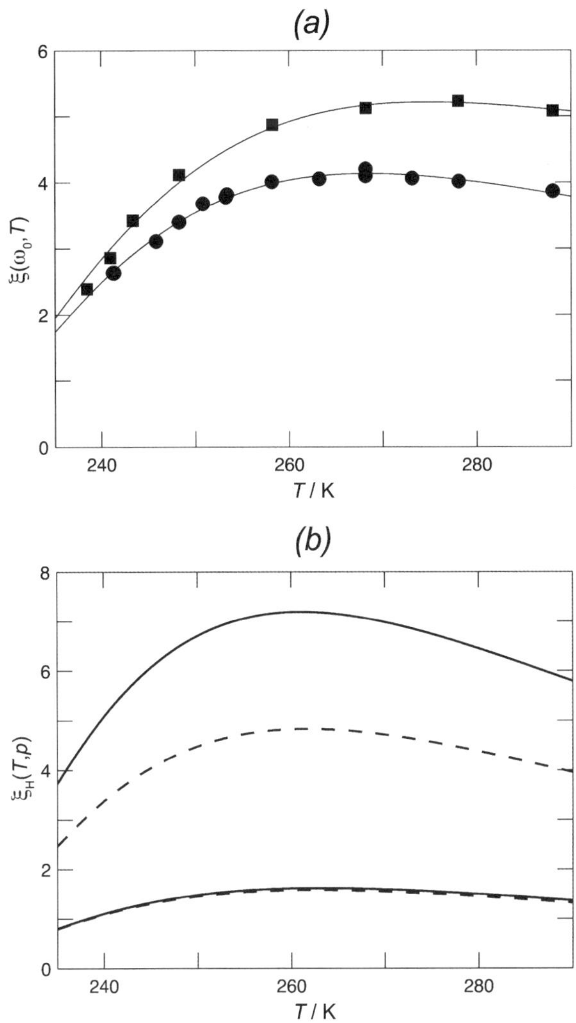

Fig. 4 (a) Temperature dependence of the apparent DPF for BLG (■) and ubiquitin (●), derived from water ^{17}O spin relaxation measurements at 81.3 MHz on dilute (0.98 and 5.1 mM respectively) aqueous solutions. The curves were obtained by fitting the two model parameters. (b) Temperature dependence of the full ($p = 1$, upper pair of curves) and partial ($p = 0.5$, lower overlapping curves) hydration-layer DPF for BLG (solid curves) and ubiquitin (dashed curves), calculated from the parameters determined in (a).

if we disregard the ~10% most strongly perturbed water molecules. In fact, the DPFs for the most mobile half of the hydration layer in the two proteins are virtually identical. Furthermore, with the exception of a small fraction of secluded hydration sites, the protein hydration layer differs little from the hydration shell of peptides and other small organic solutes. In both cases, the dynamic perturbation factor is less than 2 at room temperature and exhibits a maximum near 260 K (Table 1).

Hydration dynamics *in vivo*

In the protein solutions studied here, only 1–4% of the water molecules are in contact with the protein surface. It has been estimated that ~15% of the water in an *E. coli* cell belongs to the first hydration layer of proteins and other macromolecular

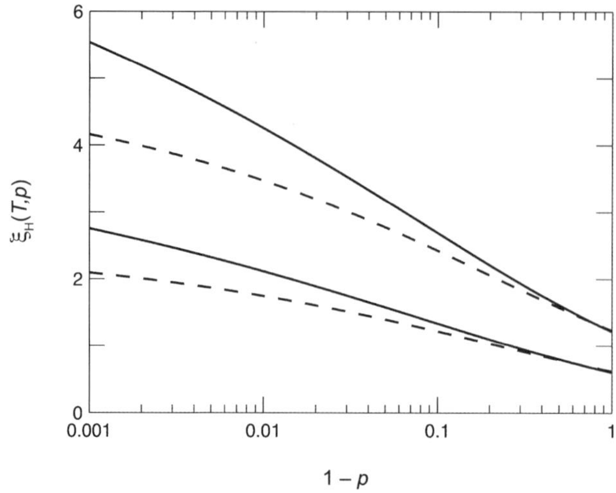

Fig. 5 Partial hydration-layer DPF at 260 K (upper pair of curves) and at 235 K (lower pair of curves) for BLG (solid curves) and ubiquitin (dashed curves). The partial DPF pertains to the most mobile fraction p of the hydration layer.

structures.[34] Most solvent-exposed surfaces in the cell are thus likely to be surrounded by multiple water layers. The long-standing question that we now address is whether the water in a living cell is similar to the water in a protein solution, with a modest dynamic perturbation in the hydration layer and bulk-water properties outside this layer. We have chosen to study the bacterium *Escherichia coli*, because of the wealth of information available about this organism, and the extreme halophilic archaeon *Haloarcula marismortui*, because of reports of unusual hydration behavior of halophilic proteins[35] and of extremely slow water diffusion in *H. marismortui* cells.[17]

Even a prokaryotic cell exhibits a high degree of spatial organization. Whether or not this structural complexity impacts on the water dynamics, it fundamentally alters the spin relaxation behavior for the biopolymers and the water molecules that are buried inside them.[13,36] The reason is that a large fraction of the biopolymers in a cell are rotationally immobilized on the NMR time scale. In fact, we have exploited the sensitivity of ^1H–^{14}N cross-relaxation to rotational immobilization to demonstrate that $50 \pm 10\%$ of the peptide groups in the two cell preparations studied here are immobilized.[34] Because of immobilization, the interpretation of water ^1H relaxation data from biological materials is a complicated task.[37] We therefore cultured the cells in D_2O and measured the water ^2H relaxation rate. To frequency-resolve contributions from different water populations, we varied the resonance frequency over five orders of magnitude.

Fig. 6 shows the water ^2H MRD profiles from living *E. coli* and *H. marismortui* cells. The MRD profile $R_1(\omega_0)$ is essentially a mapping in the frequency domain of the distribution of rotational correlation times for all water molecules in the sample. At a given frequency ω_0, R_1 reports on water motions with correlation times shorter than $1/\omega_0$. The dramatic increase in R_1 in the kHz–MHz range corresponds to water dynamics on the 0.1–10 μs time scale. Similar low-frequency ^2H relaxation dispersions are observed in biopolymer gels (but not in solutions of freely tumbling proteins) and have been quantitatively linked to the exchange of internal water molecules in rotationally immobilized proteins.[13] The R_1 dispersion below ~1 MHz is a frequency mapping of the residence time distribution of these internal water molecules. An essentially model-independent analysis[34] shows that the low-frequency dispersions in Fig. 6 are quantitatively consistent with the internal-water content

 This journal is © The Royal Society of Chemistry 2009

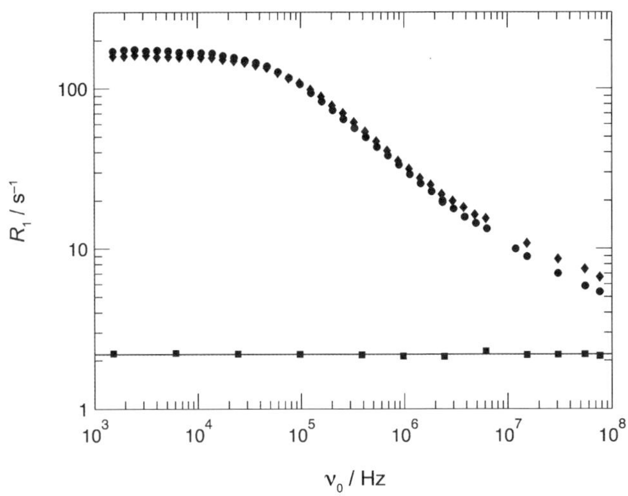

Fig. 6 Water ^{2}H MRD profiles of cells in stationary phase at 300 K: *E. coli* (●) and *H. marismortui* (◆). The results of parallel measurements on a pure-D_2O reference sample are also shown (■). Note the logarithmic R_1 scale.

of the cell samples, predicted from their molecular composition and the crystallographically-determined mean abundance of one internal water molecule per 29 amino acid residues.[38]

Our focus here is not on internal water molecules, but on water dynamics in the external hydration layers. As in the case of protein solutions (Fig. 3), this information is contained in the relaxation rate measured at high frequencies. To obtain the apparent hydration-layer DPF in eqn (5), we redefine N_W as the total number of water molecules per gram dry cell mass (DCM) in the sample and ν_H as the total number of water molecules in hydration layers per gram DCM. N_W was obtained by drying the sample at 130 °C and ν_H was calculated as A_S/a_W, as for the protein solutions. The total SASA per gram DCM, A_S, was estimated from a detailed molecular inventory of the *E. coli* cell and from elemental and amino acid analyses of our samples.[34]

The apparent hydration-layer DPFs for the two cell samples are shown in Fig. 7. At lower frequencies, $\xi_H(\omega_0,T)$ contains contributions from internal water molecules and labile hydrogens in freely tumbling biopolymers, as in a protein solution.[7] (At frequencies below 1 MHz, water molecules buried in immobilized biopolymers dominate R_1.) However, at the highest accessed ^{2}H frequency of 76.8 MHz, corresponding to a correlation time of 2 ns, the apparent DPF is expected to differ by at most 10% from the true DPF $\xi_H(T)$.[34] The convergence at high frequencies of the apparent DPFs for *E. coli* and *H. marismortui* (Fig. 7) demonstrates that the hydration water in these cells has essentially the same mobility on average. The hydration-layer DPFs are 16 ± 3 for *E. coli* and 15 ± 3 for *H. marismortui*. These results are obtained from the measured spin relaxation rates in a straightforward way, without invoking motional models. The only other input is the SASA-based estimate of the amount of hydration water (ν_H), which accounts for the ± 20% uncertainty in the DPF.

As seen from Table 2, the hydration-layer DPF for the cells is a factor of 3 larger than for small (6.5–18.4 kDa) monomeric proteins,[39] but only ~70% larger than for a set of more representative proteins (intestinal fatty-acid-binding protein, carbonic anhydrase II, β-trypsin and serum albumin),[39] which have a larger fraction of secluded hydration sites than the small proteins. (These sites are reflected directly in the MRD profiles of these proteins.) We therefore conclude that hydration layers in a living cell do not differ in any essential way from the hydration layers in a dilute

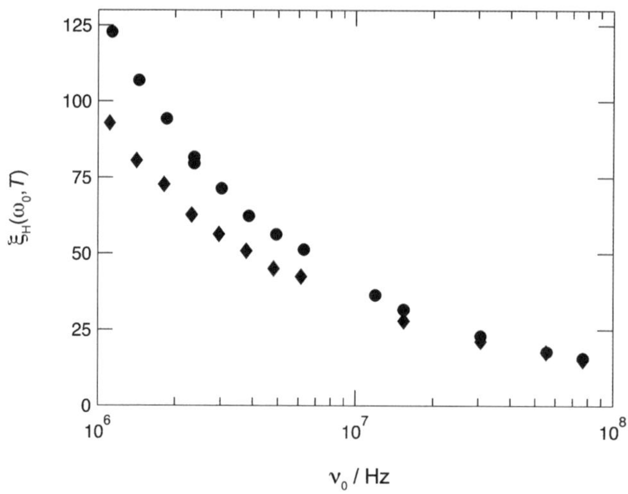

Fig. 7 Convergence at high frequencies of the apparent hydration-layer DPF for *E. coli* (●) and *H. marismortui* (◆), obtained from the data in Fig. 6, and estimates of the fraction hydration water in the cell samples.

Table 2 Hydration-layer DPF for proteins (at 298 K) and cells (at 300 K)

System	$\xi_H(T)$
Ubiquitin	3.6
BLG	5.2
11 small proteins	4.9 ± 0.6
4 larger proteins	9 ± 2
E. coli	16 ± 3
H. marismortui	15 ± 3

protein solution. The modestly larger DPF in the cells can be attributed to a larger fraction of secluded hydration sites at subunit interfaces in enzyme complexes, ribosomes, cytoskeleton and other supramolecular assemblies.

In deriving the hydration-layer DPF from the relaxation data, we tacitly assumed that the (intracellular and extracellular) water that does not belong to hydration layers has the rotational correlation time τ_0 of bulk water. This assumption is supported by the finding that it yields a hydration-layer DPF of the magnitude expected from studies of protein solutions. A more general and definite result that does not rely on our ν_H estimate can be obtained by defining a *cell-averaged* DPF by replacing ν_H in eqn (5) by the number (N_W^{cell}) of intracellular water molecules per gram DCM in the sample. The fraction of intracellular water, N_W^{cell}/N_W, is 0.74 for both samples, yielding $\xi_{cell} = 3.02 \pm 0.04$ for *E. coli* and 3.77 ± 0.04 for *H. marismortui*. Averaged over the whole cell, water rotation is thus slowed by a factor of 3 in *E. coli*. As we have seen, this average perturbation can be reasonably attributed to the 15% of the intracellular water (or 10% of the water in the sample) that belongs to hydration layers, leaving little or no room for a significant perturbation of the dominant water fraction.

The somewhat larger ξ_{cell} value for *H. marismortui* does not indicate slower water dynamics in this organism, but is a trivial dilution effect. Because of the very high salt concentration in the *H. marismortui* sample, the water/protein mass ratio is lower than in the *E. coli* sample. According to the elemental analysis, the *H. marismortui* sample contains only 12 water molecules per Na^+ or K^+ ion. Despite their

 This journal is © The Royal Society of Chemistry 2009

high concentration, these ions have a negligible effect on water dynamics, partly because of cancellation (Na^+ ions slow down, but K^+ ions speed up water rotation in the hydration shell).[40] In striking contrast to our results, a recent QENS study of a *H. marismortui* sample similar to ours concluded that 76% of the cell water is slowed down in its short-range translational diffusion by a factor of 260 at 300 K.[17] It was speculated that this extreme slowing down reflected a specific structure of cell water that is responsible for the selective accumulation of K^+ ions in the halophile cell.[17] This viewpoint, which can be traced back to the time before ion pumps had been discovered,[41] is not supported by our results. In fact, we find no significant difference in water dynamics between *E. coli* and the extreme halophile *H. marismortui*.

Conclusions

We have used ^2H and ^{17}O spin relaxation to selectively probe water dynamics in the hydration layers of peptides, proteins and cells. We report our findings in terms of the hydration-layer-averaged dynamic perturbation factor $\xi_H = \langle\tau_H\rangle/\tau_0$, a model-independent quantitative measure of the slowing of single-molecule water rotation in the hydration layer relative to bulk water. Our principal findings are as follows:

(1) Below 256 K, the activation energy for water rotation is lower in the hydration shell of a hydrophobic peptide than in bulk water. This finding challenges the classical "iceberg" view of hydrophobic hydration. The lower activation energy implies that hydration water rotates faster than bulk water at sufficiently low temperatures. Our data suggest that this is the case below 237 K for NALMA.

(2) The strong dynamical heterogeneity of the protein hydration layer can be described by a power-law distribution of rotational correlation times with an exponent close to 2.

(3) The long-τ tail of the power-law distribution is contributed by a protein-specific small population of slowly rotating water molecules, consistent with the finding from MD simulations that the most strongly perturbed water molecules reside in secluded sites.

(4) Most water molecules in the protein hydration layer exhibit a weak and generic dynamic perturbation. At room temperature, the average dynamic perturbation is a factor of ~2 for 90% of the hydration layer and only ~1.3 for the most mobile half of the layer.

(5) The hydration layer can be regarded as a defect in the H-bond network of bulk water, induced by a protein surface that provides fewer and less flexible H-bonding opportunities for the adjacent water molecules. These constraints slow down water rotation because they interfere with the cooperative mechanism that facilitates rotation in bulk water.

(6) Because the constraints are essentially temperature-independent, hydration water does not follow the strongly super-Arrhenius temperature dependence of bulk water. In this sense, hydration water is less anomalous than bulk water.

(7) With the exception of a small fraction of secluded hydration sites, the protein hydration layer differs little from the hydration shell of peptides and other small organic solutes. In both cases, the dynamic perturbation factor is less than 2 at room temperature and exhibits a maximum near 260 K.

(8) Water in the hydration layers of living cells behaves as expected from studies of simple model systems, the only difference being a somewhat larger fraction of secluded (strongly perturbed) hydration sites, associated with the supramolecular organization in the cell.

(9) Intracellular water that is not in direct contact with biopolymers has essentially the same dynamics as bulk water. This applies to >80% of the cell water.

(10) There is no significant difference in cell water dynamics between *E. coli* and *H. marismortui*, despite the high K^+ and Na^+ concentrations in the latter organism.

Acknowledgements

This work has been funded by the Swedish Research Council, the Crafoord Foundation and the Knut & Alice Wallenberg Foundation.

References

1 D. E. Woessner, *J. Magn. Reson.*, 1980, **39**, 297–308.
2 G. Carlström and B. Halle, *Langmuir*, 1988, **4**, 1346–1352.
3 B. Kim, T. Young, E. Harder, R. A. Friesner and B. J. Berne, *J. Phys. Chem. B*, 2005, **109**, 16529–16538.
4 F. Sciortino, A. Geiger and H. E. Stanley, *Nature*, 1991, **354**, 218–221.
5 D. Laage and J. T. Hynes, *Science*, 2006, **311**, 832–835.
6 J. A. R. Rodríguez Fris, G. A. Appignanesi, E. La Nave and F. Sciortino, *Phys. Rev. E*, 2007, **75**, 041501.
7 B. Halle, V. P. Denisov and K. Venu, in *Biological Magnetic Resonance*, ed. N. R. Krishna and L. J. Berliner, Kluwer Academic/Plenum, New York, 1999, pp. 419–484.
8 P. J. Rossky and M. Karplus, *J. Am. Chem. Soc.*, 1979, **101**, 1913–1937.
9 I. I. Vaisman, F. K. Brown and A. Tropsha, *J. Phys. Chem.*, 1994, **98**, 5559–5564.
10 R. Abseher, H. Schreiber and O. Steinhauser, *Proteins: Struct., Funct., Genet.*, 1996, **25**, 366–378.
11 A. E. García and G. Hummer, *Proteins: Struct., Funct., Genet.*, 2000, **38**, 261–272.
12 R. H. Henchman and J. A. McCammon, *Protein Sci.*, 2002, **11**, 2080–2090.
13 E. Persson and B. Halle, *J. Am. Chem. Soc.*, 2008, **130**, 1774–1787.
14 K. Luby-Phelps, *Int. Rev. Cytol.*, 2000, **192**, 189–221.
15 V. A. Shepherd, *Curr. Top. Dev. Biol.*, 2006, **75**, 171–223.
16 M. Chaplin, *Nat. Rev. Mol. Cell Biol.*, 2006, **7**, 861–866.
17 M. Tehei, B. Franzetti, K. Wood, F. Gabel, E. Fabiani, M. Jasnin, M. Zamponi, D. Oesterhelt, G. Zaccai, M. Ginzburg and B.-Z. Ginzburg, *Proc. Natl. Acad. Sci. U. S. A.*, 2007, **104**.
18 D. H. Rasmussen and A. P. MacKenzie, *J. Chem. Phys.*, 1973, **59**, 5003–5013.
19 J. Qvist, M. Davidovic, D. Hamelberg and B. Halle, *Proc. Natl. Acad. Sci. U. S. A.*, 2008, **105**, 6296–6301.
20 M. Ishimura and H. Uedaira, *Bull. Chem. Soc. Jpn.*, 1990, **63**, 1–5.
21 A. Bagno, G. Lovato, G. Scorrano and J. W. Wijnen, *J. Phys. Chem.*, 1993, **97**, 4601–4607.
22 Y. Ishihara, S. Okouchi and H. Uedaira, *J. Chem. Soc., Faraday Trans.*, 1997, **93**, 3337–3342.
23 A. Shimizu, K. Fumino, K. Yukiyasu and Y. Taniguchi, *J. Mol. Liq.*, 2000, **85**, 269–278.
24 E. W. Lang and H.-D. Lüdemann, *Angew. Chem., Int. Ed. Engl.*, 1982, **21**, 315–329.
25 D. Russo, G. Hura and T. Head-Gordon, *Biophys. J.*, 2004, **86**, 1852–1862.
26 D. Russo, R. K. Murarka, G. Hura, E. Verschell, J. R. D. Copley and T. Head-Gordon, *J. Phys. Chem. B*, 2004, **108**, 19885–19893.
27 D. Russo, R. K. Murarka, J. R. D. Copley and T. Head-Gordon, *J. Phys. Chem. B*, 2005, **109**, 12966–12975.
28 C. Malardier-Jugroot and T. Head-Gordon, *Phys. Chem. Chem. Phys.*, 2007, **9**, 1962–1971.
29 R. K. Murarka and T. Head-Gordon, *J. Chem. Phys.*, 2007, **126**, 215101.
30 M. Bée, *Quasielastic Neutron Scattering*, Adam Hilger, Bristol, 1988.
31 K. Modig, E. Liepinsh, G. Otting and B. Halle, *J. Am. Chem. Soc.*, 2004, **126**, 102–114.
32 V. A. Makarov, B. K. Andrews, P. E. Smith and B. M. Pettitt, *Biophys. J.*, 2000, **79**, 2966–2974.
33 A. Luise, M. Falconi and A. Desideri, *Proteins: Struct., Funct., Genet.*, 2000, **39**, 56–67.
34 E. Persson and B. Halle, *Proc. Natl. Acad. Sci. U. S. A.*, 2008, **105**, 6266–6271.
35 K. L. Britton, P. J. Baker, M. Fisher, S. Ruzheinikov, D. J. Gilmour, M.-J. Bonete, F. J,C. Pire, J. Esclapez and D. W. Rice, *Proc. Natl. Acad. Sci. U. S. A.*, 2006, **103**, 4846–4851.
36 B. Halle, *Prog. Nucl. Magn. Reson. Spectrosc.*, 1996, **28**, 137–159.
37 B. Halle, *Magn. Reson. Med.*, 2006, **56**, 60–72.
38 S. Park and J. G. Saven, *Proteins*, 2005, **60**, 450–463.
39 B. Halle, *Philos. Trans. R. Soc. London, Ser. B*, 2004, **359**, 1207–1224.
40 J. R. C. van der Maarel, D. Lankhorst, J. de Bleijser and J. C. Leyte, *J. Phys. Chem.*, 1986, **90**, 1470–1478.
41 R. Damadian, *Science*, 1976, **193**, 528–530.

 This journal is © The Royal Society of Chemistry 2009

Structure and dynamics of interfacial water in model lung surfactants

Avishek Ghosh,[ab] R. Kramer Campen,[a] Maria Sovago[a] and Mischa Bonn*[a]

Received 7th April 2008, Accepted 2nd June 2008
First published as an Advance Article on the web 18th September 2008
DOI: 10.1039/b805858j

The last decade has seen a transformation in understanding of the role of membrane-bound interfacial water. Whereas until recently water was treated principally as a continuum (primarily screening charges of lipids and proteins), it has become apparent recently that consideration of water's molecular-level properties is critical in understanding a variety of biochemical and biophysical processes. Here we investigate the structure and dynamics of water in contact with a monolayer of artificial lung surfactant, composed of four types of lipids and one protein. We probe this water using frequency-domain sum-frequency generation (SFG) spectroscopy, and a newly developed time-domain, three-pulse technique, in which the vibrational relaxation of interfacial water molecules is followed in real time. We characterize interfacial water in three systems: a monolayer of the pure lipid that is the majority of the lung surfactant mixture, a monolayer of the four lipids constituting the mixture, and a monolayer of the four lipids and the protein. We find subtle differences in the water structure and dynamics that depend on the mixture density and composition. In particular, frequency-domain measurements suggest that in the lipid mixture and the lipid mixture + protein, the relatively bulky lipids (those that have either three or unsaturated hydrocarbon tails) tend to be squeezed out at higher pressure. Measurements using the time-domain, three-pulse technique make clear that structural relaxation of interfacial water is significantly slowed down upon adding small amounts of protein to the lipids. Both results are consistent with prior measurements using other techniques in which more fluid lipids were shown to be 'squeezed out' of lung surfactant at high compression and the role of protein in the mixture was demonstrated to be a catalyst for the formation of multilayers under compression that are subsequently reintegrated into the monolayer on expansion.

1 Introduction

The 'classical' view of the cellular membrane—as a sea of lipid abutting a sea of water molecules with isolated proteins that imbue the membrane with functionality—has been revolutionized in the last 15 years.[1] During this time it has become increasingly clear that proteins are a significant percentage of natural membranes and that the various lipid components are heterogeneously distributed in the plane of the membrane.[2,3] The mechanisms by which the various components of

[a]*FOM Institute for Atomic and Molecular Physics AMOLF, Kruislaan 407, 1098 SJ Amsterdam, The Netherlands. E-mail: bonn@amolf.nl*
[b]*Leiden Institute of Chemistry, Leiden University, P.O. Box 9502, 2300 RA Leiden, The Netherlands*

membranes are organized and cellular function is maintained are known to be both passive (both thermodynamically favourable and kinetically permitted) and active (in the sense of requiring consumption of an energy source).[4] Interestingly, this new understanding of the complexity of membrane function has largely *not* extended to the role of water in membranes: water is still often thought of as 'just along for the ride', perhaps efficiently represented as a dielectric constant.[5]

Various prior studies imply that this view of water may not do justice to the complexity of its role in membrane function. For instance, studies of partially hydrated bilayers by X-ray scattering, neutron scattering and calorimetry suggest that the lipid phase—an essential parameter for membrane function—varies strongly with membrane hydration.[6] Further, NMR data from stacks of hydrated bilayers make clear that lipid head groups exercise a strong influence on local water structure: the mobility of membrane associated water is ~100 times slower than that in bulk.[7–9] While there thus appears to be bidirectional feedback between the basic biophysical properties of membranes and their associated water (water influences membrane structure and also water structure is influenced by the presence of membranes), there is also growing evidence for the importance of water in specific biochemical/biophysical functions: for example, recent computational studies have highlighted the possibility that water may mediate the interaction between some transmembrane proteins and the surrounding lipids.[10,11]

Part of the challenge in the experimental study of water in these environments has been a lack of appropriate tools: a method is required that allows the probing of water within ~1 nm of the membrane interface under ambient conditions. Because this layer is both liquid and deformable it is challenging to apply either ultra-high vacuum (using either electrons or atoms as probes) or scanning probe techniques (although the latter approach has in fact been proven insightful[12,13]). Ideally, one would like a non-invasive, label-free, analytical method that is photon-in/photon-out. As the frequency of the OH stretch is known to vary strongly as a function of hydrogen-bond strength,[14] a vibrational spectroscopy provides an attractive approach. Through measurement of the OH stretch frequency of membrane-associated interfacial water we might, therefore, gain useful insight into the interaction of this water both with itself and various other membrane components.

Measurement of the frequency of the OH stretch of membrane-bound water using conventional linear spectroscopies (*e.g.* IR) is challenging in that it requires the cancellation of a large bulk signal. This obstacle was largely overcome 20 years ago with the first measurements of interfacial water in the air/water and quartz/water systems using vibrational sum-frequency spectroscopy (VSFS).[15–17] In VSFS a pulsed visible and IR beam are overlapped at an interface and the emission of photons at the sum of the frequencies of the two incoming beams measured. In the dipole approximation the sum-frequency process is prohibited in environments with local inversion symmetry and is thus, for many systems, an interface-specific probe (inversion symmetry is always broken at the interface). The VSFS intensity is significantly enhanced when the IR frequency is resonant with a vibrational transition in the interfacial species. These two characteristics allow VSFS to measure the vibrational spectrum of water at interfaces generally, and the spectrum of water at the water/membrane interface in particular. Since the initial measurements of interfacial water at the air/water and quartz/water interfaces, such spectra have been produced for a variety of interfacial water systems: water/TiO_2, water/CaF_2, acidic water/air, basic water/air, salt solutions/air, water/surfactant and the water/lipid systems.[18–27] These spectra are typically described by vibrational resonances of specific amplitudes, centre frequencies and linewidths.

While these observables are useful, it is clear that the structure of interfacial water in general, and that of membrane-bound water in particular, is under-constrained by the few parameters that can be obtained from VSFS spectra. This situation is analogous to studies of bulk water, where linear infrared absorbance and spontaneous Raman spectral lineshapes convolute homogenous, inhomogeneous and structural

 This journal is © The Royal Society of Chemistry 2009

effects. For bulk aqueous systems pump–probe and multi-dimensional infrared spectroscopy have proven extremely useful in both describing the structure and time-dependent structural evolution of water itself, as well as the manner in which water solvates small molecules.[28–35] This pump–probe concept has recently been extended by us, and one other group independently, to the study of interfacial water by using a reflected IR pulse to excite OH stretch vibrations at an interface and tracking the dissipation of this excitation using VSFS[36–38] [hereafter we term this experimental approach time-resolved vibrational sum-frequency spectroscopy (trVSFS)].

As an initial step in understanding the role of membrane-bound water in membrane function more generally, we focus here on the role of water in model human lung surfactant (LS). Our reasons for focusing on this system are twofold. The first is pragmatic: LS is a mixture of lipids and proteins which, in its physiological role, exists at the air/water interface (the surface of the lung might usefully be thought of as air/LS/water). Useful insight into such systems may thus be gained by mono-/multilayers spread over an aqueous subphase: an easy to manipulate experimental geometry. Secondly, there is an extensive body of literature and existing therapeutic products that suggest LS function can be mimicked by a mixture of 3–4 well-defined synthetic lipids and a small, amphiphilic protein that mimics the function of lung surfactant protein B.[39,40] As such a mixture is known to be therapeutically effective, study of water in this system seems particularly poised to make a significant impact.

From the biophysical point of view LS must satisfy three requirements: it must (i) reach $\approx$ 0 mN m^{-1} surface tension near the end of exhalation, (ii) have relatively high surface activity, and (iii) adsorb/desorb from the interface rapidly.[41,42] The first requirement is met by monolayers of pure DPPC (1,2-dipalmitoyl-sn-glycero-3-phosphocholine). Unfortunately, this system is neither sufficiently surface-active, nor adsorbs sufficiently rapidly, to be useful *in situ*. Both requirements (ii) and (iii) are generally addressed by mixing lipids with DPPC that have either unsaturated bonds in their alkyl tails or tails that are particularly bulky. Pure DPPC monolayers have a relatively small lipid-condensed (LC)/ lipid-expanded (LE) phase coexistence that spans ~2–5 mN m^{-1}. The addition of unsaturated or bulky lipids expands this coexistence region over a much larger range in surface pressure, where the LC phase is then dominated by DPPC and the LE by the bulky or unsaturated lipids.[42] The pressure range of this phase coexistence can be further enhanced by the addition of a component that specifically interacts with DPPC in such a manner as to cause an increase in density; such a component might be thought of as a way of making LC domains more condensed.[43] LS films are thought to function in lungs by excluding components that form the LE domains at high pressure: the "squeeze-out hypothesis".[42,44] In the absence of any mechanism of retaining such material near the air/water interface one might expect that, over many expansion–compression cycles, loss of LS components through diffusion away from the film might be significant. Adding small amounts of a synthetic mimic of lung surfactant protein B helps overcome this problem by catalyzing the reversible formation of multilayers at high compression: instead of being lost to diffusion, LS components form multilayers at high compression and reinsert into the monolayer at low compression.[44–46]

In this study we follow prior workers and use DOPG (1,2-dioleoyl-sn-glycero-3-[phospho-*rac*-(1-glycerol)]) and tripalmitin (TP) as unsaturated and bulky components, add palmitic acid (PA) to drive the further condensation of DPPC and employ the artificial protein KL_4 (see below) as an SP-B mimic.[42,47] We take VSFS spectra over the entire CH and OH H-bonded region for first DPPC, then a 6 : 2 : 1 : 1 (by mass) mixture of DPPC : DOPG : PA : TP (called mixture) and finally a 6 : 2 : 1 : 1 : 1 mixture of DPPC : DOPG : PA : TP : KL_4 (called mixture + KL_4) at several molecular densities. In addition to the VSFS measurements, we also perform trVSFS measurements centred at 3200 cm^{-1}. Water in all three systems shows a new kind of dynamical response to vibrational excitation. The combined frequency-resolved and pump–probe study of the LS system and its constituents sheds new light on the role of water in the molecular organization of the LS monolayer.

2 Experimental

Sample preparation

1,2-Dipalmitoyl-sn-glycero-3-phosphocholine (DPPC) and 1,2-dioleoyl-sn-glycero-3-[phospho-*rac*-(1-glycerol)](sodium salt) (DOPG) were purchased from Avanti Polar Lipids. Palmitic acid (PA) and tripalmitin (TP) were purchased from Sigma-Aldrich. All were used as received. The structures of the lipids and surfactants are shown in Fig. 1. The 21-amino-acid peptide KL_4 used in our experiment consists of 4 repeating units of Lys(K)–Leu(L)–Leu–Leu–Leu terminating with K. This short-sequence peptide was custom-made by AnaSpec, Inc. (San Jose). Stock solutions ($\sim$2 mg mL^{-1}) of DPPC, PA and TP were prepared in HPLC-grade chloroform while DOPG ($\sim$2 mg mL^{-1}) and KL_4 (0.6 mg mL^{-1}) were prepared in 1 : 3 methanol–chloroform. The model lung surfactant lipid mixture without the KL_4 (*i.e.* mixture) was prepared by mixing DPPC–DOPG–PA–TP in a 6 : 2 : 1 : 1 mass ratio; and with the KL_4 (*i.e.* mixture + KL_4) a mass ratio of (DPPC–DOPG–PA–TP–KL_4) 6 : 2 : 1 : 1 : 1.[48] For the VSFS experiments, Langmuir monolayers of DPPC and the model surfactant with and without KL_4 were prepared by dropwise addition of the stock solutions onto an ultrapure (Millipore, 18 MΩ cm resistivity) water subphase in a home-built Teflon trough. A monolayer or multilayer forms on top of the aqueous phase after rapid evaporation of the chloroform and methanol. As the time-resolved measurements may take up to several hours in our set-up, a rotating trough was used in these experiments to prevent cumulative heating by the laser pulses. The surface pressures for the monolayers were monitored using a tensiometer (Kibron). Experiments were performed at two different surfactant densities for each sample type. For single component DPPC monolayers experiments were performed at surface densities of 2.9 and 4.4 $\times$ 10^{-4} mg cm^{-2} (corresponding to surface pressures of 15 and 35 mN m^{-1} respectively), for the mixture

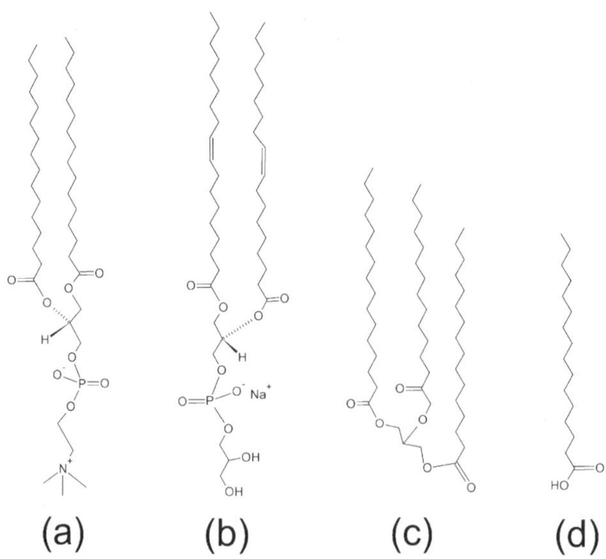

Fig. 1 Chemical structures of the model lung surfactant constituents. Molecule (a) is 1,2-dipalmitoyl-sn-glycero-3-phosphocholine (DPPC). Molecule (b) is 1,2-dioleoyl-sn-glycero-3-[phospho-*rac*-(1-glycerol)](sodium salt) (DOPG). Molecule (c) is glycerol tripalmitate or tripalmitin (TP). Molecule (d) is palmitic acid (PA).

surface densities of 1.7 and 3.1×10^{-4} mg cm^{-2} and for the mixture + KL_4 2.1 and 3.1×10^{-4} mg cm^{-2} (trVSFS results are presented only for the high-density mixture + KL_4 sample). We observed large variability in the surface pressure for the mixture and mixture + KL_4 systems as measured with our tensiometer. Prior fluorescence and scanning probe images of systems of similar composition to ours indicate lipid-phase separation. It is therefore expected to be, as we observe, difficult to extrapolate point-pressure measurements to entire monolayers in these systems[41,48]

trVSFS experimental set-up

Our implementation of frequency-domain VSFS has been described previously.[26] The optical processes excited in frequency-resolved VSFS and in our novel trVSFS, along with a qualitative illustration of the results in each case, are shown in Fig. 2. For the trVSFS experiments, tunable, intense IR pulses are generated using a two-step method:[49] first, 1 out of 3 mJ of 800 nm Ti : Sa light (100 fs pulse duration, 1 kHz repetition rate) is used to pump a commercial optical parametric amplifier (TOPAS, Light Conversion). In the second step, the idler beam generated in the TOPAS at about 2200 nm is doubled in a β-barium borate (BBO) crystal and subsequently mixed with the remaining 2 mJ of the Ti : Sa fundamental to generate the

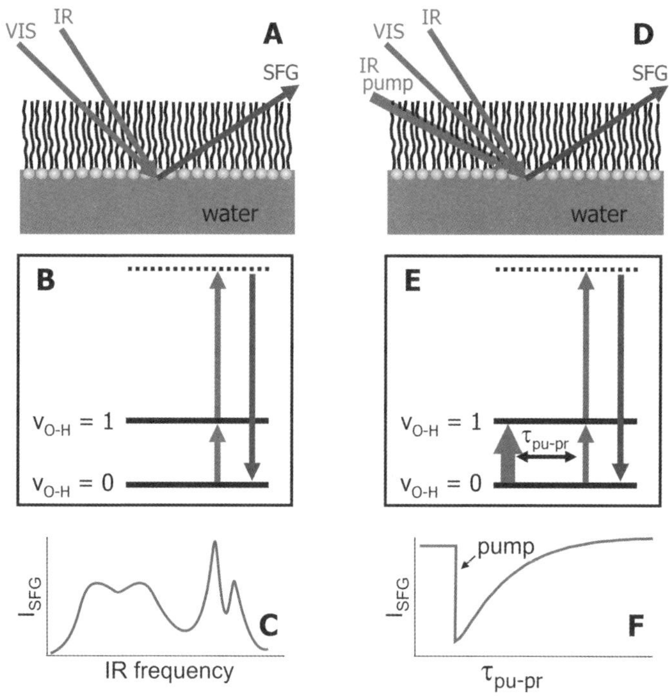

Fig. 2 Schematic representations of the two types of SFG experiments reported here: (A–C) in the spectrally resolved experiment, a weak infrared (IR) and visible (VIS) pulse impinge upon the surface at which sum-frequency generation (SFG) occurs, as depicted in the energy level scheme in (B). If the IR is resonant with a ($v = 0 \rightarrow 1$) vibrational transition, the SFG process is resonantly enhanced, and a spectrum can be recorded (C). In a time-resolved SFG experiment (D–F), a powerful infrared pump pulse which is resonant with the O–H stretch vibration excites a significant fraction of the water O–H groups to their first vibrationally excited state (E). The SFG intensity generated by a weak infrared probe pulse and a visible pulse is recorded as a function of delay time τ_{pu-pr} between pump and probe pair (F). Due to the excitation, the generated SFG intensity decreases temporarily. The recovery of the SFG signal reflects vibrational relaxation back to the ground state.

difference frequency in a 3 mm thick $KTiOPO_4$ (KTP) crystal. This results in IR pulses with an energy of ~50 μJ, tunable around 3000 nm, with a duration of ~100 fs. For the one-colour pump–probe experiments described in this study the infrared beam was split into IR pump and probe beams (intensity ratio 2.5 : 1). The visible (VIS) pulse used to generate the sum frequency at the interface (in combination with the IR probe pulse) is the residual 800 nm light of the commercial OPA, passed through a home-built pulse shaper (resulting in a bandwidth of the VIS beam of typically 10 cm^{-1}). The IR pump, IR probe and VIS probe beams lie in the same vertical plane and have incident angles of 46°, 50° and 56°, respectively, relative to the surface normal. The polarizations of all beams can be rotated independently using $\lambda/2$-plates. All beams were focused and intersected at the water interface, with focal sizes of 150, 120 and 120 μm respectively. The sum-frequency generation (SFG) signal emitting from the interface was detected in reflection geometry (see Fig. 2). Before detection the reflected light was filtered using bandpass filters, the remaining SFG signal spectrally resolved with a monochromator and detected with an electron-multiplied charge-coupled device (EMCCD, Andor Technologies). Because the SFG signal is between 0.1 and 10 detected photons per laser shot, typically 20 time scans were averaged, each consisting of 42 points (10 000 laser shots per point). To clarify vibrational relaxation dynamics, the emitted VSFS signal was normalized by the VSFS signal in the absence of the pump, by chopping the 1 kHz pump beam to 500 Hz and recording the normalized, differential SFG signal as the ratio between the signals with and without the pump. One-colour pump–probe transients are obtained in our set-up using a mechanical delay line to change the timing of the IR pump relative to the probe pulse pair. The time-zero and the instrument response for the experiments was determined by monitoring the $\chi^{(3)}$ process of infrared–infrared–visible SFG as a function of the delay between the pump and probe pulses.[50]

3 Results and analysis

Measurements were performed on 3 samples: pure DPPC, a 6 : 2 : 1 : 1 mixture of DPPC : DOPG : PA : TP (mixture) and a 6 : 2 : 1 : 1 : 1 mixture of DPPC : DOPG : PA : TP : KL_4 (mixture + KL_4). The frequency-resolved experiments were performed on a subphase of D_2O, rather than H_2O, as our set-up operates better in the OD stretching range than the OH. Without exception, experiments performed on an H_2O subphase reveal the same behaviour for H_2O as for D_2O, allowing a straightforward comparison through a renormalization of the frequency axis. Time-resolved experiments were performed on an H_2O subphase. For the conventional VSFS measurements, spectra were taken for each of the 3 systems at 2 distinct lipid densities corresponding to surface pressures of approximately ~15 and 30 mN m^{-1} for the lipids and ~7 and 15 mN m^{-1} for the lipid–protein mixture, giving a total of 6 measurements. For the trVSFS measurement, no low-density mixture + KL_4 traces were collected.

Frequency-resolved VSFS measurements

Fig. 3 shows VSFS spectra for pure DPPC, the mixture and the mixture + KL_4, at two densities each. The spectra are characterized by broad OD resonances in the 2200–2600 cm^{-1} range and narrow CH resonances around 2900 cm^{-1}. The VSFS data is fitted using a Lorentzian model, as described in literature.[51] Briefly, the sum-frequency generation (SFG) intensity is proportional to the square of the nonlinear polarization P_{SFG} generated at the surface by the visible and infrared optical fields:

$$I_{SFG} \propto |P_{SFG}|^2 \propto |\chi^{(2)}|^2 I_{VIS} I_{IR} \quad (1)$$

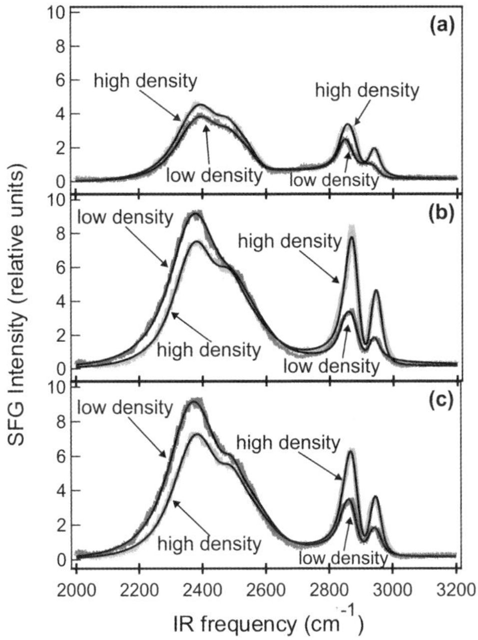

Fig. 3 Frequency-resolved VSFS spectra of the three monolayers at low (dark grey) and high (light grey) lipid densities each corresponding to surface pressures of approximately ~15 and 30 mN m^{-1}. Lipid surface densities for low and high densities are: 2.9 and 4.4 mg cm^{-2} for DPPC, 1.7 and 3.1 for the mixture, and 2.1 and 3.1 for the mixture with KL_4, all expressed in units of 10^{-4} mg cm^{-2}. Fits to the data (see text) are shown as black lines. Panel (a) is data for the DPPC monolayer, panel (b) data for the lipid mixture (no KL_4) and panel (c) data for the lipid mixture + KL_4. All static SFG data was collected in a SSP (SFG : VIS : IR) polarization configuration.

where $\chi^{(2)}$ is the second-order nonlinear susceptibility and $I_{\rm VIS}$ and $I_{\rm IR}$ are the intensities of the incident fields. As discussed above, when the frequency of the incident infrared field is resonant with a vibrational mode, n, the SFG field is strongly enhanced.

The susceptibility, $\chi^{(2)}$, contains both a nonresonant and a resonant term:

$$\chi^{(2)} = \chi^{(2)}_{\rm NR} + \chi^{(2)}_{\rm R} = A_0 e^{i\phi} + \sum_n \frac{A_n}{\omega_n - \omega_{\rm IR} - i\Gamma_n} \tag{2}$$

Here, A_0 is the real amplitude of the non-resonant susceptibility, ϕ its phase, A_n is the amplitude of the nth vibrational mode, ω_n the resonant frequency, and Γ_n the line width of the transition. We substituted eqn (2) into eqn (1) and used the resulting expression to fit the frequency-resolved VSFS spectra; the results are shown as black solid lines in Fig. 3.

The signal in the OD stretch region is significantly larger for the lipid mixture—both with and without KL_4—than for DPPC. The fits to the data reveal that the amplitudes of the susceptibility for the CH stretch are very similar for the 3 systems; the lowered SFG intensity for DPPC in the CH stretch region is caused by reduced interference with the OD stretch. As expected, for all three systems the CH resonances increase in intensity when the lipid density increases.[52] This is an effect of both an increase in the number of methyl and methylene groups contributing to the signal and the increased order among the methyl CH_3 groups. There is, however, a remarkable difference between the behaviour of the water resonances as the lipid density is varied: for DPPC the water signal *increases* weakly with increasing lipid density, while for the mixture, with and without KL_4, there is a significant *decrease* in the water signal as the lipid density is increased.

Time-resolved VSFS measurements

In addition to the static VSFS measurements, we also conducted trVSFS measurements centered at 3200 cm^{-1} (using pulses with a spectral width of ~120 cm^{-1}). This frequency corresponds to the maximum in the SFG response in Fig. 3, for H_2O in the OH stretch region. In all three systems the dynamical response of water looks qualitatively similar. Immediately after excitation the SFG intensity decreases, due to depopulation of the ground state (bleach). This is followed by a partial recovery, after which the signal decreases once more, and finally continues to recover to a significant fraction of its original value. The initial dynamics (bleach, partial recovery and second signal decrease) occur on subpicosecond timescales, whereas the final slow rise occurs over tens of picoseconds. The initial loss of signal, the bleach, occurs nearly instantaneously. After this bleach, then, inspection of the data makes clear that it is characterized by three other distinct processes: partial rapid recovery, a second signal decrease and continued slow recovery.

Such a response is most simply described by a 5-level system, as shown in Fig. 4, and is physically justified below. The relevant differential equations describing transfer of population in such a system read:

$$
\begin{aligned}
\frac{\mathrm{d}N_0(t)}{\mathrm{d}t} &= -\sigma_{01} I(t)(N_0(t) - N_1(t)) \\
\frac{\mathrm{d}N_1(t)}{\mathrm{d}t} &= \sigma_{01} I(t)(N_0(t) - N_1(t)) - \frac{N_1}{T_{12}} \\
\frac{\mathrm{d}N_2(t)}{\mathrm{d}t} &= \frac{N_1}{T_{12}} - \frac{N_2}{T_{23}} \\
\frac{\mathrm{d}N_3(t)}{\mathrm{d}t} &= \frac{N_2}{T_{23}} - \frac{N_3}{T_{34}} \\
\frac{\mathrm{d}N_4(t)}{\mathrm{d}t} &= \frac{N_3}{T_{34}} \\
\Delta I_{\mathrm{SFG}}(\mathrm{norm.}) &= \frac{[N_0(t) - N_1(t) + \sigma_{21}N_2(t) + \sigma_{32}N_3(t) + \sigma_{43}N_4(t)]^2}{[N_0(t=0)]^2}
\end{aligned}
\tag{3}
$$

Where $N_X(t)$ is the time-dependent population in state X and T_{XY} the relaxation time from the X state into the Y state. ΔI_{SFG} is the experimental observable, the normalized change in the SFG intensity, and σ_{XY} is the SFG cross section of the Y to X transition. In our modelling of the data, cross sections and relaxation times are fit parameters.

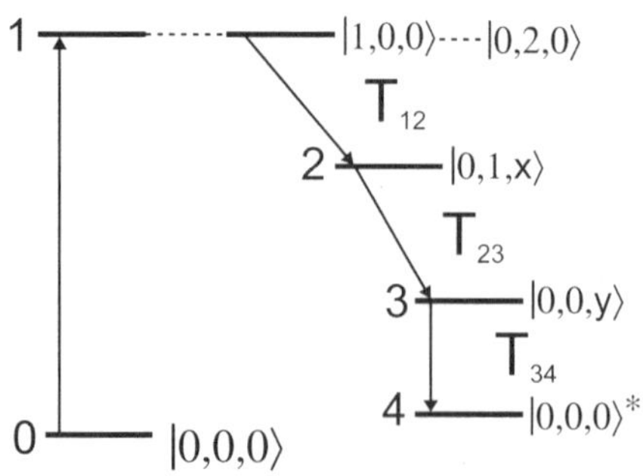

Fig. 4 5-Level scheme with three relaxation times used to describe the transients of Fig. 5.

 This journal is © The Royal Society of Chemistry 2009

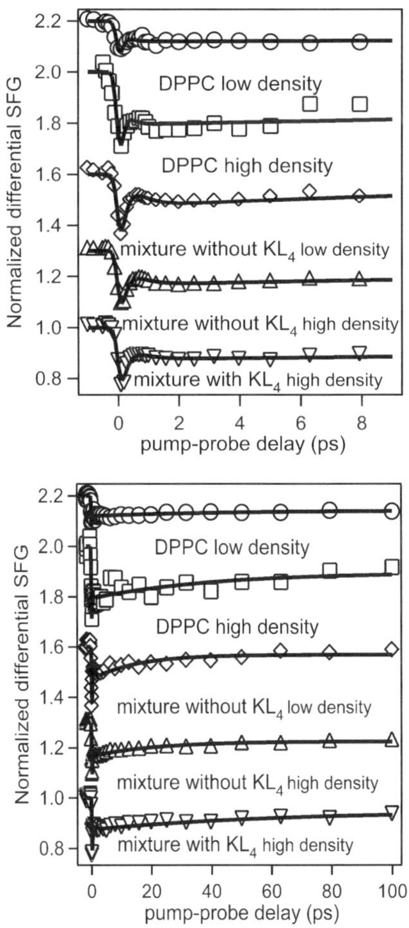

Fig. 5 Results of the trVSFS pump–probe experiments at the OH stretch frequency of water at the interface of DPPC, lung surfactant mixture with and without KL_4 on a 10 and 100 ps time-scale (upper and lower panel respectively). The pump and probe IR pulses are P-polarized, while the visible upconversion pulse is S-polarized. This polarization configuration of the incoming beams generates an SFG pulse that is S-polarized. Lipid surface densities for low and high densities are: 2.9 and 4.4 mg cm^{-2} for DPPC, 1.5 and 3.0 for the mixture, and 2.5 for the mixture with KL_4, all expressed in units of 10^{-4} mg cm^{-2}. Fits to the data (see text) are shown as black lines.

Table 1 Relaxation times (in femtoseconds) obtained from the fits to the data in Fig. 5 using the level scheme of Fig. 4 and eqn (3)[a]

	DPPC low dens.	DPPC high dens.	Mixture low dens.	Mixture high dens.	Mixture +KL_4 high dens.
T_{12}/fs	240	190	430	320	300
T_{23}/fs	80	110	230	270	310
T_{34}/fs	45 000	40 000	20 000	21 400	60 000

[a] Values are ±15%, except for T_{23} for DPPC, for which the values are ±50%.

As is evident from inspection of Fig. 5, this simple model provides an adequate description of the data. The time constants that are inferred from the fit are shown in Table 1. Two significant observations can be made. First of all, the T_{12} and T_{23} for DPPC are significantly faster than for both the mixture and the mixture + KL_4; the latter pair have similar fast dynamics. Secondly, the slow time constant, T_{34}, has a distinct dependence on monolayer composition, and is clearly slowest for the mixture with KL_4.

4 Discussion

Spectrally resolved VSFS measurements

VSFS spectra of interfacial liquid water—regardless of adjoining phase—all have in common a broad, double-peaked feature from ~3100–3550 cm^{-1} originating from OH stretch vibrations of interfacial water. The results presented in Fig. 3 are no exception. The assignment of this broad, double-peaked feature has been a subject of disagreement in the literature: on the one hand, in analogy with the IR spectrum of liquid water and ice, it has been argued that the two apparent peaks are the result of different quasi-static water types. This argument has been bolstered by a series of MD simulations that have been interpreted to suggest that each spectral feature is actually a combination of multiple, distinct, water structures.[16,17,19,22,23,53–55] On the other hand, it has also been argued based on analogy to gas-phase water molecules, other simulations and some interpretations of the spontaneous Raman spectra of bulk water, that the two peaks can be largely explained as the symmetric and asymmetric vibrational modes of the water molecule.[21,56] Recently published work, however, by showing that each peak has a similar dependence on IR and VIS polarization, has demonstrated that both peaks result from vibrational modes with the same symmetry thus ruling out the latter interpretation.[25]

We have recently demonstrated, using isotopic dilution experiments, that a third mechanism best explains the double-peaked nature of the interfacial water spectrum. At both the air/water and lipid/water interfaces the two peaks collapse to one upon isotopic dilution, showing that the two-peak structure is not the result of a distribution of H-bonded structures.[57] Owing to the frequencies and polarization dependences of the signals,[25] the interpretation in terms of symmetric and asymmetric vibrational modes can also be discarded. The variety of experimentally observable characteristics of the double-peaked spectral feature can be thus best explained if the visible 'dip' in the spectrum is the result of a Fermi resonance of the bend overtone and the stretch fundamental.[57] Such coupling has been demonstrated to exist, using two-colour IR pump–probe experiments in bulk water.[33,34]

This reasoning makes clear that a quantitative structural interpretation of the OH-stretch spectrum is complicated because OH groups are coupled both to their neighbour (intermolecular coupling) and to other modes on the same water molecule (intramolecular coupling). However, a number of qualitative statements can be made from the spectra, in particular regarding the observations that (i) the signal in the OD stretch region is significantly larger for the lipid mixture—with and without KL_4—than for DPPC, and (ii) that for DPPC the water signal *increases* weakly with increasing lipid density but for the mixture with and without KL_4 it *decreases* as the lipid density is increased.

The amplitude of the OH (OD) stretch in interfacial water as measured using VSFS is known to increase at water/X interfaces when the X phase has nominal net surface charge (see, for example, the study by Gragson and coworkers of monolayers of differently charged surfactants at the air/water interface[18]). As indicated in Fig. 1, DOPG is charged. Armed with this information, observation (i) can thus be rationalized by noting that both the mixture and the mixture + KL_4 contain significant amounts of the negatively charged DOPG and that thus the amplitude of the OH (OD) stretch in these systems might be expected to be larger than that near the

 This journal is © The Royal Society of Chemistry 2009

nominally neutral, pure DPPC monolayer. Rationalizing observation (ii) requires noting that mixtures of DPPC, DOPG, PA and TP are known to phase-separate (under a wide range of pressures) into relatively more condensed domains (enriched in DPPC and PA) and relatively more liquid domains (enriched in DOPG, TP and, if applicable, KL_4). The relatively liquid domains are thought to preferentially 'squeeze out' as the monolayer is compressed.[42,44] Therefore, as monolayer pressure is increased an increasing amount of interfacial water is expected to be in contact with the neutral DPPC, relative to that fraction of water in contact with charged DOPG, and the amplitude of the OD stretch is expected to decrease. Our observation of differential solvation of high-pressure and low-pressure model lung surfactants is thus in good agreement with prior reports of enrichment in DPPC after compression of compositionally similar model LS.[42] Before proceeding we note that it is unlikely that the contribution to the VSFS signal from the fraction that is squeezed out is significant, given the quadratic dependence of VSFS on surface population.

Time-resolved VSFS measurements

As discussed above, it is challenging to extract details of changes in water structure from the static VSF spectra. Although differences in lineshape are clearly evident, understanding the structural significance of these observations is not straightforward. As will be shown below, the dynamical response can shed additional light on local water organization.

We have previously applied trVSFS to both the air/water and 1,2-dimyristoyl-sn-glycero-3-[phospho-L-serine](DMPS)/water interfaces while others have applied it to water at the silica/water interface.[36–38] Two general types of interfacial responses were observed in these studies: at the air/water and silica/water interfaces the dissipation of the vibrational excitation was not a function of pump polarization, could be described by a biexponential in which the time constants were not a function of frequency, and were equal to those found for bulk water in IR pump–IR probe experiments.[36,37] Vibrational excitations are known to dissipate in pure H_2O or D_2O through ultrafast nonradiative energy transfer (Förster energy transfer).[29] The similarity in interfacial and bulk time constants, and the lack of frequency dependence in each case, is most easily understood as resulting from ultrafast energy transfer between bulk and interfacial water species.[37] Such energy transfer in water requires that an individual OH oscillator has, on average, a chemically equivalent OH oscillator within the Förster radius ($\sim$2.8 Å). At the DMPS/water interface, on the other hand, dissipation of vibrational energy rates varied with pump polarization and was monoexponential at all frequencies; the single time constant was strongly a function of frequency. The frequency dependence of the time constant was quantitatively described using a model of Staib and Hynes in which the OH stretch relaxes only by coupling to a hydrogen bond accepting mode.[58] This mechanism of relaxation is consistent with a scenario in which the OH groups of membrane-bound water are energetically distinct and spatially separate with, on average, similar OH groups more than 2.8 Å away.

The interfacial water studied here shows a new kind of dynamical response to vibrational excitation: unlike the air/water interface it has time constants that are frequency-dependent[59] and distinct from those that describe bulk relaxation, while unlike the DMPS/water interface it has clear coupling to an additional two states during relaxation requiring a total of three time constants to explain the data. In analogy to pump–probe measurements on bulk water[33,34] performed at 3150 cm^{-1}, we attribute our phenomenological states to quantum mechanical equivalents (see Fig. 4). Here the quantum mechanical description of each phenomenological state is given by the wave functions $|\nu,\delta,l\rangle$ in which the quantum numbers ν, δ and l describe the OH stretch, bend and water intermolecular librational modes respectively. Given this description, a typical experiment might be described in the following manner: from state $|0,0,0\rangle$ (N_0), excitation is taken to occur into state $|1,0,0\rangle$ (N_1) during the pump pulse. Due to the strong coupling between the stretch and bend overtone, $|1,0,0\rangle$ is indistinguishable from $|0,2,0\rangle$; N_1 is a superposition of those two states. From $|0,2,0\rangle$ relaxation occurs to $|0,1,x\rangle$ (N_2), a quantum mechanical state structurally interpretable as a mixture of one quantum of the water-bend and high-frequency intermolecular librations. $|0,1,x\rangle$ subsequently relaxes to $|0,0,y\rangle$ (N_3) or low-frequency inter-molecular librations. Relaxation of the low-frequency librational modes leads to a heated ground state: $|0,0,0\rangle^*$ (N_4).

Hence, we see relaxation of the excited stretch state through the overtone of the bend mode, to a mixed state bend fundamental and high-frequency inter-molecular librations, followed by low-frequency intermolecular librations and finally followed by a thermalized ground state. The rates of the first two relaxation steps (time constants T_{12} and T_{23}), prior work has demonstrated, are strongly a function of the energy overlap between intermolecular librations and the bend fundamental (relaxation is fastest in bulk water in which this overlap is significant, slows down substantially when examining HOD in D_2O and further slows for isolated water molecules in nonpolar solvents (where there are no relevant intermolecular modes)[33,34,60–63]). For bulk water, these time constants are T_{12} = 200 fs and T_{23} = 170 fs. For DPPC-bound water very similar time constants are observed (we note that the fits are more sensitive to T_{12} than to T_{23} times), but for the water bound to the lipid mixture (with and without KL_4), these times are slowed down significantly: from $T_{12} \approx$ 200 fs and $T_{23} \approx$ 100 fs for DPPC to $T_{12} \approx T_{23} \approx$ 300 fs for the mixture and mixture + KL_4. This result suggests that, for the lipid mixture and the mixture + KL_4, coupling to high-frequency inter-molecular modes is suppressed (relative both to water at the DPPC/water interface and to pure water in bulk). Such suppression is most simply explained by the reduced density of membrane-bound water in these environments. KL_4 exists in the system in relatively low concentration (<10% by mass) and its principal role is thought to lie in restraining the diffusion of DOPG and TP away from the interface after 'squeeze out'. It is, therefore, little surprise that the addition of KL_4 does not affect the fast water dynamics significantly.

In liquid water the third time constant (T_{34}) is interpreted physically as the relaxation of intermolecular librational modes and its characteristic ~1 ps timescale as the result of structural reorganization: hydrogen-bond breaking. For DPPC, the mixture and the mixture + KL_4, this timescale is absent from the data. Instead T_{34} is much longer, ≈40 ps for the DPPC layer, ≈20 ps for the mixture and ≈60 ps for the mixture + KL_4. NMR studies of partially hydrated bilayers suggest this to be a significant fraction of the timescale on which exchange of individual water molecules between those strongly associated with lipid head groups and the bulk occur (typical exchange times (τ_{exch}) reported to be ≈100 ps).[9] Hence, it seems plausible to interpret T_{34} as a restructuring of the membrane-bound water following thermalization of the excess vibrational energy. It is unsurprising that there is not a 1 : 1 correspondence between τ_{exch} and T_{34} as NMR and trVSFS are sensitive to different aspects of membrane hydration: τ_{exch} obtained from NMR provides the timescale necessary to remove water from near a particular functional group, while trVSFS responds to more collective structural changes. While molecular level details of the exchange of membrane-bound water molecules with bulk have not been described, the fact that our T_{34} is 1/2 to 1/3 τ_{exch} is consistent with a scenario in which several collective rearrangements of water molecules are necessary to extract a single water molecule from a confined, inter-headgroup environment. This scenario is the object of current research. Although the above interpretation of the 5-level system is in good agreement with the data and consistent with bulk measurements,[33,34] we cannot exclude the following, alternative interpretation: from state $|0,0,0\rangle$ (N_0), excitation is taken to occur into state N_1 in which the stretch is excited. Subsequent vibrational relaxation occurs to N_2, resulting in 'hot' water molecules. N_2 subsequently relaxes to N_3, a state where the excess energy is equilibrated over all degrees

 This journal is © The Royal Society of Chemistry 2009

of freedom. The relaxation of the system in response of the system to the quasi-instantaneous temperature increase results in relaxation into state N_4.

As described above, the addition of KL_4 to the mixture has relatively little effect on vibrational relaxation (T_{12} and T_{23}) but does have a significant influence on relatively slow structural relaxation processes (*i.e.* T_{34}). Interestingly, the addition of KL_4 to the lipid mixture does have a significant effect on the pressure isotherm:[64,65] at high surface pressures, the lipid density (normalized to the monolayer interfacial area) is lower in the presence of KL_4. This observation, in combination with scanning probe studies of model LS films, is most simply understood as the result of formation of multilayers in the lipid mixture with KL_4 present.[41] The formation of multilayers also explains the slowing down of structural rearrangement: multilayer formation is expected to slow down the exchange of water with the bulk.

From the spectrally resolved data we have already concluded that at high pressures in both the mixture and mixture + KL_4 interfacial water contact with DOPG is likely decreased. As shown in Table 1, we have found a smaller T_{12} for DPPC than for either the mixture or mixture + KL_4. The resulting density dependence of T_{12} for the mixture (at high pressure T_{12} is 90 fs less) is in qualitative agreement with this 'squeeze out' scenario: at high pressures water associated with DPPC becomes a larger fraction of the signal.

5 Conclusions

Over the last 5–10 years there has been an increasing appreciation of the necessity of describing the molecular nature of water in contact with membranes as one step in the full understanding of a variety of membrane biophysical and biochemical processes. Describing the molecular level properties of membrane-bound water has, in the past, been a significant technical challenge: membranes are soft and deformable, the description of membrane-bound water in conventional spectroscopies requires distinguishing a relatively small interfacial signal from a much larger bulk and the structure of membrane-bound water likely evolves on subnanosecond timescales.

Lung surfactant is a mixture of lipids and protein found at the air/water interface in the lung and plays an important role in its physiological function. In the current study we describe the structure and dynamics of interfacial water in contact with a monolayer of model lung surfactant, composed of four types of lipids and one protein, at various surface densities. We address the technological challenge of describing water in this environment by employing frequency- and time-domain vibrational sum-frequency spectroscopy: the latter allowing the rate and mechanism of vibrational relaxation of specifically interfacial water molecules to be followed. We deconvolve the differential hydration of each LS component by examining three systems: a monolayer of the pure lipid that is dominant in the lung surfactant mixture (DPPC), a monolayer of the four lipids (mixture), and a monolayer of the four lipids including the protein (mixture + KL_4).

While the lineshapes of the OD stretch observed in the static SFG spectra for each experimental system are similar, qualitative changes in spectral amplitudes as a function of monolayer density and composition are clearly present. Prior VSFS studies have shown that the amplitude of the OH (or OD) stretch in membrane-bound water abutting lipid monolayers is significantly higher near lipids whose head groups possess net charge (as does DOPG in our mixture). Prior work on model LS films of similar composition to ours has demonstrated that at higher pressures model lung surfactant systems form multilayers that are enriched in DOPG, TP and KL_4, while DPPC and PA preferentially remain in the surface film. Our results, that the OD stretch amplitude is larger for the mixture and the mixture + KL_4 than for DPPC, regardless of pressure, and that the amplitude of the O–D stretch increases in the DPPC monolayer as a function of pressure but decreases with pressure for the mixture and the mixture + KL_4, are consistent with these prior observations: the presence of DOPG in the LS film gives it a large O–D stretch amplitude relative to DPPC but this amplitude decreases with pressure as DOPG is preferentially excluded from the film into relatively less-ordered multilayers.

The vibrational relaxation dynamics in this system are remarkably different from those previously reported for interfacial water, and reveal a decrease in water–water interactions in increasingly realistic artificial lung surfactants (relative to bulk water and interfacial water near pure DPPC monolayers). The observed change in both the fast time constants describing vibrational relaxation (T_{12} and T_{23}) and in the slower constant describing structural relaxation (T_{34}) are consistent with the 'squeeze out' of DOPG in high-pressure monolayers and in the importance of KL_4 in multilayer formation under these circumstances. Thus our two approaches provide a consistent picture of the squeeze-out of lipids in the liquid state (DOPG and TP), through the disappearance of the signature of water interacting with the charged DOPG head-groups and the slowing down of the structural relaxation of interfacial water in systems expected to have multilayers.

Acknowledgements

This work is part of the research program of the Stichting Fundamenteel Onderzoek der Materie (Foundation for Fundamental Research on Matter) with financial support from the Nederlandse Organisatie voor Wetenschappelijk Onderzoek (Netherlands Organization for the Advancement of Research). M. B. and R. K. C. are grateful to Senter Novem for financial support.

References

1 D. M. Engelman, *Nature*, 2005, **438**, 578–580.
2 S. Mukherjee and F. R. Maxfield, *Annu. Rev. Cell Dev. Biol.*, 2004, **20**, 839–866.
3 J. F. Hancock, *Nat. Rev. Mol. Cell Biol.*, 2006, **7**, 456–462.
4 G. W. Feigenson, *Annu. Rev. Biophys. Biomol. Struct.*, 2007, **36**, 63–77.
5 B. Poolman, J. J. Spitzer and J. A. Wood, *Biochim. Biophys. Acta*, 2004, **1666**, 88–104.
6 J. Milhaud, *Biochim. Biophys. Acta*, 2004, **1663**, 19–51.
7 Z. Zhou, B. G. Sayer, D. W. Hughes, R. E. Stark and R. M. Epand, *Biophys. J.*, 1999, **76**, 387–399.
8 V. Kurze, B. Steinbauer, T. Huber and K. Beyer, *Biophys. J.*, 2000, **78**, 2441–2451.
9 K. Gawrisch, H. C. Gaede, M. Mihailescu and S. H. White, *Eur. Biophys. J. Biophys. Lett.*, 2007, **36**, 281–291.
10 J. A. Freites, D. J. Tobias, G. von Heijne and S. H. White, *Proc. Natl. Acad. Sci. U. S. A.*, 2005, **102**, 15059–15064.
11 Z. A. Sands and M. S. P. Sansom, *Structure*, 2007, **15**, 235–244.
12 M. J. Higgins, M. Polcik, T. Fukuma, J. E. Sader, Y. Nakayama and S. P. Jarvis, *Biophys. J.*, 2006, **91**, 2532–2542.
13 T. Fukuma, M. J. Higgins and S. P. Jarvis, *Biophys. J.*, 2007, **92**, 3603–3609.
14 T. Steiner, *Angew. Chem., Int. Ed.*, 2002, **41**, 48–76.
15 Q. Du, R. Superfine, E. Freysz and Y. R. Shen, *Phys. Rev. Lett.*, 1993, **70**, 2313–2316.
16 Q. Du, E. Freysz and Y. R. Shen, *Science*, 1994, **264**, 826–828.
17 Q. Du, E. Freysz and Y. R. Shen, *Phys. Rev. Lett.*, 1994, **72**, 238–241.
18 D. E. Gragson, B. M. McCarty and G. L. Richmond, *J. Am. Chem. Soc.*, 1997, **119**, 6144–6152.
19 D. E. Gragson and G. L. Richmond, *J. Phys. Chem. B*, 1998, **102**, 3847–3861.
20 S. Kataoka, M. C. Gurau, F. Albertorio, M. A. Holden, S. M. Lim, R. D. Yang and P. S. Cremer, *Langmuir*, 2004, **20**, 1662–1666.
21 D. F. Liu, G. Ma, L. M. Levering and H. C. Allen, *J. Phys. Chem. B*, 2004, **108**, 2252–2260.
22 E. A. Raymond, T. L. Tarbuck, M. G. Brown and G. L. Richmond, *J. Phys. Chem. B*, 2003, **107**, 546–556.
23 E. A. Raymond and G. L. Richmond, *J. Phys. Chem. B*, 2004, **108**, 5051–5059.
24 T. L. Tarbuck, S. T. Ota and G. L. Richmond, *J. Am. Chem. Soc.*, 2006, **128**, 14519–14527.
25 W. Gan, D. Wu, Z. Zhang, R. Feng and H. Wang, *J. Chem. Phys.*, 2006, **124**, 114705.
26 M. Sovago, G. W. H. Wurpel, M. Smits, M. Muller and M. Bonn, *J. Am. Chem. Soc.*, 2007, **129**, 11079–11084.
27 G. W. H. Wurpel, M. Sovago and M. Bonn, *J. Am. Chem. Soc.*, 2007, **129**, 8420–8421.

 This journal is © The Royal Society of Chemistry 2009

28 S. Woutersen, U. Emmerichs and H. J. Bakker, *Science*, 1997, **278**, 658–660.
29 S. Woutersen and H. J. Bakker, *Nature*, 1999, **402**, 507–509.
30 E. T. J. Nibbering and T. Elsaesser, *Chem. Rev.*, 2004, **104**, 1887–1914.
31 S. Park and M. D. Fayer, *Proc. Natl. Acad. Sci. U. S. A.*, 2007, **104**, 16731–16738.
32 N. Huse, S. Ashihara, E. T. J. Nibbering and T. Elsaesser, *Chem. Phys. Lett.*, 2005, **404**, 389–393.
33 S. Ashihara, N. Huse, A. Espagne, E. T. J. Nibbering and T. Elsaesser, *Chem. Phys. Lett.*, 2006, **424**, 66–70.
34 S. Ashihara, N. Huse, A. Espagne, E. T. J. Nibbering and T. Elsaesser, *J. Phys. Chem. A*, 2007, **111**, 743–746.
35 H. J. Bakker, *Chem. Rev.*, 2008, **108**, 1456–1473.
36 J. A. McGuire and Y. R. Shen, *Science*, 2006, **313**, 1945–1948.
37 M. Smits, A. Ghosh, M. Sterrer, M. Muller and M. Bonn, *Phys. Rev. Lett.*, 2007, **98**, 098302.
38 A. Ghosh, M. Smits, J. Bredenbeck and M. Bonn, *J. Am. Chem. Soc.*, 2007, **129**, 9608–9609.
39 C. G. Cochrane and S. D. Revak, *Science*, 1991, **254**, 566–568.
40 I. Frerking, A. Gunther, W. Seeger and U. Pison, *Intensive Care Med.*, 2001, **27**, 1699–1717.
41 J. A. Zasadzinski, J. Ding, H. E. Warriner, F. Bringezu and A. J. Waring, *Curr. Opin. Colloid Interface Sci.*, 2001, **6**, 506–513.
42 G. Ma and H. C. Allen, *Photochem. Photobiol.*, 2006, **82**, 1517–1529.
43 G. Ma and H. C. Allen, *Langmuir*, 2007, **23**, 589–597.
44 G. Mao, J. Desai, C. R. Flach and R. Mendelsohn, *Langmuir*, 2008, **24**, 2025–2034.
45 P. Cai, C. R. Flach and R. Mendelsohn, *Biochemistry*, 2003, **42**, 9446–9452.
46 L. Wang, P. Cai, H. J. Galla, H. X. He, C. R. Flach and R. Mendelsohn, *Eur. Biophys. J. Biophys. Lett.*, 2005, **34**, 243–254.
47 G. Ma and H. C. Allen, *Langmuir*, 2006, **22**, 11267–11274.
48 A. Saenz, O. Canadas, L. A. Bagatolli, M. E. Johnson and C. Casals, *FEBS J.*, 2006, **273**, 2515–2527.
49 A. M. Dokter, S. Woutersen and H. J. Bakker, *Phys. Rev. Lett.*, 2005, **94**.
50 M. Bonn, C. Hess, J. H. Miners, T. F. Heinz, H. J. Bakker and M. Cho, *Phys. Rev. Lett.*, 2001, **86**, 1566–1569.
51 C. D. Bain, P. B. Davies, T. H. Ong, R. N. Ward and M. A. Brown, *Langmuir*, 1991, **7**, 1563–1566.
52 X. Y. Chen and Z. Chen, *Biochim. Biophys. Acta*, 2006, **1758**, 1257–1273.
53 M. J. Shultz, S. Baldelli, C. Schnitzer and D. Simonelli, *J. Phys. Chem. B*, 2002, **106**, 5313–5324.
54 V. Buch, *J. Phys. Chem. B*, 2005, **109**, 17771–17774.
55 D. S. Walker, D. K. Hore and G. L. Richmond, *J. Phys. Chem. B*, 2006, **110**, 20451–20459.
56 A. Morita and J. T. Hynes, *Chem. Phys.*, 2000, **258**, 371–390.
57 M. Sovago, R. K. Campen, G. W. H. Wurpel, M. Müller, H. J. Bakker and M. Bonn, *Phys. Rev. Lett.*, 2008, **100**, 173901.
58 A. Staib and J. T. Hynes, *Chem. Phys. Lett.*, 1993, **204**, 197–205.
59 A. Ghosh, R. K. Campen, M. Sovago and M. Bonn, 2008, in preparation.
60 O. F. A. Larsen and S. Woutersen, *J. Chem. Phys.*, 2004, **121**, 12143–12145.
61 G. Seifert, T. Patzlaff and H. Graener, *J. Chem. Phys.*, 2004, **120**, 8866–8867.
62 P. Bodis, O. F. A. Larsen and S. Woutersen, *J. Phys. Chem. A*, 2005, **109**, 5303–5306.
63 G. Seifert and H. Graener, *J. Chem. Phys.*, 2007, **127**, 224505.
64 F. Bringezu, J. Q. Ding, G. Brezesinski and J. A. Zasadzinski, *Langmuir*, 2001, **17**, 4641 4648.
65 F. Bringezu, J. Q. Ding, G. Brezesinski, A. J. Waring and J. A. Zasadzinski, *Langmuir*, 2002, **18**, 2319–2325.

The terahertz dance of water with the proteins: the effect of protein flexibility on the dynamical hydration shell of ubiquitin

Benjamin Born,[a] Seung Joong Kim,[b] Simon Ebbinghaus,[a] Martin Gruebele[bc] and Martina Havenith*[a]

Received 20th March 2008, Accepted 2nd June 2008
First published as an Advance Article on the web 1st October 2008
DOI: 10.1039/b804734k

The role of water in the functioning of proteins has been a hot topic over the years. We use terahertz (THz) spectroscopy as an experimental tool to probe the protein-induced fast solvation dynamics of ubiquitin. In order to investigate the effect of protein flexibility on the changes in the solvation dynamics, we have measured the concentration-dependent THz absorption of several site-specific ubiquitin mutants. The observed non-linear dependence of absorption on concentration is a signature of a long-range hydration shell with properties distinct from bulk water. We determined a dynamical hydration shell of a thickness of at least 18 Å on the protein surface. This exceeds the static hydration layer as it is typically observed by scattering methods (3 Å) by far. We also conclude that any increase in flexibility obtained by side-chain truncations that decrease the structural rigidity of the protein results in more bulk-like behaviour of the dynamical hydration shell. Furthermore, our THz measurements show that a single phenylalanine-to-tryptophan substitution to introduce a fluorescent marker leads to measurable changes in the solvation dynamics.

Introduction

Solvation water plays an integral role in the folding and function of proteins. For example, the expulsion of solvation water sheets from the hydrophobic core has been implicated as a major cause of the final folding barrier leading up to the native state.[1–3] Specific functional water molecules have been resolved by NMR spectroscopy and X-ray crystallography, for example mediating water transport through pores.[4,5]

Solvation water and proteins also interact in more subtle ways. For instance, water mediates and screens interactions between proteins, and between proteins and small substrates, facilitating docking processes through a multiplicity of protein–water contacts.[6] Dynamical and structural fluctuations of water molecules contribute to the flexibility of the polypeptide chain, which is often critical for the function of proteins. The classic example is myoglobin, whose heme binding site is accessible to substrates only because of structural fluctuations of protein and

[a]*Lehrstuhl für Physikalische Chemie II, Ruhr-Universität Bochum, 44780 Bochum, Germany. E-mail: Martina.Havenith@rub.de; Fax: +49 (0)234 32 14183; Tel: +49 (0)234 32 28249*
[b]*Department of Physics, University of Illinois, Urbana, IL, 61801, USA*
[c]*Department of Chemistry and Center for Biophysics and Computational Biology, University of Illinois, Urbana, IL, 61801, USA*

solvent.[7] The solvent and protein modes can be highly correlated over many timescales, and this effect is referred to as 'slaving'.[8]

Zhong and coworkers have combined tryptophan mutant scans with time-gated fluorescence up-conversion to monitor site-specific hydration dynamics with sub-picosecond time resolution. They observed local (1–8 ps) and global (20–200 ps) relaxation of water networks around the protein surface.[9] Whenever the sample has increased flexibility and is exposed to more water, the relaxation becomes faster which implies that these water molecules are less structured.

Recently, terahertz (THz) spectroscopy has been demonstrated as a new probe of the coupling between biomolecules and their hydration shells.[10–13] This is possible because key large-amplitude motions of water and biomolecules occur on the ps timescale characteristic of THz spectroscopy, and new powerful THz sources are capable of penetrating water films. The THz absorption coefficient is even more sensitive to fast water dynamics than dielectric spectroscopy or IR spectral changes.[14] We have been able to quantify the size of the dynamical hydration shell around a small protein by concentration-tuning studies in aqueous solution and measuring the non-linear absorption dependence on concentration.[13] We also found the absorption coefficient to be very sensitive to pH-induced increased flexibility of the protein.[15]

Recent molecular dynamics simulations showed that the heterogeneous rigidity of the water network and its coupling to the protein surface both influence the vibrational density of the low frequency modes.[12,15,16] Changes in the THz absorption can therefore also be used as a probe of the local response of the hydration dynamics to point mutations, and we have observed large changes in the THz absorption as a result of surface mutations of the small protein λ-repressor fragment.[15]

In this paper, we use THz spectroscopy of the small globular β-sheet protein ubiquitin to study the correlation between protein structural flexibility and the absorption properties of the dynamical aqueous solvation shell around the protein. To tune protein flexibility, we introduce two single-point and one pair mutation known to decrease the structural rigidity of the protein, as probed by fluorescence anisotropy.[17] To deconvolve protein and hydration contributions to the spectra, we systematically vary the protein-to-buffer concentration ratio. Such studies can now be carried out systematically in the laboratory thanks to the advent of table-top THz radiation sources with sufficient output power to penetrate aqueous solutions of proteins.[18]

Terahertz spectroscopy of solvated proteins

A small but growing number of researchers, including the groups of Durbin, Markelz and Plaxco, the groups of Plusquellic, Heilweil and Siegrist at the National Institute of Standards and Technology, and ours, are now looking at the terahertz response of hydrated to fully solvated large biomolecules.[10,11,13,15,19–24]

First measurements showed rather featureless THz absorption spectra. Proteins have a high density of vibrations at these low frequencies, such that it makes sense to speak of a spectral density $\rho(\omega)$, rather than accounting for what would be called a fingerprint, *e.g.* detection of single vibrational absorption bands. Whereas at the highest frequency found in the IR, mostly localized vibrational modes are found, below 10 THz the modes get delocalized and include the motion of a large number of atoms. When approaching zero frequency (*i.e.* in the dielectric regime), the absorption declines towards zero as some power of the frequency, or $A \sim \omega^a$.[25–27] In the region between 1–5 THz, water absorption *also* plays a significant role due to its high absorption coefficient in the THz range.

Although the high density of delocalized modes involved does not allow the assignment of each contributing mode, important information on the coupled

 This journal is © The Royal Society of Chemistry 2009

protein–water network motions can be extracted from a measurement of the THz absorption with protein concentration.

Previous studies showed that hydration of the protein can substantially enhance the THz absorbance of protein systems.[13,20] For a thin film of hen-egg-white lysozyme a sharp increase of the THz absorption was found when the hydration exceeded 23 gram water per gram protein.[28]

The addition of proteins to a solution can result in either an increase or a decrease of the THz absorption.[13] This is a result of the coupling of the protein with the water network motion which is shown to depend sensitively on protein configuration and flexibility.[15,22,29]

In our previous work we introduced two key concepts to describe the experimental results of terahertz spectroscopy of biomolecules in water:

The first is the "terahertz defect". The protein absorbs less THz light than water over part of the THz frequency range, therefore the absorption coefficient of the solution is found to decrease with increasing protein concentration.[11] In some cases, the experimental data resembled a simple model assuming that the protein behaves like a THz-transparent cavity in the water.[15] The missing absorption is explained by the assumption that the protein, although large and extended, does not contribute in the same way to the THz absorption as the full water network. Being bounded in size, they lack some of the lowest-frequency motions that can be found in the bulk aqueous solvent. Thus, at sufficiently low frequencies, water will absorb much more than proteins. The difference can range from a few percent to percentages close enough to 100% that the biomolecule-as-hole-in-water (or volume-exclusion concept) can be used as a first approximation.

A second key concept is the "terahertz excess". Despite the fact that pure biomolecule solids or films or high concentrated protein solutions generally absorb less than bulk water between 1–3 THz, we have found conditions where the protein–water mixture does absorb more than the bulk water sample alone. This can be explained only by the contribution from a third substance—the hydration water. If the presence of the proteins affects nearby water molecules, this could have an effect on any of the experimentally accessible properties of water, especially the relaxation rates and reorientation rates within the network.

This reorientation arises as water molecules within the hydrogen bonding network tumble around and diffuse, constantly making and breaking hydrogen bonds. Measuring the terahertz absorption is an effective tool to probe this since the absorption coefficient depends sensitively on the dynamical reorientation of dipole moments on the picosecond timescale. THz absorption studies demonstrated that the dynamical reorientation of the water dipole moment can be affected over rather long distances (up to 20–30 Å from the surface of the protein)[13] which includes up to 1000 water molecules. Our improved picture of a protein in water thus has to include the protein (causing a terahertz defect), bulk water (if far enough away) and hydration water with new physical properties, which might include an increase enhanced terahertz absorption—the terahertz excess. The hydration water within this so-called dynamical hydration shell and the protein are strongly coupled with the motions extending over the barriers of the protein surface.

We further note that the power of THz spectroscopy to detect any changes upon structural reorientation lies in the fact that at frequencies above 1 THz we begin to approach the resonance of the coupled diffusional–rotational motions.

It is important to note that this dynamical hydration shell is not identical to the static hydration shell as probed by X-ray crystallography or NMR which includes sterically bound water molecules. It is defined to include all water molecules that show water network dynamics distinct from the bulk, and thus a distinct THz absorbance. The influence of the protein can reach much further than the static hydration radius since it involves only a change in the motions and not a fixed H-bond to the protein.

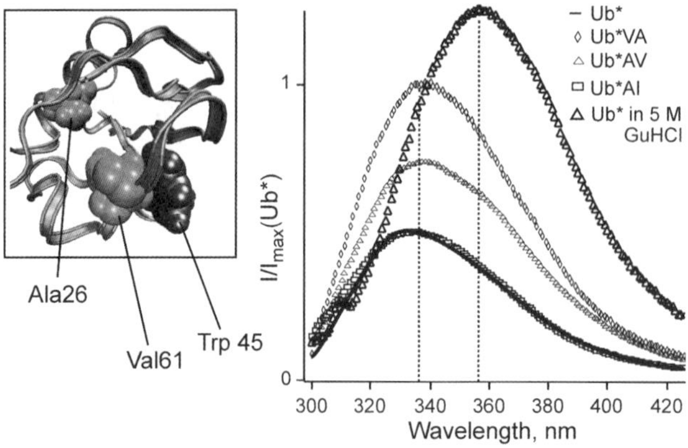

Fig. 1 Left: structural model of Ub*AV, obtained by making side-chain substitutions with XPLOR, using the SOLVATE feature of VMD to add TIP3P water, then relaxing the structure at 298 K for 8 ps in an *NVT* ensemble.[39] Right: fluorescence intensities of mutants compared to Ub* and denatured Ub*. Note that despite different fluorescence intensity maxima, all mutants have peak shifts close to native Ub*. Data from ref. 17 and 38.

Materials and methods

Ubiquitin mutants

Ubiquitin is a small α/β protein with one α-helical segment and a short 3_{10} helical segment.[30] We studied the wild-type human sequence, obtained from Sigma. In addition, we introduced a Phe45 Trp mutation to have a strongly fluorescent residue as an independent probe of protein flexibility.[31] We call this ubiquitin pseudo-wild-type Ub*, and the ubiquitin wild-type Ub. Fig. 1 shows the two sites we chose for single- and double-point mutations to increase protein flexibility. Ile61 is in van der Waals contact with Trp45, and Val26 has at least one intervening residue to Trp45.[17] The three mutants of Ub* represent two single truncations and one double, and are summarized in Table 1. All mutations truncated aliphatic side-chain residues to shorter aliphatic side chains, to avoid direct effects on the tryptophan fluorescence quenching *via* introduction or elimination of electron transfer, proton transfer or Förster mechanisms. We use the abbreviation Ub*XY, where * indicates the tryptophan, X is the residue in position 26, and Y is in position 61.

The plasmids for Ub*VA, Ub*AI and Ub*AV were obtained by single-point mutations of the original Ub* plasmid (provided by Tracy Handel) using site-directed mutagenesis (Stratagene). Proteins were over-expressed in *E. coli* (BL21) and purified as indicated elsewhere.[32] Purity was checked by gel electrophoresis (SDS-PAGE) and protein identity by low-resolution mass spectrometry (by electro-spray ionisation). Samples were lyophilized and kept at −20 °C before use.

Table 1 Parameters for the ubiquitin mutants used in this study. ΔG is the folding free energy at 0 M GuHCl, 25 °C, pH 5.9 in 40 mM phosphate buffer from ref. 17. f_{ratio} is the peak ratio of fluorescence intensity compared to Ub*

Mutant	Abbreviation	ΔG/kJ/mol	f_{ratio}
Wild-type (WT)	Ub		1.0
Pseudo-WT F45W	Ub*	−34 ± 1.5	1.0
Ile61Ala	Ub*VA	−18 ± 1	2.0
Val26Ala	Ub*AI	−20 ± 1.5	1.0
Val26Ala/Ile61Val	Ub*AV	−14 ± 1	1.4

 This journal is © The Royal Society of Chemistry 2009

Ubiquitin mutants were resuspended in buffer for terahertz (THz) absorption studies at (20 ± 0.5) °C. Unless otherwise indicated, all solvents were buffered with 50 mM magnesium acetate at pH 4.8. Protein concentration was varied between 0 and 3.6 mM. The actual concentration was measured with an uncertainty of 3% using UV absorption at 280 nm, and assuming an extinction coefficient of 6970 M^{-1} cm^{-1}.[33] We did not observe any aggregation below a concentration of 3.8 mM.

Fig. 1 summarizes the effect of the side-chain truncations studied here on protein flexibility, relative to Ub*. As shown in ref. 17, local flexibility of the ubiquitin structure is directly correlated with tryptophan fluorescence intensity. The Ub*VA and Ub*AV mutants with a truncation adjacent to Trp45 have greatly increased fluorescence intensities compared to Ub*. The truncation mutant Ub*AI looks very similar to Ub* because tryptophan is not in contact with residue 26 and only probes flexibility locally, but molecular dynamics (MD) simulations showed that the Val26Ala core truncation induces a local increase in flexibility greater than the Ile61Ala near-surface truncation. Such truncations typically decreased the anisotropy parameter $\langle \cos^2 \theta \rangle$ of the tryptophan side chain from 0.95 to 0.85, with excursions as low as 0.55 (isotropic: 0.5) for the Ub*AV double mutant.

Experimental results: terahertz (THz) measurements of ubiquitin

Using our THz p-type Germanium laser spectrometer[18] we measured the change of the absorption coefficient in the spectral range from 2.1–2.8 THz. In Fig. 2 we show the THz absorption spectrum for the buffer and for 0.59 mM ubiquitin dissolved in buffer. As was found in other cases, a featureless linear increase of THz absorption with increasing frequency is observed. However, significant changes in the integrated THz absorption upon changing the protein concentration are found.

The transmitted intensities were measured at a fixed layer thickness using a standard Bruker liquid-sample cell with PTFE spacers and z-cut quartz windows. The layer thickness of the aqueous sample was determined to be (52.6 ± 0.3) μm using FTIR spectroscopy. The temperature of the sample was kept at (20 ± 0.5) °C by using a Peltier element. The measured humidity near the purged sample cell was below 8%.

To measure the difference in absorption between protein solution and buffer blanks accurately, we set up a THz difference spectrometer (see also ref. 34). This approach also minimized any additional systematic errors due to temperature drifts or changes in the air humidity which would be present in the case of subsequent measurements. Specifically, we determine:

$$\alpha(c) = \alpha_{\text{sample}}(c) - \alpha_{\text{buffer}}(c), \quad (1)$$

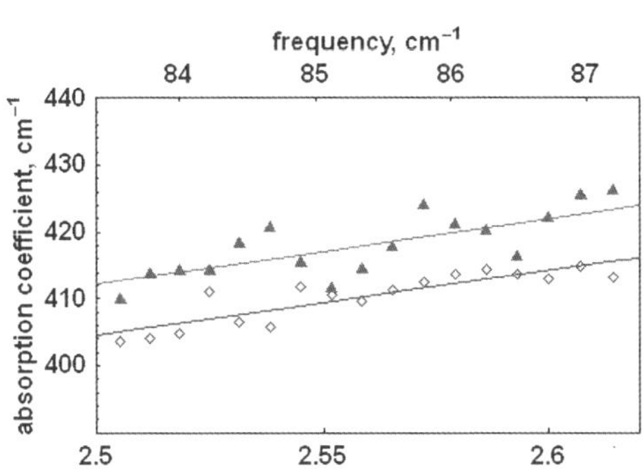

Fig. 2 Measured THz absorption for buffer (empty diamonds) and for a solution of 0.59 mM ubiquitin wild-type dissolved in buffer (filled triangles) in the frequency range from 2.5–2.62 THz.

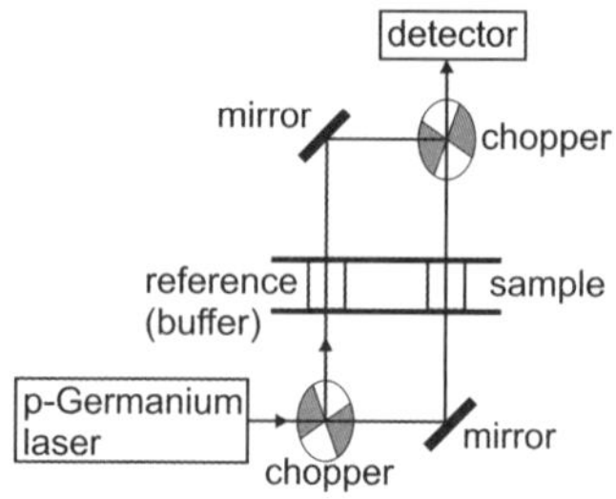

Fig. 3 Experimental set-up. The path of the THz radiation is displayed. The mirror–chopper splits the radiation into two distinct beams. One part probes the reference (buffer), the alternating beam probes the sample. Both are focused onto the detector.

with $\alpha_{\text{sample}}(c)$ and $\alpha_{\text{buffer}}(c)$ being the integrated absorption coefficients (2.1–2.8 THz) of the sample and buffer at a given concentration c. The experimental set-up is displayed in Fig. 3.

The THz-pulse train was split into two beams using a mirror–chopper. One beam was directed into the reference, the alternating beam into the sample. Each signal was detected by a gated integrator. In order to further minimize systematic errors, we interchanged the sample and reference channel at each concentration. Measurements were repeated five times at each concentration to provide an error estimate for the absorbance difference. Each point corresponds then to the average of 10 000 pulses. The main error source was found to be the manual refilling of the sample cell, which leads to slight sample-to-sample, cell positioning or path-length variations.

By data evaluation with Beer's law the absorption coefficients of the sample and the reference in the two separate channels were determined. We determined the ubiquitin absorption relative to the buffer absorption, showing a non-linear behaviour of the integrated THz absorption with increasing ubiquitin concentrations for all samples.

Results

We summarize the results of the measured changes in the THz absorption of all five protein variants in Fig. 4. For reference, the dotted line in Fig. 4 also shows what would be expected for a simple two-component model with protein and bulk water only. In that case, the following model would fit the absorption coefficient as a function of concentration:

$$\begin{aligned}\alpha = {} & \alpha_{\text{protein}} \frac{V_{\text{protein}}}{V} + \alpha_{\text{buffer}} \frac{V - V_{\text{protein}}}{V} \\ & \approx \alpha_{\text{buffer}} \left(1 - \rho_{\text{protein}}\, c_{\text{protein}}\right)\end{aligned} \tag{2}$$

In this equation, the protein has concentration c_{protein} in total solution volume V and ρ_{protein} is the protein density of *ca.* 1.4 g cm^{-3}.[35] The approximation in the second line, plotted in Fig. 4, corresponds to the limit where protein absorption is negligible compared to the buffer absorption (described previously as the THz defect: $\alpha_{\text{protein}} = 0$). Note that each ubiquitin molecule, with a bare radius of gyration of $R_g = 11.7$ Å estimated from the X-ray crystal structure and different MD simulations,[36] has a surface radius of $\sqrt{5/3}R_g \approx 15.1$ Å and displaces *ca.* 14 400 Å^3 of buffer.

Any two-component model which considers only the absorption of the buffer and the ubiquitin wild-type would lead to a linear concentration dependence, although the slope would differ from eqn (2) in a more sophisticated dielectric cavity model. The measured THz absorption of all five proteins deviates strongly from a linear

 This journal is © The Royal Society of Chemistry 2009

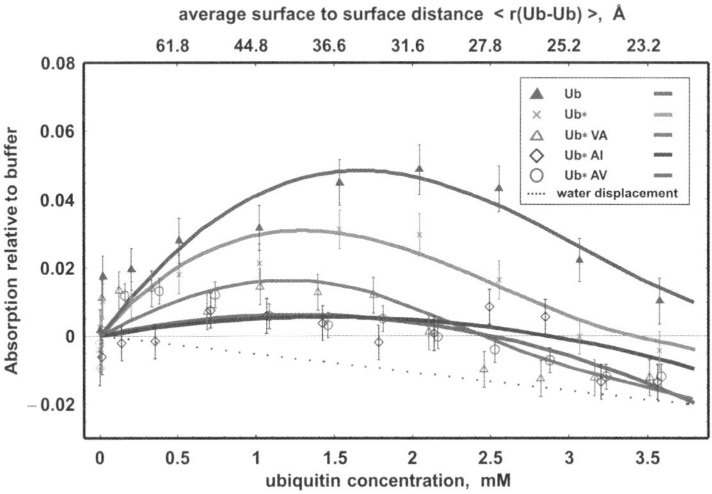

Fig. 4 Integrated THz absorption coefficient (between 2.1–2.8 THz) of the protein as a function of protein concentration. The results for ubiquitin wild-type, ubiquitin pseudo-wild-type and three ubiquitin mutants relative to the buffer are displayed. The measurements ware carried out at (20 ± 0.5) °C and at pH 4.8. The dotted line shows the predicted decrease in the case where the protein does not contribute to the total THz absorption, but is just displacing water molecules (excluded-volume model).

decrease which is predicted in eqn (2). The wild-type and pseudo-wild-type (containing a tryptophan) in particular deviate strongly from a linear concentration dependence. The mutants whose fluorescence indicates higher flexibility (Fig. 1) deviate less from the bulk buffer absorption, but still significantly outside the measurement uncertainty shown by the error bars in Fig. 4. Note that the ranking from highest to lowest flexibility based on our fluorescence measurements is Ub*VA > Ub*AV > Ub*AI ≈ Ub*, whereas the ranking from lowest to highest deviation in the THz absorption is Ub*AI ≈ Ub*AV < Ub*VA < Ub* < Ub.

When extrapolating the THz absorption coefficient towards very high protein concentrations we find slight differences between the different mutants. Whereas the wild-type shows the highest net absorption in this spectral range, Ub* and Ub*AI are found to have a similar but slightly smaller THz absorption coefficient at the highest protein concentration measured. Both exceed the coefficients of Ub*AV and Ub*VA, which approach the value expected for a completely transparent sphere of the protein volume.

All protein variants proceed through an absorption maximum at concentrations between 1 and 1.5 mM. The minimum fitting model required to explain this deviation even qualitatively must incorporate at least a third component, attributed to water in the dynamical hydration shell around the protein, whose absorption coefficient is increased compared to bulk water by the presence of the protein:

$$\alpha = \alpha_{\text{protein}} \frac{V_{\text{protein}}}{V} + \alpha_{\text{shell}} \frac{V_{\text{shell}}}{V} + \alpha_{\text{buffer}} \frac{V - V_{\text{protein}} - V_{\text{shell}}}{V} \tag{3}$$

For disaccharides this three-component model provided an excellent description and allowed us to fit the size of the dynamical hydration shells, since the solvation is rather homogeneous.[34] For proteins in general the situation is more complicated.

For small concentrations the different contributions from the protein and the water in the coupled dynamical hydration shell both contribute linearly to the total absorption since the volume of the dynamical solvation shell increases linearly with protein concentration. In order to make clear how changes in the dynamical

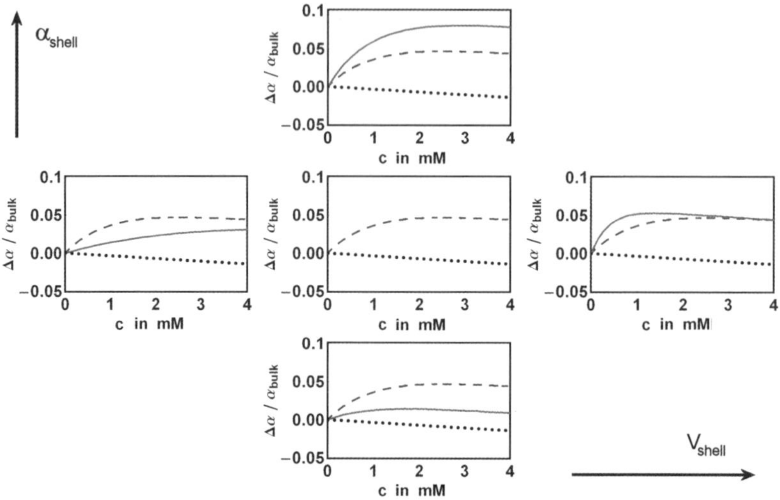

Fig. 5 Predicted dependence of the THz absorption for given values of α_{shell} and V_{shell}, both for ubiquitin wild-type (Ub). The solid line reflects the predicted THz absorption in each case, assuming an empty cavity of the size of the protein (α_{solute} is set to zero). The dotted curve shows the THz defect. The dashed curve (middle panel) corresponds to the measured concentration dependence of a ubiquitin–water solution and remains fixed for comparison. The bending of the dashed curve is attributed to an increased absorption for water within a given dynamical hydration shell. For comparison: the solid curve was obtained when varying α_{shell} (upper and lower) and V_{shell} (left and right).

hydration shell affect the measured THz absorbance, we have plotted in Fig. 5 the predicted THz absorption of a protein–water solution for various values of α_{shell} and V_{shell}. The dotted line in Fig. 5 corresponds to the pure THz defect (*i.e.* α_{shell} and V_{shell} are both set to zero). If the absorption coefficient of the water in the hydration shell is smaller than that of buffer ($\alpha_{shell} < \alpha_{buffer}$), the predicted curve (represented by the solid line) will lie below the dashed line (lower picture). If the absorbance of the dynamical hydration shell exceeds the absorbance of the bulk water displaced by the shell and protein (THz excess: $\alpha_{shell} > \alpha_{buffer}$), the overall absorption will at first increase linearly with protein concentration. Eventually, the dynamical hydration shells overlap, and their volume actually decreases relative to the increasing volume of protein. In the extreme limit of hexagonal packing of proteins and negligible protein absorption compared to the solvent, $\alpha \rightarrow 0.29\ \alpha_{shell}$ if the shell is wide enough to displace all bulk water in the interstitial spaces. Thus unless the shell absorption coefficient exceeds that of the buffer by a factor of at least three, absorption will drop below the bulk value if we assume $\alpha_{protein}$ much smaller than α_{buffer}. As a result, there is a turnover in the absorption coefficient. Thus a further linear increase in concentration does not lead to a linear increase of the contribution from the water within the dynamical hydration shell any more.

From left to right in Fig. 5 the extension of the dynamical hydration shell (and thereby the volume V_{shell}) is increased. As we can see, the onset of non-linearity (the turning point) is shifting towards smaller and smaller concentrations. This is a consequence of the fact that for larger dynamical hydration shells an overlap of adjunct hydration shells occurs already at lower concentrations as for very small dynamical hydration shells. The concentration at which the onset of non-linearity occurs is directly correlated with the smallest concentration at which the dynamical hydration shells start to overlap. We can also distinguish between the case where the water in the dynamical hydration shell has an increased (THz excess) or decreased THz absorption compared to the bulk.

 This journal is © The Royal Society of Chemistry 2009

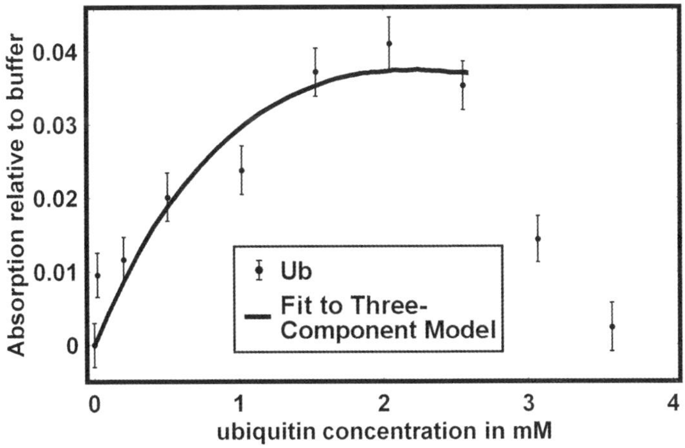

Fig. 6 Wild-type ubiquitin THz absorption fitted to a three-component Monte Carlo model that takes into account overlapping hydration water shells at higher concentrations.

Fig. 6 shows a fit of the ubiquitin data to eqn (3). We have restricted the fit to concentrations below 2.5 mM because the three-component model cannot account for the full decrease of absorption observed at higher concentrations. The truncated bulk solvent and hydration-sphere volumes required by eqn (3) were simulated by a Monte Carlo distribution of globular proteins with spherical hydration shells whose diameter can be adjusted. The fit of a three-component model is too simple to describe a more complex system such as proteins in solution. Instead of fitting the size of the dynamical hydration shell we take the point of turnaround as a measure of the size of the hydration shell.

For each concentration we can deduce an average protein - protein distance when assuming a random (Monte Carlo) distribution of proteins. We show these calculated average protein surface–protein surface distances at the x-axis of Fig. 4. For ubiquitin the molar concentration at the maximum of the THz absorption is around 1.5 mM, which corresponds to an average protein - protein center of mass distance of approx. 66.6 Å. If we take a surface radius of 15 Å for bare ubiquitin, the average distance between the two protein surfaces at 1.5 mM is 36.6 Å (which corresponds to the distance $\langle r(\text{Ub–Ub})\rangle$ as shown in the additional x-axis of Fig. 4). For comparison, 3.6 mM was the highest concentration measured because we found an onset of aggregation. At this concentration the distance between the two ubiquitin surfaces is about 23 Å.

Following our previous argument, the precise measurement of the onset of non-linearity allows us to estimate the average size of the dynamical hydration shell: The concentration at the point of turnaround in the THz absorption corresponds to $\langle r(\text{Ub–Ub})\rangle/2$ of *ca.* 18 Å, which gives the size of the dynamical hydration shell.

However, it must be noted that for ubiquitin the three-component model oversimplifies the situation, therefore it is not possible to obtain a quantitative fit to the three-component model over the full concentration range as was possible for the carbohydrates.[34] As illustrated in Fig. 6, the fitted function does not drop off rapidly enough at higher concentrations. Proteins are rather large molecules compared to disaccharides and at high enough concentration a large fraction of the solvation water lies in the hydration shells around two or more proteins. Such multiple hydrated waters may differ from solvation water around a single protein. The overestimate of absorbance in Fig. 6 at higher concentrations would in fact indicate that water interacting with multiple proteins absorbs less than water interacting with one protein. Our previous study of λ-repressor also showed that the assumption of a single hydration shell with a distinct, but constant absorption coefficient is too simple for proteins. The accompanying molecular dynamics simulations supported a more complex THz absorption which depends on the protein–protein distance.[13]

Discussion

We observe a correlation between increased protein flexibility and a smaller deviation from the linear concentration absorption of the terahertz (THz) absorption expected from a two-component model. The double mutant Ub*AV shows the smallest deviation from two-component behaviour, whereas the least flexible (according to fluorescence in Fig. 1) Ub* pseudo-wild-type has the highest deviation from the two-component model.

However, fluorescence and THz spectroscopy report differently on the flexibility of the protein. Tryptophan is a local probe, and one would expect it to be most sensitive to the environment near the side chain. THz spectroscopy of the dynamical hydration shell is a global probe which averages over the entire protein surface. This is borne out by the ranking reported in the results. The fluorescence measurements show large deviations from the pseudo-wild-type when residue isoleucine 61, adjacent to the tryptophan, is mutated to alanine or valine. On the other hand, mutation of the remote residue valine 26 to alanine has almost no effect on the fluorescence, even though this mutation is more destabilizing to the protein overall. In contrast, THz spectroscopy shows the largest deviation from Ub* for the more disruptive core mutation Val26Ala (Ub*AI) and for the corresponding double mutant (Ub*AV), but a smaller effect for the less disruptive Ile61Ala mutation of a near-surface residue (Ub*VA). Thus tryptophan detection emphasizes the side-chain truncation near the tryptophan residue, while THz spectroscopy emphasizes the more destabilizing truncation of a core residue.

The tryptophan probe itself can be evaluated further by our THz measurements. The interesting question is: how much does insertion of a tryptophan side chain modify the solvation dynamics? As we can see from Fig. 4, the introduction of the fluorescent tryptophan in the Phe45 Trp variant Ub* shows a peak at a slightly lower concentration (1.25 mM) than the wild-type Ub (1.5 mM), and we obtain a statistically significant reduction of the THz absorption at the maximum concentration. Thus the tryptophan probe has an impact on the absorbance of the hydration water network THz vibrations, and thereby on the fast hydration dynamics. This is an important consideration for fluorescence studies that depend on the insertion of tryptophan probes at various sites, and which generally assume unperturbed solvation dynamics upon insertion. Although Fig. 4 warrants some caution, the good news for fluorescence studies of solvation water is that phenylalanine-to-tryptophan replacements cause smaller changes in the absorbance characteristics of the hydration water than any of the other mutations, *e.g.* the side-chain truncations.

In our recent study of λ-repressor we found a similar non-linear behavior (*e.g.* a maximum at low concentrations followed by a subsequent decrease), which was attributed to the hydration shell absorption. As mentioned in the results section, the simple three-component model does not provide a completely quantitative description of the hydration layer thickness because it neglects that hydration water near multiple proteins contributes in a non-trivial way to the total THz absorption coefficient. However, using more extensive molecular dynamics simulations, we showed that this model provides a qualitative estimate of the point where, statistically, the hydration shells start to overlap significantly. The concentration at which the maximum in the THz absorption for Ub and Ub* is found lies around 1.25–1.5 mM, which is higher than the maximum in the THz absorption for the five-helix bundle λ^*_{6-85} at a concentration of approximately 0.6 mM. This indicates that ubiquitin has a smaller dynamical hydration shell than λ^*_{6-85}.

From the Monte Carlo simulation which does not consider aggregation, 1.5 mM corresponds to an average distance between the protein centers of *ca.* 67 Å, which corresponds to an average distance between the protein surfaces of *ca.* 37 Å. The

 This journal is © The Royal Society of Chemistry 2009

THz spectrum thus monitors a dynamical hydration shell whose absorbance differs from bulk water out to distances of 18 Å. This number still exceeds by far the estimated size of the sterically bound first hydration shell (~3 Å).

Whether the solvent should be considered bulk or not of course depends on the probe used. THz spectroscopy measures the dipole–dipole autocorrelation function of water molecules on the approximately 1 ps timescale, and so this is the criterion by which the ubiquitin dynamical hydration shell extends up to 18 Å from the protein surface. The 18 Å dynamical hydration shell corresponds to approximately 6 water layers, a significant range beyond just solvation waters interacting directly with the protein surface, and similar to values that have been reported for λ-repressor by THz spectroscopy. Ubiquitin and the λ^*_{6-85} protein that we studied previously[13,15] are thus capable of reorganizing hydration water well beyond solvent molecules in direct or second-shell exchange with the protein. Such long-range interactions imply that cytoplasmic water, at concentrations of protein, RNA and carbohydrates in the 300 mg ml^{-1} range, is mostly 'biological water', and not bulk water, at least by the THz criterion.

The differences between the mutants with different flexibility and the non-linearity due to solvation are remarkably large. It will be very interesting to explore these results by molecular dynamics (MD) simulation. Because of the 1 ps timescale on which THz experiments report, MD simulations are ideally suited for the purpose, and extensive sampling in full solvent can be made.[12,13] A heterogeneous position-dependent solvation dynamics has been proposed by MD studies.[37] This heterogeneity in the solvation dynamics was found to correspond to distinct shifts in the absorption band of the O···O···O bending or restricted translations of the water below 100 cm^{-1}. A comparison of the experimentally observed THz absorption coefficient for specifically selected mutants of ubiquitin therefore yields valuable information about the relation between changes in the solvation dynamics and site-specific mutations, and such simulations should be repeated with protein mutants.

Although simulations will be required to determine the mechanism by which side-chain truncations that increase protein flexibility bring the THz absorbance closer to the bulk water value in Fig. 4, we can rule out some mechanisms and speculate on others.

In general, increased flexibility of a ubiquitin mutant results in a shift of protein normal modes towards lower frequency. This would lead to a different value for the THz absorption of the protein itself. In our 2.1–2.8 THz range, the absorbance would decrease if modes were shifted below that range and not replaced by modes above that range. Simulations could determine whether such an asymmetric shift takes place, but preliminary analysis shows only a very small shift.[15] So the protein by itself can probably not account for our observations.

A complete absence of the maximum was found for denatured λ-repressor at pH 2, which is more flexible than the native structure.[15] 1-Anilino-8-naphthalene-sulfonate (ANS) binding has indicated that this reduction might be associated with the increased exposure of more hydrophopic sites which affect the water in its hydration shell. Thus increased surface hydrophobicity of the mutants is a candidate for changes in hydration water structure that lead to the smaller bulk-water-like THz absorption, as compared to the protein ubiquitin and its Ub* mutants. Based upon MD simulations[37] we propose that the solvent-exposed hydrophobic side chains induce a negative THz absorption coefficient in their surrounding, whereas hydrophilic parts lead to an increase in the hydration water compared to bulk water. The simultaneous exposure of both partially hydrophobic and partially hydrophilic parts of the protein to water could therefore lead to a kind of cancellation which means that the net increase of THz absorption can partially be compensated by the contributions with a decreased THz absorption in the dynamical hydration layer.

To test this idea further, we measured the absorption coefficient of denatured ubiquitin as a function of concentration. Fig. 7 shows that ubiquitin, like

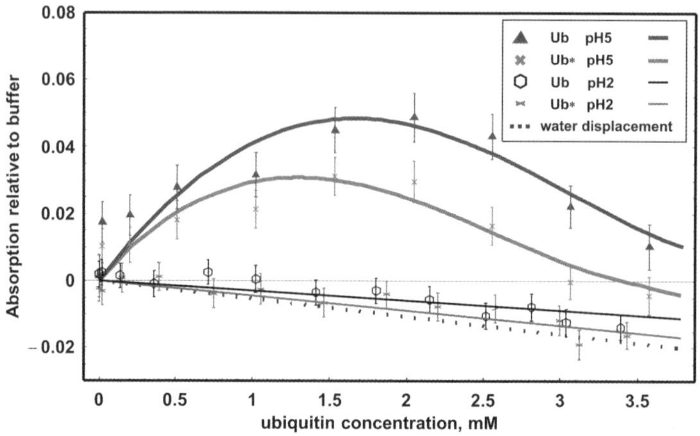

Fig. 7 Integrated THz absorption coefficient (between 2.1–2.8 THz) of Ub and Ub* as a function of protein concentration at pH 2 and pH 4.8. The measurements were carried out at (20 ± 0.5) °C. The dotted line shows the predicted decrease in the case where the protein does not contribute to the total THz absorption, but is just displacing water molecules (two-component excluded-volume model).

λ-repressor, has a signature very close to the two-component bulk water model of eqn (2) once it has been denatured under acidic conditions at pH 2. This supports our previous hypothesis that in case of partial unfolding the increased exposure of the hydrophobic core leads to a decrease of the initial maximum in THz absorption at 1.5 mM. The resulting curve now resembles that of bulk water with the protein displacing a water volume of 14 400 $Å^3$ × number of proteins.

At the highest protein concentration which was probed (3.6 mM), the average distance between the protein surfaces amounts to about 23 Å. This means that for higher concentrations the THz absorption monitors only the collective motions of the protein itself and its first hydration shell. For concentrations of 3.6 mM we find the following ranking from highest to lowest integrated THz absorption: Ub > Ub* ≈ Ub*AI > Ub*VA ≈ Ub*AV. Interestingly, this resembles the same order as was found for the fluorescence measurements which are sensitive to surface mutations. Based upon these results we may suggest that the water dynamics of the closely bound first hydration shell seems to be determined by the protein surface.

In summary we find that THz absorption spectroscopy provides a sensitive tool for probing the fast hydration dynamics around proteins. At low concentrations we find a non-linear absorption dependence on concentration. The non-linearity indicates a long-range (up to 18 Å from the protein surface) influence on the hydration dynamics, corresponding to 6 water layers. This long-range influence is sensitive to changes in the overall flexibility. The THz absorption at low concentrations is significantly altered when the protein is partially unfolded. In addition, the THz absorption at high concentrations which probes the coupled protein–hydration dynamics up to the first shell seems to be mainly determined by the protein surface.

Acknowledgements

We wish to thank E. Bründermann and G. Schwaab for the initial instrument development and many helpful discussions. K. Schröck is acknowledged for help at the THz spectrometer and M. Heyden for programming the Monte Carlo simulation. B. B. thanks the DAAD and the Ruhr-University Research School for financial support. M. H. and M. G. gratefully acknowledge the Human Frontier Science Program.

 This journal is © The Royal Society of Chemistry 2009

References

1 G. J. A. Vidugiris, J. L. Markley and C. A. Royer, *Biochemistry*, 1995, **34**, 4909–4912.
2 S. Khorasanizadeh, I. Peters and H. Roder, *Nat. Struct. Biol.*, 1996, **3**, 193–205.
3 M. Jäger, H. Nguyen, J. C. Crane, J. W. Kelly and M. Gruebele, *J. Mol. Biol.*, 2002, **311**, 373–393.
4 Y. Fujiyoshi, K. Mitsuoka, B. L. de Groot, A. Philippsen, H. Grubmuller, P. Agre and A. Engel, *Curr. Opin. Struct. Biol.*, 2002, **12**, 509–515.
5 J. A. Ernst, R. T. Clubb, H.-X. Zhou, A. M. Gronenborn and G. M. Clore, *Science*, 1995, **267**, 1813–1817.
6 M. Karplus and J. A. McCammon, *Nat. Struct. Biol.*, 2002, **9**, 646–652.
7 A. R. Panchenko, J. Wang, G. U. Nienhaus and P. G. Wolynes, *J. Phys. Chem.*, 1995, **99**, 9278–9282.
8 P. W. Fenimore, H. Frauenfelder, B. H. McMahon and F. G. Parak, *Proc. Natl. Acad. Sci. U. S. A.*, 2002, **99**, 16047–16051.
9 L. Zhang, L. Wang, Y.-T. Kao, W. Quia, Y. Yang, O. Okobiah and D. Zhong, *Proc. Natl. Acad. Sci. U. S. A.*, 2007, **104**, 18461–18466.
10 J.-Y. Chen, J. R. Knab, J. Cerne and A. G. Markelz, *Phys. Rev. E*, 2005, **72**, 040901–040904.
11 J. Xu, K. W. Plaxco and S. J. Allen, *Protein Sci.*, 2006, **15**, 1175–1181.
12 D. M. Leitner, M. Havenith and M. Gruebele, *Int. Rev. Phys. Chem.*, 2006, **25**, 553–582.
13 S. Ebbinghaus, S. J. Kim, M. Heyden, X. Yu, U. Heugen, M. Gruebele, D. Leitner and M. Havenith, *Proc. Natl. Acad. Sci. U. S. A.*, 2007, **104**, 20749–20752.
14 C. Ronne, L. Thrane, P.-O. Astrand, A. Wallqvist, K. V. Mikkelsen and S. R. Keiding, *J. Chem. Phys.*, 1997, **107**, 5319–5330.
15 S. Ebbinghaus, S. J. Kim, M. Heyden, X. Yu, M. Gruebele, D. Leitner and M. Havenith, *J. Am. Chem. Soc.*, 2008, **130**, 2374–2375.
16 S. Chakraborty, S. K. Sinha and S. Bandyopadhyay, *J. Phys. Chem. B*, 2007, **111**, 13626–13631.
17 J. Ervin, E. Larios, S. Osvath, K. Schulten and M. Gruebele, *Biophys. J.*, 2002, **83**, 473–483.
18 A. Bergner, U. Heugen, E. Brundermann, G. Schwaab, M. Havenith, D. R. Chamberlin and E. E. Haller, *Rev. Sci. Instrum.*, 2005, **76**, 063110.
19 U. Heugen, G. Schwaab, E. Brundermann, M. Heyden, X. Yu, D. M. Leitner and M. Havenith, *Proc. Natl. Acad. Sci. U. S. A.*, 2006, **103**, 12301–12306.
20 C. F. Zhang and S. M. Durbin, *J. Phys. Chem. B*, 2006, **110**, 23607–23613.
21 D. F. Plusquellic, K. Siegrist, E. J. Heilweil and O. Esenturk, *ChemPhysChem*, 2007, **8**, 2412–2431.
22 A. G. Markelz, *IEEE J. Sel. Top. Quantum Electron.*, 2008, **14**, 180–190.
23 J. Xu, K. W. Plaxco and S. J. Allen, *J. Phys. Chem. B*, 2006, **110**, 24255–24259.
24 J. R. Knab, J. Y. Chen, Y. He and A. G. Markelz, *Proc. IEEE*, 2007, **95**, 1605–1610.
25 R. Elber and M. Karplus, *Phys. Rev. Lett.*, 1986, **56**, 394–397.
26 D. M. Leitner, *Annu. Rev. Phys. Chem.*, 2008, **59**, 233–259.
27 R. C. Herrick and H. J. Stapleton, *J. Chem. Phys.*, 1976, **65**, 4778–4785.
28 J.-Y. C. Knab and A. G. Markelz, *Biophys. J.*, 2006, **90**, 2576–2581.
29 S. E. Whitmire, D. Wolpert, A. G. Markelz, J. R. Hillebrecht, J. Galan and R. R. Birge, *Biophys. J.*, 2003, **85**, 1269–1277.
30 D. L. DiStefano and A. J. Wand, *Biochemistry*, 1987, **26**, 7272–7281.
31 P. Laub, S. Khorisanizadeh and H. Roder, *Protein Sci.*, 1995, **4**, 973–982.
32 G. A. Lazar, J. R. Desjarlais and T. M. Handel, *Protein Sci.*, 1997, **6**, 1167–1178.
33 H. Edelhoch, *Biochemistry*, 1967, **6**, 1948–1954.
34 M. Heyden, E. Bründermann, U. Heugen, G. Niehues, D. M. Leitner and M. Havenith, *J. Am. Chem. Soc.*, 2008, **130**, 5773–5779.
35 H. Fischer, I. Polikarpov and A. F. Craievich, *Protein Sci.*, 2004, **13**, 2825–2828.
36 H. Fan, A. E. Mark, J. Zhu and B. Honig, *Proc. Natl. Acad. Sci. U. S. A.*, 2005, **102**, 6760.
37 A. A. Golosov and M. J. Karplus, *J. Phys. Chem. B*, 2007, **111**, 1482–1490.
38 Z. Qin, J. Ervin, E. Larios, M. Gruebele and H. Kihara, *J. Phys. Chem. B*, 2002, **106**, 13040–13046.
39 W. F. Humphrey, A. Dalke and K. Schulten, *J. Mol. Graphics*, 1996, **14**, 33.

General discussion

Professor Luzar opened the discussion of the paper by Professor Halle: We should not question the validity of QENS in comparison with NMR. The results are good in both cases and one must find what is the best description that is compatible with *both* techniques. In your present careful analysis, what is measured is the relative change of rotational relaxation times of water molecules, which is *not* measured directly by QENS. Also the time window in your case (Fig. 6 and 7 in your paper) is centered on long times: microseconds and nanoseconds, while in QENS the time scales are picoseconds. QENS measures exactly, and directly. Nothing more. The value of this correlation function is normally dominated by (i) diffusion (not seen in your case), and (ii) rotations of hydrogen around the center of mass. That is why this method is also more local and less dependent on concentration.

Professor Halle replied: No-one is questioning the "validity of QENS"; it is a powerful technique for probing single-particle dynamics on relatively short time and length scales. What is at issue here is rather the *interpretation* of QENS data from a very complex system and, in particular, the inference that 76% of the cell water in *H. marismortui* (55% of the water in the sample) has a translational diffusion coefficient that is 250 times smaller than the bulk-water value.[1] As you know, translation and rotation in liquid water take place on the same time scale (because they are strongly coupled by the H-bond network). Therefore, if the QENS interpretation is correct, also the water rotation that we measure should be 250 times slower. At the highest frequency (76.8 MHz), we should then have $R_1 = 0.45 \times 2.2 + 0.55 \times 2.2 \times 250 \approx 300\ \mathrm{s}^{-1}$, whereas the measurement[2] gives $R_1 = 6.66 \pm 0.03\ \mathrm{s}^{-1}$. On the other hand, the interpretation of QENS data from *E. coli* suggested that the visible cell water does not differ significantly from bulk water.[3] Yet, the primary NMR data (Fig. 6 of our paper) from *E. coli* and *H. marismortui* are virtually superimposable (and the small difference is removed when the data are concentration-normalized). Complacency in the face of such order-of-magnitude discrepancies is simply bad science.

1 M. Tehei, B. Franzetti, K. Wood, F. Gabel, E. Fabiani, M. Jasnin, M. Zamponi, D. Oesterhelt, G. Zaccai, M. Ginzburg and B.-Z. Ginzburg, *Proc. Natl. Acad. Sci. U. S. A.*, 2007, **104**, 766.
2 E. Persson and B. Halle, *Proc. Natl. Acad. Sci. U. S. A.*, 2008, **105**, 6266.
3 M. Jasnin, M. Moulin, M. Härtlein, G. Zaccai and M. Tehei, *EMBO Rep.*, 2008, **9**, 543.

Dr Zaccai commented: You stated, with a certain sense of humour that since it is the parameter that you measure with your technique, water rotational diffusion being a more localised motion reports more accurately on the hydration layer. Certainly, when measured by NMR, translational motions are on a long length scale (~μm), which is not informative on local mobility. With neutron scattering, however, translational and rotational diffusion are measured on an atomic length scale (~Å–nm) on different time scales (~ps–ns). The primary data from QENS are line widths, which provide information on time scale, measured as a function of scattering vector Q, which informs us on the length scale. At small Q values, the Q^2 dependence of the line-width yields the atomic scale translational diffusion. This is a direct visualisation of a trajectory according to a space–time correlation function. Note that such trajectories are what is calculated in MD simulations, so that a direct comparison with experiment is possible. See for example the paper of Professor Tobias *et al.* in this volume. Because of this, contrary to another of your assertions, the primary interpretation of QENS data is not model-dependent.

Dr Zaccai then said: You have several times pointed out the contradiction between neutron scattering and NMR results. The two methods measure different parameters or effects and I cannot see how their results can be in contradiction. On the contrary, assuming that the data has been collected and treated correctly, an interpretation that is compatible with both sets can only enrich our knowledge of a complex system. In your paper you compare rotational correlation times from NMR and neutrons. Incoherent neutron scattering measures the single particle space-time correlation of protons. What does NMR measure?

Professor Halle replied: I think we agree that translation diffusion coefficients derived from water displacements on the micrometre length scale, whether determined by pulsed-gradient spin echo NMR or any other technique, tells us little about cell water dynamics. Indeed, the clinical value of the apparent diffusion coefficients measured in MRI examinations lies in their dependence on tissue morphology and cell ultrastructure.

Like QENS, NMR relaxation probes single-particle dynamics on a molecular length scale. But there are important differences in how the dynamical information is extracted from the primary data. In both NMR and QENS this is a two-step process, which can be schematically illustrated as follows:

$$\text{NMR: } M(t) \rightarrow R(\omega) \rightarrow \upsilon_{\mathrm{R}}$$

$$\text{QENS: } S(Q,\omega) \rightarrow \Gamma(Q) \rightarrow (D,\, \upsilon_0,\, \upsilon_{\mathrm{R}},\, \ldots)$$

In NMR, the first step is a single-exponential fit of the magnetization curve $M(t)$ as a function of evolution time t. The exponential form is found in all water-rich systems, even if they are dynamically heterogeneous, and the resulting frequency-dependent spin relaxation rate $R(\omega)$ is a model-independent quantity. In QENS, the first step is to fit the spectrum $S(Q,\omega)$ to an empirical function that may include one or more Lorentzians in addition to an elastic peak (due to unresolved and/or confined motions) and a uniform background (which includes motions faster than the dynamic range of the instrument). Whereas the exponential form of $M(t)$ follows from the general theory of nuclear spin relaxation, the precise shape of $S(Q,\omega)$ cannot be predicted without specifying a model for the proton motions in the particular sample that is being investigated. Therefore, already the choice of fitting function introduces a model-dependence in the quasielastic linewidths $\Gamma(Q)$.

The second step in NMR depends strongly on the nuclear isotope used and the frequency range covered. For isotopes with a nuclear electric quadrupole moment (like ^{2}H and ^{17}O), the spin relaxation behavior is particularly simple. The relaxation dispersion $R(\omega)$ is then a direct mapping of the spectral density function, which is the Fourier transform of an orientational time correlation function (TCF). In the case of ^{2}H relaxation, the TCF describes the rotational motion of the water O–H bond. With current magnet technology, the highest accessible ^{2}H or ^{17}O frequency is ~100 MHz. This means that rotational motions faster than ~1 ns do not produce a frequency dependence in $R(\omega)$. The relaxation rate then yields the time integral of the TCF, which defines a model-independent rotational correlation time υ_{R} (equal to $1/(6D_{\mathrm{R}})$ for a rotational diffusion model).

In QENS, the interpretation of the empirically fitted linewidths $\Gamma(Q)$ is an open problem even for bulk water. QENS probes displacements of protons, not water molecules. This distinction can be ignored on large length scales (small momentum transfer, Q), but not in the Q range 0.2–2 Å^{-1} usually probed in QENS studies. Water translation and rotation are certainly not statistically independent processes (since both are governed by H-bond rearrangements), as invariably assumed to simplify the interpretation of QENS data. As a result of these complications, the physical significance of the correlation times (υ_0 and υ_{R}) deduced with the aid of the standard QENS model remains obscure even for homogeneous systems.

 This journal is © The Royal Society of Chemistry 2009

In complex systems (like cells), water motions take place on multiple time scales. Such dynamic heterogeneity is manifested in quite different ways in NMR and QENS: while $R(\omega)$ is a fast-exchange population-weighted average, $S(Q,\omega)$ is a slow-exchange superposition of Lorentzians. In QENS, a small population of slow water produces a narrow peak of small amplitude that may easily escape detection. But in the NMR relaxation rate $R(\omega)$, the small population is multiplied by a long correlation time. It is therefore possible, for example, to detect a single water molecule buried inside a protein against a background of 10^5 bulk water molecules. In dynamically heterogeneous systems, a model is needed to describe exchange averaging of $R(\omega)$, but because the exchange is invariably fast on the spin relaxation time scale, there is no need to model the exchange dynamics. In most cases, a two-state model is sufficient, where the single parameter is the water fraction f in hydration shells, the remaining fraction $(1 - f)$ being considered as bulk water. The models that you used to interpret QENS data from cells ignore this dynamic heterogeneity.

Professor Finney addressed Professor Halle and Dr Zaccai: It is often claimed that interpreting neutron scattering data on the dynamics of water—particularly in biological systems—requires the use of models which may themselves force an interpretation that depends on the model used. In contrast, Halle claims his NMR results are essentially model-independent. Can Dr Zaccai and Professor Halle explore a little the way in which the models used to interpret their respective data could be influencing the physical conclusions being drawn from their data, both qualitatively and quantitatively?

In this context of model dependency we might recall the conventional wisdom in the early 1970s that water close to protein was "slowed down" by a factor of about 10^6 compared to the bulk. I believe that this conclusion was the result of the model(s) used to interpret NMR data, and these were shown (I believe by Professor Halle and others) to be internally inconsistent. As interpretation of data improved (removing model dependence?) the conventional wisdom changed dramatically from "slowing downs" of 10^6 to "a few".

Professor Halle responded: Regarding the first, general part of your question, I addressed it in my response to the previous question by Dr Zaccai. The historical literature on NMR relaxation studies of protein (and DNA) hydration makes fascinating and sobering reading. Much of the confusion resulted from incomplete understanding of the spin relaxation mechanism, rather than from the use of inappropriate models. Thus, the earliest NMR studies (1950s) measured proton linewidths, which may be influenced by a variety of relaxation mechanisms, in addition to dipolar relaxation. Many NMR spectroscopists were also lead astray by the confounding effects of labile protons, that "looked" like long-lived water molecules. These problems were sorted out in the mid 1990s by measuring the longitudinal relaxation rate of the ^2H and ^{17}O isotopes over a wide frequency range (magnetic relaxation dispersion, MRD). It was thus demonstrated that the only long-lived (more than a few ns) water molecules that exist in protein solutions are buried in cavities inside the protein and that water molecules in the external hydration layer are highly mobile.[1] This general picture has withstood the test of time and subsequent MRD studies have refined the dynamical characterization of internal water molecules[2] and the external hydration layer.[3]

The theory of NMR relaxation involves dynamics in the nuclear spin system, which must be described quantum-mechanically, and dynamics in the molecular system, and the coupling of the two *via* nuclear interactions rendered time-dependent by molecular motions. There are many different spin couplings and a universe of NMR experiments used for different purposes. All of this may deter the 'outsider' who just wants to know if a particular NMR study of, say, cell water dynamics, is trustworthy. But if we don't make an effort to understand other methods than our own, the scientific enterprise will degenerate into a collection of subcultures which can neither communicate with nor benefit from each other. We will increasingly attend conferences catering exclusively to NMR spectroscopists, neutron scatterers, computational chemists and so on. Perhaps we are already there...

1 V. D. Denisov and B. Halle, *J. Am. Chem. Soc.*, 1994, **116**, 10324.
2 E. Persson and B. Halle, *J. Am. Chem. Soc.*, 2008, **130**, 1774.
3 C. Mattea, J. Qvist and B. Halle, *Biophys. J.*, 2008, **95**, 2951.

Dr Zaccai answered: I agree that we should look carefully at the NMR and neutron scattering data (which essentially measure different phenomena) in a model-independent way, in order to reach a better understanding of water behaviour in these complex systems.

Professor Halle opened the discussion of the paper by Dr Zaccai: In your QENS study of *E. coli*,[1] you find that the observed cell water is indistinguishable (within experimental accuracy) from bulk water as regards translational and rotational dynamics. However, on the basis of another parameter (υ_0) in the model that you used to establish the bulk-like dynamics, you infer that "water molecules spend longer times in the first hydration shell of macromolecular structures than in the bulk phase". Can you explain how you can extract information about hydration water from a model that does not take into account the dynamical heterogeneity of cell water?

In our NMR study of *E. coli*,[2] we found that hydration water (estimated to 15% of total cell water) is rotationally slowed down by a factor 15 on average. This substantial retardation is directly manifested as an increase (relative to bulk water) of the measured spin relaxation rate R_1 by a factor that is >250 times larger than our measurement error. If the NMR results are accepted, one must conclude that the QENS data[1] indicate bulk-like behavior for the trivial reason that they cannot resolve the hydration water. Do you share this view?

From your QENS study of *H. marismortui*,[3] you conclude that 45% of the water molecules in your cell pellet have bulk-like dynamics, while the remaining 55% (corresponding to 76% of the cell water) are slowed down 250-fold. According to your interpretation, this extremely slow motion occurs in confined spaces of 3.3 Å radius. This seems to imply that the ultrastructure of this halophile differs radically from that of other microorganisms. Is this view consistent with current cell biology?

Our NMR data[2] show that cell water dynamics are very similar in *E. coli* and *H. marismortui*. This model-independent result cannot be reconciled with your QENS interpretation.[3] We therefore question your assignment of the quasielastic broadening observed on the high-resolution instrument (IN16) to 55% of the water molecules in the sample. If this were so, would you not expect the elastic peak on the low-resolution instrument (IN6) to account for >55% of the integrated spectral intensity on IN6? But in your fit (Fig. 1a in ref. 3), the elastic intensity appears to be less than 10%.

1 M. Jasnin, M. Moulin, M. Härtlein, G. Zaccai and M. Tehei, *EMBO Rep.*, 2008, **9**, 543.
2 E. Persson and B. Halle, *Proc. Natl. Acad. Sci. U. S. A.*, 2008, **105**, 6266.
3 M. Tehei, B. Franzetti, K. Wood, F. Gabel, E. Fabiani, M. Jasnin, M. Zamponi, D. Oesterhelt, G. Zaccai, M. Ginzburg and B.-Z. Ginzburg, *Proc. Natl. Acad. Sci. U. S. A.*, 2007, **104**, 766.

Dr Zaccai answered: Regarding the QENS study of *E. coli*, as discussed earlier in this session, neutron scattering and NMR measure different rotational relaxation processes and there is reason to expect qualitative but not quantitative agreement between the two parameter sets. There have been various neutron studies of hydration water. In 1996 Bellissent-Funel *et al.*[1] found a proton residence time longer by a factor of 6 compared to bulk water in a 0.4 g per g hydrated powder of phycocyanin. Your factor of 15 for NMR rotational diffusion was found in solution in

 This journal is © The Royal Society of Chemistry 2009

which, because of rapid exchange, one might expect a smaller rather than larger slowing-down factor, but then again I believe these factors cannot be compared quantitatively. In the *E. coli* neutron scattering study, because of rapid exchange one expects to measure a mean residence time between the protons in hydration sites and the protons in the bulk. The published value of longer residence time by a factor of 2 compared to bulk, is in quantitative agreement with 20% of the cell water behaving as in the paper by Bellissent-Funel *et al.*[1]—like hydration water in rapid exchange with 80% of the cell water, which behaves like bulk -water.

Regarding the QENS study of *H. marismortui*, indeed it is consistent with current cell biology, with respect to the extreme halophilic archaeon *Haloarcula marismortui* and to the particular salt–ionic and hydration interactions of its proteins. Please see ref. 2–11.

With respect to your question about Fig. 1a in our *PNAS* paper, please note that the (integrated elastic)/(integrated quasi-elastic) ratio measured with a given energy resolution is dependent on the scattering vector, Q. The data shown in Fig 1a of our paper are for a large Q value (1.65 Å^{-1}). The ratio at a given Q value is not equal to the ratio of respective proton populations. Formally, the population ratio could be obtained from that ratio as Q tends to zero. We estimated the ratio for a Q value of 0.5 Å^{-1} to be between 0.7 and 0.8, above the 55%, as expected.

1 M.-C. Bellissent-Funel, J.-M. Zanotti and S. H. Chen, *Faraday Discuss.*, 1996, **103**, 281.
2 M. Ginzburg, L. Sachs and B.-Z. Ginzburg, *J. Gen. Physiol.*, 1970, **55**, 187.
3 B.-Z. Ginzburg and M. Ginzburg, in *Biophysics of Water*, ed. F. Franks and S. F. Mathias, Interscience, London, 1982, pp. 340–343.
4 B.-Z. Ginzburg, *Thermochim. Acta*, 1981, **46**, 249.
5 H. Morgan, M. Ginzburg and B.-Z. Ginzburg, *Biochim. Biophys. Acta*, 1987, **924**, 54.
6 S. Bone, B.-Z. Ginzburg, H. Morgan, G. Wilson and B. Zaba, *Phys. Med. Biol.*, 1996, **41**, 45.
7 B.-Z. Ginzburg and M. Ginzburg, in *Biomembranes*, ed. H. Eisenberg, E. Katchalski-Katzir and L. A. Manson, Plenum, New York, 1975, vol. 7, pp. 219–251.
8 C. Ebel, L. Costenaro, M. Pascu, P. Faou, B. Kernel, F. Proust De Martin and G. Zaccai, *Biochemistry*, 2002, **41**, 13234.
9 D. Madern, C. Ebel and G. Zaccai, *Extremophiles*, 2000, **4**, 91.
10 C. Ebel, P. Faou, B. Kernel and G. Zaccai, *Biochemistry*, 1999, **38**, 9039.
11 S. B. Richard, D. Madern, E. Garcin and G. Zaccai, *Biochemistry*, 2000, **39**, 992.

Dr Patel continued the discussion of the paper by Professor Halle: For halophiles, neutron scattering studies seem to indicate a 2 orders of magnitude slowing down of the hydration water dynamics. In your work, the dynamic perturbation factor is inferred from the high-frequency data in order to avoid contribution from internal water molecules that relax on the time-scale that the protein relaxes. My question is: if the 2 orders of magnitude slowing down in the hydration water dynamics seen by neutron scattering were indeed present, wouldn't its effect be manifest at lower frequencies and be indistinguishable from the signal due to the relaxation of the internal water molecules?

Professor Halle responded: This is a valid concern and it is the reason why it was important to measure the spin relaxation rate down to very low frequencies, even though it turned out that the strong frequency dependence comes from internal water molecules that make up merely ~0.1% of the cell water.[1] If the interpretation of the QENS data[2] were correct, the rotational correlation time for 76% of the cell water (55% of the water in the sample) would be $\upsilon = 250 \times 2.8 = 700$ ps. Even at our highest frequency of 76.8 MHz, $(\omega_0\upsilon)^2 = 0.11 << 1$, which means that these putative water molecules would not give rise to a frequency dependence but would increase R_1 to the same extent over the entire frequency range. At the highest frequency, where internal water molecules do not contribute, we measured $R_1/R_1^o = 3.06 \pm 0.02$ (where R_1^o is the spin relaxation rate in pure D_2O). According to the QENS interpretation, we should have measured $R_1/R_1^o = 0.45 + 0.55 \times 250 = 138$ at this frequency. We therefore conclude that the QENS interpretation is incorrect.

It should be said that the QENS data that led to the proclamation of "extremely slow cell water" in *H. marismortui* were acquired on an instrument that cannot resolve water rotation.[2] The dynamical information was thus extracted in the form of a translational diffusion coefficient in a particular model of spherically confined diffusion. It might be argued that water translation is slowed 250-fold, while water rotation is bulk-like. However, because of the extensive H-bonding in liquid water, translation and rotation are strongly coupled and invariably occur on the same time scale.

1 E. Persson and B. Halle, *Proc. Natl. Acad. Sci. U. S. A.*, 2008, **105**, 6266.
2 M. Tehei, B. Franzetti, K. Wood, F. Gabel, E. Fabiani, M. Jasnin, M. Zamponi, D. Oesterhelt, G. Zaccai, M. Ginzburg and B.-Z. Ginzburg, *Proc. Natl. Acad. Sci. U. S. A.*, 2007, **104**, 766.

Dr Zaccai replied: The frequency ranges used in the *H. marismortui* experiments corresponded to time scales from the picosecond to the nanosecond on different spectrometers. The "slow" water was "seen" in the longer time scale. With respect to translational diffusion, internal water molecules would probably appear as immobile on this time scale. Bulk-like water was observed on the spectrometer with the shorter time scale. In *E. coli* there was no "slow" water component observed on the longer time scale spectrometers.[1,2]

1 M. Tehei, B. Franzetti, K. Wood, F. Gabel, E. Fabiani, M. Jasnin, M. Zamponi, D. Oesterhelt, G. Zaccai, M. Ginzburg and B.-Z. Ginzburg, *Proc. Natl. Acad. Sci. U. S. A.*, 2007, **104**, 766.
2 M. Jasnin, M. Moulin, M. Härtlein, G. Zaccai and M. Tehei, *EMBO Rep.*, 2008, **9**, 590.

Dr Nutt remarked: How do you unify the two apparently opposite effects described in your conclusions (1) and (5), whereby rotation of hydration water is faster at low T around NALMA, but disruption of the tetrahedrality of hydration water interferes with the cooperative rotation of water molecules, slowing rotation?

Professor Halle responded: The picture of water rotation, inspired by computer simulations of bulk water,[1,2] that we use to rationalize our experimental results is as follows. Rotation in bulk water at ambient temperature is fast because high-energy configurations with one or more completely broken H-bonds can be avoided by concerted H-bond interchanges made possible by cooperative rearrangements of several neighboring water molecules. At the interface of an inert solute, a water molecule has fewer water neighbors with which to swap H-bonds and so the cooperative mechanism is impeded. This is the case even if the solute can H-bond with the water molecule, because the bulky solute is less mobile than a water molecule and thus cannot participate in the cooperative rearrangement. Therefore, water rotates more slowly at an interface. This is what we try to summarize in conclusion (5) in our paper. We do not mean to imply that "disruption of tetrahedrality" leads to slower water rotation. On the contrary: when bulk water is (super)cooled, water rotation slows down drastically (in a super-Arrhenius fashion) because the tetrahedral order increases. This ordering interferes with the cooperative rotation mechanism, which requires that a fifth water molecule can approach closely enough to replace one of the four nearest neighbors. For an interfacial water molecule, rotation is also slowed down on cooling. But below a certain temperature, which is 256 K for NALMA and ~260 K for the proteins that we studied, it slows down less than it does in bulk water. The reason for this difference, we believe is that the structure of the solute, which constitutes part of the environment of the interfacial water molecule, hardly changes with temperature. If water rotation in the hydration layer slows down less on cooling than it does in bulk water, there will come a temperature where hydration water actually rotates faster than bulk water. For NALMA, our data indicate that this temperature is 237 K. This is the substance of our conclusion (1).

 This journal is © The Royal Society of Chemistry 2009

1 F. Sciortino, A. Geiger and H. E. Stanley, *Nature*, 1991, **354**, 218.
2 D. Laage and J. T. Hynes, *Science*, 2006, **311**, 832.

Dr Ricci remarked: There is evidence from both simulations and experiments that structural and dynamical properties of water are influenced by the presence of solutes[1,2] or the vicinity of a substrate[3,4] well beyond the first hydration shell. Most of these studies are performed on solutions of small electrolites or in confinement within a matrix formed of molecules as simple as SiO_2. How is it possible that, as you say: "the perturbation induced by a biological interface is of short range, essentially limited to the first shell"? In particular this contrasts with the results reported in Professor Havenith's paper in this same volume.[5] Could you please tell us on which evidence your statement is based?

1 R. Mancinelli, A. Botti, F. Bruni, M. A. Ricci and A. K. Soper, *Phys. Chem. Chem. Phys.*, 2007, **9**, 2959.
2 J. Holzmann, R. Ludwig, A. Geiger and D. Paschek, *Angew. Chem., Int. Ed.*, 2007, **46**, 8907.
3 H. Thompson, A. K. Soper, M. A. Ricci, F. Bruni and N. T. Skipper, *J. Phys. Chem. B*, 2007, **111**, 5610.
4 P. Gallo, M. Rovere and E. Spohr, *J. Chem. Phys.*, 2000, **113**, 11324.
5 B. Born, S. J. Kim, S. Ebbinghaus, M. Gruebele and M. Havenith, *Faraday Discuss.*, 2009, DOI: 10.1039/b804734k.

Professor Halle answered: The solute-induced perturbations of water structure and dynamics surely decay gradually and do not vanish identically at any finite distance. A simple Born model estimate indicates that some 200 kJ mol^{-1} of the hydration free energy of a monovalent ion is contributed by water molecules beyond the first hydration shell. Yet, NMR relaxation measurements[1] show that the average rotational correlation time of the water molecules in a 4 molal KCl solution (with 14 water molecules per K^+Cl^- pair) differ by merely a few percent from that in pure water. When discussing the range of solute-induced perturbations, it is thus necessary to specify which solvent property is being considered and at what quantitative level the perturbation can be deemed negligibly small. A 0.2 Å shift of the second peak in the O–O RDF for water molecules outside the first shell of a Na^+ ion, as you found,[2] does not necessarily imply a significant dynamic perturbation. And, unlike the extended silica surfaces examined in two of the other studies that you cite, biological interfaces tend to induce less pronounced solvent layering. Around nonpolar groups, in particular, the boundary between the first and second hydration shells is quite fuzzy (that is, the first minimum in the RDF is shallow and very broad).

We do not imagine that the dynamical perturbation outside the first hydration shell is zero, but we believe that it is much smaller than in the first shell. Thus, if the slowing down is 50% (a typical value) in the first shell, it might be 5% in the second and 0.5% in the third shell. It is then a reasonable approximation to attribute the entire dynamical perturbation to the first shell. This viewpoint is strongly supported by experiments and simulations, see for example ref. 1, 2 and 8–10 in our paper.

1 J. R. C. van der Maarel, D. Lankhorst, J. de Bleijser and J. C. Leyte, *J. Phys. Chem.*, 1986, **90**, 1470.
2 R. Mancinelli, A. Botti, F. Bruni, M. A. Ricci and A. K. Soper, *Phys. Chem. Chem. Phys.*, 2007, **9**, 2959.

Dr Russo commented: In the Fig. 2 in your paper, you show NAGMA and NALMA DPF values as arising from spin relaxation measurement together to QENS data. You find a discrepancy which you attribute to the strong model dependence in the interpretation of the QENS data generalizing this idea to the all QENS technique.

It is true that the QENS NALMA data that you report in Fig. 2 have a strange behavior. The DPF is smaller than 1, which would suggest that pure bulk water

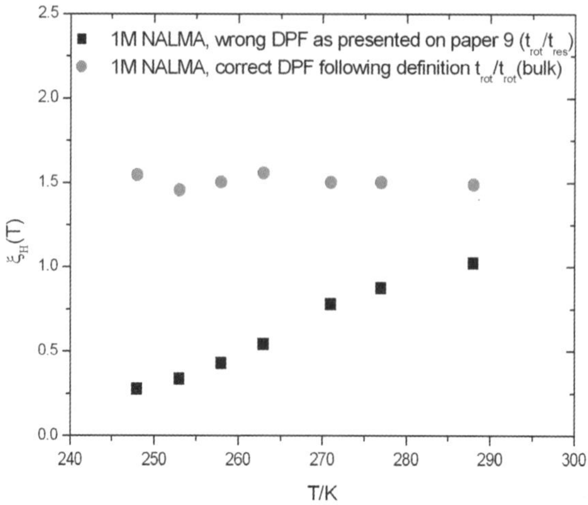

Fig. 1

has a rotational dynamics slower than hydration water. Most likely the problem on the Figure arises from a bad interpretation of the experimental data, from your side, instead of a model-dependence interpretation. Indeed you calculated the DPF factor using the wrong "t_0". Instead to normalize the t_{rot} data reported in the Table 1 of ref. 28 in your paper, for the t_{rot} of bulk water in the same range of temperature ($t_{bulk} = 0.0485 \exp(E_a/kT)$, $E_a = 1.85$ kcal mol^{-1} [1]) it seems that you normalized for the residential time t_0 reported on the same Table. Residential times arise from translational dynamics. In Fig. 1, I plot the QENS DPF as you reported on the paper and the DPF as should have been calculated following the information given in Table 1 of ref. 28 and the paper from Teixeira *et al.*[1] Correlation times are calculated from the following relation: $tE = 0.6582$ (where t is in ps and E in meV), and if we speak in term of the lifetime of the hydrogen bond we divide this result by a factor of 3.[2] Therefore it is not very clear in ref. 28 what the authors calculate and what they call t_{rot}. There is a discrepancy between their Fig. 4 and Table 1, in addition to the fact that the sample has an "undefined" concentration as a consequence of some experimental problems.

In Fig. 2, I report the DPF calculated in a smaller range of temperature for three different concentrations: 1 M, 2 M and 0.5 M.[3,4] The DPF dependence on the NALMA concentration and temperature suggests that only the "really first" hydration water layer has a rotational correlation which is affected from the presence of the biomolecule and that the outer layers (~2–3 shells) are bulk-like. The value reported in Fig. 2 for 2 M NALMA are similar to those that you account for the same peptide on the same range of temperature. No discrepancy seems to appear and most importantly the data presented do not arise from a model.

Can you clearly define what you mean by hydration shell and what you measure for it in your paper? Are your NALMA DPF data comparable with the 2 M NALMA (where there is only one hydration layer is in place), or with the 1–0.5 M NALMA (where ~2–3 hydration layers are in place, and the pair correlation function $g(r)$ is available)? Is your rotational time (NMR) directly comparable with our rotational correlation time (QENS)?

1 J. Teixeira, M.-C. Bellisent-Funel, S. H. Chen and A. J. Dianoux, *Phys. Rev. A*, 1985, **31**, 1913.
2 V. F. Sears, *Can. J. Phys.*, 1966, **44**, 1299.
3 D. Russo, R. K. Murarka, G. Hura, E. Verschnell, J. R. D. Copley and T. Head-Gordon, *J. Phys. Chem. B*, 2004, **108**, 19885.
4 D. Russo, data not yet published.

 This journal is © The Royal Society of Chemistry 2009

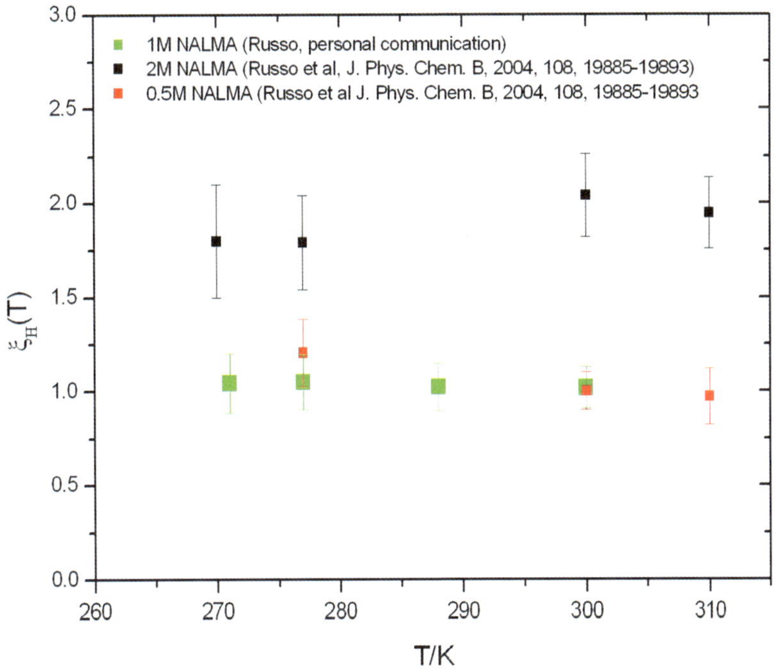

Fig. 2

Professor Halle responded: Your questions relate to Fig. 2 in our paper, which shows the temperature dependence of the dynamic perturbation factor (DPF) ξ_H for the hydration shell of the peptide NALMA. The DPF is obtained from the spin relaxation rate R_1 measured in dilute (<0.2 M) NALMA solutions and that (R_1^o) measured in pure water. In this concentration range R_1 increases linearly with the NALMA concentration, which means that the relaxation enhancement $R_1 - R_1^o$ can be attributed to the solute-induced perturbation of water rotation in the hydration shell of the solute at effectively infinite dilution of solute. These measurements do not allow us to separately determine the degree of rotational slowing down and the spatial range of this perturbation. However, the range of the linear dependence of R_1 on solute concentration in this and many other systems as well as many MD simulations demonstrate that the dynamic perturbation is essentially confined to the first water layer as I explained earlier to Dr Ricci. As a reasonable approximation, we therefore attribute the entire perturbation to the first layer of water molecules. The number of water molecules in the first layer, $\nu_S = 43$ for NALMA, is obtained from MD simulations. The DPF is then obtained directly from the measured data according to eqn (2) in our paper.

The interpretation of the DPF does not rely on a dynamical model (such as rotational diffusion or large-angle jumps), but follows from the rigorous link[1] between R_1 and the integral of the time correlation function $\langle P_2(\cos\theta(t))\rangle$, where $\theta(t)$ specifies the orientation of the O–H bond at time t. MD simulations show that this function drops from 1 to $S_{lib}^2 \approx 0.8$ as a result of subpicosecond O–H bond librations and thereafter decays exponentially with a rotational correlation time υ_R. The integral correlation time, which we denote by υ, is thus given by $S_{lib}^2\upsilon_R$ to an excellent approximation. The DPF is the ratio of the integral correlation time in the hydration shell (υ_H), averaged over the 43 molecules in the shell, and that in pure bulk water (υ_0). Since S_{lib} should be nearly the same in hydration shell and bulk, the DPF can also be regarded as the ratio of υ_R values in the two regions. Further details about our NMR study of hydration dynamics for NALMA and three other solutes can be found in ref. 2.

Our Fig. 2 also includes a DPF derived from QENS data for a 1 M NALMA solution.[3] As in previous QENS studies[4,5] of peptide solutions, ref. 3 reported a quantity called υ_{rot}, defined as $1/(6D_{rot})$, and deduced from a model of unrestricted, torque-free, isotropic rotational diffusion (eqn (6) in ref. 3). If this model is valid, and if the analysis of the QENS data is correct also in other respects, then the QENS-derived quantity υ_{rot} should be equal, apart from the librational factor $S_{lib}{}^2$, to the integral rotational correlation time υ derived from NMR. To calculate a DPF, we need to divide υ_{rot} by the bulk rotational correlation time, υ_{rot}(bulk). Since the latter quantity was not reported in ref. 3 and since both methods should give the same rotational correlation time for bulk water (if the data are properly analyzed), we used the NMR-derived rotational correlation for bulk water, which we denote by υ_0. Unfortunately, the same symbol υ_0 is used for a parameter in the Singwi–Sjölander model[6] that was used to describe the translational part of the scattering function.[3–5] Thus, we did not, as you assume, use υ_0 values from Table 1 in ref. 3 to compute the DPF.

As seen from Fig. 2 in our paper, the DPFs derived from NMR and QENS depend on temperature in opposite ways and differ by an order of magnitude at low temperatures. This discrepancy might be caused by the >5-fold higher solute concentration used in the QENS study, but MD simulations at this concentration agree rather well with the NMR results. Therefore, we conclude that the quantity υ_{rot} derived from QENS data is, in fact, not the rotational correlation time of water. Even for bulk water,[8] the parameter υ_{rot} extracted from QENS data with the same model assumptions as in the peptide work, differs substantially from the rotational correlation time determined by NMR and other methods. In particular, υ_{rot} is found to obey the Arrhenius law with a small activation energy of ~8 kJ mol^{-1},[8] whereas the NMR-derived rotational correlation time has a much stronger temperature dependence with the (apparent) activation energy increasing from 19 to 33 kJ mol^{-1} between +20 and −20 °C. This discrepancy has been recognized for some time,[9,10] but the QENS model has persisted because "it provides an excellent fit to the data".[8] It has been suggested that υ_{rot} reflects "large-amplitude librations" or "hindered rotational diffusion",[9,10] but this is not what the model describes. Furthermore, librations in water occur on time scales <100 fs and are presumably taken into account *via* the Debye–Waller factor. It seems clear, therefore, that the interpretation of QENS data on bulk water remains an open problem. And the models that don't work for bulk water cannot be expected to work for hydration water.

As you point out, it would have been more consistent to calculate the QENS-DPF with υ_{rot}(bulk) values measured by QENS.[8] When this is done, the DPF is 1.5 with no significant temperature dependence. So again there is a qualitative discrepancy with the NMR-DPF. But the important point is that NMR measures the rotational correlation time of water molecules, whereas the standard analysis of QENS data yields a quantity (υ_{rot}) that is model-dependent and so far has not been convincingly linked to anything that can be measured by other experiments or obtained from MD simulations. Other QENS models have been proposed,[11,12] but they are of an empirical nature and therefore cannot resolve the issue.

The strong model-dependence in the interpretation of QENS data from systems as simple as peptide solutions is further illustrated by the conclusion that a large fraction (of order 50%) of the water molecules in 0.5–3 M solutions of NAGMA or NALMA are immobile on time scales up to 40 ps (the time scale corresponding to an energy resolution of 35 μeV).[5] If this were the case, the DPF would be at least an order of magnitude larger than what we find by NMR (where no water is outside the "time window").[2] Furthermore, the immobilized fraction was found to decrease with increasing solute concentration.[5] This unphysical behavior was deduced from small deviations of the (apparent) elastic incoherent structure factor from the Q-dependence of the rotational structure factor. These deviations may be caused by inadequacies of the QENS model at high Q values.

 This journal is © The Royal Society of Chemistry 2009

1 A. Abragam, *The Principles of Nuclear Magnetism*, Clarendon Press, Oxford, 1961.
2 J. Qvist and B. Halle, *J. Am. Chem. Soc.*, 2008, **130**, 10345.
3 C. Malardier-Jugroot and T. Head-Gordon, *Phys. Chem. Chem. Phys.*, 2007, **9**, 1962.
4 D. Russo, G. Hura and T. Head-Gordon, *Biophys. J.*, 2004, **86**, 1852.
5 D. Russo, R. K. Murarka, J. R. D. Copley and T. Head-Gordon, *J. Phys. Chem. B*, 2005, **109**, 12966.
6 K. S. Singwi and A. Sjölander, *Phys. Rev.*, 1960, **119**, 863.
7 R. K. Murarka and T. Head-Gordon, *J. Chem. Phys.*, 2007, **126**, 215101.
8 J. Teixeira, M.-C. Bellisent-Funel, S. H. Chen and A. J. Dianoux, *Phys. Rev. A*, 1985, **31**, 1913.
9 M.-C. Bellisent-Funel and J. Teixeira, *J. Mol. Struct.*, 1991, **250**, 213.
10 J. Teixeira, M.-C. Bellisent-Funel and S.-H. Chen, *J. Mol. Liq.*, 1991, **48**, 123.
11 S. H. Chen, C. Liao, F. Sciortino, P. Gallo and P. Tartaglia, *Phys. Rev. E*, 1999, **59**, 6708.
12 L. Liu, A. Faraone and S. H. Chen, *Phys. Rev. E*, 2002, **65**, 041506.

Professor Klein opened the discussion of the paper by Professor Tobias, addressing Professor Halle, Dr Zaccai and Professor Tobias: I have two questions for the three speakers.

Firstly, what are the implications of the rapid exchange dynamics of the water-of-hydration molecules together with their large dehydration energies? These can result in large hydration repulsion effects, as well as in interesting hydration lubrication phenomena (see *e.g.* Raviv and Klein[1]).

Secondly, were large variations in the relaxation times of hydration water molecules observed in the vicinity of multivalent ions?

1 U. Raviv and J. Klein, *Science*, 2002, **297**, 1540.

Professor Halle replied: In response to your first question, just as you contrast the fast hydration-to-bulk water exchange with the large dehydration energy, one could contrast the fast dynamics in bulk water with the large heat of vaporization. In both cases, the implication is that dynamics do not require dehydration. When a water molecule in the bulk liquid or in a hydration layer moves, it does not leave behind a cavity. Rather, the movement is a cooperative process where another water molecule enters the vacated space in a concerted manner. In this way, high (dehydration) barriers are avoided and the motion is therefore fast. However, dehydration energies are relevant to dynamics in two cases. The first is hydration/dehydration of a protein surface from the vapor phase (of limited biological relevance), where the water residence times are very long. The other case concerns secluded hydration sites, such a deep pocket or pore on the surface of a protein.[1] The geometry then prevents the new water molecule from entering the site until the old one has left, giving rise to a high barrier and long residence time. For this reason, the residence time distribution in protein hydration layer is strongly influenced by the surface topography.

Secondly, the water residence times in the hydration shell of monatomic ions (in bulk electrolyte solutions) span some 20 orders of magnitude.[2] For ions at surfaces, less is known. The residence time of water molecules coordinating the protein-bound Ca^{2+} ions in calbindin D_{9k} is in the ns–μs range, much longer than for Ca^{2+} in bulk solution.[3]

1 B. Halle, *Philos. Trans. R. Soc. London, Ser. B*, 2004, **359**, 1207.
2 L. Helm and A. E. Merbach, *Chem. Rev.*, 2005, **105**, 1923.
3 V. P. Denisov and B. Halle, *J. Am. Chem. Soc.*, 1995, **117**, 8456.

Dr Zaccai responded: Regarding your first question, indeed the rapid exchange dynamics of hydration water associated with their large dehydration energies has direct implications for neutron scattering observations. The dehydration energy leads to a time-averaged structural modification of the hydration layer with respect to bulk water that was observed by small angle neutron and X-ray scattering.[1] Rapid exchange leads to an averaging of the times of residence observed in quasi-elastic neutron scattering.[2]

As to your second, a recent review on water around various ions, measured by neutron scattering, is by Collins *et al.*[3] A QENS paper on water diffusion on the presence of multivalent ions was published in 1987.[4] I note that the biological aspects in this discussion are dominated by proteins. Hydrated multivalent ions play an important role in the stabilisation of nucleic acid structures, and it would certainly be interesting to study the interactions involved in detail.

1 D. I. Svergun, M. H. J. Koch, S. Kuprin, S. Richard and G. Zaccai, *Proc. Natl. Acad. Sci. U. S. A.*, 1998, **95**, 2267.
2 M. Jasnin, M. Moulin, M. Härtlein, G. Zaccai and M. Tehei, *EMBO Rep.*, 2008, **9**, 543.
3 K. D. Collins, G. W. Neilson and J. E. Enderby, *Biophys. Chem.*, 2007, **128**(2–3), 95.
4 P. S. Salmon, W. S. Howells and R. Mills, *J. Phys. C.: Solid State Phys.*, 1987, **20**, 5727.

Professor Tobias answered: By large dehydration energies I assume you mean that a water molecule near a biomolecule has a much lower free energy than a water molecule in the bulk. Rapid exchange dynamics implies a low free energy barrier to move from next to the molecule to the bulk. I don't know what are the implications for hydration repulsion forces and lubrication phenomena, but I imagine it could be quite different for biomolecules from the case for the flat and rigid mica surfaces employed in the study you referred to, because biomolecular surfaces are rough on the length scale of a water molecule, and mobile on the time scale of water dynamics. We have not included multivalent ions in our simulations, but I would expect large variations based on experimental data (*e.g.* water mobility near calcium dications is orders of magnitude greater than near magnesium dications).

Professor Halle addressed Professor Tobias and Dr Zaccai: In the papers presented by Professor Tobias and Dr Zaccai, a large number of so-called "dynamical transitions" are identified from the temperature dependence of atomic mean-square displacements in hydrated biological systems. These transitions, variously referred to as "inflections" (Professor Tobias' paper) or "breaks in slope" (Dr Zaccai's paper), are not always apparent to the untrained eye. Could you therefore describe the objective quantitative procedure that you used to determine the transition temperature and its associated error, taking the data for nonexchangeable protons in maltose binding protein as an example (shown in Fig. 1A of both papers)?

Professor Tobias replied: So-called dynamical transitions in many proteins have been identified by various experimental techniques (most commonly neutron scattering) and discussed for more than two decades by numerous authors. As far as I am aware, there is no objective quantitative procedure for determining the temperature at which the transition(s) occur. In most cases it is clearly manifested (at least to a trained eye) by a change of slope in a plot of the mean-squared fluctuations *vs.* temperature. The maltose binding protein data plotted in our papers is, of all cases that I have seen, the one where the transition in the protein dynamics is least evident (although a corresponding transition in the water dynamics is more clear). In any case, the dynamical transition is not expected to be sharp, in the sense of a thermodynamic phase transition.

Dr Zaccai responded: We used these terms because of the discussion within the community concerning the nature of the transition. In the simplest approximation, the quantitative definition of the "transition" is where there is an observable deviation from harmonic behaviour. This happens at various temperatures depending on the protein, the energy resolution of and scattering vector range of the spectrometer.

Professor Jungwirth continued the discussion of the paper by Dr Zaccai: Extreme halophiles fill the intracellular region with potassium and the surface of intracellular proteins has many carboxylate anionic groups. A single COO^- group prefers sodium over potassium. So, is the situation for halophiles a result of massive active pumping of sodium out of the cell or of a special arrangement of carboxylates to prefer potassium (like on the potassium channel) or both of these effects?

Dr Zaccai responded: The work of the Ginzburgs,[1–6] which was at the origin of our neutron study,[7] presented evidence that the high K^+ intracellular concentration was maintained somehow, even when the membrane barrier was diminished in *H. marismortui*. This would be in favour of a special structural arrangements of water and potassium ion "binding" rather than massive pumping.

1 M. Ginzburg, L. Sachs and B.-Z. Ginzburg, *J. Gen. Physiol.*, 1970, **55**, 187.
2 B.-Z. Ginzburg and M. Ginzburg, in *Biophysics of Water*, ed. F. Franks and S. F. Mathias, Interscience, London, 1982, pp. 340–343.
3 B.-Z. Ginzburg, *Thermochim. Acta*, 1981, **46**, 249.
4 H. Morgan, M. Ginzburg and B.-Z. Ginzburg, *Biochim. Biophys. Acta*, 1987, **924**, 54.
5 S. Bone, B.-Z. Ginzburg, H. Morgan, G. Wilson and B. Zaba, *Phys. Med. Biol.*, 1996, **41**, 45.
6 B.-Z. Ginzburg and M. Ginzburg, in *Biomembranes*, ed. H. Eisenberg, E. Katchalski-Katzir and L. A. Manson, Plenum, New York, 1975, vol. 7, pp. 219–251.
7 M. Tehei, B. Franzetti, K. Wood, F. Gabel, E. Fabiani, M. Jasnin, M. Zamponi, D. Oesterhelt, G. Zaccai, M. Ginzburg and B.-Z. Ginzburg, *Proc. Natl. Acad. Sci. U. S. A.*, 2007, **104**, 766.

Professor Lyashchenko commented: I am a specialist in high frequency dielectric properties of electrolyte and nonelectrolyte solutions so my comment will be from a solution point of view. We have a timescale break between the elementary dynamic processes in bulk water (less than 1 ps) and the water in the biological gel and glass structures (more than 1 ns). I think that we can use the dielectric data of simple aqueous systems for our understanding of this functional difference. There is a very interesting phenomenon in the simple water–electrolyte systems (maybe it is a common case). In the first concentration region with bulk tetrahedral water we have one Debye relaxation process with one relaxation time (about 2–8 ps). There is a concentration boundary around the first zone.[1] In the second concentration zone we have the hydration shells and clusters with a small number of water molecules. This cluster formation is connected with specific concentrations. In this case we have two relaxation processes. The time of relaxation therefore has two values: not only 2 ps (as in hydration shell) but also 20 or 40 ps (for the second process)![2] In addition, it is only in this concentration region with heterogeneous dynamics of water that we have gel and glass formation in water–electrolyte systems. It can be assumed that the concentration of bulk water is the main functional parameter in the case of biological aqueous systems. The absence of bulk water can determine the possibility of gel formation. What do you think about it?

1 A. Lyashchenko, *J. Mol. Liq.*, 2001, **91**, 21.
2 A. Lyashchenko and A. Zasetsky, *J. Mol. Liq.*, 1998, **77**, 61.

Dr Zaccai answered: Certainly, in our current understanding, the specific hydrogen bonding properties of bulk water that for example lead to the so-called hydrophobic effect, are essential for many fundamental biological processes such as protein folding and stabilisation, protein–protein and protein–nucleic acid interactions.

Professor Finney remarked: I'd like to make a further comment on the so-called "dynamical transition", perhaps suggesting it might have served its function as a concept and that it might be more productive to move on to something more sophisticated. From my own experience in obtaining plots of mean square deviation with temperature, locating a temperature at which a "transition" is clearly seen can involve almost an act of faith. In most systems there is a change in gradient as temperature is increased. However, in many cases, this looks like a gradual change in gradient, and to assign a specific temperature at which a gradient change occurs may not only be unhelpful but could be misleading, putting the results in a conceptual straitjacket. As more measurements of MSD *vs.* T are made on a wider range of systems and over a wider range of conditions, we increasingly see that, if we do try to assign temperatures at which a "dynamical transition" may occur, these can be dependent on sample and timescale probed and more than one "transition" is sometimes found.[1] We should perhaps go beyond the two-well model used to explain the early hydrated myoglobin powder work.[2]

1 R. M. Daniel, R. V. Dunn, J. L. Finney and J. C. Smith, *Annu. Rev. Biophys. Biomol. Struct.*, 2003, **32**, 69.
2 W. Doster, S. Cusack and W. Petry, *Nature*, 1989, **37**, 754.

Professor Finney then continued the discussion of the paper by Professor Halle: The conclusion that water rotation might be faster than in the bulk is interesting but the apparent extrapolation to lower temperature of the DPFs for NAGMA and NALMA in your Fig. 2 are not convincing. Is this conclusion based on other arguments, and if so what?

On another topic, some of the results are dependent on the use of surface accessible area to obtain estimates of the number of water molecules "covering" a solute. In my experience, calculations of numbers of hydration waters based on surface accessible areas can vary over a wide range depending on the assumptions made. Can you say why you think the numbers you obtain for hydration shell water molecules are realistic, and also estimate the likely maximum error in these numbers? And how would this error propagate to affect your later conclusions?

Professor Halle replied: The dynamic perturbation factor (DPF) ξ_H shown in Fig. 2 is the ratio of the average correlation time in the hydration layer $\langle \upsilon_H \rangle$ and the bulk water correlation time υ_0, each of which was measured as a function of temperature. It is evident from these data (only their ratio is shown in Fig. 2) that, at low temperatures, the activation energy is higher in bulk water than in the hydration shell, which means that the DPF decreases on cooling in this range. Since this trend is not likely to be overturned a few degrees below the lowest measured temperature, we can confidently predict that $\xi_H < 1$ at temperatures below ~237 K. As briefly indicated in the Figure legend, we fitted the temperature dependence of $\langle \upsilon_H \rangle$ and υ_0 separately (and over a somewhat wider temperature range). These fits are displayed as Arrhenius plots in Fig. 3 of ref. 1. The curves shown in Fig. 2 of our paper were obtained by dividing these fits. Because $\langle \upsilon_H \rangle$ and υ_0 vary monotonically with temperature, we were brave enough to extrapolate the fitted curves 5–10 K below the 65 K data range.

Regarding the dependence on surface accessible area, the answer is yes: for small molecules (but not for proteins) the solvent accessible surface area (SASA) is rather sensitive to the choice of van der Waals radii. Partly for this reason, we determined the number of water molecules ν_H in the first hydration shell of each solute from MD simulations (1 ns trajectory with ~1700 SPCE water molecules per solute molecule). From simulations of four small solutes,[1] we found that the mean SASA occupied per water molecule, obtained by dividing the SASA (calculated with a standard set of vdW radii and a probe radius of 1.7 Å) by the simulation-based hydration number ν_H, were all within the narrow range 10.65–10.87 Å^2. Moreover, nearly the same value (10.61 Å^2) was obtained from simulations of the protein β-lactoglobulin.[2] It thus appears that accurate hydration numbers can be obtained for small organic solutes as well as for proteins with a "water area" of 10.75 Å^2 and with the SASA calculated as described. These hydration numbers include water oxygens that are inside the first minimum of the RDF for O–O_W (3.3 Å), N–O_W (3.5 Å) and C–O_W (5.0 Å) pairs. As seen from Fig. 1 in our paper, this prescription corresponds closely to the intuitive picture of a first water layer. The simulation-based hydration numbers depend mostly on the C–O_W cutoff, which is not very well-defined since the

 This journal is © The Royal Society of Chemistry 2009

RDF minimum is wide and shallow. An uncertainty of ±0.2 Å in this cutoff corresponds to ±10% in ν_H and to ±(3–5)% in ξ_H. This error is probably smaller than the one incurred by attributing the total dynamical perturbation to the first layer as I said earlier in response to Dr Ricci's question

1 J. Qvist and B. Halle, *J. Am. Chem. Soc.*, 2008, **130**, 10345.
2 C. Mattea, J. Qvist and B. Halle, *Biophys. J.*, 2008, **95**, 2951.

Professor Debenedetti continued the discussion of the paper by Professor Tobias: Fig. 6 of your paper and the accompanying discussion refers to "anomalous" diffusion, whereby the mean-squared displacement increases sub-linearly with time. In simulations in simple geometries (*e.g.* water between parallel slabs) one finds that the diffusion is normal in the direction parallel to the slabs (MSD ~ t) but not in the direction normal to the slabs. Have you tried a similar "decomposition" in your simulations?

Professor Tobias answered: The interlamellar space between membranes in purple membrane stacks seems at first glance to share some similarities with the "simple geometries" you refer to, namely the space between parallel, essentially planar slabs. One might therefore expect similarities in the decomposition of the water mean-squared displacement (and corresponding diffusion constant). However, the "slabs" in our case are very rough and chemically heterogeneous on the length scale of a water molecule, and the protein protrusions from the approximately planar membrane matrix presents barriers to lateral diffusion (see Fig. 3 of our paper). Thus, it is indeed very interesting to compare and contrast water dynamics between planar slabs and rough membrane surfaces. We have not done a systematic analysis as a function of temperature and hydration, but we have performed the decomposition of the water mean-square displacement into in-plane and out-of-plane (normal to the membrane surface) components for the purple membrane system with 1924 water molecules at 296 K. The plot shown here (Fig. 3) reveals that the water in the interlamellar space of purple membranes displays anomalous diffusion on the 100 ps time scale (sub-linear temporal evolution of mean-squared displacement) in both the in-plane and out-of-plane directions. Moreover, the plot shows that the diffusion is somewhat anisotropic, because the in-plane component is more than 2/3 of the total.

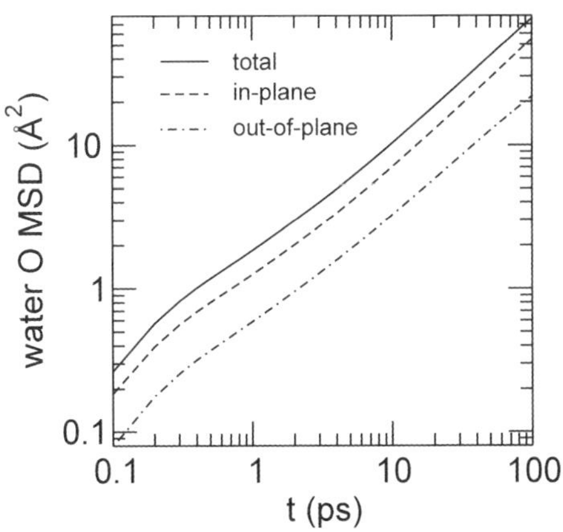

Fig. 3 Mean-squared displacements (MSD) computed from molecular dynamics trajectories of oxygen atoms of hydration water molecules between purple membrane stacks at 296 K. Solid: total MSD (three-dimensional). Dashed: components of the MSD in the plane of the membranes (two-dimensional). Dot-dashed: component of the MSD normal to the plane of the membranes.

Professor Debenedetti addressed Professor Tobias, Professor Halle and Dr Zaccai: In recent years the view has been advocated (by simulations from Stanley's group in Boston,[1] and in experiments by Chen at MIT[1]) that the so-called dynamical transition is associated with a change in water dynamics from super-Arrhenius at high temperature to Arrhenius at low temperature. Do you see evidence for this in your simulations/experiments?

1 P. Kumar, Z. Yan, L. Xu, M. G. Mazza, S. V. Buldyrev, S. H. Chen, S. Sastry and H. E. Stanley, *Phys. Rev. Lett.*, 2006, **97**, 177802.

Professor Halle responded: To make progress in this area, it is first of all necessary to distinguish between (1) bulk water, (2) water in "free" hydration layers of *e.g.* proteins in dilute aqueous solution, and (3) water molecules trapped/confined in (sub-)nanometer spaces in *e.g.* rehydrated solid protein powders or nanoporous silica matrices. The free hydration layer, being in contact with bulk water on one side, is expected to have properties intermediate between those of bulk and confined water. Some authors, including the ones you mention, argue that the physical mechanism underlying the anomalously strong temperature dependence of bulk-water dynamics (with an apparent divergence at 220–230 K) also operate in confined water. We would then reasonably expect this mechanism to prevail also in free hydration layers, where water dynamics differ much less from bulk water than in *e.g.* protein powders. We have recently studied water dynamics over a wide temperature range in the free hydration layers of several proteins (at millimolar concentration).[1] We find Arrhenius-like dynamics in the hydration layer down to at least 238 K (our limit of supercooling), while bulk water dynamics is highly super-Arrhenius in this temperature range. A similar behavior is found for the hydration shells of small organic solutes.[2] Going from bulk to confined water *via* "semiconfined" water (free hydration layer), the temperature dependence thus goes from super-Arrhenius to Arrhenius and back to super-Arrhenius. This nonmonotonic trend suggests that the underlying causes of the super-Arrhenius behavior in bulk and confined water are different. Deviations of the Arrhenius law can occur for many reasons, such as changes in the confining potential or structure. One should also bear in mind that low-temperature confined water is unfreezable rather than supercooled.

1 C. Mattea, J. Qvist and B. Halle, *Biophys. J.*, 2008, **95**, 2951.
2 J. Qvist and B. Halle, *J. Am. Chem. Soc.*, 2008, **130**, 10345.

Dr Zaccai replied: Our experimental work has shown the coincidence of protein and water "dynamical transitions" at about 200 K for the soluble maltose binding protein from *E. coli*.[1] We have to look more carefully at the data with respect to the Stanley and Chen results to which you refer.

1 K. Wood, A. Fröhlich, P. Paciaroni, M. Moulin, M. Härtlein, G. Zaccai, D. J. Tobias and M. Weik, *J. Am. Chem. Soc.*, 2008, **130**, 4586.

Professor Tobias replied: The behavior you refer to has been observed in a plot of the reciprocal of the diffusion constant of hydration water *vs.* the reciprocal of the temperature (see *e.g.* Fig. 3 in ref. 37 of our paper). The calculation of the diffusion constant from the MD trajectories was based on the Einstein relation. Given that the mean-square displacements of hydration water in our purple membrane systems displays a sublinear time-dependence (indicative of anomalous diffusion), at least up to the 100 ps over which we have calculated them, I am hesitant to extract a diffusion constant using the Einstein relation. Another dynamical quantity that could be expected to show the strong/fragile crossover is the relaxation time of the protein–water hydrogen-bonds (see Fig. 13 of our paper). The temperature dependence of this quantity (and the corresponding quantity for protein–lipid hydrogen bonds) reveals a change in behavior at ~240 K, which is close to the temperature of the

 This journal is © The Royal Society of Chemistry 2009

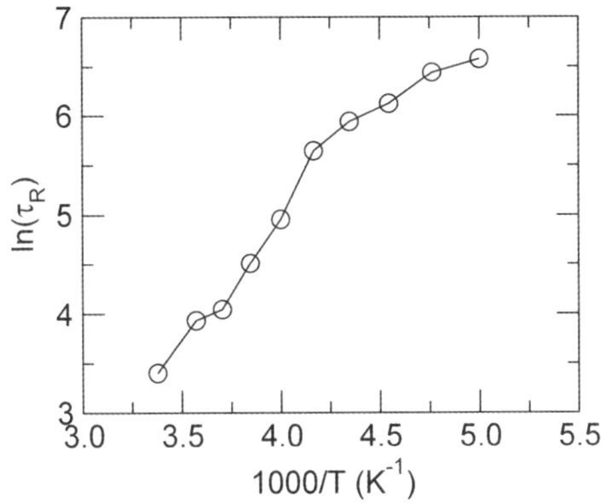

Fig. 4

dynamical transition in the purple membrane (protein and lipids). In Fig. 4 I have plotted the logarithm of the protein–water hydrogen-bond relaxation times *vs.* 1/T. There is a kink in the curve at $T = 240$ K, and the shape of the curve looks qualitatively similar to the diffusion constant data in ref. 37 of our paper, but I cannot confidently say whether or not the low T behavior is well-described by the Arrhenius law and the high T by a Vogel–Fulcher–Tamman form.

Dr Nutt said: Since bR (bacteriorhodopsin) is an asymmetric protein in that protons are pumped across the membrane in one specific direction, have you observed, or would you expect to see, some difference in the hydration properties at the intra- and extra-cellular sides of the membrane, which could be linked to protein function?

Professor Tobias replied: In addition to the protein asymmetry, there is an asymmetric distribution of lipids in the purple membrane. The simulations we report in our paper consisted of a single membrane with only a few layers of hydration that could be influenced by both sides of the membrane *via* periodic boundary conditions (mimicking a multilamellar sample), so it might be difficult to discern differences in water behavior on each side of the membrane that arise from asymmetry. It is certainly worth looking into, and I thank you for the suggestion.

Dr Zaccai asked: The C-terminus of bR is missing from the crystal structures because it is disordered—it is strongly negatively charged and could lead to different hydration behaviour on the two surfaces of the membranes. Did you include it in your model?

Professor Tobias answered: The all-atom purple membrane model that we used to initiate our simulations was kindly provided to us by Professor Emad Tajkhorshid at the University of Illinois. The model includes the complete bR sequence. Details on the construction of the model are reported in ref. 50 of our paper. Prompted by Dr Nutt's question and yours, we will investigate the possibility of differences in hydration dynamics on the two sides of the purple membrane.

Professor Havenith remarked: You stated that the hydration dynamics as probed by dynamical transitions is just determined by the topography of proteins in terms of caging and trapping them on chemical composition as seen in the simulations. We see from results of THz spectroscopic measurements that hydrophobic and hydrophilic amino acids influence the hydration dynamics in different ways. If the dynamical transition were seen experimentally to be independent of tertiary structure† would this be in contradiction to this statement?

Professor Tobias answered: I did not say that hydration water dynamics is only determined by the topography of proteins. I did say that the topography of the protein surface plays an important role, and that it is difficult to clearly dissect the role of the different types of functional groups (acidic, basic, neutral polar, hydrophobic) in determining water dynamics because most water molecules on the surfaces of proteins are interacting with more than one functional group. I also mentioned that there is a substantial body of literature that has explored possible correlations between protein surface chemistry and topology on hydration water dynamics and, in my impression, no clear conclusion has emerged except that both factors are important.

The existence of dynamical transitions in hydrated proteins seems to be universal, *i.e.* it occurs in soluble and membrane proteins with a wide variety of secondary and tertiary structures (it is also worth noting that it also occurs in nucleic acids). However, in detail, the temperature-dependence of the mean-squared displacements does depend to some degree on identity of the protein. Regarding the role of water dynamics in the transition, as discussed in our paper and references therein (ref. 17, 24 and 26), it appears that the transition is connected with the relaxation of the protein–water hydrogen-bond network which is, in turn, correlated with the onset of water translational diffusion. The details of the behavior of the protein–water hydrogen-bond network will certainly depend on the chemistry and topology of the protein surface.

Ms Mohana Sundaram communicated to Professor Tobias and Professor Havenith: There appears to be some disagreement as to whether the dynamics of hydration water around proteins is dictated primarily by the protein surface topology (according to Professor Tobias) or by the chemical composition of the protein surface (according to Professor Havenith). I would like to mention our recent observations[1] that neither surface topology nor chemical heterogeneity alone is sufficient enough to describe the water dynamics near protein surfaces. We showed that hydration waters reside for longer periods of time in regions of protein surface that are both chemically heterogeneous and topologically rough. Can you describe your observations from experimental or simulations leading to your assertions that water dynamics is dictated by either chemical heterogeneities or surface roughness?

1 M. Hamsa Priya, J. K. Shah, D. Asthagiri and M. E. Paulaitis, *Biophys. J.*, 2008, **95**, 2219.

Professor Havenith communicated in reply: Neither point of view, whether hydration dynamics is dictated by topology or chemical composition, should be taken to the extreme. I assume that both contribute and speculate that topology will have a larger effect on water molecules very close to the surface whereas surface composition, which affects the long range electrostatics, will have a larger effect on water molecules at greater distances.

Using THz spectroscopy we probe the fast (sub-ps) collective network oscillations over a longer range, beyond 10 Å for proteins, and 5–6 Å for carbohydrates.[1,2] This implies a change in the fast solvent motions over 4–5 hydration layers for proteins. Most of the 1000 water molecules which are shown to have a distinct water network motion from that of bulk water are not in the first hydration shell. Whereas the first hydration shell is dictated by sterical hindering and the formation of hydrogen bonds, surely this will not be the case for the third or fourth hydration shell, which will contribute most to the our signals.

† After the discussion a paper on this subject was published: Y. He, P. I. Ku, J. R. Knab, J. Y. Chen and A. G. Markelz, *Phys. Rev. Lett.*, 2008, **101**, 178103.

 This journal is © The Royal Society of Chemistry 2009

The effects are also the subject of recent modeling studies by us and by other groups. The group of Bandyopadhyay has carried out an atomistic molecular dynamics study of the solvated protein villin headpiece subdomain.[3] He could also deduce an influence on the water properties beyond the first hydration shell. Whereas in the first hydration shell they find distinct hydration dynamics depending on the hydrophobic or hydrophilic property of the specific segment, the heterogeneous nature of the hydration water diminishes as the thickness is increased. Thus, in the range where we preferentially probe the hydration dynamics, the hydration dynamics is mostly homogeneous and not dictated by topological roughness, but topology is surely of importance for the hydration dynamics in the first hydration layer.

1 U. Heugen, G. Schwaab, E. Bründermann, M. Heyden, X. Yu, D. M. Leitner and M. Havenith, *Proc. Natl. Acad. Sci. U. S. A.*, 2006, **103**, 12301.
2 S. Ebbinghaus, S. J. Kim, M. Heyden, X. Yu, U. Heugen, M. Gruebele, D. M. Leitner and M. Havenith, *Proc. Natl. Acad. Sci. U. S. A.*, 2007, **104**, 20749.
3 S. K. Sinha, S. Chakraborty and S. Bandyopadhyay, *J. Phys. Chem. B*, 2008, **112**, 8203.

Professor Tobias also communicated in response: I essentially agree with your statement that neither surface roughness nor chemical heterogeneity alone are sufficient to explain the changes in water dynamics near protein surfaces compared to in bulk. The issue of surface topography is important, but it is not the whole story. I would like to point out that, contrary to expectations, correlations of water dynamics with the types of functional groups (charged, polar neutral, hydrophobic) that they interact with on the surface of the protein are difficult to establish. A simulation paper that examines this issue closely is Luise *et al.*[1]

1 A. Luise, M. Falconi and A. Desideri, *Proteins: Struct., Funct., Genet.*, 2000, **39**, 56.

Ms Mohana Sundaram also communicated: In your paper, you show that the dynamical transition of proteins is coupled to the translation diffusion of water (Fig. 1). During your presentation, you referred to similar results for proteins in aqueous solutions containing different concentrations of glycerol. Does the dependence of this dynamical transition on glycerol concentration provide any insight into the mechanism of protein stabilization by glycerol? Further, does glycerol influence protein dynamics directly or indirectly through its impact on water dynamics? Have you also studied the effect of denaturants, such as urea, on this protein dynamical transition?

Professor Tobias communicated in reply: The focus of our paper and my presentation was on the differences between the protein–water dynamical coupling in the context of the so-called dynamical transition in soluble and membrane proteins. In my presentation, I included results of measurements of the dynamical transition of soluble proteins in the presence of a glycerol co-solvent to demonstrate the profound effect of changing the solvent environment of the protein on protein dynamics. The results I emphasized are that the temperature of the transition increases, and the amplitudes of protein motion at temperatures above the transition temperature are decreased, with the addition of glycerol. We have suggested (see ref. 24 and 26 of our paper), based on MD simulations, that the nature of protein–solvent coupling, and its role in the dynamical transition, are similar in water and in glycerol. The basic idea is that the dynamical transition is correlated with the relaxation of the protein–solvent hydrogen-bond network which is, in turn, associated with the onset of translational diffusion of the solvent. The connection of protein dynamics to protein stability is not very well established, as far as I am aware. The amplitude of motion is a reflection of the width or volume of a free energy basin, whereas the stability of a protein is a reflection of the relative free energy of the basins corresponding to folded and unfolded states. Although it is not obvious that a correlation between the solvent dependence of protein dynamics and protein stability should exist, such a correlation has been noted in a study of lysozyme in water and glycerol (see ref. 22 of our paper).

Our simulations of glycerol effects on protein-solvent coupling (ref. 26 of our paper) were carried out in neat glycerol, so they only provide insight into the effects of "direct" protein–glycerol interactions *vs.* protein–water interactions on protein dynamics. We have not investigated the effects of urea on protein dynamics, but it is an interesting proposition. If the hypothesis (ref. 22 of our paper) concerning the correlation between protein dynamics and stability is correct, one would expect increased motion (corresponding to decreased stability) at a given temperature above the dynamical transition in the presence of urea *vs.* in neat water.

Professor Madden communicated: Did you measure the distribution of rotational correlation times in this system?

Professor Tobias communicated in response: We have not yet analyzed the rotational dynamics in this system. In his paper, Professor Halle showed that it is possible to characterize the distribution of rotational correlation times in protein hydration water based on magnetic resonance relaxation measurements on soluble peptides and proteins. It therefore seems timely to compare distributions of rotational times from simulation and experiment. It will also be interesting to compare distributions for protein hydration water in solution to those for water confined between purple membrane stacks.

Professor Finney remarked: Professor Tobias commented (a) that there was no enzyme activity at temperatures below the dynamic transition and (b) that enzymes were also inactive below a certain critical hydration level.

Both these statements are commonly assumed to be correct. In particular, the assertion that enzyme activity requires a minimum water content tends to be based on the work of Rupley and colleagues on lysozyme which failed to find significant activity below a hydration level of 0.2 g water per g protein (see for example ref. 19 of Professor Tobias' paper).

More recent work has demonstrated that neither of these accepted assertions are correct. Enzyme activity has been found below the dynamical transition for a number of enzymes, for example glutamate dehydrogenase.[1] Furthermore, enzyme activity has been observed at very low hydrations indeed, with an implication that for certain enzymes, activity may be possible at essentially zero hydration (see Dunn and Daniel[2]). See also Daniel *et al.*[3] for a review of the experimental work on the relationship between dynamics and enzyme activity up to 2003, which shows that the enzyme dynamics–hydration–activity relationship is more complex than is implied by the often accepted conventional wisdom of statements (a) and (b) above. That review also discusses problems of identification of a dynamic transition that were also raised in this discussion by Professor Halle.

1 R. M. Daniel, J. C. Smith, M. Ferrand, S. Hery, R. V. Dunn and J. L. Finney, *Biophys. J.*, 1998, **75**, 2504.
2 R. V. Dunn and R. M. Daniel, *Philos. Trans. R. Soc. London, Ser. B*, 2004, **359**, 1309.
3 R. M. Daniel, R. V. Dunn, J. L. Finney and J. C. Smith, *Annu. Rev. Biophys. Biomol. Struct.*, 2003, **32**, 69.

Professor Tobias responded: I do not recall making the two statements that you attributed to me, and I doubt I would have made such strong statements because I am aware of the observations (and corresponding papers) you refer to in your comment. I do recall saying that the onset of aspects of the function of some proteins (*e.g.* myoglobin, bacteriorhodopsin) appears to be correlated with dynamical transitions that do not occur in dry samples. As we say in our paper, the connection (or lack thereof) between the fast (ps–ns) dynamics probed in most studies of dynamical

 This journal is © The Royal Society of Chemistry 2009

transitions remains to be established. From the few cases for which the temperature/solvent dependence of fast dynamics has been measured in conjunction with functional assays, a consensus has not emerged. I agree that, as you say, the dynamics–hydration–activity relationship is more complex than is implied by the conventional wisdom drawn from early studies.

Dr Ricci opened the discussion of the paper by Professor Bonn: I have really appreciated your discussion about the lineshape of the OH stretching band of water. Can you give more details about these "isotopic dilution" experiments, both under confinement and in the bulk? Is it possible to project an experiment in bulk water, aimed at evidencing the role of the Fermi resonance in determining the OH spectral lineshape?

Professor Bonn responded: Thank you. Many of the details of this experiment can be found in ref. 1. One measurement that would be useful to 'nail down' the Fermi resonance scenario directly is a pump OH/OD stretch, probe H_2O/D_2O bend. If the vibrational excitation leaves the stretch mode at the same rate that it enters the bend these states must be directly coupled. It is a 2D-IR experiment of this type, in fact, that has demonstrated the importance of the Fermi resonance in bulk liquid water.[2] Performing the equivalent experiment in our IR pump–SFG probe set up is technically demanding but an object of current focus in our group.

1 M. Sovago, R. K. Campen, G. W. H. Wurpel, M. Müller, H. J. Bakker and M. Bonn, *Phys. Rev. Lett.*, 2008, **100**, 173901.
2 S. Ashihara, N. Huse, A. Espagne, E. T. J. Nibbering and T. Elsaesser, *Chem. Phys. Lett.*, 2006, **424**, 66.

Professor Kim said: As we know, the origin of two peaks (or more) in the O–H stretching vibration region is not clearly understood yet; so your H_2O–D_2O mixture data are very interesting. Although it is known that the bulk liquid spectra of an H_2O and D_2O mixture can easily be deconvoluted with the pure H_2O and D_2O spectra, can the same approach be applicable to the interfacial water at the water/air interface? I wonder if mixing H_2O and D_2O may introduce a new degree of freedom in analysis of vibrational features of water at the interface. For example, it is known that protonated and deuterated polyethylene are not completely miscible although they are initially expected to be miscible. If the 3200 and 3400 cm^{-1} peaks are representing two different states (or two different distributions/populations of multiple states), it would be expected to see that in your pump–probe experiment, the recovery time of the depopulated vibration peak would be different whether you probe the system with 3200 or 3400 cm^{-1}, but your paper does not describe these issues. Have you done the frequency-dependent pump–probe experiments? If so, what are the results and implications?

Professor Bonn answered: There is a very small difference in H-bonding strength between the OH and OD groups of H_2O and D_2O but for bulk H_2O and D_2O; this does not have a significant effect on the properties. Most importantly from our point of view, the vibrational spectrum of D_2O is exactly equivalent to that of H_2O after a scaling for mass.

The limited miscibility of protonated and deuterated polyethylene is due to entropic contributions to the free energy, which become significant for very large molecules, but are not expected to play any significant role for water.

We have indeed performed pump–probe experiments as a function of frequency, and find that the vibrational lifetime depends strongly on frequency at the lipid/water interface (see ref. 1 and a forthcoming paper). This indicates that the absorption lines for water in contact with lipid layers are strongly inhomogeneously broadened, and that there exists a continuum of conformations that water molecules may adopt—not just two. Spectral hole-burning experiments will allow us to quantify the degree of inhomogeneity.

1 A. Ghosh, M. Smits, J. Bredenbeck and M. Bonn, *J. Am. Chem. Soc.*, 2007, **129**, 9608.

Professor Yarwood commented: There are at least two ways of interpreting the broad OH stretching bands shown in your Fig. 3 (apart from the well-known interference of the Fermi resonance with the bending mode).[1] One interpretation is that the band profile is broadened by very fast inhomogeneous processes (resulting from dephasing of the molecular vibrators on a very fast time scale).[2] The other possible interpretation is that the profile is made up of a number of subspectra[1,2] representing water 'species' which have lifetimes which are long compared with the IR timescale (these might be, for example, species with different strengths of hydrogen bonding). Such species have been examined both in the bulk[1] and in interfacial, *e.g.* polymer, systems[2] (as well as in many other systems).

So my question is whether a more detailed examination of the interactions and dynamics implied by your Fig. 3 might be useful for the interpretation of your pump–probe experiments?

1 F. O. Libnau, A. A. Christy and O. M. Kvalheim, *Appl. Spectrosc.*, 1995, **49**, 1431.
2 C. Sammon, C. Mura, J. Yarwood, N. Everall, R. Swart and D. Hodge, *J. Phys. Chem. B*, 1998, **102**, 3402.

Professor Bonn responded: We can imagine three different conceptual models for describing the double peaks present in the hydrogen-bonded OH and OD stretching spectra (of H_2O and D_2O respectively).

Firstly, in the fast, or homogenously broadened, limit any exchange between different subensembles of molecular vibrators is rapid relative to vibrational relaxation and dephasing occurs through elastic interactions of the oscillators with bath modes. This process will result only in motional narrowing: it would only lead to a 'smearing out' of two peaks not to their presence and thus cannot explain these well-known spectral features.

Secondly, in the slow, or inhomogenously broadened, limit any exchange between different subensembles of molecular vibrators is slow relative to vibrational relaxation and dephasing occurs as a result of different subensembles having different fundamental frequencies. This interpretation (*i.e.* sub-spectra corresponding to different H-bonding strengths) is equivalent to the 'ice-like' and 'liquid-like' interpretation suggested initially.[1]

Thirdly, points 1 and 2 can be applied to isolated oscillators. However, neither H_2O nor D_2O are isolated oscillators: OH or ODs on the same molecule may couple to each other as well as to OH/ODs on other molecules. In prior work we observed that the two peaks in the H_2O and D_2O spectrum merge into one peak upon isotopic substitution of the water molecules (*i.e.* going from H_2O to HDO, or from D_2O to HDO).[2] If differently H-bonded substates did exist for H_2O and D_2O, these should also be present for HDO. The observation that only one peak remains for HDO thus suggests that the double-peaked spectra present in the H_2O and D_2O spectra are likely the result of intramolecular coupling. Note, though, that this argument only suggests that the double-peaked structure is not the result of inhomogenous broadening. Other frequency-dependent pump–probe data,[3] suggests strongly that the absorption lines for water in contact with lipid layers are strongly inhomogeneously broadened, as the vibrational lifetimes depends strongly on frequency. Spectral hole-burning experiments will allow us to quantify the degree of inhomogeneity.

1 Q. Du, R. Superfine, E. Freysz and Y. R. Shen, *Phys. Rev. Lett.*, 1993, **70**, 2313.
2 M. Sovago, R. K. Campen, G. W. H. Wurpel, M. Müller, H. J. Bakker and M. Bonn, *Phys. Rev. Lett.*, 2008, **100**, 173901.
3 A. Ghosh, M. Smits, J. Bredenbeck and M. Bonn, *J. Am. Chem. Soc.*, 2007, **129**, 9608.

Professor Buch noted: Professor Shen does not seem to be convinced lately that "ice-like" is a good description of a low-frequency peak.

 This journal is © The Royal Society of Chemistry 2009

In addition, the low-frequency SFG peak of neat water seems to be an analogue of a Raman peak of bulk water and ice, contributed by collective in-phase OH stretch vibrations of strongly H-bonded OH, coupled intermolecularly by dipole–dipole coupling.[1]

1 V. Buch, T. Tarbuck, G. L. Richmond, H. Groenzin, I. Li and M. J. Schultz, *J. Chem. Phys.*, 2007, **127**, 204710.

Professor Bonn responded: Firstly, the argument that we have made in our prior work does not rest on the particular notion of whether the two peaks in OH or OD stretch (of H_2O or D_2O) are in fact "ice-like" or "liquid water-like" and, if so, what those words would mean in that context. Rather, our conclusion, based on the collapsing of the two peaks to one on isotopic dilution and both peaks having the same symmetry, was that the notion of these peaks representing different types of quasi-static water structure, whatever these were called, was highly unlikely.[1] While Professor Shen may no longer be comfortable using the term "ice-like" to describe interfacial water structure, publications from his group within the past couple of years do assign different features in the interfacial water spectrum to different, quasi-static water structure types.[2–4]

Secondly, our prior isotopic dilution measurements do not rule out the possibility that the "3200 cm^{-1} peak" apparent in the OH stretch spectrum is the result of a delocalization of the OH excitation over multiple water molecules. It is important to note, however, that in this interpretation of the OH stretch SFG spectrum the difference between the 3200 and 3400 cm^{-1} peaks is not that each is the result of different, quasi-static types of water, but rather that in the former case the excitation is delocalized while in the later it is local. Current work in our group is focused on both understanding how different aspects of the existing static SFG data for interfacial water fit into a possible delocalization model (*e.g.* the identical symmetry of the double peaks and the different relationship between the central frequency of the HOD peak and the observed doublet in the D_2O (H_2O) spectrum for water at different interfaces) and on performing multidimensional SFG experiments that can more directly address this issue.

1 M. Sovago, R. K. Campen, G. W. H. Wurpel, M. Müller, H. J. Bakker and M. Bonn, *Phys. Rev. Lett.*, 2008, **100**, 173901.
2 Y. R. Shen and V. Ostroverkhov, *Chem. Rev.*, 2006, **106**, 1140.
3 N. Ji, V. Ostroverkhov, C. Y. Chen and Y. R. Shen, *J. Am. Chem. Soc.*, 2007, **129**, 10056.
4 N. Ji, V. Ostroverkhov, C. S. Tian and Y. R. Shen, *Phys. Rev. Lett.*, 2008, **100**, 096102.

Dr Tyrode communicated: Though not discussed in the paper, in your presentation you showed how the spectral features of the air/D_2O interface vary upon isotopic dilution with H_2O. The two bonded OD stretches, clearly defined in the pure D_2O spectrum, collapse into a single band at a 1 : 2 molar D_2O : H_2O ratio.[1] This led you to the conclusion that the simplistic assumption frequently employed in the literature, assigning those two bonded OH (or OD) bands to "ice-like" and "liquid-like" environments is not appropriate. Other researchers have previously carried out similar isotopic dilution experiments[2,3] in the OH stretching region, and have not observed the obvious collapse that you have showed in OD region. This is in spite of their significant better signal-to-noise ratio and overall bandwidth (less than 10 cm^{-1} compared to ~50 cm^{-1} in your paper). How can you explain this difference? Could it be due to a stronger coupling between the bending overtone and the stretching modes in the OD region when compared to the OH region? How much of the observed effect can be attributed to experimental errors?

1 M. Sovago, R. K. Campen, G. W. H. Wurpel, M. Müller, H. J. Bakker and M. Bonn, *Phys. Rev. Lett.*, 2008, **100**, 173901.
2 T. L. Tarbuck, S. T. Ota and G. L. Richmond, *J. Am. Chem. Soc.*, 2006, **128**, 14519.
3 E. A. Raymond, T. L. Tarbuck and G. L. Richmond, *J. Phys. Chem. B*, 2002, **106**, 2817.

Professor Bonn communicated in reply: We were indeed aware of these references when we were analyzing our earlier work. On inspection of the data and fit in Fig. 1 and 2 of the 2002 paper it seems clear that Raymond, Tarbuck and Richmond do, in fact, see the collapse of two peaks to one peaks that we have observed. In general terms, these authors justify their analysis of the vapor/HOD interface by noting that "isotopic dilution studies such as these are ideally suited for deconvolving complicated H_2O spectral features due to the energetically uncoupled nature of the OH and OD stretching modes in HOD". We have argued that the nature of the decoupling is not what they proposed: rather the peak collapse on isotopic dilution implies the two peaks are likely the result of intramolecular coupling. It is worth reemphasizing that our interpretation is fully consistent with both peaks having the same symmetry: a result unavailable to Raymond *et al.* in 2002.

While the data are less clearly plotted, the situation is much the same in the 2006 paper by Tarbuck, Ota and Raymond. Here in their Fig. 2d, the H-bonded portion of the OH in HOD can be virtually completely fitted with a single resonance. As for their 2002 results, however, we believe the observation of this peak collapse, in conjunction with the two peaks in either the H_2O or D_2O spectrum having the same symmetry, is most easily explained by the two peaks being the result of Fermi resonance with the bend overtone.

Quantitative descriptions of the coupling strength between the bend overtone and the stretch in either H_2O or D_2O are non-trivial. While important to determine in the long term, such an effort is not necessary for our point. We only wish to note that in both the interfacial D_2O and interfacial H_2O spectrum (for water abutting a variety of other phases) a double-peaked feature is observed to collapse to a single peak on conversion of H_2O (or D_2O) to HOD. This fact, along with the identical symmetry both peaks are observed to have in either of the two isotopically pure phases, suggests strongly that the double-peaked feature is the result of intramolecular coupling: most probably a Fermi resonance with the bend overtone.

While the details of the quality of the experimental data are less clear to us than Dr Tyrode, for example it seems clear that our signal-to-noise ratio (see Fig. 2 in our paper) is better than that in Fig. 2 of the 2002 paper by Raymond *et al.*, these differences are not germane. In our reading both the 2002 paper and the 2006 paper by Tarbuck *et al.* see the same spectral evolution we have observed on isotopic dilution. Based on other data we have simply proposed a different interpretation of this observation.

Professor Uosaki asked: I have three questions. Firstly, have you measured the SFG spectra by changing the H/D ratio?

Secondly, you showed the spectra in the OD stretching region. What about those in the OH stretching region?

Thirdly, can you explain the significant change in the spectra by surface modification of quartz by octadecyltrichlorosilane (OTS), namely the almost equal intensity of two ("ice-like" and "liquid-like") peaks at bare quartz and the "ice-like" peak-dominated spectra at the OTS modified quartz.[1]

1 S. Ye, S. Nihonyanagi and K. Uosaki, *Phys. Chem. Chem. Phys.*, 2001, **3**(16), 3463.

Professor Bonn responded: Firstly, we have indeed measured the OD spectrum at a variety of isotopic dilutions. Amplitudes from the fits to these spectra appear in our prior work on this topic[1] and are consistent with our interpretation.

Secondly, as alluded to in this paper and discussed in our prior work we did not, for technical reasons, take static spectra of the OH stretch spectral region. We believe this is justified both because the static SFG spectrum of the OD stretch in pure D_2O is precisely the same as that of the static SFG spectrum of the OH stretch in pure H_2O (after scaling by the mass difference) and because, as I mentioned in another comment, such spectra have been taken by others,[2,3] who observed that the double-peaked feature in the OH stretch also collapses to a single peak on isotopic dilution.

Thirdly, the central frequency of the OH stretch is known, from a variety of prior studies in both solids and liquids, to shift as a function of hydrogen bond strength (see, for example, ref. 4). In contrast, the central frequency of the H_2O bend mode is known to be relatively insensitive to hydrogen bonding strength (for example, upon weakening the H-bonds by increasing the temperature, the OH stretch frequency is known to change ~1 cm^{-1} K^{-1} while the bend changes only ~0.05 cm^{-1} K^{-1} [5,6]). If the Fermi resonance model of the bonded OH stretch spectral region is correct one would expect that the position of the 'dip' between the two peaks should be roughly the same (this is determined by the frequency of the bending mode) but that the position of the underlying OH stretch mode should shift as hydrogen bond strength changes (thus changing the relative intensity of the "ice-like" and "liquid-like" peaks). This mechanism can explain both the observation alluded to above as well as others of the change in the OH stretch spectrum with temperature or with the addition of acids, bases or salts to the aqueous phase near various solids.

1 M. Sovago, R. K. Campen, G. W. H. Wurpel, M. Müller, H. J. Bakker and M. Bonn, *Phys. Rev. Lett.*, 2008, **100**, 173901.
2 E. A. Raymond, T. L. Tarbuck and G. L. Richmond, *J. Phys. Chem. B*, 2002, **106**, 2817.
3 T. L. Tarbuck, S. T. Ota and G. L. Richmond, *J. Am. Chem. Soc.*, 2006, **128**, 14519.
4 T. Steiner, *Angew. Chem., Int. Ed.*, 2002, **41**, 48.
5 M. Falk, *J. Raman Spectrosc.*, 1990, **21**, 563.
6 A. J. Lock and H. J. Bakker, *J. Chem. Phys.*, 2002, **117**, 1708.

Professor Tobias said: In Fig. 3 of your paper you report sum frequency spectra that were measured using the ssp polarization configuration. Have you tried different polarizations that might provide additional insight into the orientation of water molecules at the lipid/water interface?

Professor Bonn responded: We reported both ssp and pss spectra of the OD stretch in prior work[1] and have, since that time, measured such spectra below lipid monolayers of varying types. We have made little effort, however, to use this information to infer water orientation for two reasons. Firstly, at least in the case of the air/H_2O interface, this has been previously done by others both using vibrationally resonant SFG and nonresonant second harmonic generation.[2,3] Second, and perhaps more importantly, because there are few independent constraints on the density of interfacial water molecules, at either the air/water interface or the substantially more complicated lipid/water interface, inferring molecular orientation from the polarization dependence of SFG intensity requires the assumption of a particular angular distribution (generally a δ function). Both in published and forthcoming work we have observed substantial frequency dependence of the vibrational relaxation of interfacial water at the lipid/water interface where relaxation pathways appear qualitatively different with changes in lipid type.[4–6] These observations strongly suggest the importance of inhomogeneous broadening mechanisms in determining the OH stretch lineshape. If these quasi-static differences in OH stretch frequency reflect different hydrogen bonding environments, it seems likely that assuming a δ function orientation of the ensemble of water molecules in the laser spot may be misleading.

1 M. Sovago, R. K. Campen, G. W. H. Wurpel, M. Müller, H. J. Bakker and M. Bonn, *Phys. Rev. Lett.*, 2008, **100**, 173901.
2 W. Gan, D. Wu, Z. Zhang, R. Feng and H.-f. Wang, *J. Chem. Phys.*, 2006, **124**, 114705.
3 W. Zhang, D. Zheng, Y. Xu, H. Bian, Y. Guo and H.-f. Wang, *J. Chem. Phys.*, 2005, **123**, 224713.
4 A. Ghosh, M. Smits, J. Bredenbeck and M. Bonn, *J. Am. Chem. Soc.*, 2007, **129**, 9608.
5 A. Ghosh, R. K. Campen, M. Sovago and M. Bonn, *Faraday Discuss.*, 2009, DOI: 10.1039/b805858j.
6 A. Ghosh, R. K. Campen, M. Sovago and M. Bonn, 2008, in preparation.

Professor Lyashchenko opened the discussion of the paper by Professor Havenith: My first question and comment are about the comparison of the data of dielectric spectroscopy in all orientational frequency ranges (up to 800 cm^{-1} in the case of water[1]) and the THz spectroscopy data. Is it possible to compare them? It would be an additional piece of information. In the case of dielectric spectroscopy we have the theoretical analytical models for the analysis of elementary processes. Is that so in your case?

Secondly, about long distance hydration effects: are they connected with hydrophobic or hydrophilic parts of hydration? In our experimental data[1,2] from dielectric spectroscopy of small molecules in water (DMSO) the long distance effect is connected only with hydrophobic part of polar molecules. The hydrophilic part is constant and its realized only in the first hydration sphere.

1 A. Lyashchenko and T. Novskova, *J. Mol. Liq.*, 2006, **125**, 130.
2 Z. Klemenkova, T. Novskova and A. Lyashchenko, *Russ. J. Phys. Chem. A*, 2008, **82**, 571.

Professor Havenith answered: Dielectric spectroscopy (in the GHz range) and THz spectroscopy (here: 2.7 THz) access different time scales of water motions. However, if dielectric spectroscopy is extended experimentally towards the high frequency limit and THz spectroscopy towards the low frequency limit, we find of course a smooth transition between the results of these experimental techniques. Extrapolation based upon on an analytical model always carries the risk of missing unexpected modes of the protein–water mixture in the upper frequency range. However, molecular dynamics has evolved in the meantime, so that we can now deduce a complete microscopic (instead of macroscopic) description of the dynamics based on vibrational, librational, rotational and diffusional motions of the protein and solvent. By Fourier-transforming the auto-correlation function of the derivative of the dipole function and including a harmonic quantum correction factor, the spectrum is then directly be calculated as:

$$\alpha(\omega) = \frac{1}{\omega^2}\frac{4\pi^2\omega\left(1 - e^{-\frac{\hbar\omega}{k_B T}}\right)}{3\hbar nc}\int\left\langle\frac{\partial M}{\partial t}_{t=0}\frac{\partial M}{\partial t}_{t=\tau}\right\rangle e^{i\omega t}dt$$

where $\hbar$ is the reduced Planck's constant, k_B is the Boltzmann factor, T the system's temperature, n the index of refraction, c the speed of light, and M the dipole moment of the system. The frequency dependence of the index of refraction will be neglected. The time derivative of the dipole moment can be expressed as:

$$\frac{\partial M}{\partial t} = \sum_i^n q_i \frac{\partial \boldsymbol{x}_i}{\partial t} = \sum_i^n q_i \boldsymbol{v}_i$$

where q_i, $\boldsymbol{x}_i$ and $\boldsymbol{v}_i$ are the partial charge, coordinates and velocity of atom i respectively. The collective motions of the water network are clearly influenced far beyond the first hydration shell. When probing at lower frequencies (GHz) the fast sub-ps motions which are probed by THz spectroscopy will be averaged over a longer time. Any fast exchange between hydration water and bulk water will thus lead to an overall averaged result, which then lacks the distinction which can be made when probing faster motions.

Professor Halle remarked: On the sub-100 fs time scale probed by your THz absorption measurements, water dynamics are generally believed to be governed by the local intermolecular structure.[1,2] Hence, it is difficult to understand how the THz dynamics of ~5000 water molecules (corresponding to a "dynamical hydration shell" of 18 Å thickness) can be significantly altered by the removal of two methyl groups from the deeply buried hydrophobic core of ubiquitin (the Val26Ala mutant *versus* pseudo-WT in your Fig. 4). Moreover, the proposed correlation between "dynamical hydration" and protein flexibility does not appear to be supported by your data since the "anomalous" THz absorption excess is (i) greatly reduced for a mutant (Ub-AI in Fig. 4) that is not more flexible than pseudo-WT[3] and (ii) is abolished by reducing pH from 5 to 2 (your Fig. 7) even though the protein is in the native state at both pH values.[4]

These considerations suggest that one should look for alternative interpretations of the THz absorption excess, such as protein self-association or differential heating effects. Self-association would be promoted by your use of a magnesium acetate buffer in this and previous studies[5] of protein hydration (>30 Mg^{2+} ions per protein at the THz absorption maximum in Fig. 4), whereas the increased net charge (from +3 to +13) of ubiquitin at pH 2 (Fig. 7) should prevent self-association. Did you measure the THz absorption coefficient as a function of ubiquitin concentration in the absence of buffer?

1 B. Guillot, *J. Chem. Phys.*, 1991, **95**, 1543.
2 C. J. Fecko, J. J. Loparo, S. T. Roberts and A. Tokmakoff, *J. Chem. Phys.*, 2005, **122**, 054506.
3 J. Ervin, E. Larios, S. Osváth, K. Schulten and M. Gruebele, *Biophys. J.*, 2002, **83**, 473.
4 M. M. Harding, D. H. Williams and D. N. Woolfson, *Biochemistry*, 1991, **30**, 3120.
5 S. Ebbinghaus, S. J. Kim, M. Heyden, X. Yu, U. Heugen, M. Gruebele, D. M. Leitner and M. Havenith, *Proc. Natl. Acad. Sci. U. S. A.*, 2007, **104**, 20749.

Professor Havenith answered: THz radiation does not probe the fs time scale only. Our experiments at 2.5 THz extend to 100s of fs. This is the time scale of librational and diffusion motions within the water network as has been shown based upon MD simulations in my talk.

To state that at pH 2 ubiquitin is in the native state is not true. The protein is found to be partially unfolded as can be proven by CD spectroscopy. Note that in Ervin *et al.* (your ref. 3), the data was taken at pH 5.9, not pH 2; a reference for non-native ubiquitin at pH 2 is Stockman *et al.*[1] We have used acid denaturation and not methanol denaturation (as in Harding *et al.*, your ref. 4) in our experiment. However, in both cases the CD spectra change considerably, proving considerable denaturation of the native secondary structure. The corresponding CD spectra for a λ-repressor are displayed in the supplementary information of our previous paper[2] and show a clear change in the CD spectra between pH 5.0 and pH 2.0, again due to partial unfolding.

The MD data in your ref. 3 showed that the VV mutation increases flexibility of the Trp more than the AI mutation because it is adjacent to the tryptophan. However, the MD simulation showed the same increase in average flexibility for both mutants because of truncation of Val26 to Ala as truncation of Ile61 to Val (unpublished results, Larios *et al.*). One has to keep in mind that Trp fluorescence and also NMR are local probes, whereas THz provides a method to probe global changes.

In Ervin *et al.*, some mutations investigated by molecular dynamics cause relatively small changes in flexibility of a nearby tryptophan at low temperature (*e.g.* their Fig. 6, Ile61Val mutant), but large changes at higher temperature (*e.g.* their Fig. 8, Ile61Val mutant). In their Fig. 7, the Val26Ala mutation greatly enhances the flexibility induced by the Ile61Val mutation in Fig. 6, showing that a truncation of two methyl groups can have a large effect.

When doing site-directed mutagenesis, such a truncation in some cases even causes these proteins to unfold (Gruebele and coworkers, unpublished results), although those unstable mutants are of course rarely published in the literature because they failed to fold. Thus it is not surprising that even small truncations can affect protein flexibility quite dramatically. The typical stability of proteins is only 10s of kJ mol^{-1} (equivalent to a few H-bonds) and even single point mutants could cause drastic changes in flexibility, since they unfold a protein entirely in some cases. We must remember how marginally stable proteins are to begin with.

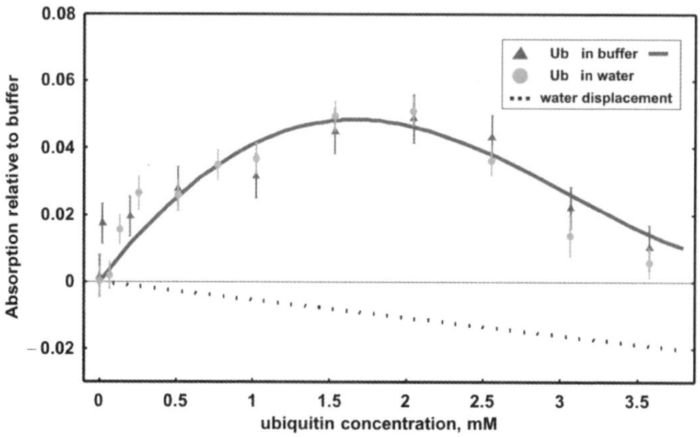

Fig. 5 Measured THz absorption of a protein–buffer and a protein–water mixture as a function of ubiquitin concentration. Both show a pronounced THz excess at 2 mM.

The probe is temperature controlled (within ±0.1 K), we use a broad beam waist and small energies; therefore differential heating can be excluded. You made the statement that the observed THz signal is significantly influenced by the added buffer, 50 mM magnesium acetate, which was suspected by you to cause aggregation. We have repeated the measurement using proteins solvated by water. As can be seen in Fig. 5 here, we find no difference between the protein solvated in buffer and the protein solvated in water. All data were reproducible within less than 1% over several month and with different spectrometers. Thus there is no doubt that the observed change in THz absorption has to be attributed to fast (sub-ps) long range hydration dynamics caused solely by the protein. In fact, the data below show that the effect of the macromolecules dominates completely over any small effects salts may have.

1 B. J. Stockman, A. Euvrard and T. A. Scahill, *J. Biomol. NMR*, 1993, **3**, 285.
2 S. Ebbinghaus, S. J. Kim, M. Heyden, X. Yu, M. Gruebele, D. M. Leitner and M. Havenith, *J. Am. Chem. Soc.*, 2008, **130**, 2374.

Professor Tobias asked: One of the conclusions of your paper is that hydrophobic and hydrophilic sidechains have opposite effects on the terahertz absorption. Specifically, you claim that hydrophobic sidechains decrease the absorption due to hydration water while hydrophilic sidechains increase it. This conclusion is based in part on your Fig. 7, which shows decreased absorption of denatured ubiquitin relative to native ubiquitin, and the assumption that hydrophobic and hydrophilic sidechains have similar solvent exposure in the unfolded state. This assumption seems to imply that the denatured state is fully unfolded. However, there are a number of experimental studies that have demonstrated the existence of substantial residual structure in the unfolded state, even under strongly denaturing conditions (see, for example, Plaxco and Gross,[1] and references therein). Thus, additional measurements on proteins of known structure with varying amounts of solvent exposed hydrophobic/hydrophilic surface are in order before a definitive conclusion can be reached concerning the potentially different effects of hydrophobic and hydrophilic groups on the hydration water dynamics probed by terahertz spectroscopy. Molecular dynamics simulations of proteins have suggested that the topography of the surface has a stronger effect on hydration water dynamics than does the chemical identity of solvent exposed groups (see, for example, Luise *et al.*[2]).

1 K. W. Plaxco and M. Gross, *Nat. Struct. Biol.*, 2001, **8**, 659.
2 A. Luise, M. Falconi and A. Desideri, *Proteins: Struct., Funct., Genet.*, 2000, **39**, 56.

 This journal is © The Royal Society of Chemistry 2009

Professor Havenith answered: I agree with the argument that based on the small number of measurements so far we are not able to come to a conclusive answer on this topic. However, our argument does not rely on the assumption that hydrophobic and hydrophilic sidechains have similar solvent exposure in the unfolded states, nor on the idea that the denatured state is a fully unfolded random coil. The argument relies only on the notion that the unfolded state has relatively more hydrophobic exposure than the native state. This has been proved by ANS binding experiments over and over again, and we performed such experiments for ubiquitin also, showing that hydrophobic surface exposure is increased in the unfolded state. Thus we have clearly established a correlation between the degree of hydrophobic exposure and decrease of absorbance in the spectral range we measured. Furthermore, we have also shown a complete absence of a maximum in the THz absorption for a denatured λ-repressor at pH 2, which is more flexible than the native structure.[1] ANS binding again indicated that this reduction is correlated with the increased exposure of more hydrophobic sites which affect the water in its hydration shell. That said, a correlation for two proteins is of course not proof of a causal relationship in general at this stage. Thus we have stated in our paper that "increased surface hydrophobicity of the mutants is a candidate for changes in hydration water structure that leads to the smaller bulk water-like THz absorption, as compared to the protein ubiquitin and its Ub* mutants". There are, however, two more lines of inquiry that support our proposed correlation. Recent MD simulations proposed that the solvent exposed hydrophobic sidechains induce a negative THz absorption coefficient in their surrounding, whereas hydrophilic parts lead to an increase in the hydration water compared to bulk water.[2] Very recent systematic studies by us (which are not yet published), where we measured the THz absorption coefficient of the hydration waters for several amino acids, showed a systematic trend with an increase in THz absorption for hydrophilic amino acid and a decrease for hydrophobic amino acids. This further supports the idea that chemical identity of solvent exposed groups is important for the fast long range hydration dynamics as probed by THz absorption spectroscopy.

1 S. Ebbinghaus, S. J. Kim, M. Heyden, X. Yu, U. Heugen, M. Gruebele, D. M. Leitner and M. Havenith, *Proc. Natl. Acad. Sci.*, 2007, **104**, 20749.
2 A. A. Golosov and M. J. Karplus, *J. Phys. Chem. B*, 2007, **111**, 1482.

Professor Klein continued the discussion of the paper by Professor Bonn: Is it known why the hydration level of phosphorylcholine groups (as at lipid headgroups) is so high, typically 15–25 water molecules in the primary hydration shell (depending on the method used)?

Professor Bonn replied: The hydration level of phosphatidylcholine groups, as in for example DPPC, is known to be high relative to that of phosphoethanolamine in DPPE for example, because the phosphatidylcholine group is both relatively bulky and is not a hydrogen bond donor (so does not interact strongly with either other lipid head groups or water). The combination of being both bulky and relatively poorly packed allows significant numbers of relatively weak interactions with water molecules. These issues have been addressed in a variety of prior studies by other authors.[1]

1 See *e.g.* S. Leekumjorn and A. K. Sum, *Biophys. J.*, 2006, **90**, 3951.

Dr Henchman asked: About the interpretation of the double peak of the OH stretch frequencies at the interface, a single water molecule has two OH stretches that may be coupled as they are in the gas phase. At the interface, it may not be possible for both OH groups to form strong hydrogen bonds simultaneously and could this therefore be responsible for the splitting?

Professor Bonn responded: If both OH(OD) groups in $H_2O(D_2O)$ do not form strong hydrogen bonds one might expect that the two groups would couple with each other resulting in the symmetric/asymmetric splitting seen in the gas phase. As I described in more detail in my earlier response to Professor Kim's comment, the interpretation of the double-peaked feature present in the OH(OD) SFG spectrum of interfacial $H_2O(D_2O)$ as the result of symmetric/asymmetric splitting is not consistent with our observation of the symmetry of OD in D_2O or prior work on OH in H_2O. If only one of the OH(OD) groups forms a strong hydrogen bond while the other does not one might expect that the absorption line associated with each mode should have different symmetry. This effect is observed at the air/water interface when comparing the symmetry of the higher frequency, free OH peak, with that of the two peaks visible in the hydrogen bonded region.[1] Unfortunately the two peaks in the hydrogen bonded portion of the spectrum (our area of focus in the current contribution and the only OH (or OD) stretching modes observable at the lipid/water interface) do have the same symmetry, an observation most easily rationalized if the hydrogen bonding environments of the local oscillators are not drastically different.

1 W. Gan, D. Wu *et al.*, *J. Chem. Phys.*, 2006, **124**, 114705.

Dr Bakó continued the discussion of the paper by Professor Havenith: If I understand properly, you explain your results as a significant change in the long range dynamical structure around protein in the range 40–120 cm^{-1}! In this range you have different local mode too (libration, wagging, frustrated translation). Can you explain your results as a change in the first shell of water around the protein?

Professor Havenith replied: No, it is not possible to explain the experimental observations as a change in the first hydration shell. We observe here a clear deviation from a nonlinear concentration dependence of the absorption even for very low concentrations. For each concentration we can calculate the average separation between two protein surfaces at that given concentration. A concentration of 4 mM corresponds to average distances of more than 20 Å between the protein surfaces. If only the first hydration shell would be affected we would expect a linear change in the total absorption up to a concentration where the first hydration shells start to overlap. Assuming that only the first hydration shell is affected we would expect a linear behaviour over the entire concentration range. This is in clearly in disagreement with our experimental result! The only way out of this would be some 'action at a distance' where the first shell is highly perturbed as the 5th shells come into contact, but the 5th shell is somehow not affected, which seems like a contradiction in itself.

Dr Patel commented: Professor Kim invoked the relevance of the finding that the hydrogenous and deuterated polyethylene phases separate, with respect to the current work that employs H_2O and D_2O. I would like to point out that in polymeric systems entropic factors are inversely proportional to the molecular weight. This means that small enthalpic differences between deuterated and hydrogenous polymers can have a significant impact on the free energy differences between them. This argument, however, does not apply in the case of small molecules like water.

Professor Kim answered: You are right about the polymers. I just used the $-(CH_2)_n-$ and $-(CD_2)_n-$ system as an example where almost identical molecules with the only difference is H/D do not see each other identically. In the case of H_2O/D_2O at the interface, I am questioning if the structure or hydrogen bonding dynamics at the interface will not be influenced by the presence of molecules that have a different symmetry (HOD) than their pure counter parts (H_2O and D_2O).

 This journal is © The Royal Society of Chemistry 2009

In several talks presented in this conference, we see many examples in which H_2O at the surface behaves differently from H_2O in the bulk. In this situation, can we assume that the H_2O–D_2O mixture effect at the interface will be the same as in the bulk and interpret the SFG spectra using the knowledge of the bulk behavior? I do not know the answer.

Professor Kim then continued the discussion of the paper by Professor Bonn: Based on your data, it appears that the doublet features at 3200 cm^{-1} (often called "ice-like" or "solid-like" water) and 3400 cm^{-1} (called "liquid-like" or "disordered" water) are observed only when pure H_2O or pure D_2O are probed with SFG; but it is not observed for the case of a mixture of H_2O and D_2O which contains HOD molecules *via* H–D exchange reactions. In that case, the HOD molecule is not in the C_{2v} symmetry. Could this be the reason that the H_2O–D_2O mixture shows only a single broad peak, instead of two peaks at 3200 and 3400 cm^{-1}? The tetrahedral arrangement of HOD (and its mixtures with H_2O and D_2O) would be different from the tetrahedral arrangement of pure H_2O or pure D_2O. I wonder if there is any possibility that this mixture effect at the interface might be different from that in the bulk.

Professor Bonn replied: The isotope-induced symmetry breaking is a very good point, and what we, in fact, initially considered to be the correct explanation of our data: that the 3200 and 3400 cm^{-1} peaks would be caused by the symmetric and antisymmetric stretch vibrations of water.[1] However, this explanation is inconsistent with several observations, the most important one being that SFG experiments at different polarization combinations of the IR, VIS and SFG light show that the two peaks have the same symmetry.[2]

More generally, there has been no evidence that the behavior of (partially) isotopically substituted water is significantly different at the interface than in the bulk.

1 M. Sovago, R. K. Campen *et al.*, *Phys. Rev. Lett.*, 2008, **100**, 173901.
2 W. Gan, D. Wu *et al.*, *J. Chem. Phys.*, 2006, **124**, 114705.

Professor Wu asked: I would like to ask a question about spectroscopic intensities—which kind of water molecules have you observed in your SFG spectrum? For my studies$_{1,2}$ we found that the Raman intensity of water becomes too strong when the water species interact with the polarised ground. We know the SFG signal is a product of Raman and infrared intensities.

1 D. Y. Wu, S. Duan, X. M. Liu, Y. C. Xu, Y. X. Jiang, B. Ren, X. Xu, S. H. Lin and Z. Q. Tian, *J. Phys. Chem. A*, 2008, **112**, 1313.
2 Y. X. Jiang and J. F. Li, *Chem. Commun.*, 2007, 4608.

Professor Bonn answered: As we noted in this contribution, and many others have noted before, the amplitude of the OH stretch of interfacial water in the intensity SFG spectrum is dramatically larger for water at charged interfaces. This increase in intensity is usually attributed to, at least, two effects: interfacial charge leads to an increase in ordering of interfacial water (*i.e.* increases the interfacial thickness) and to a $\chi(3)$ enhancement of the $\chi(2)$ signal. It is challenging to discriminate between the two effects experimentally.

Professor Bain asked: The slow relaxation time, T_{34}, that you measure in your experiments is associated with the thermal equilibration of the water structure on the timescale of tens of picoseconds after the pump pulse. In his paper, Professor Tobias posed the question of whether water dynamics drives the melting of the hydrocarbon chains in lipids or whether melting of the chains drives the changes in water dynamics. Perhaps I could ask the same question here? Do the lipid chains melt in your experiments and, if so, does the relaxation of the interfacial water occur on a faster or slower timescale?

Professor Bonn responded: This is indeed a very important issue. We know from previous experiments[1] that energy equilibrates very rapidly within a lipid monolayer: heat can be transported from the water phase to the lipid terminal methyl groups on a timescale of ~1 ps. So the molecular lipid response is not what determines T_{34}. One would expect chain melting to occur on a similar timescale, *i.e.* appreciably faster than the T_{34} times reported here. Moreover, in our experience, both in the current paper and previous work,[2] the water dynamics are also very fast, occurring on sub-picosecond timescales, so it is also not the response of the water. We think that the T_{34} times correspond to the timescale of exchange of membrane-bound water and bulk water. This exchange is necessary to achieve the new thermal equilibrium after vibrational relaxation is complete.

1 M. Smits, A. Ghosh *et al.*, *New J. Phys.*, 2007, **9**, 390.
2 A. Ghosh, M. Smits *et al.*, *J. Am. Chem. Soc.*, 2007, **129**, 9608.

Dr Ricci communicated: It is certainly true that in order to extract information on the diffusion coefficient of the system from a QENS experiment a model, such as the jump diffusion, is needed. On the other hand, the MSD data shown by Dr Zaccai are almost raw EISF data, and these bring into evidence clear difference between different biological materials and different solvation conditions, without any background model. There is no doubt that when a soluble protein is dry or wet the behavior with temperature of its MSD is different and that a sensible change of the slope of this function is correlated with a similar change visible for the MSD of hydration water. Where does this change take place? It is a gently change which takes place between 200 and 240 K. The situation is more complicated in the case of membrane proteins, where the membrane lipids also play a role and possibly two transitions can be identified. Nevertheless the merit of these experiments is that of evidencing clear changes of the dynamics of proteins at given hydration/temperature conditions along with peculiarities of different classes of proteins, which may be of biological relevance.

Dr Zaccai answered: I thank Dr Ricci for her comment. I agree fully, and would add that primary data from QENS can also provide model-independent information on diffusion coefficients, for events taking place within the length-time scales of the spectrometer. The primary data are line widths, which provide information on time scale measured as a function of scattering vector modulus, Q, which informs us on the length scale. At small Q values, the Q^2 dependence of the line-width yields the atomic scale translational diffusion.

Professor Tobias responded: I agree completely with your comments. Regardless of whether or not one sees a clear "transition" in a plot of mean-squared amplitudes *vs.* temperature, there are clear changes in such plots when solvent/environmental conditions are varied, and there seem to be (at least from limited number of examples currently available) clear differences between soluble and membrane proteins. Thus, experiments and simulations of dynamical "transitions" are providing insight into the dynamical coupling between proteins and their environments.

Professor Yarwood continued the discussion of the paper by Professor Havenith: One of the disadvantages (it seems to me, as someone who for many years looked routinely at the 400–20 cm^{-1} region) of using THz technology is that one is looking at a very narrow frequency range—in your case at only one frequency. So how, for example, do you know that some of your intensity data variations are not caused by slight shifts in the band profile to high or low frequency?

Professor Havenith answered: We do in fact know that some of the intensity changes are associated with shifts in broad absorption bands. The conclusions are

 This journal is © The Royal Society of Chemistry 2009

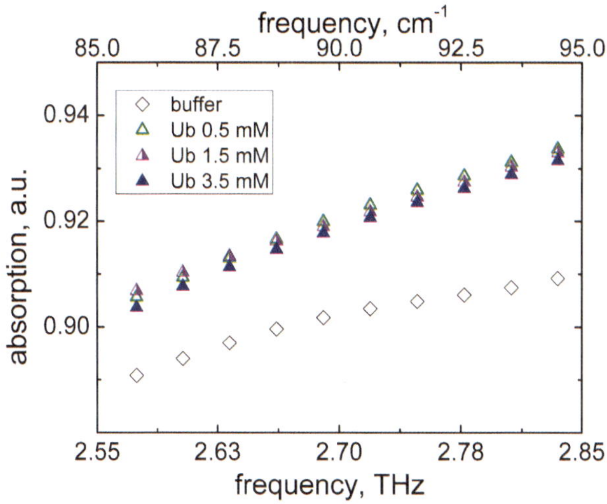

Fig. 6

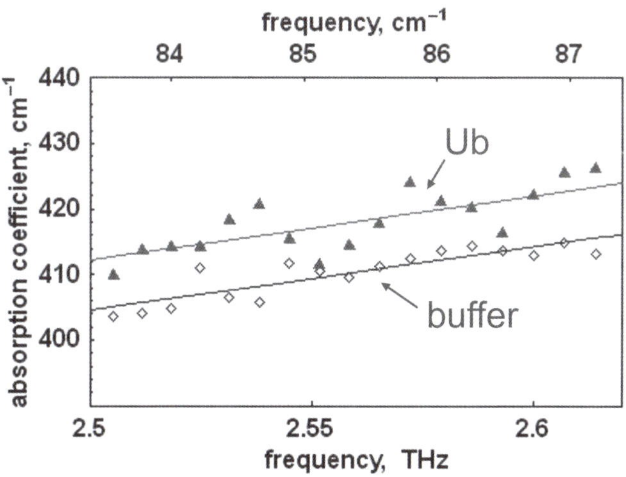

Fig. 7

valid independent of whether the intensity change is caused by a band shift, an overall intensity change, or a combination of both. We only use the fact that the concentration dependence is nonlinear. If we assume a two component model with a shift in the band center, we would—for a given frequency—still observe a linear change with increasing solute concentration. Only when taking into account a third component (the hydration shell) can we expect a nonlinear variation with linear increasing protein concentration. We have also measured the frequency dependence of the bands using a THz-FT spectrometer (Fig. 6). Although the results are not as precise as our laser measurements, they provide a general trend. If we restrict the scale to the investigated frequency range we find a very good agreement between both methods (see Fig. 6 and 7).

Coarse-grained modeling of the interface between water and heterogeneous surfaces

Adam P. Willard and David Chandler*

Received 7th April 2008, Accepted 28th May 2008
First published as an Advance Article on the web 9th October 2008
DOI: 10.1039/b805786a

Using coarse-grained models we investigate the behavior of water adjacent to an extended hydrophobic surface peppered with various fractions of hydrophilic patches of different sizes. We study the spatial dependence of the mean interface height, the solvent density fluctuations related to drying the patchy substrate, and the spatial dependence of interfacial fluctuations. We find that adding small uniform attractive interactions between the substrate and solvent cause the mean position of the interface to be very close to the substrate. Nevertheless, the interfacial fluctuations are large and spatially heterogeneous in response to the underlying patchy substrate. We discuss the implications of these findings for the assembly of heterogeneous surfaces.

1 Introduction: interfaces and hydrophobicity

The interaction of liquid water with oily components in aqueous solution is central to many phenomena.[1] These phenomena are called "hydrophobic effects." The hydrophobic effects leading to robust assembly, such as micelle formation and protein aggregation, follow from the nucleation of water/vapor-like interfaces adjacent to sufficiently extended hydrophobic surfaces.[2] This article reports on computer simulation studies of the behavior of such interfaces when the surfaces contain hydrophilic regions.

Our analysis focuses on probability distributions for density fluctuations in water, distributions like $P_v(N)$, which stands for the probability that the centers of N water molecules are found in a volume v. We consider such distributions because arrangements of water molecules near hydrophobic solutes are similar to those of water near voids in the liquid. The connection between voids and hydrophobicity is found in Stillinger's[3] proposal that a water/oil interface is similar to a water/vapor interface. This proposal is supported by recent experiments.[4–6]

Hummer, Pratt and their coworkers[7] were the first to call attention to the behavior of $P_v(N)$. For small microscopic volumes in water, they used computer simulation to show that this probability is almost exactly Gaussian. Some theories of homogeneous liquids and hydrophobic effects are based upon the assumption that microscopic density fluctuations obey Gaussian statistics,[8] so the finding of ref. 7 provides support for those theories. But more important is a remarkably simple yet quantitatively accurate theory for solvation free energies of small hydrophobic species.[7,9] This consequence follows from the fact that the solvation free energy or excess chemical potential for a solvent excluding volume v, $\Delta\mu_v$, is given by[10]

$$\beta\Delta\mu_v = -\ln P_v(0),$$

Department of Chemistry, University of California, Berkeley, California 94720, USA. E-mail: chandler@cchem.berkeley.edu where $1/\beta = k_B T$ is Boltzmann's constant times temperature. Therefore, to the extent that $P_v(N)$ is Gaussian, $\Delta\mu_v$ can be expressed entirely in terms of bulk water's mean density and mean-square density fluctuations, both of which can be determined from experimentally known quantities. With this solvation free energy in hand, hydration free energies of real apolar molecules can be estimated, using perturbation theory to account for solvent–solute forces beyond those of excluded volume interactions.[2,11]

But for large volumes v, or for small volumes in proximity to an extended hydrophobic surface, the Gaussian approximation to the probability $P_v(N)$ ceases to be accurate. Here, probabilities for fluctuations are not the same as those for the bulk liquid because the chemical potentials of liquid water and vapor differ by very little compared to $k_B T$. As such, a large void in water can nucleate vapor-like configurations,[3] and vapor-like configurations increase the likelihood of large density fluctuations. Thus, for large v and N much smaller than its mean, $\langle N \rangle_v$, $P_v(N)$ is much larger than the probability predicted by the Gaussian approximation. In other words, for large enough v, $P_v(N)$ possesses a fat non-Gaussian tail at small values of N. Understanding the nature of this tail is important for estimating values of solvation free energies of sufficiently large clusters and extended surfaces of hydrophobic species. It is also important for estimating likely pathways by which water molecules are displaced during assembly of hydrophobic clusters.

A theory for the fat tail or large length scale hydrophobicity must account for the presence of both liquid and vapor and the interface between them. The theory of Lum, Chandler and Weeks (LCW)[12] does so by partitioning the density field of water into two components. One is the field $n_s(\mathbf{r})$, which describes two-phase coexistence and is presumed to vary slowly in space (hence the subscript "s"). The other, $\delta\rho(\mathbf{r})$, is a Gaussian field, which is assumed to include all variation not captured by $n_s(\mathbf{r})$. The theory predicts the onset of large length scale hydrophobicity at a size of roughly 1 nm. Variations of $n_s(\mathbf{r})$ occur on length scales greater than 0.2 or 0.3 nm. The equations governing the behavior of $n_s(\mathbf{r})$ are continuum versions of equations governing a binary field (*i.e.* the density field of a lattice gas or the equivalent spin field of an Ising model) on a three-dimensional cubic lattice.[13,14] To the extent that a spatially coarse resolution of the liquid density is of interest, a lattice-gas field suffices to describe the fluctuations of water. The role of $\delta\rho(\mathbf{r})$ in that case is implicit, affecting the values of parameters of the lattice-gas Hamiltonian. We have adopted this approach in earlier work modeling dynamics of a hydrophobic chain in water,[15] dynamics of water in and around nanometer-scale tubes,[16] and the dynamics of dimerization for two nanometer-scale spheres in water.[17] Some additional justification for the approach can be found in theoretical analysis[18] and explicit atomistic modeling.[19] Examples of other applications of the lattice-gas model to treat water fluctuations near extended hydrophobic surfaces are found in ref. 20.

The models we employ are defined in the next section. We then present results of this general coarse-grained approach, examining density fluctuations adjacent to extended complex surfaces in a way that elucidates recent atomistic modeling.[21,22] Some of our calculations focus directly on the fluid interface, *i.e.* a $d = 2$ dimensional manifold in $d = 3$ dimensional space. Our principal results demonstrate the variability of probabilities for dewetting fluctuations, and further show that the behaviors of these probabilities are richer and more physically pertinent than might be guessed from the behaviors of their mean values alone.

2 Models

2.1 Lattice gas

As noted, we use the lattice-gas model as the discrete version of the slowly varying field of the LCW theory,[12] with the parameterization of Ref. [18]. In particular the density field is taken as a binary field on a three dimensional cubic lattice with lattice

 This journal is © The Royal Society of Chemistry 2009

spacing l. Positions of a lattice site are specified with $\boldsymbol{r} = x\hat{\mathbf{x}} + y\hat{\mathbf{y}} + z\hat{\mathbf{z}}$ where x, y and z are Cartesian components. In units of l these components have integer values. The lattice site at position $\boldsymbol{r}$ has occupation number $n_{\boldsymbol{r}}$ equal to 1 if the volume within that site is liquid-like and equal to 0 if the volume within that site is vapor-like. This lattice-gas model is capable of supporting liquid/vapor phase coexistence and its interfaces. The energy for a given binary density field $\{n_{\boldsymbol{r}}\}$ is given by,

$$E_{\mathrm{L}}(\{n_{\boldsymbol{r}}\}) = -\varepsilon \sum_{\boldsymbol{r},\boldsymbol{r}'}{}' n_{\boldsymbol{r}} n_{\boldsymbol{r}'} - \sum_{\boldsymbol{r}} \mu n_{\boldsymbol{r}}, \tag{1}$$

where the primed summation is taken over nearest-neighbor pairs of lattice sites, and at coexistence the chemical potential, μ, has value 3ε. To match the surface tension, compressibility and proximity of water to coexistence, we use $\varepsilon = 1.51k_{\mathrm{B}}T$, $l = 0.21$ nm and $\mu = 3\varepsilon + 2.25 \times 10^{-4}k_{\mathrm{B}}T$. In the simulation of this model, we use cells that are periodically replicated in the x and y directions, while the upper z boundaries are populated with liquid-like lattice sites and the lower z boundaries serve to characterize patchy surfaces described below.

Fluctuations across a liquid/vapor interface can be correlated over very long distances while correlation lengths in the bulk liquid are very small. Thus, fluctuations of the interface can be interesting even while those in the bulk are uninteresting. For this reason we choose to simulate systems with large interfacial area (in the xy plane) and relatively little bulk liquid. Taking the perspective to the extreme would disregard the bulk entirely, as is done with an interface model, which we turn to now.

2.2 Fluid interface

Here we adapt the two-dimensional model of a liquid/vapor interface described by Weeks.[23] In this model, a liquid/vapor interface is treated as a two-dimensional manifold that is coarse-grained onto a square lattice in a reference plane with lattice spacing approximately equal to the bulk correlation length of water, $\xi = 0.42$ nm. This two-dimensional interface is periodically replicated in each of the two dimensions of the reference xy plane. The interfacial profile is characterized by the set of height variables $\{h_{\boldsymbol{a}}\}$, where $h_{\boldsymbol{a}}$ is the distance of the interface normal to the reference xy plane, and $\boldsymbol{a} = x\hat{\mathbf{x}} + y\hat{\mathbf{y}}$ refers to a lattice point on that plane. In this model, which we call the "fluid interface model," the free interface has the energy function

$$E_{\mathrm{W}}(\{h_{\boldsymbol{a}}\}) = \frac{\Gamma}{2} \sum_{\boldsymbol{a},\boldsymbol{a}'}{}' (h_{\boldsymbol{a}} - h_{\boldsymbol{a}'})^2, \tag{2}$$

where the primed summation is taken over nearest-neighbor pairs, and to be consistent with the surface tension of water, we use $\Gamma\xi^2 = 0.1k_{\mathrm{B}}T$.

2.3 Patchy substrate

We couple the models described above to a patchy substrate. The substrate interacts with the solvent through excluded volume interactions as well as attractive potentials. The substrate lies in the $z = 0$ plane, and is taken to be effectively infinite in the xy plane, and it prevents the liquid (and its interface) from existing at $z < 0$. In addition to excluding volume, the substrate surface contains regions which attract the liquid or its interface. These are called "hydrophilic" regions. The other portions of the substrate, which simply exclude the liquid, are the "hydrophobic" regions. The surface is partitioned with the same length scale as the solvent and the state (hydrophilic or hydrophobic) is specified with a variable $\sigma_{\boldsymbol{a}}$ (1 or 0 respectively). For the lattice-gas model, the interaction energy between the patchy substrate $\{\sigma_{\boldsymbol{a}}\}$ and the first (*i.e.* $z = 1$) layer of lattice sites is

$$\Delta E_{\mathrm{L}}(\{n_r\};\{\sigma_a\}) = -\sum_a (\varepsilon\sigma_a n_{a,1} + \Delta\gamma n_{a,1}). \tag{3}$$

Here, $n_{a,z}$ refers to n_r for $\boldsymbol{r} = \boldsymbol{a} + z\hat{z}$, and $\boldsymbol{a}$ specifies x and y. The strong adhesive forces used to mimic the hydrophilic interactions come from the first term of eqn (3). The quantity $\Delta\gamma$ is the surface adhesive interaction which is included to capture small attractive interactions such as van der Waals attractions between oil and water. In the absence of hydrophilic regions ($\sigma_a = 0$ for all $\boldsymbol{a}$), $\Delta\gamma\langle n_{a,1}\rangle/l^2$ is the mean field estimate of the difference between water/oil surface tension and water/vapor surface tension. In this case (all σ_a set to zero), the choice $\Delta\gamma \approx 0.05\varepsilon$, gives an average density profile $\langle n_{a,z}\rangle$ consistent with that found experimentally for water/oil interfaces.[4]

For the fluid interface model, the interaction energy between the patchy substrate $\{\sigma_a\}$ and the interfacial profile is

$$\Delta E_{\mathrm{W}}(\{h_a\};\{\sigma_a\}) = W(\{h_a\}) - \sum_a \Theta(h_{\mathrm{c}} - h_a)\left(\sigma_a\mathscr{E}\xi^2 + \Delta\Gamma\xi^2\right), \tag{4}$$

where $\Theta(x)$ is the Heaviside step function which is equal to 1 for $x > 0$ and zero otherwise, h_{c} is a cutoff distance equal to the bulk correlation length, the hydrophilic interaction energy $\mathscr{E}\xi^2$ is taken to be $3k_{\mathrm{B}}T$ (which amounts to about half the attractive energy per surface unit of ε for the lattice gas), and the function $W(\{h_a\})$ excludes the interface from crossing the $z = 0$ plane (it is infinite if any $h_a < 0$, and it is zero if all $h_a \geq 0$). The surface adhesion $\Delta\Gamma$ is included to capture small attractive interactions (as above). For the case of no hydrophilic patches, the choice of $\Delta\Gamma\xi^2 = 0.7k_{\mathrm{B}}T$ gives an interfacial profile which is consistent with experimental observations for water adjacent to oil.[4]

2.4 Monte Carlo

For the lattice-gas model, we carry out Monte Carlo trajectories for $\{n_r\}$. Acceptance rejection obeys a detailed balance for the grand canonical ensemble with net energy function

$$E(\{n_r\},\{\sigma_a\}) = E_{\mathrm{L}}(\{n_r\}) + \Delta E_{\mathrm{L}}(\{n_r\};\{\sigma_a\}). \tag{5}$$

To specify hydrophilic patches, *i.e.* to specify $\{\sigma_a\}$, we tile the substrate with $d \times d$ squares thus creating a square lattice with lattice spacing d, where d is an integer multiple of l. At random, a fraction f of the $d \times d$ squares are made hydrophilic (*i.e.* $\sigma_a = 1$ for each substrate lattice site in the square). The pattern so formed is then held fixed while the Monte Carlo trajectory is carried out for all n_r values. Averages over these trajectories are then recorded, the results of which are described in the next section.

For the fluid interface model, we carry out Monte Carlo trajectories for $\{h_a\}$. Acceptance and rejection obeys a detailed balance with the net energy function

$$E(\{h_a\},\{\sigma_a\}) = E_{\mathrm{W}}(\{h_a\}) + \Delta E_{\mathrm{W}}(\{h_a\};\{\sigma_a\}). \tag{6}$$

Here, the hydrophilic patterns are created as above, but now with underlying lattice spacing ξ rather than l. Monte Carlo trajectories for h_a values are performed and analyzed with $\{\sigma_a\}$ fixed. The h_a values evolve continuously, unlike the n_r values, which change discontinuously between 0 and 1.

3 Mean interfacial height for various substrates

We have chosen to focus on surfaces which can be characterized by two parameters. These parameters are the overall fraction f of hydrophilic sites and the size d of the

 This journal is © The Royal Society of Chemistry 2009

hydrophilic patches. The interface height of the lattice gas, $h_{\boldsymbol{a}}$, is defined such that $h_{\boldsymbol{a}} + l$ is the smallest value of z, for $z > 0$, where $n_{\boldsymbol{a},z}$ is not zero. For example, if only the cell immediately adjacent to the surface at $\boldsymbol{a}$ is empty, $h_{\boldsymbol{a}} = l$. If none were empty, $h_{\boldsymbol{a}} = 0$. In other words, $h_{\boldsymbol{a}}$ is the value of z for the occupied lattice facet closest to the patchy surface at $\boldsymbol{a} = (x,y)$. Consider first the average interfacial height $\langle h \rangle_{f,d}$, where $\langle \cdots \rangle_{f,d}$ denotes the equilibrium average with patchy surfaces characterized by f and d. Specifically,

$$\langle h \rangle_{f,d} = \frac{1}{N_{\text{rep}}} \sum_{\alpha=1}^{N_{\text{rep}}} \frac{1}{N_{\text{surf}}} \sum_{\boldsymbol{a}} \bar{h}_{\boldsymbol{a}}^{(\alpha)}, \tag{7}$$

where the first summation is over N_{rep} different realizations of the surface patterns, α refers to a specific realization, *i.e.* $\alpha = \{\sigma_{\boldsymbol{a}}\}$, consistent with f and d, and the second summation is over the N_{surf} lattice sites in the patchy surface (xy plane). The quantity $\bar{h}_{\boldsymbol{a}}^{(\alpha)}$ is the mean height of the interface over the surface site $\boldsymbol{a} = (x,y)$, for a specific realization

$$\bar{h}_{\boldsymbol{a}}^{(\alpha)} = \frac{1}{N_{\text{obs}}} \sum_{\tau=1}^{N_{\text{obs}}} h_{\boldsymbol{a}}^{(\alpha)}(\tau), \tag{8}$$

where the summation here is over the N_{obs} averaging time steps in a single Monte Carlo trajectory.

In the absence of weak surface adhesive interactions, *i.e.* $\Delta\gamma = \Delta\Gamma = 0$, for some fraction of hydrophilic coverages, f, the mean interface height $\langle h \rangle_{f,d}$ is non-monotonic in the hydrophilic patch size d. The results for these cases are shown in Fig. 1. This non-monotonic behavior, which is observed in both the fluid interface model as well as the lattice-gas model, demonstrates that for fixed surface composition f there is an optimal patch size d for attracting the fluid.

The mean interface height $\langle h \rangle_{f,d}$ is an average over a surface which is locally heterogeneous. The influence of the local surface structure on the mean $\langle h \rangle_{f,d}$ can be visualized for a surface realization α with the spatially resolved average interface height $\bar{h}_{\boldsymbol{a}}^{(\alpha)}$. Fig. 2 shows $\bar{h}_{\boldsymbol{a}}^{(\alpha)}$ for representative patchy surfaces with $f = 0.03$ and $d = l$, $2l$ and $3l$. The Figure shows that a hydrophilic patch affects interface fluctuations beyond the region immediately above the patch. In ref. 22, atomistic simulations were used to study water confined between nano-scale heterogeneous surfaces. Among their findings are that the solvent density within the first hydration layer over a nano-scale hydrophobic region is increased significantly by the introduction of a border of hydrophilic sites. This effect is not symmetric with respect to the exchange of hydrophilic and hydrophobic material as the hydration layer over a nano-scale hydrophilic region is affected very little by the introduction of a hydrophobic border. We have found (not shown here) that these results are duplicated with the lattice models used here.

The non-monotonic behavior seen in Fig. 1 can be understood qualitatively through the mean interfacial profile $\bar{h}_{\boldsymbol{a}}^{(\alpha)}$ (shown in Fig. 2). For a surface with $f = 0.03$ and $d = l$ in the absence of surface adhesive interactions the mean interfacial profile is only slightly perturbed by the underlying hydrophilic patches. Locally, regions of the surface with a relatively high density of hydrophilic patches lower the average height of the interface. There are however, many isolated hydrophilic patches for which the mean interface height shows little response. For $d = 2l$, however, the hydrophilic patches pin the local interface, and because d is relatively small there are many such patches so that the distance between a patch and its neighbors is not very large. At $d = 3l$ there are fewer hydrophilic patches to pin the interface and thus the hydrophobic domains are generally larger than when $d = 2l$. The effect of adding a small attractive interaction between the surface and solute is significant as seen in Fig. 1 and 2. The mean interface is pulled much closer to the patchy

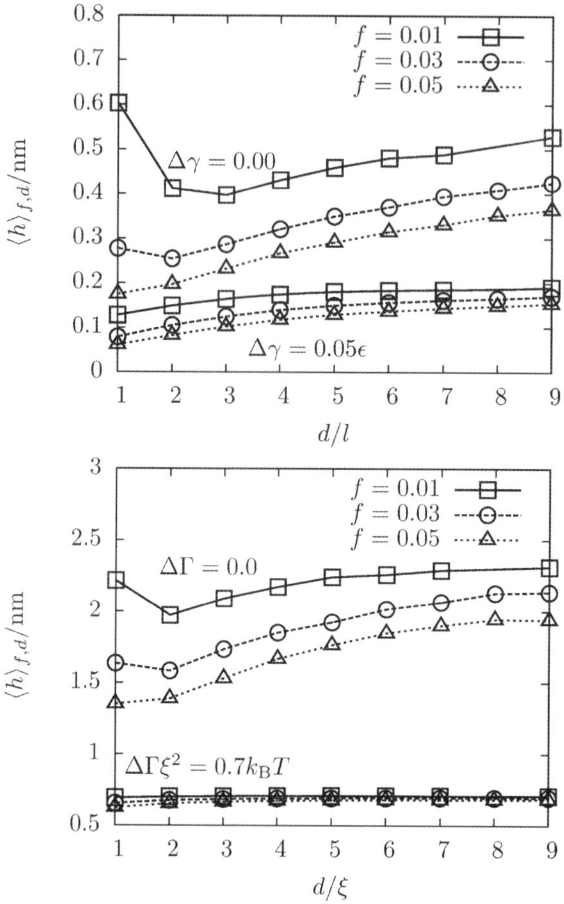

Fig. 1 The equilibrium height, $\langle h \rangle_{f,d}$, of the liquid vapor interface over the surface layer as a function of patch size d for the lattice-gas model (top) and the fluid interface model (bottom) at different hydrophilic fractions f. In each panel, the top three curves are computed in the absence of weak water–substrate adhesive interactions, *i.e.* $\Delta\gamma = \Delta\Gamma = 0$, and the bottom three curves are computed with weak water–substrate adhesive interactions.

surface when weak attractions are present. This result is consistent with the role of weak attractions uncovered in earlier work.[2,24–26] The non-monotonic behavior seen in the average interfacial height is eliminated and the fluctuations are reduced.

4 Density fluctuations and dewetting

To explore the onset of drying-like phenomena we have considered a finite patchy substrate. The specific substrate size is $32 \times 32\ l^2$ for which we have studied density fluctuations in the adjacent lattice gas. In particular, we have computed the probability distribution, $\mathscr{P}_\alpha(\rho)$ for various substrate realizations, α, corresponding to different hydrophilic fractions f and patch sizes d. The variable ρ refers to the density in the first two solvent layers above the substrate:

$$\mathscr{P}_\alpha(\rho) = \left\langle \delta(\rho - \frac{1}{2N_S}\sum_a [n_{a,1} + n_{a,2}]) \right\rangle_\alpha, \tag{9}$$

where $\delta(\cdots)$ stands for Dirac's delta function, N_S is the number of lattice sites in a single layer of the lattice gas (in this case $N_S = 32^2$), and the averaging implied

 This journal is © The Royal Society of Chemistry 2009

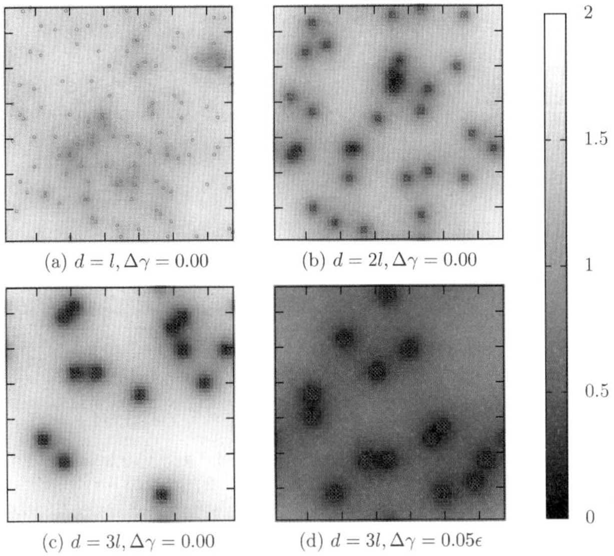

Fig. 2 Mean interfacial height, $\bar{h}_a^{(\alpha)}$ projected onto the xy plane for hydrophilic fraction $f = 0.03$ and $d = l$ (a), $d = 2l$ (b), and $d = 3l$ (c and d). The height h is indicated by the shading (scale at right) and the location of hydrophilic surface sites (substrate sites with $\sigma_a = 1$) are shown with red circles. Panels (a), (b) and (c) correspond to averages in the absence of surface adhesive interactions ($\Delta\gamma = 0.0$) and panel (d) corresponds to an average with surface adhesive interactions ($\Delta\gamma = 0.05\varepsilon$). The surfaces pictured are $64 \times 64\ l^2$ in size.

by $\langle\cdots\rangle_\alpha$ is carried out by umbrella sampling[29] with the surface realization α fixed. This distribution coincides with $P_v(N)$ discussed in the introduction, in this case where v is the combined volume of the first and second layers of cells above the substrate with pattern α, and ρ multiplied by that volume is N.

4.1 Without weak surface adhesive interactions, $\Delta\gamma = 0$

The inset to Fig. 3 shows $\mathscr{P}_\alpha(\rho)$ for the case of $\Delta\gamma = 0$ (plotted with open symbols) for representative surfaces with $f = 0.03$ and $d = l$, $3l$ and $5l$. These distributions are very broad, showing significant fluctuations over a large range of densities. These broad distributions arises because the liquid/vapor interface wanders in and out of the observed volume. The distribution with $d - l$ shows an additional narrow peak near $\rho = 0$, which corresponds to the dewetted state where the interface has completely pulled away from the patchy surface.

The bimodal character of $\mathscr{P}_\alpha(\rho)$ is related to the non-monotonic behavior seen in Fig. 1. That is, the metastable "dry" state can emerge when the hydrophilic patch sizes are small, but this state is absent in the case of larger patch sizes. The loss of the metastability occurs when hydrophilic patch sizes are large enough to pin the adjacent interface. Consider the effect of excluding volume from a fluctuating liquid/vapor interface, as the substrate does. This imposes constraints on the configurations accessible to the interface, and the closer the interface is to the substrate the more severe the constraints. Thus, for a purely hydrophobic substrate there is an entropic force that drives the liquid/vapor interface away from the surface, thereby drying the substrate. By adding hydrophilic patches, the entropic driving force is overcome by energetically favorable interactions between the interface and the substrate, but just barely for the case of $f = 0.03$ and $d = l$, as is evident in Fig. 3.

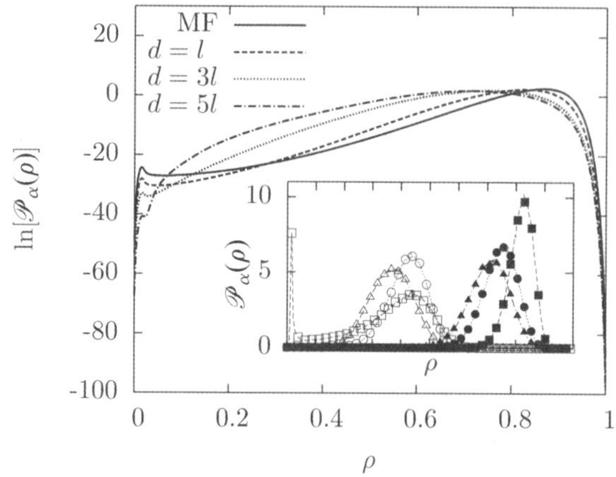

Fig. 3 The probability distribution for density within the first two layers of lattice sites with weak adhesive interactions ($\Delta\gamma = 0.05\varepsilon$) over the substrate surface for specific realizations of patchy substrates with hydrophilic fraction $f = 0.03$ and patch sizes $d = l$, $3l$ and $5l$. The curve label "MF" is the result for the mean field distribution, where the substrate is uniformly attracted to the solvent with an attractive strength equal to $f\varepsilon$.[33] The inset compares the distributions for $d = l$ (squares), $d = 3l$ (circles), and $d = 5l$ (triangles) for substrates with (hollow symbols) and without (filled symbols) weak adhesive interactions.

4.2 With weak surface adhesive interactions

The inset of Fig. 3 also shows $\mathscr{P}_\alpha(\rho)$ for the same surface parameters, $f = 0.03$ and $d = l$, $3l$ and $5l$, but with the interface adhesion turned on, *i.e.* $\Delta\gamma = 0.05\varepsilon$ (plotted with filled symbols). With the interface pulled close to the surface the distributions shift to larger values of ρ (more liquid-like). For $d = l$ the addition of adhesive interactions destabilizes the "dewetted" state and as a result the distribution is unimodal and considerably more narrow than when $\Delta\gamma = 0$. The main panel in Fig. 3 shows $\ln[\mathscr{P}_\alpha(\rho)]$ for $f = 0.03$, and $d = l$, $3l$ and $5l$. The distributions exhibit non-Gaussian tails which are a characteristic of density distributions near phase coexistence.[27,28] The departure from Gaussian behavior arises because density fluctuations in the presence of a liquid/vapor interface occur through the translation of the interface. Fluctuations without the interface require compressing a nearly incompressible fluid. The non-Gaussian tails imply a far greater probability for large density fluctuations near the substrate than for those in the bulk.

The quantity $\ln[\mathscr{P}_\alpha(\rho)]$ is proportional to the free energy to alter the solvent density in the volume adjacent to the substrate, and thus reflects the cost of drying the patchy substrate. There is a trivial dependence of the free-energy difference between the $\rho = 0$ and $\rho = 1$ state on the hydrophilic fraction f. For fixed f, however, the shapes of the distributions depend on the hydrophilic patch size d, and this feature is pertinent to the kinetic pathways to drying. Moving from the wet state to the dry state as a function of ρ, initially near $\rho \approx 0.9$, a surface with larger d incurs less of a free-energy cost to dry than for a surface with a relatively small d. As ρ approaches $\rho = 0$, however, for the system with smaller d, fluctuations to lower ρ have lower free energy than for systems with larger d. As the solvent density, ρ, over the patchy substrate is reduced, the substrate and the hydrophilic patches dry. The solvent density over the hydrophilic patches, the patches where $\sigma_{\mathbf{a}} = 1$, is given by

$$m_1 = \frac{1}{fN_S} \sum_{\mathbf{a}} \sigma_{\mathbf{a}} n_{\mathbf{a},1}. \tag{10}$$

Fig. 4 shows the average value of m_1 for substrate realization α, $\langle m_1 \rangle_{\rho,\alpha}$, as a function of ρ, where the average implied by $\langle\cdots\rangle_{\rho,\alpha}$ is an average over configurations with

 This journal is © The Royal Society of Chemistry 2009

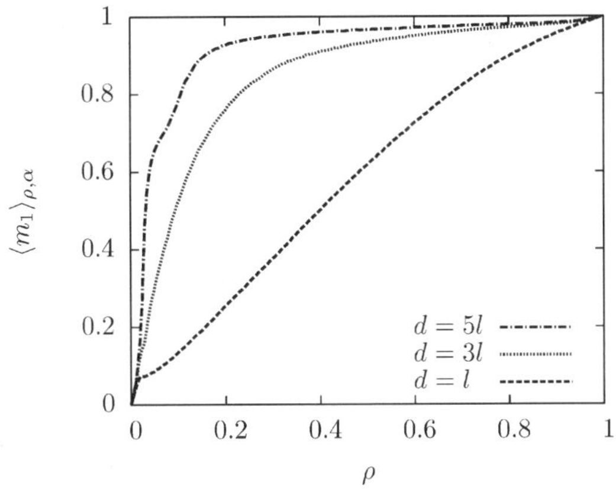

Fig. 4 The mean solvent density over the hydrophilic patches, $\langle m_1 \rangle_{\rho,\alpha}$ (see text for definition), as a function of the total solvent density in the first two solvent layers, ρ. Each curve is averaged over a fixed substrate realization, α, with $f = 0.03$ and weak surface adhesive interactions ($\Delta\gamma = 0.05\varepsilon$).

fixed ρ, and substrate realization α. In Fig. 4 we see that when $d = l$, the hydrophilic patches dry essentially uniformly as the solvent density, ρ, is decreased. When the hydrophilic patches are large, however, $\langle m_1 \rangle_{\rho,\alpha}$ does not show a significant response to ρ until ρ is quite small, indicating that when d is large the hydrophilic patches are among the last regions to dry. Qualitatively, therefore, the initial stages of drying are easier for systems with larger d because there are relatively large hydrophobic domains from which to pull the interface. The final stages of drying are difficult when large hydrophilic patches exist, and comparatively easy for a surface with poorly pinning small patches.

5 Spatial dependence of height fluctuations

The distribution $\mathscr{P}_\alpha(h;\boldsymbol{a})$ is the probability of observing interface height h over surface site $\boldsymbol{a} = (x,y)$ for surface pattern α. To visualize these distributions we define the free energy $A_\alpha(h;\boldsymbol{a})$,

$$A_\alpha(h;\boldsymbol{a}) = -k_B T \ \log\left[\frac{\mathscr{P}_\alpha(h;\boldsymbol{a})}{\mathscr{P}_\alpha\left(\tilde{h}_{\boldsymbol{a}}^{(\alpha)};\boldsymbol{a}\right)}\right], \tag{11}$$

where $\tilde{h}_{\boldsymbol{a}}^{(\alpha)}$ is the most likely value of $h_{\boldsymbol{a}}$ over the substrate at position $\boldsymbol{a}$. Therefore, $A_\alpha(h;\boldsymbol{a})$ is the free energy required to move the interface at equilibrium from its most likely position $\tilde{h}_{\boldsymbol{a}}$ to a height h. Fig. 5 shows $A_\alpha(h;\boldsymbol{a})$ for a surface α with $f = 0.05$ and $d = l$. The weak adhesive interaction, $\Delta\gamma = 0.05\varepsilon$, causes the $h = 0$ interfacial configuration to be favored by the interface. Fluctuations into $h = l$ are thermally accessible to the interface ($A(h;\boldsymbol{a}) \sim k_B T$) and for many regions, configurations with $h = 2l$ (interface is 0.4 nm from the surface) are accessible through fluctuations of less than $3k_BT$. Recall that for $d = 3l$ with weak adhesions turned on, the interface is on average pulled very close to the substrate (Fig. 1), and solvent volume adjacent to the substrate has a more liquid-like density (Fig. 3). But the distribution $A_\alpha(h;\boldsymbol{a})$ shows that local fluctuations in the interface through the second layer are within the range of thermal fluctuations. On the other hand, the free energy to pull the interface off of a hydrophilic patch is quite large. These features lead to the contrasting shapes

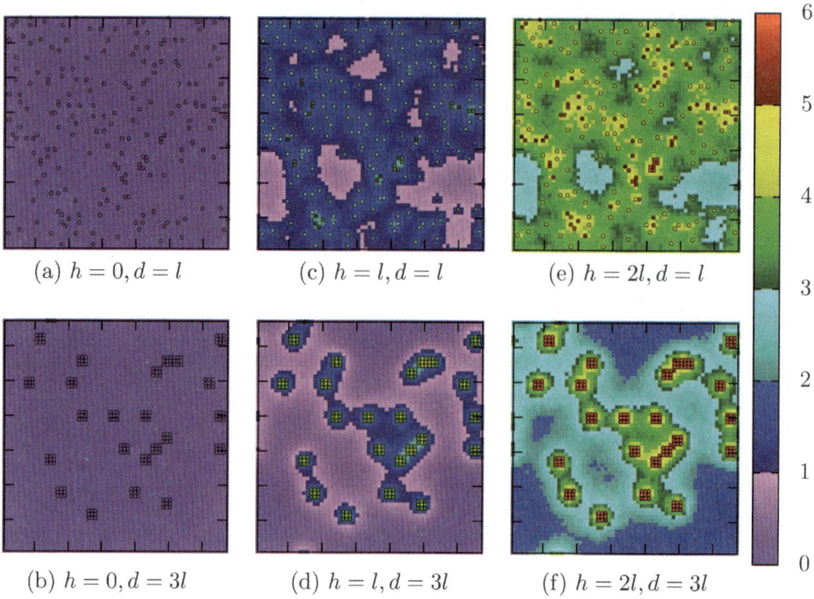

(a) $h = 0, d = l$ (c) $h = l, d = l$ (e) $h = 2l, d = l$

(b) $h = 0, d = 3l$ (d) $h = l, d = 3l$ (f) $h = 2l, d = 3l$

Fig. 5 The spatial variation of the free energy, $\beta A_\alpha(h;\boldsymbol{a})$ to displace the interface at $\boldsymbol{a}$ from its most likely height to the indicated height h for a surface realization with $f = 0.05$ and with $d = l$ and $3l$ in the presence of weak adhesive interactions ($\Delta\gamma = 0.05\varepsilon$). The surfaces pictured are $64 \times 64\ l^2$ in size.

of the $\mathscr{P}_\alpha(\rho)$ values shown in Fig. 3, as well as the response of $\langle m_1 \rangle_{\rho,\alpha}$ to ρ shown in Fig. 4.

6 Implications

Many meso-scale solutes in nature are patchy. Hua *et al.*[30] characterized the distributions of hydrophobic and hydrophilic subunits on an assembling surface of proteins by coarse-graining the hydrophobicity on protein surfaces over 5×5 Å squares. The resulting hydrophobic surface distributions look like hydrophilic sites distributed on a hydrophobic background, not unlike the surface models we are considering. The relative fraction of hydrophilic sites depends upon the protein considered and coincide qualitatively with our model with $f \approx 0.25$–0.50 and hydrophilic patch size $d \approx 2l$–$3l$.

The protein surfaces studied in ref. 30 and 31 are finite in size, extending over several nanometers. Our model surfaces are effectively infinite and thus lack a boundary term which may be significant in the actual kinetics of assembly. Nonetheless, for solutes with hydrophobic surfaces extending over about 1 nm, the mechanism for assembly can involve the drying of the solvent volume between the two assembling proteins.[17,20,32] This drying event and subsequent solute aggregation is preempted by the formation of a vapor tunnel which arises when the interfaces surrounding both solutes come close to contact. The height fluctuations of the interface next to these solutes describe this phenomenon. Specifically, these interfacial fluctuations set the range over which solutes can broadcast their presence into the bulk solvent. Fig. 6 shows the free energy $A_\alpha(h;\boldsymbol{a})$ for a substrate α with $f = 0.25$ and $d = 3l$ chosen to resemble the assembling surfaces of the proteins in ref. 30. For surfaces consistent with these parameters, the typical range which is thermally accessible to fluctuations of the interface height is 0–$2l$ (0.0–0.4 nm), implying that two solutes with similar surface distributions in water can form a vapor tunnel at

 This journal is © The Royal Society of Chemistry 2009

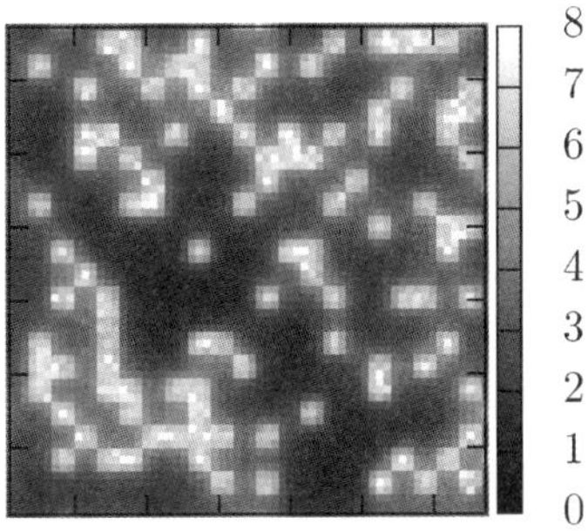

Fig. 6 The spatial variation of the free energy $\beta A_\alpha(h;\boldsymbol{a})$ for $h = 2l$, and a patch pattern realization with $f = 0.25$ and $d = 3l$. The pattern of the patchy substrate $\{\sigma_{\boldsymbol{a}}\}$ is not explicitly displayed but is evident in the pattern of $\beta A_\alpha(h;\boldsymbol{a})$. The surface pictured is $64 \times 64\ l^2$ in size.

separations smaller than about 0.8 nm. Not surprisingly, therefore, Hua *et al.*[30] find spontaneous dewetting at surface separations of 0.6 nm or less in a reasonably short period of time (within 100 ps) using atomistic simuations with patchy hydrophobic protein pairs.

The expulsion of water over buried hydrophilic patches presents a barrier to the water-mediated assembly of patchy solutes. For an isolated patchy surface, large hydrophilic patches remain "wet" until the final stages of drying. This is because drying these large patches is very costly, as manifest in the downward curving tails for $d = 5$ in Fig. 3 near $\rho = 0$. These downward tails do not exist for $f = 0$ (not shown), which is consistent with the finding of ref. 22 that the addition of a hydrophilic site to an otherwise hydrophobic surface significantly slows drying. The free energy associated with fluctuations of the interface to $h = 2l$ (0.41 nm) over the hydrophilic patches are close to $10k_BT$ (Fig. 6). With free-energy barriers this large it is unlikely that the first stage of assembly occurs through a complete drying of the assembling surfaces. Thus, during assembly, water molecules over large hydrophilic regions are likely expelled in the latter stages of assembly through a mechanism different from those which are responsible for drying the hydrophobic regions. In fact, retaining some amount of buried water until the final stages of assembly can be advantageous. Specifically, water retained between assembling protein surfaces can aid in the final stages of successful aggregation.[34,35] Indeed, the grouping of hydrophilic regions can allow for much larger fluctuations of the interface over the substrate. This is demonstrated in Fig. 3 through the comparison of the distribution $\mathscr{P}_\alpha(\rho)$ for homogeneous and patchy surfaces. Along with results for patchy substrates, Fig. 3 has density distributions for systems with substrate layers of uniform attractive interactions with magnitude f (these are the "mean field" lines in Fig. 3).[33] Large fluctuations in solvent density towards the vapor-like state are less likely in the case of the homogenous substrate. We conclude that as the geometry of a patchy hydrophobic surface considerably affects the mean behavior of the adjacent solvent, it also significantly affects the interfacial fluctuations and thus can strongly influence the kinetics of assembly.

Acknowledgements

This work was supported in its initial stages by the Director, Office of Science, Office of Basic Energy Sciences, Chemical Sciences, Geosciences and Biosciences Division, U. S. Department of Energy under Contract No. DE-AC02-05CH11231, and then by the National Institutes of Health. We are grateful to Lan Hua, Bruce Berne and their co-workers for sharing with us their unpublished data[31] relating to the surface distributions of self-assembling proteins.

References

1 C. Tanford, *The Hydrophobic Effect: Formation of Micelles and Biological Membranes*, Wiley, New York, 1980.
2 D. Chandler, *Nature*, 2005, **437**, 640.
3 F. H. Stillinger, *J. Solution Chem.*, 1973, **2**, 141.
4 A. Poynor, L. Hong, I. K. Robinson, S. Granick, Z. Zhang and P. A. Fenter, *Phys. Rev. Lett.*, 2006, **97**, 266101.
5 M. Maccarini, R. Steitz, M. Himmelhaus, J. Fick, S. Tatur, M. Wolff, M. Grunze, J. Janecek and R. R. Netz, *Langmuir*, 2007, **23**, 598.
6 M. Mezger, H. Reichert, S. Schoder, J. Okasinski, H. Schroder, H. Dosch, D. Palms, J. Ralston and V. Honkimaki, *Proc. Natl. Acad. Sci. U. S. A.*, 2006, **103**, 18401.
7 G. Hummer, S. Garde, A. Pohorille, A. E. Garcia and L. Pratt, *Proc. Natl. Acad. Sci. U. S. A.*, 1996, **93**, 8951.
8 D. Chandler, *Phys. Rev. E*, 1993, **48**, 2898.
9 S. Garde, G. Hummer, A. E. Garcia, M. E. Paulaitis and L. R. Pratt, *Phys. Rev. Lett.*, 1996, **77**, 4966.
10 B. Widom, *J. Chem. Phys.*, 1963, **39**, 2808.
11 L. R. Pratt and D. Chandler, *J. Chem. Phys.*, 1977, **67**, 3683.
12 K. Lum, D. Chandler and J. D. Weeks, *J. Phys. Chem. B*, 1999, **103**, 4570.
13 D. Chandler, *Introduction to Modern Statistical Mechanics*, Oxford University Press, New York, 1987.
14 P. M. Chaikin and T. C. Lubensky, *Principals of Condensed Matter Physics*, Cambridge University Press, Cambridge, 2000.
15 P. R. ten Wolde and D. Chandler, *Proc. Natl. Acad. Sci. U. S. A.*, 2002, **99**, 6539.
16 L. Maibaum, A. R. Dinner and D. Chandler, *J. Phys. Chem. B*, 2004, **108**, 6778.
17 A. P. Willard and D. Chandler, *J. Phys. Chem. B*, 2008, **112**, 6187.
18 P. R. ten Wolde, S. X. Sun and D. Chandler, *Phys. Rev. E*, 2001, **65**, 011201.
19 T. F. Miller, E. Vanden-Eijnden and D. Chandler, *Proc. Natl. Acad. Sci. U. S. A.*, 2007, **104**, 14559.
20 For example: A. Luzar and K. Leung, *J. Chem. Phys.*, 2000, **113**, 5836; K. Leung and A. Luzar, *J. Chem. Phys.*, 2000, **113**, 5845; T. A. McCormick, *Phys. Rev. E*, 2003, **68**, 061601; K. Lum and A. Luzar, *Phys. Rev. E*, 1997, **56**, R6283.
21 N. Giovambattista, C. F. Lopez, P. J. Rossky and P. G. Debenedetti, *Proc. Natl. Acad. Sci. U. S. A.*, 2008, **105**, 2274.
22 N. Giovambattista, P. J. Rossky and P. G. Debenedetti, *J. Phys. Chem. C*, 2007, **111**, 1323.
23 J. D. Weeks, *J. Chem. Phys.*, 1977, **67**, 3106.
24 L. Maibaum and D. Chandler, *J. Phys. Chem. B*, 2007, **111**, 9025.
25 D. M. Huang and D. Chandler, *J. Phys. Chem. B*, 2002, **106**, 2047.
26 H. S. Ashbaugh and M. E. Paulaitis, *J. Am. Chem. Soc.*, 2001, **123**, 10721.
27 S. T. Bramwell, J.-Y. Fortin, P. C. W. Holdsworth, S. Peysson, J. F. Pinton, B. Portelli and M. Sellitto, *Phys. Rev. E*, 2001, **63**, 041106.
28 D. M. Huang and D. Chandler, *Phys. Rev. E*, 2000, **61**, 1501.
29 D. Frenkel, B. Smit, *Understanding Molecular Simulation*, Academic Press, London, 2nd edn, 2002.
30 L. Hua, X. H. Huang, P. Liu, R. H. Zhou and B. J. Berne, *J. Phys. Chem. B*, 2007, **111**, 9069.
31 L. Hua, private communication, 2008.
32 X. Huang, R. Zhou and B. J. Berne, *J. Phys. Chem. B*, 2005, **109**, 3546.
33 The uniform mean field substrate with hydrophilic fraction f interacts with the adjacent lattice gas with the energy $\Delta E_{\mathrm{L}}^{(\mathrm{MF})}(\{n_r\}) = -\sum_{a}(f\varepsilon + \Delta\gamma)n_{a,1}$.
34 Y. Levy and J. N. Onuchic, *Proc. Natl. Acad. Sci. U. S. A.*, 2004, **101**, 3325.
35 F. B. Sheinerman and C. L. Brooks III, *J. Mol. Biol.*, 1998, **278**, 439.

 This journal is © The Royal Society of Chemistry 2009

Water growth on metals and oxides: binding, dissociation and role of hydroxyl groups

M. Salmeron,[ab] H. Bluhm,[a] M. Tatarkhanov,[ac] G. Ketteler,[a] T. K. Shimizu,†[a] A. Mugarza,‡*[a] Xingyi Deng,[a] T. Herranz,[a] S. Yamamoto[d] and A. Nilsson[d]

Received 16th April 2008, Accepted 19th June 2008
First published as an Advance Article on the web 22nd September 2008
DOI: 10.1039/b806516k

We discuss the role of the presence of dangling H-bonds from water or from surface hydroxyl species on the wetting behavior of surfaces. Using scanning tunneling and atomic force microscopies and photoelectron spectroscopy, we have examined a variety of surfaces, including mica, oxides and pure metals. We find that in all cases, the availability of free, dangling H-bonds at the surface is crucial for the subsequent growth of wetting water films. In the case of mica, electrostatic forces and H-bonding to surface O atoms determine the water orientation in the first layer and also in subsequent layers with a strong influence in its wetting characteristics. In the case of oxides like TiO_2, Cu_2O, SiO_2 and Al_2O_3, surface hydroxyls form readily on defects upon exposure to water vapor and help nucleate the subsequent growth of molecular water films. On pure metals, such as Pt, Pd and Ru, the structure of the first water layer and whether or not it exhibits dangling H-bonds is again crucial. Dangling H-bonds are provided by molecules with their plane oriented vertically, or by OH groups formed by the partial dissociation of water. By tying the two H atoms of the water molecules into strong H-bonds with pre-adsorbed O on Ru can also quench the wettability of the surface.

Introduction

Understanding wetting at the molecular scale implies knowing the structure of water near interfaces. This is exceedingly important for a large number of applications, from atmospheric and environmental phenomena, to biology, catalysis and electrochemistry.[1,2] Only a good knowledge of the water structure at the molecular level can bring understanding and predictability to these important phenomena. The discussion in this paper will therefore be centered on the study of the adsorption, reactions and growth of water on surfaces. In our laboratory we have developed and applied scanning tunneling and atomic force microscopies (STM, AFM) as

[a]*Lawrence Berkeley National Laboratory, Berkeley, CA, 94720, USA*
[b]*Materials Science and Engineering Department, University of California, Berkeley, CA, 94720, USA*
[c]*Physics Department, University of California, Berkeley, CA, 94720, USA*
[d]*Stanford Synchrotron Radiation Laboratory, SLAC, 2575 Sand Hill Road, Menlo Park, CA, 94025, USA*

† Present address: Surface Chemistry Laboratory, RIKEN (The Institute of Physical and Chemical Research), 2-1, Hirosawa, Wako, Saitama, 351-0198, Japan
‡ Present address: Centre d'Investigació en Nanociència i Nanotecnologia CIN2 (ICN-CSIC), Campus UAB, 08193 Bellaterra, Spain

well as photoelectron spectroscopy techniques (XPS and XAS) to study water adsorption on pure metals (Pd, Ru, Cu), oxides (mica, TiO_2, Cu_2O, SiO_2, Al_2O_3) and metal surfaces modified by pre-adsorbed oxygen (Ru). In all cases we find that the formation of structures that expose dangling H-bonds from water or hydroxyl groups is crucial to the subsequent adsorption and 2-dimensional growth of water films.

Orientation and wetting properties of water on mica

Atomic force microscopy was used in the non-contact electrostatic force mode called scanning polarization force microscopy (SPFM), to image the structure of water in equilibrium with vapor.[3,4] We found that at room temperature water forms a complete monolayer when the relative humidity (RH) reaches between 70 and 80%, although the film is nearly complete at humidities as low as 30–40%. The SPFM images reveal that the first and second monolayers form hexagonal structures in good orientational epitaxy with the mica lattice.[4] Sum-frequency generation (SFG) vibrational spectra, shown in Fig. 1, revealed that the bonding and orientation of the water molecules is such that no free or dangling H-bonds exist in the complete first monolayer,[5] as shown by the absence of the characteristic O–H stretch vibration mode at 2740 cm^{-1} (for D_2O, 3600 cm^{-1} for H_2O). However, above the first monolayer the free or dangling H became visible, as shown in the Figure. In these experiments, due to the small local heating from the laser beam, the ambient humidity is larger than on the surface. As we show later, in the absence of laser

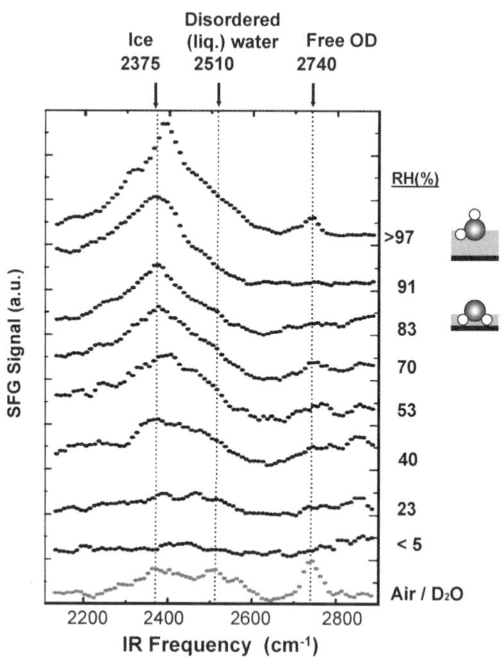

Fig. 1 Sum-frequency generation spectra of water on mica as a function of relative humidity. The spectrum for the neat water/air surface (deuterated) is shown in grey at the bottom for reference. Due to a small local heating the ambient relative humidity (RH) is 91% when the monolayer is saturated, which under normal conditions occurs around 80% RH. The most important finding of this experiment is that at the completion of the monolayer the peak due to stretching of the free H-bonds, at 2740 cm^{-1}, is absent. The peak becomes visible when the water coverage exceeds 1 monolayer.

heating the monolayer is completed around 70% RH. This result indicates that in the second and subsequent layers, water molecules exhibit free dangling H at the vapor interface.

Supporting these findings, contact potential measurements of the mica surface as a function of RH with SPFM operating in the Kelvin probe (KP) mode revealed a decrease of surface potential, which indicates that on average the water dipole is pointing towards the surface.[6] After the monolayer was completed, additional water adsorption changed the potential to positive values. These findings provide additional support for a model of the growth of water on mica at room temperature, where the molecular dipole points down on average and where no free H-bonds are present in the first layer. Reversal of the average dipole direction occurs in the second and following layers.

We propose that the hydrophilicity of mica is related to the fact that dangling H from the water molecules are formed easily above the monolayer upon adsorption of additional water.

A recent result by Michaelides and Hu is relevant and should be mentioned here.[7] These authors concluded using DFT calculations that on kaolinite, water does not form a stable layer on the basal plane termination that exposes O only, while a 2-dimensional film is stable on the opposite surface that exposes OH groups with dangling H-bonds. The authors point out the relevant fact that these groups can reorient on that surface termination to expose the H-bonds.[7]

Two-dimensional ice rules

We switch now to the behaviour of water on metal surfaces. Using low-temperature STM we studied the adsorption of water on Pd(111) and Ru(0001); both surfaces have hexagonal symmetry and lattice parameters that are within a few percent of a good match with those of the basal plane of Ih ice. At low coverage, water molecules form small clusters made of side-sharing hexamer units of 6 water molecules. We found that the flat honeycomb structures formed by side-sharing hexamers are limited laterally to no more than two or three hexagonal cells. This is shown in Fig. 2 for Pd. The surface temperature is 100 K and the honeycomb clusters can be seen to have widths of only a few cells.[8] A similar observation was made for Ru(0001): when the temperature is below 140 K, where water also forms limited-size clusters with the molecule oriented flat and parallel to the surface maximizing

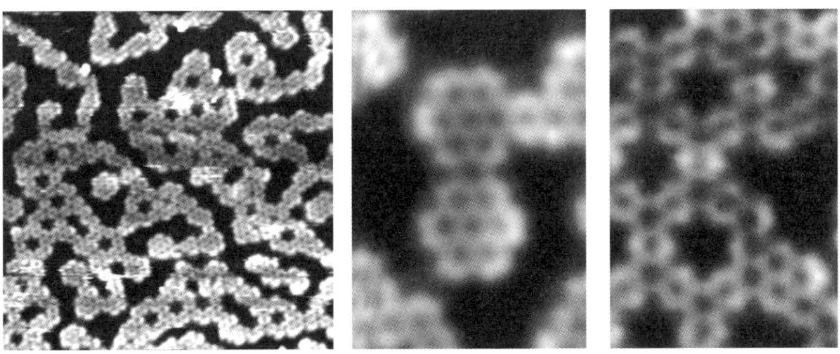

Fig. 2 STM images of a partial monolayer of water on Pd(111) at 100 K. Hexagonal clusters of ice form in epitaxial relationship with the substrate, with ($\sqrt{3} \times \sqrt{3}$) R30 structure. The Pd atoms in the dark region are not visible due to their very weak contrast (~2 pm), compared to that of the water clusters (~50 pm). The expanded views on the right show the limited size of the clusters, which are due to the saturation of the H-bonding by the water molecules laying flat (molecular plane parallel to the surface). The clusters stop growing laterally when dangling H-bonds form at the periphery.

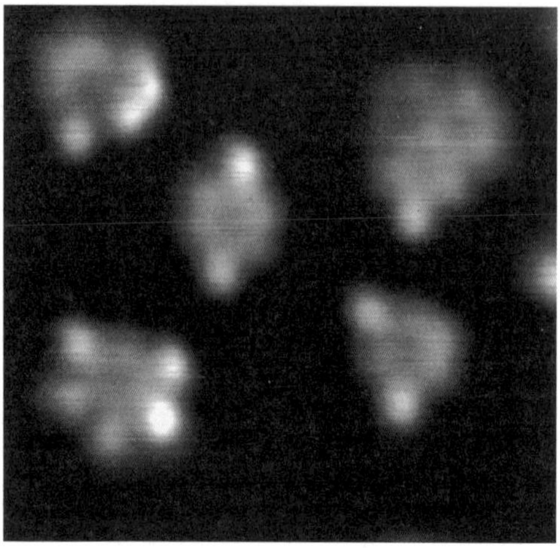

Fig. 3 Hexagonal clusters of water on Ru(0001) formed by exposure of the sample to 2×10^{-8} Torr of water at 50 K followed by heating to 130 K and cooling back to the imaging temperature of 50K. In this image 4 hexamers are visible that have additional water molecules attached to the edges (3, 5, 2 and 2). In subsequent images taken at 1 min intervals, the position of these molecules changes as they visit each periphery site without any preference. A larger cluster of 3 fused hexamers with one additional water molecule at the edges is also visible. The corners of the hexagons in the images are at the position of the water molecules and are located on top of Ru atoms. The Ru atoms are not visible due to their weak contrast of ~2 pm, as compared to the 40 pm of the water in the clusters and the 100 pm of the additional water molecules at the periphery.

in-plane H-bonds and terminating with dangling H at the peripheries. In the example shown in Fig. 3, we can see a few hexamers with additional water molecules attached at the edges. Even at 50 K these additional molecules are moving along the periphery. When 2 or more molecules are attached to the edges they can occupy contiguous or opposed corners, with no observed change in the STM contrast of the hexamer ring. This indicates that the water molecules in the hexamer are equivalent, *i.e.* have the same bonding configuration with the metal atoms underneath.

These results can be interpreted by postulating a set of "2D ice rules". According to these rules: (a) water molecules bind to metal atoms on a top position through their lone-pair orbital in a flat-lying geometry, as predicted theoretically;[9,10] (b) the water molecules form as many H-bonds with other molecules as possible, *i.e.* 3 inside the cluster, two being donors and one acceptor; (c) cluster growth terminates when a H-bond cannot be satisfied in this way. The unsatisfied or non-donor bond is therefore confined to the periphery of the cluster. It is obvious from geometrical considerations that the rules can only be satisfied when the cluster size is limited in at least one dimension. Our experiments indicate that the 2D ice rules are followed remarkably well on both the Pd and Ru hexagonal planes at temperatures below 130 K, although they break down above that temperature where extensive honeycomb structures are formed. Clearly the 2D ice rules describe a metastable growth pattern with an important local energy minimum. Growth beyond the limited size requires energy to overcome barriers, which trigger the formation of extended hexagonal structures where alternating molecules are flat, with 2 donor and 1 acceptor bonds, and vertical, with 2 acceptor and 1 donor H-bonds. The other H is not involved in a H-bond and points either towards the surface in Pt [11] and Pd, or away towards the vacuum.

If the temperature is maintained below 100 K for Pd and Ru, the addition of more water to the surface results in the formation of 3D structures, even if the metal surface is not yet covered with a continuous film. The formation of a second layer

 This journal is © The Royal Society of Chemistry 2009

before the first one is completed indicates again that the surface is hydrophobic. This result could be interpreted as due to the lack of dangling H atoms in the first water layer and to the difficulty of "turning around" the molecules in the first later to provide donating H-bonds to the second layer. This process might be either energetically unfavorable or require and activation energy, as in the mica example discussed above. Interestingly, theoretical calculations by Michaelides *et al.*[12] indicate that the energy difference between H-up and H-down extended structures is very small, which would suggest that a second layer could favorably bond with a modified first layer. Experiments to explore the structure of the second layer of water are thus highly desirable and are being planned in our laboratory.

We should mention here the work of Kimmel *et al.*[13] on the growth of water on Pt(111). In their experiments using thermal programmed desorption of Xe on the water films they observed that water grows in a 3D manner beyond the first monolayer after annealing to 135 K, a result that they interpret as due to the lack of dangling H-bonds. Similar observations were made by Hodgson and Haq[14] on Ru(0001) where, according to these authors, the strong binding of the first layer rendered the surface hydrophobic. We will show below that this behavior is also followed by a modified Ru(0001) surface by the preadsorption of 0.25 monolayers of oxygen.

Wetting in the presence of OH groups

We now turn our attention to oxide surfaces where hydroxyl groups can form easily. The experiments were performed using a special X-ray photoelectron spectrometer designed in our laboratory that operates under ambient pressures of gases (a few Torr),[15] sufficient to reach humidities near 100% at temperatures close to the triple point of water. In this manner we can study surfaces in thermodynamic equilibrium with water vapor, which is extremely important in environmental studies.[16]

We describe two experiments here involving single-crystal $TiO_2(110)$ and a thin (less than 1.5 nm) film of Cu_2O formed by thermal oxidation of a polycrystalline Cu foil. In both cases XPS reveals the surface composition clearly differentiating between lattice O in the oxide substrate, O in OH groups and H_2O. In addition a peak due to the gas phase above the surface can also observed. We follow the growth of water by measuring the areas under the peaks and after calibration of the XPS signals. Fig. 4 shows the results for TiO_2. The results can be summarized as follows: after preparation of the TiO_2 surface by sputtering and annealing in O_2, the surface contains a number of O vacancies that previous investigations have identified as located in the bridge O-rows.[17–19] Under our conditions the concentration of these vacancies is ≤ 0.125 ML. Water adsorbs efficiently on these sites where it dissociates giving rise to 2 OH groups. Molecular water always grows after saturation of the vacancies, forming strong bonds with the OH groups.[20]

Similarly, on Cu_2O the presence of a small amount of O vacancies gives rise to the formation of OH groups, which nucleate the subsequent adsorption of molecular water.[21] This is shown in Fig. 5.

The clean Cu metal is a particularly interesting case in our discussion. We found out that the (111) and (110) surfaces of this metal behave in surprisingly different ways. On the (111) face no water adsorption was observed using XPS in the presence of 1 Torr of water at room temperature. This clearly indicates that the clean metal surface is hydrophobic.[22] On the (110) surface of the same metal however, partial water dissociation into OH and H occurred readily, which was then followed by molecular water adsorption.

Although not shown here, similar results were obtained on two other oxide surfaces: SiO_2 that contain OH groups as a result of the etching and oxidation procedure,[23] and Al_2O_3 films on Al foil.[21] Other examples, including Fe_2O_3 and MgO, are currently being investigated in our laboratory.

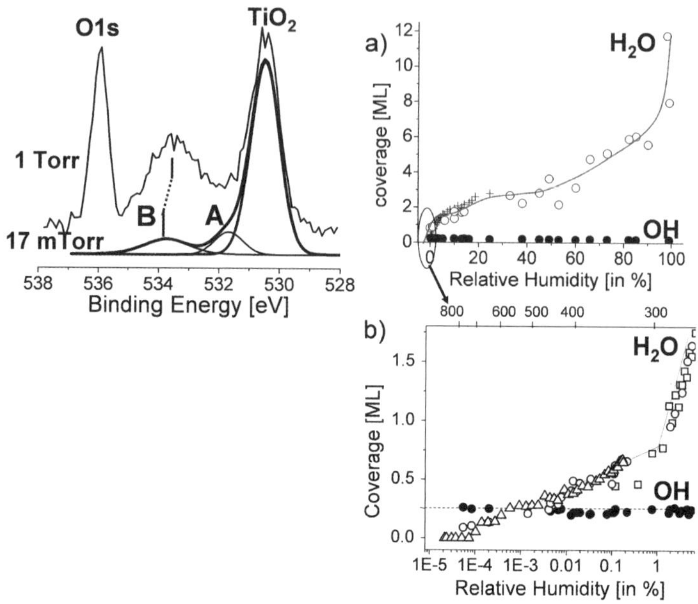

Fig. 4 Left: X-ray photoelectron spectra of water on TiO_2(110) showing peaks due to excitation of the O1s electrons by 690 eV photons. Various O species can be distinguished that correspond to lattice O in TiO_2 (lowest binding energy), hydroxyls, adsorbed molecular water (near 534 eV) and gas-phase H_2O (536 eV). By measuring the areas under the peaks, isotherms and isobars of the amount of surface water and hydroxyls can be determined and are shown on the right. Panel (b) shows an expansion of the circled area near 0% RL in panel (a). Isotherms and isobars collapse into a single curve when plotted against relative humidity. OH forms by dissociation of water on O-vacancies in the bridge O rows. Two OH groups form for every defect site. Subsequently water molecules bind to the OH groups forming several layers well before reaching the saturation humidity of 100%, *i.e.* with good wetting.

Adsorption of water on an O-covered Ru(0001) surface

In this last example we explore the effect of blocking the formation of dangling H-bonds on the wetting properties of a Ru(0001) surface. Such blocking can be accomplished by pre-adsorbing 0.25 monolayer of oxygen on the surface. The oxygen atoms form a 2×2 structure on the Ru surface with each O atom sitting on an hexagonal close-packed (hcp) site.[24] On this surface water adsorbs readily forming also a 2×2 structure and does not dissociate.[25] We found that water adsorbs on the top Ru site inside the 2×2 cell, forming a covalent bond with Ru through the lone-pair orbital and two additional H-bonds with the chemisorbed O of the 2×2 structure. This is shown by the STM images acquired at 6 K, in Fig. 6. The drawing on the right shows the DFT-calculated structure by Cabrera-Sanfelix *et al.*[26] The result of interest for our discussion is that the H-bonds to the chemisorbed O atoms are strong enough to prevent H-bonding between water molecules, which then remain separated from each other in their 2×2 unit cells. As the coverage of water increases beyond 0.25 monolayers, the molecules in excess of 0.25 ML are weakly bound and are easily moved about by interaction with the tip, *i.e.* no stable second layer is formed. This result indicates that once the H atoms of the water molecule are tied strongly and prevented from presenting dangling free on the surface, the surface becomes hydrophobic. We found also that heating beyond 140 K results in desorption of the water with no dissociation, unless defects (missing O atoms) in the initial O(2×2) structure are present. These results are again in line with the previous results presented in this paper. This time we have shown that if water is prevented from reorienting to present dangling H atoms, no wetting takes place.

 This journal is © The Royal Society of Chemistry 2009

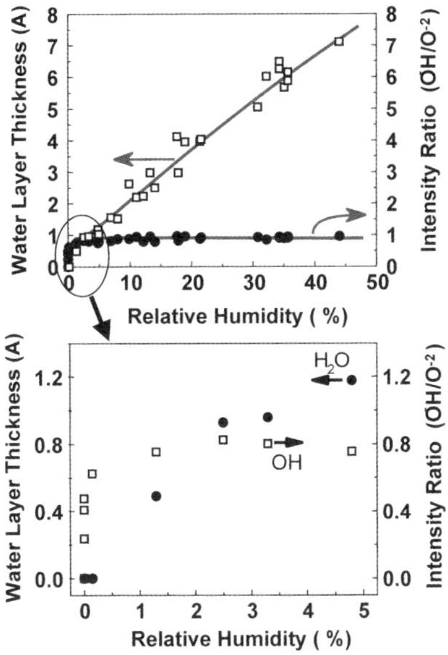

Fig. 5 Adsorption isobars and isotherms of water adsorption on polycrystalline Cu_2O measured from the areas under the XPS peaks of the corresponding O species, similar to those shown in Fig. 4 for TiO_2. Both isobars and isotherms collapse into the same curve when plotted against relative humidity. The lower panel shows an expansion of the circled area near 0% RL in the upper panel. As in the case of TiO_2, formation of OH precedes the formation of molecular water films.

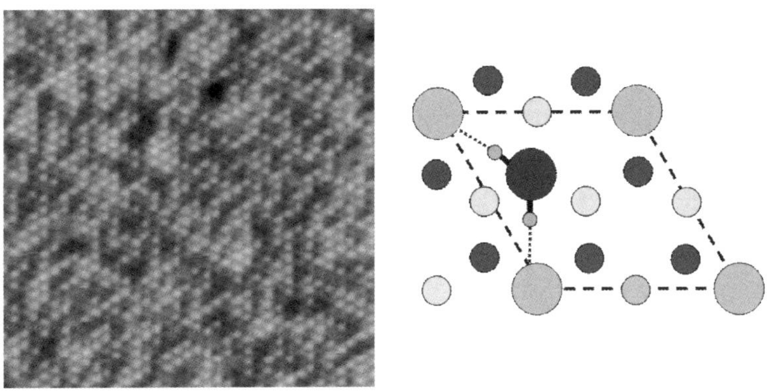

Fig. 6 Left: STM image (18 nm × 18 nm) of an O(2 × 2)-Ru(0001) surface with a water layer on top with a coverage of 0.18 ML, acquired at 6 K. The bright spots correspond to water molecules. The underlying 2 × 2 structure (dark) is from an O overlayer on Ru. The molecules adsorb as shown in the schematic drawing on the right, with the O of the water molecule (large dark circle) on top of a Ru atom (small dark circles are from Ru atoms in the first layer, small light gray circles are from second layer Ru atoms). The 2 H atoms from the water molecule form H-bonds with the O atoms of the 2 × 2 cell. Image acquired at a constant current of 95 pA and an applied bias voltage of −220 mV, after annealing to 140 K.

Conclusions and outlook

All the examples discussed here support a mechanism for molecular-scale wetting by water that is strongly dependent on the ability of the surface to produce dangling H-bonds, from either molecular water (mica, Pd, Ru), or OH groups (oxides with defects). We have also shown how pre-adsorbed oxygen on Ru(0001) forms strong H-bonds with water, blocking the formation of new H-bonds with other water molecules, making the surface hydrophobic. Although a firmly established mechanism cannot yet be presented, the observations clearly point to the importance of dangling H-bonds. Several experiments are suggested, with some underway in our laboratory, to clarify some of the questions raised by the results and discussion presented in this paper. One is the study of water orientation on mica at temperatures below 0 °C using SFG beyond the first monolayer. Another is STM (and/or AFM) experiments to determine the molecular-level structure of multilayer water films as a function of layer thickness at various temperatures. Finally, it would be important to explore the adsorption of water on perfect oxide surfaces, with no O-vacancy defects, to see if OH groups can still form and the effect on wetting.

Acknowledgements

This work was supported by the Director, Office of Energy Research, Office of Basic Energy Sciences, Materials Sciences Division, of the U. S. Department of Energy under Contract No. DE-AC02-05CH11231. A. M. was financed by the Marie Curie Outgoing International Foundation, Project No. 514412. T. H. also acknowledges financial support from the Ramon Areces Foundation.

References

1 M. A. Henderson, *Surf. Sci. Rep.*, 2002, **46**, 5, and references therein.
2 A. Verdaguer, G. M. Sacha, H. Bluhm and M. Salmeron, *Chem. Rev.*, 2006, **106**, 1478–1510.
3 J. Hu, X.-d. Xiao, D. F. Ogletree and M. Salmeron, *Science*, 1995, **268**, 267. Note: in this paper the water structures formed are due to contact with the tip and correspond to a second water layer, not the first (see ref. 4).
4 L. Xu and M. Salmeron, *Nano-Surface Chemistry*, ed. M. Rosoff, Marcel Dekker, New York, 2001, vol. 6, pp. 243–287.
5 P. B. Miranda, L. Xu, Y. R. Shen and M. Salmeron, *Phys. Rev. Lett.*, 1998, **81**, 5876.
6 H. Bluhm, T. Inoue and M. Salmeron, *Surf. Sci.*, 2000, **462**, 1–L599.
7 Xiao Liang Hu and Angelos Michaelides, *Surf. Sci.*, 2008, **602**, 960.
8 J. Cerda, A. Michaelides, M. L. Bocquet, P. J. Feibelman, T. Mitsui, M. Rose, E. Fomin and M. Salmeron, *Phys. Rev. Lett.*, 2004, **93**, 6101.
9 A. Michaelides, V. A. Ranea, P. L. de Andres and D. A. King, *Phys. Rev. Lett.*, 2003, **90**, 216102.
10 S. Meng, E. G. Wang and S. Gao, *Phys. Rev. B*, 2004, **69**, 195404.
11 H. Ogasawara, B. Brena, D. Nordlund, M. Nyberg, A. Pelmenschikov, L. G. M. Petterson and A. Nilsson, *Phys. Rev. Lett.*, 2002, **89**, 276102.
12 A. Michaelides, A. Alavi and D. A. King, *J. Am. Chem. Soc.*, 2003, **125**, 2746.
13 G. A. Kimmel, N. G. Petrik, Z. Dohna'lek and B. D. Kay, *Phys. Rev. Lett.*, 2005, **95**, 166102.
14 S. Haq and A. Hodgson, *J. Phys. Chem. C*, 2007, **111**, 5946.
15 D. Frank Ogletree, H. Bluhm, G. Lebedev, C. Fadley, Z. Hussain and M. Salmeron, *Rev. Sci. Instrum.*, 2002, **73**, 3872.
16 M. Salmeron and R. Schlögl, *Surf. Sci. Rep.*, 2008, **63**, 169.
17 S. Wendt, J. Matthiesen, R. Schaub, E. K. Vestergaard, E. Lægsgaard, F. Besenbacher and B. Hammer, *Phys. Rev. Lett.*, 2006, **96**, 066107.
18 S. Wendt, R. Schaub, J. Matthiesen, E. K. Vestergaard, E. Wahlstrom, M. D. Rasmussen, P. Thostrup, L. M. Molina, E. Lægsgaard, I. Stensgaard, B. Hammer and F. Besenbacher, *Surf. Sci.*, 2005, **598**, 226.
19 O. Bikonda, C. L. Pang, R. Ithnin, C. A. Muryn, H. Onishi and G. Thornton, *Nat. Mater.*, 2005, **5**, 189.
20 G. Ketteler, S. Yamamoto, H. Bluhm, K. Andersson, D. E. Starr, D. F. Ogletree, H. Ogasawara, A. Nilsson and M. Salmeron, *J. Phys. Chem. C*, 2007, **111**, 8278.
21 Xingyi Deng, Tirma Herranz, Christoph Weis, Hendrik Bluhm and Miquel Salmeron, *J. Phys. Chem. C*, 2008, **112**, 9668.

 This journal is © The Royal Society of Chemistry 2009

22 Susumu Yamamoto, Klas Andersson, Hendrik Bluhm, Guido Ketteler, David E. Starr, Theanne Schiros, Hirohito Ogasawara, Lars G. M. Pettersson, Miquel Salmeron and Anders Nilsson, *J. Phys. Chem. C*, 2007, **111**, 7848.
23 A. Verdaguer, C. Weis, G. Oncins, G. Ketteler, H. Bluhm and M. Salmeron, *Langmuir*, 2007, **23**, 9699.
24 M. Lindroos, H. Pfnür, G. Held and D. Menzel, *Surf. Sci.*, 1989, **222**, 451.
25 M. J. Gladys, A. Mikkelsen, J. N. Andersen and G. Held, *Chem. Phys. Lett.*, 2005, **414**, 311.
26 Pepa Cabrera-Sanfelix, Daniel Sanchez-Portal, Aitor Mugarza, Tomoko K. Shimizu, Miquel Salmeron and Andres Arnau, *Phys. Rev. B*, 2007, **76**, 205438.

Order and disorder in the wetting layer on Ru(0001)

Mark Gallagher, Ahmed Omer, George R. Darling and Andrew Hodgson*

Received 9th May 2008, Accepted 11th June 2008
First published as an Advance Article on the web 30th September 2008
DOI: 10.1039/b807809b

The growth of an intact water monolayer on Ru(0001) has been investigated by comparing the ordering of O and Ru, determined by low-energy electron diffraction (LEED), with that of the top layer of O and H, as probed by He atom scattering (HAS). Although LEED shows that water forms an extended commensurate ($\sqrt{3} \times \sqrt{3}$) R30° structure as the coverage approaches 0.67 monolayer, the HAS distributions are insensitive to the exact water coverage and show a very low specular reflectivity, indicating a disordered water layer. The angular profile from a D_2O monolayer shows a broad diffuse peak in the angular scattering distribution at a momentum exchange similar to the position of the second-order (1/3,1/3) peaks, but the maxima show little variation with scattering azimuth. H_2O shows a slightly higher He reflectivity and more clearly resolved angular structure, with broad, faint peaks appearing close to the first-order diffraction positions. The origin of this disorder is discussed based on density functional calculations for the monolayer which find that water forms chains of flat and H-down molecules within a hexagonal hydrogen-bonding network, rather than the ice bilayer usually assumed. This arrangement leads to long-range order in the O location, but disorder in the O height and the proton orientation. We discuss how this combination of lateral order in the adsorption site, but disorder in the water orientation, is reflected in the sharp $\sqrt{3}$ LEED pattern but diffuse, broad peaks in He scattering.

1. Introduction

The wetting of metal surfaces represents an important test of our understanding of how water adapts its hydrogen-bonding network in response to adsorption at a surface.[1,2] Clearly this is a problem with many different scientific ramifications, from understanding how water nucleates in the atmosphere[3] to how the structure of water near interfaces affects other properties, such as electrochemical reactivity.[4] Despite its apparent simplicity, water adsorption has proved particularly difficult to study and a systematic understanding of the factors influencing wetting and film growth is emerging only slowly.[5]

The wetting behaviour of Ru(0001) has become a particular test for our understanding of water adsorption at metal surfaces. It was largely based on the early studies of this system that the well-known bilayer model for water adsorption became established[1] and, perhaps as a result, this surface has received more interest than any other. Electron-stimulated desorption ion angular distributions (ESDIAD),[6–8] vibrational spectroscopy using electron energy-loss spectroscopy (EELS)[7,9,10] and

The University of Liverpool, Surface Science Research Centre, Oxford Street, Liverpool, UK L69 3BX. E-mail: ahodgson@liverpool.ac.uk; Tel: +44 (0)151 794 3536

reflection–absorption IR (RAIRS),[11,12] low-energy electron diffraction (LEED)[7,8] and temperature-programmed desorption (TPD)[6–8] have all been used to investigate the monolayer structure and the influence of O coadsorption. Vibrational spectroscopy[7,12] shows that water forms a hydrogen-bonded network even at low temperatures (85 K). At higher temperatures water orders into a structure containing two thirds of a monolayer (ML) water which shows a $(\sqrt{3} \times \sqrt{3})$ R30° LEED pattern.[7,8] Based on the LEED and ESDIAD results, Doering and Madey proposed that water was adsorbed in a commensurate buckled bilayer, similar to the (0001) plane of ice Ih.[8] In this model water is adsorbed in a hexagonal, honeycomb network, with alternate water molecules adsorbed O-down atop Ru with the other water hydrogen-bonded between these in the upper half of the bilayer, without any direct bond to Ru. Each water has three hydrogen bonds, leaving the waters in the upper half of the layer with one uncoordinated H atom, assumed to be dangling out into the vacuum.[1] This has become the standard model for water adsorption,[1,2] sometimes modified by allowing the uncoordinated H atoms to point 'H-down' towards the metal surface.[13]

Held and Menzel made a detailed analysis of the D_2O structure based on LEED IV simulations of two O atoms in a $(\sqrt{3} \times \sqrt{3})$ R30° unit cell.[14] The O atoms of water in the two halves of the bilayer were found to be adsorbed atop Ru, almost co-planar ($\Delta z_{(O-O)} = 0.10 \pm 0.02$ Å), with the first Ru layer buckled in anti-correlation to the O height. They concluded that half the waters do indeed have a chemical bond to Ru atoms, with the other half hydrogen-bonded in a second layer between them with the uncoordinated H pointing 'H-up' into the vacuum, but that this structure is considerably flattened compared to an ideal ice bilayer ($\Delta z_{(O-O)} = 0.96$ Å).[15] This study still represents the only detailed structural analysis of a wetting layer on a metal surface.

Feibelman[16] re-ignited interest in this problem when he showed that the calculated binding energy of water, based on the established bilayer model, is too low for water to wet the surface—it would prefer to form 3D clusters and bare Ru. A more stable, partially dissociated $H/OH/H_2O$ structure was found and proposed for the wetting structure. Although density functional theory (DFT) finds that the mixed OH/H_2O phase has O approximately coplanar, a reanalysis of the IV simulations concluded this structure was probably not responsible for the $\sqrt{3}$ layer seen in the LEED IV study.[15] This controversy has stimulated a number of experimental studies that show the water/Ru system is more complex than was originally assumed. Two different water structures are formed, depending on the isotope and the adsorption conditions. At low temperatures (below *ca.* 150 K) a metastable intact water layer is formed, which wets the surface.[17,18] This structure shows a dramatic kinetic isotope effect; heating the monolayer causes D_2O to desorb intact, whereas H_2O decomposes to form a partially dissociated structure that is stable up to 210 K.[17] The same mixed OH/H_2O (OD/D_2O) structures can be formed by reacting a low coverage of O ($\theta <$ 0.09 ML) with water. The confirmation that a partially dissociated $H/OH/H_2O$ structure is formed on Ru(0001)[19] provides further support for the validity of using DFT to investigate water adsorption on metal surfaces.[16]

The conclusion that an intact water bilayer is not expected to wet Ru(0001)[16] is supported by a number of calculations,[20–22] all of which find the bilayer binding energy is less than in bulk ice. A comparison with Ar adsorption indicates that this discrepancy cannot be explained by the incomplete inclusion of dispersion interactions in the DFT calculations,[23] leaving the structure of the intact film unclear. Using RAIRS and LEED, we have previously shown that water forms small, flat lying clusters on Ru(0001), only aggregating into an extended hydrogen-bonded network as the coverage increases.[24] The $(\sqrt{3} \times \sqrt{3})$ R30° structure is observed in LEED as the coverage approaches 0.67 ML, but then the diffraction pattern disappears as the structure compresses to accommodate more water.[8,24] The flattened 'H-up' water layer proposed from LEED IV simulations[14,15] is inconsistent with the absence of a free-OH stretching signal in sum-frequency generation (SFG)[25] and RAIRS,[17,24] or with the calculated O corrugation.[16] Based on DFT calculations

 This journal is © The Royal Society of Chemistry 2009

we have proposed a wetting structure consisting of chains of water lying flat, linked by H-down chains into a hexagonal hydrogen-bonded network. This structure has a binding energy 20% greater than the bilayer structure, sufficient to wet the Ru(0001) surface.[24]

Here we compare He atom scattering (HAS) and LEED measurements on water films and discuss the ordering and structure of the intact water layer. Our results show that, even when LEED finds a well-defined ($\sqrt{3} \times \sqrt{3}$) R30° ordering associated with the O and Ru atoms, HAS reveals the structure is highly disordered. We discuss the structure of the wetting layer, based on DFT calculations of the intact 0.67 ML water layer. These calculations find water adsorbs atop Ru, forming a hexagonal hydrogen-bonding network containing flat-lying chains of water, interspaced by chains of water oriented H-down.[24] This structure results in a well-defined lateral site for O, but does not have the simple alternation in O height that characterises a water bilayer. The water chains are intrinsically disordered, with no long-range correlation in the height of the O atoms or the location of the protons. We argue that these structures are consistent with the disorder seen in He scattering[26] and the well-ordered ($\sqrt{3} \times \sqrt{3}$) R30° pattern seen in LEED.[14,15]

2. Experimental

Experiments were performed in an ultrahigh vacuum (UHV) chamber, with a base pressure less than 5×10^{-11} Torr. In addition to the HAS apparatus described below, the main chamber was equipped with a low-current LEED system, based around a dual microchannel plate (MCP) amplifier. Sustained exposure to the electron beam during LEED measurements can cause significant restructuring of water films adsorbed on the Ru surface.[24,27] To protect against this, great care was taken to reduce the exposure time to the absolute minimum necessary and the beam current was kept low (<3 nA) at all times. These conditions allow us to maintain stable ice structures for the duration of our experiments, as shown in this paper and previously for Pt(111)[28] and Ni(111).[29]

The Ru crystal was aligned to ±0.25° of the (0001) face and was cleaned by repeated Ar^+ ion sputter and thermal anneal cycles (1100 K). The crystal was attached *via* Ta heating wires to liquid-nitrogen-cooled Ta support posts, allowing the crystal to be heated to 1400 K and cooled to 90 K within a few minutes. Sample temperature was measured by a K-type thermocouple spot welded to the back of the crystal and controlled *via* computer feedback to a DC heating supply, allowing the temperature to be ramped linearly for TPD measurements. Cleanliness was judged primarily from the D_2O desorption behaviour and LEED patterns of the ice monolayer, which are very sensitive to surface contamination. Mounting the Ru crystal directly on the heating wires ensures that nothing except the sample desorbs gas during heating measurements, allowing accurate assessment of the quality of the water layer. We have been cautious when using specular He reflectivity as a further measure of the surface quality, either for the clean metal surface or the water layer, for the reasons detailed in the next section.

Water layers have been grown using two separate experimental arrangements, to highlight the critical role of film preparation on the resultant ice structure. The first method involves dosing with a molecular beam (Fig. 1). The beam flux is typically in the order of 0.005 layers s^{-1} and creates a pressure increase of $\leq 2 \times 10^{-11}$ Torr in the chamber during adsorption. A retractable flag is placed in front of the crystal, to accurately control the surface exposure to the molecular beam. For D_2O the water coverage was obtained by measuring directly the uptake required to form a saturated ice monolayer at the surface, using the King and Wells direct-reflection technique.[30] TPD measurements provided a second measure of the water coverage by calibration against the saturated monolayer (defined here as 1 layer of water). For H_2O, dissociation of the layer during heating makes accurate measurement of the coverage more complicated. Typically the saturation coverage is determined from the dose

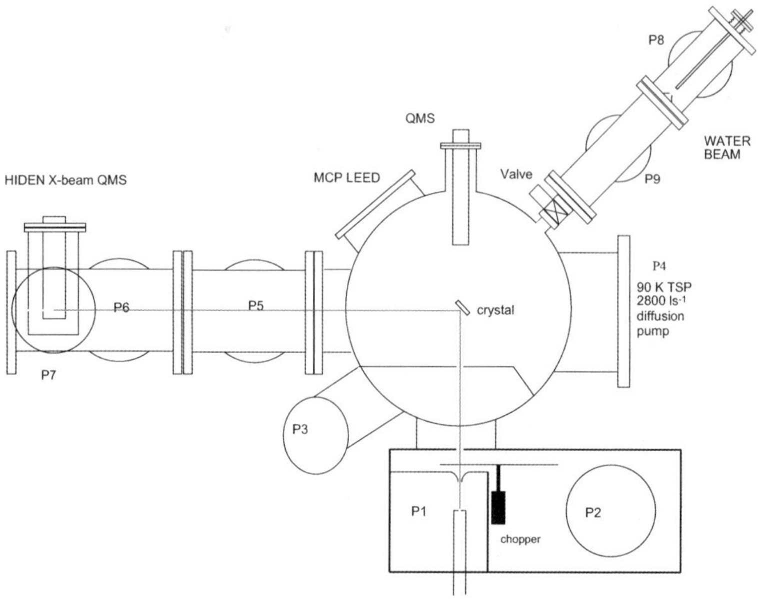

Fig. 1 Schematic diagram showing the experimental chamber (not to scale). The chamber is separated into four main sections: the helium source and chopper chambers, the main experimental/scattering chamber, the detector path chambers and the molecular beam section. The helium atom source consists of a temperature-controlled nozzle with differential pumping (P1, P2 and P3) by 9″, 6″ and 6″ diffusion pumps respectively. The detector arm is at a scattering angle of 90° and contains a cross-beam QMS, triply differentially pumped by turbomolecular pumps (P5, P6), themselves backed by diffusion pumps, and an ion pump (P7), giving base pressures in the low 10^{-11} Torr range. The water beam consists of an effusive glass nozzle, pinhole skimmer and collimating aperture, differentially pumped by 240 l s^{-1} turbo pumps (P8 and P9), and is arranged so that the water beam just fills the crystal surface. The main chamber is also equipped with a variable leak valve and collimated tube doser (not shown). Further details can be found in the text.

time required before appearance of a multilayer peak in the TPD spectra, which is found to be consistent with the D_2O dosing conditions. The detector used for TPD of beam-dosed layers, was an unshielded VG Smart-IQ quadrupole mass spectrometer (QMS), located near the surface in the main experimental chamber.

The second dosing method used was background dosing *via* a collimating tube, supplied by a variable leak valve (not shown in Fig. 1). In contrast to the beam-dosed layers, this leads to additional adsorption on the surface surroundings (support posts, heating wires, edge and sides of crystal *etc.*). In order to obtain TPD spectra characteristic of desorption from the surface alone, the QMS detector is used in a stagnation configuration, housed inside a retractable gas-collection tube. Water desorbing from the sample surroundings is shielded from entering the detector housing, whereas gas from the surface enters through a 4 mm hole placed directly in front of the sample. The TPD spectra obtained from this method are consistent with those from beam-dosed layers, with an additional time constant due to the pump out of the detector volume. This time constant can be simply modelled, and the tail removed from the TPD, allowing direct comparison with beam-dosed layers.

Although the TPD spectra are broadly consistent between the two dosing methods, there is a significant difference in the intensity of the partially dissociated TPD peak and in the LEED and HAS data for water layers prepared using the beam or collimated doser, as we will discuss later. We consider this to be due to the higher purity achievable with the molecular beam and is particularly important when dosing H_2O, where dissociation can occur. The LEED and TPD results, for both beam- and background-dosed water films, have been repeated in a number of separate experimental chambers, using different Ru(0001) crystals and a conventional (VG) LEED unit, confirming the reproducibility of the data for this system.[17,24,31]

The HAS apparatus consists of seven independently pumped sections; the first chamber containing the He beam source. High-purity helium (BOC) is supplied to the nozzle at a stable pressure of up to 10 bar, passing first through a nitrogen-cooled coil to remove any remaining condensable impurities. The nozzle assembly is enclosed within a liquid nitrogen dewar, which, coupled with dc resistive heating of the tip, gives a working temperature range of approximately 110 to 380 K. The nozzle temperature is stable to within ±0.1 K, as measured by a K-type thermocouple spot welded to the tip, and corresponds to beam energies of 26–80 meV, as confirmed by time-of-flight (TOF) measurements using the retractable chopper (Fig. 1). The He beam is incident at the centre of the crystal, mounted on a VG (x,y,z,θ) manipulator whose axis can also be tilted relative to the scattering plane. The sample is mounted at the centre of the azimuthal rotation of the manipulator (θ), with the surface perpendicular to the incident beam. The incident and reflected beam paths are fixed at 90° to each other and the crystal is rotated about θ by a stepper motor under computer control. The reflected beam leaves the experimental chamber through a 3 mm aperture, passing through two stages of differential pumping before entering the detector (Hiden HAL 301 X beam QMS). The entrance to the mass spectrometer housing is a 3 mm aperture with the beam exiting through a 4 mm hole, in the time-of-flight configuration. The final aperture can be blocked off to use the QMS as a stagnation detector, as is the case for the scattering data shown here. In stagnation mode the clean Ru surface gives a maximum reflected intensity of around 1.5×10^6 cps in the specular beam. The HAS measurements presented here were recorded at 300 K corresponding to an energy, incident wavevector and energy resolution of $E = 65$ meV, $k_i = 11.1$ Å^{-1} and $\Delta E/E = 7.0\%$ respectively. The azimuthal orientation of the sample was initially set using LEED and then optimised using He diffraction from a (2×2) O overlayer.

3. Results and discussion

3.1 Preparation of intact H_2O and D_2O layers

The dashed black line in Fig. 2 shows the TPD of D_2O recorded at a ramp rate of 0.5 Ks^{-1}. The simple two-peak desorption profile observed is consistent with an intact water layer, as previously reported.[17] The onset of multilayer desorption begins at ~142 K, with the peak maximising at ~160 K. The high-temperature peak near 180 K originates from desorption of the intact wetting layer that is in contact with the surface. Dissociation of this layer, either due to contamination during adsorption or induced during structural measurements, stabilises water on the surface and gives rise to a desorption peak near 210 K.[17] The weak tail to the QMS signal above 190 K is caused by D_2O adsorption/desorption at the chamber walls; this interpretation is supported by the He reflectivity measurements described later. The absence of any high-temperature peak near 210 K confirms the quality of the ice layer grown and, as such, this TPD profile is used as the benchmark for a clean non-dissociated layer, to which we attach great importance. Subsequent to any scattering experiment, recording the TPD provides both an accurate coverage measurement and an indication of any dissociation in the structure being studied. We find that it is relatively simple to avoid damaging the ice layer with the low-current LEED optics and, as expected, the helium measurements have no impact on the TPD. This allows us to optimise the dosing conditions and then record the He profile and LEED for H_2O and D_2O layers formed at different coverage and growth temperatures.

A number of experimental factors were found to deteriorate the quality of the TPD profile. Firstly, although the general TPD characteristics are similar for layers

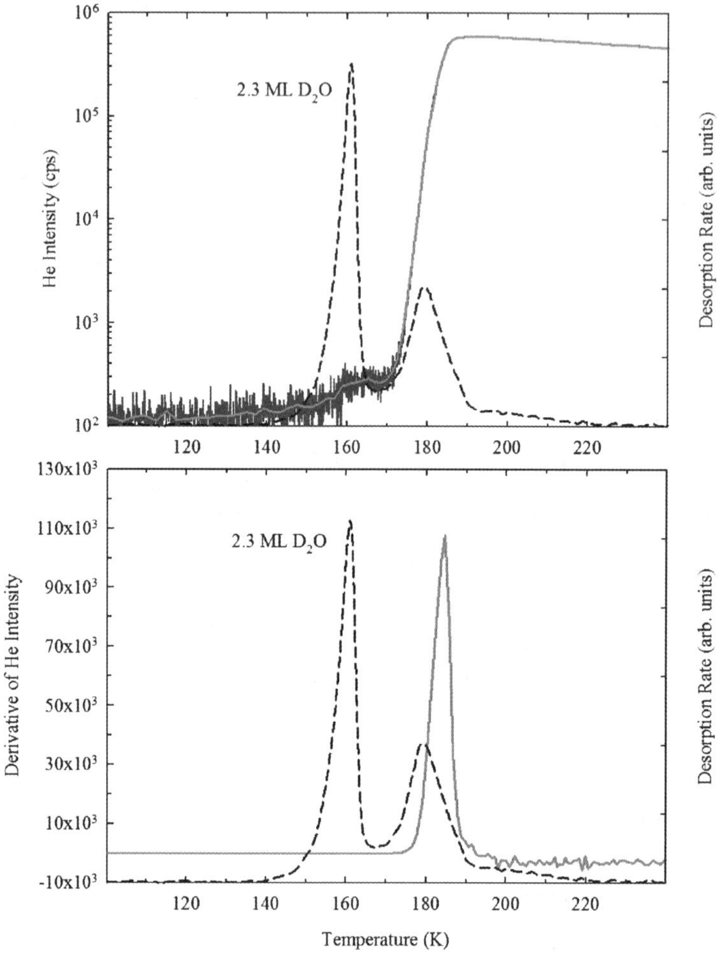

Fig. 2 Top: comparison of the He reflectivity (solid lines) and TPD (dashed black line) for a D_2O film deposited at 100 K showing their simultaneous measurement as the temperature is ramped at ~0.5 K s^{-1} to desorb the intact film. He beam energy = 65 meV. Bottom: the solid line corresponds to the derivative of the He reflectivity signal shown in the top panel as a function of temperature, showing that all the D_2O has desorbed from the surface by 190 K.

grown by beam or background dosing (not shown here) the area around 200 K was only completely featureless when water was dosed using the beam. We associate this with the reduced impurity concentration achievable with a molecular beam source, which transfers the source water purity directly to the sample. Whereas beam adsorption causes a negligible pressure increase in the main chamber, using the effusive source causes the pressure to increase to *ca.* 6×10^{-10} Torr and takes 5–10 min to recover to the base level. During this time water displaces other gases from the walls and contributes to contamination of the surface. Reducing the pressure increase in the chamber during adsorption by cooling the titanium sublimation pump to 80 K reduced the desorption intensity above 200 K, supporting this interpretation. This conclusion is also consistent with LEED data, especially for the case of H_2O where we found it impossible to obtain an intact $\sqrt{3}$ LEED pattern using the background-dosing method,[32] instead finding the streaked LEED pattern associated with domains of partially dissociated $OH/H_2O/H$.[24] Using the molecular beam to dose H_2O it is relatively simple to obtain a $(\sqrt{3} \times \sqrt{3})$ R30° LEED pattern, implying there

 This journal is © The Royal Society of Chemistry 2009

is always a small amount of contamination whilst background dosing. Although this leads to dissociation of H_2O layers, there is a strong kinetic isotope effect[17] and the impact on the D_2O TPD and LEED is less noticeable. Dissociation of the water layer also has significant implications for He scattering, as we discuss below.

Because of the extremely low reflectivity of the water monolayer relative to He, discussed below, any minor but ordered contamination becomes significant. By annealing the crystal at a higher temperature (1400 K) during the cleaning process, we induce an ordered impurity structure on the "clean" surface. The exact origin of this structure is unknown, but we associate it with segregation of impurities, probably C, from the bulk to form small regions of ordered superstructure on the surface. This structure is observed as sharp satellite peaks either side of the specular signal at a momentum transfer $\Delta k_{\|} = \pm 0.23$ Å^{-1}, which were stable over the entire water cover range from submonolayer to multilayer in He scattering. These peaks are only visible when the bare Ru reflectivity is suppressed by water adsorption (see below), otherwise they disappear underneath the specular peak. This structure also gives rise to a weak, multiple-spot pattern in the LEED at 10–40 eV, with the same $\Delta k_{\|}$, and is observed for both the bare metal and the water-covered surface.[33] The specular He reflectivity from the bare metal surface prepared in this way is very high (comparable to clean (1 × 1) Ru) and therefore the He reflectivity is not considered a good measure of surface cleanliness. As well as generating these satellite peaks, the high anneal temperature also resulted in a high temperature shoulder to the monolayer D_2O TPD peak, indicative of partial dissociation or contamination. To avoid this problem the anneal temperature was reduced to 1100 K, which is adequate for producing a good (1 × 1) surface, as evidenced by the LEED, HAS and water TPD.

3.2 HAS from H_2O and D_2O layers

The top half of Fig. 2 shows a simultaneous measurement of the change in He specular reflectivity (solid line) and TPD for 2.3 layers of D_2O (dashed black line). During heating the reflectivity is unchanged until ~150 K, where there is a small step as the multilayer desorbs. The intensity plateaus until the onset of monolayer desorption, where there is a sharp increase as the bare Ru surface becomes exposed. This is highlighted by the solid line in the lower pane of Fig. 2, which shows the derivative of the He intensity as the temperature is ramped. By comparison with the water TPD, the large change in He reflectivity clearly occurs as the monolayer desorbs, with no evidence for any increase above 190 K. The helium reflectivity on the plateau, corresponding to completion of the water monolayer, is very low, ~0.05% of the clean surface reflectivity, and does not depend on the preparation temperature (135 to 160 K). This reflectivity is lower than previous reports in the literature, where the reflectivity of the D_2O monolayer was found to be 0.2–0.3%[26] or 3–6%[33] of the bare metal signal, 5 and 100 times higher than we report. We find that an increased He reflectivity at saturation coverage occurs only when we have a partially dissociated layer, which we can induce either by deliberately exposing the water layer to electrons or by heating a structure containing H_2O above 150 K. Water layers grown using background dosing, or the collimated gas doser, also showed a higher reflectivity and a high-temperature tail to the TPD peak, indicating partial dissociation. In contrast the pure D_2O layer, prepared using the molecular beam, shows only a slow decrease in reflectivity above 190 K (Fig. 2), associated with Debye–Waller effects, without the increase near 200 K that is observed for a partially dissociated structure.

In order to investigate the structure of the ice layer we measured the angular scattering profile as a function of the water coverage and preparation conditions. Fig. 3 shows HAS from an intact D_2O layer whose coverage (0.67 ML) has been chosen to optimise the $(\sqrt{3} \times \sqrt{3})$ R30° LEED pattern. This LEED pattern has been widely reported on Ru(0001) and detailed measurements of the coverage dependence

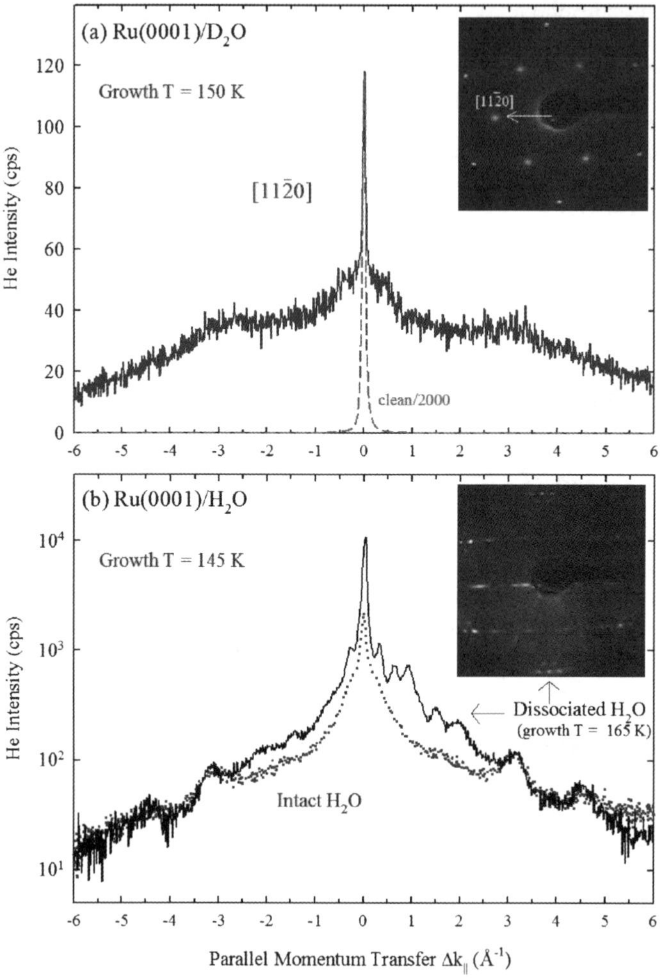

Fig. 3 (a) He atom scattering for clean Ru(0001) (dashed line) and a 0.67 ML D_2O layer, grown at 150 K (solid line). No classical elastic diffraction peaks corresponding to a $\sqrt{3}$ structure are observed. Inset: LEED (55 eV) from the D_2O layer. (b) HAS of an intact 0.67 ML H_2O layer grown at 145 K (dotted line) and a deliberately dissociated structure (solid line) grown at 165 K. Inset: LEED (56 eV) corresponding to the dissociated H_2O layer grown at 165 K. The HAS measurements shown were recorded for a 65 meV beam, directed along the $[11\bar{2}0]$ azimuth, at a surface temperature of 100 K.

show that this forms at a coverage of ~0.6 to 0.67 ML, slightly before the monolayer saturates with a coverage of 0.75 ML and a diffuse LEED pattern.[8,24] The solid line in Fig. 3a shows a wide-angle (20–70°) He scattering profile, corresponding to a parallel momentum transfer $\Delta k_{\|}$ of up to ± 6 Å^{-1} along the $[11\bar{2}0]$ direction. The dashed line corresponds to the angular scattering profile from the clean Ru surface (divided by a factor of 2000). Despite the well-defined LEED pattern, we find no indication of sharp diffraction peaks in our helium data. Rather we have weak broad maxima centred close to (although not precisely at) the position expected for second-order $\sqrt{3}$ diffraction. It is clear from Fig. 3 that we do not see any satellite peaks near to the specular reflection, as observed by Traeger *et al.*,[33] although they also report diffuse broad maxima at wide angle. The angular distribution shown here is similar to the findings of Kondo *et al.*[26] who interpret the pattern as originating from a ($\sqrt{3}$ ×

 This journal is © The Royal Society of Chemistry 2009

$\sqrt{3}$) R30° ice bilayer. They explained the absence of the first-order diffraction in terms of the large corrugation in the interaction potential between He and the surface, although we have observed an intensity maximum near to the first order position in HAS from a saturated H_2O layer (as shown in Fig. 3b). The broad maximum in intensity is quite unlike the sharp diffraction peaks seen by LEED and that might be expected in HAS based on the bilayer model. In fact we find the HAS intensity maximum is insensitive to the scattering azimuth, also appearing (at slightly different $\Delta k_{\|}$) following rotation of the sample to scatter perpendicular to the Ru rows.

The HAS patterns shown in Fig. 3 for D_2O (and H_2O) are essentially independent of coverage from approx 1.5 to 0.5 layers of water, the only change being an increase in intensity close to the specular peak as the coverage departs from 0.67 ML and a steady increase in reflectivity as multilayer clusters disappear and bare Ru is exposed. In contrast the degree of ordering found in LEED changes significantly.[8,24] At low coverage, water appears to form flat-lying clusters of limited size, which order to give a sharp LEED pattern near 0.67 ML before compressing to incorporate more water as adsorption saturates. The absence of any change in He scattering implies that disorder in the water structure exists irrespective of the degree of long-range lateral order found in LEED. For higher coverage the specular reflectivity of the multilayer, which forms clusters on top of the monolayer,[26,31] is so low that the intensity is rapidly reduced to background levels.

The angular HAS profile from an intact saturated H_2O layer is shown as the dotted line in Fig. 3b. With the adsorption temperature maintained below 150 K, the water layer is not dissociated and the general features of the HAS data are similar to that obtained for D_2O, the main differences being that the specular reflectivity is slightly higher and the diffuse peaks near the second-order positions appear sharper than for D_2O. Additionally, there is a weak intensity rise close to $\Delta k_{\|} = \pm 1.5$ Å^{-1} which cannot be resolved in the D_2O distribution shown in Fig. 3a. If the H_2O structure is grown at high temperature (above 152 K) dissociation occurs. The solid black line in Fig. 3b corresponds to an ice layer grown at 165 K, at which temperature multilayer H_2O is unstable and a complete, partially dissociated layer is formed. This gives rise to a series of sharper diffraction peaks along the [11$\bar{2}$0] azimuth and the specular reflectivity is also increased by a factor of ~5 compared to intact H_2O. The additional structure is consistent with LEED patterns obtained from this layer, shown as an inset to Fig. 3b, and has been interpreted as domains of a mixed OH/H_2O structure separated by H covered Ru.[17,24] This LEED pattern shows sharp features around the central beam, oriented to the $\sqrt{3}$ direction, the spacing of which corresponds to the HAS peaks.

The absence of sharp HAS diffraction features corresponding to an ordered ($\sqrt{3} \times \sqrt{3}$) R30° ice bilayer indicates that such structures can be ruled out. Instead, the HAS, which samples the repulsive wall some way above the surface, is consistent with the He-surface potential being highly corrugated, but disordered. Since H/D has a low electron-scattering cross section, the LEED pattern reflects an ordering of the Ru and O atoms. At the correct water coverage (*ca.* 0.67 ML) this shows sharp $\sqrt{3}$ spots, indicating formation of well-ordered domains of O but the HAS distribution is not sensitive to the water coverage (*i.e.* whether the LEED pattern shows long-range $\sqrt{3}$ order). We argue below that this is consistent with the ($\sqrt{3} \times \sqrt{3}$) R30° layer having a well-defined, periodic water adsorption site within an intrinsically disordered hydrogen-bonding structure. We note that the remainder of this discussion is based on formation of an idealised 0.67 ML water structure, something that requires careful control of the coverage.[8,24]

3.3 Calculated 0.67 ML water structures

DFT calculations were performed to explore structural disorder in the 2/3 ML water phase using the plane-wave basis set VASP code.[34,35] Standard ultra-soft pseudopotentials were used for all atoms and exchange and correlation were treated with the

PW91 generalised gradient approximation.[36] The plane-wave cutoff energy was 396 eV. The surface was represented by a three-layer slab of Ru, with the experimental lattice constants and with the bottom two layers fixed in position. A vacuum gap of 12 Å was used (measured from the H atom furthest from the surface). Reciprocal space was sampled with a 3 × 3 × 1 Monkhorst–Pack k-point set.[37] For supercells larger than ($2\surd3 \times 2\surd3$), tests were performed varying the k-point set to maintain the same density of k-points in each dimension, but the influence on the final binding energies was found to be negligible. In previous work we found that the binding energies were also unaffected by using thicker slabs and including dipole corrections normal to the surface.[24]

Fig. 4a shows the 0.67 ML structure with the highest binding energy (0.655 eV per molecule) we have found on this surface.[24] In this model, half of the water molecules lie flat atop Ru sites, while the other half adopt an H-down orientation. As we shall show below, the key to obtaining favourable binding in this structure is the existence of extended chains in which water adsorbs flat, with a short hydrogen bond of ~1.67 ± 0.02 Å. To maximise this bonding, the adlayer has to sacrifice the strict ($\surd3 \times \surd3$) R30° symmetry imposed in many earlier models of water adsorption, increasing the binding energy by ~25% compared to that calculated for the H-down ($\surd3 \times \surd3$) R30° bilayer on Ru(0001). Even in its simplest form, as shown in Fig. 4a, there is an element of disorder in the positions of the H-down molecules. For the structure shown, the O of the flat-lying water deviates from the atop-Ru positions by only 0.06 Å (RMS), while the H-down molecules have an RMS lateral displacement of 0.22 Å. The H-down chains have long, weak hydrogen bonds between water and do not form a specific bond to Ru. In other optimisation runs we have observed the H-down lateral displacement increase as high as 0.58 Å due to a displacement of the complete H-down chain. This indicates that the binding energy of this structure is insensitive to the exact position of the H-down chain above Ru, so long as these waters are present to complete the hydrogen-bonding network. As in bulk water, we can also introduce an element of disorder by changing the direction of the OH bonds along the water chains, without disturbing the overall hydrogen-bonding network.

More significant changes in the structure can be imposed by noticing that the direction of the water chains is not constrained. Thus in Fig. 4b, we introduce a bend into the chain by changing the placement of the flat-lying molecules to create chains running along the close-packed direction. Every flat-lying molecule still donates

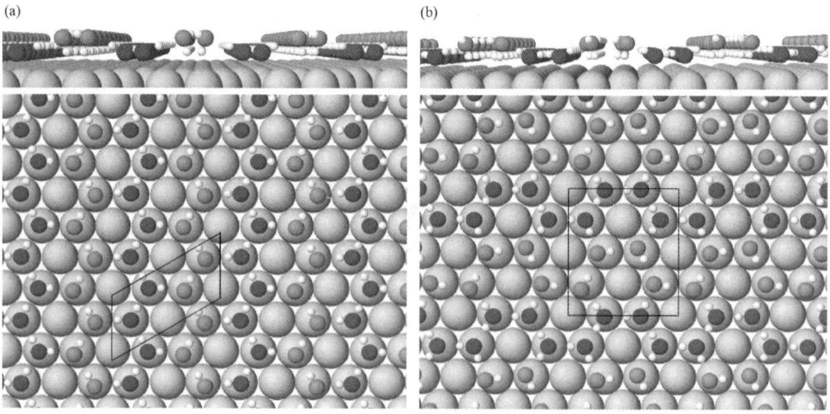

Fig. 4 Minimum-energy structures for an intact water layer calculated in (a) a ($\surd3 \times 2\surd3$) and (b) a ($3 \times 2\surd3$) unit cell. Water forms chains of flat water (O indicated by small dark circles) and H-down water (O indicated as small lighter circles) embedded within the overall hydrogen-bonded hexagonal network.

 This journal is © The Royal Society of Chemistry 2009

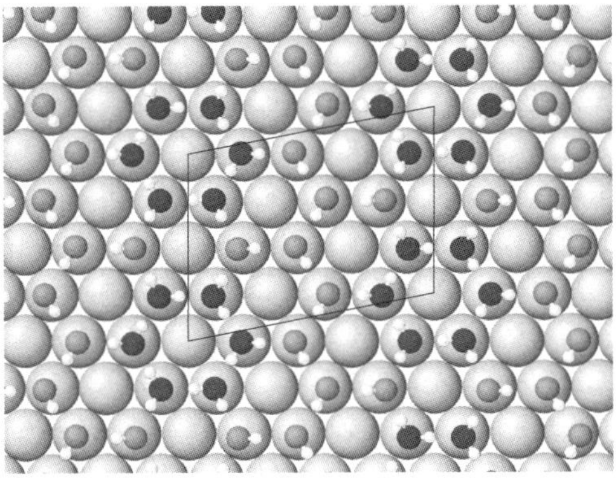

Fig. 5 Calculated structure for a water network containing cyclic rings of flat water, arranged in a $(2\sqrt{3} \times \sqrt{21})$ unit cell.

and accepts one hydrogen bond within the flat chain, donating the remaining hydrogen bond to the H-down water chain. The binding energy computed for this structure is 0.648 eV per molecule, essentially the same as for the original chains, indicating that the direction of the flat-lying chains is likely to be completely disordered within the overall hydrogen-bonded honeycomb structure. An alternative way to make every flat-lying molecule hydrogen bond with two other flat-lying molecules is to have closed rings. Fig. 5 shows a structure composed of hexagons of flat-lying water molecules, interspaced by branched chains of H-down molecules that complete the network. This structure maintains the bonding motif of unbroken flat-lying "chains" and consequently still has a high binding energy of 0.658 eV per molecule.

Since we associate the stability of these structures with the formation of chains of flat water, we have examined the energy cost of terminating these chains by calculating binding energies for structures with a range of chain lengths, as shown in Fig. 6. All of these structures have half of the waters lying flat and half H-down, but the number of hydrogen bonds between water molecules with different geometry increases for the shorter chains. For chains of length 2, shown in Fig. 6a, the binding energy is 0.596 eV per molecule, changing only marginally to 0.604 eV per molecule for a chain length of 3. Two structures with chain lengths of 4 have been examined, formed of short chains related to those of Fig. 4a and 4b respectively. The linear chain shown in Fig. 6c has a binding energy of 0.631 eV per molecule while the structure composed of molecules arranged in simple arcs is slightly less favourable, having a binding energy of 0.621 eV (Fig. 6d). The 5- and 6-unit chain structures shown in Fig. 6e and 6f have a binding energy of 0.621 and 0.627 eV per molecule respectively. The structures shown for different chain lengths are representative; in most cases there are other possible arrangements of the waters that give the same chain length, but broadly support the idea that terminating the water chains decreases the stability of the water layer.

The structures shown in Fig. 6 are still well-ordered chain-like arrangements of the water molecules. We have also used the DFT calculations to estimate the costs of introducing single errors into the chains, and of introducing chain branches. Fig. 7 shows four of the models considered. Introducing a single swap of a flat-lying and H-down water (Fig. 7a) has a substantial cost, reducing the binding energy from 0.655 to 0.615 eV per molecule. This is because we have produced two particularly unfavourable configurations: a flat-lying molecule bonded to 3 H-down molecules (*i.e.* with no bond to another flat-lying molecule) and an H-down molecule bound between 3 flat-lying molecules. These units are the building blocks of the conventional H-down bilayer[13] and are evidently not favoured. Although there are two medium-length (1.77–1.87 Å) bonds with hydrogen donated to the H-down

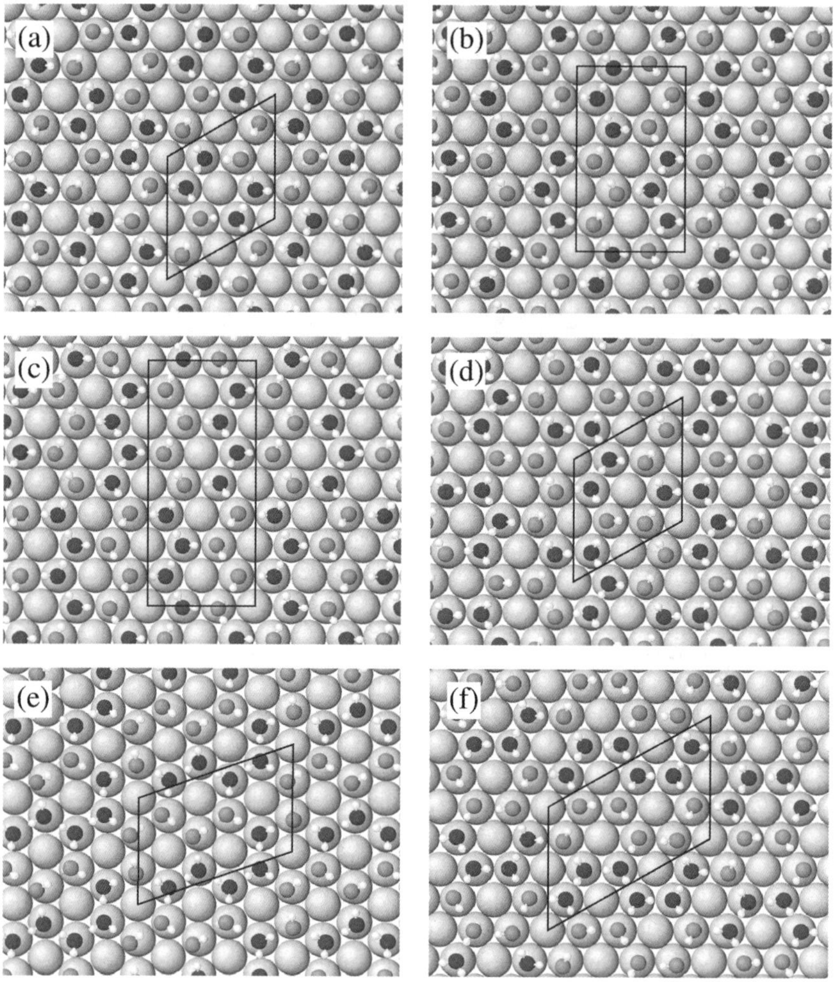

Fig. 6 Calculated structures for water arranged in chains of different lengths within the overall hydrogen-bonding network. Each water always has three hydrogen bonds to neighbouring molecules, but is arranged with chains of (a) two, (b) three, (c) and (d) four, (e) five and (f) six flat-lying waters.

molecule, the last bond donated by the H-down molecule is much longer at ~2.37 Å (this should be compared with a hydrogen bond length of 1.93–2.0 Å in complete chains of H-down molecules).

We have also considered branching structures in chains of finite length. Fig. 7b shows a chain of flat-lying molecules with a branch at the right-hand side, while Fig. 7c shows the structure with the last molecule at the left-hand side moved down to form another branch at the left-hand side of the chain. These structures have identical binding energies, 0.634 eV per molecule, indicating that there is no clear preference for or against branching a chain—provided the number of flat–flat hydrogen bonds remains unchanged. Finally, we have considered the energy cost of orienting one water molecule H-up in one of these large-scale structures. Altering Fig. 7b to obtain Fig. 7d, we find only a small decrease in binding energy to 0.631 eV per molecule, a destabilisation (increase in total energy) of ~70 meV for the unit cell indicated, suggesting that H-up water may form as a minor species within disordered chain structures. (Note that replacing complete

 This journal is © The Royal Society of Chemistry 2009

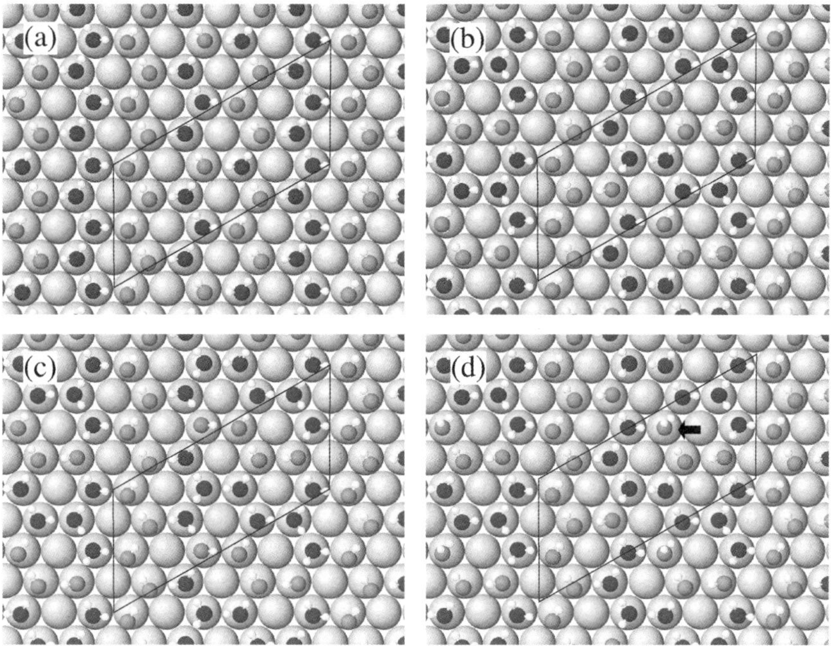

Fig. 7 Water networks containing disordered chains, calculated in a $(2\surd3 \times 4\surd3)$ unit cell. (a) Exchange of one flat and one H-down water to break half of the infinite chains into units of three waters. Branched chain structures containing (b) one flat and one H-down water and (c) two flat waters each coordinated by three waters of the same geometry. (d) Structure (b) but with one H-down water replaced by an H-up species (arrowed). The binding energies were calculated as 0.615, 0.634, 0.634 and 0.631 eV respectively.

H-down chains with H-up chains in Fig. 7 reduces the binding energy substantially to 0.564 eV.)

Examining all of the structures calculated, we find certain common elements. The flat-lying molecules adsorb almost exactly atop Ru, with O–Ru bond lengths of ~2.36 to ~2.40 Å. The RMS lateral deviation of O ranges from 0.06 Å for the optimum structure shown in Fig. 4a to 0.154 Å for the unfavourable 2-unit chains (Fig. 6a). The H-down molecules have a much greater deviation from exact atop adsorption, the RMS lateral deviation being ~0.15 Å for the 4-unit chains of Fig 6c, increasing to ~0.58 Å for the wiggles in Fig. 4b. The greater lateral displacement of O in the H-down chains is a direct consequence of the chain structure itself. Molecules in these chains are not symmetrically bound to flat-lying molecules, as is the case for the standard $(\surd3 \times \surd3)$ R30° bilayer structures; consequently they can tilt and displace to improve the hydrogen bonding to their flat-lying neighbours. On average, the lateral deviation of O within the H-down chains is more than three times that of the flat-lying water and, combined with a very soft translational mode, is expected to lead to a large Debye–Waller factor for these water molecules.

The importance of the short hydrogen bonds within the flat-lying chains can be clearly demonstrated by graphing the binding energy *versus* the fraction, f_s, of hydrogen bonds that are short (*i.e.* flat water–flat water) within the computational unit cell. The fraction f_s varies from 0 for the standard bilayer structures (we have considered 3 versions—H-down, H-up and alternating up and down) to 1/3 for the perfect chain structures shown in Fig. 4. As Fig. 8 shows, there is an excellent correlation across all the structures considered here. Although there are a range of different binding energies for structures with the same f_s, the precise details of

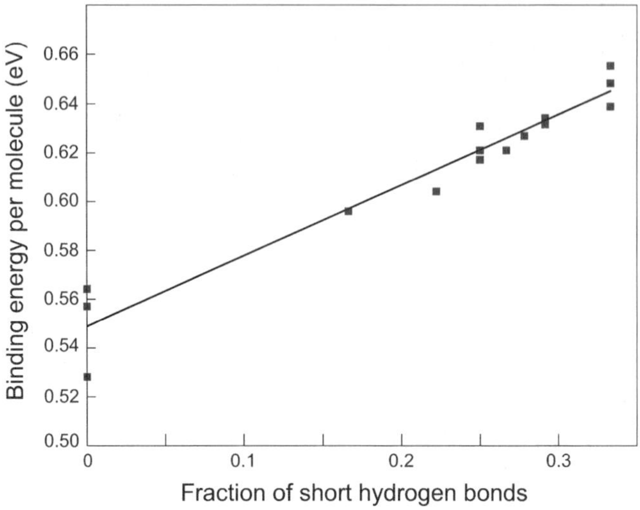

Fig. 8 Calculated binding energy as a function of the number of hydrogen bonds between flat-lying waters.

the bonding of flat chains to H-down molecules and within the H-down chains influence the total binding energy to a lesser extent than the completion of the flat chains. For example, there are 3 structures with $f_s = 1/3$ (Fig. 4 and 5), with a range of binding energies, similarly there are a range of binding energies for $f_s = 1/4$ (Fig. 6c, 6d and 7a). This is the clearest correlation in our data, although there is also a weaker correlation between the binding energy and the average O–Ru bond length.

The infinite chain structures described here are expected to be intrinsically disordered, with no long-range correlation in the direction of the chains and hence in the O height. We have shown that terminating the chains adds to this disorder, at the expense of a reduction in binding energy. Based on the trend in Fig. 8, and a thermal excitation $kT = 0.013$ eV at 150 K, we estimate f_s to be reduced by ~13% compared to the ideal infinite chain structure, indicating that chain termination will be significant in disordering these water structures. Apart from the energetic stability of different water structures, entropy effects will also influence the relative free energy. Feibelman and Alavi[38] showed that the residual Pauling-type entropy of a completely disordered surface water layer is $0.5k_B\ln[9/2]$, reducing the free energy by 0.010 eV at 150 K. Although this value is relatively small compared to the difference in binding energy of different chain structures, it is comparable to the thermal energy available at 150 K and will favour disordering the layer by terminating the water chains. The vibrational contribution to the entropy is somewhat larger and will be enhanced by any softening of the low-frequency acoustic modes.[38] Since we find the binding energy is insensitive to the lateral displacement of the H-down water chains, we expect these modes to be soft, contributing to the vibrational entropy and to disordering of the top layer.

The effect that disorder in the chain length and direction has on the (static) corrugation of the ice surface is illustrated in Fig. 9, which shows a side view of the $(2\sqrt{3} \times 4\sqrt{3})$ structure in Fig. 7b. The corrugation of the surface is large, the O atoms of flat-lying water (dark spheres) lie ~1 Å below the upper H and O atoms of H-down water (lighter spheres), but because the site and orientation of the proton of H-down molecules follows no simple pattern, the outer surface appears highly disordered, consistent with the absence of diffraction intensity in HAS.

 This journal is © The Royal Society of Chemistry 2009

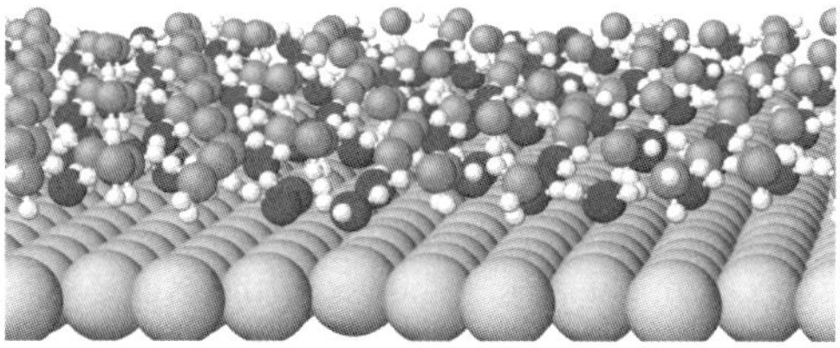

Fig. 9 Side view of the $(2\sqrt{3} \times 4\sqrt{3})$ structure shown in Fig. 7b showing the high corrugation and disordered nature of the top layer of water (lighter O atoms).

3.4 LEED patterns for the chain structure

The model for the water layer proposed above has been criticised because it does not follow the simple $(\sqrt{3} \times \sqrt{3})$ R30° unit cell that was expected based on the observation of (1/3,1/3) LEED spots for the 0.67 ML water layer on Ru(0001).[14] Describing the chain structures requires a $(\sqrt{3} \times 2\sqrt{3})$ unit cell, or larger, containing at least 4 different water molecules. The contention is that disorder in the lateral site of flat and H-down water molecules will lead to different phase factors, causing the intensity of the $\sqrt{3}$ superstructure LEED beams to average to zero, whereas formation of more ordered chain structures would give rise to additional diffraction beams that are not observed in experiment. Here we discuss this picture, which underpins the original LEED IV analysis of the water structure on Ru(0001)[14] and resulted in the conclusion that the O atoms of the water layer are nearly coplanar.

Based on the DFT calculations described in section 3.3, the O atoms are assumed to be arranged in a commensurate hexagonal network, atop Ru, with two different heights depending on whether the molecule is adsorbed flat or H-down. Considering a water island containing 2N water molecules in an $(n\sqrt{3} \times m\sqrt{3})$ island, where m and n are integers and $N = (n \times m)$, the diffraction beams will have intensity provided the structure factor $F_{h,k}$ does not go to zero. The overall structure factor for the island is given by

$$F_{hk} = \sum f_n e^{2\pi i(\bar{G}.\bar{r}_n)} = \sum f_n e^{2\pi i(hu_n + kv_n + \Delta k_z z_n)}$$

where the sum runs over all the atoms in the structure, with positions defined by $\boldsymbol{r}_n = u_n\boldsymbol{a}_1 + v_n\boldsymbol{a}_2 + z_n\boldsymbol{j}$, surface reciprocal lattice vectors $\boldsymbol{G}_{h,k} = h\boldsymbol{b}_1 + k\mathbf{b}_2$ and $\Delta k_z = (k_{zi} - k_{zf})$ is the change in wavevector perpendicular to the surface. The intensity of the $\sqrt{3}$ beams can be obtained by calculating the structure factor $|F_{n,m}|^2$ or equivalent,

$$F_{n,m} = \sum_{p,q=0}^{n-1,m-1} f_O \Big[e^{2\pi i(pn/n + qm/m + \Delta k_z z_{O_A}^{p,q})} + e^{2\pi i(pn/n + n/3n + qm/m + m/3m + \Delta k_z z_{O_B}^{p,q})} \Big]$$

$$F_{n,m} = \sum_{p,q=0}^{n-1,m-1} f_O \Big[e^{2\pi i(\Delta k_z z_{O_A}^{p,q})} + e^{4\pi i/3} e^{2\pi i \Delta k_z z_{O_B}^{p,q}} \Big]$$

where $z_{O_A}^{p,q}$ and $z_{O_B}^{p,q}$ are the heights of the O atoms in sub-cell (p,q). For different arrangements of the O atoms within the hydrogen-bonded superstructure we obtain different structure factors, although for none of them will the structure factor (and hence the intensity of the diffraction peaks) be uniquely zero. Assuming that the height of the O atoms within the network is random, with no correlation whatsoever, leads to a structure factor

$$F_{n,m} = Nf_{\mathrm{O}}\left\langle \mathrm{e}^{2\pi\mathrm{i}(\Delta k_z z_{\mathrm{O}})} \right\rangle \left[1 + \mathrm{e}^{4\pi\mathrm{i}/3}\right]$$

where $\langle \mathrm{e}^{2\pi\mathrm{i}(\Delta k_z z_{\mathrm{O}})} \rangle$ is just the expectation value for the O phase factor, averaged across the entire $(n\sqrt{3} \times m\sqrt{3})$ network. For two different types of water, flat and 'H-down' with heights z_{OA} and z_{OB} this just corresponds to the average,

$$\left\langle \mathrm{e}^{2\pi\mathrm{i}(\Delta k_z z_{\mathrm{O}})} \right\rangle = \left(\mathrm{e}^{2\pi\mathrm{i}(\Delta k_z z_{\mathrm{O_A}})} + \mathrm{e}^{2\pi\mathrm{i}(\Delta k_z z_{\mathrm{O_B}})}\right) \Big/ 2$$

so even for an entirely random arrangement of water orientations within the overall hexagonal network, the structure factor will not be zero and we expect to see $\sqrt{3}$ LEED beams. A coplanar array of O, with $z_{\mathrm{OA}}^{p,q} = z_{\mathrm{OB}}^{p,q}$, and an ordered bilayer, with $z_{\mathrm{OA}}^{p,q} = z_{\mathrm{OA}}$ and $z_{\mathrm{OB}}^{p,q} = z_{\mathrm{OB}}$, are just special cases of this, again with non-zero structure factors.

A similar argument can be applied to the displacement of the top layer Ru atoms caused by water adsorbed in a hexagonal commensurate array. The DFT calculations find that adsorption of water slightly displaces Ru vertically, depending whether the Ru site is bare or has water adsorbed flat or H-down. The Ru atoms beneath flat-lying water are pulled out of the surface by 0.054 ± 0.014 Å compared to Ru that is bare or has H-down water above. This buckling of the Ru surface has the same well-defined lateral period as the location of the flat-lying waters and can be included in the scattering factor, providing a contribution to the $\sqrt{3}$ LEED beams. An anti-correlation of O and Ru heights was also found in the LEED IV simulations.[14,15]

So far this simple model has assumed that the O is adsorbed exactly atop Ru, with a vertical height that depends only on whether water is adsorbed in a flat or H-down adsorption geometry. In practice the DFT calculations predict that the lateral and vertical O location also depends on the local coordination of neighbouring water molecules. This disorder will broaden the distribution of O positions and contribute to the averaging, reducing the scattering factor and hence the diffraction intensity. The importance of this can be tested using the DFT calculations to see how the O positions vary depending on the local environment of the water. As discussed earlier, the two different types of water, flat and H-down, behave very differently. Flat-lying water forms a well-defined bond to the Ru and the O position is tightly constrained. For a typical disordered $(2\sqrt{3} \times 4\sqrt{3})$ chain structure (*e.g.* structure 7b) the average Ru–O separation is 2.38 ± 0.07 Å, with a comparable variation in absolute O height (σ_z = 0.08 Å). Similarly, the RMS lateral displacement of these O atoms from atop the Ru is small, 0.09 Å. The O position is still more tightly constrained for the ordered chains shown in Fig. 4, the O displacement increasing only for the shortest chain structures shown in Fig. 6. In fact, the variation in the (equilibrium) O position for the flat waters in different structures is rather less than the 0.18 Å RMS displacement used in the recent re-analysis of the LEED IV data.[15]

The H-down waters that complete the hexagonal hydrogen-bonded network are very different, often being displaced considerably further from atop the Ru site, depending on the local hydrogen-bonding environment. For example, in Fig. 7b the O atoms of the H-down waters have a typical variation in height of σ_z = 0.15 Å, an RMS lateral displacement of σ_{xy} = 0.34 Å and a Ru–O separation of 3.44 ± 0.22 Å. Even for the more regular chain structures of Fig. 4, the binding energy is rather insensitive to the lateral location of the H-down water O, the RMS displacement increasing to as much as 0.6 Å with little change in the binding energy. The softness of these vibrational modes is expected to result in a large Debye–Waller factor for these O atoms. We conclude that scattering from sites associated with flat-lying water is likely to suffer limited averaging as a result of disorder in the hydrogen-bonding network, whereas the O atoms in the upper layer adopt

 This journal is © The Royal Society of Chemistry 2009

a much greater range of geometries and are expected to contribute less to the scattering intensity when averaged across the structure.

Obviously this qualitative discussion says nothing about the intensity or IV characteristics of the LEED spots; these require a full multiple scattering calculation for the disordered water network that may well prove impractical. However, multiple scattering cannot, *per se*, remove the diffraction intensity and it is useful to address the relationship between the previous LEED IV analysis by Held and Menzel[14,15] and the structures proposed here. Using the low current MCP LEED system, we have re-measured LEED IV data for the intact D_2O ($\surd 3 \times \surd 3$) R30° structure. The data was taken in several different series of experiments, varying the current, spot position and overlayer preparation. No changes to the IV profiles were observed which could be attributed directly to electron-induced dissociation and the TPD curves recorded after rastering the electron beam across the surface indicated that D_2O was still intact. The IV curves we obtain are similar to those reported previously, with slightly more intensity and structure at high energy than was originally observed. The agreement between the two data sets is consistent with the same, intact structure being formed in both studies, with negligible electron damage. Held and Menzel did observe electron-induced damage but took steps to minimise its effect when collecting the IV data. This agreement tends to rule out Feibelman's suggestion that the flattened O layer found in the LEED IV analysis[14,15] is caused by partial dissociation to form the mixed O/H_2O structure[16] (although we note that the two structures have similar IV characteristics, reflecting a similar atop adsorption site[14,32]).

Having ruled out electron damage in the original LEED IV data, we should consider the effect of the disorder found in HAS on the original LEED IV analysis of the structure of the intact D_2O layer. This analysis assumed that there were two distinct O atoms in a ($\surd 3 \times \surd 3$) R30° unit cell, with a well-defined periodic corrugation in their height, and fitted the site and position of the O, as well as the relaxation of the Ru layer.[14] The original simulation was unable to rule out a buckled bilayer structure with Δz_O of 0.7 Å, but a subsequent reanalysis, optimising the O RMS displacements and adding a H atom above the upper O, concluded the O layer could not be buckled.[15] The discussion presented here suggests that any re-analysis of this system should take into account lateral disorder in the O height within the hydrogen-bonding network, as well as the much greater RMS displacement of the upper water molecules; this can not be done in a simple ($\surd 3 \times \surd 3$) R30° unit cell. Attempts to fit the new IV data are in progress.

3.5 Discussion

Two other pieces of indirect evidence support the disorder evidenced by He scattering. The vibrational spectrum shows broad stretching bands,[17,24] quite unlike small clusters[39] or proton-ordered OH/H_2O layers,[40] but consistent with an inhomogeneous broadening of the vibrational spectrum. Disorder in the water layer results in a distribution of O–O distances, which is strongly correlated to the OH stretching frequency.[41] Secondly, calculations for a disordered hexagonal network can reproduce the experimental workfunction for the water film (1.3 eV[32]), whereas that for the ordered H-up, H-down or partially dissociated structures do not. Meng *et al.*[42] proposed a bilayer structure containing a mixture of H-up and H-down water, the correct proportion of which (between 1 : 2 and 1 : 3) can reproduce the observed workfunction. Such a structure relies on the H-up and H-down water bilayers having a very similar binding energy, and would be qualitatively consistent with the disorder seen in HAS, but appears to be inconsistent with the absence of a free OH stretching band in the vibrational SFG[25] and RAIR spectra.[17,24] The water chain structures described here give a calculated workfunction change of −1.15 eV, close to that observed, without any need to make an arbitrary assumption about a particular H-up to H-down ratio.[24] Unlike the bilayer structures, where the site associated with second layer water is ordered, the chain model results in a stable 2D wetting layer that is intrinsically disordered, with nothing to constrain the long-range lateral registry of water to be flat or H-down.

4. Conclusion

He atom scattering from intact H_2O/D_2O layers finds a very low He reflectivity and diffuse features in scattering, with no well-defined diffraction peaks. Partial dissociation of the water layer increases the He reflectivity and can be induced by contamination during adsorption. HAS is insensitive to the exact coverage of water, or the degree of lateral ordering indicated by LEED, indicating that the $(\sqrt{3} \times \sqrt{3})$ R30° layer is disordered. We interpret this behaviour in terms of a well-defined lateral order in the location of the O atoms, combined with disorder in the height of the O atoms and the positions of the protons. DFT calculations find that water prefers to form hexagonal networks containing chains of flat lying molecules, bound atop Ru, interspaced by H-down chains. The stability of the network correlates closely to the fraction of the hydrogen bonds that are formed between flat waters. Disorder in these structures originates from directional disorder of the chains and from the low energy cost of chain termination, driven by thermal excitation and entropy effects. We argue that the long-range periodicity of the O adsorption site is consistent with an ordered $\sqrt{3}$ LEED pattern, the large variation in O position likely contributing to make the O atoms of H-down water in the upper part of the layer invisible to LEED.

Acknowledgements

We would like to acknowledge the EPSRC for support of this work and Sam Haq and Takahiro Kondo for valuable comments.

References

1 P. A. Thiel and T. E. Madey, *Surf. Sci. Rep.*, 1987, **7**, 211.
2 M. A. Henderson, *Surf. Sci. Rep.*, 2002, **46**, 5.
3 X. L. Hu and A. Michaelides, *Surf. Sci.*, 2007, **601**, 5378.
4 G. S. Karlberg, J. Rossmeisl and J. K. Norskov, *Phys. Chem. Chem. Phys.*, 2007, **9**, 5158.
5 C. Clay and A. Hodgson, *Curr. Opin. Solid State Mater. Sci.*, 2005, **9**, 11.
6 T. E. Madey and J. T. Yates, *Chem. Phys. Lett.*, 1977, **51**, 77.
7 P. A. Thiel, F. M. Hoffmann and W. H. Weinberg, *J. Chem. Phys.*, 1981, **75**, 5556.
8 D. L. Doering and T. E. Madey, *Surf. Sci.*, 1982, **123**, 305.
9 P. A. Thiel, F. M. Hoffmann and W. H. Weinberg, *Phys. Rev. Lett.*, 1982, **49**, 501.
10 P. A. Thiel, R. A. Depaola and F. M. Hoffmann, *J. Chem. Phys.*, 1984, **80**, 5326.
11 K. Kretzschmar, J. K. Sass, P. Hofmann, A. Ortega, A. M. Bradshaw and S. Holloway, *Chem. Phys. Lett.*, 1981, **78**, 410.
12 K. Kretzschmar, J. K. Sass, A. M. Bradshaw and S. Holloway, *Surf. Sci.*, 1982, **115**, 183.
13 H. Ogasawara, B. Brena, D. Nordlund, M. Nyberg, A. Pelmenschikov, L. G. M. Pettersson and A. Nilsson, *Phys. Rev. Lett.*, 2002, **89**, 276102.
14 G. Held and D. Menzel, *Surf. Sci.*, 1994, **316**, 92.
15 S. R. Puisto, T. J. Lerotholi, G. Held and D. Menzel, *Surf. Rev. Lett.*, 2003, **10**, 487.
16 P. J. Feibelman, *Science*, 2002, **295**, 99.
17 C. Clay, S. Haq and A. Hodgson, *Chem. Phys. Lett.*, 2004, **388**, 89.
18 K. Andersson, A. Nikitin, L. G. M. Pettersson, A. Nilsson and H. Ogasawara, *Phys. Rev. Lett.*, 2004, **93**, 196101.
19 J. Weissenrieder, A. Mikkelsen, J. N. Andersen, P. J. Feibelman and G. Held, *Phys. Rev. Lett.*, 2004, **93**, 196102.
20 A. Michaelides, A. Alavi and D. A. King, *J. Am. Chem. Soc.*, 2003, **125**, 2746.
21 A. Michaelides, A. Alavi and D. A. King, *Phys. Rev. B*, 2004, **69**, 113404.
22 S. Meng, E. G. Wang and S. W. Gao, *Phys. Rev. B*, 2004, **69**, 195404.
23 P. J. Feibelman, *Phys. Rev. B*, 2005, **72**, 113405.
24 S. Haq, C. Clay, G. R. Darling, G. Zimbitas and A. Hodgson, *Phys. Rev. B*, 2006, **73**, 115414.

 This journal is © The Royal Society of Chemistry 2009

25 D. N. Denzler, C. Hess, R. Dudek, S. Wagner, C. Frischkorn, M. Wolf and G. Ertl, *Chem. Phys. Lett.*, 2003, **376**, 618.
26 T. Kondo, S. Mae, H. S. Kato and M. Kawai, *Surf. Sci.*, 2006, **600**, 3570.
27 N. S. Faradzhev, K. L. Kostov, P. Feulner, T. E. Madey and D. Menzel, *Chem. Phys. Lett.*, 2005, **415**, 165.
28 G. Zimbitas, M. Gallagher, G. R. Darling and A. Hodgson, *J. Chem. Phys.*, 2008, **128**, 074701.
29 M. Gallagher, A. Omer, S. Haq and A. Hodgson, *Surf. Sci.*, 2007, **601**, 268.
30 D. A. King and M. G. Wells, *Surf. Sci.*, 1972, **29**, 454–482.
31 S. Haq and A. Hodgson, *J. Phys. Chem. C*, 2007, **111**, 5946.
32 G. Held and D. Menzel, *Surf. Sci.*, 1995, **327**, 301.
33 F. Traeger, D. Langenberg, Y. K. Gao and C. Wöll, *Phys. Rev. B*, 2007, **76**, 033410.
34 G. Kresse and J. Furthmuller, *Phys. Rev. B*, 1996, **54**, 11169.
35 G. Kresse and J. Hafner, *Phys. Rev. B*, 1993, **47**, 558.
36 J. P. Perdew, J. A. Chevary, S. H. Vosko, K. A. Jackson, M. R. Pederson, D. J. Singh and C. Fiolhais, *Phys. Rev. B*, 1992, **46**, 6671.
37 H. J. Monkhorst and J. D. Pack, *Phys. Rev. B*, 1976, **13**, 5188.
38 P. J. Feibelman and A. Alavi, *J. Phys. Chem. B*, 2004, **108**, 14362.
39 M. Nakamura and M. Ito, *Chem. Phys. Lett.*, 2005, **404**, 346.
40 C. Clay, S. Haq and A. Hodgson, *Phys. Rev. Lett.*, 2004, **92**, 046102.
41 P. J. Feibelman, *Chem. Phys. Lett.*, 2004, **389**, 92.
42 S. Meng, E. G. Wang, C. Frischkorn, M. Wolf and S. W. Gao, *Chem. Phys. Lett.*, 2005, **402**, 384.

What ice can teach us about water interactions: a critical comparison of the performance of different water models

C. Vega,* J. L. F. Abascal, M. M. Conde and J. L. Aragones

Received 1st April 2008, Accepted 9th May 2008
First published as an Advance Article on the web 19th September 2008
DOI: 10.1039/b805531a

The performance of several popular water models (TIP3P, TIP4P, TIP5P and TIP4P/2005) is analyzed. For that purpose the predictions for ten different properties of water are investigated, namely: 1. vapour–liquid equilibria (VLE) and critical temperature; 2. surface tension; 3. densities of the different solid structures of water (ices); 4. phase diagram; 5. melting-point properties; 6. maximum in the density of water at room pressure and thermal coefficients α and κ_T; 7. structure of liquid water and ice; 8. equation of state at high pressures; 9. self-diffusion coefficient; 10. dielectric constant. For each property, the performance of each model is analyzed in detail with a critical discussion of the possible reason of the success or failure of the model. A final judgement on the quality of these models is provided. TIP4P/2005 provides the best description of almost all properties of the list, the only exception being the dielectric constant. In second position, TIP5P and TIP4P yield a similar performance overall, and the last place with the poorest description of the water properties is provided by TIP3P. The ideas leading to the proposal and design of the TIP4P/2005 are also discussed in detail. TIP4P/2005 is probably close to the best description of water that can be achieved with a non-polarizable model described by a single Lennard-Jones (LJ) site and three charges.

I. Introduction

Water is probably the most important molecule in our relation to nature. It forms the matrix of life,[1] it is the most common solvent for chemical processes, it plays a major role in the determination of the climate on earth, and it also appears on planets, moons and comets.[2] Water is interesting not only from a practical point of view, but also from a fundamental point of view. In the liquid phase water presents a number of anomalies when compared to other liquids.[3–7] In the solid phase it exhibits one of the most complex phase diagrams, having fifteen different solid structures.[3] Due to its importance and its complexity, understanding the properties of water from a molecular point of view is of considerable interest. The experimental study of the phase diagram of water has spanned the entire 20th century, starting with the pioneering works of Tammann[8] and Bridgman[9] up to the recent discovery of ices XII, XIII and XIV.[10,11] The existence of several types of amorphous phase at low temperatures,[12–14] the possible existence of a liquid–liquid phase transition in water,[15,16] the properties of ice at a free surface[17–23] and the interaction with hydrophobic molecules[24,25] have also been the focus of much interest in the last two decades.

Departamento de Química Física, Facultad de Ciencias Químicas, Universidad Complutense de Madrid, 28040 Madrid, Spain. E-mail: cvega@quim.ucm.es

Computer simulations of water started their road with the pioneering papers by Watts and Barker[26] and by Rahman and Stillinger[27] about forty years ago. A key issue when performing simulations of water is the choice of the potential model used to describe the interaction between molecules.[28–32] A number of different potential models have been proposed. An excellent survey of the predictions of the different models proposed up to 2002 was made by Guillot.[33] Probably the general feeling is that no water potential model is totally satisfactory and that there are no significant differences between water models.

In the recent years the increase in computing power has allowed the calculation of new properties which can be used as new target quantities in the fit of the potential parameters. More importantly, some of these properties have been revealed as a stringent test of the water models. In particular, recently we have determined the phase diagram for different water models and we have found that the performance of the models is quite different.[34–36] As a consequence, new potentials have been proposed. In this paper we want to perform a detailed analysis of the performance of several popular water models including the recently proposed TIP4P/2005, and to include in the comparison some properties of the solid phases, thus extending the scope of the comparison performed by Guillot.[33] The importance of including solid-phase properties in the test of water models was advocated by Morse and Rice[37] and Whalley,[38] among others, and Monson suggested that the same should be true for other substances.[39,40]

It must be recognised that water is flexible and polarizable. That should be taken into account in the next generation of water models.[41,42] However, it is of interest at this stage to analyse how far it is possible to go in the description of real water with simple (rigid, non-polarizable) models. For this reason we will restrict our study to rigid, non-polarizable models. Besides, in case a certain model performs better than others, it would be of interest to understand why is this so since this information can be useful in the development of future models. As commented above, the study of the phase diagram of water can be extremely useful to discriminate among different water models. Thus, in the comparison between water models, we incorporate not only properties of liquid water but also properties of the solid phases of water. The scheme of this paper is as follows. In Section II, the potential models used in this paper are described. In Section III, we present the properties that will be used in the comparison between the different water models. Section IV gives some details of the calculations of this work. Section V presents the results of this paper. Finally, in Section VI, the main conclusions will be discussed.

II. Water potential models

In this work we shall focus on rigid, non-polarizable models of water. All the models we are considering locate the positive charge on the hydrogen atoms and a Lennard-Jones interaction site on the position of the oxygen. What are the differences between them? They differ in three significant aspects:

• The bond geometry. By bond geometry we mean the choice of the OH bond length and H–O–H bond angle of the model.

• The charge distribution. All the models place one positive charge at the hydrogens but differ in the location of the negative charge(s).

• The target properties. By target properties we mean the properties of real water that were used to fit the potential parameters (forcing the model to reproduce the experimental properties).

In this paper we compare the performance of the following potential models: TIP3P,[43] TIP4P,[43] TIP5P[44] and TIP4P/2005.[36] This selection presents an advantage. All these models exhibit the same bond geometry (the d_{OH} distance and H–O–H angle in all these models are just those of the molecule in the gas phase, namely 0.9572 Å and 104.52° respectively). Therefore, any difference in their performance

 This journal is © The Royal Society of Chemistry 2009

is due to the difference in their charge geometry and/or target properties. Let us now present each of these models.

A. TIP3P

TIP3P was proposed by Jorgensen *et al.*[43] In the TIP3P model the negative charge is located on the oxygen atom and the positive charge on the hydrogen atoms. The parameters of the model (*i.e.* the values of the Lennard-Jones (LJ) potential parameters σ and ε and the value of the charge on the hydrogen atom) were obtained by reproducing the vaporization enthalpy and liquid density of water at ambient conditions. TIP3P is the model used commonly in certain force fields of biological molecules. A model similar in spirit to TIP3P is SPC.[45] We shall not discuss the SPC model so extensively in this paper, since we want to keep the discussion within TIP*n*P geometry (by TIP*n*P geometry we mean that d_{OH} adopts the value 0.9572 Å and the H–O–H angle is 104.52°). In the SPC model the OH bond length is 1 Å and the H–O–H is 109.47° (the tetrahedral value). The parameters of SPC were obtained in the same way as those of TIP3P. Thus TIP3P and SPC are very similar models. Nine distances must be computed to evaluate the energy between two TIP3P (or SPC) water models so their computational cost is proportional to 9.

B. TIP4P

The key feature of this model is that the site carrying the negative charge (usually denoted as the M site) is not located at the oxygen atom but on the H–O–H bisector at a distance of 0.15 Å. This geometry was already proposed in 1933 by Bernal and Fowler.[46] TIP4P was proposed by Jorgensen *et al.*[43] who determined the parameters of the potential in order to also reproduce the vaporization enthalpy and liquid density of liquid water at room temperature. It is fair to say that TIP4P, although quite popular, is probably less often used than TIP3P or SPC/E. The reader may be surprised to learn that the reasons for that are the appearance of a massless site (the M center) and the apparently higher computational cost. To deal with a massless site within a MD package one must either solve the orientational equations of motion for a rigid body (for instance, using quaternions) or using constraints after distributing the force acting on the M center among the rest of the atoms of the system. These options should be incorporated in the MD package and this is not always the case. Also the computational cost of TIP4P is proportional to 16 when no trick is made (due to the presence of four interaction sites) but it can be reduced to 10 by realizing that one must only compute the O–O distance to compute the LJ contribution, plus 9 distances to compute the Coulombic interaction. Thus, for some users TIP4P is not a good choice as a water model either because it is slower than TIP3P in their molecular dynamics package or because the package may have problems in dealing with a massless site. These are methodological rather than scientific reasons but they appear quite often. Some modern codes such as GROMACS[47] (just to mention one example) solve these two technical aspects quite efficiently.

C. TIP5P

TIP5P is a relatively recent potential: it was proposed in 2000 by Mahoney and Jorgensen.[44] This model is the modern version of the models used in the seventies (ST2) in which the negative charge was located at the position of the lone-pair electrons.[48] Thus instead of having one negative charge at the center M, this model has two negative charges at the L centers. Concerning the target properties this model reproduces the vaporization enthalpy and density of water at ambient conditions. This is in common with TIP3P and TIP4P. However, TIP5P incorporates a new target property: the density maximum of liquid water. The existence of a maximum in density at approximately 277 K is one of the fingerprint properties of liquid water.

Obtaining density maxima in constant temperature and constant pressure (NpT) simulations of water models is not difficult from a methodological point of view.[31,49,50] However, very long runs are required [of the order of several millions of time steps (MD) or cycles (MC)] to determine accurately the location of the density maximum. Thus, it is not surprising that the very first model able to reproduce the density maximum was proposed in 2000 when the computing power allowed an accurate calculation of the temperature and density at the maximum. Since TIP5P consists of five interaction sites, it apparently requires the evaluation of 25 distances. However, using the trick described for TIP4P, it is possible to show that one just requires the evaluation of 17 distances.

D. SPC/E

In this paper we shall focus mainly on TIP*n*P-like models. However, it is worth introducing SPC/E,[51] considered by many as one of the best water models. SPC/E has the same bond geometry as SPC. Concerning the charge distribution, it locates the negative charge at the oxygen atom, as do SPC and TIP3P. The target properties of SPC/E were the density and the vaporization enthalpy at room temperature. So far everything seems identical to SPC. The key issue is that SPC/E only reproduces the vaporization enthalpy of real water when a polarization energy correction is included. Berendsen *et al.*[51] pointed out that when using non-polarizable models one should include a polarization correction before comparing the vaporization enthalpy of the model to the experimental value. This is because the dipole moment is enhanced in this type of model in order to account approximately for the neglected many-body polarization forces. Thus, when the vaporization enthalpy is calculated one compares the energy of the liquid with that of a gas with the enhanced dipole moment. It is then necessary to correct for the effect of the difference between the dipole moment of the isolated molecule and that of the effective dipole moment used for the condensed phase. The correction term is given by:

$$\frac{E_{\mathrm{pol}}}{N} = \frac{(\mu - \mu_{\mathrm{gas}})^2}{2\alpha_{\mathrm{p}}} \tag{1}$$

where α_{p} is the polarizability of the water molecule, μ_{gas} is the dipole moment of the molecule in the gas phase and μ is the dipole moment of the model. SPC/E reproduces the vaporization enthalpy of water only if the correction given by eqn (1) is used. The introduction of the polarization correction is the essential feature of SPC/E.

E. TIP4P/2005

The TIP4P/2005 was recently proposed by Abacal and Vega[36] after evaluating the phase diagram for the TIP4P and SPC/E models of water.[34,52] It was found that TIP4P provided a much better description of the phase diagram of water than SPC/E. Both models yielded rather low melting points.[53] For this reason, it was clear that the TIP4P could be slightly modified to still predict a correct phase diagram but improve the description of the melting point. The TIP4P/2005 has the same bond geometry as the TIP4P family. Also, TIP4P/2005 has the same charge distribution as TIP4P (although the distance from the M site to the oxygen atom is slightly modified). The main difference between TIP4P and TIP4P/2005 comes from the choice of the target properties used to fit the potential parameters. TIP4P/2005 not only uses a larger number of target properties than any of the water models previously proposed but also they represent a wider range of thermodynamic states. As has become traditional, the first of the target properties is the density of water at room temperature and pressure (but in TIP4P/2005 we have also tried to account for the densities of the solid phases). More significantly, TIP4P/2005 incorporated

 This journal is © The Royal Society of Chemistry 2009

those target properties used in the successful SPC/E and TIP5P models. As in SPC/E, the polarization correction is included in the calculation of the vaporization enthalpy. Secondly (as in TIP5P) the maximum in the density of liquid water at room pressure (TMD) was used as a target property. Finally, completely new target properties are the melting temperature of the hexagonal ice (Ih) and a satisfactory description of the complex region of the phase diagram involving different ice polymorphs.

In Table 1 the parameters of the models SPC, SPC/E, TIP4P, TIP5P and TIP4P/2005 are shown. The dipole moment and the three values of the traceless quadrupolar tensor are given in Table 2. There, we also present the values of the quadrupole moment Q_T which is defined as $Q_T = (Q_{xx} - Q_{yy})/2$. We have chosen the H–O–H bisector as the z axis, the x axis in the direction of the line joining the hydrogen atoms so that the y axis is perpendicular to the molecular plane. It can be shown that, for models consisting of just three charges, the value of Q_T is independent of the origin (when the z axis is located along the H–O–H bisector).[54–56] An interesting feature appears in Table 2. Although for most of water models the dipole moment is close to 2.3 Debye, they differ significantly in their values of the quadrupole moment. Notice that neither the dipole nor the quadrupole moment were used as target properties by any of the water models discussed here. Thus the charge distribution determines the aspect of the quadrupolar tensor. One may suspect that, since quadrupolar forces induce a strong orientational dependence, the differences in the quadrupolar tensor between different water models will be manifested significantly

Table 1 Potential parameters of the water potential models. The distance between the oxygen and hydrogen sites is d_{OH}. The angle formed by hydrogen, oxygen and the other hydrogen atom is denoted as H–O–H. The LJ site is located at the oxygen with parameters σ and (ε/k_B). Proton charge q_H. All the models (except TIP5P) place the negative charge in a point M at a distance d_{OM} from the oxygen along the H–O–H bisector. For TIP5P, d_{OL} is the distance between the oxygen and the L sites placed at the lone electron pairs

Model	d_{OH}/Å	H–O–H/°	σ/Å	(ε/k_B)/K	q_H/e	d_{OM}/Å	d_{OL}/Å
SPC	1.0	109.47	3.1656	78.20	0.41	0	—
SPC/E	1.0	109.47	3.1656	78.20	0.423	0	—
TIP3P	0.9572	104.52	3.1506	76.54	0.417	0	—
TIP4P	0.9572	104.52	3.1540	78.02	0.52	0.15	—
TIP4P/2005	0.9572	104.52	3.1589	93.2	0.5564	0.1546	—
TIP5P	0.9572	104.52	3.1200	80.51	0.241	—	0.70

Table 2 Dipole moment and components of the quadrupole moment. Debye units are used for the dipole moment μ while the components of the quadrupole moment are given in DÅ. Q_T is defined as $Q_T = (Q_{xx} - Q_{yy})/2$. The center of mass is used as origin, the z axis being that of the H–O–H bisector and the x axis located in the direction of the vector joining the two hydrogen atoms

Model	μ	Q_{xx}	Q_{yy}	Q_{zz}	Q_T	μ/Q_T
SPC	2.274	2.12	−1.82	−0.29	1.969	1.155
SPC/E	2.350	2.19	−1.88	−0.30	2.035	1.155
TIP3P	2.350	1.76	−1.68	−0.08	1.721	1.363
TIP4P	2.177	2.20	−2.09	−0.11	2.147	1.014
TIP4P/2005	2.305	2.36	−2.23	−0.13	2.297	1.004
TIP5P	2.290	1.65	−1.48	−0.17	1.560	1.460
Gas (expt.)	1.850	2.63	−2.50	−0.13	2.565	0.721

in the solid phases where the relative orientation of the molecules is more or less fixed by the structure of the solid. The coupling between dipolar and quadrupolar interactions is well known since some time ago due to early studies about the behavior of hard spheres with dipole and quadrupole moments.[57] However, the role of the quadrupole in the properties of water has probably been overlooked although there were some clear warnings about its importance.[58–60]

III. An exam for water models

In 2002 Guillot presented[33] a study of the performance of water models to describe several properties of water. Although this is only six years ago there are several reasons to perform this study once again. At that time TIP5P had just been released. TIP4P/2005 was proposed three years later. In the intervening years a more precise determination of some properties of the water models (surface tension, temperature of maximum density) has been obtained. More importantly, the calculation in recent years of water properties that were almost completely unknown before (as, for instance, predictions for the solid phases of water, the determination of the melting points and the phase diagram calculations) makes a new comparison of the performance of the models interesting. It is necessary to select some properties to establish the comparison. The selected properties should be as many as possible but representative of the different research communities with an interest in water. We will not include properties of the gas phase (second virial coefficients, vapour densities) in the comparison because only a polarizable model can be successful in describing all the phases of water. Non-polarizable models cannot describe the vapour phase and the condensed phases simultaneously. Thus, the models described above fail in describing vapour properties because they ignore the existence of the molecular polarizability.[61,62] Therefore, here we will focus only on the properties of the condensed phases (liquid and solids). The ten properties of the test will be the following ones:

- 1. Vapour–liquid equilibria (VLE) and critical point.
- 2. Surface tension.
- 3. Densities of the different ice polymorphs.
- 4. Phase diagram calculations.
- 5. Melting temperature and properties at the melting point.
- 6. Maximum in the density of water at room pressure (TMD). Values of the thermal expansion coefficient, α, and the isothermal compressibility, κ_T.
- 7. Structure of water and ice Ih.
- 8. Equation of state at high pressures.
- 9. Self-diffusion coefficient.
- 10. Dielectric constant.

In order to give an assessment of the quality of the predictions we will assign a score to each of properties depending on the predictions of the model. We recognise that any score is rather arbitrary. We do not pretend to give an absolute test but rather to give a qualitative idea of the relative performance of these water models. So we have devised a simple scoring scheme: for each property we shall assign 0 points to the model with the poorest performance, 1 to the second, 2 to the third and 3 points to the model showing the best performance.

IV. Calculation details

In this work we compare the performance of TIP3P, TIP4P, TIP5P and TIP4P/2005. To perform the comparison we shall use data taken from the literature, either from our previous works or from other authors. For the cases where no results are available we have carried out new simulations to determine them. In all the calculations of this work the LJ part of the potential has been truncated and standard long-range corrections are used. Unless stated otherwise all calculations of this work were

 This journal is © The Royal Society of Chemistry 2009

obtained by truncating the LJ part of the potential at 8.5 Å. The importance of an adequate treatment of the long-range Coulombic forces when dealing with water simulations has been pointed out in recent studies.[63–66] In fact, in the case of water, the simple truncation of the Coulombic part of the potential causes a number of artifacts.[63–66] In our view, for water the simple truncation of the Coulombic part of the potential should be avoided and one should use a technique treating the long-range Coulombic interactions adequately (as, for instance, Ewald sums or reaction field[67–69]). The Ewald sums are especially adequate for phase diagram calculations since they can be used not only for the fluid phase but also for the solid phase. Because of this, the Ewald summation technique[69] has been employed in this work for the calculation of the long-range electrostatic forces. NpT and phase diagram calculations were done with our Monte Carlo code. Transport properties were determined by using GROMACS 3.3.[47] Isotropic NpT simulations were used for the liquid phase (and cubic solids) while anisotropic Monte Carlo simulations (Parrinello–Rahman-like)[70,71] were used for the solid phases.

Recently, we have computed the phase diagram for the TIP4P model by means of free-energy calculations. For the solid phases, the Einstein crystal methodology proposed by Frenkel and Ladd was used.[72] Further details about these free-energy calculations can be found elsewhere.[73] For the fluid phase, the free energy was computed by switching off the charges of the water model to arrive at a Lennard-Jones model for which the free energy is well known.[74] The free-energy calculations for the fluid and solid phases lead to the determination of a single coexistence point for each coexistence line. Starting at this coexistence point, the complete coexistence lines were obtained by using Gibbs–Duhem integration.[75] The Gibbs–Duhem integration (first proposed by Kofke) is just the numerical integration of the Clapeyron equation:

$$\frac{\mathrm{d}p}{\mathrm{d}T} = \frac{s_{\mathrm{II}} - s_{\mathrm{I}}}{v_{\mathrm{II}} - v_{\mathrm{I}}} = \frac{h_{\mathrm{II}} - h_{\mathrm{I}}}{T(v_{\mathrm{II}} - v_{\mathrm{I}})} \tag{2}$$

where we use lower case for thermodynamic properties per particle, and the two coexistence phases are labelled as I and II respectively. Since the difference in enthalpy and volume between two phases can be easily determined in computer simulations, the equation can be integrated numerically. Therefore, a combination of free-energy calculations and Gibbs–Duhem integration allowed us to determine the phase diagram of the TIP4P model.

In this work we have also computed the phase diagram for the TIP3P and TIP5P models. Instead of using the free-energy route to obtain the initial coexistence point we have used Hamiltonian Gibbs–Duhem integration[76] which we briefly summarize. Let us introduce a coupling parameter, λ, which transforms a potential model A into a potential model B (by changing λ from zero to one):

$$U(\lambda) = \lambda U_{\mathrm{B}} + (1 - \lambda)U_{\mathrm{A}}. \tag{3}$$

It is then possible to write a generalized Clapeyron equation as

$$\frac{\mathrm{d}T}{\mathrm{d}\lambda} = \frac{T\left(\langle u_{\mathrm{B}} - u_{\mathrm{A}}\rangle^{\mathrm{II}}_{N,p,T,\lambda} - \langle u_{\mathrm{B}} - u_{\mathrm{A}}\rangle^{\mathrm{I}}_{N,p,T,\lambda}\right)}{h_{\mathrm{II}} - h_{\mathrm{I}}} \tag{4}$$

where u_{B} is the internal energy per molecule when the interaction between particles is described by U_{B} (and a similar definition for u_{A}). If a coexistence point is known for the system with potential A, it is possible to determine the corresponding coexistence point for the system with potential B (by integrating the previous equations changing λ from zero to one). In this way, the task of determining an initial point of the coexistence lines of the phase diagram of system B is considerably simplified. The initial coexistence properties for the system A (*i.e.* the reference system) must be known. We have used TIP4P as reference system because its coexistence lines are now well known.[34,52] The phase diagram for TIP3P and TIP5P reported in this work has been obtained by means of the Hamiltonian Gibbs–Duhem integration starting from the known coexistence points of the TIP4P model. Again, the rest of the coexistence lines have been calculating using the usual Gibbs–Duhem [eqn (2)] integration.

V. Results

1. Vapour–liquid equilibria (VLE) and critical point

The vapour–liquid equilibria of TIP4P and TIP5P have been reported by Nezbeda *et al.*[64,77,78] Besides, we have recently calculated the VLE of TIP4P/2005.[79] However, predictions of the vapour–liquid equilibria for the rigid TIP3P model were not found after a literature search (although it was possible to find results[80] for a flexible TIP3P model). For this reason we proceeded to evaluate the VLE of the TIP3P model. Firstly, we used Hamiltonian Gibbs–Duhem integration to obtain an initial coexistence point for the TIP3P model at 450 K. As initial reference point, we used the coexistence data reported by Lisal *et al.*[77] of the TIP4P model at 450 K. The LJ part of the potential was truncated at a distance slightly smaller than half the box length (both for the liquid and for the vapour phase). Once the coexistence point at 450 K was obtained for the TIP3P model, the rest of the coexistence line was obtained from Gibbs–Duhem simulations. The coexistence points are presented in Table 3. The critical properties of all TIP*n*P models are given in Table 4 and Fig. 1 shows the vapour–liquid equilibria for TIP3P, TIP4P, TIP4P/2005 and TIP5P.

The lowest critical temperature corresponds to that of the TIP5P model followed by that of the TIP3P and TIP4P. TIP4P/2005 provides a critical point in very good agreement with experiment. As it can be seen in Fig. 1, TIP4P/2005 also provides an excellent description of the coexistence densities along the orthobaric curve. The description of the critical pressure is not good for any of the models which suggests that the inclusion of polarizability is probably required to reproduce the

Table 3 Vapour liquid equilibria for the TIP3P model as obtained from the computer simulations of this work

T/K	p/bar	ρ_{gas}/g cm^{-3}	ρ_{liquid}/g cm^{-3}
550	81.32	0.074(3)	0.543(8)
545	74.74	0.066(2)	0.561(7)
540	68.64	0.058(2)	0.585(5)
530	57.83	0.043(1)	0.613(5)
520	48.44	0.0350(9)	0.643(7)
510	40.48	0.0278(8)	0.668(4)
500	33.58	0.0224(5)	0.697(4)
490	27.62	0.0183(5)	0.720(4)
470	18.30	0.0112(3)	0.760(5)
450	11.72	0.0070(1)	0.799(3)
425	6.313	0.00375(8)	0.836(3)
400	3.142	0.00187(4)	0.873(2)
375	1.412	0.00088(1)	0.904(2)
350	0.5587	0.000361(6)	0.934(2)
325	0.1890	0.000129(2)	0.959(2)
300	0.05203	0.0000380(4)	0.984(2)
275	0.01101	0.00000827(7)	1.004(2)

 This journal is © The Royal Society of Chemistry 2009

Table 4 Critical properties of TIP3P, TIP4P, TIP4P/2005 and TIP5P. Values taken from ref. 77 (TIP4P), ref. 79 (TIP4P/2005), ref. 64 (TIP5P) and this work (TIP3P)

Model	T_c/K	p_c/bar	ρ_c/g cm^{-3}
TIP3P	578	126	0.272
TIP4P	588	149	0.315
TIP4P/2005	640	146	0.31
TIP5P	521	86	0.337
Experiment	647.1	220.64	0.322

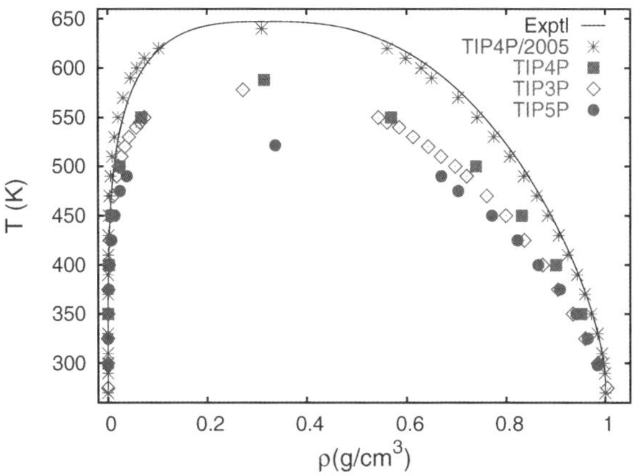

Fig. 1 Vapour–liquid equilibria for TIP3P, TIP4P, TIP5P and TIP4P/2005 models.

experimental value of the critical pressure. Notice that models reproducing the vaporization enthalpy of water without the addition of the polarization term (TIP3P, TIP4P and TIP5P) tend to predict critical temperatures which are too low. The difference between TIP3P and TIP4P is quite small suggesting that three charge models yield a critical point around 580 K when forced to reproduce the vaporization enthalpy. This idea is further reinforced by the critical temperature of SPC, which is of about 590 K.[81] The results for TIP5P indicate that the location of the negative charge at the lone pairs does not solve the problem and this model gives the worst predictions for both the critical temperature and pressure. The success of TIP4P/2005 in estimating the critical point seems to be related to the fact that this model reproduces the vaporization enthalpy only after the inclusion of the polarization correction [as given by eqn (1)]. Further evidence of this is obtained from the fact that SPC/E (which also incorporates the polarization correction) also reproduces the critical temperature of water reasonably well.

The correlation between the critical point and the vaporization enthalpy was first pointed out by Guillot.[33] It is obvious that the value of the vaporization enthalpy is related to the strength of the hydrogen bond in the model. One may try to relate the strength of the hydrogen bond to the magnitude of the dipole and quadrupole moments. The dipole moments of TIP3P, TIP4P, TIP4P/2005 and TIP5P are rather similar but their quadrupole moments differ significantly. TIP5P has the lowest quadrupole moment and TIP4P/2005 has the highest and this is also true for the critical temperatures. However the SPC/E model, which has a relatively low value of the quadrupole moment, yields a satisfactory critical temperature. Obviously the strength of the hydrogen bond depends not only on the multipole moments but also on the parameters of the LJ potential and on the bond geometry. From the results of the vapour–liquid equilibria, we assign 3 points to TIP4P/2005, 2 points to TIP4P, 1 point to TIP3P and 0 points to TIP5P.

2. Surface tension

The values of the surface tension, γ, for TIP4P and TIP4P/2005 have been reported recently.[82] For these models the surface tension was obtained using the mechanical route[83] (from the normal and perpendicular components of the pressure tensor) and the test area method[84] (where the Boltzmann factor of a perturbation that changes the interface area but keeps the total volume constant is evaluated). Results obtained from these two methodologies were in good agreement. For TIP5P, results of the surface tension has been reported recently by Chen and Smith.[85] In this work we have calculated the surface tension for TIP3P using both the mechanical route and the test-area method. Simulations were performed with GROMACS 3.3[47] for 1024 water molecules with an interface area of about ten by ten molecular diameters. The length of the runs was 2 ns. The LJ part of the potential was truncated at 13 Å. Long-range corrections to the surface tension were included as described in ref. 82. The results for TIP3P are presented in Table 5. The surface tension of TIP3P can be described quite well by the following expression:[86]

$$\gamma = c_1(1 - T/T_c)^{11/9}[1 - c_2(1 - T/T_c)] \tag{5}$$

This expression is used by the IAPWS (International Association for Properties of Water and Steam) to describe the experimental values of the surface tension of water.[6,87] A fit of the surface tension results for TIP3P furnishes the parameters $T_c = 578.17$ K, $c_1 = 176.42$ mN m^{-1}, $c_2 = 0.57857$. The critical temperature obtained in this way is in good agreement with that obtained from Gibbs–Duhem calculations. Fig. 2 shows the surface tension for TIP3P, TIP4P, TIP4P/2005 and TIP5P. The lowest values of the surface tension correspond to the TIP5P model followed by TIP3P and TIP4P. Again, models reproducing the vaporization enthalpy of water give rather low values of the surface tension. This is consistent with the lower critical temperature of these models. It seems that the TIP4P charge distribution provides a slightly higher value of the surface tension when compared to TIP3P and TIP5P. The predictions of the surface tension of the TIP4P/2005 are in excellent agreement with experiment.[82,88] How to explain the performance of the different models? Since the surface tension of SPC is similar to that of TIP3P, and that of

Table 5 Surface tension (in mN m^{-1}) for the TIP3P model of water at different temperatures as obtained from computer simulation. ρ_l and ρ_v are the densities (in g cm^{-3}) of the liquid and vapour phases at coexistence. $\bar{p}_N$ and $\bar{p}_T$ are the macroscopic values of the normal and tangential components of the pressure tensor (in bar units). t is the thickness of the vapour–liquid interface (in Å). γ_v^* and γ_{ta}^* are the values of the surface tension obtained from the virial route and the test-area method respectively, without including long-range corrections. γ_v and γ_{ta} are the corresponding values of the surface tension, but including long-range corrections which correct for the truncation of the LJ part of the potential at $r_c = 13$ Å (see ref. 82 for details). The values of the surface tension as estimated from this work (γ_{sim}) correspond to the arithmetic average $(\gamma_v + \gamma_{ta})/2$. γ_{exp} are the experimental values of the surface tension

T/K	ρ_l	ρ_v	$\bar{p}_N$	$\bar{p}_T$	γ_v^*	γ_{ta}^*	t	γ_v	γ_{ta}	γ_{sim}	γ_{exp}
300	0.980	0.000024	−0.11	−98.44	49.2	49.8	3.87	52.4	52.2	52.3(2.2)	71.73
350	0.930	0.00035	0.22	−80.63	40.4	40.7	4.91	43.3	43.1	43.2(2.0)	63.22
400	0.867	0.0018	2.86	−61.64	32.3	32.9	6.12	34.7	34.5	34.6(1.8)	53.33
450	0.790	0.0069	11.95	−33.81	22.9	23.3	7.82	24.8	24.6	24.7(1.5)	42.88
500	0.689	0.025	33.01	7.61	12.7	12.0	11.53	13.9	13.8	13.9(1.8)	31.61

 This journal is © The Royal Society of Chemistry 2009

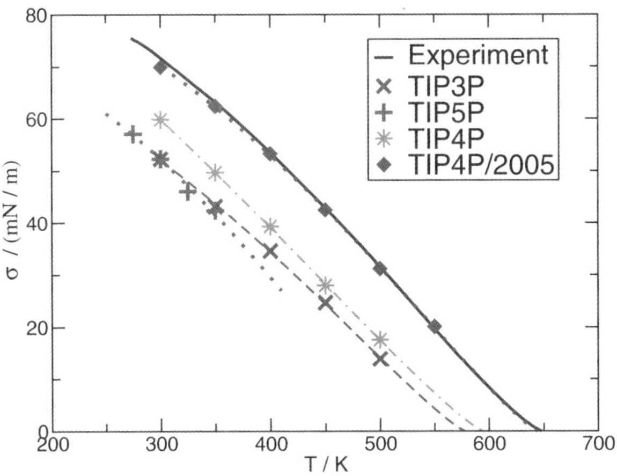

Fig. 2 Surface tension for different water models.

SPC/E is similar to that of TIP4P/2005 (although the predictions of TIP4P/2005 are even better[82,89] than those of SPC/E), the use of the polarization correction for the vaporization enthalpy seems to be responsible for the improvement of TIP4P/2005 with respect to the TIP3P, TIP4P and TIP5P models. Thus, the use of the polarization correction of Berendsen *et al.*[51] seems to be a prerequisite to obtain models with a good description of the surface tension of water (in non-polarizable models). From the results for the surface tension, we give 3 points to TIP4P/2005, 2 points to TIP4P, 1 point to TIP3P and 0 points to TIP5P. Not surprisingly, the scores of the models are identical to those obtained for the vapour–liquid equilibria.

3. Densities of the different ice polymorphs

In the book *Physics of Ice*,[4] Petrenko and Whitworth reported the density for several solid phases of water at a certain thermodynamic state. These states are used often to test the performance of water models.[34,36,90] Simulation results[34–36,91] indicate that TIP5P overestimates the density of the ices by about seven per cent whereas TIP4P overestimates the densities by less than three per cent and TIP4P/2005 gives an error of less than one per cent. The failure of TIP5P is probably a consequence of a too short distance between the oxygens when the hydrogen bond is formed.[92] This might also be related to the fact that the negative charge is too far away from the oxygens in TIP5P. In this work we have obtained the predictions for the TIP3P and SPC models which have not been reported yet. We have carried out NpT simulations with anisotropic scaling. For the initial configurations we used the structural data obtained from diffraction experiments. This is all that is needed for ices in which the protons are ordered (II, IX, XI, VIII, XIII, XIV). For ices with proton disorder the oxygens were located using crystallographic information and a proton-disordered configuration (with no net dipole moment and satisfying the Bernal–Fowler rules[46,93,94]) was generated with the algorithm proposed by Buch *et al.*[95] Notice that, although we use experimental information to generate the initial solid configuration, the system can relax since we are using the anisotropic NpT scaling in the simulations.

The ice densities for the TIP3P and SPC models are given in Table 6. SPC yields errors of about 2.2%. In general all solid phases were mechanically stable with the SPC model, the only exception being ice XII which melted spontaneously at the simulated temperature. The situation is much worse for the TIP3P model for which several of the solid phases melted spontaneously (in particular ice Ih, III, V and XII). As will be discussed later, this is related to the extremely low melting points of solid

Table 6 Densities and residual internal energies of the different ice phases from NpT simulations for TIP3P and SPC. The experimental data of ices are taken from ref. 4, except those for ice VII (ref. 119) and ice II (ref. 98). The numbers in parenthesis correspond to the cases where the ices are not mechanically stable and melt into liquid water (the reported densities and internal energies then correspond to those of the liquid)

			ρ/g cm^{-3}			U/kcal mol^{-1}	
Phase	T/K	p/bar	Exptl	SPC	TIP3P	SPC	TIP3P
Liquid	300	1	0.996	0.975	0.982	−9.96	−9.66
Ih	250	0	0.920	0.923	(1.021)	−11.72	(−10.23)
Ic	78	0	0.931	0.951	0.959	−13.04	−12.49
II	123	0	1.190	1.219	1.219	−12.94	−12.66
III	250	2800	1.165	1.150	(1.130)	−11.32	(−10.34)
IV	110	0	1.272	1.298	1.286	−12.23	−11.75
V	223	5300	1.283	1.270	(1.226)	−11.60	(−10.73)
VI	225	11000	1.373	1.379	1.366	−11.47	−10.91
VII	300	100000	1.880	1.809	1.826	−8.46	−8.16
VIII	10	24000	1.628	1.661	1.683	−11.41	−11.02
IX	165	2800	1.194	1.198	1.194	−12.55	−12.16
XII	260	5000	1.292	(1.216)	(1.183)	(−10.78)	(−10.26)
XI	5	0	0.934	0.967	0.972	−13.54	−13.01
XIII	80	1	1.251	1.282	1.289	−12.86	−12.48
XIV	80	1	1.294	1.33	1.338	−12.65	−12.22

phases for the TIP3P model.[53] Table 6 also reports the internal energies for the different crystalline phases of TIP3P and SPC (those for other models can be found elsewhere[36,96,97]).

Fig. 3 represents the average deviation from experiment for several ice polymorphs. For the experimental density of ice II we are using the value recently reported by Fortes *et al.*,[98] which evidenced that the value of Kamb[99] could be distorted by the presence of helium inside the ice II structure (see also the discussion

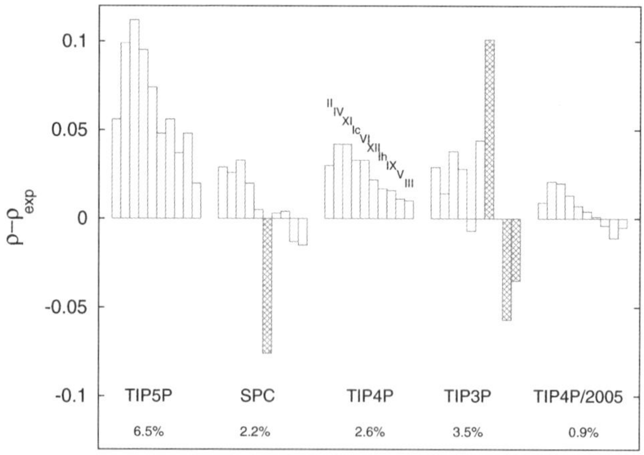

Fig. 3 Deviations in the density predictions for several ice polymorphs at the thermodynamic states reported by Petrenko and Whitworth.[4] The experimental value of the density for ice II is taken from Fortes *et al.*[98,121] (instead of the value of Kamb *et al.*[99,122] used in our previous work). For TIP3P and SPC, the filled rectangles indicate that the corresponding solid phase melts and that the density of the liquid was used to compute the error.

 This journal is © The Royal Society of Chemistry 2009

in ref. 97). In fact, the simulation results agree much better with the new reported data of Fortes *et al.*, reinforcing the idea that the value reported by Kamb is probably incorrect.

According to the results of Fig. 3, we assign 3 points to the performance of TIP4P/2005, 2 points to TIP4P, 1 point to TIP5P and 0 points to TIP3P (the low score of TIP3P is due to the fact that several ices melt well below their experimental melting temperatures so that this model is not useful to study the behaviour of the solid phases of water).

4. Phase diagram calculations

The phase diagram of TIP4P, SPC/E and TIP4P/2005 has been determined in previous work.[34,36,52,100] The phase diagram for TIP3P has not been determined so far. In this work we have used Hamiltonian Gibbs–Duhem integration to compute it and to complete the previous calculations for TIP5P. The results are presented in Fig. 4. The predictions of TIP3P and TIP5P are quite poor. Ice Ih is thermodynamically stable only at negative pressures. The stable phase at the normal melting point for both models is ice II. This surprising result has been confirmed recently by

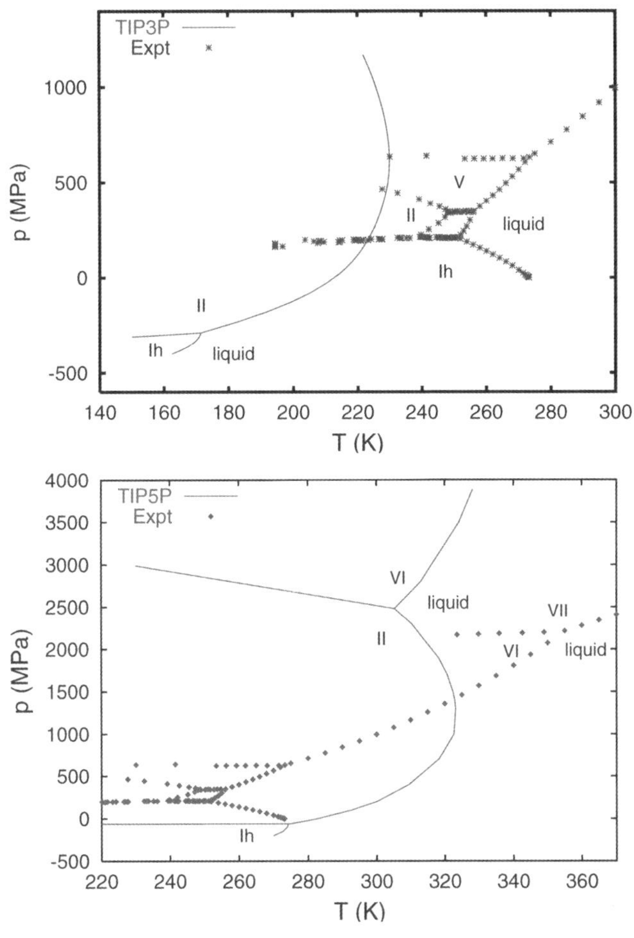

Fig. 4 Phase diagram as obtained in this work for TIP3P and TIP5P. Symbols: experimental phase diagram, lines: computed phase diagram. Top: results for the TIP3P model, bottom: results for the TIP5P model.

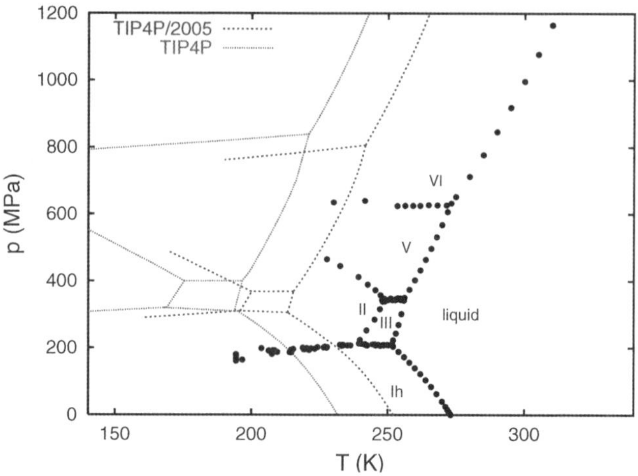

Fig. 5 Phase diagram of TIP4P and TIP4P/2005 models. Symbols: experimental phase diagram, lines: computed phase diagram.

evaluating the properties of ices at zero temperature and pressure.[96] The performance of TIP4P/2005 is quite good (Fig. 5). The overall performance of TIP4P is also quite good although somewhat shifted to lower temperatures as compared to TIP4P/2005.

What is the origin of the success of TIP4P/2005 and TIP4P? Certainly, it cannot be related to the value of the vaporization enthalpy. This is clear since TIP4P/2005 and TIP4P give good phase diagram predictions and the first uses the polarization correction whereas the second does not. Since the bond geometry of all models is the same, the different performance between them should be related to the differences in the charge distribution. It is not surprising that the solid phases provide information about the orientational dependence of the pair potential in the molecule of water. After all, in the solid phase the molecules adopt certain relative orientations which define the structure of the crystal. Polar forces are strongly dependent on the orientation. Since the dipole moment is similar for TIP3P, TIP4P, TIP4P/2005 and TIP5P, the different predictions found in Fig. 4 and 5 must be related to differences in higher multipolar moments. In fact, as was discussed above, the quadrupole moments for these water models are quite different. We have recently reported that the ratio of the dipole to quadrupole moment seems to play a crucial role in the quality of the phase diagram predicted by the different water models.[55] As can be seen in Table 2, the ratio μ/Q_T for TIP4P and TIP4P/2005 is about 1 whereas for TIP3P it increases to 1.33. For the TIP5P model the ratio is even larger. In summary, a qualitatively good description of the phase diagram of water requires a reasonable balance between dipolar and quadrupolar forces, and the factor affecting this ratio is just the charge distribution. Not surprisingly, the phase diagram prediction for SPC/E (with a charge distribution similar to that of TIP3P) is also quite poor. Therefore, we assign 3 points to TIP4P/2005, 2 points to TIP4P, 1 point to TIP5P and 0 points to TIP3P.

5. Melting temperature. Properties at melting

Obtaining the melting point of water models is not as obvious as it may appear. The simplest approach (which works fine in the lab[9]) of heating the ice within the simulation box and determining the temperature at which it melts, does not provide the true melting point. In fact, when NpT simulations are performed under periodical boundary conditions, ices usually melt at a temperature about 80–90 K above the

 This journal is © The Royal Society of Chemistry 2009

true melting point[101] (*i.e.* the temperature at which the chemical potential of the liquid and solid phases are identical). But, in real experiments, ices can not be superheated (at least for a reasonable time). The absence of a free surface is responsible for the superheating of ices in NpT simulations. In fact, when the simulations are performed with a free surface, superheating is suppressed.[102] For this reason the evaluation of the melting point of water models requires special techniques. The melting point of TIP4P was obtained from free-energy calculations of both the fluid and the solid phase. Details of these free-energy calculations have been given recently.[73] The melting temperatures of TIP5P, TIP3P and TIP4P/2005 were obtained with Hamiltonian Gibbs–Duhem integration using TIP4P as a reference model. The melting points obtained *via* the free-energy route are in complete agreement with those obtained from direct coexistence simulations[103–106] where the fluid and solid phases are kept in contact within the simulation box.

We now discuss the melting temperatures of ice Ih for TIP3P, TIP4P, TIP5P, TIP4P/2005 at the standard pressure of 1 bar. It should be mentioned once again that for TIP3P and TIP5P, ice II is more stable than ice Ih at ambient pressure. It is possible to locate the melting point of ice Ih, even though is metastable, provided that it is mechanically stable. For TIP5P the melting point of ice II is about 2 K above that of ice Ih but the melting point of ice II for TIP3P is about 60 K above that of ice Ih. In Table 7, the melting temperature of ice Ih (at 1 bar) and the properties of the solid and the liquid phases at coexistence are given for these models. Concerning the melting temperature, it is as low as 146 K for TIP3P. Thus, TIP3P is probably the poorest model for studying solid phases of water. Since it is possible to simulate ices up to temperatures about 80 K above the melting point, the temperature of 230 K is, roughly speaking, the highest one that can be used to study ice Ih with the TIP3P model. At higher temperatures the TIP3P ice Ih melts. Our estimates for the melting temperature of TIP4P, TIP4P/2005 and TIP5P are approximately 232, 252 and 274 K respectively. The result for TIP4P is in good agreement with the value reported by Gao *et al.*[107] and Koyama *et al.*[108] Another interesting quantity is the ratio of the normal melting temperature to the critical temperature, T_m/T_c. Since the difference between the normal melting temperature and the triple point temperature is only about 0.01 K, this ratio determines the range of existence of the liquid phase for the considered model. Experimentally, $T_m/T_c = 0.42$. The value of the ratio for TIP4P models (TIP4P and TIP4P/2005) is essentially the same, 0.394, but it increases to 0.525 for TIP5P. Thus TIP4P models describe significantly better the experimental range of existence of the liquid phase for water. For TIP3P the ratio is considerably lower (0.25) when the melting temperature of ice Ih (146 K) is considered. But it increases to 0.36 using 210 K as the melting temperature for the stable phase at normal melting for the TIP3P model, which is ice II.

Table 7 Melting properties of ice Ih at $p = 1$ bar for different models. T_m, melting temperatures; ρ_l and ρ_{Ih}, coexistence densities of liquid water and ice; ΔH_m, melting enthalpy; dp/dT, slope of the coexistence curve (between ice Ih and water at the normal melting temperature of the model). We also include for comparison the ratio T_m/T_c. Melting properties are taken from ref. 53 (TIP3P, SPC, SPC/E, TIP4P) and ref. 35 (TIP4P/2005)

Model	TIP3P	SPC	SPC/E	TIP4P	TIP4P/2005	TIP5P	Exptl
T_m/K	146	190	215	232	252	274	273.15
ρ_l/g cm^{-3}	1.017	0.991	1.011	1.002	0.993	0.987	0.999
ρ_{Ih}/g cm^{-3}	0.947	0.934	0.950	0.940	0.921	0.967	0.917
ΔH_m/kcal mol^{-1}	0.30	0.58	0.74	1.05	1.16	1.75	1.44
dp/dT/bar K^{-1}	−66	−115	−126	−160	−135	−708	−137
T_m/T_c	0.25	0.321	0.337	0.394	0.394	0.525	0.422

Table 7 also presents the coexistence properties (densities at coexistence, slope of the melting curve dp/dT and enthalpy of melting). TIP4P/2005 provides the best estimates of the coexistence densities while TIP5P gives a poor estimate of the density of ice Ih and a completely wrong prediction of the slope dp/dT. This is because, for TIP5P, the densities of ice Ih and of the liquid are quite similar. On the other hand, no model is able to reproduce the melting enthalpy. It is too small (three times lower) for TIP3P. The TIP4P and TIP4P/2005 models also underestimate the melting enthalpy but the errors are much smaller, about 30 and 20% respectively. Finally, TIP5P overestimates the melting enthalpy by approximately 20%.

Let us now try to provide a rational basis for the previous results. For three charge models (TIP3P, TIP4P and TIP4P/2005) there is a clear correlation between the melting temperature of ice Ih and the quadrupole moment of the molecule.[54,56] Our conclusion is that models locating the negative charge on the oxygen atom have low melting points for ice Ih. Locating the negative charge shifted from the oxygen along the H–O–H bisector increases the melting temperature. The improvement is better if the polarization correction is used in the calculation of the vaporization enthalpy as a target property. When this is done (as in TIP4P/2005) the melting point is about 20 K below the experimental value. This is the closest you can get for the melting point of ice Ih with a three charge model while still describing the vaporization enthalpy with the polarization correction of Berendsen *et al.*[51] The only way of reproducing the experimental melting point of ice Ih within three charge models is to sacrifice the vaporization enthalpy as a target property. In fact, we have developed a model (denoted as TIP4P/Ice)[35] which reproduces the melting point of ice Ih but overestimates the vaporization enthalpy. The behaviour of TIP5P is different; it yields a good prediction for the melting temperature in spite of having a low quadrupole moment. Probably the fact that there are two negative charges instead of just one is the responsible for this different behaviour. It is likely that locating the negative charge on the lone-pair electrons enhances the formation of hydrogen bonds in the solid phases provoking an increase in the melting point.

TIP5P reproduces the melting temperature of water. But this is not for free, since the estimate of the coexistence density of ice and of the dp/dT slope then becomes quite poor. Conversely, TIP4P/2005 predicts a rather low melting temperature but it gives quite reasonable estimates for the properties at melting. For this reason, there is no clear winner in this test so we have decided to assign 2.5 points to both TIP5P and TIP4P/2005, 1 point to TIP4P and 0 points to TIP3P.

6. Temperature of maximum density. Thermal coefficients α and κ_T

The density of liquid water for the room pressure isobar as a function of the temperature for TIP3P, SPC and TIP4P with a simple truncation of the Coulombic forces have been reported by Jorgensen and Jenson.[109] Recently, we have also calculated the liquid densities at normal pressure[36,50] for TIP3P, TIP4P, TIP5P and TIP4P/2005 using Ewald sums to deal with long-range Coulombic forces. Good agreement between our results and those reported by Paschek was found.[31] Our results are shown in Fig. 6. All models do exhibit a maximum in the density of water along the isobar. For some time it was believed that TIP3P did not exhibit this maximum but it is now clear that the maximum also exists for this model (although located at very low temperatures). In Table 8, the location of the TMD and the values of the thermal expansion coefficient (α) and of the isothermal compressibility (κ_T) at room temperature and pressure are presented. It is clear that the poorest description of the TMD is provided by TIP3P. The behaviour of TIP4P is noticeably better. TIP4P/2005 reproduces nicely the density of water (and its maximum) for all the temperatures along the room pressure isobar. We have shown recently[50] that—for three charge models like TIP3P, TIP4P and TIP4P/2005—the difference between the TMD and the melting temperature of ice is around 25 K (the experimental difference is 4 K). For this reason, the location of the TMD correlates very well with the

 This journal is © The Royal Society of Chemistry 2009

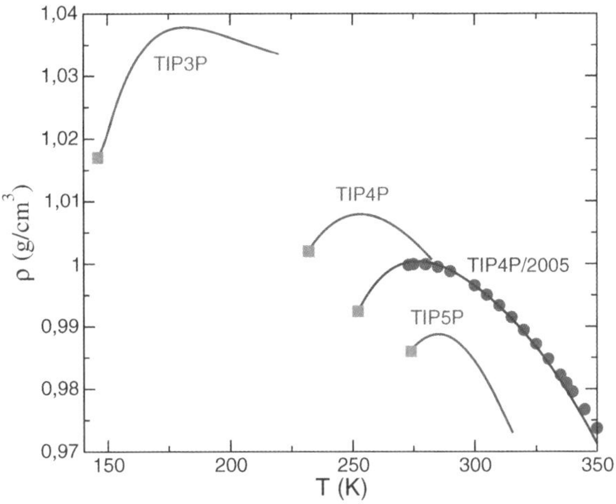

Fig. 6 Maximum in density for several water models at room pressure. Filled circles: experimental results, lines: simulation results. For each model the square represents the location of the melting temperature of the model for ice Ih.

Table 8 Temperature at which the maximum in density occurs T_{TMD} (at room pressure) for different water models. The value of the coefficient of thermal expansion α and of the isothermal compressibility κ_T at room temperature and pressure are also given. The values of α and κ_T for TIP3P, TIP4P and TIP5P as reported in ref. 120. The values for TIP4P/2005 were taken from ref. 36. The value of the maximum in density is as given in ref. 50, except those of the TIP5P that were taken from the original ref. 44

Model	T_{TMD}/K	$10^5\kappa_T$/MPa^{-1}	$10^5\alpha$/K^{-1}
TIP3P	182	64	92
TIP4P	253	59	44
TIP5P	277	41	63
TIP4P/2005	278	46	28
Experiment	277	45.3	25.6

melting temperature. TIP5P also reproduces the temperature at which the maximum density in water occurs. For the TIP5P the difference between the melting point of ice Ih and the temperature of the maximum in density is about 11 K. In Fig. 6 the densities of TIP5P as obtained from simulations using Ewald summation are presented.

Notice that TIP5P does not reproduce the experimental densities when Ewald summation is used. In the original paper where the TIP5P model was proposed by Mahoney and Jorgensen[44] the potential was truncated at 9 Å. In these conditions TIP5P reproduces the density of water at the maximum. It has been noticed by other authors that, in general, lower densities are obtained when Ewald sums are implemented than when the electrostatic potential is merely truncated at a given distance. The decrease in density is particularly noticeable in the case of the TIP5P model. For this reason the maximum in density of the TIP5P model occurs at 285 K when Ewald sums are used[50,64] whereas the maximum takes place at 277 K when the potential is truncated at 9 Å. Fig. 6 shows that for the TIP5P model when the whole normal pressure isobar is considered (not just the maxima), the curvature is not correct. In other words, the dependence of the density on temperature at constant pressure (given by the thermal expansion coefficient α) is not properly predicted by TIP5P (see Table 8). Importantly, this is also true when the potential is truncated at 9 Å. Therefore, TIP5P does not reproduce α correctly, neither with the potential truncated at 9 Å nor with Ewald sums.

In summary, both TIP4P/2005 and TIP5P reproduce the temperature at which the density maximum occurs. This is by design since the TMD was used as a target property in both models. However, the description of the complete room pressure isobar is much better in the TIP4P/2005 which nicely reproduces the whole curve providing a very good estimate of the coefficient α. Also the estimate of κ_T is better for TIP4P/2005. The excellent description of the densities along the isobar has an interesting consequence. Since the vapour pressure of water is quite small up to relatively high temperatures, orthobaric densities are essentially identical to those obtained from the room pressure isobar. Therefore, a model correctly describing the equation of state along the room pressure isobar will also provide reliable estimates of the orthobaric densities (at least, at temperatures not too close to the critical one). Thus the good description of the ambient pressure isobar of TIP4P/2005 explains in part the good description of the coexistence curve presented previously. Although TIP5P estimates the location of the TMD correctly (when using a cutoff for the electrostatic interactions), it does not yield satisfactory densities for the rest of the room pressure isobar and this explains the poor performance in describing orthobaric coexistence densities. Thus, using the TMD as a target property is a good idea, but it is far better to use the complete room pressure isobar (of course including the maximum) as a target property. For future developments of water potentials we do really recommend using the complete room pressure isobar as a target property. According to the discussion above, we give 3 points to TIP4P/2005, 2 points to TIP5P, 1 point to TIP4P and 0 points to TIP3P.

7. Structure of liquid water and ice

The experimental oxygen–oxygen radial distribution function for liquid water and for ice Ih will be compared to the predictions from the simulations. In Fig. 7 the comparison is made for liquid water. Experimental results are taken from Soper.[110] TIP5P provides the best estimate of the radial distribution function. The predictions of TIP4P/2005 are good but the first peak is too high. TIP4P gives quite acceptable predictions but not as good as the previous models. Once again, the discrepancy between the results for TIP3P and experiment is notorious. In Fig. 8 the comparison is made for ice Ih at 77 K and 1 bar. Experimental results are taken from Narten *et al.*[111] The predictions of TIP4P/2005 are now the best (although it again overestimates the height of the first peak). The results for TIP4P are also very good. Both TIP5P and TIP3P fail in the description of the structure of the solid beyond the first

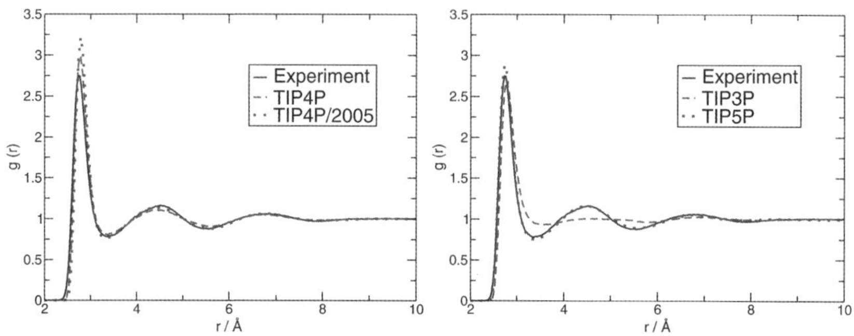

Fig. 7 Oxygen–oxygen radial distribution function for liquid water at $T = 298$ K and $p = 1$ bar. Experimental results were taken from Soper.[110] Left: results for TIP4P and TIP4P/2005, right: results for models TIP3P and TIP5P.

 This journal is © The Royal Society of Chemistry 2009

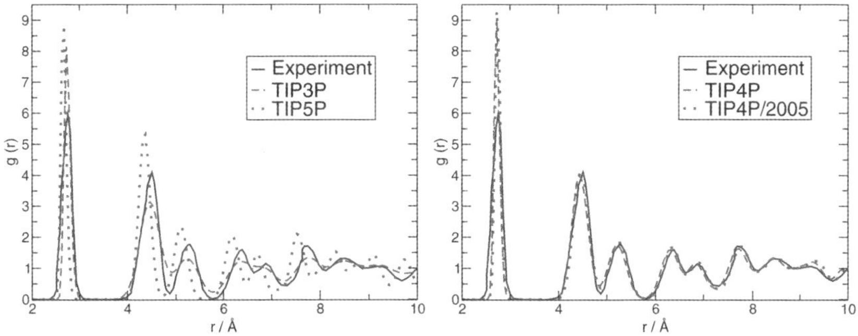

Fig. 8 Oxygen–oxygen radial distribution function for ice Ih at $T = 77$ K and $p = 1$ bar. Experimental results as reported by Narten *et al.*[111] Left: results for models TIP3P and TIP5P, right: results for TIP4P, TIP4P/2005.

coordination shell. TIP4P/2005 provides good estimates of the radial distribution functions but it overestimates the height of the first peak. It is likely that quantum effects (which could be determined by using path integral simulations) may be significant to understand the amplitude of the first maximum. Indeed, this has been shown to be the case not only for liquid water[112,113] but also for ice Ih as reported by Kusalik and de la Peña.[114,115] Further work on this point is probably needed. There is no clear winner concerning structural predictions (TIP5P performs better for water and TIP4P/2005 for ice Ih). For this reason we have decided to give 2.5 points to both TIP5P and TIP4P/2005, one point to TIP4P and zero points to TIP3P.

8. Equation of state of liquid water at high pressures

In a number of geological applications the equation of state (EOS) of water at high temperatures and pressures is needed. Therefore it is of interest to analyze the capacity of these models to predict the behaviour of liquid water at high pressure. Fig. 9 displays the EOS for TIP3P, TIP4P, TIP4P/2005 and TIP5P along the 373 K isotherm for pressures up to 24 000 bar (obtained with 360 water molecules). Experimental results are taken from the EOS of Wagner *et al.*[116] and from the experimental measurements of Abramson and Brown.[117] TIP4P/2005 provides an

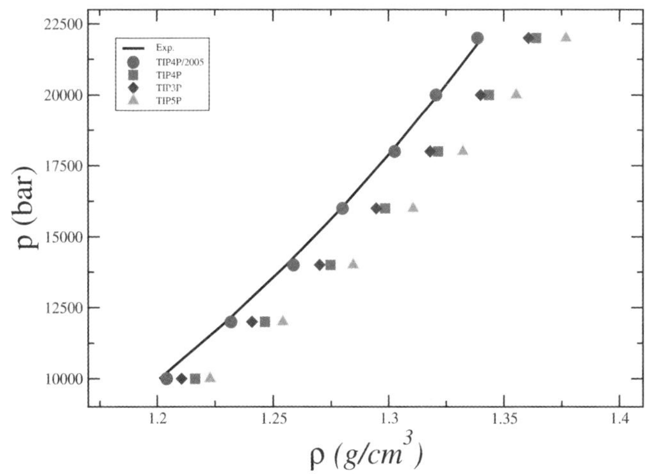

Fig. 9 Equation of state at $T = 373$ K for several water models.

excellent description of the EOS. The differences with experimental data increase for TIP3P followed by TIP4P. The performance of TIP5P is quite poor. We thus assign 3 points to TIP4P/2005, 2 points to TIP3P, 1 point to TIP4P and 0 points to TIP5P.

9. Self-diffusion coefficient

We have computed the self-diffusion coefficient as a function of temperature (at 1 bar) for TIP3P, TIP5P, TIP4P and TIP4P/2005. Simulations were performed with the GROMACS[47] package and the diffusion coefficients were determined from the slope of the mean square displacement *versus* time (using 360 molecules). A relaxation time of 5 ps was used for the thermostat and for the barostat. Results are presented in Table 9. The diffusion coefficient of TIP3P is too high at all the temperatures investigated. This suggests that the hydrogen bonding for this model is probably too weak. Likely, the weakness of the hydrogen bond is also responsible for the low melting temperature of ice Ih and the loss of structure for the liquid phase beyond the first coordination shell. This may be a big concern in simulation studies of proteins where there must be a competition between intramolecular and intermolecular hydrogen bonds. The diffusion coefficients of TIP4P are closer to experiment but they are still too high, reflecting the fact that using the vaporization enthalpy as a target property leads to highly diffusive water models. TIP5P yields results in agreement with experimental data in the vicinity of 300 K but the departures from experiment greatly increase as the temperature moves away from the ambient one. In fact, at 318 K the result furnished by TIP5P is almost the same as that for TIP4P. Fig. 10 shows that the slope of the line log D *vs.* $1/T$ is quite poor (TIP4P

Table 9 Self-diffusion coefficient 10^9D (in m^2 s^{-1}) for liquid water at 1 bar as a function of temperature

T/K	TIP3P	TIP4P	TIP5P	TIP4P/2005	Experiment
278	3.71	2.08	1.11	1.27	1.313
288	4.34	2.71	1.74	1.57	1.777
298	5.51	3.22	2.77	2.07	2.299
308	6.21	4.12	3.68	2.60	2.919
318	6.32	4.90	4.81	3.07	3.575

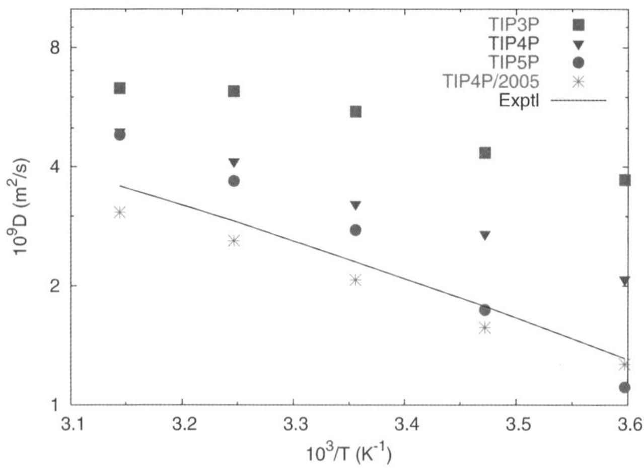

Fig. 10 Logarithmic plot of the self-diffusion coefficients *versus* the inverse of the temperature.

and even TIP3P are superior in this respect). This seems to reflect the overall results of TIP5P: good results at ambient conditions but becoming increasingly bad as one moves away from that point. The best predictions are those of TIP4P/2005. Although a little bit below the experimental values they show the correct trend in its dependence with temperature. Within three charge models, the use of the polarization correction to describe the vaporization enthalpy leads to water models with improved diffusion properties. That was already true for SPC/E (which provides much better diffusion coefficients than SPC) and seems also to be true for TIP4P/2005 (which provides much better estimates of the diffusion coefficients than TIP4P). We therefore give 3 points to TIP4P/2005, 2 points to TIP5P, 1 point to TIP4P and zero points to TIP3P.

10. Dielectric constant

Let us finish by presenting results for the dielectric constant of liquid water at room temperature and pressure. Again, simulations were performed with the package GROMACS[47] at room temperature and pressure. The simulations lasted 8 ns, and the system consisted of 360 molecules. The value of the dielectric constant is presented in Table 10. TIP5P provides the best estimate of the dielectric constant followed by TIP3P. TIP4P yields the worst value for the dielectric constant. TIP4P/2005 predicts a better dielectric constant than TIP4P but it is still far from the experimental value. It is obvious that the TIP4P charge distribution tends to give low dielectric constants. This is the only property for which the TIP4P charge distribution is in trouble against charge distributions with the negative charge located at the oxygen atom. In fact, the dielectric constants of SPC and SPC/E are better than those of TIP4P and TIP4P/2005, respectively, indicating that locating the charge on the oxygen tends to give better predictions for the dielectric constant. Recently, Rick[65] has studied the behaviour of the dielectric constant in detail as a function of the dipole and quadrupole moments for different water models. He has shown that a larger dipole increases the value of the dielectric constant whereas a larger quadrupole decreases it. He proposed an equation to correlate the dielectric constant of water models with these multipole moments:

$$\varepsilon = -85 + 98\mu - 35.7Q_T \tag{6}$$

where the dipole moment is given in Debye and Q_T in (DÅ). The dipole moments of all water models are quite similar. However they differ significantly in the value of the quadrupole moment. Thus, the lower value of the dielectric constant for TIP4P models is a direct consequence of the higher quadrupole moment of these type of models. The higher quadrupole moment of TIP4P/2005 was quite good for improving predictions for the melting point and phase diagram. Unfortunately, it also seems responsible for the low dielectric constant of the model.

The dielectric constant is given by the fluctuations of the dipole moment of the sample. It is thus related to the instantaneous values of the polarization of the sample. Non-polarizable models attempt to incorporate (in a mean field way)

Table 10 Computed static dielectric constant ε at 298 K and 1 bar

Model	ε
TIP3P	94
TIP4P	50
TIP4P/2005	58
TIP5P	91
Experiment	78.4

the effect of the polarizability. Thus, the dipole moments of non-polarizable models are higher than those of the gas phase. It is likely that a property like the dielectric constant, which depends so dramatically on just the dipole moment fluctuations, can hardly be reproduced by an effective potential in which the molecular charge cannot fluctuate. In fact, polarizable models tend to have higher dielectric constants[41] than their non-polarizable counterparts. If this is the case, the good agreement of TIP3P and TIP5P may be somewhat forced. Probably, polarizable versions of TIP3P and TIP5P will tend to give too high dielectric constants whereas the introduction of polarizability will improve the predictions of models based on the TIP4P charge distribution. Further work is required to analyse this issue in more detail since it is difficult at this stage to establish definitive conclusions. Concerning the dielectric constant predictions, we give 3 points to TIP5P, 2 points to TIP3P, 1 point to TIP4P/2005 and 0 points to TIP4P.

VI. Conclusions

Table 11 summarizes the scores obtained by the models for each of the test properties. The final result is that TIP4P/2005 has obtained 27 points, TIP5P 14, TIP4P 13 and TIP3P 6 out of a maximum possible number of 30 points. For most of the properties TIP4P/2005 yielded the best performance. The main exception is the dielectric constant for which TIP4P/2005 yields too low a value. For the second position, TIP5P and TIP4P obtained very similar scores. TIP5P improves the melting point predictions, TMD, dielectric constant and diffusivity with respect to TIP4P, but it is clearly worse in phase diagram predictions, critical point, density of ices and high pressure behaviour. In that respect TIP5P and TIP4P yielded a similar performance and the choice of one or the other potential may depend on the property to be studied. The least satisfactory model, well below any of the others, is TIP3P (see Table 11). Somewhat surprisingly TIP3P is probably the most used model in simulations of biomolecules. In our opinion the only reason to continue using TIP3P is that certain force fields were optimized to be used with TIP3P water. It is not fully obvious whether the force fields must be used with a given water model. Some researchers have challenged this idea.[118] In any case, it is clear that new force fields should also be built around better water models.

We have not included the SPC or SPC/E models in the comparison. The performance of SPC is certainly better than that of TIP3P. This is due to the fact that if the negative charge is located on the oxygen, the larger OH bond length of SPC and the tetrahedral bond angle increases the values of the quadrupole moment, and this improves the performance of the model. However, SPC/E yields an overall better performance than SPC. In fact, it improves the prediction of almost all of the

Table 11 Scores obtained by each model for the ten properties considered in this work

Property	TIP3P	TIP4P	TIP4P/2005	TIP5P
1. VLE, T_c	1	2	3	0
2. Surface tension	1	2	3	0
3. ρ ices	0	2	3	1
4. Phase diagram	0	2	3	1
5. T_m melting prop.	0	1	2.5	2.5
6. T_{TMD}, α, κ_T	0	1	3	2
7. Structure	0	1	2.5	2.5
8. EOS (high p)	2	1	3	0
9. D	0	1	3	2
10. ε	2	0	1	3
Total	6	13	27	14

 This journal is © The Royal Society of Chemistry 2009

properties. The performance of SPC/E, if it had been included in the test, would have been better than that of TIP5P and TIP4P but well below that of TIP4P/2005. The number of points obtained by SPC/E would have been about 21 points (3 for vapour–liquid equilibria, 2 points for surface tension, 2 points for the density of ices, 1 point for the phase diagram, 1 point for the melting properties, 1 point for the TMD, α and κ_T, 2 points for structure predictions, 3 points for the equation of state at high pressures, 3 points for the diffusion coefficient and 3 points for the dielectric constant). Thus, overall SPC/E improves the predictions with respect to TIP4P and TIP5P but is well below the number of points obtained by TIP4P/2005.

Why is the performance of TIP4P/2005 so good in general? TIP4P/2005 is just a TIP4P model with the hydrogen charge augmented from 0.52 *e* to 0.556 *e* and with the value of ε/k_B increased from 73 to 93 K (the values of σ and d_{OM} are quite similar in both models). It has taken more than twenty years to realize that these two small changes dramatically improve the performance of the model. The steps leading to TIP4P/2005 are basically three. First, the realization of the fact that TIP4P provided a better phase diagram than SPC/E. Secondly, the attempt to improve the prediction of the melting temperature could not be done without accepting the idea of introducing a correction term in the calculation of the vaporization enthalpy. The idea, first introduced by Berendsen *et al.*,[51] enabled them to transform SPC into SPC/E which, in our opinion, is a much better model than SPC. The same could also be useful within the TIP4P geometry, and the final result is TIP4P/2005, but there was an additional refinement. Using the TMD as a target property (as first done by TIP5P) or, even better, fitting the whole room pressure isobar could improve the overall performance. Thus, TIP4P/2005 incorporated this property as a target property. In summary, TIP4P/2005 takes the charge distribution from TIP4P (which yields a reasonable prediction of the phase diagram of water), the use of the correction of Berendsen *et al.* to the vaporization enthalpy from SPC/E, and the use of the TMD as a target property from TIP5P. The resulting model predicts quite nicely the orthobaric densities, critical temperature, surface tension, densities of the different solid phases of water, phase diagram, melting properties (with a reasonable prediction of the melting point about 20 K below the experimental value), TMD, isothermal compressibility, coefficient of thermal expansion and the structure of water and ice. The properties analyzed here cover a temperature range from 120 to 640 K and pressures up to 30 000 bar.

Obviously TIP4P/2005 is not the last word in water potentials and it does have some deficiencies. It fails in the prediction of the vapour properties (vapour pressure, dew densities, critical pressure, second virial coefficient) and in the prediction of the dielectric constant in the liquid phase. However, it really points out clearly the limits that can be achieved by rigid, non-polarizable models. We believe that inclusion of polarizability within a TIP4P/2005-like model (with an adequate fitting of the parameters) would yield further improvement. The extreme sensitivity of phase diagram to the water potential model may help in developing new potential models for water. Now that it is possible to evaluate phase diagrams (both vapour–liquid and fluid–solid) or the TMD on relatively routine basis, it sounds a good idea to extend the incorporation of the phase diagram and the room pressure isobar as target properties in the fitting (and/or checking) of new water models. Although there may be a current of opinion suggesting that polarizable models have not yet improved the performance of non-polarizable models, it is our belief that we have not worked hard enough to refine a reliable polarizable model for water. For the time being, TIP4P/2005 can be considered as a reliable and cheap model of water offering good performance over a wide range of conditions. Since TIP4P/2005 is a minor modification of the model proposed by Bernal and Fowler[46] in 1933, one may say that we have spend seventy-five years refining their parameters to finally obtain a reliable model of water for the condensed phases.

Acknowledgements

It is a pleasure to acknowledge to L. G. MacDowell, E. G. Noya, C. McBride and R. G. Fernandez for many helpful discussions and for their contribution to different parts of this work. Discussions with E. de Miguel (Huelva), A. Patrykiejew (Lublin) and I. Nezbeda (Prague) are also gratefully acknowledged. This work has been funded by grants FIS2007-66079-C02-01 from the DGI (Spain), S-0505/ESP/0229 from the CAM, MTKD-CT-2004-509249 from the European Union and 910570 from the UCM.

References

1 P. Ball, *Life's Matrix. A Biography of Water*, University of California Press, Berkeley, 2001.
2 J. P. Poirier, *Nature*, 1982, **299**, 683.
3 D. Eisenberg and W. Kauzmann, *The Structure and Properties of Water*, Oxford University Press, London, 1969.
4 V. F. Petrenko and R. W. Whitworth, *Physics of Ice*, Oxford University Press, Oxford, 1999.
5 J. L. Finney, *Philos. Trans. R. Soc. London, Ser. B*, 2004, **359**, 1145.
6 M. Chaplin, http://www.lsbu.ac.uk/water/, 2005.
7 J. L. Finney, *J. Mol. Liq.*, 2001, **90**, 303.
8 G. Tammann, *Kristallisieren und Schmelzen*, Johann Ambrosius Barth, Leipzig, 1903.
9 P. W. Bridgman, *Proc. Am. Acad. Arts. Sci.*, 1912, **47**, 441.
10 C. Lobban, J. L. Finney and W. F. Kuhs, *Nature*, 1998, **391**, 268.
11 C. G. Salzmann, P. G. Radaelli, A. Hallbrucker, E. Mayer and J. L. Finney, *Science*, 2006, **311**, 1758.
12 C. G. Venkatesh, S. A. Rice and A. H. Narten, *Science*, 1974, **186**, 927.
13 P. G. Debenedetti, *J. Phys.: Condens. Matter*, 2003, **15**, R1669.
14 O. Mishima, L. D. Calvert and E. Whalley, *Nature*, 1984, **310**, 393.
15 P. H. Poole, F. Sciortino, U. Essmann and H. E. Stanley, *Nature*, 1992, **360**, 324.
16 O. Mishima and H. E. Stanley, *Nature*, 1998, **396**, 329.
17 M. T. Suter, P. U. Andersson and J. B. C. Petterson, *J. Chem. Phys.*, 2006, **125**, 174704.
18 J. G. Dash, A. W. Rempel and J. S. Wettlaufer, *Rev. Mod. Phys.*, 2006, **78**, 695.
19 H. Bluhm, D. F. Ogletree, C. S. Fadley, Z. Hussain and M. Salmeron, *J. Phys.: Condens. Matter*, 2002, **14**, L227.
20 F. Paesani and G. A. Voth, *J. Phys. Chem. C*, 2008, **112**, 324.
21 V. Buch, H. Groenzin, I. Li, M. J. Shultz and E. Tosatti, *Proc. Natl. Acad. Sci. U. S. A.*, 2008, **105**, 5969.
22 P. Jedlovszky, L. Partay, P. N. M. Hoang, S. Picaud, P. von Hessberg and J. N. Crowley, *J. Am. Chem. Soc.*, 2006, **128**, 15300.
23 G. J. Kroes, *Surf. Sci.*, 1992, **275**, 365.
24 D. Chandler, *Nature*, 2005, **437**, 640.
25 S. J. Wierzchowski and P. A. Monson, *J. Phys. Chem. B*, 2007, **111**, 7274.
26 J. A. Barker and R. O. Watts, *Chem. Phys. Lett.*, 1969, **3**, 144.
27 A. Rahman and F. H. Stillinger, *J. Chem. Phys.*, 1971, **55**, 3336.
28 R. M. Lynden-Bell, J. C. Rasaiah and J. P. Noworyta, *Pure Appl. Chem.*, 2001, **73**, 1721.
29 K. M. Åberg, A. P. Lyubartsev, S. P. Jacobsson and A. Laaksonen, *J. Chem. Phys.*, 2004, **120**, 3770.
30 M. Ferrario, G. Ciccotti, E. Spohr, T. Cartailler and P. Turq, *J. Chem. Phys.*, 2002, **117**, 4947.
31 D. Paschek, *J. Chem. Phys.*, 2004, **120**, 6674.
32 C. R. W. Guimaraes, G. Barreiro, C. A. F. de Oliveria and R. B. de Alencastro, *Brazil. J. Phys.*, 2004, **34**, 126.
33 B. Guillot, *J. Mol. Liq.*, 2002, **101**, 219.
34 E. Sanz, C. Vega, J. L. F. Abascal and L. G. MacDowell, *Phys. Rev. Lett.*, 2004, **92**, 255701.
35 J. L. F. Abascal, E. Sanz, R. G. Fernández and C. Vega, *J. Chem. Phys.*, 2005, **122**, 234511.
36 J. L. F. Abascal and C. Vega, *J. Chem. Phys.*, 2005, **123**, 234505.
37 M. D. Morse and S. A. Rice, *J. Chem. Phys.*, 1982, **76**, 650.
38 E. Whalley, *J. Chem. Phys.*, 1984, **81**, 4087.
39 P. A. Monson and D. A. Kofke, in *Advances in Chemical Physics*, ed. I. Prigogine and S. A. Rice, John Wiley and Sons, New York, 2000, vol. 115, p. 113.

 This journal is © The Royal Society of Chemistry 2009

40 P. A. Monson, *AIChE J.*, 2008, **54**, 1122.
41 P. Paricaud, M. Predota, A. A. Chialvo and P. T. Cummings, *J. Chem. Phys.*, 2005, **122**, 244511.
42 B. Chen, J. Xing and J. I. Siepmann, *J. Phys. Chem. B*, 2000, **104**, 2378.
43 W. L. Jorgensen, J. Chandrasekhar, J. D. Madura, R. W. Impey and M. L. Klein, *J. Chem. Phys.*, 1983, **79**, 926.
44 M. W. Mahoney and W. L. Jorgensen, *J. Chem. Phys.*, 2000, **112**, 8910.
45 H. J. C. Berendsen, J. P. M. Postma, W. F. van Gunsteren, and J. Hermans, *Intermolecular Forces*, ed. B. Pullman, Reidel, Dordrecht, 1982, p. 331.
46 J. D. Bernal and R. H. Fowler, *J. Chem. Phys.*, 1933, **1**, 515.
47 E. Lindahl, B. Hess and D. van der Spoel, *J. Mol. Model.*, 2001, **7**, 306.
48 F. H. Stillinger and A. Rahman, *J. Chem. Phys.*, 1974, **60**, 1545.
49 L. A. Báez and P. Clancy, *J. Chem. Phys.*, 1994, **101**, 9837.
50 C. Vega and J. L. F. Abascal, *J. Chem. Phys.*, 2005, **123**, 144504.
51 H. J. C. Berendsen, J. R. Grigera and T. P. Straatsma, *J. Phys. Chem.*, 1987, **91**, 6269.
52 E. Sanz, C. Vega, J. L. F. Abascal and L. G. MacDowell, *J. Chem. Phys.*, 2004, **121**, 1165.
53 C. Vega, E. Sanz and J. L. F. Abascal, *J. Chem. Phys.*, 2005, **122**, 114507.
54 J. L. F. Abascal and C. Vega, *Phys. Chem. Chem. Phys.*, 2007, **9**, 2775.
55 J. L. F. Abascal and C. Vega, *Phys. Rev. Lett.*, 2007, **98**, 237801.
56 J. L. F. Abascal and C. Vega, *J. Phys. Chem. C*, 2007, **111**, 15811.
57 G. N. Patey and J. P. Valleau, *J. Chem. Phys.*, 1976, **64**, 170.
58 J. L. Finney, J. E. Quinn, and J. O. Baum, in *Water Science Reviews 1*, ed. F. Franks, Cambridge University Press, Cambridge, 1985.
59 K. Watanabe and M. L. Klein, *Chem. Phys.*, 1989, **131**, 157.
60 G. A. Tribello and B. Slater, *Chem. Phys. Lett.*, 2006, **425**, 246.
61 K. M. Benjamin, A. J. Schultz and D. A. Kofke, *J. Phys. Chem. C*, 2007, **111**, 16021.
62 K. M. Benjamin, J. K. Singh, A. J. Schultz and D. A. Kofke, *J. Phys. Chem. B*, 2007, **111**, 11463.
63 D. van der Spoel, P. J. van Maaren and H. J. C. Berendsen, *J. Chem. Phys.*, 1998, **108**, 10220.
64 M. Lísal, J. Kolafa and I. Nezbeda, *J. Chem. Phys.*, 2002, **117**, 8892.
65 S. W. Rick, *J. Chem. Phys.*, 2004, **120**, 6085.
66 Y. Yonetani, *J. Chem. Phys.*, 2006, **124**, 204501.
67 M. P. Allen and D. J. Tildesley, *Computer Simulation of Liquids*, Oxford University Press, Oxford, 1987.
68 B. Garzon, S. Lago and C. Vega, *Chem. Phys. Lett.*, 1994, **231**, 366.
69 D. Frenkel and B. Smit, *Understanding Molecular Simulation*, Academic Press, London, 2002.
70 M. Parrinello and A. Rahman, *J. Appl. Phys.*, 1981, **52**, 7182.
71 S. Yashonath and C. N. R. Rao, *Mol. Phys.*, 1985, **54**, 245.
72 D. Frenkel and A. J. C. Ladd, *J. Chem. Phys.*, 1984, **81**, 3188.
73 C. Vega, E. Sanz, E. G. Noya and J. L. F. Abascal, *J. Phys.: Condens. Matter*, 2008, **20**, 153101.
74 J. K. Johnson, J. A. Zollweg and K. E. Gubbins, *Mol. Phys.*, 1993, **78**, 591.
75 D. A. Kofke, *J. Chem. Phys.*, 1993, **98**, 4149.
76 D. A. Kofke, *Phys. Rev. Lett.*, 1995, **74**, 122.
77 M. Lisal, W. R. Smith and I. Nezbeda, *Fluid Phase Equilib.*, 2001, **181**, 127.
78 M. Lisal, I. Nezbeda and W. R. Smith, *J. Phys. Chem. B*, 2004, **108**, 7412.
79 C. Vega, J. L. F. Abascal and I. Nezbeda, *J. Chem. Phys.*, 2006, **125**, 034503.
80 C. C. Liew, H. Inomata and K. Arai, *Fluid Phase Equilib.*, 1998, **144**, 287.
81 J. R. Errington and A. Z. Panagiotopoulos, *J. Phys. Chem. B*, 1998, **102**, 7470.
82 C. Vega and E. de Miguel, *J. Chem. Phys.*, 2007, **126**, 154707.
83 J. S. Rowlinson and B. Widom, *Molecular Theory of Capillarity*, Clarendon Press, Oxford, 1982.
84 G. J. Gloor, G. Jackson, F. J. Blas and E. de Miguel, *J. Chem. Phys.*, 2005, **123**, 134703.
85 F. Chen and P. E. Smith, *J. Chem. Phys.*, 2007, **126**, 221101.
86 S.-B. Zhu and M. W. Evans, *Water in Biology, Chemistry and Physics: Experimental Overviews and Computational Methodologies*, ed. G. W. Robinson, World Scientific, Singapore, 1996.
87 H. J. White, J. V. Sengers, D. B. Neumann and J. C. Bellows, *IAPWS Release on the Surface Tension of Ordinary Water Substance*, 1995, available from http://www.iapws.org.
88 A. Ghoufi, F. Goujon, V. Lachet and P. Malfreyt, *J. Chem. Phys.*, 2008, **128**, 154716.
89 J. Lopez-Lemus, G. A. Chapela and J. Alejandre, *J. Chem. Phys.*, 2008, **128**, 174703.
90 A. Baranyai, A. Bartok and A. A. Chialvo, *J. Chem. Phys.*, 2005, **123**, 54502.

91 C. Vega, C. McBride, E. Sanz and J. L. Abascal, *Phys. Chem. Chem. Phys.*, 2005, **7**, 1450.
92 T. James, D. J. Wales and J. Hernandez-Rojas, *Chem. Phys. Lett.*, 2005, **415**, 302.
93 L. Pauling, *J. Am. Chem. Soc.*, 1935, **57**, 2680.
94 L. G. MacDowell, E. Sanz, C. Vega and L. F. Abascal, *J. Chem. Phys.*, 2004, **121**, 10145.
95 V. Buch, P. Sandler and J. Sadlej, *J. Phys. Chem. B*, 1998, **102**, 8641.
96 J. L. Aragones, E. G. Noya, J. L. F. Abascal and C. Vega, *J. Chem. Phys.*, 2007, **127**, 154518.
97 E. G. Noya, C. Menduina, J. L. Aragonés and C. Vega, *J. Phys. Chem. C*, 2007, **111**, 15877.
98 A. D. Fortes, I. G. Wood, M. Alfredsson, L. Voèadlo and K. S. Knight, *J. Appl. Crystallogr.*, 2005, **38**, 612.
99 B. Kamb, *Acta Crystallogr.*, 1964, **17**, 1437.
100 C. Vega, J. L. F. Abascal, E. Sanz, L. G. MacDowell and C. McBride, *J. Phys.: Condens. Matter*, 2005, **17**, S3283.
101 C. McBride, C. Vega, E. Sanz, L. G. MacDowell and J. L. F. Abascal, *Mol. Phys.*, 2005, **103**, 1.
102 C. Vega, M. Martin-Conde and A. Patrykiejew, *Mol. Phys.*, 2006, **104**, 3583.
103 R. G. Fernandez, J. L. F. Abascal and C. Vega, *J. Chem. Phys.*, 2006, **124**, 144506.
104 J. Wang, S. Yoo, J. Bai, J. R. Morris and X. C. Zeng, *J. Chem. Phys.*, 2005, **123**, 036101.
105 O. A. Karim and A. D. J. Haymet, *J. Chem. Phys.*, 1988, **89**, 6889.
106 D. Quigley and P. M. Rodger, *J. Chem. Phys.*, 2008, **128**, 154518.
107 G. T. Gao, X. C. Zeng and H. Tanaka, *J. Chem. Phys.*, 2000, **112**, 8534.
108 Y. Koyama, H. Tanaka, G. Gao and X. C. Zeng, *J. Chem. Phys.*, 2004, **121**, 7926.
109 W. L. Jorgensen and C. Jenson, *J. Comput. Chem.*, 1998, **19**, 1179.
110 A. K. Soper, *Chem. Phys.*, 2000, **258**, 121.
111 A. H. Narten, C. G. Venkatesh and S. A. Rice, *J. Chem. Phys.*, 1976, **64**, 1106.
112 H. A. Stern and B. J. Berne, *J. Chem. Phys.*, 2001, **115**, 7622.
113 L. Hernandez de la Peña and P. G. Kusalik, *J. Chem. Phys.*, 2004, **121**, 5992.
114 L. Hernandez de la Peña and P. G. Kusalik, *J. Am. Chem. Soc.*, 2005, **127**, 5246.
115 L. Hernandez de la Peña, M. S. Gulam Razul and P. G. Kusalik, *J. Chem. Phys.*, 2005, **123**, 144506.
116 W. Wagner, A. Saul and A. Pruss, *J. Phys. Chem. Ref. Data*, 1994, **23**, 515.
117 E. H. Abramson and J. M. Brown, *Geochim. Cosmochim. Acta*, 2004, **68**, 1827.
118 D. R. Nutt and J. C. Smith, *J. Chem. Theory Comput.*, 2007, **3**, 1550.
119 R. J. Hemley, A. P. Jephcoat, H. K. Mao, C. S. Zha, L. W. Finger and D. E. Cox, *Nature*, 1987, **330**, 737.
120 W. L. Jorgensen and J. Tirado-Rives, *Proc. Natl. Acad. Sci. U. S. A.*, 2005, **102**, 6665.
121 A. D. Fortes, I. G. Wood, J. P. Brodholt and L. Voèadlo, *J. Chem. Phys.*, 2003, **119**, 4567.
122 B. Kamb, W. C. Hamilton, S. J. LaPlaca and A. Prakash, *J. Chem. Phys.*, 1971, **55**, 1934.

 This journal is © The Royal Society of Chemistry 2009

On thin ice: surface order and disorder during pre-melting

C. L. Bishop,[ab] D. Pan,[c] L. M. Liu,[bd] G. A. Tribello,†[abe] A. Michaelides,[abd] E. G. Wang[c] and B. Slater*[abe]

Received 30th April 2008, Accepted 24th June 2008
First published as an Advance Article on the web 9th October 2008
DOI: 10.1039/b807377p

The effect of temperature on the structure of the ice Ih (0001) surface is considered through a series of molecular dynamics simulations on an ice slab. At relatively low temperatures (200 K) a small fraction of surface self-interstitials (*i.e.* admolecules) appear that are formed exclusively from molecules leaving the outermost bilayer. At higher temperatures (*ca.* 250 K), vacancies start to appear in the inner part of the outermost bilayer exposing the underlying bilayer and providing sites with a high concentration of dangling hydrogen bonds. Around 250–260 K aggregates of molecules formed on top of the outermost bilayer from self-interstitials become more mobile and have diffusivities approaching that of liquid water. At ~270–280 K the inner bilayer of one surface noticeably de-structures and it appears that at above 285 K both surfaces are melting. The observed disparity in the onset of melting between the two sides of the slab is rationalised by considering the relationship between surface energy and the spatial distribution of protons at the surface; thermodynamic stability is conferred on the surface by maximising separations between dangling protons at the crystal exterior. Local hotspots associated with a high dangling proton density are suggested to be susceptible to pre-melting and may be more efficient at trapping species at the external surface than regions with low concentrations of protons thus potentially helping ice particles to catalyse reactions. A preliminary conclusion of this work is that only about 10–20 K below the melting temperature of the particular water potential employed is major disruption of the crystalline lattice noted which could be interpreted as being "liquid", the thickness of this film being about a nanometre.

Introduction and background

Despite the importance of ice in regulating the earth's temperature *via* albedo and within atmospheric chemistry (where it serves as a catalyst for many trace gas reactions[1]), a key unknown in these two roles is the physicochemical nature of the ice surface as a function of temperature. Faraday[2] first mooted the idea that the surface

[a]*Department of Chemistry, Christopher Ingold Building, University College London, 20 Gordon Street, London, UK WC1H 0AJ. E-mail: b.slater@ucl.ac.uk*
[b]*Materials Simulation Laboratory, University College London, London, UK WC1H 0AJ*
[c]*Institute of Physics, Chinese Academy of Sciences, Box 603, Beijing, 100080, China*
[d]*London Centre for Nanotechnology, University College London, London, UK WC1H 0AJ*
[e]*Davy Faraday Research Laboratory, Kathleen Lonsdale Building, University College London, Gower Street, London, UK WC1E 6BT*

† Present address: Computational Science, Department of Chemistry and Applied Biosciences, ETH, Zurich, USI-Campus, Via Giuseppe Buffi 13, C-6900, Lugano, Switzerland

of ice could have liquid-like properties, a quasi-liquid-layer (QLL), but the nature of this layer, in particular its thickness approaching the melting temperature, is disputed—where more than an order of magnitude separates the estimates of thickness obtained using proton channelling and ellipsometry.[3] A review of experimental and computer simulation studies pertinent to the pre-melting phenomenon has recently been compiled, which provides a useful introduction to this topic.[4] Computer modelling approaches could potentially play a role in resolving the crystallinity of the surface as a function of temperature, and the atomic-scale mechanisms together with the role of defects in the pre-melting phenomenon in ice. Naturally, the verity of any such model and approach hinges upon accurately describing the relationship between free energy and temperature. In this work we examine one particular model, the six-site model due to Nada and van der Eerden (NvdE)[5] which was developed specifically to reproduce the melting temperature of ice Ih. In analysing the surface structure we focus upon (i) the evolution of the surface structure in the temperature regime above 200K, (ii) the identification of defects at the surface and (iii) the spatial distribution of protons on the external surface of ice and how this influences pre-melting and stability.

Methods

Here we use a combination of first-principles calculations adjunct to empirical potential simulations to probe the detailed nature of structures evolving from long-timescale pre-melting studies. Despite the incredible versatility of the TIP4P[6] model, indicated by its ability to describe large portions of the water–ice phase diagram, a shortcoming is its ability to predict the melting temperature of water (not withstanding recent reparameterisation of this potential due to Abascal, Vega and co-workers[7–10]). In previous simulation work due to Kroes[11] and others, the melting of an ice slab with the TIP4P potential was considered in the temperature range of 190–250 K. Although melting was observed, the bulk melting temperature for TIP4P has been found to range from around 190–230 K (the most recent estimate being 232 K[12]) and hence the liquefaction mechanism at the superheated temperatures of >230 K could be distinct from pre-melting. More recently, Ikeda-Fukazawa[13] and others have addressed pre-melting using the Kawamura potential which has a reported melting temperature of 290 K. In their work, an ice slab was heated to 270 K for 60 ps and the mean-square displacement of water molecules in the outermost bilayer was noted to be very strongly enhanced in the *c* axis, perpendicular to the crystal surface rather than the *a* or *b* axis parallel to the crystal surface. At this temperature, only a very slight loss of structure is noted at the external surface evidenced by a broadening of the density at the outermost layer. A very relevant study is that of Carignano *et al.*[14] who looked at ice growth up to 275 K using the same potential used here. Their study monitored the liquid-like layer on either side of a slab of ice, which was in contact with both vacuum and water; comparisons between our studies are drawn later. Bolton and Pettersson[15] have also studied pre-melting and their work highlighted rapid transport of water between outer bilayers at temperatures in the region 180–210 K (TIP4P model), which was associated with pre-melting. However, in this paper, we present the first preliminary account of our ongoing study that attempts to shed further light on the pre-melting phenomenon, by monitoring systematic changes in surface structure up to the melting temperature.

In this work, we use the potential due to NvdE, which is a rigid-body, six-site model. In the original paper of NvdE the melting temperature was estimated to be 271 ± 9 K but we note the most recent study[16] with this potential puts the melting temperature at 289 K. In addition to the oxygen and hydrogen sites, two sites represent the lone pairs, whilst a further site sits along the bisector of the two hydrogen atoms and attempts to mimic the concentration of electron density close to the centre of mass of the water molecule. We have used the $6 \times 3 \times 3$ cell due to Hayward and Reimers,[17] containing 768 molecules and 8 bilayers (96 molecules per bilayer).

 This journal is © The Royal Society of Chemistry 2009

Hayward and Reimers developed ice cell geometries in which the protons are maximally disordered and the cell has no net dipole or quadrupole. The latter characteristics are an important pre-requisite for surface simulation; in this work we use a periodic slab approach and any significant dipole across the slab can potentially interact through the vacuum with its periodic image, which would lead to spurious artefacts. The bulk cell was initially geometry optimised at 0.1 K using NPT molecular dynamics‡ (employing the DL_POLY[18] code, version 2.18) to give cell parameters of a = 26.92, b = 31.08, c = 29.30 Å, which compare favourably with the experimental geometry of a = 26.98, b = 31.48, c = 29.28 Å taken from Röttger *et al.*[19] at 10 K (note that these parameters relate to an ice sample which contains some fraction of ice XI. Ice XI displays small negative thermal expansion below ~70 K, hence the maximum error of 1.3% in b is an upper bound. At 85 K, above the Ih → XI transition temperature, the maximum error in b reduces to 1.2%). A vacuum gap of 30 Å was introduced along the Cartesian z axis to create two (0001) surfaces and the slab was geometry-optimised. We determined that the surface energy was converged at 6 bilayers but use 8 bilayers in this work to screen interactions across the slab arising from melting at the external surface. In a contiguous series of MD runs, using a 1 fs timestep and starting at 200 K, the slab was equilibrated for 5 ps and then production runs of between 0.5 and 5 ns were conducted for 210, 220, 230, 240, 250, 255, 260, 265, 270, 273, 275, 280, 285 and 290 K. Trajectory samples were taken at 0.5 ps intervals and analysis was performed over all the configurations sampled and over sections of the longer MD runs. Conservation of kinetic energy was reasonable: at 285K the RMS fluctuation in the temperature over a 1 ns run was 2.4 K. Due care was exercised in ensuring the statistical samples taken over MD runs of unequal length were not directly compared. However, it is noteworthy that for the 285 K run, the coordination numbers extracted from 0–0.25 ns and 0.75–1.0 ns are essentially indistinguishable from one another. Additionally, the gradient of the MSD at this temperature is converged within <5 ps of the production phase.

The DFT calculations have been performed with the CP2K/Quickstep program,[20] which employs a hybrid Gaussian and plane-wave basis set. The Perdew, Burke and Ernzerhof (PBE)[21] exchange–correlation functional has been used throughout and the core electrons are described with norm-conserving Goedecker, Teter and Hutter (GTH) pseudopotentials.[22,23] The valence electrons are expanded in terms of Gaussian functions with a triple-ζ doubly polarised basis set (TZV2P) and for the auxiliary basis set of plane waves a 340 Ry cutoff is used. Due to the larger computational demands of the DFT calculations compared to the empirical potentials, a considerably smaller (but still moderately large) unit cell with 48 water molecules per bilayer was employed. Calculations on slabs of between 2 and 6 bilayers thickness were performed and during the DFT geometry optimizations all atoms were fully relaxed. A vacuum gap of 20 Å was used throughout.

Results

In the following sections, we address the response of the thin ice slab to increasing temperature in the range 200 to 290 K. At all temperatures up to about 285 K the energies throughout the course of the MD simulations are well conserved. This can be seen for three high temperatures (275, 280 and 285 K) from Fig. 1. However at 290 K it can be seen that the total energy is no longer conserved and the crystal appears to be in a non-equilibrium melting state. Towards the later part of the simulation

‡ A timestep of 1 fs was used throughout, integration was performed using the velocity Verlet algorithm. The van der Waals cutoff used was 10.0 Å, as used by NvdE, and the Berendsen thermostat/barostat was used to control the temperature and pressure (1 atm.). The frequency of scaling was 0.005 ps which was necessary in order to minimise fluctuations in the temperature. The NVT ensemble was used for the surface simulation runs.

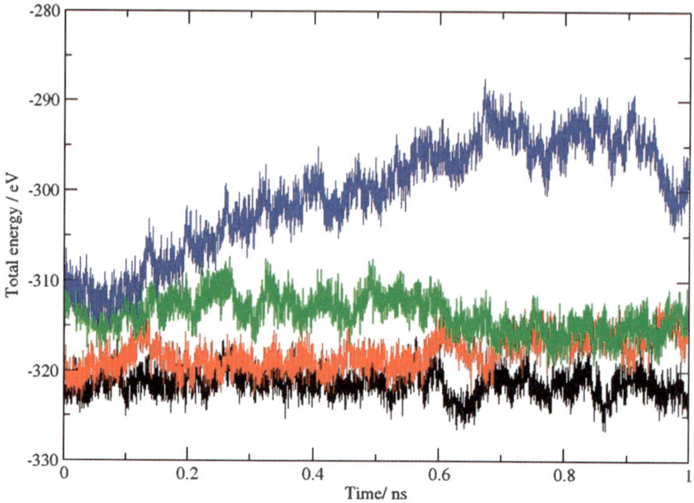

Fig. 1 Evolution of the total energy over the duration of the simulation. Total energy at 275 K denoted in black, 280 K data is displayed in red, 285 K in green and 290 K in blue.

at 290 K, it appears that the energy is reducing and this may indicate a very long timescale restructuring process. However, we crudely infer from this that the melting temperature for the entire slab with this potential is at around 285–290 K. This is consistent with Abascal and Vega's estimate and hence the temperatures reported here should be reduced by ~15 K to be comparable to experiment.

Firstly we examine the density across the slab to monitor the crystallinity of each bilayer in a qualitative way and then address the degree of order in each layer. We then comment on the incidence and type of defects formed and then attempt to rationalise the disparity in melting rates that are observed for the two surfaces.

(i) Density profile as a function of temperature

Fig. 2 shows a composite of the density profile, generated with 0.1 Å bins, for four-coordinated oxygen atoms across the slab taken for two temperatures, 200 and 285 K. The coordination number of each molecule was evaluated within a cutoff radius of 2.27 Å, around 0.5 Å larger than the equilibrium hydrogen-bond lengths. The number density peaks are unequal for the sub-surface slabs which is caused by the vibration and relaxation of the bilayers which leads to a broadening of the peaks that is most pronounced at the surface. The surfaces of the slab are the extrema of the coordinates plotted along the *x* axis. At 200 K, the outermost part of the surface bilayer is well crystalline but shows slight broadening in comparison to the inner part of the bilayer. The peak heights of the outermost peaks are noticeably smaller than the corresponding peaks of the inner bilayer because a fraction of the water molecules (the water molecules in the upper bilayer) are three-coordinated. There is evidence of surface relaxation, principally manifested in the reduced peak heights of the outer layers compared to the inner layers, which reflects a tendency to deviate from the ideal lattice position. At 200 K, the combined width of the outermost bilayer peaks are 3.00 and 2.90 Å for the lower and upper surfaces respectively, which compares to 2.40 and 2.50 Å for the central bilayers. The central bilayers are 2.70 Å apart whilst the surface layer to second bilayer distance (measured from the inner bilayer surface peak to the outer peak of the second bilayer) is 2.80 Å. For comparison, at 285 K, the central bilayers have peak widths of 2.90 and 3.0 Å. A final point upon which we comment on in more detail later is that the profile at 200 K already contains defects in the outermost layers. However,

 This journal is © The Royal Society of Chemistry 2009

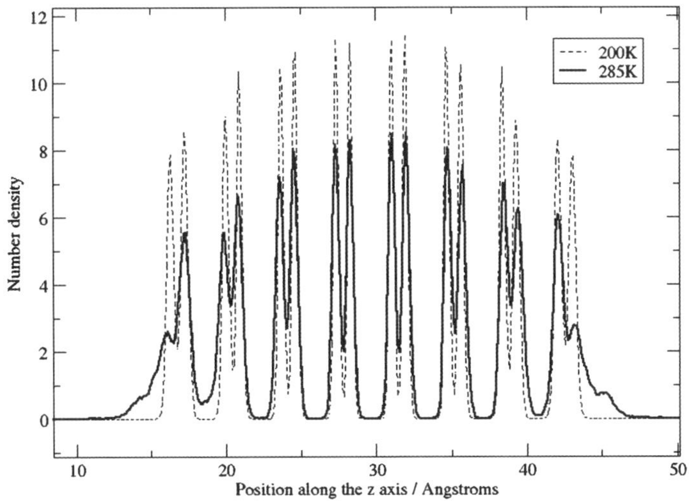

Fig. 2 Number density of four-coordinated oxygen atoms at 200K (depicted with the dotted line) and at 285K (displayed with a continuous line) as a function of the slab height. Note the asymmetry between the two outer surfaces, where the lower surface is more disordered in the region between the first and second bilayers.

the lifetime and incidence of these defects is extremely small and therefore the density plot is essentially unchanged from the defect-free state.

Upon increasing the temperature, the peak heights and widths for the first bilayers reduce and increase respectively. For the sake of clarity, only the 285 K plot is superimposed upon the 200 K data. Upon close inspection of Fig. 2, it can be seen that all the peaks have broadened but most noticeably at the surface. More significantly, the two surfaces on either side of the slab have disordered to different extents, the lower surface (smaller values of x) showing less structure than the upper surface. In the lower surface, there is a build-up of density between the first two bilayers indicating pre-melting or melting which is less pronounced for the upper surface. The inner part of the bilayer for the upper surface is noticeably more well-ordered than the analogous peak for the lower surface and the lower surface shows significant loss of order in the second bilayer whilst the corresponding peak in the upper part of the slab is relatively well ordered. The new overlayer on top of the original external surface has a depth of approximately 5 Å at 285 K for both the upper and lower layer.

Since the density profiles reflect the averaged position of species, the spatial relationship of species and crystallinity of each layer is open to question. Fig. 3a and 3b show a snapshot taken from the end of the dynamics run for the oxygen sub-lattice in the first three bilayers of the lower surface. In Fig. 3a, the surface is shown at 250 K and shows a vacancy in the outermost bilayer (blue), which exposes the second bilayer, but otherwise the hexagonal oxygen sub-lattice is intact. The situation at 285 K shown in Fig. 3b contrasts strongly with that at lower temperatures; all characteristic hexagonal ordering in the oxygen sub-lattice of the outer bilayer has been lost, the second layer is noticeably disordered exposing the third bilayer but the third layer is still strongly crystalline. The inferences made about ordering from the density of four-coordinated oxygen atoms in Fig. 2 appear to be supported by the visualisation of the slab. From viewing the evolution of the sub-lattice networks at 285 K, it is evident that the outermost layer depicted in blue undergoes very rapid rearrangement and from frame to frame (0.5 ps) the structure changes. Molecules furthest away (the highest molecules) from the outer bilayer are particularly mobile as they are typically one- or two-coordinated and these contribute strongly to the apparent height of the surface in the density profile as they fleetingly

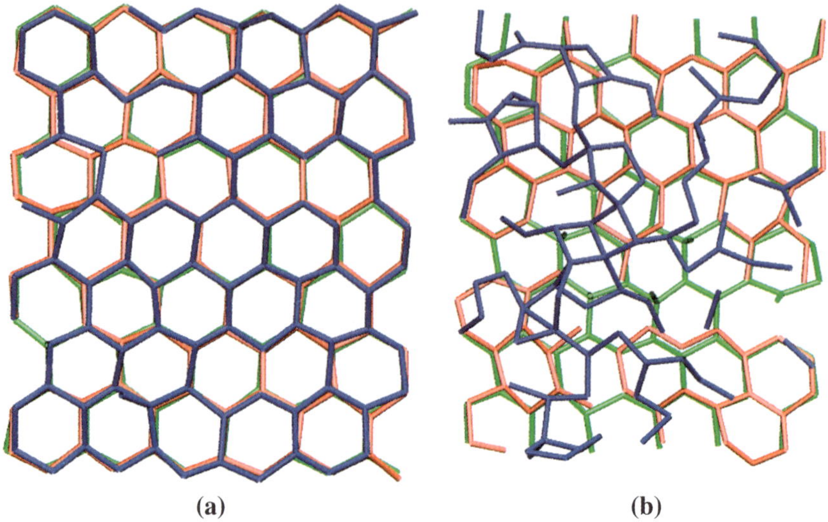

Fig. 3 Three bilayers of water ice viewed from the [001] direction, where the oxygen sub-lattice is depicted by creating bonds between oxygen atoms within 3.0 Å of one another. The external bilayer is in blue, the second bilayer is shown in red and the third bilayer towards the centre of the crystal slab is shown in green. (a) This snapshot was taken at 250 K. Only the outermost bilayer is interrupted whilst the sub-surface bilayers display a fully connected hexagonal sub-lattice. (b) This snapshot was taken at 285 K. Obvious hexagonal structure has been lost in the outermost layer and the second bilayer has also lost significant order.

leave the surface. Underlying layers have a more rigid sub-lattice although the second layer shows significant fluctuation in the 2D structure in comparison to the third layer. The external surface layer is rather rough at this temperature, from the maximum height of the external surface to the lowest exposed point in the topography of the ice surface is some ~10 Å.§ From the perspective of catalysis, it is readily conceivable that these crevasses may trap or even encapsulate species which could then participate in reactions. The walls and base of the crevasse are lined with dangling hydrogen bonds which could presumably strongly bind foreign or natant water species.

(ii) Ordering and disordering in the external surface layer

We now turn to consideration of the density of under-coordinated water ice molecules, in particular the three-coordinated water molecules, to understand more of the nature of the quasi-liquid layer and its intrinsic order. In ice, each molecule is surrounded by four neighbouring molecules except at the surface where one dangling hydrogen bond is exposed. In water, using this potential model approximately half of the water molecules are three-coordinated and half are four-coordinated. Hence the proportion of four- to three-coordinate H_2O molecules could be used to monitor how liquid-like such a layer is. In Fig. 4a and 4b, the number of three-coordinated species on the lower and upper surfaces, respectively, are shown for selected temperatures. There is some anisotropy between the two plots. In particular, there is a qualitative change in between 230 and 260 K when three-coordinate species appear in the inner-part of the bilayer. For the upper surface inner bilayer, under-coordination is less pronounced than for the lower surface upon heating from 260 to 273 K. The number of 3-coordinate species above

§ The estimate of QLL thickness made here is slightly larger than the value of ~0.6 nm from Carignano *et al.*[14] which was obtained at a lower temperature of 275 K.

 This journal is © The Royal Society of Chemistry 2009

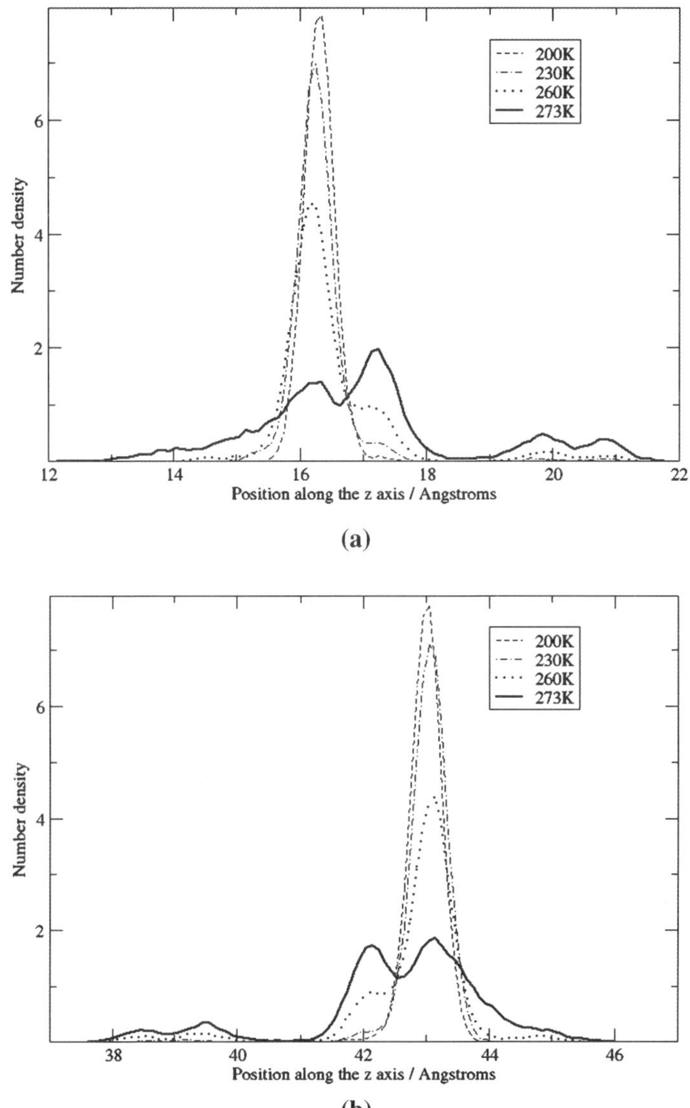

Fig. 4 Number density of three-coordinated oxygen atoms at selected temperatures for two bilayers on (a) the lower and (b) the upper part of the slab. (a) The peak at around 16 Å indicates the outer bilayer of the lower surface. (b) The peak around 43 Å indicates the upper part of the bilayer *i.e.* the external surface.

the external surface remains constant for both surfaces, potentially suggesting that these structures are in (dynamic) equilibrium. Notionally one can compare the number densities of four- and three-coordinate species and in the outer bilayers, the ratio of four- to three-coordinated species is approximately 1, suggesting this region is liquid-like. However, the structure is far from homogenous as Fig. 3 attests and hence such a measure of liquid-like nature is possibly misleading. We note that the number of one- and two-coordinated molecules increases with temperature and comprises the majority of the new overlayer. At 285 K the relative proportions of coordination are: 1-coordinate (0.1%), 2-coordinate (4.9%), 3-coordinate (12.8%), 4-coordinate (80.7%) and 5-coordinate (0.6%).

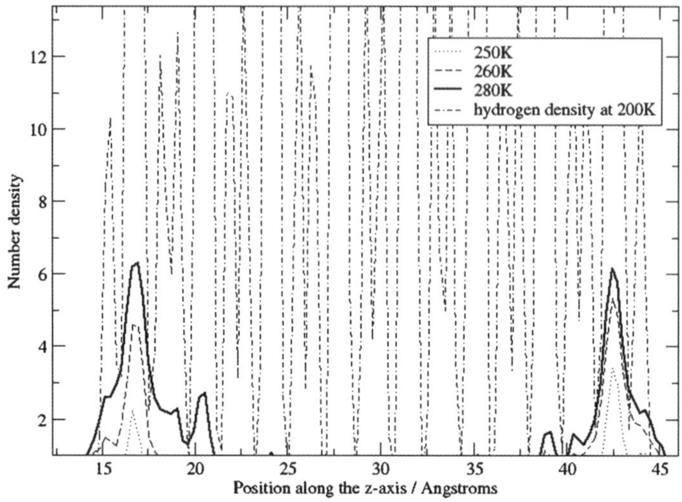

Fig. 5 Against a backdrop of the total density of hydrogen atoms in the slab as function of slab depth, the density of hydrogen atoms that have exchanged which oxygen they were hydrogen bonded to is displayed, for selected temperatures.

Another guide to the transience of the hydrogen-bond network that one attributes to water-like behaviour is exchange of hydrogen bonds. Fig. 5 shows the density of hydrogen atoms that have exchanged which oxygen they are bonded to. Again the asymmetry in the pre-melting of the ice slab is visible with slightly more exchange in the first and second bilayer for the lower surface than the upper surface. The upper surface is somewhat more resistant to the effects of increasing temperature. For both surfaces, there is evidence that both the first and second bilayer contain fractions of H_2O molecules that are freely exchanging hydrogen bonds in a way that is characteristic of liquid-like water rather than crystalline water. In the outermost bilayer approximately 40% of H_2O molecules are 3-coordinated compared to 25% in the ideal surface and around 5% are two-coordinated. One interesting feature is that at the lower temperatures (250–260 K) the upper surface has a higher concentration of waters that have changed their hydrogen-bonding neighbours than the lower surface but at much higher temperatures of *circa* 280 K the lower surface has a more mobile network. This observation may be explained by the upper surface somehow auto-passivating or reconstructing itself. In passing we note that provided the external water-like layer is thick enough, a radial distribution function could be generated to help establish whether it possesses long-range order and the degree of short-range order, although for the run described here we suspect the statistics are not plentiful enough for such an analysis to be unequivocal.

Interestingly, even at 280 K as displayed in Fig. 6, the ratio of H_2O molecules which exchange over the final 0.25 ns of a run is relatively small (9.5%) in comparison to the total number of molecules. The ratio of waters with a rigid hydrogen-bond network *versus* a flexible hydrogen-bond network peaks between the first and second bilayers over the lifetime of the run, with the largest ratio being found just inside the surface. This observation suggests that a proportion of the sub-surface H_2O molecules are more mobile than those above them, a situation analogous to a frozen lake with liquid water underneath. A potentially important observation in the context of distinguishing between melting and pre-melting is that the density and exchange plots averaged over the first 0.25 ns and final 0.25 ns (for 260–280 K) were essentially unchanged and suggest the crystal is in equilibrium *i.e.* not melting.

 This journal is © The Royal Society of Chemistry 2009

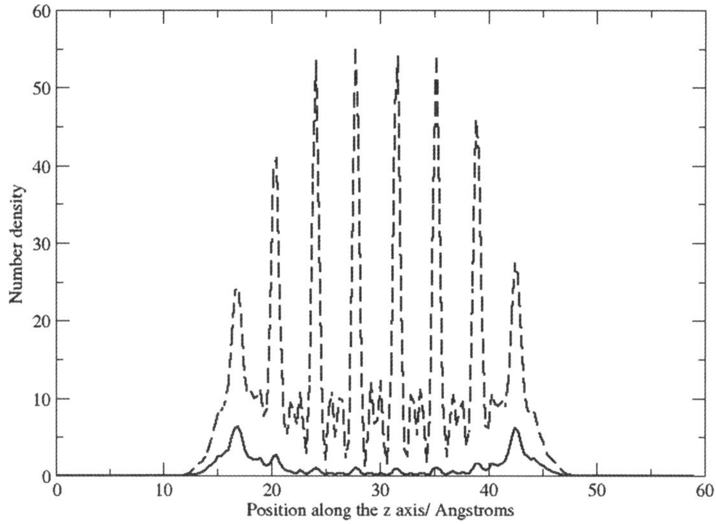

Fig. 6 The dashed line represents total hydrogen density at 280 K averaged over the final 0.25 ns of a MD run. The solid line represents the density of hydrogen atoms that exchange hydrogen bonds with other molecules. The ratio of exchangeable hydrogen bonds is maximal at the external surfaces, decreases at the inner bilayer and increases again below this layer before becoming negligible at the bulk interior.

Although not reported here, we have analysed the mean-square displacement (MSD) profile of H_2O molecules throughout the slab. What is particularly noteworthy is that the MSD monotonically increases with temperature and by 285 K the measured diffusion coefficient is 2.37×10^{-9} m^2 s^{-1} for the outer layers (0.23×10^{-9} m^2 s^{-1} averaged over the entire slab) which is considerably larger than the experimental values of 0.7×10^{-9} m^2 s^{-1} at 263.7 K[24] (the approximate temperature this corresponds to in real terms) but close to the experimental value of 2.23×10^{-9} m^2 s^{-1} at 298 K. Carignano's data[14] suggest that the NvdE potential overestimates the true diffusion of water below 298 K but not to the degree noted here. However, if the MSDs are deconvolved to show their individual diffusivities, it is clear that a small fraction of surface bound species have extraordinary diffusivity that is more than one order of magnitude faster than the QLL. The presence of these highly mobile molecules skews the apparent diffusivity considerably and highlights that collective diffusivity is not a reliable gauge of the liquid-like nature of the surface.

Analysis of structural changes during pre-melting

To aid interpretation of the structural changes occurring during pre-melting, the trajectories have been analysed visually to try and untangle the local events that occur which trigger the collective loss of structural order in a layer associated with melting. Three distinct temperature regimes are discussed here.

(i) 200–230 K

According to experiments on dislocation loops in ice crystals,[3] evidence has been presented that interstitial defects are the dominant defect in ice. Previous simulation work by Itoh *et al.*[25] using the TIP4P potential produced reasonable pathways for self-interstitial migration in the crystal bulk, but the underlying energetics were deemed to be qualitative at best by the authors. More recently de Koning *et al.*[26] have used generalised gradient approximation (GGA) DFT approaches to examine interstitial defect formation in the crystal bulk and arrived at a value of ~0.7 eV, which is considerably greater than the experimental estimate of 0.4 eV.[3] The significance of the defect energies is that concentration of defects is used to rationalise the proton conductivity as well as other properties. The evidence from de Koning's studies on these defects and other more complex defects suggests that interstitials may be less populous relative to other defects than previously thought. In the light of these insights, it is of interest to monitor the type and frequency of defect formation at the surface since the energy of formation of these species remain unknown at this time (though we note work is underway within our groups to establish surface defect energies from DFT calculations).

At 200 K self-interstitials form relatively rarely—to the extent that just one surface interstitial is seen over the duration of a 0.25 ns MD run. The mobility of the interstitials is low and migration was only noted within a radius equivalent to the next nearest neighbour at this temperature. Recent DFT calculations due to Thierfelder *et al.*[27] showed four distinct positions for surface molecular adsorption of water on the ice (0001) face, with energies ranging between 0.55 and 0.58 eV (computed using a periodic slab, plane-wave approach and the PW91 functional). Essentially Thierfelder notes two adsorption modes, one where a water molecule binds towards the centre of the hexamer ring which is the most stable site (0.58 eV), and a second where the sorbed water sits atop an ideal lattice site (~0.55 eV) corresponding to the inner part of the outermost bilayer. In our simulations from above 200 K the interstitials are readily identified and it is heartening that despite the fact that a rigid dipole is used to describe the water in these pair-potential calculations, the centre site mode of adsorption is observed with increased frequency to the atop mode. Within this temperature range, the interstitials still undergo relatively slow diffusion across the surface such that probability of two interstitials being proximate is rare enough that no dimerisation of the interstitials is seen. A final observation is that in this temperature regime, only the outermost, undercoordinated molecules of the external bilayer form self-interstitials and there is no visual evidence that the lower part of the outermost bilayer is disrupted.

(ii) 230–250 K

Above 230 K, there is a marked increase in the number of self-interstitials formed and consequently the number of vacancies in the outermost bilayer. The interstitials begin to aggregate from around 230 K onwards, initially forming dimer pairs. We note that Thierfelder also found that dimers formed readily with an adsorption energy which is fractionally more favourable (0.576 eV per molecule) than forming two isolated interstitials (0.575 eV per molecule). However, whether DFT with the PW91 exchange–correlation functional is reliable enough to accurately compare the stability of adsorbed monomers and dimers remains open to question.[28,29] The majority of the dimer pairs formed in our MD simulations remain as a complex which is either static or which diffuses very slowly across the surface. From around 240 to 250 K there is a qualitative change in the surface topography where molecules diffuse from the lower part of the outermost bilayer to either the outer part of the bilayer, occupying vacant lattice sites or directly above the surface as an interstitial. At around 240 K temporary holes appear in the outermost bilayers that provide access to the sub-surface bilayer but these have very short lifetimes of just a few ps and are quickly repaired by interstitial molecules and molecules from the outer part of the bilayer. Approaching 250 K, the holes become more permanent and have lifetimes of >50 ps. Over the temperature range 230–250 K, dimer clusters on the external surface begin to aggregate with one another and lone interstitials to form islands of quasi-crystalline water atop the original external surface. In this temperature regime, particularly at around 250 K the surface becomes considerably rougher as a consequence of forming vacancies in the outer bilayer and a new but incomplete quasi-crystalline layer.

 This journal is © The Royal Society of Chemistry 2009

(iii) Above 250 K

At temperatures above approximately 250 K, it becomes more difficult to distinguish distinct events occurring on the external surface. Analysis of the oxygen sub-lattice at ≥250 K shows that the characteristic regular hexagonal structure is interrupted by defects and dynamically evolves, presumably in response to melting events on the surface and complex surface structural rearrangements. A qualitative ring analysis shows that the surface consists of largely 6-membered rings, with smaller strained 4- and 5-membered rings compensated by 7- and 8-membered rings at 260 K. However, evidently the network still retains both 2D and 3D structure and therefore appreciable long-range order is still present. At around 270 K, hexagonal order is more difficult to detect and from the disruption observed in the top two bilayers of each surface at 275–280 K the lower and upper surfaces have lost any semblance of hexagonal order. Raising the temperature to 290 K noticeably accelerates the loss of order and which is accompanied by a drift in the total energy, suggesting that over the lifetime of 1 ns the (melting) ice slab has not equilibrated.

Before leaving this qualitative assessment of structural changes we highlight two other observations, which will possibly influence the mechanism and the rate of melting at ice surfaces.

Proton disordering events

The commonly accepted mechanism of proton conduction was put forward by Bjerrum (see ref. 3 for a more complete explanation), who highlighted the role of orientational defects in proton conduction. The two so-called Bjerrum defects violate the ice rules: (i) an *L-defect* is formed when a molecule rotates around from a water to which it donates a hydrogen bond, leaving no proton between two oxygen sites; (ii) a *D-defect* is created when a water rotates resulting in two protons between two oxygen atoms. Taking the instance of the D-defect, the two protons are in an unfavourable interaction and in the Bjerrum model, one of the waters rotates away forming a new D-defect and the process continues until a series of molecules change orientation and in effect a proton has migrated. Detailed analysis of the 230 K dynamics run revealed two distinct events which impact on the proton disordering and proton conduction behaviour of ice which are distinct from the Bjerrum orientational defect. We call the two mechanisms (i) relay and (ii) exchange or dosey-do. In the relay mechanism shown in Fig. 7a, water molecules rotate about their centre of mass in a process reminiscent of the Grotthus mechanism of proton transport. By contrast, in the exchange mechanism (Fig. 7b), the water molecules physically pass one another in a concerted movement. The exchange mechanism is slightly more complex than the Figure implies in that the participating molecules do not lie in the same plane during the transition. One molecule goes sub-surface whilst the other migrates *via* a supra-surface pathway.¶ The two mechanisms also contrast in that in the relay mechanism the dangling proton is passed from one molecule to its next nearest neighbour *via* a water in the inner part of the bilayer. In the dosey-do mechanism the dangling proton is passed to its nearest neighbour from the outer bilayer to the inner bilayer. The duration of both processes differ considerably; the relay mechanism involving just concerted flipping of the water occurred over ~2.5 ± 0.5 ps whilst the dosey-do takes considerably longer (~17.5 ± 0.5 ps) suggesting approximately an order of magnitude difference in the proton conduction rates However, both mechanisms occurred just twice in a 0.5 ns 230 K run and hence more extensive sampling is needed to validate the relative rates of these processes.

¶ Both the processes referred to were only ever seen at the crystal surface though there is no obvious reason why such events could not happen in the crystal bulk; however, we presume the barrier is much higher for steric and electrostatic reasons.

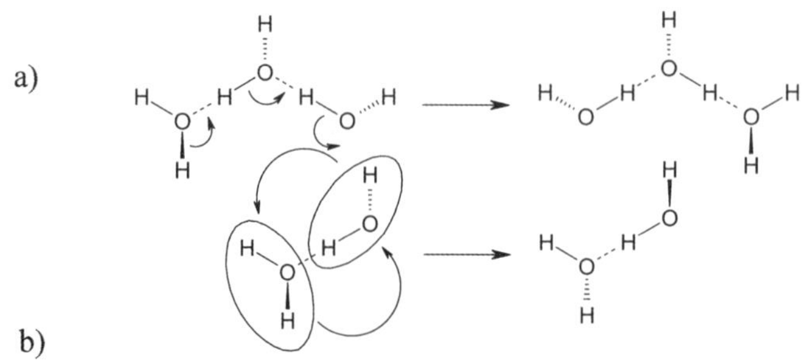

Fig. 7 The dashed lines represent protons below the plane of the paper, whilst the emboldened lines indicate protons above the plane of the paper. (a) Proton relay mechanism by orientational reordering. In this example, the net effect is for one proton to be "passed" from the left to the right in the plane of the paper. In the simulations, we only observed a mechanism in which the central molecule is within the inner bilayer whilst the other two molecules are in the outer bilayer. (b) The exchange or dosey-do mechanism in which two water molecules exchange position within the lattice in a concerted event which is only observed to occur between a nearest neighbour inner bilayer and outer bilayer water molecule. The molecules avoid steric clashes by one water following a trajectory sub-surface whilst the other molecule goes supra-surface. An important distinction is that these mechanisms allow for a change in polarisation but are not proton conduction mechanisms in and of themselves.

General discussion

An interesting result of this paper is the observation of slightly different disordering behaviour and onset of melting for the two sides of the slab. Since each surface of our slab is representative of the (0001) surface of ice Ih being, as we mentioned, obtained with the procedure of Hayward and Reimers, it is somewhat surprising to observe two slightly different melting temperatures. However, at the outset each surfaces does not have identical proton distributions and so we looked in to this issue in detail to see if we could understand if a relation between the onset of melting and surface order existed. We sought to consider precisely how the stability of a variety of defect-free (0001) ice surfaces differs with respect to the arrangement of the surface protons. Conscious that surfaces which differ only in the spatial arrangement of the protons are likely to have very similar energies, these calculations were performed with DFT. However, the general conclusion that we reach from the DFT calculations is not altered if we use the classical water potential described here (with an additional term to better reproduce proton ordering energetics).

The intrinsic stability of the various surfaces at zero Kelvin was compared by computing their total surface energies, γ, which is defined here per water molecule as:

$$\gamma = \frac{E_{\text{tot}}^{\text{slab}}(n) - nE_{\text{tot}}^{\text{bulk}}}{m} \tag{1}$$

where $E_{\text{tot}}^{\text{slab}}$ is the total energy of the ice slab obtained from our first principles DFT calculations, n is the number of bilayers in the slabs, and m is the number of water molecules on the two surfaces of the slab. The bulk reference per molecular bilayer, $E_{\text{tot}}^{\text{bulk}}$, is extracted from calculations on ice slabs from the relation:

$$E_{\text{tot}}^{\text{bulk}} = E_{\text{tot}}^{\text{slab}}(n) - E_{\text{tot}}^{\text{slab}}(\text{n} - 1) \tag{2}$$

where, as before, n is the number of bilayers in the slab. The surface energy was computed for more than 20 perfect (0001) ice Ih surfaces that were identical except for the spatial arrangement of the protons in the slab. For each slab a different value of the surface energy was obtained. The fact that we do not get a unique value for ice

 This journal is © The Royal Society of Chemistry 2009

Ih is partly to be expected since it is already known, for example, that the cohesive energy of bulk ice depends on the degree of proton disorder within the simulation cell.[30–33] However, the interesting finding here is that the variation of the surface energy (~50 meV per H_2O) with the degree of proton order by far exceeds the typically small range (~4 meV per H_2O) seen in bulk ice Ih. After exploring in detail the proton distributions at each of the surfaces, we find that the spatial distribution of the upright OH groups at the surface, those which "dangle" out of the surface, are key. In order to distinguish different distributions of these dangling OH bonds, we define an order parameter:

$$C_{OH} = \frac{1}{N_{OH}} \sum_{i=1}^{N_{OH}} c_i \tag{3}$$

where N_{OH} is the total number of dangling OH bonds on the two surfaces of our simulation slab and C_i is the number of the next nearest-neighbour dangling OH bonds around the ith dangling OH bond. The order parameter, C_{OH}, quantifies the uniformity in the distribution of the dangling OH bonds at the surface. The smaller the value of C_{OH}, the more uniform the distribution. The larger the value of C_{OH}, the more inhomogeneous or disordered the distribution of the dangling OH groups is, with the important consequence that on average the dangling OH groups at the surface are closer together. The smallest possible value of C_{OH} for a non-polar surface is 2, which is in accord with the structure proposed by Fletcher[34] for low-temperature ice. Remarkably, we find that when all the surface energies of the many surfaces considered are plotted as a function of C_{OH} a rather clear linear correlation is observed, as shown in Fig. 8. Moreover, for certain values of C_{OH} we have computed more than one structure and despite having different proton arrangements, surfaces with the same value of C_{OH} always come within ~4 meV per H_2O.

Having observed a correlation between the intrinsic stability of a surface, measured here by means of the total surface energy, we now revisit the slab used

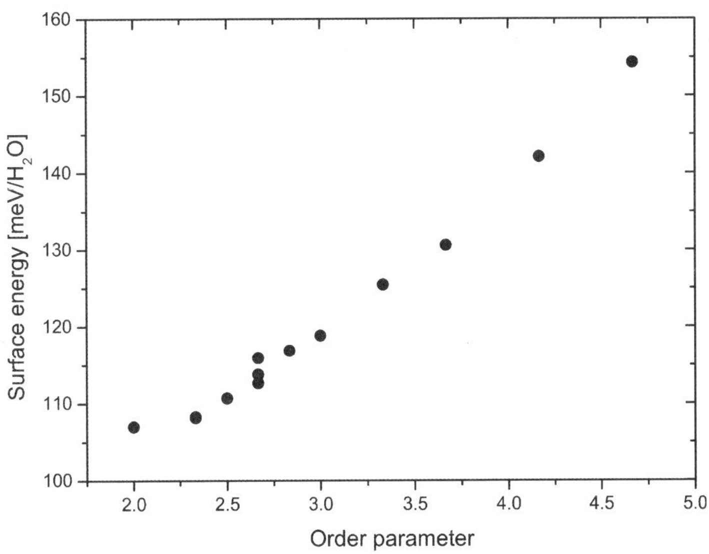

Fig. 8 The surface total energy of ice (0001) as a function of the order parameter defined in the text.

in the molecular dynamics simulations, in a way anticipating that this might explain the different melting behaviour on the two sides of the slab. However, employing the order parameter defined in eqn (3) we find that the two surfaces of the slab have precisely the same value: 2.92. Thus, although we have observed a correlation between surface order and surface energy, it appears that even surfaces with the same average extent of surface order may start to melt at different temperatures. This leads us then to examine proton ordering in more detail and consider whether local, as opposed to global, proton arrangements have an impact on the susceptibility of a surface to pre-melting. Indeed from the relation displayed in Fig. 8 we get an indication that molecules will likely be unstable, or "hot" when three or more protons surround them. It turns out that the lower surface has sixteen molecules with highly unfavourable sites with clusters of three or more adjacent dangling protons whilst the upper surface has only twelve. Thus we suggest that the greater proportion of the local hot spots with high local concentrations of dangling OHs may be responsible for the slightly lower onset of melting on the lower surface of the slab. We can imagine figuratively speaking that at the hot-spots on the real ice surface associated with clusters of protons, these regions are prone to pre-melting and form 'puddles', which may also act as traps for trace gases in atmospheric reactions. The previous discussion is somewhat speculative at this stage and it could be that the slightly different melting behaviour at the two surfaces is an artefact of having reasonably small surface areas with a relatively low variance in the local C_{OH}. In future work, a systematic study of the melting temperature of various ice surfaces as a function of average and local C_{OH} is planned. An additional point in critique of the results presented here which also acts as a word of caution for all simulations of ice-surface pre-melting relates to the rather long timescales required. Indeed, despite the structures observed at the external surface appearing to be in equilibrium at around $\leq$285 K, there is still the possibility that the surface has not always attained its equilibrium structure. In Carignano *et al.*,[14] the QLL was found to equilibrate in just 200 ps at 275 K and this concurs with our observations. However, in particular, by monitoring the structure visually, it can be seen that the outermost water molecules very slowly organise themselves to form chains after 1 ns at temperatures above 275 K. Possibly, more extensive simulations are required to establish whether more ordered structures develop on a very long timescale.

The final issue worthy of note is the melting temperature; our results indicate that the surface, despite being manifestly disordered at the surface for two bilayers at the crystal surface, does not melt until 285–290 K, in agreement with the estimate of Abascal *et al.*[16] However, it is somewhat surprising that given the surface was in contact with free vacuum, the surface does not melt at a lower temperature. Our model of the surface does not contain any ionic defects which are known to be present in real ice samples and perhaps this prevents us from observing disordering at relatively low temperatures (observed at just above 200 K by Wei *et al.*[35]). We might expect ionic defects and extrinsic impurities to surface segregate which may well affect the response of the surface to temperature and perhaps the thickness of the QLL reported by experiment.

Conclusions

In collating the conclusions, in this section we subtract 15K from the temperatures at which we observe phenomena in order to make contact with experiment. We find that:

• The ionic defect-free pure ice (0001) surface shows anisotropy in its melting which we attribute in part to the spatial arrangement of protons on the surface.

• The outer surface is liquid-like at temperatures above ~250 K but this lies in a crust of interrupted ice immediately below which the sub-surface H_2O is rather mobile (and can diffuse to the outer surface).

 This journal is © The Royal Society of Chemistry 2009

• Surface interstitials (ad-molecules) are seen at 200 K. These interstitials are not evidence of melting or pre-melting since they occur at temperatures well below the formation of the QLL. At temperatures of around 240–250 K, these *single* interstitials are extremely mobile, diffusing faster than liquid water, hence the collective diffusion coefficient should not be used to gauge the extent of water-like structures at the surface or the identification of water at the surface. When single interstitials dimerise their diffusion is markedly slower.

• Large holes of width ~0.5–1.0 nm appear in the outermost bilayer at temperatures below 250 K. These 'crevasses' are at least 1.0 nm deep at temperatures around 10 K below the melting temperature and have the capability of trapping foreign species. The roughness of the surface is considerably larger than expected, suggesting even at temperatures typical of arctic and antarctic lower and upper atmospheres, the sticking probability of species to the surface will deviate significantly from the classic picture of the Kossel crystal ideal surface.

In current work, we are extending the timescales of the MD runs described here and we are exploring in detail how proton arrangement influences the melting of the ice surface. We are also investigating more quantitative schemes for discriminating between liquid and crystalline water, such as that of Carignano *et al.*[14]

Acknowledgements

The work of L. M. L. and A. M. is supported by the EURYI scheme (see http://www.esf.org/euryi) and the EPSRC. D. P. and E. G. W. are supported by the NSFC. Partial computational support from the London Centre for Nanotechnology is warmly acknowledged.

References

1 M. J. Molina, *Angew. Chem., Int. Ed. Engl.*, 1996, **35**, 1778–1785.
2 M. Faraday, *The Athenaeum*, 1850, **1181**, 640–641.
3 V. F. Petrenko and R. W. Whitworth, *Physics of Ice*, Oxford University Press, USA, 2002.
4 Y. Li and G. A. Somorjai, *J. Phys. Chem. C*, 2007, **111**, 9631–9637.
5 H. Nada and J. P. van der Eerden, *J. Chem. Phys.*, 2003, **118**, 7401–7413.
6 W. L. Jorgensen, J. Chandraeskhar, J. D. Madura, R. W. Impey and M. L. Klein, *J. Chem. Phys.*, 1983, **79**, 926–935.
7 E. Sanz, C. Vega, J. L. F. Abascal and L. G. McDowell, *J. Chem. Phys.*, 2004, **121**, 1165–1166.
8 E. Sanz, C. Vega, J. L. F. Abascal and L. G. McDowell, *Phys. Rev. Lett.*, 2004, **92**, 255701.
9 C. Vega, J. L. F. Abascal, E. Sanz, L. G. McDowell and C. McBride, *J. Phys.: Condens. Matter*, 2005, **17**, S3283–S3288.
10 J. L. F. Abascal and C. Vega, *J. Chem. Phys.*, 2005, **123**, 234505.
11 G. J. Kroes, *Surf. Sci.*, 1992, **275**, 365–382.
12 J. F. Abascal and C. Vega, *Phys. Chem. Chem. Phys.*, 2007, **9**, 2775–2778.
13 T. Ikeda-Fukazawa, *J. Chem. Phys.*, 2004, **120**, 1395–1401.
14 M. A. Carignano, P. B. Shepson and I. Szleifer, *Mol. Phys.*, 2005, **103**, 2957–2967.
15 K. Bolton and J. B. C. Petterson, *J. Phys. Chem. B*, 2000, **104**, 1590–1595.
16 J. F. Abascal, R. G. Fernandez, C. Vega and M. A. Carignano, *J. Chem. Phys.*, 2006, **125**, 16601.
17 J. A. Hayward and J. R. Reimers, *J. Chem. Phys.*, 1997, **106**, 1518–1529.
18 W. Smith and T. R. Forester, *J. Mol. Graphics*, 1996, **14**, 136–141.
19 K. Röttger, A. Endriss, J. Ihringer, S. Doyle and W. F. Kuhs, *Acta Crystallogr., Sect. B*, 1994, **50**, 644–648.
20 J. VandeVondele, M. Krack, F. Mohamed, M. Parrinello, T. Chassaing and J. Hutter, *J. Comput. Phys. Commun.*, 2005, **167**, 103.
21 J. P. Perdew, K. Burke and M. Ernzerhof, *Phys. Rev. Lett.*, 1996, **77**, 3865; J. P. Perdew, K. Burke and M. Ernzerhof, *Phys. Rev. Lett.*, 1997, **78**, 1396.
22 S. Goedecker, M. Teter and J. Hutter, *Phys. Rev. B*, 1996, **58**, 3641.
23 M. Krack, *Theor. Chem. Acc.*, 2005, **114**, 145.
24 K. T. Gillen, D. C. Douglass and M. J. R. Hoch, *J. Chem. Phys.*, 1972, **57**, 5117.
25 H. Itoh, K. Kawamura, T. Hondoh and S. Mae, *J. Chem. Phys.*, 1996, **105**, 2408–2413.

26 M. de Koning, A. Antonelli, A. J. R. de Silva and A. Fazzio, *Phys. Rev. Lett.*, 2006, **97**, 155501.
27 C. Thierfelder, A. Hermann, P. Schwerdtfeger and W. G. Schmidt, *Phys. Rev. B*, 2006, **74**, 045422.
28 B. Santra, A. Michaelides and M. Scheffler, *J. Chem. Phys.*, 2007, **127**, 184104.
29 X. L. Hu and A. Michaelides, *Surf. Sci.*, 2008, **602**, 960–974.
30 G. A. Tribello and B. Slater, *Chem. Phys. Lett.*, 2005, **425**, 246–250.
31 T. K. Hirsch and L. Ojamäe, *J. Phys. Chem. B*, 2004, **108**(40), 15856–15864.
32 J.-L. Kuo, *Phys. Chem. Chem. Phys.*, 2005, **7**, 3733–3737.
33 S. J. Singer, J.-L. Kuo, T. K. Hirsch, C. Knight, L. Ojamäe and M. L. Klein, *Phys. Rev. Lett.*, 2004, **94**, 135701.
34 N. H. Fletcher, *Philos. Mag.*, 1992, **B36**, 109–115.
35 X. Wei, P. B. Miranda, C. Zhang and Y. R. Shen, *Phys. Rev. B*, 2002, **66**, 085401.

 This journal is © The Royal Society of Chemistry 2009

Reactivity of water–electron complexes on crystalline ice surfaces

Mathieu Bertin, Michael Meyer, Julia Stähler, Cornelius Gahl,† Martin Wolf and Uwe Bovensiepen*

Received 1st April 2008, Accepted 2nd May 2008
First published as an Advance Article on the web 19th September 2008
DOI: 10.1039/b805198d

The interactions between long-living electrons trapped in defects of crystalline D_2O and electronegative molecules have been investigated using two-photon photoemission spectroscopy. When covered by a Xe adlayer, the spectroscopic signature of the trapped electrons vanishes, which provides evidence that the trapping sites are located on the surface of the crystalline ice. The reactive character of these surface-trapped electrons with molecules has been studied. In the case of $CFCl_3$ adsorbed on top of the ice, we show that the trapped electrons induce the dissociation of the molecules, *via* a dissociative electron attachment process, resulting in $\bullet CFCl_2$ and Cl^- formation. The latter species are responsible for the observed increase of the work function and presumably for the deactivation of the surface trapping sites with respect to subsequent light-induced population by excited electrons. This process is thought to be of high efficiency since it is observed for a very low $CFCl_3$ coverage of only ~0.004 monolayer (ML). In the case of exposure of the crystalline ice to a partial pressure of gaseous O_2, the deactivation of the trapping site has also been observed. The mechanism is thought to involve the formation of the O_2^{*-} transient anion by electron attachment, followed by its reactive interaction with the water molecules of the defect. In both cases, the mechanisms are triggered by negative ion resonances which are known from experiments using a primary electron beam to be effective for isolated molecules for ballistic electrons of ~0 eV. We thereby demonstrate a similarity between the processes induced by these primary, very low kinetic-energy electrons and by the long-living surface electrons on the crystalline ice surface. These results suggest that the photoexcited trapped electrons can play an important role in the heterogeneous chemical processes on condensed water surfaces and could be relevant in the polar stratosphere chemistry.

1. Introduction

Electrons in an environment of dipolar molecules are known to form electron–solvent complexes.[1,2] Well-established examples are solvated electrons in ammonia[3] or water,[4] which are referred to as ammoniated or hydrated electrons respectively. These solvated electrons gain binding energy with respect to electrons in continuum states due to molecular rearrangement of the adjacent polar molecules. The resulting electron–solvent complex screens the excess charge which leads to the increase in binding energy as revealed from photodetachment spectroscopy.[5–7] In the case of

Freie Universität Berlin, Fachbereich Physik, Arnimallee 14, 14195 Berlin, Germany. E-mail: uwe.bovensiepen@fu-berlin.de

† Present address: Max-Born-Institute für Nichtlineare Optik und Kurzzeitspektroskopie, Max-Born-Str. 2A, 12489 Berlin, Germany

bulk water the energy gain is estimated to be 3.2 eV from extrapolation of the vertical binding energies determined in water-cluster anions to infinite cluster size.[5,6] Essential for this extrapolation is the solvation site of the excess electron in such clusters. Currently, it is controversially discussed whether the excess electron resides in the bulk or at the surface of water clusters[6,8] and the actual structure of the investigated clusters could be a key ingredient to settle this question.[9]

Beside these quasi-stationary properties of excess electrons in an environment of dipolar molecules, considerable effort has been devoted to the investigation of the ultrafast dynamics related to molecular rearrangement in water driven by a photo-excited excess electron in the liquid,[10–13] in clusters[14] and at amorphous ice/metal interfaces.[7] Early molecular dynamics simulations established that an excess electron attached to the surface of a water cluster gains binding energy and is transferred from a surface site to an interior site within a few picoseconds.[15] This behavior depends on the confinement effects for different cluster sizes and configurations, as shown by Barnett *et al.*,[15] highlighting the correlation between dynamic and confined-water structures. Recently, related questions have been addressed by a surface science approach investigating well-characterized amorphous ice structures adsorbed on metal surfaces by femtosecond (fs) time-resolved two-photon photo-electron spectroscopy (2PPE) and scanning tunneling microscopy.[9,16] The lifetime of hydrated electrons at amorphous ice/metal interfaces is governed by the rate of the electron transfer back to the metal substrate, which is the final decay channel of the hot electrons that are photoinjected from the metal substrate into the ice. For Cu(111) and Ru(001) substrates, initial lifetimes of 140 fs and 34 fs respectively have been reported.[17] Using "titration" of the excess electron by Xe overlayers it was shown that for small clusters on a Cu(111) surface the excess electron resides at the ice/vacuum interface while for wetting layers of several bilayer (BL) height the excess electron was found to be solvated in the bulk of the ice.[18,19] This implies that in the case of clusters the electron density may reside in the vacuum in front of the ice surface and thus these particular excess electrons could drive a chemical reaction at the ice surface. Due to the short electron lifetime in the fs to ps regime their reactivity might be limited. However, Lu *et al.* reported on an enhanced Cl^- yield produced by dissociative electron attachment (DEA) on a submonolayer of chlorofluorocarbons (CFCs) when they are coadsorbed with water or ammonia.[20,21] More recently, Ryu *et al.* concluded that photodissociation of $CFCl_3$ adsorbed on an *amorphous* ice surface could be enhanced by the photoinjection of excess electrons into the ice layer.[22]

Recently, these studies have been extended to crystalline ice structures on metal surfaces[23,24] where the lifetimes of trapped excess electrons can be enhanced by many orders of magnitude. We first provide evidence for the long-living character of these electrons and show by the Xe titration method that such electrons reside at the ice/vacuum interface. We investigate electron-induced dissociation of coadsorbed CFCs and find that only minute amounts of adsorbed CFC are needed to quench the spectroscopic signature of the trapped electrons in 2PPE. From the concomitant increase in workfunction we conclude a production of negative charge due to electron-induced dissociation of the CFC molecules. By a similar approach we study the interaction of O_2 in the gas phase with the long-living electrons and conclude that the interaction leads to the formation of a transient negative ion, O_2^{*-}, that can dissipate its excess energy through vibrational excitation and presumably through chemical modification of the ice surface.

Thereby we clearly identify two analogue chemical reaction pathways driven by bound, optically excited electrons. Our findings are closely related to chemical processes induced at surfaces by ballistic electron irradiation, and allow us to point out a similarity between surface-residing excess electrons and ballistic electrons having a null kinetic energy. The chemical processes induced by the latter particles are indeed of great interest since very low-energy electrons ($E < 5$ eV) are known to be able to trigger complex chemical reactions in condensed molecules on

 This journal is © The Royal Society of Chemistry 2009

surfaces.[25,26] Thereby we demonstrate a close correlation of photoinduced processes driven by optically excited electrons with processes driven by primary low-energy electrons. These findings suggest that the bound photoexcited electrons play an important role in the chemistry of water-ice surfaces, and may for this reason also be of high relevance in stratospheric chemistry.

2. Experiment

The studies were performed in an ultrahigh vacuum (UHV) chamber (base pressure $< 1 \times 10^{-10}$ mbar). The Ru(001) substrate is mounted on a liquid helium cryostat which can be cooled down to 30 K. Its preparation involves ion sputtering and thermal treatment following standard procedures.[27] The surface quality is probed by low-energy electron diffraction (LEED) as well as by workfunction measurement using 2PPE spectroscopy.

Crystalline D_2O is grown as follows. D_2O, purified by several freeze–pump–thaw cycles, is dosed *in situ* on the Ru surface, which is kept at 155 K, through a pinhole. The resulting film is subsequently annealed to 165 K to complete its crystallization, monitored by the isothermal water desorption yield with a quadrupolar mass spectrometer (QMS).[28] The annealing at 165 K is stopped when a constant desorption rate is observed, resulting in crystalline D_2O. The coverage in the experiments discussed below is determined to be 4.5 BL from thermal desorption spectroscopy (TDS) measurements. Following this procedure, the first bilayer of crystalline water is a well-ordered wetting layer exhibiting a $\sqrt{3} \times \sqrt{3}$ structure on the Ru(001) surface.[29] At higher coverage, crystalline water clusters are formed on the top of the wetting layer according to findings on Cu(111), Pt(111) and Ru(001).[23,30,31] The overlayer depositions of Xe, $CFCl_3$ and O_2 have been performed by exposing the crystalline ice to a background pressure of the atom/molecule during a given time. In the case of Xe and $CFCl_3$, the resulting coverage at 30 K has been calibrated by means of TDS experiments.

One-color 2PPE spectroscopy has been performed using a femtosecond (fs) laser system, operating at a repetition rate of 200 to 300 kHz. The amplified laser pulses at 800 nm are converted to tunable laser pulses in the visible range with an optical parametric amplifier (OPA), and frequency doubled to obtain ultraviolet pulses used in the present study. In the experiments presented here, photon energies $h\nu$ between 3.6 and 3.9 eV have been used. In two-photon photoemission, the photon energy $h\nu$ is kept below the work function Φ of the system. As depicted in Fig. 1, the first photon excites electrons from the metal substrate to intermediate states. A second photon, with the same energy $h\nu$, is then used to photoemit electrons from the intermediate state to the vacuum, where they are analyzed regarding their kinetic energy $E_{\rm kin}$ using a time-of-flight spectrometer (TOF). Thus, the energy E of the intermediate state, with respect to the Fermi energy $E_{\rm F}$, can be obtained from the kinetic energy of the electrons by:

$$E - E_{\rm F} = E_{\rm kin} + \Phi - h\nu. \tag{1}$$

In principle, time-resolved experiments can also be carried out with two time-delayed laser pulses. However, in the present study, only steady-state experiments are performed using 2PPE spectroscopy.

3. Results and discussion

3.1 2PPE spectra of crystalline D_2O deposited on Ru(001)

Single-color (UV) 2PPE spectra of crystalline D_2O/Ru(001) are shown in Fig. 1 and exhibit distinct features: the low energy cut-off, a *d*-band surface resonance of Ru(001) at $E - E_{\rm F} = 0.9$ eV, a pronounced peak around 2 eV above $E_{\rm F}$, termed $e_{\rm T}$

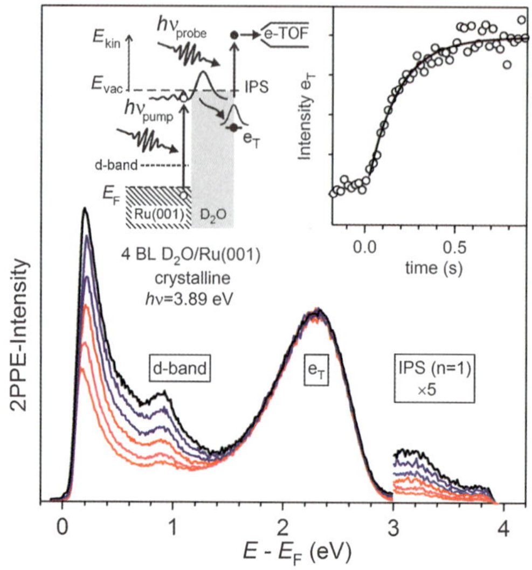

Fig. 1 One-color two-photon photoemission (2PPE) using one photon from an ultrashort laser pulse with a photon energy $h\nu$ smaller than the work function $\Phi = E_{vac} - E_F$ to excite an electron from an occupied state into a bound excited state. A second photon of the same color photoemits the excited electron, which is analyzed according to its kinetic energy by a time-of-flight spectrometer. Thereby the spectrum of the excited, intermediate states is probed. In the main panel 2PPE spectra realized at a repetition rate of 200 kHz on a 4 BL crystalline D_2O coverage on Ru(001) at $T = 43$ K are shown for laser pulse durations between 100 fs (black, top) and 300 fs (red, bottom). The spectra exhibit three features (low energy cut-off, *d*-band, and $n = 1$ image potential state) that vary with the inverse square of the laser pulse duration and one (e_T) that is independent of the pulse duration. The intensity of the latter builds up with a time constant of 0.17 s, as shown in the inset, forming a photostationary state with population and depopulation in equilibrium (solid curve).

for trapped electrons, that only occurs for crystalline ice films, and the first image potential state (IPS) of the D_2O/Ru interface at 3.2 eV. The *d*-band resonance and the IPS are intermediate states in the 2PPE process with sub-10 fs lifetimes.[32] Thus, 2PPE from these states is a second-order (non-linear) process and the respective 2PPE intensity depends on the power density of the UV laser pulse. 2PPE spectra shown in Fig. 1 have been recorded at $T = 43$ K for different UV pulse durations between 100 and 300 fs, *i.e.* for different power densities, while keeping the photon energy constant. As expected, an increase of the 2PPE yield is observed for decreasing pulse duration (*i.e.* increasing instantaneous power) in the case of the IPS, the *d*-band resonance and the low energy cut-off. However, the intensity of the e_T feature is unaffected by the change of pulse duration. This effect can be explained exclusively by lifetimes of e_T electrons which exceed the inverse laser repetition rate of 5 μs. The e_T state is thus populated by a first laser pulse and probed by subsequent laser pulses so that excitation of and photoemission from that state are not correlated. In other words, the e_T spectrum represents a photostationary state, where population of the long-living intermediate state ~2 eV above E_F and its depopulation are balanced.

Under the employed absorbed fluence of 0.02 mJ cm^{-2}, formation of the photostationary state—measured by the 2PPE intensity build-up of e_T in real time—occurs with a time constant of 0.17(1) s, as shown in the inset of Fig. 1. The rise time—following a simple rate equation analysis—is equivalent to the average time of an electron residing in the trap. In ref. 24 and 33 we provide detailed information on these long-living electrons e_T, where we show that the lifetime of these species extends up to minutes. Theoretical predictions suggest that they are formed and

 This journal is © The Royal Society of Chemistry 2009

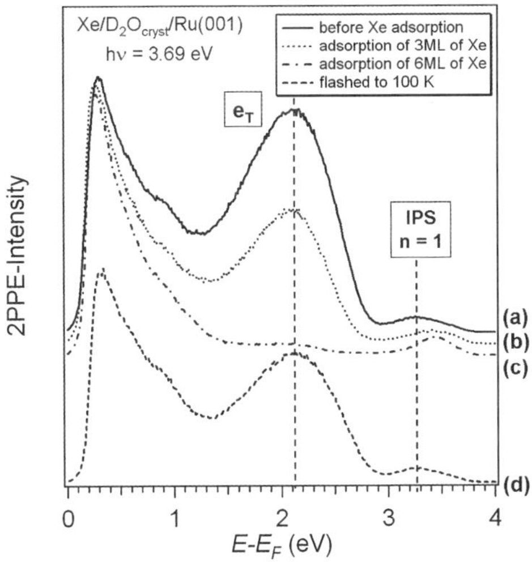

Fig. 2 2PPE spectra at ~30 K of 4.5 BL of crystalline D_2O deposited on Ru(001) (a), after deposition of 3 ML of Xe (b), after deposition of 6 ML of Xe (c) and after flashing the surface to ~100 K (d).

stabilized at structural defects, located on the surface of the crystalline ice. From fluence-dependent studies‡, the surface density of these surface solvation sites is estimated to be $\sim 2 \times 10^{12}$ cm^{-2}, corresponding to $\sim 10^{-3}$ sites per surface water molecule. A detailed discussion on the dynamics of the long-living electrons is out of the scope of the present paper and we focus here on the surface character and reactivity of these excess electrons e_T.

3.2 Xe titration experiment on crystalline D_2O

Fig. 2 compares 2PPE spectra of the crystalline ice layer, after deposition of 3 and 6 monolayers (ML) of Xe and after flashing the sample to 100 K respectively. Adsorption of Xe results in a dielectric wetting layer, and is therefore expected to modify the energetic environment only for electrons residing at the surface.[9] Moreover, due to its closed-shell structure, no spontaneous reaction is expected to take place upon Xe adsorption. Therefore, any modification observed has to be related to a purely electrostatic perturbation of the solvated electron due to (i) the polarizability of the Xe adatoms and (ii) a confinement effect, limiting the spatial extension of the e_T electron wave function.[9,34] As observed from the 2PPE spectra, the adsorption of 6 ML of Xe results in a total quenching of the long-living electrons signal, which unambiguously shows that the e_T electrons are localized on the surface of the ice layer. A coverage of 3 ML results in a relative decrease of the e_T peak only.

The observed quenching can be understood as an electrostatic perturbation of the trapping sites (defects) by the surrounding Xe atoms, therefore inducing the quenching of the e_T signal in the 2PPE spectra. The relatively high Xe coverage (6 ML) needed for quenching can be explained by (i) the fact that each trapping site has to be surrounded by Xe atoms to generate the required electrostatic perturbation to induce the change in the 2PPE spectra and (ii) the structure of the crystalline D_2O layer. Indeed, as mentioned in the experimental part, for coverages of more than one

‡ The surface density of the trapping site has been estimated from a fluence-dependent study on the e_T peak intensity using several photon energies. Rate equations, involving the equilibrium between population (excitation probability) and depopulation (probing probability) of the trapping sites, have been used, assuming a photoelectron detection probability of 10^{-4}.

bilayer, the D_2O molecules form clusters on top of the first wetting bilayer, thus increasing the effective interface area between D_2O and Xe atoms.

Beside quenching the long-living electrons signal, the Xe coverage does not have a significant effect on the system work function. The work function change can be directly accessed by the shift of the low energy ($E - E_F$) cut-off in the 2PPE spectra (see eqn (1)), which is very small in the case of Xe adsorption as observed in Fig. 2. It is noteworthy that the interaction of the Xe atoms with the trapping sites is not destructive, as the e_T peak is recovered after desorption of the Xe adlayer at 100 K. Nevertheless, the e_T peak is less intensive after the flashing than in the case of freshly prepared crystalline ice. This intensity difference might be related to thermal-induced modification of the surface trapping sites upon the annealing to 100 K.

3.3 $CFCl_3$ adsorption on crystalline D_2O

Surface photochemistry involving condensed chlorofluorocarbons (CFC) has attracted a lot of attention because of their role in atmospheric ozone depletion.[35,36] A large part of the polar stratospheric chemistry occurs mainly heterogeneously, on the surface of water ice particles.[37] Thus, the photoinduced dissociation of these adsorbed molecules, releasing chloride as an ozone-scavenger, is of great interest in the understanding of the chemical mechanisms in the stratospheric medium. In addition to direct photodissociation, the chlorofluoromethanes, such as $CFCl_3$, present a high cross section for dissociative electron attachment (DEA) of ballistic electrons with zero kinetic energy.[38–40] In this resonant process, such electrons can attach to the molecule, resulting in the formation of a transient negative ion, which can spontaneously decay either *via* the autodetachment of the electron, or *via* a C–Cl bond cleavage and the formation of a Cl^- ion, following the equation:

$$CFCl_3 + e^- \text{ (0 eV)} \leftrightarrows [CFCl_3]^{*-} \rightarrow \bullet CFCl_2 + Cl^- \quad (2)$$

Some other DEA resonances exist at higher incident electron energy (>3 eV) in the chlorofluoromethanes, leading to the formation of other negative ions, but the process at 0 eV presents a much higher cross section.[41] Hence, this barrier-free DEA process has to be taken into account in photoinduced reactions, since UV photons can also generate low-energy secondary electrons (hot electrons) which may dissociate the CFC molecules. The role played by these very low-energy electrons in condensed CFC reactivity has indeed motivated many experimental studies conducted with low-energy electron beams.[26,42]

In the present study, we aim to study the relevance of these dissociative attachment processes to reactions involving photoexcited solvated electrons, which present a finite binding energy with respect to the vacuum level. In the case of electron attachment to condensed molecules, it is well known that the energy of the resonant negative ion state is lowered compared to the one in the gas phase because of polarization effects (image-charge potential due to the metal substrate and polarizability of the surrounding molecules).[43] Therefore, solvated electrons, with a binding energy below the vacuum level, may attach to a condensed molecule *via* the stabilized negative ion state, which is accessible with ~0 eV electrons for gas-phase molecules. Moreover, these bound electrons might access negative ion states at lower-lying energy, below the vacuum energy, and thus trigger electron attachment processes which would not be accessible using ballistic electrons. Finally, these sub-vacuum electrons are not coupled to the continuum, and therefore cannot be detached into the vacuum after their attachment to a molecule. This implies that the decay channel of electron autodetachment after the negative ion formation depends solely on the electron transfer rate back to the metal substrate. Previous studies of CF_2Cl_2 and $CFCl_3$ condensed on top of water layers showed that the DEA cross section can be greatly enhanced due to the presence and attachment of solvated electrons in the ice layer.[20–22,44] Even if these last results were concerned with water

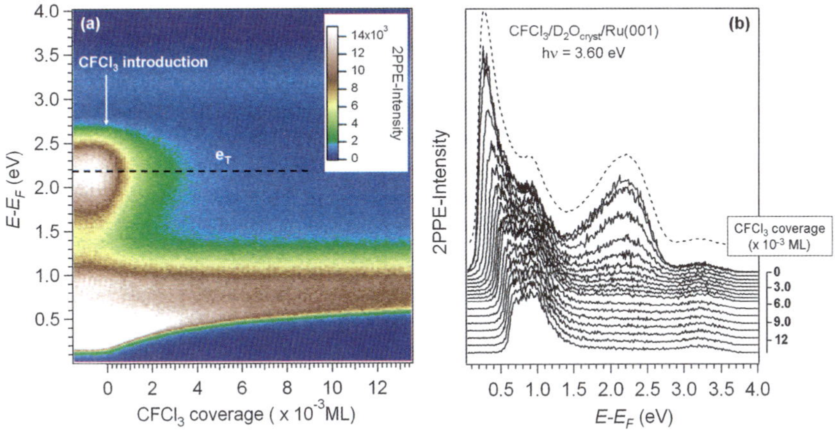

Fig. 3 2PPE spectroscopy performed on crystalline D_2O (coverage ~4.5 BL) deposited on Ru(001) during its exposition at ~30 K to 1×10^{-9} mbar of $CFCl_3$. Figure (a) represents a 2D plot of the 2PPE intensity in false color as a function of both $E - E_F$ and the coverage of $CFCl_3$, as calibrated from the exposition time by TDS experiments. Figure (b) shows vertical cuts of the 2D plot (2PPE intensity versus $E - E_F$) for different $CFCl_3$ coverages. The dashed spectrum is a 2PPE spectrum realized on 4.5 BL of freshly deposited crystalline D_2O before $CFCl_3$ adsorption.

ice in the amorphous state, the capability of the CFC molecules to react efficiently with hydrated electrons makes them good candidates for probing the reactivity of the long-living electrons created on the crystalline ice surface during UV irradiation. Moreover, the DEA cross section is expected to be greatly enhanced on crystalline ice because of (i) the surface character of the trapped electron, (ii) their very long lifetime, which is 14 orders of magnitude longer than in amorphous ice and (iii) their very weak coupling with the metal substrate, which nearly closes the decay channel of autodetachment of the electron and therefore favors the dissociative decay channel.[24]

Fig. 3 presents 2PPE spectra of 4.5 BL of crystalline D_2O, grown on Ru(001), acquired during an *in situ* exposure at ~30 K to a partial pressure of $CFCl_3$ (1×10^{-9} mbar). The $CFCl_3$ coverage (in ML) on top of the surface has been determined by means of TDS calibration in a higher coverage regime (0.05–1.50 ML), and by assuming a constant sticking coefficient in the lower coverage regime. Fig. 3a shows a 2D plot of the 2PPE intensity (false color scale) as a function of both intermediate state energy, $E - E_F$, and $CFCl_3$ coverage. The broad and intensive feature observed at $E - E_F \sim 2.2$ eV before $CFCl_3$ introduction is associated with the e_T peak of the pure crystalline ice surface (Fig. 1). Right after the start of $CFCl_3$ exposure, this peak's intensity vanishes. Simultaneously, one can clearly identify a shift of the low energy cut-off towards higher $E - E_F$, which is attributed to an increase of the surface work function Φ. These observations can be clearly distinguished in the Fig. 3b, which presents horizontal cuts of the 2D plot (2PPE spectra as a function of $E - E_F$) for several $CFCl_3$ coverages. These effects are very remarkable, since they occur at an extremely small coverage of $CFCl_3$. The e_T peak becomes totally quenched for a coverage as small as ~0.004 ML, and the concomitant work function increase corresponds to $\Delta\Phi = +250$ meV. Such an increase of the system work function has to be related to the presence of an effective negative charge density on the surface, inducing a surface dipolar potential that virtually lowers the vacuum energy with respect to the Fermi level.[45]

To explain the vanishing of the long-living electrons' peak intensity, a simple electrostatic perturbation by the surrounding $CFCl_3$ molecules of the trapping sites, as invoked in the case of Xe overlayer, can be ruled out. As learnt from the Xe overlayer experiment, a relatively high coverage (~6 ML) is needed to deactivate the e_T signal by purely electrostatic effects. Therefore, the low $CFCl_3$ surface density (<10^{-3} ML) cannot induce such a strong electrostatic effect on the surface defects. Moreover, the work function increase, observed simultaneously with the e_T signal quenching, cannot be explained by the dipole moment of intact $CFCl_3$ molecules at such low coverage. Indeed, the $CFCl_3$ dipole moment is 0.45 Debye,[46] *i.e.* lower than the D_2O dipole moment of 1.86 Debye,[47] and is therefore expected to have little impact on the system work function due to the low density of molecules and screening by the underlying 4.5 BL of crystalline ice. In addition, the UV irradiation performed on crystalline ice alone leads to a work function change of ~+50 meV for a coverage of 4.5 BL, due to the charging of the layer which is five times smaller than the one observed in the present case. Finally, any process triggered by the direct photodissociation of the $CFCl_3$ can not be operative in our case, since the photon energy employed (~3.6–3.8 eV) is below the dissociation threshold of the intact molecule (~5 eV from ref. 22).

Hence, another interpretation has to be considered to explain the effects observed in the 2PPE spectra, involving a reactive interaction between the $CFCl_3$ molecules and the long-living electrons trapped at the ice/vacuum interface. As mentioned above, $CFCl_3$ molecules present a high cross section of DEA at ~0 eV, resulting (eqn (2)) in the formation of the radical $\bullet CFCl_2$ and the Cl^- anion formation. Moreover, it has already been concluded that solvated electrons in amorphous water ice could also perform the DEA process on $CFCl_3$.[20,22] Thus, in the present case, the concomitant quenching of the e_T peak and increasing of the work function can be understood as follows. The lowering and quenching of the e_T peak's intensity in the 2PPE spectra can be related to the dissociative attachment of the long-living electrons on the $CFCl_3$ molecules. This dissociation is then accompanied by the formation of stable Cl^- anions at the surface of the ice. This density of negative charge present on the surface is then responsible for the observed increase of the work function by ~250 meV. As discussed above, we observe a total quenching of the e_T peak for a $CFCl_3$ coverage of 0.004 ML, which corresponds to a surface density of $\sim 2.3 \times 10^{12}$ cm^{-2} (considering a surface density of 5.8×10^{13} cm^{-2} for 0.1 ML [44]). It appears that this molecule density is in the same order of magnitude as the estimated surface density of trapping sites. This observation shows the high reactive character of the long-living surface electrons.

In order to get a better insight into the interactions involved between the e_T electrons and the $CFCl_3$ molecules, 2PPE has been performed on crystalline D_2O covered by 0.01 ML of $CFCl_3$ and subsequently annealed to different temperatures below the multilayer desorption temperature of crystalline ice. The resulting spectra, presented in Fig. 4, have all been taken after a cooling to 30 K following each flash to the indicated temperature. Compared to the spectrum of the crystalline D_2O, and as described before, deposition of $CFCl_3$ results in the quenching of the e_T peak and in the increase of the work function of the system. For flashing temperatures of 60 and 140 K, Fig. 4 shows spectra acquired on a fresh spot, *i.e.* where no UV irradiation has lead to the trapping sites population by e_T electrons prior to the flashing (dashed curve), and pre-irradiated spot, *i.e.* where UV irradiation has already been realized before the annealing, and therefore where the reaction could already have taken place (solid curve). After flashing the system to ~60 K, we observe that the spectrum taken in the pre-irradiated spot (solid curve) presents a slight increase in the work function, whereas the spectrum taken on a fresh spot (dashed curve) is basically identical to the one before flashing. After a temperature flash to 140 K, the 2PPE spectra acquired on a pre-irradiated spot (solid curve) and fresh spot (dashed curve) present the same work function, which is now decreased to a value close to the one of the pure crystalline ice. Nevertheless, no clear signal related to the long-living electrons is observed in the pre-irradiated spot, whereas the fresh spot spectrum exhibits a structure at $E - E_F \sim 2.2$ eV associated with the e_T signal. Finally, one can

 This journal is © The Royal Society of Chemistry 2009

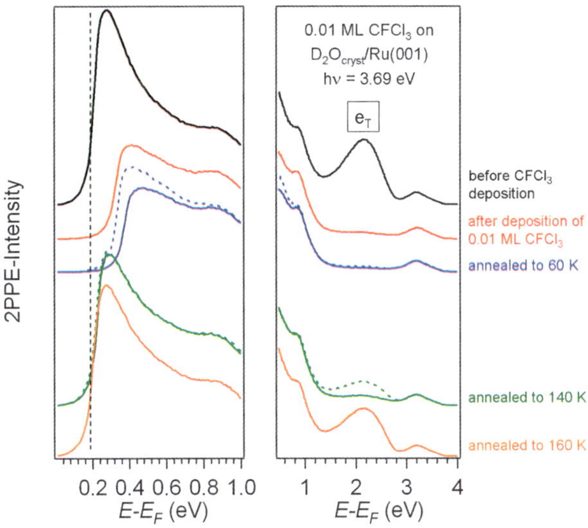

Fig. 4 2PPE spectra acquired at 30 K on 4.5 BL of crystalline D_2O before and after deposition of 0.01 ML of $CFCl_3$. The resulting surface has then been flashed to 60, 140 and 160 K, the spectra being acquired each time after a cooling back to 30 K. For the flashing at 60 and 140 K, the solid curves denote spectra performed on already-irradiated spots before the flash, whereas the dashed curves are associated with spectra acquired on fresh spots. A vertical dashed line represents the low energy cut-off, associated with the work function, of the pure crystalline ice at 30 K.

observe that a flashing to 160 K leads to a total recovery of the e_T peak in the 2PPE spectrum, which is observed independently for a fresh and a pre-irradiated spot.

We believe that the work function increase of the system and the vanishing of the e_T electron signal are the clear signature of a reactive process triggered by the dissociative attachment of the solvated electrons to the $CFCl_3$ molecules. Nevertheless, the exact nature of the subsequent reaction mechanisms is difficult to establish. Several options can be considered to explain the decrease of the e_T peak intensity as well as the flashing-temperature dependency of the 2PPE spectra. Indeed, the quenching of the e_T peak can be associated with (i) a pronounced reduction of the lifetime of the solvated electrons because of their attachment to intact molecules whereas the trapping sites can still be populated by excited electrons during the illumination or (ii) an energetic or chemical modification of the trapping sites that prevent further photoexcited electrons from populating them during UV irradiation. These scenarios are discussed in the following.

We observe in our 2PPE experiments at a constant $CFCl_3$ coverage that the e_T signal is not recovered with time of illumination. If this signal quenching were only due to electron attachment to intact CFC molecules, which reduces the electron lifetimes (case (i) described above), the e_T peak would be observed again after all $CFCl_3$ molecules around the site were dissociated. As this is not the case in our experiment, we believe that the DEA process also leads to a chemical modification and a deactivation of the trapping site, *i.e.* it cannot be populated any more. Several propositions can be envisaged to explain such a site deactivation. A first scenario considers a constant negative charge density on the ice surface. After one *unique* cycle of population of the site and attachment of the electron to the $CFCl_3$, the resulting Cl^- is trapped at the site and therefore blocks it with regard to further population. In this case, the global surface charge of the surface remains the same before and after dissociation, since the chlorine anion has the same charge as the solvated electron. Therefore, the observed increase of the work function would originate from the difference in the screening efficiency of the extra negative charges on the surface by the water molecules. Indeed, it is expected that the screening of the electron in the site is much more efficient than the one of the more voluminous Cl^-. Finally, the fact that we observe a total quenching of the e_T peak for a number of $CFCl_3$ molecules which is of a similar order of magnitude to the number of trapping sites fits with the present interpretation.

However, other mechanisms, triggered by the DEA process on $CFCl_3$, could also be invoked to explain the e_T signal quenching and the work function increase. In particular, another scenario could involve an increase of the surface negative charge density. In this case, *several* dissociation processes occur at a unique trapping site before it becomes deactivated. Its blocking could in this case be due to a high density of Cl^- formed in the vicinity of the defect, resulting in a strong electrostatic perturbation leading to its deactivation. In this case, the work function increase would not be caused by the difference in the screening efficiency by the water molecule but by a higher number of negative charges present on the surface. To explain the work function change due to the Cl^- formation ($\Delta\Phi = +250$ meV) in comparison to the one due to the charging of the pure crystalline D_2O layer by the e_T electrons ($\sim$50 meV), at least five Cl^- would need to be formed per trapping site. Although in the case of a total quenching of the e_T peak we estimated the number of $CFCl_3$ molecules to be comparable to the number of e_T electron sites, the uncertainty inherent to the determination of the density of the sites—principally the determination of the detection sensitivity in the 2PPE experiments—does not allow us to exclude unambiguously the possibility that several dissociation processes can occur per surface defect.

Considering the two possible scenarios, the observation of the flashing-temperature dependency (Fig. 4) of the 2PPE spectra can be explained as follows. The flashing to 60 K promotes the diffusion of intact $CFCl_3$ molecules to available sites. In the pre-irradiated spot, where a certain number of sites are already blocked by the irradiation before the flashing, the temperature flash may activate the diffusion of intact $CFCl_3$ to some remaining available sites. The reaction is therefore enhanced and results in a larger amount of trapped Cl^- compared to the one present before flashing. The higher amount of Cl^- increases the local work function of the system. In the non-irradiated spot, as no dissociation took place before the flash, the diffusion of $CFCl_3$ prior to the reaction has no effect on the 2PPE spectrum.

The flashing of the system to 140 K is thought to desorb the majority of the intact $CFCl_3$,[22] which is seen on the non-irradiated spot where the e_T signal is again observed after the flash. For the pre-irradiated spot, we conclude that the trapping sites are still blocked as no e_T signal is present in the 2PPE spectrum. However, the work function of the system is now very close to the one of the pure crystalline ice. If we consider the first scenario, the work function increase is due to a less effective screening of the Cl^- by the water molecules of the e_T electron sites. The decrease of the work function upon flashing to 140 K means then that the thermal energy brought to the system allows the D_2O molecules to re-orient around the Cl^-, therefore enhancing the charge screening, and thus decreasing the work function. This scenario is thus capable of explaining the flashing-temperature dependence of the 2PPE spectra. If we consider the second scenario, where the work function increase and the deactivation of the trapping site is due to the accumulation of several Cl^- ions, the interpretation of the behavior observed after the annealing at 140 K becomes more complex. Indeed, the work function reduction in the pre-irradiated spot could be explained by the thermal-induced diffusion of the Cl^- ions on the surface which would then decrease the local density of surface negative charge. However, the blocking of the trapping sites cannot be explained by the anions' density any more and other mechanisms, like a thermal-induced reaction (*e.g.* involving $\bullet CFCl_2$ radicals) leading to the chemical modification of the sites, have to be invoked. After flashing to 160 K, the 2PPE spectra are identical to the one of the pure crystalline ice. Indeed, as annealing to this temperature results in considerable D_2O desorption, in both scenarios we expect the Cl^- to diffuse on the whole surface or in the bulk of the ice, therefore drastically decreasing its local density on the surface due to its weak concentration, and the surface defects to be regenerated.

Finally, as pointed out above, other explanations for the site blocking could involve a chemical modification of the trapping site after dissociation of the $CFCl_3$ molecule. For instance, the other product of dissociation, the radical $\bullet CFCl_2$, is a very reactive species and could perform a reactive interaction with the water molecules in the vicinity of the e_T electron sites, leading to their chemical modification. Indeed, it has been showed by Perry *et al.*[48] that radicals $\bullet CF_xCl_{4-x}$ ($x = 0, 2$), formed from electron dissociative attachment on CF_2Cl_2 or CCl_4 coadsorbed with water, are at the origin of subsequent chemical reactions with H_2O molecules, leading in particular to CO_2 formation. Nevertheless, this finding has been realized on amorphous water ices under X-ray and 200 eV irradiation, and the invoked mechanisms involve other electron-induced dissociations, in particular of the water molecules, by hot electrons having a broad kinetic energy distribution. Therefore, a direct comparison of this chemistry induced by higher energy irradiation to our case should be considered with caution. One should also consider that two neighboring $\bullet CFCl_2$ radicals could recombine to form the stable $C_2F_2Cl_4$, as has been observed in the case of amorphous ice.[22] Nevertheless, such a recombination is not likely in our case due to the very small concentration of $CFCl_3$. Unfortunately, because of the very small coverage of $CFCl_3$ used here, attempts to find direct evidence of any product of a possible chemical reaction by TDS were unsuccessful since the small amount desorbed is below our sensitivity threshold.

Summarizing, UV irradiation on a crystalline ice surface leads to the formation of long-living e_T electrons, trapped on surface defects. These electrons exhibit a reactive interaction with adlayers of $CFCl_3$ molecules *via* a dissociative electron attachment process, leading to the formation of the radical $\bullet CFCl_2$ and the Cl^- anion. As observed from our 2PPE experiments, this process results in the deactivation of the trapping sites, preventing them from being populated by further electrons, and in the clear increase of the system work function ($\Delta\Phi \sim$+250 meV), which is attributed to the presence of surface Cl^-. The exact mechanism involved in the deactivation of the site is difficult to establish in our case. A reasonable explanation suggests that the Cl^- formed upon dissociation remains in the trapping site, therefore deactivating it, which provides an interpretation of the flashing-temperature dependence of the 2PPE spectra. Finally, we note that such strong effects on the photoelectron spectra have been observed for very low coverages of $CFCl_3$ molecules ($\sim$0.004 ML), which correspond to a density of molecules comparable to the density of e_T electron sites. This remarkable result shows the high efficiency of the DEA process, which has to be related to the extreme lifetime and the surface location of the electrons excited by UV light and residing in pre-existing defects on crystalline ice.

3.4 Gas-phase O_2 exposure to crystalline D_2O

While CFCs exhibit a high dissociative electron attachment cross section for electrons with no kinetic energy, no equivalent for this dissociative interaction exists in the case of dioxygen O_2 at very low energy (<3 eV). Nevertheless, it is well known from gas- and condensed-phase experiments[49,50] that a $\sim$0 eV electron can temporarily attach to an O_2 molecule in its fundamental state $^3\Sigma_g^-$, leading to the formation of the vibrationally excited O_2^{*-} ($^2\Pi_g$) anion. Since this anionic state is associative, the only decay channels are either the autodetachment of the excess electron or a transfer of the excess vibrational energy to a third body, leaving the O_2^- in its fundamental state, as has been shown in the case of O_2 condensed on a Kr layer[49] or in O_2 clusters.[51]

The study of the interaction between O_2 molecules and the crystalline D_2O surface during UV irradiation has been performed by exposing the ice layer to a partial pressure of dioxygen ($\sim 3 \times 10^{-9}$ mbar) while keeping it at 68 K. It is well known that

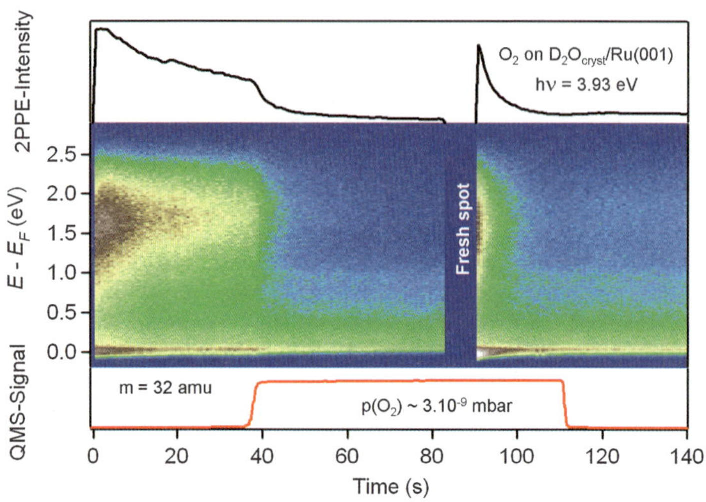

Fig. 5 2D plot of the 2PPE intensity of 9 BL of crystalline D_2O layer at 68 K, as a function of both $E - E_F$ and time, together with the integrated 2PPE intensity of the e_T peak (above) and the partial pressure of O_2 measured by a QMS (below) plotted versus time. At $t \sim 90$ s, the position of the laser spot on the sample was changed in order to probe a fresh region of the surface.

physisorbed multilayers of O_2 desorb below 30 K. Hence, exposure at 68 K does not lead to the formation of an O_2 adlayer on the ice surface with time. Moreover, it has been reported from electron-stimulated desorption experiments (ESD)[52] that no penetration or diffusion of O_2 inside the crystalline ice was observed. Thus, any effects induced by dioxygen exposure can only be attributed to gas–surface interactions. Fig. 5 presents the 2PPE experiment performed on 9 BL of crystalline D_2O at 68 K. The 2D plot of 2PPE intensity as a function of time and $E - E_F$ is shown together with the evolution of the integrated e_T peak intensity (above) and the partial pressure of O_2 (below) *versus* time. Note that at 68 K the e_T peak is observed for $E - E_F \sim 1.8$ eV, instead of ~2.2 eV in the case of measurement at 30 K. This effect has already been reported from temperature-dependent measurements of the 2PPE spectrum of the crystalline D_2O.[53] Before exposing the sample to O_2 (for $0 < t < 40$ s), a slow decrease in the e_T intensity is observed with irradiation time. This effect has been studied before in the case of the higher temperature of the ice (68 K) and coverage (9 BL).[53] However, starting with the O_2 exposition ($t \sim 40$ s), a quick quenching of the intensity of the e_T peak occurs. The same observation is made when another region of the surface is probed ($t \sim 90$ s). Finally, after 120 s, the O_2 exposure is stopped, but no e_T feature is recovered.

Similarly to the $CFCl_3$ overlayer experiment, we associate the observed changes in the 2PPE spectra to be the signature of the attachment of the trapped electron to gas-phase O_2 during collisions with the ice surface, which leads to the deactivation of the e_T electron sites. At very low energy this attachment is not dissociative and leads to the formation of the negative ion O_2^{*-}. After its formation in the vicinity of the trapping site, the excited O_2^{*-} species can decay (i) by auto-detachment of the extra-electron, leading to the formation of O_2 which is subsequently desorbed, (ii) by a transfer of vibrational energy to the surrounding medium, in this case to the water molecules forming the trapping site, leaving O_2^- in its fundamental state or (iii) by a chemical reaction with the surrounding water molecules. Whereas in the $CFCl_3$ experiment the deactivation of the sites were accompanied by a clear increase of the work function, associated with the presence of negative charges on the crystalline ice surface (Cl^-), no such strong effect is observed in the case of O_2, meaning that the O_2^- species is not expected to remain on the surface, stabilized in the trapping defects

 This journal is © The Royal Society of Chemistry 2009

and therefore blocking them. In this case, the loss of surface charge is counter-balanced by the repopulation of the site by electrons. Hence, the deactivation of the trapping sites has to be realized *via* a reactive interaction between the resonant state O_2^{*-} and the D_2O molecules of the site, leading to their chemical modification and thus to their deactivation with respect to further population by excited electrons. The exact nature of this reaction is not accessible in our case, but the capability of the resonant state O_2^{*-} formed by electron attachment to induce chemical reaction has previously been reported in the case of surface oxidation by primary low-energy electrons[54] or X-ray processing.[55] Such equivalent reactions with molecular media have to our knowledge never been observed following non-dissociative electron attachment to O_2 molecules. Nevertheless we believe that the reactive resonant state O_2^{*-} is involved in the chemical modification of the trapping sites in the present experiment, resulting in its deactivation.

4. Summary

We studied the interactions between trapped electrons, photoinjected in crystalline D_2O ice from a Ru(001) substrate, and atomic and molecular adlayers. The trapped photoexcited electrons in crystalline ice present a very long lifetime, that extends up to minutes, and are stabilized in defects in the ice structure.[24,33] Titration by 6 ML of Xe atoms results in the vanishing of the spectroscopic signature of these long-living electrons in 2PPE spectra, due to an electrostatic perturbation of the trapping defects by the so-formed dielectric adlayer. This observation shows clearly that the long-living electrons are trapped in defects located on the surface of the crystalline D_2O.

The reactive character of these long-living surface electrons with molecules was then studied using (i) $CFCl_3$ adsorbed on the surface of the ice and (ii) exposure of the ice to a partial pressure of gas-phase O_2. In both cases, the presence of these molecules leads to a quenching of the long-living electrons signal in 2PPE experiments. In the case of $CFCl_3$ adlayers, the quenching occurs at an extremely small CFC coverage ($\sim$ 0.004 ML) simultaneously with an increase of the system work function ($\Delta\Phi = +250$ meV). These effects are attributed to the dissociative electron attachment (DEA) of the long-living electrons to intact $CFCl_3$ molecules, leading to the formation of the $\bullet CFCl_2$ radical and the Cl^- anion. The formation of the Cl^- results in the increase of the system work function because of an enhancement in the effective surface negative charge density. Based on the flashing-temperature dependence of the 2PPE spectra, we proposed that after the dissociation of the $CFCl_3$ molecule the trapping site of the photoelectrons is blocked by a Cl^- ion. As the surface density of molecules needed to quench the trapped electrons signal totally is of the same order of magnitude as the surface density of trapping sites, we conclude that the DEA mechanism is a very effective process.

In the case of gaseous O_2 exposure to crystalline ice, the quenching of the trapped electrons signal in 2PPE spectra is also related to the deactivation of the trapping sites. The experiments were performed at a temperature above O_2 desorption, and the work function of the system was only slightly modified by the exposure. Hence, we believe the blocking of the sites to be a consequence of the formation of the transient negative ion O_2^{*-} by non-dissociative attachment of the long-living electrons to gas-phase O_2 during a collision with the surface. Thus, the deactivation of the trapping site would be related to a chemical reaction driven by this activated O_2^{*-} species with the surface water molecules, therefore chemically modifying the defect and preventing its further population by photoexcited electrons.

In these studies, we showed that the quenching of the long-living electrons signal in the 2PPE spectra is triggered by their attachment to the molecular adlayer: dissociative electron attachment in the case of $CFCl_3$ and non-dissociative attachment in the case of O_2. It is interesting to note that these two processes are resonant and effective for $\sim$0 eV ballistic electrons as has been studied in gas-phase experiments using primary electron beams. Therefore, we demonstrate a strong correlation between processes induced by primary very low-energy electrons (~0 eV) and photoexcited bound electrons trapped at the surface of the crystalline ice under UV irradiation. The efficiency of the processes in this latter case has to be related to the very long lifetime, the surface location and the weak coupling with the metal substrate, which favors the reactive pathway by preventing the autodetachment of the electron. Thus, our results suggest that the processes induced by these surface-trapped electrons play an important role in the surface chemistry of water ice subjected to photon irradiation and may be considered, for example, in the study of heterogeneous chemistry in stratospheric media.

Acknowledgements

The authors acknowledge funding by the Deutsche Forschungsgemeinschaft (through Sfb 450) and the German Israeli Foundation (GIF).

References

1 T. R. J. Tuttle and S. Golden, *J. Phys. Chem.*, 1991, **95**, 5725.
2 A. L. Sobolewski and W. Domcke, *Phys. Chem. Chem. Phys.*, 2002, **4**, 4.
3 C. A. Kraus, *J. Am. Chem. Soc.*, 1908, **30**, 1323.
4 E. J. Hart and J. W. Boag, *J. Am. Chem. Soc.*, 1962, **84**, 4090.
5 J. V. Coe, G. H. Lee, J. G. Eaton, S. T. Arnold, H. W. Sarkas, K. H. Bowen, C. Ludewigt, H. Haberland and D. R. Worsnop, *J. Chem. Phys.*, 1990, **92**, 3980–3982.
6 J. R. R. Verlet, A. E. Bragg, A. Kammrath, O. Cheshnovsky and D. M. Neumark, *Science*, 2005, **307**, 93.
7 C. Gahl, U. Bovensiepen, C. Frischkorn and M. Wolf, *Phys. Rev. Lett.*, 2002, **89**, 107402.
8 L. Turi, W.-S. Sheu and P. J. Rossky, *Science*, 2005, **309**, 914.
9 J. Stähler, M. Mehlhorn, U. Bovensiepen, M. Meyer, D. O. Kusmierek, K. Morgenstern and M. Wolf, *Phys. Rev. Lett.*, 2007, **98**, 206105.
10 F. H. Long, H. Lu and K. B. Eisenthal, *Phys. Rev. Lett.*, 1990, **64**, 1469.
11 C. Silva, P. K. Walhout, K. Yokoyama and P. F. Barbara, *Phys. Rev. Lett.*, 1998, **80**, 1086.
12 M. F. Emde, A. Batuska, A. Kummrow, M. S. Pshenichnikov and D. A. Wiersma, *Phys. Rev. Lett.*, 1998, **80**, 4645.
13 R. Laenen, T. Roth and A. Lauberau, *Phys. Rev. Lett.*, 2000, **85**, 50.
14 A. E. Bragg, J. R. R. Verlet, A. Kammrath, O. Cheshnovsky and D. M. Neumark, *Science*, 2004, **306**, 669.
15 R. N. Barnett, U. Landman and A. Nitzan, *Phys. Rev. Lett.*, 1989, **62**, 106.
16 U. Bovensiepen, *Prog. Surf. Sci.*, 2005, **78**, 87–100.
17 J. Stähler, C. Gahl, U. Bovensiepen and M. Wolf, *J. Phys. Chem. B*, 2006, **110**, 9637–9644.
18 M. Meyer, J. Stähler, D. O. Kusmierek, M. Wolf and U. Bovensiepen, *Phys. Chem. Chem. Phys*, **10**, 4932–4938.
19 C. Gahl, U. Bovensiepen, C. Frischkorn, K. Morgenstern, K.-H. Rieder and M. Wolf, *Surf. Sci.*, 2003, **108**, 532–535.
20 Q. B. Lu and T. E. Madey, *J. Chem. Phys.*, 1999, **111**, 2861–2864.
21 Q. B. Lu and L. Sanche, *Phys. Rev. Lett.*, 2001, **87**, 078501.
22 S. Ryu, J. Chang, H. Kwon and S. K. Kim, *J. Am. Chem. Soc.*, 2006, **128**, 3500–3501.
23 M. Mehlhorn and K. Morgenstern, *Phys. Rev. Lett.*, 2007, **99**, 246101.
24 U. Bovensiepen, C. Gahl, J. Stähler, M. Bockstedte, M. Meyer, X.-Y. Zhu, A. Rubio and M. Wolf, *J. Chem. Phys. C*, submitted.
25 S. Lacombe, K. Jacobi, M. N. Hedhili, Y. L. Coat, R. Azria and M. Tronc, *Phys. Rev. Lett.*, 1997, **79**, 1146.
26 R. Balog and E. Illenberger, *Phys. Rev. Lett.*, 2003, **91**, 213201.
27 S. Funk, M. Bonn, D. N. Denzler, C. Hess, M. Wolf and G. Ertl, *J. Chem. Phys.*, 2000, **112**, 9888–9897.
28 R. S. Smith and B. D. Kay, *Surf. Rev. Lett.*, 1997, **4**, 781–797.
29 G. Held and D. Menzel, *Surf. Sci.*, 1995, **327**, 301–320.
30 G. A. Kimmel, N. G. Petrik, Z. Dohnalek and B. D. Kay, *Phys. Rev. Lett.*, 2005, **95**, 166102.
31 S. Haq and A. Hodgson, *J. Phys. Chem. C*, 2007, **111**, 5946–5953.
32 U. Bovensiepen, C. Gahl, J. Stähler and M. Wolf, *Surf. Sci.*, 2005, **584**, 90–97.

 This journal is © The Royal Society of Chemistry 2009

33 J. Stähler, PhD thesis, Freie Universität, Berlin, 2007, http://www.diss.fu-berlin.de/2007/479/.
34 M. Wolf, E. Knoesel and T. Hertel, *Phys. Rev. B*, 1996, **54**, R5295.
35 S. F. Rowland, *Annu. Rev. Phys. Chem.*, 1991, **42**, 731–768.
36 M. J. Prather and R. T. Watson, *Nature*, 1990, **344**, 729–734.
37 A. R. Ravishankara, *Science*, 1997, **276**, 1058–1065.
38 K. Aflatooni and P. D. Burrow, *Int. J. Mass Spectrom.*, 2001, **205**, 149–161.
39 D. Klar, M. W. Ruf, I. Fabrikant and H. Hotop, *J. Phys. B*, 2001, **34**, 3855–3878.
40 C. D. Finch, R. Parthasarathy, H. C. Akpati, P. Nordlander and F. B. Dunning, *J. Chem. Phys.*, 1997, **106**, 9594–9601.
41 T. Oster, A. Kuhn and E. Illenberger, *Int. J. Mass Spectrom. Ion Processes*, 1989, **89**, 1–72.
42 M. N. Hedhili, M. Lachgar, Y. Le Coat, R. Azria, M. Tronc, Q. B. Lu and T. E. Madey, *J. Chem. Phys.*, 2001, **114**, 1844–1850.
43 H. Sambe, D. E. Ramaker, L. Parenteau and L. Sanche, *Phys. Rev. Lett.*, 1987, **59**, 505–508.
44 Q. B. Lu and L. Sanche, *J. Chem. Phys.*, 2004, **120**, 2434–2438.
45 A. Zangwill, *Physics at Surfaces*, Cambridge University Press, Cambridge, 1988.
46 C.-P. Chai Kao, M. E. Paulaitis, G. A. Sweany and M. Yokozeki, *Fluid Phase Equilib.*, 1995, **108**, 27–46.
47 S. A. Clough, Y. Beers, G. P. Klein and L. S. Rothman, *J. Chem. Phys.*, 1973, **59**, 2254–2259.
48 C. C. Perry, N. S. Faradzhev, D. H. Fairbrother and T. E. Madey, *Int. Rev. Phys. Chem.*, 2004, **23**, 289–340.
49 L. Sanche and M. Deschênes, *Phys. Rev. Lett.*, 1988, **61**, 2096–2098.
50 G. J. Schulz, *Rev. Mod. Phys.*, 1973, **45**, 423–486.
51 T. D. Märk, K. Leiter, W. Ritter and A. Stamatovic, *Int. J. Mass Spectrom. Ion Processes*, 1986, **74**, 265–279.
52 R. Azria, Y. Le Coat, M. Lachgar, M. Tronc, L. Parenteau and L. Sanche, *Surf. Sci. Lett.*, 1999, **436**, L671–L676.
53 C. Gahl, PhD thesis, Freie Universität, Berlin, 2004, http://www.diss.fu-berlin.de/2004/338/.
54 Y. Chen, Y.-S. Luo, J. M. Seo and J. H. Weaver, *Phys. Rev. B*, 1991, **43**, 4527–4530.
55 J. M. Seo, S. E. Harvey, Y. Chen and J. H. Weaver, *Phys. Rev. B*, 1991, **43**, 11893–11902.

General discussion

Professor Luzar opened the discussion of the paper by Dr Patel and Mr Willard: Several years ago we carried out a rather similar study,[1,2] at that time the first that quantitatively and systematically examined the effect of chemical inhomogeneity (surface heterogeneity) on water film density fluctuations (local interface fluctuation amplitudes). We used almost identical coarse-grained models. We further explicitly computed the evaporation rates from between heterogeneous surfaces, a rate-determining factor in kinetics of the assembly of such surfaces. Our conclusions are similar to yours: "as the geometry of the patchy hydrophobic surface considerably affects the mean behavior of the adjacent water, it also significantly affects the interfacial fluctuations and therefore strongly influences the kinetics of assembly". We explicitly put forward the quantitative results in our kinetic studies of water expulsion, to corroborate our conclusions. It is of interest to summarize here some of our findings that complement yours and thus could give a more complete picture of the subject. Following the same notion that virtually all surfaces are chemically heterogeneous, we altered surface properties by varying (a) its texture and (b) its size/area.

With respect to its texture we made our weakly attractive hydrophobic surfaces sprinkled with various fractions of hydrophilic patches of different sizes, and different coverages (a low one, useful for materials science and tuning of surface properties; and high one, useful for biology where coverage is prescribed in advance). At constant surface coverage we found the larger the hydrophilic patch size, the easier it was to find voids in the liquid. Larger voids resulted in larger local vapor film thickness. One of the interesting aspects we considered was a distinction between random and ordered surface patterns. We found that for random distribution of patches there was a finite fraction of voids of above average size always present, in contrast to regular distribution, with no fluctuations of void size. This contributed to evaporation rates (expulsion of water between such surfaces) up to 2 orders of magnitude faster in a random case. The result may be significant because the compositional heterogeneity of surfaces in biological systems can be regularly patterned. Apart from connecting our results to protein assembly (aggregation), our results also provide guidelines for efficient manipulation of surface properties in materials science, *e.g.* how to prevent adhesion with little change in surface composition. In order to drastically slow down, or even prevent, water evaporation from confining surfaces, a very low hydrophilic coverage is sufficient provided the patches are small and evenly distributed.

1 A. Luzar and K. Leung, *J. Chem. Phys.*, 2000, **113**, 5836.
2 K. Leung and A. Luzar, *J. Chem. Phys.*, 2000, **113**, 5845.

Mr Willard answered: The mesoscopic physics of water near hydrophobic surfaces is captured to a significant extent by that of an Ising or lattice gas field. This is the perspective that underlies the Lum–Chandler–Weeks theory of hydrophobicity;[1,2] it is with this perspective that the Chandler group has used the Ising model to study de-wetting and related phenomena. Further, recent work in the Chandler group has verified that this perspective is consistent with the coarse graining of atomistic simulations.[3] Your 1997 paper with Lum[4] reports on a part of that effort, and your subsequent work with Leung[5,6] is similarly motivated, as is a host of other works from the Chandler group including our contribution to this discussion.[2,7–11] All these papers bear similar intellectual roots, as well as distinctions. In the case of our current contribution,[11] we have considered the lattice model influenced by a single surface while your papers considered the model confined by two surfaces. We consider the effect of surface heterogeneity on the spatially resolved mean interface height, whereas your papers consider the effect of heterogeneities on the surface-averaged mean interface height. We measure the free-energy profile associated with drying the entire heterogeneous surface, whereas your work measured the free energy to nucleate a vapor tube bridging the two heterogeneous surfaces. Finally, we measured the spatially resolved free energy associated with interface displacements of different magnitudes, whereas your papers measure the mean squared fluctuations in the vapor film thickness averaged over the surface heterogeneity.

1 K. Lum, D. Chandler and J. D. Weeks, *J. Phys. Chem. B*, 1999, **103**, 4570.
2 P. R. tenWolde, S. X. Sun and D. Chandler, *Phys. Rev. E*, 2001, **65**, 011201.
3 T. F. Miller, E. Vanden-Eijnden and D. Chandler, *Proc. Natl. Acad. Sci. U. S. A.*, 2007, **104**, 14559.
4 K. Lum and A. Luzar, *Phys. Rev. E*, 1997, **56**, R6283.
5 A. Luzar and K. Leung, *J. Chem. Phys.*, 2000, **113**, 5836.
6 K. Leung and A. Luzar, *J. Chem. Phys.*, 2000, **113**, 5845.
7 P. R. tenWolde and D. Chandler, *Proc. Natl. Acad. Sci. U. S. A.*, 2002, **99**, 6539.
8 L. Maibaum and D. Chandler, *J. Phys. Chem. B*, 2003, **107**, 1189.
9 T. A. McCormick, *Phys. Rev. E*, 2003, **68**, 061601.
10 A. P. Willard and D. Chandler, *J. Phys. Chem. B*, 2008, **112**, 6187.
11 A. P. Willard and D. Chandler, *Faraday Discuss.*, 2009, DOI: 10.1039/b805786a.

Mr Willard went on to say: In adopting an Ising or lattice gas model to treat the mesoscopic physics of water adjacent to a hydrophobic surface, we implicitly assume that the water/oil interface is much like that of a water/vapor interface. In response to queries requesting evidence in support of this assumption, let me describe a series of atomistic simulation results I have recently obtained.

The system I have considered consists of two different kinds of hydrophobic solutes: an extended planar solute and a spherical solute with a radius of 0.5 nm. These solutes are fixed in space and solvated by a slab of SPC/E water molecules,

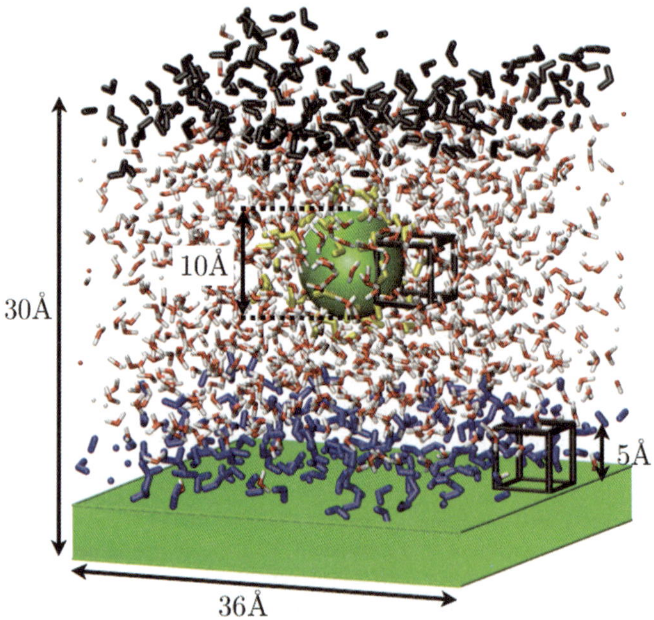

Fig. 1 A snapshot of the molecular dynamics simulation used in the analysis of the solute/solvent interface. The planar solute (bottom) and the spherical solute (middle) are colored green. Water molecules are colored red/white to indicate their belonging to the bulk liquid or are colored solid black, blue, or yellow to indicate their belonging to the liquid/vapor, planar solute/solvent or spherical solute/solvent interface respectively. The cubic boxes represent the volumes within which $P_v(N)$ was computed.

 This journal is © The Royal Society of Chemistry 2009

see for example Fig. 1. The upper boundary of the system is set so that the overall density of the system is well within the two-phase region of the temperature–density phase diagram so that there is a fluctuating liquid/vapor interface at the top of the solvent slab. The solutes interact with the oxygen centers of water molecules through the potential shown in Fig. 2, where the parameter λ sets the magnitude of the dispersive type solute–solvent attractions.

First consider the average solvent density profile as a function of the distance z or r from the center of the planar solute (which is located at the $z = 0$ plane) or spherical solute respectively (see Fig. 3). I find that for the case of the small length-scale

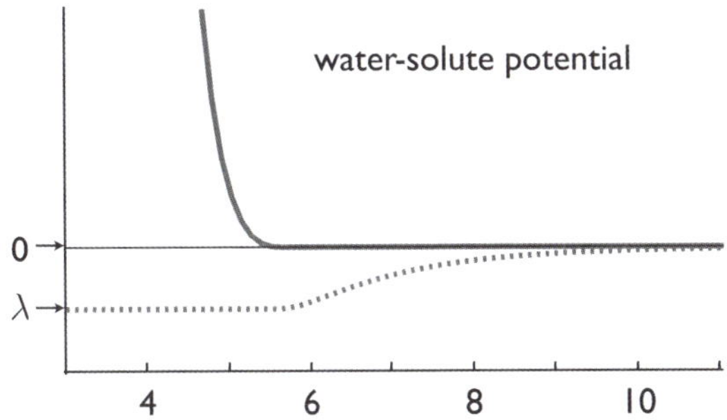

Fig. 2 The tunable solute–solvent interaction potential used in the atomistic simulations. The potential is the sum of the repulsive (solid red) and attractive (dotted blue) branches of the Lennard-Jones potential, the latter of which is parameterized by λ.

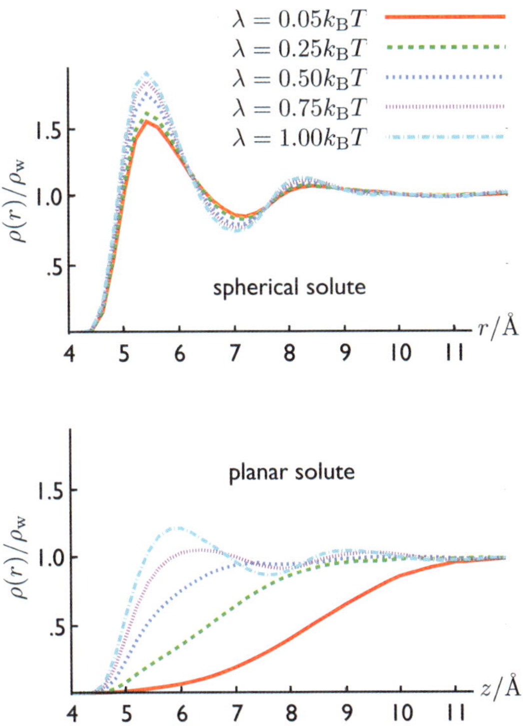

Fig. 3 The mean solvent density profile at a distance r or z from the center of the spherical or planar solute respectively. The different lines correspond to different values of the strength of the solute–solvent adhesive interactions. Density profiles are measured relative to the density of the bulk liquid, ρ_w.

spherical solute the solvent structure shows a distinct layering associated with the strain imposed on the local hydrogen bonding network by the presence of the solute. For the case of the large length-scale hydrophobic solute, at small values of λ when the planar solute is most like an ideal hydrophobic surface there is a pronounced depletion layer. When the solute–solvent attraction is increased the width of the depletion layer narrows and the solvent density profile develops layering. This layering and considerable narrowing of the depletion layer is often interpreted as evidence that the solute/solvent interface is not liquid/vapor-like.

Next consider the density fluctuations adjacent to either solute, as measured by the probability $P_v(N)$ of finding the centers of N water molecules within the $5 \times 5 \times 5$ Å cubic subvolume v. I find that the density fluctuations are bulk-like for all but the nearly ideal hydrophobic solute (see Fig. 4). If one were to stop looking here it would be easy to conclude that the solute/solvent interface is unlike that of an ordinary liquid/vapor interface, but the picture changes as one digs deeper.

Now consider the distribution of interface heights, as measured from the center of the solute to the center of water molecules considered to be members of the solute/solvent interface. As shown in Fig. 5 the height fluctuations of the planar solute/solvent interface are liquid/vapor-like when λ is small but have different behavior around the mean when λ is made larger. What is significant in Fig. 5 is that even for larger λ, the fluctuations in the wings of the distributions continue to be characteristic of a liquid/vapor interface. For the small length-scale spherical solute, fluctuations of the interface to larger values of h come at a large free energy cost. Finally, the orientational distribution of water molecules belonging to the planar solute/solvent interface (as measured by the angle θ_b of the OH bond vector relative to the solute surface) are indistinguishable from that of a liquid/vapor interface when λ is small. The features of the distribution narrow as λ is increased (see Fig. 6). The qualitative picture I draw from both the distribution of interface heights (Fig. 5) and the distributions of interfacial molecules (Fig. 6) is that for small values of there exists a liquid/vapor-like interface whose entropic fluctuations result in a layer of depleted solvent density next to a large hydrophobic solute. When λ is increased the entropic 'capillary wave' fluctuations are damped away in favor of the enthalpic adhesive interactions with the solute surface. In this case the interface is pulled up against the solute and spatial fluctuations usually associated with a liquid/vapor

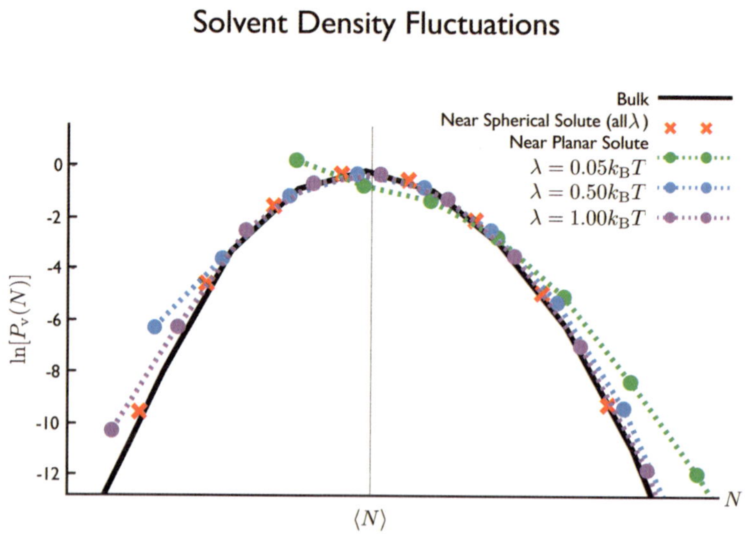

Fig. 4 The probability distribution, $P_v(N)$, for finding the centers of N water molecules in a $5 \times 5 \times 5$ Å cubic subvolume v, positioned adjacent to the surface of the solute.

 This journal is © The Royal Society of Chemistry 2009

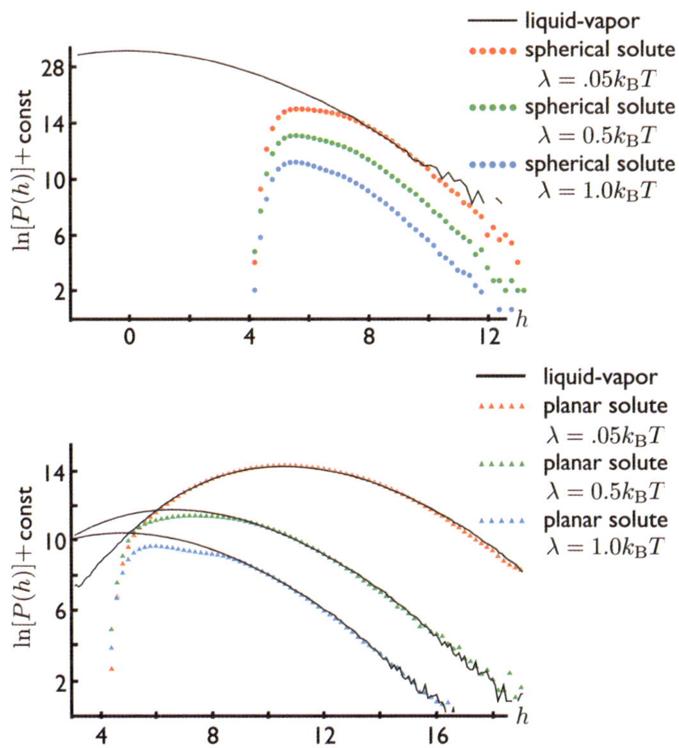

Fig. 5 The probability distribution for finding the center of a water molecule a distance h from the center of a solute. The solid line corresponds to the height distribution of molecules belonging to the liquid/vapor interface as measured from an arbitrary reference plane.

interface are not easily accessed. This implies typical density fluctuations near the solute are bulk-like and do not occur through translation of the localized interface. The intrinsic structure of the liquid/vapor interface is not changed by the addition of these adhesive forces as is evident in the distributions of molecular orientations. Specifically, the peaks in orientational distribution functions of molecules at the planar solute/solvent interface correspond to hydrogen bonds made in the plane of the interface ($\cos\theta_b \approx 0$) and into the bulk liquid ($\cos\theta_b \approx 1$). These peaks sharpen as λ is increased which is consistent with the idea that the effect of the adhesive interactions is to damp out the capillary wave fluctuations that are responsible for the width of the free interface. With a more localized interface the distribution of interface orientations relative to the planar solute narrow.

Dr Patel remarked: With further reference to questions asking for evidence that water near hydrophobic surfaces behaves like water near a water/vapor interface, I would like to add the following:

The ease with which assembly proceeds depends directly on the ease with which water molecules can be vacated from the region between two assembling hydrophobic surfaces.[1–3] The probability of density depletion near a hydrophobic surface is like that near a water/vapor interface and significantly larger than that in bulk, thus prompting the analogy. In order to highlight this point, I would like to describe some results from a simulation study of water near idealized hydrophilic and hydrophobic surfaces: In Fig. 7, a snapshot of the system that is simulated is shown. An idealized hydrophilic surface is constructed with fixed water molecules, whose positions were taken from a typical bulk water configuration. The corresponding

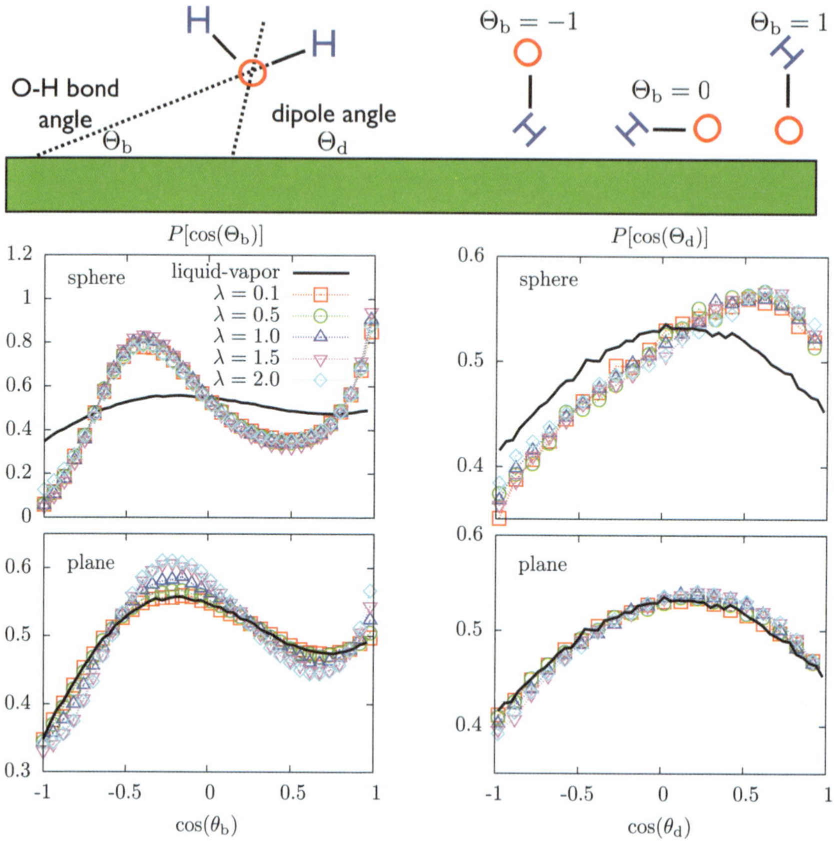

Fig. 6 The distribution of OH bond angles, *b*, and dipole angles, *d*, for interfacial water molecules. Solid lines correspond to the distribution of orientations of water molecules in the liquid/vapor interface.

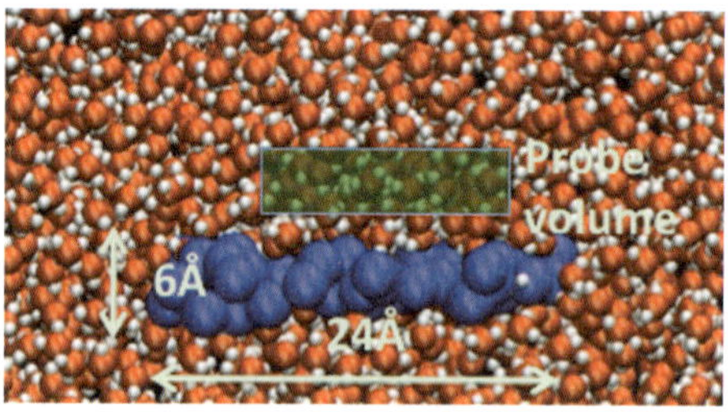

Fig. 7 System

hydrophobic surface is made by replacing the fixed waters with oily particles that interact with water *via* a Lennard-Jones potential, typical of alkanes. The quantity of interest is the particle number distribution, $P_v(N)$ of water molecules, within a probe volume, v, of dimensions $3 \times 12.3 \times 12.3$ Å near the surface being considered.

In Fig. 8, we show $P_v(N)$ for a probe volume near a hydrophobic surface, near a hydrophilic surface, near a water/vapor interface and in the bulk. The $P_v(N)$ for the probe volume near a hydrophilic surface is nearly the same as that for the bulk $P_v(N)$. However, near the hydrophobic surface, it is significantly easier to empty out the probe volume than it is in the bulk. This is manifest in the higher

 This journal is © The Royal Society of Chemistry 2009

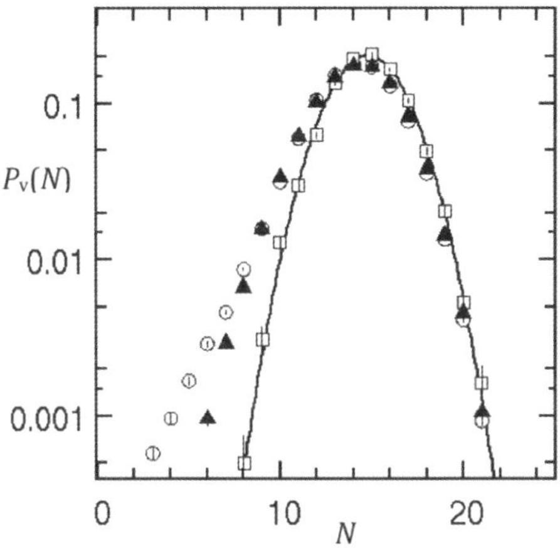

Fig. 8 Particle number distribution of water molecules in a probe volume near a hydrophobic surface (filled triangles), a hydrophilic surface (open squares), in the bulk (solid line) and near a water/vapor interface (open circles).

values of $P_v(N)$ as $N \to 0$ for the hydrophobic surface. This is akin to what is observed for the probe volume near a water/vapor interface, where the effect is even more pronounced.

Furthermore, I would like to invoke evidence found in the numerous simulation studies of water confined by hydrophobic surfaces.[4,5] These studies have observed bistability (or bimodal distributions for $P_v(N)$) with respect to the density of water in the confined region. The theory of Lum, Chandler and Weeks[6] also predicts bistability.[7] This bistability is a demonstration of coexistence between a vapor-like phase and a liquid-like phase. Thus, the water in the confined region has a character which is like that of interfacial water at liquid-vapor coexistence.

1 D. Chandler, *Nature*, 2005, **437**, 640.
2 P. R. tenWolde and D. Chandler, *Proc. Natl. Acad. Sci. U. S. A.*, 2002, **99**, 6539.
3 A. P. Willard and D. Chandler, *J. Phys. Chem. B*, 2008, **112**, 6187.
4 X. Huang, C. J. Margulis and B. J. Berne, *Proc. Natl. Acad. Sci. U. S. A.*, 2003, **100**, 11953.
5 N. Choudhury, *J. Phys. Chem. B*, 2008, **112**, 6296.
6 K. Lum, D. Chandler and J. D. Weeks, *J. Phys. Chem. B*, 1999, **103**, 4570.
7 K. Lum, PhD thesis, University of California, Berkeley, 1998.

Mr Willard then commented: Further results derived from the atomistic simulations described earlier are plotted in Fig. 9. These graphs show the free energy to empty a 5 × 5 × 5 Å cubic subvolume as a function of the distance z or r between the nearest edge of the cube and the $z = 0$ center of the planar solute or center of the spherical solute respectively. The parameter λ sets the depth of the attractive well in the solute/solvent potential which is shown in Fig. 2 (above). Looking first at the planar solute, the free energy cost to solvate a cubic subvolume is greater in the bulk liquid than it is adjacent to the planar solute. This is true even at values of λ large enough to induce a layering in the local solvent density (see Fig. 3 above). At low values of λ, the lowering of the solvation free energy near the planar solute arises because the density fluctuations associated with emptying the cubic subvolume are dominated by the fluctuations of the liquid/vapor-like interface induced at the planar solute surface. Inhibiting the fluctuations in this interface, as increasing λ does, thus effects the solvation free energy of the subvolume in the vicinity of the planar solute. Notice, however, that even at the larger values of λ the planar solute exerts a thermodynamic attractive force on small hydrophobic solutes. The range of this effective attraction is determined by the amplitude of interfacial fluctuations. For the case of the spherical solute, the solvation free energy of the cubic subvolume

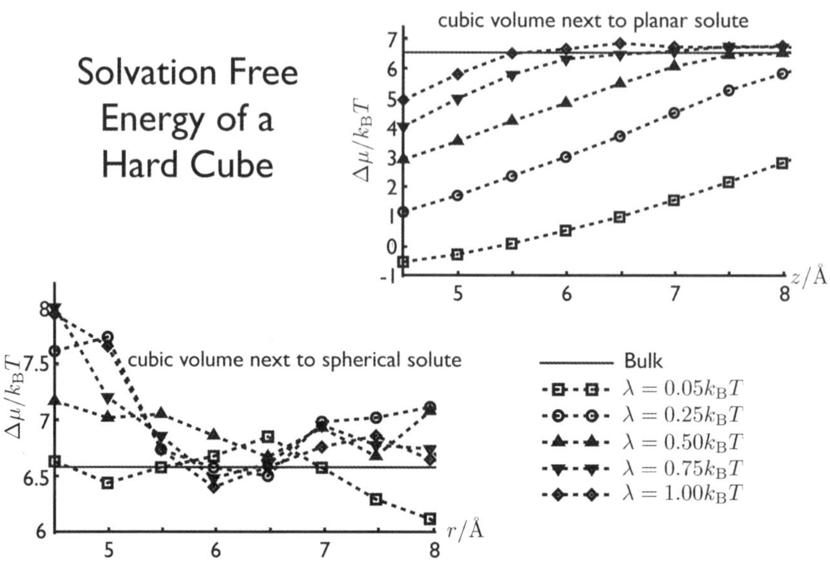

Fig. 9 The free energy $\Delta\mu$ to empty a 5 × 5 × 5 Å volume with its nearest surface a distance z or r from an extended planar or spherical solute with radius 5 Å respectively. The different lines for the cubic subvolume near a planar or spherical solute refer to different values of the solute–solvent attractive potential λ. The solid black line indicates the value of $\Delta\mu$ where the volume is in bulk liquid far from the solute.

is larger very near the solute than it is in the bulk. Furthermore, the ordering of the solvent induced by the presence of the spherical solute (see Fig. 3 in my previous comment) is reflected in the spatial dependence of the solvation free energies. These curves, which reflect the nature of the solvent density fluctuations in the vicinity of the solutes serve to highlight the difference between small and large length-scale hydrophobic solvation. They also demonstrate that the behavior of the mean solvent density does not necessarily reflect the nature of the fluctuations.

Dr Paulaitis addressed Mr Willard and Dr Patel: The conclusion I draw from this paper is that your model in its current state has limited applicability to the hydration of protein surfaces. What direct validation of the model do you anticipate in order to describe hydrated protein systems of experimental interest? The following observations lead to this conclusion and motivate my question.

First, you state that the fraction of hydrophilic patches, f, ranges from 0.25 to 0.50 for proteins, and you show results in Fig. 6 of your paper for $f = 0.25$. However, your earlier analysis of depletion, which you characterize by the mean height of the vapor/liquid interface, $\langle h \rangle$, over a heterogeneous surface, considers much lower values of f ~0.01–0.05. It appears therefore that your earlier analysis of depletion does not have specific relevance to the hydration of protein surfaces. Note, in particular, the results in Fig. 1 of your paper for hydrophilic patch sizes, d, between $2l$ and $3l$ that correspond to the range of patch sizes you identify as appropriate for protein surfaces. Here $\langle h \rangle$ is on the order of 2 Å for $f = 0.05$ when there are no surface adhesive interactions—*i.e.* the mean height of the vapor/liquid interface is already less than the diameter of a water molecule at this f value. Increasing f to 0.25 to obtain the lowest f value applicable to protein surfaces, would reduce $\langle h \rangle$ even further. Thus, your model predicts a depletion layer adjacent to protein surfaces that is at most a small fraction of the diameter of a water molecule for the reasonable values of f and d that characterize the proteins considered. Further, since the formation of a larger depletion layer is central to your proposed mechanism of water-mediated assembly of proteins, it does not appear that your model can describe the role of hydration in protein assembly.

 This journal is © The Royal Society of Chemistry 2009

Second, including the surface adhesion interaction reduces $\langle h \rangle$ even further, as shown in Fig. 1 in your paper. Specifically, $\langle h \rangle$ is reduced to less than 1 Å for f = 0.05 when considering a surface adhesive interaction consistent with that found experimentally for water/oil interfaces. The surface adhesive interaction you used for the water/protein interface is not specified, and it is not obvious how one would estimate a reasonable value for this parameter in your model when the interface is both chemically heterogeneous and molecularly rough. Could a reasonable estimate of the surface adhesion interaction be obtained, for example, from experimentally determined osmotic second virial coefficients of globular proteins? Would you expect typical values so obtained to be significantly larger than the value corresponding to water/oil interfaces that you used in your earlier analysis of depletion over a heterogeneous surface?

Mr Willard answered: In response to your first observation, your claim that our proposed mechanism of collapse depends on the development of a large depletion layer is only true of instantaneous configurations. As you note, the quantities plotted in our Fig. 1 and 2 demonstrate very small depletion layers, but these are time-averaged quantities. Fig. 5 shows that although the time averaged interface height is very small, the spatially resolved fluctuations in the height of the interface is appreciable and depends significantly on size and relative position of the underlying heterogeneities. In Fig. 6 we demonstrate that the above analysis holds true even at more biologically relevant values of the hydrophilic coverage f.

In response to your second observation, chemical details of the solvent and the surface at length scales smaller than the coarse graining length are irrelevant to the phenomena we treat, except to the extent that these details determine the values of the interaction parameters in our model. It would be useful to have a prescription to determine these parameters starting from the chemical details of the system one wishes to treat. This prescription is not provided in our paper, as your comment correctly notes. It is possible that the ideas you put forth in your two questions would lead to a satisfactory prescription.

Professor Luzar commented: This is a comment with regard to Dr Patel's remark about using John Weeks' local molecular theory (LMF)[1] on proteins in water. LFM theory has been very successfully applied[2] to nonuniform Lennard-Jones fluids,[3,4] to ionic solutions,[5–7] and very recently to bulk and confined water as described by the SPC/E model.[8] Results for solvation, dielectric and other properties of water are very promising. However, another study by the same group[9] makes the point that all charges should be treated equally if one wants to get that nice result of a Poisson's equation defined with a smoothed charge distribution.[10] Therefore, when having a macroion (*e.g.* colloid or protein) in solution using this method will become much more challenging.

1 J. D. Weeks, *Annu. Rev. Phys. Chem.*, 2002, **53**, 533.
2 For a very recent review see, for example, B. J. Berne, J. D. Weeks and R. Zhou, *Annu. Rev. Phys. Chem.*, 2009, **60**, 85.
3 J. D. Weeks, K. Katsov and K. Vollmayr, *Phys. Rev. Lett.*, 1998, **81**, 4400.
4 K. Katsov and J. D. Weeks, *J. Phys. Chem. B*, 2001, **105**, 6738.
5 Y. G. Chem, C. Kaur and J. D. Weeks, *J. Phys. Chem. B*, 2004, **108**, 19874.
6 Y. G. Chen and J. D. Weeks, *Proc. Natl. Acad. Sci. U. S. A.*, 103, 7560-7565 (2006).
7 J. M. Rodgers, C. Kaur, Y. G. Chen and J. D. Weeks, *Phys. Rev. Lett.*, 2006, **97**, 097801.
8 J. M. Rodgers and J. D. Weeks, *Proc. Natl. Acad. Sci. U. S. A.*, 2008, in press.
9 J. M. Rodgers and J. D. Weeks, *J. Phys.: Condens. Matter*, 2008, **20**, 494206.
10 J. D. Weeks, personal communication, 2008.

Dr Feibelman opened the discussion of the paper by Dr Salmeron: Does wetting = layer by layer growth? Granting that energy must be expended to flip down-pointing OH bonds of the first water adlayer on mica to an up-pointing geometry that (according to Salmeron) allows bonds to form a second layer of H_2O molecules, what physics underlies the extra binding energy relative to H-bonds in bulk ice? If there is no extra binding, should we not attribute layer by layer growth, if it occurs, to kinetics rather than thermodynamics?

Dr Salmeron responded: Good question. Wetting is a macroscopic concept (the spreading of a macroscopic amount of liquid deposited on a surface). Below the vapor saturation pressure, the formation of water multilayers observed in our *in situ* experiments on oxides, implies that the forces between first and second and between second and third *etc.* are larger (albeit progressively smaller) than in bulk ice or water. The extra binding must originate from film–substrate interactions, be it short range forces of the first layer with the substrate, and from the first to the second due to the modified structure of the first, *etc.* with decreasing strength, or from long range forces such as van der Waals and electrostatic forces. Understanding these interactions is the key to a proper microscopic description of wetting.

As to the physics underlying the flipping of downward oriented first-layer water molecules, on mica or on metal substrates such as Pt(111), this is more complicated. Still, we know that water molecules at the liquid surface exhibit dangling H-bonds. So, if the surfaces are wetting the transition from H-down to H-up should occur at the early stages of growth on hydrophilic substrates.

Assuming that this flipping costs some energy, thermodynamic equilibrium might not be reached in vacuum at low temperatures, and the structures observed there might be determined by kinetics.

Professor Hodgson also answered: Requiring layer by layer growth seems too restrictive to me, albeit this is the usual definition of solid on solid wetting. It is a precondition for a surface to wet that water should preferentially form a stable first monolayer, otherwise the surface will be hydrophobic and prefer to form 3D clusters. However, at least on a metal surface, I can't envisage how bonding of the first to a second water layer is likely be more favourable than within crystalline ice. Optimisation of the water–metal bonding substantially distorts the wetting monolayer from the ideal bulk ice bilayer structure, preventing layer by layer growth of bulk ice on top of this monolayer. Certainly on all the metal surfaces we have looked at so far the ice multilayer does not grow layer by layer, but either by forming non-wetting clusters on top of the first layer if this is tightly bound (*e.g.* on Ru(0001), see ref. 1) or else by reconstructing the first layer to allow an incommensurate bulk ice like film to form (*e.g.* on Ni(111), see ref. 2, and on Pt(111), see ref. 3). This behaviour is determined by the energy cost of reconstructing the monolayer to form the metal/bulk ice interface, something that appears to be relatively easy on Ni(111) but requires formation of thicker ice clusters on Pt(111) (as discussed by Thürmer and Bartelt[4,5]).

1 S. Haq and A. Hodgson, *J. Phys. Chem. C*, 2007, **111**, 5946.
2 M. Gallagher, A. Omer, S. Haq and A. Hodgson, *Surf. Sci.*, 2007, **601**, 268.
3 G. Zimbitas, S. Haq and A. Hodgson, *J. Chem. Phys.*, 2005, **123**, 174701.
4 K. Thürmer and N. C. Bartelt, *Phys. Rev. B*, 2008, **77**, 195425.
5 K. Thürmer and N. C. Bartelt, *Phys. Rev. Lett.*, 2008, **100**, 186101.

Professor Klein continued the discussion of the paper by Dr Patel and Mr Willard: Could your approach make definite predictions about the surface forces expected between two such heterogeneous surfaces across water? We have carried out measurements between charge-patched surfaces across water (*e.g.* Perkin *et al.*[1,2]) and these could be extended to hydrophobic/hydrophilic patches.

I would also like to comment that there has been some very recent discussion concerning recent results on depletion of water near hydrophobic surfaces cited in your manuscript, and their re-interpretation, see *e.g.* Ocko *et al.*[3]

1 S. Perkin, N. Kampf and J. Klein, *J. Phys. Chem.*, 2005, **109**, 3832.
2 S. Perkin, N. Kampf and J. Klein, *Phys. Rev. Lett.*, 2006, **96**, 038301.
3 B. M. Ocko, A. Dhinojwala and J. Daillant, *Phys. Rev. Lett.*, 2008, **101**, 039601.

 This journal is © The Royal Society of Chemistry 2009

Mr Willard replied: In response to your question, the answer is yes. Closely related models have been used to study the solvent mediated forces and dynamics between different manifolds.[1,2] With regard to Ocko *et al.*'s comment on the work of Granick and co-workers (ref. 4 in our paper), I note that Granick and co-workers have also responded to the comment.[3]

1 P. R. tenWolde and D. Chandler, *Proc. Natl. Acad. Sci. U. S. A.*, 2002, **99**, 6539.
2 A. P. Willard and D. Chandler, *J. Phys. Chem. B*, 2008, **112**, 6187.
3 A. Poynor, L. Hong, I. K. Robinson, S. Granick, P. A. Fenter and Z. Zhang, *Phys. Rev. Lett.*, 2008, **101**, 039602.

Professor Lynden-Bell said: Would it be useful to consider the density depletion near a hydrophobic surface in terms of free energy changes? Consider a slab of liquid next to a solid hydrophilic surface. Its density is enhanced near the surface by attraction to the solid and by packing effects. Now turn down the solid–liquid attraction to change the surface into a hydrophobic one. Then the liquid surface can increase its entropy at little cost in energy by moving away from the solid so that its fluctuations increase. If the loss of solid–liquid attractive energy is smaller than $T\Delta S$, the liquid will move away from the solid with a resulting depletion of energy.

Mr Willard answered: I think this is a good way to think about water at hydrophobic interfaces. When confronted with an extended ideal hydrophobic surface, liquid water forms what is akin to an ordinary liquid/vapor interface. Entropic fluctuations of that interface lead to a blurring-out of the solvent density profile near the solute. This is the origin of the depletion layer. Even small dispersive-type attractions

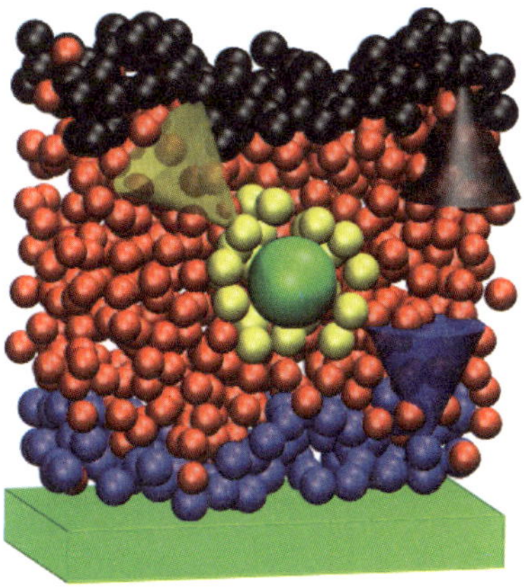

Fig. 10 A snapshot taken from an atomistic simulation where the front half of the molecules have been omitted for clarity. The atomistic system consitst of a slab of SPC/E water molecules that is approximately 3.6 × 3.6 × 3.6 nm and periodically replicated in the x and y Cartesian directions. The liquid water solvates an extended planar solute at the lower z boundary and a spherical solute with radius 0.5 nm which is fixed in space near the center of the solvent slab. The solutes interact with the solvent through a potential $w_\lambda(r)$, where r is the distance of the oxygen center of a water molecule from either the center of the spherical solute or the $z = 0$ center of the planar solute. The potential $w_\lambda(r)$ is the sum of Weeks–Chandler–Andersen[1] repulsive and attractive branches of the Lennard-Jones potential, the latter of which is scaled by the parameter λ. There are two different hydrophobic solutes (green), an extended planar solute (bottom of image) and a spherical solute with radius 0.5 nm (middle of image). The smaller spheres correspond to the positions of the centers of oxygen atoms and are colored to indicate their belonging to either the bulk liquid (red), the planar solute interface (blue), the spherical solute interface (yellow) or the liquid/vapor interface (black). The constrained radial distribution functions in Fig. 11 are computed for the centers of oxygen atoms which are a distance r from an interfacial water molecule and within the cones pictured above. Each cone extends approximately 25° from its normal.

between the hydrophobic surface and water can overcome the entropically-driven spatial fluctuations in the liquid/vapor-like interface, pulling it against the substrate surface. With the spatial fluctuations in the interface damped-out by enthalpic dispersive attractions a wide depletion layer does not develop, but the intrinsic structure of the liquid/vapor-like interface persists. As such, one expects that the solvent properties which are inherent to the structure of the interface to be independent of the magnitude of the dispersive attractions.

For example, I have considered an atomistic model and computed a constrained radial distribution function which measures the mean solvent density seen by interfacial water molecules in the direction of the bulk liquid (see Fig. 10). I find that for water at the planar solute interface, as well as for water near an ordinary liquid/vapor interface, the solvent structure propagating from the interface into the bulk

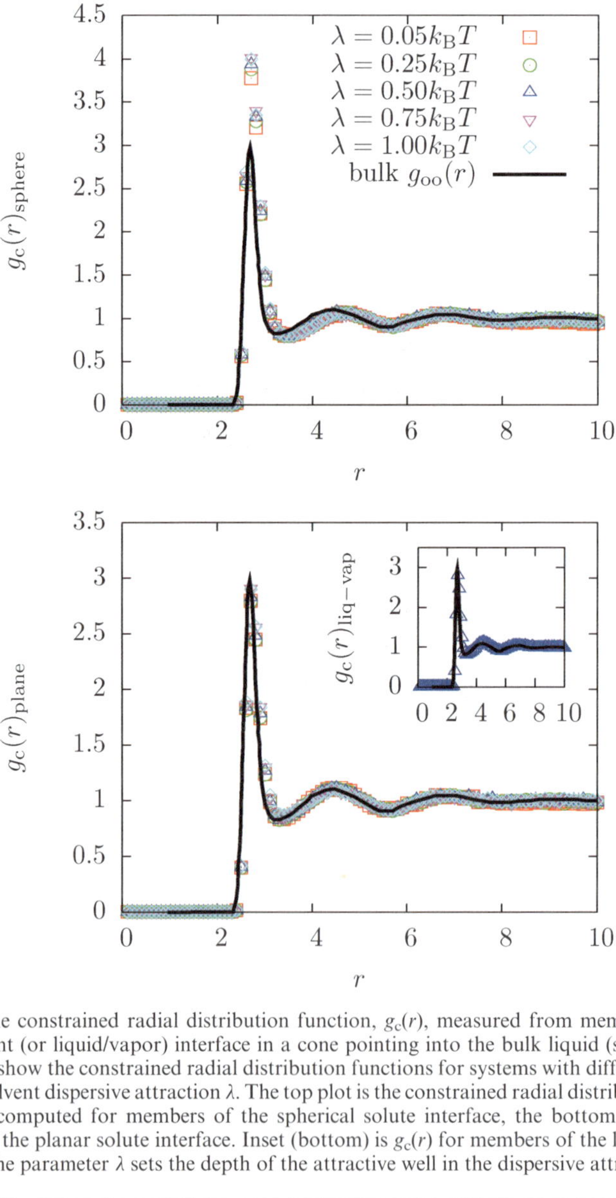

Fig. 11 The constrained radial distribution function, $g_c(r)$, measured from members of the solute/solvent (or liquid/vapor) interface in a cone pointing into the bulk liquid (see Fig. 10). The graphs show the constrained radial distribution functions for systems with different values of solute–solvent dispersive attraction λ. The top plot is the constrained radial distribution function, $g_c(r)$, computed for members of the spherical solute interface, the bottom is $g_c(r)$ for members of the planar solute interface. Inset (bottom) is $g_c(r)$ for members of the liquid/vapor interface. The parameter λ sets the depth of the attractive well in the dispersive attractions.

 This journal is © The Royal Society of Chemistry 2009

is the same as that for bulk liquid (see Fig. 11). This result remains true as I vary the strength of the solute–solvent dispersive attractions over a range that begins with a nearly ideal hydrophobic surface (demonstrating a pronounced depletion layer) and ends with a hydrophobic surface with dispersive attractions strong enough to induce a layering of the adjacent solvent density profile. The same is not true, however, for the case of a small spherical hydrophobic solute. In this case the solute/solvent interface is not like that of a water/vapor interface, but rather consists of water molecules engaged in a strained hydrogen bond network encompassing the solute. One interesting fact gleaned from the inset of Fig. 11 is that the intrinsic width of the free interface is roughly one molecular diameter, which in turn is roughly the bulk correlation length of the liquid.

1 J. D. Weeks, D. Chandler and H. C. Andersen, *J. Chem. Phys.*, 1971, **54**, 5237.

Professor Jungwirth asked: Can you include into your model a varying rigidity/softness of the hydrophobic wall as this has been shown to be an important factor for interfacial water structuring (see simulations by Vácha *et al.*[1])?

1 R. Vácha, R. Zangi, J. B. F. N. Engberts and P. Jungwirth, *J. Phys. Chem. C*, 2008, **112**, 7689.

Mr Willard responded: The model used in our paper is coarse-grained. Short length-scale details such as those you refer to are not treated in the model. The general perspective from which our model is built is that of Lum, Chandler and Weeks, where net liquid density field is an Ising or lattice gas field plus a Gaussian field.[1,2] The Gaussian field would account for the effects your refer to.

1 P. R. tenWolde, S. X. Sun and D. Chandler, *Phys. Rev. E*, 2001, **65**, 011201.
2 K. Lum, D. Chandler and J. D. Weeks, *J. Phys. Chem. B*, 1999, **103**, 4570.

Dr Christenson continued the discussion of the paper by Dr Salmeron: Any discussion of water adsorption to mica must take into consideration the K^+ ions that are present on air-cleaved mica. On average there is one K^+ per 0.5 nm^2 so the mean separation between these ions is only 7 Å! The K^+ may be exchanged for H^+ by dipping the

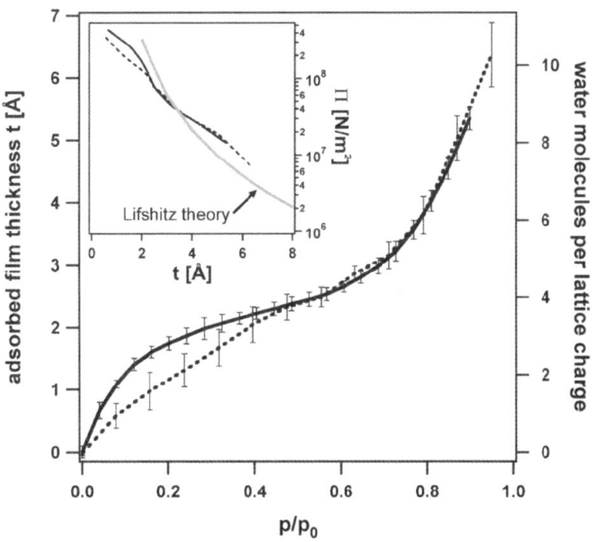

Fig. 12 Adsorption isotherms of water on mica with K^+ on the surface (solid lines) and on mica with H^+ on the surface (dashed lines). The temperature was 22.7 °C. Reproduced with permission from *Langmuir*, 2008, **24**, 1566.[1] Copyright 2008 American Chemical Society.

mica in acidic solution and this changes the surface properties. The water vapour adsorption isotherm changes in the submonolayer regime (see Fig. 12 from Balmer *et al.*[1]), the properties of water films on mica and water condensed between mica surfaces becomes very different. Water films on air-cleaved mica behave very much as films of a concentrated salt solution. By contrast, experiments with H^+ mica should give a much better idea of the wetting behaviour *etc.* of water on a smooth oxide surface.

1 T. E. Balmer, H. K. Christenson, N. D. Spencer and M. Heuberger, *Langmuir*, 2008, **24**, 1566.

Dr Salmeron replied: I agree completely. We have shown spectroscopically in the past (by XPS) that the ions on mica can indeed be easily exchanged by other ions in solution.[1]

1 L. Xu and M. Salmeron, *Langmuir*, 1998, **14**(20), 5841.

Miss Van Der Niet opened the discussion of the paper by Professor Hodgson: In the experimental section of the paper you state that the required amount of D_2O to form a saturated ice monolayer was determined using the King and Wells technique. I was wondering how this was done in practice, since I thought that the sticking probability for water (at least on platinum) is approximately 1 for both the monolayer as well as the multilayer.

Professor Hodgson answered: The accommodation coefficient for water trapping on top of either the monolayer or multilayer ice is certainly roughly unity, but this precursor must then find a stable binding site or else it will desorb. For a particular temperature the net adsorption rate depends on the availability and stability of binding sites that stabilise this precursor. Since the monolayer is significantly more stable than the multilayer, it has a higher desorption temperature and lower desorption rate than the multilayer. This difference is clear when following the net uptake from the molecular beam as a function of temperature and coverage using the King and Wells technique (see for example Fig. 2 in ref. 1). To determine the saturation coverage of the first monolayer one simply increases the surface temperature until adsorption terminates with completion of the monolayer, the multilayer being kinetically unstable to desorption. This temperature depends on the flux of water and the thermal stability of the multilayer ice. In some cases, for example on Pt(111),[1] multilayer nucleation requires a critical precursor density and there is a pronounced minimum in the adsorption probability as the monolayer completes (rather than just a permanent decrease as we go from monolayer to multilayer), this is caused by kinetic competition between desorption and stabilisation of the precursor at multilayer clusters. In this case the uptake of water increases again as multilayer clusters nucleate, providing favourable adsorption sites,[1] although growth on Ru(0001) does not show any obvious inhibition of multilayer nucleation.[2] The uptake determined from the adsorption measurements can also be cross checked against the coverage dependence of the temperature-programmed desorption (TPD) traces to confirm saturation of the monolayer peak.[3] Typical cross agreement is about 3%, a far better precision than can be achieved by any other approach. Of course an absolute coverage requires calibration of the beam flux against some ordered structure; we use the highly ordered (3×3) ($OH + H_2O$) layer on Pt(111)[4] as an absolute calibration or, as a less accurate relative reference, the ordered $(\sqrt{3} \times \sqrt{3})R30^\circ$ water layer on Ru(0001).[2]

1 S. Haq, J. Harnett and A. Hodgson, *Surf. Sci.*, 2002, **505**, 171.
2 S. Haq and A. Hodgson, *J. Phys. Chem. C*, 2007, **111**, 5946.
3 C. Clay, S. Haq and A. Hodgson, *Chem. Phys. Lett.*, 2004, **388**, 89.
4 C. Clay, S. Haq and A. Hodgson, *Phys. Rev. Lett.*, 2004, **92**, 046102.

Professor Debenedetti continued the discussion of the paper by Dr Patel and Mr Willard: In atomistic simulations (*e.g.* Giovambattista *et al.*[1]) one finds non-trivial dependence of "local" hydrophobicity (as measured by the ability to exclude water from the first layer on the surface) on the size and geometry of patches. The rules that govern this behaviour are still incompletely understood. In a coarse-grained model such as you presented, you seem to have already washed out such effects.

1 N. Giovambattista, P. Debenedetti and P. J. Rossky, *J. Phys. Chem. C*, 2007, **111**, 1323.

Mr Willard answered: Professor Debenedetti's comment possibly suggests that the coarse-grained Ising field used in our modeling cannot capture the essential features of results of atomistic modeling that he and his co-workers have reported.[1] This suggestion is not correct, as I illustrate with Fig. 13 presented here. The atomistic modeling considered a large bath of water with two parallel planar 9 nm^2 surfaces separated by various distances, with various compositions, and focused on water adjacent to the inner surfaces of the plates. The representative case illustrated in the Figure uses unhydroxylated SiO_2 groups for hydrophobic units and hydroxylated SiO_2 groups for hydrophilic units, each on a rectangular lattice with lattice spacing of about 0.2 nm. To capture the essential elements of such surfaces with a coarse grained model I have used a square lattice, also with lattice spacing of about 0.2 nm, treating surface sites around modeled SiO_2 as being vapor-like for hydrophobic units or being liquid-like for hydrophilic units, and assigning a volume excluding energetic penalty to the top sites of the SiO_2 groups and a uniform adhesive interaction to all other surface sites. Specifically, in addition to the energetics of the lattice gas solvent (eqn (1) in our paper), the surface couples to the adjacent lattice gas through the energy function,

$$E_{\mathrm{surf}}[\{n_i\},\{\sigma_i\}] = \sum_{i=1}^{N_x \times N_y} -\varepsilon n_i \sigma_i - \mu_i n_i \,,$$

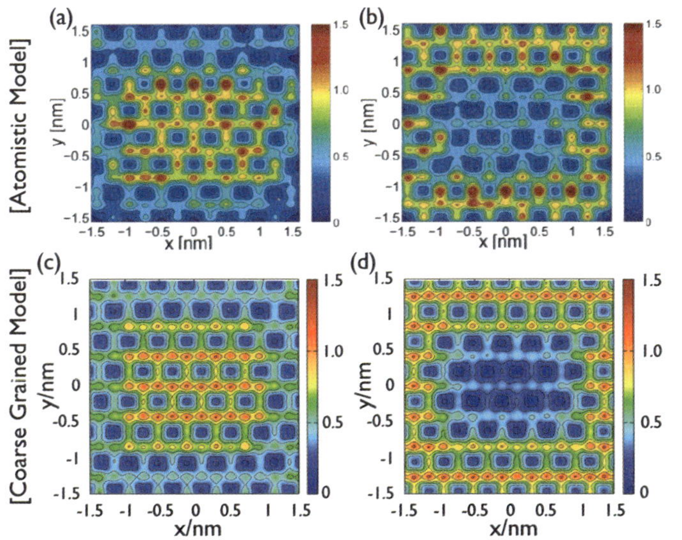

Fig. 13 The spatially resolved surface density of water molecules adjacent to the inner surface of a hydrophilic plate with hydrophobic borders (left) or a hydrophobic plate with hydrophilic borders (right). The upper panels (a) and (b) are taken from Fig. 3 of ref. 1 by Debenedetti and co-workers (reprinted with permission from *J. Phys. Chem. C*, 2007, **111**, 1323; copyright 2007 American Chemical Society) and corresponds to the surface density of water molecules in a 0.266 nm slab next to the surface. The bottom panels (c) and (d) correspond to the average surface density of the lattice gas solvent in a 0.21 nm slab adjacent to the coarse grained surface. Contour lines are linear in $\rho(x,y)$.

where the sum is over all $N_x \times N_y$ lattice sites adjacent to the surface, the n_i values are the occupation numbers of the lattice gas adjacent to the surface, the σ_i values specify the surface disorder ($\sigma_i = 0$ for hydrophobic sites or 1 for hydrophilic sites), and μ_i is the potential field of the surface layer which has a value of $\mu_i = -4.4k_BT$ for sites adjacent to the volume excluding SiO_2 groups or $\mu_i = 0.8k_BT$ for all other sites reflecting the uniform surface–solvent adhesive interactions. The Figure compares the equilibrium-averaged solvent density obtained from a Monte Carlo simulation of this lattice model with the average density reported for the atomistic modeling. Adopting the same convention used in the atomistic modeling[1] and pictured in panels (a) and (b) of Fig. 13, the density distribution for panels (c) and (d) have been made continuous by means of Gaussian interpolation, so that the density field pictured in (c) and (d) is

$$\rho(x,y) = \sum_{i=1}^{N_x \times N_y} \bar{n}_i e^{(-x-x_i)^2/2\sigma_x} e^{(-y-y_i)^2/2\sigma_y}$$

where $\bar{n}_i$ is the time-averaged occupation number in cell i, and $\sigma_x = \sigma_y = 0.8 \times (0.21$ nm/2). One notices some differences, primarily due to evident statistical errors in the atomistic simulation results. Nevertheless, except for details on length scales smaller than the coarse-graining length, 0.21 nm, the atomistic modeling and the lattice modeling give essentially the same results.

1 N. Giovambattista, P. Debenedetti and P. J. Rossky, *J. Phys. Chem. C*, 2007, **111**, 1323.

Dr Perkin asked: Have you considered the effect of ions in the water adjacent to the patchy surface? Firstly, how do the ions (and their hydration) behave adjacent to the hydrophobic patch compared to the hydrophilic patch? Secondly, I would be interested to know about the structure of the electrical double layer—for example if the hydrophilic patch was negatively charged with a diffuse layer of cations—when the surface has hydrophobic patches. Are there fluctuations in the electrical double layer, relevant to the longer range interactions between heterogeneous surfaces?

Mr Willard responded: It is possible to include the effects of ions on the coarse-grained solvent implicitly by reparameterizing the interaction parameters of the lattice gas to account for the change in surface tension, *etc.* A complete answer to the questions you pose, however, would require augmenting our coarse-grained treatment with shorter length-scale detail. A modified or extended version of a method that exists for providing this detail[1] might be applied to your questions.

1 P. R. tenWolde, S. X. Sun and D. Chandler, *Phys. Rev. E*, 2001, **65**, 011201.

Professor Klein continued the discussion of the paper by Dr Salmeron: Concerning the issue of contamination of freshly-cleaved mica, of course it can influence matters. Other than for cleavage carried out in UHV, it has been shown by careful dynamic SIMS that a layer of a few Ångstroms of a water–carbon complex forms very rapidly on the mica surface in ambient air; this dissolves under water, but care must be taken to ensure—*via* suitable controls—that it has in fact been removed. See *e.g.* Perkin *et al.*,[1] where it is shown in mica surface force balance experiments, that—even within a minute of cleaving—mica is contaminated by some 5 Å of this ambient-adsorbed layer, but that in water this may be removed. Unfortunately, AFM and other scanning-probe methods cannot measure absolute surface separations, and so cannot reveal the presence or absence of this layer. Often, therefore, AFM-type studies measure 'solid-like water' behaviour when what is in fact being measured are the properties of this contaminant layer; this can be particularly true when capillary condensation occurs about the probe tip, as that 'sucks up' contamination from the surrounding areas about the tip.

1 S. Perkin, L. Chai, N. Kampf, U. Raviv, W. Briscoe, I. Dunlop, S. Titmuss, M. Seo, E. Kumacheva and J. Klein, *Langmuir*, 2006, **22**, 6142.

Dr Christenson replied: I completely agree with Professor Klein regarding the danger of attributing unusual properties to water condensates studied with an AFM. I think it highly likely that a great many of these studies are actually of fairly concentrated solutions, not pure water. With the SFA it is easier to tell when artefacts due to contamination are involved.

Dr Held continued the discussion of the paper by Professor Hodgson: In recent experiments electron and X-ray stimulated dissociation of water on Ru(0001) has been quantified[1,2] and cross sections between 10^{-16} and 10^{-15} cm^2 were found. This means that less than 1 electron per molecule is necessary to convert an intact water layer into a partially dissociated one.

Our earlier LEED-IV analysis of the ($\sqrt{3} \times \sqrt{3}$) structure of water on Ru(0001)[3] was carried out with electron doses of 3–9 electrons per molecule at 150 K. It is now undisputed that under these conditions partial dissociation of the water layer would definitely have occurred (in contrast to the assumptions in our article, which were based on the spectroscopic evidence available at the time of the experiments).

The finding, reported in the paper by Gallagher *et al.*,[4] that the new IV-curves agree with those from our earlier experiments is therefore evidence that the surface layer under investigation was also partially dissociated. If this dissociation was induced by the electron beam on a very small fraction of the surface, one would not necessarily detect any significant change in TPD. Although there was a second *R*-factor minimum for a buckled arrangement of oxygen atoms in the ($\sqrt{3} \times \sqrt{3}$) structure reported in our article of 1994,[3] the coplanar arrangement of the oxygen atoms has been confirmed in a later, more precise study.[5] (It should be noted that LEED can only determine the position of heavy atoms like oxygen and ruthenium unambiguously and can, therefore, not make a reliable distinction between hydroxyl and water molecules.)

I would also like to ask a question regarding diffraction spots: the crucial argument, which is not explicitly mentioned in the discussion of whether sharp diffraction spots can be expected (section 3.4, pp. 14–15 of Gallagher *et al.*[4]), is the assumption of a long-range periodic ($\sqrt{3} \times \sqrt{3}$) vacancy structure (Ru atoms which are not covered by a water molecule, neither 'up' nor 'down'). If complete randomness would be allowed (*i.e.* if each site atop a Ru surface atom could either be unoccupied or occupied by an 'up' or 'down' molecule with equal probability) only a sharp (1×1) diffraction pattern should be observed. What is the justification for assuming this long-range order of vacancy sites?

The same arguments as above would also apply if only very small domains (<20 Å or so) of local order existed.[6] In this case, however, broad maxima in *k*-space should be observed near the points corresponding to the local periodicity (*i.e.* ($2\sqrt{3} \times 2\sqrt{3}$)). The calculations presented in your paper[4] would suggest that such short-range correlations with periodicities larger than ($\sqrt{3} \times \sqrt{3}$) exist, as the chain structures are clearly lower in energy than an ice-like structure. Given the high mobility of water molecules even at 100 K, the lowest free energy chain structure should dominate on the surface. To my knowledge, however, no extra maxima have been observed, neither in LEED nor in HAS (see also Haq *et al.*[7]) for low-temperature water structures on Ru(0001).

It is also not obvious why the arguments in the first part of section 3.4 in your paper[4] do not apply to He diffraction. The derivation uses single-scattering theory, which is—strictly speaking—not applicable to structure factors in LEED, but should describe X-ray and He diffraction very well. Compared to 'up' or 'down' water molecules vacancy sites should represent a large variation in the scattering potential for He atoms. Therefore, a long-range periodicity of vacancy sites should lead to sharp diffraction spots in HAS following the same arguments as I have outlined here.

1 N. S. Faradzhev, K. L. Kostov, P. Feulner, T. E. Madey and D. Menzel, *Chem. Phys. Lett.*, 2005, **415**, 165.
2 Gladys *et al.*, unpublished results.
3 G. Held and D. Menzel, *Surf. Sci.*, 1994, **316**, 92.
4 M. Gallagher, A. Omer, G. R. Darling and A. Hodgson, *Faraday Discuss.*, 2009, DOI: 10.1039/b807809b.
5 S. R. Puisto, T. J. Lerotholi, G. Held and D. Menzel, *Surf. Rev. Lett.*, 2003, **10**, 487.
6 J. B. Pendry, *Low Energy Electron Diffraction*, Academic Press, London, 1974, pp. 228–236.
7 S. Haq, C. Clay, G. R. Darling, G. Zimbitas and A. Hodgson, *Phys. Rev. B*, 2006, **73**, 115414.

Professor Hodgson replied: With regard to your comment on LEED IV, since it is not so long since you published a re-analysis of your original D_2O LEED IV data and concluded that the water layer was intact,[1] it is ironic to hear that you would now take similar LEED IV data as "evidence that the surface layer under investigation was also partially dissociated". Certainly the conditions used in your original experiment make it likely that there was some dissociation, but that does not constitute evidence that the observation of a similar LEED IV profile implies the film is dissociated.

Having originally demonstrated that electron damage to the intact film creates the LEED patterns you observed for H_2O,[2] we are well aware of the electron damage issues. However, we have shown that TPD provides a very simple, definitive quantitative test for dissociation of D_2O,[3] and have used this to check that the LEED IV data we have taken for D_2O is free from dissociation (we work with a LEED spot size of *ca.* 0.5 mm and a beam dosed spot of just 3 mm, allowing us to raster across the spot and see any damage, if it occurs). We can also deliberately monitor the effect of dissociation on the LEED IV behaviour, this is minimal but does reduce the structure observed at high energy to make it closer to that observed in your study. Thus however much damage occurred in your original study, we are convinced the LEED IV data we have recorded is consistent with intact D_2O adsorption/desorption. A consideration of the intact and partially dissociated structures found in calculations perhaps suggests why the IV curves do not change very much between the two structures. O has the same adsorption site (atop Ru), the same long-range $\sqrt{3}$ order, while we argue that only the lower O atoms will appear in the IV, the large Debye–Waller factor of the upper O atoms making these essentially invisible (see section 3.4 of our paper). In fact all we would expect is a small change in the O–Ru distance. Hopefully an IV analysis that takes into account the disorder in the O height with location may provide some resolution to this issue.

As for diffraction spots, firstly the simple answer is that hydrogen bonding provides the long-range registry in the O position (see Fig. 4–7 in our paper), and hence the vacancy sites. In order for each water in the 0.67 ML structure to form three hydrogen bonds we end up with a hexagonal honeycomb network of water molecules. The key is to understand the evidence available for this model. Low current LEED experiments show that long-range $\sqrt{3}$ order is observed at, and only at, 0.67 ML coverage.[2,4] DFT calculations find the minimum energy structure for this coverage has the water arranged in chains of flat and H-down water within this network. It is this long-range periodicity in the O adsorption sites that creates the sharp diffraction peaks. To disrupt this arrangement and form a (1×1) diffraction pattern would require water molecules to break hydrogen bonds, something that is energetically unfavourable. This registry is also reflected in the distortion of the top layer Ru, which contributes to the diffraction intensity as discussed in paragraph 3 of section 3.4 in our paper. This long-range registry in the O location is only observed near 0.67 ML. At lower coverage water prefers to form small

 This journal is © The Royal Society of Chemistry 2009

clusters of limited size, allowing more of the water molecules to lie flat, optimising their bonding to the metal. As a result the long-range order is lost and the LEED spots broaden (see Fig. 2 and 5 in ref. 2). At higher coverage the water layer compresses to incorporate more water next to the metal. The monolayer eventually saturates with a coverage of ~0.75 ML, possibly by the formation of domain boundaries,[4] having lost the long-range $\sqrt{3}$ order. So to describe this as the "assumption of a long-range periodic ($\sqrt{3} \times \sqrt{3}$) vacancy structure" misses the point, the evidence for long-range order in the O location for a 0.67 ML structure comes from LEED while DFT calculations show that this is caused by hydrogen bonding between the water molecules that order the water into flat chains within a 2D honeycomb network. The issue is whether water necessarily has a simple ($\sqrt{3} \times \sqrt{3}$) unit cell, which we argue it doesn't, invalidating the original LEED IV analysis.[1,5]

Secondly, you are trying to impose some locally ordered superstructure on the water that simply does not exist. The model for this structure is of a long-range ($\sqrt{3} \times \sqrt{3}$) honeycomb network of water locations, atop Ru, but with no ordering of the proton positions or the O height, just a preference to form chains of flat water. Just because we choose to model the surface using a DFT slab calculation (which has a periodic form) does not imply that any specific local order will be present in practice; in fact the evidence is quite the reverse. We find that chains and branched chains all have similar energy, provided we make the chains of flat water long and avoid the rigid ice bilayer alternation in O height. So there is nothing special about the $2\sqrt{3}$ repeat, or any other; we can model the structure using any unit cell consistent with the ($\sqrt{3} \times \sqrt{3}$) repeat in the O location. The surface is genuinely disordered—so there will be no extra spots of the type suggested. The only periodicity that survives this averaging is the location of the O atoms and, as we argue in the paper, this gives rise to sharp spots irrespective of the local correlation in height. In principle the intensity of the $\sqrt{3}$ LEED spots will carry information on the local correlation in O height, although whether it is possible to distinguish this above the disorder is not clear. As we argue in the paper, the LEED intensity will come primarily from the O atoms of flat water molecules and the Ru atoms beneath them which are displaced out of the surface, the upper O atoms having a large mean square displacement.

To answer your final point, you have to understand the very different nature of electron and He scattering. Whereas low energy electrons scatter off the electron density of the top few layers of atoms (*i.e.* off the full 3D electron distribution of the O atoms and top few Ru atoms), He atoms scatter from several ångstrom above the surface, sampling only the outermost tail of the surface electron density. The electron density here is not simply dependent on the identity of the atom beneath that particular point, it is also sensitive to this atom's local environment. Ironically this is perhaps most obvious in the vacancy site that you suggest will form a regular array of minima and hence give a well-defined diffraction feature. For example, the vertical turning point directly over the vacant sites for the structure shown in our Fig. 5 is calculated to vary by ~0.9 Å, almost as large as the corrugation in the O height! As a result there is no well-defined periodicity to the He scattering potential. The scattering potential is highly corrugated, giving rise to large angle scattering with diffuse elastic scattering peaks observed at large Δk. In contrast, electron scattering in LEED, with its relatively long coherence length and deep penetration into the surface, picks out the long-range order in the location of the Ru atoms and O atoms of flat lying water within this disordered structure.

Finally it is misleading to suggest that single scattering should describe He diffraction very well, but not LEED. Single scattering only describes He scattering in the limit of low corrugation, which is certainly not the case here. In fact the key point is that the surface shows large angle scattering, consistent with the high corrugation expected for an intact water structure and not the flat O network implied by your interpretation of the LEED IV data. With regard to the discussion of which spots will appear in LEED, a single scattering analysis tells us definitively which LEED spots will appear in the diffraction pattern, multiple scattering effects can change the intensity of spots but cannot cause them to disappear (except at some particular energy).

1 S. R. Puisto, T. J. Lerotholi, G. Held and D. Menzel, *Surf. Rev. Lett.*, 2003, **10**, 487.
2 S. Haq, C. Clay, G. R. Darling, G. Zimbitas and A. Hodgson, *Phys. Rev. B*, 2006, **73**, 115414.
3 C. Clay, S. Haq and A. Hodgson, *Chem. Phys. Lett.*, 2004, **388**, 89.
4 D. L. Doering and T. E. Madey, *Surf. Sci.*, 1982, **123**, 305.
5 G. Held and D. Menzel, *Surf. Sci.*, 1994, **316**, 92.

Dr Feibelman said: Note that He atoms see the electron density not nuclei. Therefore mostly you are looking at O atoms not H atoms.

Professor Hodgson responded: He atom scattering probes low electron density regions several ångstroms above the surface, so it is sensitive to whatever influences this. I agree that for the H-down water chain structures we find on Ru(0001), where the H atoms of the upper water molecules are only slightly higher than the O atoms, the H atom electron density is barely visible in the scattering potential. In this case we are primarily sensitive to the disorder in the repulsive wall induced by the O height and location. It is a different matter if uncoordinated H atoms point out from the surface as free OH.

Dr Salmeron commented: The build up of contamination can be followed spectroscopically, which we have done using XPS *in situ*[1] and with SFG (for the water on mica system). Contamination in the form of hydrocarbons is fast and prominent in "dry" mica (following C–H vibrational stretch modes). On mica cleaned under high water humidity, contamination builds up much more slowly than in dry atmospheres. Contamination can also be "soft" and "hard" using our nomenclature. "Soft" means that water can easily displace the layer of weakly bound contaminants.[2] "Hard" means strongly bound contaminates not removable by water. The important point is that contamination can be followed spectroscopically and its effects assessed quantitatively.

1 H. Bluhm, D. F. Ogletree, C. S. Fadley, Z. Hussain and M. Salmeron, *J. Phys.: Condens. Matter*, 2002, **14**, L227.
2 L. Xu, A. Lio, J. Hu, D. F. Ogletree and M. Salmeron, *J. Phys. Chem. B*, 1998, **102**, 540.

Dr Nutt continued the discussion of the paper by Dr Salmeron: Another example of a closely-related surface which might be of interest is $BaF_2(111)$, with a hexagonal lattice parameter similar to that of ice. George Ewing's experiments showed that, after initial growth of thin film water around 273 K under ambient pressure, the water layer appeared to dewet, forming isolated islands of ice.[1,2] Subsequent modelling suggested that this was due to the orientation of the initial water monolayer on the surface, which would lead to the growth of a proton-ordered film in the first instance, which then presumably dewets to form proton-disordered ice, perhaps localised at defect sites.[2,3]

This can be contrasted with Langmuir films of long-chain alcohols which have OH groups available for hydrogen-bonding at the water interface.[4] The C_{31} alcohol ($C_{31}H_{63}OH$) is a good ice-nucleator, whereas the C_{30} alcohol ($C_{30}H_{61}OH$) is not so good. A potential explanation is that the C_{30} hydroxyl groups are not oriented in a favourable way for the nucleation and stabilisation of ice. This suggests that the availability of –OH for hydrogen bonding is not a sufficient condition for stable ice growth.

1 V. Sadtchenko, P. Conrad and G. E. Ewing, *J. Chem. Phys.*, 2002, **116**, 4293.
2 V. Sadtchenko, G. E. Ewing, D. R. Nutt and A. J. Stone, *Langmuir*, 2002, **18**, 4632.
3 D. R. Nutt and A. J. Stone, *J. Chem. Phys.*, 2002, **117**, 800.
4 R. Popovitz-Biro, J. L. Wang, J. Majewski, E. Shavit, L. Leiserowitz and M. Lahav, *J. Am. Chem. Soc.*, 1994, **116**, 1179.

 This journal is © The Royal Society of Chemistry 2009

Dr Salmeron responded: The explanation maybe that the first water layer is autophobic, in the same line that in the case of kaolinite. See the paper by Hu and Michaelides[1] who studied water adsorption on the two polar faces of kaolinite. They report on the ability of the H-terminated surface of this material to bind water, but equally interesting is the finding that this water monolayer is autophobic and subsequent growth is not wetting. The reason is that the first water layer is too strongly bound and water cannot re-orient to present H bonds to the second and subsequent layers. We contend the same in our paper on water on Ru(0001) covered by an O(2×2) layer.[2,3]

1 X. L. Hu and A. Michaelides, *Surf. Sci.*, 2008, **602**, 960.
2 P. Cabrera-Sanfelix, D. Sánchez-Portal, A. Mugarza, T. K. Shimizu, M. Salmeron and A. Arnau, *Phys. Rev. B*, 2007, **76**, 205438.
3 A. Mugarza, T. K. Shimizu, P. Cabrera-Sanfelix, D. Sánchez-Portal, A. Arnau and M. Salmeron, *J. Phys. Chem. C*, 2008, **112**(36), 14052.

Professor Dr von Klitzing asked: With respect to the results at Cu (110) surfaces, is it possible to measure the potential at the surface after dissociation of H_2O—is there a change?

Dr Salmeron responded: It is possible indeed, but we have not done this experiment yet.

Dr Christenson remarked: With reference to several comments on contamination of mica, I believe that three separate issues are involved: firstly, the water that adsorbs immediately on cleavage, even in a very dry atmosphere; secondly, adsorption of spurious contaminants, *e.g.* hydrocarbons from the atmosphere, and thirdly, the reaction and diffusion of potassium ions to form a concentrated salt solution or crystals, depending on the humidity. We do not know how much water adsorbs directly on cleavage, and film thicknesses derived from adsorption measurements are additional to this. As Dr Salmeron has mentioned, hydrocarbons may be removed (at least partially) by dipping in water and are not irreversibly adsorbed. The salt problem can be eliminated by acid dipping. These matters were discussed in a 1993 paper,[1] and even though Salmeron has since done much better adhesion measurements with mica, I think that the conclusions in the paper still hold.

1 H. K. Christenson, *J. Phys. Chem.*, 1993, **97**, 12034.

Dr Lützenkirchen commented: In a recent paper Shen's group[1] reported sum-frequency data on the basal plane of sapphire (α-Al_2O_3). These authors observe dangling (free) OH even in the presence of bulk water. These dangling bonds are interpreted as pertaining to surface hydroxyls. The basal plane, prepared in a similar way as the samples for SFG measurements, has been characterized previously by X-ray reflectivity measurements[2] and is therefore well-defined. The basal plane of sapphire has also been claimed to be extremely hydrophilic.[3] It is interesting now how the two observations of dangling OH (here they come from surface hydroxyls) and extreme hydrophilicity can be reconciled in the idea of wetting as described in Dr Salmeron's paper (here the dangling OH come from adsorbed water).

An alternative point of view might be that Daum's group[4] reported the sum-frequency data on the "real" sapphire basal plane or the one that would more closely correspond to Ducker's preparation (after annealing), since in their spectra on annealed samples the dangling bonds are not present. Still the presence of dangling bonds on an oxide surface in the presence of aqueous solutions are somewhat difficult to understand.

1 L. Zhang, C. Tian, G. A. Waychunas and Y. R. Shen, *J. Am. Chem. Soc.*, 2008, **130**(24), 7686.
2 P. J. Eng, T. P. Trainor, G. E. Brown, Jr, G. A. Waychunas, M. Newville, S. R. Sutton and M. L. Rivers, *Surf. Sci.*, 2000, **12**(288), 1029.

3 W. A. Ducker, Z. Xu, D. R. Clarke and J. N. Israelachvili, *J. Am. Ceram. Soc.*, 1994, **77**(2), 437.
4 B. Braunschweig, S. Eissner, W. Daum, *J. Phys. Chem. C*, 2008, **112**(6), 1751.

Dr Salmeron replied: Thanks for bringing these references up, which as you say support our model of dangling H-bonds playing a crucial role in wetting and hydrophilicity. I will add simply that even if no dangling bonds are present, as in the case of the water monolayer on mica (in our paper), it is the ability of the next water layer to "turn around" molecules in the first layer to expose their H bonds and thus continue the wetting growth that is still important in the same line of argument.

(13:226) **Dr Rao** commented: You mentioned using resonant photons for excitation using the O–H stretch frequency. Resonant photons (even if very weak in their energy) can cause dramatic effects on the 'structure' of water to the extent that they can lead to dissociation. Please refer to the dissociation of salt water using radio frequency by Roy *et al.* Did you consider any dissociation in your technique?

1 R. Roy, M. L. Rao and J. Kanzius, *Mater. Res. Innov.*, 2008, **12**, 3.

Dr Salmeron replied: Infrared spectroscopy is a highly popular, well-seasoned technique that people have been using for many years. I don't know of any report of resonant IR radiation absorption that mentions molecular dissociation effects. The power in the IR source is quite small indeed. Regarding our sum frequency generation spectra, where IR and visible light are mixed, we never observed any effect attributable to damage from the beam, like decay of the O–H stretch peaks, with time. We can safely exclude damage from that source.

Dr Feibelman continued the discussion of the paper by Professor Hodgson: Is disorder in the first H_2O layer maybe a precondition for forming a second layer by providing energetic nucleation sites?

Professor Hodgson answered: It may well help. Certainly if the first layer is highly ordered and has an unfavourable arrangement for hydrogen bonding a second layer, for example if the first layer water has all the non-hydrogen-bonded H atoms arranged in chains pointing down, there will be no suitable template to nucleate multilayer ice (see Zimbitas *et al.*[1]). Disorder in the first layer may create favourable multilayer nucleation sites, or make it less costly to locally reorient first layer water to form sites which stabilise multilayer water. Certainly we don't observe the dramatic kinetic minimum associated with second layer water nucleation on Ru(0001) (where we know the water layer is disordered) that we find on Pt(111). However, this may be due to the formation of favourable nucleation sites on Ru(0001) as the water film compresses beyond 0.67 ML and creates free OH groups,[2] rather than reflecting the intrinsic local structure of the ordered ($\sqrt{3} \times \sqrt{3}$) layer.

1 G. Zimbitas, M. E. Gallagher, G. R. Darling and A. Hodgson, *J. Chem. Phys.*, 2008, **128**, 074701.
2 S. Haq, C. Clay, G. R. Darling, G. Zimbitas and A. Hodgson, *Phys. Rev. B*, 2006, **73**, 115414.

Dr Thürmer addressed Dr Salmeron and Professor Hodgson: In the Stranski–Krastanov growth mode the film material first covers a solid substrate with one or a few monolayers before 3D clusters are formed. If the Stranski–Krastanov mode represents equilibrium, the formation of these 3D clusters raises the chemical potential; these 3D clusters are then a source of atoms or, in the case of water films, molecules that try to squeeze into the wetting layer, possibly increasing its density. We observed this in Ag/Ru(0001) films, for example. But 3D clusters could also serve as material sinks depleting the wetting layer. For example, if Stranski–Krastanov does not represent equilibrium, a metastable 2D wetting layer could first form,

 This journal is © The Royal Society of Chemistry 2009

and then, after nucleating 3D clusters, these clusters could then feed from the wetting layer, which becomes diluted or even disappears completely. Have you ever observed a 2D water layer becoming less dense upon nucleating 3D water (ice) clusters?

Dr Salmeron responded: In the case of mica we have observed the formation of 3D clusters on top of an incomplete 2D layer.[1] The 3D clusters formed at temperatures around −36 °C (in equilibrium with vapor) spread and wet after heating to higher temperature (−16 °C). Perhaps this observation can be related to your question.

1 H. Bluhm and M. Salmeron, *J. Chem. Phys.*, 1999, **111**(15), 6947.

Professor Hodgson answered: We believe we observed something similar on Ru(0001) when multilayer clusters form on top of the first layer of D_2O.[1] In this case the first layer of water compresses during growth to increase the density of water in contact with the metal beyond the 0.67 ML associated with the ($\surd 3 \times \surd 3$) structure, removing the $\surd 3$ diffraction spots in LEED. This destabilises the monolayer slightly, as evidenced by the lowered desorption temperature.[1,2] When a water multilayer is grown on top of the first layer, the monolayer $\surd 3$ spots remain visible even when there is significant coverage of multilayer water, indicating that the density of the first layer water is lower. However, because the temperature is critical for establishing this equilibrium, and also dramatically effects the cluster size distribution, we can't be entirely sure whether the true thermodynamic equilibrium is with the $\surd 3$ layer or the compressed water layer.

1 S. Haq and A. Hodgson, *J. Phys. Chem. C*, 2007, **111**, 5946.
2 S. Haq, C. Clay, G. R. Darling, G. Zimbitas and A. Hodgson, *Phys. Rev. B*, 2006, **73**, 115414.

Professor Buch opened the discussion of the paper by Dr Slater: Three points: firstly, perhaps a more complete citation of computational work on surface premelting is in order?[1] The dangling atom pattern on surface is not random—the order parameter is 2 not ~3 (see Buch *et al.*[2]). Also, what kind of thermostat was used in production runs?

1 K. Bolton and J. B. C. Petersson, *J. Phys. Chem. B*, 2000, **104**, 1590.
2 V. Buch, H. Groenzin, I. Li, M. J. Shultz and E. Tosatti, *Proc. Natl. Acad. Sci. U. S. A.*, 2008, **105**, 5969.

Dr Slater answered: In regard to the first point, this is certainly important work and we will add a citation in the proofs.

We are aware of the issue of proton ordering and our recent work[1] addresses the same problem as discussed by Buch *et al.*,[2] rationalising the proton ordering by relating surface stability to the average spatial density of protons in the outer layer. Indeed we address much of this within this paper and recognise that our surface is too disordered. Further work is needed to ascertain whether proton order influences the mechanism of pre-melting. Since the energy scale associated with ordering the protons is rather small, being ~50 meV, it is conceivable that proton order does not exert a great influence on the processes we have observed.

We used the Berendsen thermostat, which appeared in one of the footnotes in the original manuscript.

1 D. Pan, L.-M. Liu, G. A. Tribello, B. Slater, A. Michaelides and E. Wang, *Phys. Rev. Lett.*, 2008, **101**, 155703.
2 V. Buch, H. Groenzin, I. Li, M. J. Shultz and E. Tosatti, *Proc. Natl. Acad. Sci. U. S. A.*, 2008, **105**, 5969.

Professor Finney opened the discussion of the paper by Professor Vega: The ice phase diagrams you calculate generally do not cover the (p,T) range where ices

VII and VIII occur. Presumably there is a good and hopefully interesting reason for this, from which we can learn something about the potentials. Please can you comment on this?

Also, in your Table 6, you choose for each phase a single (p,T) point. (a) Why choose ambient pressure state points for ices II and IV when under-pressure data is available? (b) For most of the high pressure phases, isothermal compressibility data is available. Testing water models against such data could be particularly informative about the repulsive core especially so for the very high pressure phases such as VI, VII and VIII which all have interpenetrating networks. Such tests would be particularly interesting in the context of earlier suggestions that an isotropic repulsion[1,2] might be a significant improvement over a spherical Lennard-Jones.

1 J. L. Finney and H. F. J. Savage, *Nature*, 1986, **322**, 717.
2 H. F. J. Savage, *Water Science Reviews*, ed. F. Franks, Cambridge University Press, Cambridge, 1986, vol. 2, p. 67.

Professor Vega replied: When the phase diagram of water was computed by the first time for TIP4P we realized that the description of ices appearing at high pressures (ice VII, ice VIII) was not as good as that of the rest of the ices (Ih, II, III, IV, V, VI) which appear at lower pressures. Although TIP4P provides a qualitatively correct description of the phase diagram of water, it fails in predicting quantitatively the equation of state (EOS) of ices VII and VIII. The same is true for other water models such as SPC/E or TIP4P/2005. This suggests that all these models present a common feature that prevents a correct description of the high pressure region.

Possible reasons are the inadequate shape of the repulsive part of the Lennard-Jones potential, or the fact that these models do not present anisotropy in the repulsive part of the potential which of course is not consistent with the non-spherical shape of the electronic cloud of the molecule. It would be of interest to identify the reason of the failure since it would allow us to design changes in water models so they are also able to describe correctly the very high pressure region.

Regarding your question (a), for each ice we took one state (p,T). They were taken from the excellent book of Petrenko and Whitworth,[1] which is easy to find and with a number of references on experimental data. In summary, the selection of p,T states was taken for convenience from this book. In fact, many of the p,T states correspond to liquid nitrogen at room temperature, so that the ice is metastable under these conditions (ice Ih should be the stable phase).

It would have been possible for some of the ices (but not for all) to select an state within the region of the phase diagram where it is thermodynamically stable. Some of these points have been include in a recent paper of our group dealing explicitly with the EOS and structure of ices. In that paper we have also computed the isothermal compressibility for some of the ices (II, III), and it was found that TIP4P/2005 was able to describe in a satisfactory way the compressibility of these ices.[2]

As to your question (b), your suggestion about computing the compressibility of high pressure ices (VI or VII, VIII) and trying to learn why water models fail seems to be a very good idea. The possibility of an incorrect description of the repulsive part of the potential (either because it is too repulsive or because the model is too spherical neglecting the anisotropy of the molecule) could be tested.

1 V. F. Petrenko and R. W. Whitworth, *Physics of Ice*, Oxford University Press, Oxford, 1999.
2 E. G. Noya, C. Menduina, J. L. Aragones and C. Vega, *J. Phys. Chem. C*, 2007, **111**, 15877.

Dr Roeselova addressed all attendees: I wish to join the authors of Professor Vega's paper in their call for a simple, but reliable *polarizable* water model for simulations of the ice phase and, in particular, the ice/vapor interface. We have been carrying out molecular dynamics simulations, following the methodology of Vega *et al.*,[1] to determine the melting point of one of the most commonly used polarizable water models, POL3.[2] Our results indicate that the melting temperature of the POL3 model is between 180 and 200 K.[3] The POL3 model with its explicit treatment of polarization interaction is considered to be an improvement of the SPC/E model that only takes into account the water polarization in an effective way. However, the low melting point of the POL3 model is comparable to that of the original SPC model (T_m = 190 K) which, unlike SPC/E, does not even have the "average" polarization of the water molecule included.

1 C. Vega, M. Martin-Conde and A. Patrykiejew, *Mol. Phys.*, 2006, **104**, 3583.
2 J. W. Caldwell and P. A. Collman, *J. Phys. Chem.*, 1995, **99**, 6208.
3 E. Muchova, P. N. M. Hoang, S. Picaud and M. Roeselova, *J. Chem. Phys.*, 2008, to be submitted.

Professor Vega responded: There are very few calculations of the melting point of polarisable models of water. One is your evaluation of the melting point of POL3. Another one is the evaluation of the melting point of TIP4P-FQ which has a melting point of 303 K.[1] Further work is needed to understand why the introduction of polarisability does not necessarily improve the prediction of the melting point of ice. Taking into account that changing the parameters of TIP4P slightly to yield the improved TIP4P/2005 has required 22 years, it is clear that it may require some time, effort and imagination to propose a polarisable model which performs better than TIP4P/2005 for the ten properties considered in this work. My feeling is that such a model exists and that we should continue working in this direction. The model will come. What I do not know is how long it may take to find such a model.

1 B. F. Nicholson, P. Clancy and S. W. Rick, *J. Cryst. Growth*, 2006, **293**, 78.

Dr Roeselova continued the discussion of the paper by Dr Slater: Your paper presents an important contribution to the advancement of microscopic, molecular-level understanding of the structure and dynamics of the ice/vapor interface, which is a topic of prime significance in atmospheric science. We have recently carried out a molecular dynamics simulation[1] of an ice slab at 250 K employing the same water model (NvdE) as you used. While our main focus is on processes associated with water vapor deposition and sublimation as an essential step towards unraveling the microscopic mechanism of vapor-deposited ice crystal growth, we have also studied the structure and dynamics of the ice surface. The rather long (70 ns) trajectory gives us the opportunity to observe dynamics on a longer time scale than you reported. Here we communicate our first results.

We have used a 10-bilayer thick slab of hexagonal ice consisting of 1280 molecules, within a unit cell measuring approx. $36(x) \times 31(y) \times 100(z)$ Å^3. The thickness of the ice slab in the z dimension was ~34 Å. The simulation cell was extended to 100 Å along the z axis in order to introduce additional space above and below the ice slab. The ice/vapor interfaces were at the basal (0001) plane. The initial (proton-disordered) configuration of the ice slab was generated using the algorithm of Buch *et al.*[2] Upon annealing the slab from 0 to 250 K, constant temperature (NVT) molecular dynamics at T = 250 K was carried out for 70 ns.

First of all, we wish to point out that in order to obtain reliable characteristics of the ice/vapor interface, it is important to perform rather long production runs, often of the order of tens of nanoseconds. In addition, it is essential to use a thick enough ice slab to screen interactions across the slab arising from (pre)melting at the two interfaces. This becomes particularly important in simulations at temperatures close to the melting point as under such conditions the surface disorder propagates relatively deep into the ice slab. The thickness of the slab should ensure that even at the melting point the slab has a distinct core layer of ice surrounded by surface molecules. However, caution should be exercised already at temperatures well below the melting point of the water model. As an illustration, Fig. 14 shows that a 5-bilayer-thick model is not thick enough to withstand melting at T = 250 K. While averaging over the first few nanoseconds of the trajectory would yield a density

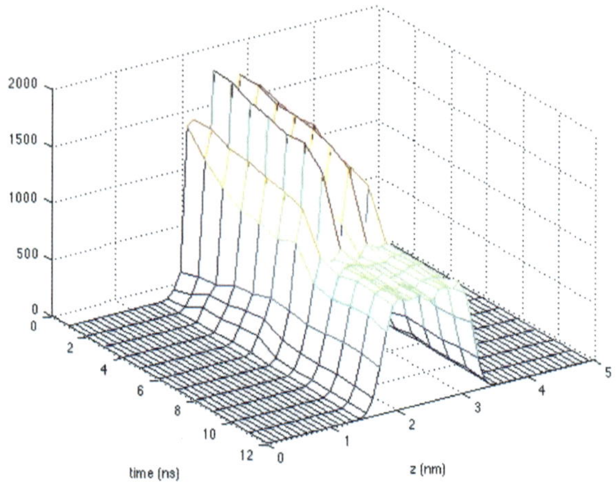

Fig. 14 Time-resolved density profile of oxygen atoms in a 5-bilayer slab, showing 5 distinct maxima each corresponding to one bilayer of the ice slab at $t = 1$ ns, and melting of the slab at $t = 8$ ns.

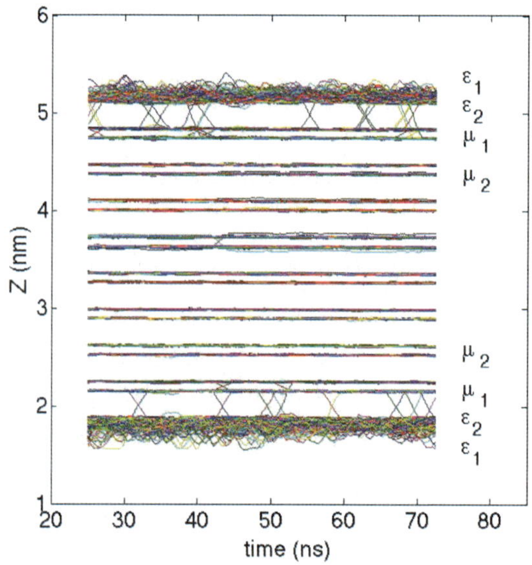

Fig. 15 Trajectories of all 1280 water molecules in the ice slab at 250 K depicted as oxygen z-coordinates using 2.5 ns running averages. For definition of ε_i and μ_i layers see text.

profile corresponding to a (seemingly) stable ice slab; the slab actually melts after ~8 ns. In contrast, the 10-bilayer-thick ice slab is stable over (at least) 70 ns.

In agreement with your results, the density profile of the oxygen atoms across the 10-bilayer slab at 250 K exhibits fairly symmetric double-maxima peaks typical of the inner bilayers (denoted below as μ_i) while the peaks corresponding to the surface bilayer (ε_2) are rather asymmetric and broadened, with an additional small peak corresponding to an (incomplete) overlayer (ε_1) of water molecules on top of the original surface. We find that individual molecular exchange between ε_2 and ε_1 occurs on a time scale of 10s to 100s of picoseconds. Fig. 15 shows that there are also exchanges, albeit much less frequent (10s of nanoseconds), of molecules

 This journal is © The Royal Society of Chemistry 2009

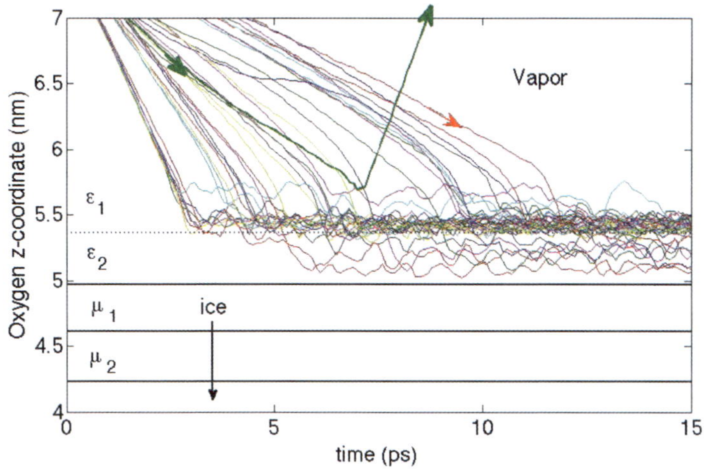

Fig. 16 Trajectories (depicted as oxygen z coordinates) of water vapor molecules sticking to ice surface at 250 K. For definition of ε_i and μ_i layers see text.

between the surface bilayer ε_2 and the adjacent crystalline bilayer μ_1 as well as within the two layers constituting the μ_1 bilayer. Consistent with your results, we find that both traffic between different layers and intra-bilayer traffic occur as a series of pair-wise exchanges. In addition to the permanent inter- and intra-layer exchanges depicted in Fig. 15, we observed a lot of exchanges between ε_2 and μ_1 layers that turned out to be transient (these events do not show up in Fig. 15 in which 2.5 ns running averages were used to depict the z coordinates of the oxygen atoms). The above observations raise some intriguing questions: are the permanent exchanges dosey-do, and the fleeting ones some other mechanism? Do intra-bilayer exchanges occur by a different mechanism than inter-bilayer ones?

Our final comment concerns adsorption of species to the ice surface. You speculate that 'crevasses' which form on the surface of ice at temperatures around 250 K may be responsible for trapping foreign species. While these 'crevasses' could presumably strongly bind some impurities, they do not seem to play a significant role in trapping of water itself. We have carried out a set of simulations in which a single water molecule was initially placed ~10 Å above the surface of the ice slab and randomly assigned an initial velocity according to the Maxwell–Boltzman distribution for $T = 250$ K. Fig. 16 shows that most of the water molecules that impinge onto the ice surface and stick to it do so into the outermost (ε_1) overlayer.

1 S. Neshyba, E. Nugent, M. Roeselova and P. Jungwirth, *J. Phys. Chem. C*, 2008, submitted.
2 V. Buch, P. Sandler and J. Sadlej, *J. Phys. Chem. B*, 1998, **102**, 8641.

Dr Slater replied: Your results are extremely interesting and it is somewhat reassuring to see that inferences drawn from our comparatively brief simulations of a few ns are consistent with the findings taken from simulations that are approximately an order of magnitude greater in duration. We have also observed the migration of water from one bilayer to another as well as intra-bilayer albeit with far fewer samples than you collected. However, we can offer some speculations on the driving forces which will likely influence the migration pathways of water molecules within and between layers. Recent work using first-principles approaches shows a surprising variance in the vacancy energy within bilayers close to the surface and in particular, at the surface (the vacancy energy relates to the self-interstitial formation energy at the surface).[1] We find the inner part of the first few bilayers have a much greater vacancy energies than those in the outer parts of the bilayer. This is consistent with density profile results in our paper where the outer bilayer shoulder height is always lower than its accompanying inner bilayer peak. Crudely, we imagine that each bilayer has an outer part in which molecules are released from their ideal lattice positions more easily than the underlying "stiff" inner bilayer. This picture is a little simplistic as a very small number of the water molecules in the outer bilayer are actually held more strongly than a few of those in the inner bilayer. Nevertheless, upon heating, more water molecules are liberated from their lattice sites in the outer bilayer than the inner bilayer; however, occasionally one of the released surface interstitials probably migrates to vacant sites in the inner bilayer. In addition, molecules from either the inner or outer bilayer of the outermost layer occasionally will diffuse into vacancies arising in the outer bilayer of the second layer. We didn't analyse the pathways of the migrations too closely because of the lack of statistics but your new data offers an exciting opportunity to look into this further.

Regarding the crevasses we reported, it is very interesting that water vapour seems not to penetrate deeply into the ice slab. According to our results, the crevasses are not too prevalent or deep at 250 K so perhaps these are not so important at these temperatures (we note again that 250 K maps on to around ~235 K in reality taking into account the true melting temperature of the NvDE model). Another consideration is that the surface structure is rather dynamic and that the crevasses are lined with dangling hydrogen bonds close to the surface and hence the probability of a molecule impinging directly on a site at the base of the crevasse is perhaps low. We do know that migration of water from one surface site to another is fairly slow being on the 10s of ps timescale and hence perhaps some the waters in Fig. 15 might migrate from the mouth to the base of the crevasse. It would be worthwhile to investigate this topic more fully.

1 M. Watkins and B. Slater, 2008, in preparation.

Dr Feibelman continued the discussion of the paper by Professor Vega: A calculation of the viscosity of the model premelted layer is wanted for comparison to experiment.

Professor Vega replied: We have determined recently the thickness of the liquid layer on the free surface of ice for several water models,[1] using both a geometrical and a dynamic criterion (related to the diffusivity of the water molecules). The analysis of the viscosity of the liquid layer present on the free surface of ice would be quite interesting.

But it is important to stress that the calculation of dynamical properties is costly in computer time. The calculation of viscosity of the liquid water would be desirable not only for TIP4P/2005 but for many water models (for which it is still unknown).

1 M. M. Conde, C. Vega and A. Patrykiejew, *J. Chem. Phys.*, 2008, **129**, 014702.

Dr Feibelman remarked: Can you provide a force field that is fitted to the water phase diagram in the universe described by a particular functional *e.g.* PBE? That would make for useful QM/MM calculations of water at solids.

Professor Vega answered: The evaluation of the phase diagram of water from first principles would be a major achievement. Before this is achieved, it would of interest to analyse if current functionals used in DFT are able to predict the relative energy between solid phases of water (ices) at 0 K. The experimental value of the relative energy between ices were estimated by Whalley.[1] We have shown recently that models predicting correctly the phase diagram of water also predict the relative energies between ices at 0 K.[2]

Therefore, before attempting the evaluation of the phase diagram from first principles, it would be of interest to determine from first principles the relative energies

 This journal is © The Royal Society of Chemistry 2009

between ices at 0 K, and to analyse the performance of different functionals. This is indeed a difficult test for any functional as the relative energy between ices at 0 K is typically between 0.1–0.4 kcal (increasing up to 1 kcal mol^{-1} when the high pressure ices VII and VIII are considered). The first principles calculations should be extremely precise to determine such a small difference. Some preliminary calculations show that this may be quite difficult.[3] At the moment there have been some attempts to parametrize the water force field to reproduce *ab initio* forces obtained from condensed phase QM/MM calculations.[4] But this force field has only been used to check the results for the radial distribution function and the diffusion coefficient.

1 E. Whalley, *J. Chem. Phys.*, 1984, **81**, 4087.
2 J. L.Aragones, J. L. F. Abascal, E. G. Noya and C. Vega, *J. Chem. Phys.*, 2007, **127**, 154518.
3 B. Slater, private communication, 2007.
4 O. Akin-Ojo, Y. Song and F.Wang, J. Chem. Phys., 2008, 129, 064108.

Dr Feibelman said: Can you account for the equality of the lattice constants of D_2O and H_2O ice Ih in a force field model?

Professor Vega replied: TIP4P/2005 and other water models have been developed using classical statistical mechanics, and therefore incorporate quantum effects implicitly through the values of the potential parameters.

It would be possible to use different potential parameters (σ, ε *etc.*) for D_2O and H_2O so that certain target properties of these compounds were reproduced (since properties like melting point, temperature of maximum density *etc.* do differ, one would obtain different parameters).

However, the value of the electronic density is hardly affected by the value of the nuclear masses. Thus, it seems more adequate to use the same potential model for D_2O and H_2O, and to introduce quantum effects on nuclear motion *via* path integral simulations.

It could be of interest to develop a water potential model in which these effects were not incorporated in the parameters. The model should then be able to reproduce water properties when dealing explicitly with nuclear motion quantum effects (*i.e.* path integral simulations).

Dr Slater remarked: Further to Dr Feibelman's comment about the potential role of DFT to train or re(de)fine water interatomic potentials; we have critically evaluated DFT functionals and their ability to reproduce calorimetric data from ice phase transitions. In general, we find that GGA DFT typically overestimates enthalpy differences by a factor of 10 with respect to experimental data. Interestingly, as Vega has shown, the modified forms of TIP4P have relatively small errors with respect to experiment. Hybrid functionals appear to substantially improve the relative energetic of ice phases.

One area where the potentials are less successful is the description of proton ordering energetics where many potential models predict the wrong structure for ice XI. Here GGA DFT is rather good and in several cases—see the works of Singer *et al.*[1,2]—the order–disorder transition has been predicted with remarkable accuracy.

So, potentials would need to be fitted to extremely expensive DFT calculations—given the emergence of periodic MP2 capabilities in codes like CRYSTAL and VASP, the latter might be a more fruitful path to pursue.

1 C. Knight, S. J. Singer, J.-L. Kuo, T. K. Hirsch, L. Ojamäe and M. L. Klein, *Phys. Rev. E*, 2006, **73**, 056113.
2 C. Knight and S. J. Singer, *J. Chem. Phys.*, 2006, **125**, 064506.

Professor Fuchs commented: I enjoyed your presentation of the TIP4P/2005 force-field and was amazed by its preformance. Among the 10 physical properties of water that you have investigated, only one is a transport property. I suggest, for future works by your group and others, that it would be useful to test TIP4P/2005 for some other transport properties, such as viscosity. One of the most important issue in the field of properties prediction is the ability of a given forcefield to do a good job for predicting both static equilibrium and transport coefficients.

Professor Vega responded: We have recently computed the viscosity at room temperature and pressure of the model TIP4P/2005. The model also provides an excellent description of the viscosity of water.

Professor Fuchs continued the discussion of the paper by Dr Slater: I was very much interested by the fact that different ice surfaces behaved differently with respect to premelting. I was wondering whether it would be possible to find an ice I surface that would NOT premelt?

Dr Slater answered: It's an interesting question. I suspect the answer is no. We have found that the energy to create surface self-interstitials is very small (~0.2 eV) and this is true of both highly proton ordered and disordered surfaces. Since pre-melting seems to be initially driven by the formation of these self-interstitials we presume that all (0001) surfaces would pre-melt. Recent work by Conde *et al.*[1] on the pre-melting of other crystal faces of ice indicates that the prism faces are actually more resilient to pre-melting but nonetheless exhibit pre-melting.

1 M. M. Conde, C. Vega and A. Patrykiejew, *J. Chem. Phys.*, 2008, **129**, 014702.

Professor Bonn opened the discussion of the paper by Dr Bovensiepen: Do you have any ideas that might explain the relatively large cross section for the reaction of the surface electron with $CFCl_3$? In particular, could it be due to mobility of the electron given its undoubtedly shallow potential well?

Dr Bovensiepen replied: $CFCl_3$ is known in general for its large dissociative electron attachment cross section. We have no indication that the mobility of the trapped electron is decisive, but report in the paper evidence that the mobility of the $CFCl_3$ contributes. As we discuss in our paper[1] that is currently being reviewed, the binding energy of the trapped electron is temperature-dependent. Actually, the temporal evolution and the temperature dependence is described simultaneously with a stretched exponential function. However, this occurs due to the electron transfer back to the metal substrate and not because of diffusion of the trapped electron at the surface. In Fig. 4 of our Faraday Discussions paper we report on the effect of temperature on the work function. We find that annealing up to 60 K enhances the work function further which suggests that thermally activated diffusion enhances the cross section. Since this is observed after the annealing process when the temperature is at 30 K it is not the trapped electrons are diffusing, but the $CFCl_3$ molecules. (The electron population is refreshed with every laser shot occurring at 200 kHz laser repetition rate. Hence, after annealing the electrons carry no information regarding their diffusion. However, the identical molecules are adsorbed before and after annealing.)

1 U. Bovensiepen, C. Gahl, J. Stähler, M. Bockstedte, M. Meyer, F. Baletto, S. Scandolo, X.-Y. Zhu, A. Rubio and M. Wolf, *J. Phys. Chem. C*, 2008, submitted.

Professor Rovere continued the discussion of the paper by Professor Vega: Did you understand the reason for which all the site models for water underestimate the critical pressure?

Professor Vega replied: Long ago Barker and Monaghan[1] showed that a reasonable prediction of the critical properties of a Lennard-Jones fluid could be obtained

 This journal is © The Royal Society of Chemistry 2009

with a virial expansion truncated at the level of the fourth virial coefficient (B4). We have recently shown that the same is true for molecular fluids.[2]

The TIP4P/2005 model predicts a critical temperature and density in good agreement with the experimental value. However the critical pressure deviates significantly from experiment. That strongly suggests that TIP4P/2005 (and the rest of non polarisable models) do not describe correctly the experimental values of the virial coefficients of water.

Kofke and coworkers have shown that this indeed the case.[3] The failure of TIP4P/2005 (and the rest of non-polarisable models) in describing the virial coefficients of water is related to the fact that these models are using an effective dipole moment (of about 2.3 D) which is not convenient to evaluate the properties of just two (B2), three (B3) or four (B4) molecules in the gas phase.

Not surprisingly, water models predict a value of B2 well below the experimental value, which is probably the reason why the critical pressure of all models is too low. The absence of polarisability prevents an accurate description of the virial coefficients and of the critical pressure.

1 J. A. Barker and J. J. Monaghan, *J. Chem. Phys.*, 1962, **36**, 2564.
2 L. G. MacDowell, C. Menduina, C. Vega and E. de Miguel, *J. Chem. Phys.*, 2003, **119**, 11367.
3 K. M. Benjamin, J. K. Singh, A. J. Schultz and D. A. Kofke, *J. Phys. Chem. B*, 2007, **111**, 11463.

Dr Nutt asked: I find the success of the TIP4P/2005 model quite impressive, and it's definitely a model which I would consider using in my work on biomolecular hydration. I have a couple of comments and questions. Firstly, whenever the TIP3P model is criticised for not reproducing a particular property, it's worth remembering that it was almost certainly not parameterised to include this property, and I find the continued criticism of TIP3P in this respect a little unfair. However, testing a range of models for their ability to reproduce a range of properties provides useful insight, as you have clearly demonstrated. I do use TIP3P water in my simulations, but only because it is the default water model for the biomolecular force-field in question. You touch on the question of transferability at the end of your paper, and I agree that this is an important issue for future work.

My question concerns the use of Ewald summation *vs.* cut-off in the treatment of long-range electrostatic interactions. In your opinion, can a given water model be used with any non-bonded scheme, or should an effort be made to use the scheme used in the original parameterisation? For example, since TIP3P was parameterised with a cut-off, is it appropriate, in your view, to use an alternative scheme with this water model?

A final comment is on future water models. You mention the number of interaction distances that need to be calculated for a given water model (9 for two TIP3P water molecules, *etc.*). There's a recent soft-sticky-dipole-quadrupole-octupole (SSDQO) model from Ichiye *et al.*,[1] in which they condense the electrostatic charges to a single point with multipoles up to order 3. This is based on a Stockmayer potential, an old idea (a single point water model with dipole and quadrupole moments was published[2] in 1951) which seems to have previously been neglected, presumably because it's easier to visualise discrete charge sites than molecular multipoles. Personally, this seems to be a useful direction in which to take future work.

1 T. Ichiye and M.-L. Tan, *J. Chem. Phys.*, 2006, **124**, 134504.
2 J. S. Rowlinson, *Trans. Faraday Soc.*, 1951, **47**, 120.

Professor Vega answered: Most of the importance of developing accurate water models is that they may serve to investigate other properties different from those used in the fitting of their potential parameters. The development of TIP3P was a major breakthrough in our knowledge of the water force field. But it is now clear that it has serious limitations and the advances in the field have allowed to develop new models. This is a natural consequence of the advance of science and, surely, the TIP4P/2005 model will be surpassed by yet another improved water model. The goal of this work was essentially to investigate to what extent some of the more popular models are able to predict a wide set of important water properties (thus extending the scope of properties used to fit any of the models). Certainly, TIP4P/2005 has a small advantage because the set of fitting properties was a bit larger than for the other models (though these are still a minor fraction of the total set studied in this work). But that is not the main question. More important is that, according to the results of this work, if one is interested in investigating a water property which is not easily accessible from experiment, he or she may have more confidence in the predictions of TIP4P/2005 than those obtained using TIP3P.

It is finally important to note that some of the TIP4P/2005 properties not included in the determination of the parameters (vapor–liquid equilibria, critical temperature, surface tension, dielectric constant, diffusion coefficient) have been reproduced by independent calculations of different groups.[1,2] It is known that, for liquid water, the use of the simple cut-off cause several artifacts, and makes the results dependent on the value of the distance at which the potential is truncated.[3,4] A change in the cut-off changes the Hamiltonian. Contrarily, when Ewald sums are used, the value at which the real space term contribution is truncated does not affect the value of the energy due to the presence of the additional Fourier term. In summary, with Ewald sums, the Coulombic energy is evaluated almost exactly and there is no truncation of the potential.

If these remarks are true for the liquid, they are even more relevant in the study of the solid phases of water. Now that Ewald sums have reduced dramatically its computational cost (due to the use of the particle mesh Ewald method, PME), and that this option is incorporated into many MD packages, we really recommend the use of Ewald sums when dealing with water simulations. That may well be the point of view of many groups since the traditional TIP3P and TIP5P have been reparametrized to be used with Ewald sums.[5,6] Notice however that these two models are essentially a minor modification of the original TIP3P and TIP5P since the same target properties were used in the fit. The score obtained by the reparametrized models would be essentially the same. Although TIP4P/2005 has been designed to be used with Ewald sums, it is not just a minor modification of TIP4P to be used with Ewald sums. The reason for that is that the target properties of real water used to determine the parameters of the potential were different in TIP4P and TIP4P/2005. For this reason, the value of ε and the hydrogen charge differ significantly in TIP4P/2005 with respect to the original TIP4P. Concerning the last comment, the use of model with a Lennard-Jones center and multipole expansion (including, dipole, quadrupole, octupole and hexadecapole terms) is appealing. It would be of interest to know more precisely the computational efficiency of the expansion (is it cheaper to evaluate 9–10 distances or to calculate the multipole–multipole terms which present a complex dependence of the relative orientation of the molecule?). On the other hand, the electrostatic field generated by a charge distribution can be approximated by the multipolar expansion at large distances. However, at small distances there may be differences between the electric field created by discrete charges and multipoles. It would be important to investigate the structure generated by those models at small distances.

1 J. Kolafa, F. Moucka and I. Nezbeda, *Coll. Czech. Chem. Commun.*, 2008, **73**, 481.
2 A. Ghoufi, F. Goujon, V. Lachet and P. Malfreyt, *J. Chem. Phys.*, 2008, **128**, 154716.
3 Y. Yonetani, *J. Chem. Phys.*, 2006, **124**, 204501.
4 M. Lisal, J. Kolafa and I. Nezbeda, *J. Chem. Phys.*, 2002, **117**, 8892.
5 S. W. Rick, *J. Chem. Phys.*, 2004, **120**, 6085.
6 D. J. Price and C. L. Brooks, *J. Chem. Phys.*, 2004, **121**, 10096.

Dr Bakó asked: Comparison of simulated and experimental data on O···O partial radial distributional level is not the best method. O···O partial radial distributional

 This journal is © The Royal Society of Chemistry 2009

level is not a direct quantity; it is determined through a complicated data processing from neutron diffraction experiment. It is significantly better, if you show the comparison on total RDF (especially H_2O) level. If you do comparison with Narten's neutron diffraction data,[1] you need to take into account the modification function used to avoid the Fourier truncation error in experiment. This factor caused significant changes on the peak height and peak width, especially at short distances.

1 A. H. Narten, C. G. Venkatesh and S. A. Rice, *J. Chem. Phys.*, 1976, **64**, 1106.

Professor Vega replied: The extraction of $g(r)$ from diffraction data is indeed a non-trivial process and requires, as you mention, some data processing. Maybe for this reason the experimental result of g_{OO} (and g_{OH}, g_{HH}) for liquid water has ben modified several times along the years. We found difficulties when comparing g_{OO} of ice Ih as obtained from simulation (TIP4P or TIP4P/2005) with experimental results.

To our knowledge, only two set of experimental data are available. Whereas we found good agreement (see our paper presented at this discussion meeting) with X-ray diffraction results,[1] the agreement was not so good[2] with the results of neutron diffraction.[3] We believe that a good water model should provide a reasonable prediction of the $g(r)$ of water (and ice as well), but forcing a water model to reproduce exactly the "experimental" values of $g(r)$ may not always be a good idea. The comparison of simulation results for the total structure factor to the corresponding experimental results always seems adequate. For liquid water, such a comparison has recently been performed[4] by Pusztai *et al.* for seven different water models. The authors conclude that the predictions "from the TIP4P/2005 potential proved to be the most consistent with the experimental neutron-weighted total scattering structure factor of heavy water".

1 A. H. Narten, C. G. Venkatesh and S. A. Rice, *J. Chem. Phys.*, 1976, **64**, 1106.
2 C. Vega, C. McBride, E. Sanz and J. L. Abascal, *Phys. Chem. Chem. Phys.*, 2005, **7**, 1450.
3 A. K. Soper, *Chem. Phys.*, 2000, **258**, 121.
4 L. Pusztai, O. Pizio and S. Sokolowski, *J. Chem. Phys.*, 2008, **129**, 184103.

Dr Michaelides asked: Your efforts to evaluate the performance of the TIP4P series of potentials is very nice and I like your analogy with the Olympic games. However, it reminds me somewhat of the cold war with only four countries (potentials) invited. There are countless other potentials in the literature, in particular those developed by comparison to explicitly correlated quantum chemistry methods. Can you say anything about how many points these potentials would get in your Olympics?

Professor Vega answered: I cannot provide an answer. The reason is that when water potential models are proposed (especially those obtained to correlate quantum chemistry calculations) only a few properties are considered.

Typically the density, internal energy, structure (*i.e.* radial distribution function) and diffusion coefficient are computed. Getting phase equilibria is quite involved and for this reason, the vapor–liquid equilibria, critical properties, melting point and the dense region of the phase diagram are usually not computed. The same is true for other properties as surface tension, or density predictions for solid polymorphs.

One of the purposes of our paper is to illustrate that the comparison of 3–4 properties does not provide a clear view of the performance of the models.

In fact, the choice of potential models for this work is basically the list of models for which we have results about the performance of the ten considered properties (SPC/E was not included in the discussion although a rough idea of its performance is given at the end of the paper). I should just mention that the vapor–liquid equilibrium has been obtained quite recently by Siepmann *et al.*[1] by using first principles simulations (using the BLYP and PBE functionals). The calculations predicted a critical temperature for water of 550 K (BLYP) and 700 K (PBE), which are far from the experimental value.

Therefore the answer is that I can not assess the number of points obtained by a large number of potential models (empirical or from correlations of quantum chemistry calculations) because nobody has calculated the ten properties considered in this work. We do hope that, when new potential models are proposed in the future, an extensive number of properties are computed and not just a few.

1 M. J. McGrath, J. I. Siepmann, I.-F. W. Kuo and C. J. Mundi, *Mol. Phys.*, 2006, **104**, 3619.

Professor Rovere commented: Concerning the derivation of an empirical potential from first principles computer simulation, as far as I know there were attempts to do it, but the potentials extracted gave good results for clusters of water but not for liquid water.

Professor Vega answered: First principles (or DFT) calculations should be quite useful in the development of water potential models. Although the general agreement is that first-principles-based potentials should make empirical potentials disappear, it is probably fair to say that empirical potentials still yield better results.

There are several reasons for that. First, since the potential energy represents only a small fraction of the total energy of the system, the precision of the calculations may compromise the results. Secondly calculations based on the water dimer or more complex water clusters do not necessarily give an accurate representation of water in condensed phases. That will change in the future, no doubt, but it is difficult to establish whether it will take a few years, a decade or several decades.

Will the interaction between water molecules be the topic of a Faraday Discussion in the future or will the problem be essentially solved so that there will be no need for such a conference? It is difficult to assess that.

Professor Jungwirth continued the discussion of the paper by Dr Bovensiepen: What is the effect of the water structure—crystalline *vs.* amorphous—on the behaviour (*e.g.* surface *vs.* interior position) of the solvated electron? (Simulations by Turi and Rossky[1] show that lowering the temperature strongly enhances the surface propensity of the electron and the amorphous phase may somewhat mimic the liquid state.)

1 A. Madarasz, P. J. Rossky and L. Turi, *J. Chem. Phys.*, 2007, **126**, 234707.

Dr Bovensiepen replied: We have carried out detailed analysis of the electron dynamics at amorphous ice/metal interfaces (see our paper for references) and find typical lifetimes of 100–300 fs while the binding energy (referenced to the vacuum level) continuously increases from the bottom of the conduction band. For crystalline structures we identify a trap state that is indirectly populated through the first image potential state at the ice-metal interface and has a binding energy which is 1 eV larger than the one of the image potential state. The lifetime of this trap state is 14 orders of magnitude larger and the electron population decays on a time scale of minutes. We explain this difference in lifetimes by an electron density residing in the trap that is extremely well screened from the metal substrate. This scenario is in agreement with density functional calculations of excess electrons at a particular reorientational defects of crystalline ice surfaces, which illustrate the molecular configuration of the long-living trap site. This work has been submitted to *J. Phys. Chem. C*.[1]

1 U. Bovensiepen, C. Gahl, J. Stähler, M. Bockstedte, M. Meyer, F. Baletto, S. Scandolo, X.-Y. Zhu, A. Rubio and M. Wolf, *J. Phys. Chem. C*, 2008, submitted.

 This journal is © The Royal Society of Chemistry 2009

Professor Wu asked: For charge transfer from metal to water molecules, could you provide any evidence to prove the CT process is an indirect or direct one from metal to water molecules? In the indirect CT process, there is a state of hot electrons.

Dr Bovensiepen responded: We carried out photon-energy-dependent studies of the electron transfer from the metal into the ice layer and observed a threshold of 3.2 eV photon energy that is required to transfer the electron into the adlayer. This energy is identical with the energy difference between the Fermi level of the metal substrate and the first image potential state at the ice/metal interface. We conclude that for electron trapping first the electron has to be excited into the laterally delocalized image potential state from where preexisting traps can be populated. The relevant work is reported in a manuscript which is currently under review in *J. Phys. Chem. C*.[1]

1 U. Bovensiepen, C. Gahl, J. Stähler, M. Bockstedte, M. Meyer, F. Baletto, S. Scandolo, X.-Y. Zhu, A. Rubio and M. Wolf, *J. Phys. Chem. C*, 2008, submitted.

Dr Fraxedas asked: Do trapped electrons on crystalline ice form regular structures, *i.e.* 2D Wigner crystals?

Dr Bovensiepen responded: The dispersion of the trapped electron state with electron momentum parallel to the surface has been analyzed and the data are part of a publication submitted to *J. Phys. Chem.*[1] The results suggest a trapped, solvated state and bears no sign of a periodicity of the trapped electron wave function. This is in agreement with our femtosecond time-resolved study which led to the conclusion that these trapped electrons form at preexisting reorientational defects on crystalline ice surfaces. Also the estimated density of trapped electrons, which is on the order of 10^{-4} electrons per lattice site argues against a long range order of trapped electrons.

1 U. Bovensiepen, C. Gahl, J. Stähler, M. Bockstedte, M. Meyer, F. Baletto, S. Scandolo, X.-Y. Zhu, A. Rubio and M. Wolf, *J. Phys. Chem. C*, 2008, submitted.

Professor Buch continued the discussion of the paper by Dr Slater: There is an experimental demonstration of orientational defects originating from the surface[1]—is this consistent with your paper?

1 J. P. Devlin and V. Buch, *J. Chem. Phys.*, 2007, **127**, 091101.

Dr Slater responded: Another interesting point. We have additional data from DFT studies which suggests that these and other orientational defects do prefer to surface segregate. A manuscript on this topic is in preparation.[1]

1 M. Watkins and B. Slater, in preparation, 2008

Professor Buch then commented on Professor Vega's paper: Vaporization energy should be also affected by QM effects in addition to polarizability—because of large QM contribution (zero point energy of a few kcal mol^{-1} per molecule) to the kinetic energy.

Professor Vega replied: Certainly, besides polarizability, nuclear quantum effects (both intramolecular and intermolecular) should be taken into account for a quantitative description of the vaporization energy of real water.

Professor Luzar said: It would be really helpful to include the SPC/E model in your Table 11 for all the properties listed, because it seems that this potential does pretty well in overall performance (structure and thermodynamics and dynamics), especially if we take into account that 3-site models should be cheaper to run. Since the submission of your Faraday Discussions paper, have you calculated any other dynamic property with your model, in addition to diffusion coefficient, and if so, can you share your preliminary results here with us?

Professor Vega answered: SPC/E would obtain 21 points (to be compared with 27 obtained by TIP4P/2005, and 13 and 14 by TIP4P and TIP5P respectively). The justification of this number of points is the following: 3 for vapour–liquid equilibria, 2 points for surface tension, 2 points for the density of ices, 1 point for the phase diagram, 1 point for the melting properties, 1 point for the TMD, α and κ_T, 2 points for structure predictions, 3 points for the equation of state at high pressures, 3 points for the diffusion coefficient and 3 points for the dielectric constant. Overall, SPC/E improves the predictions with respect to TIP4P and TIP5P but it is well below the score obtained by TIP4P/2005. The introduction of the polarization correction is responsible for the overall improvement of SPC/E with respect to TIP4P and TIP5P. However, when the polarization correction is included within the TIP4P geometry the resulting model (TIP4P/2005) seems to be superior.

The computational time of three-site models (as SPC/E) and four-site models (as TIP4P/2005) is almost the same. Nine distances are required to evaluate the energy in SPC/E whereas TIP4P/2005 requires the evaluation of ten distances: 9 distances between the charges and one O–O distance to compute the Lennard-Jones energy. Notice that the computation of the six oxygen–charge distances is not required. For this reason, using SPC/E instead of TIP4P/2005 only reduces the computer time by 10%. Notice that some MD codes may (unnecessarily) evaluate $4 \times 4 = 16$ distances. The trick to reduce it to 10 distances requires just three lines of code. Check whether the trick is implemented in your program or MD package. Gromacs for instance incorporates this trick, so with Gromacs the computational gain of using SPC/E is only 10%.

The interest manifested by several attendees at the Faraday Discussion has prompted us to evaluate the viscosity of several water models. Preliminary results for the viscosity at 298.15 K (calculated using the Green–Kubo equation with 20 ns runs) are in line with the findings of this work. Again, TIP4P/2005 yields the better predictions, 0.86 mPa s, much closer to the experimental value (0.893) than SPC/E (0.73) and TIP5P (0.70). The departure from experiment of the result for TIP4P (0.51) is already considerable although it is a relatively acceptable prediction when compared to the poor performance of TIP3P (0.33 mPa s), which is only almost one third of the experimental value.

Professor Tobias said: During the discussion of the paper by Professor Vega, several critical comments were made concerning the use of outdated/inferior water models (*e.g.* TIP3P) in simulations of biological molecules, and it was suggested that a different, presumably superior (as judged by ability to reproduce properties of bulk liquid water and water ice) water model could simply be substituted. I wish to point out that the biomolecular force fields are typically refined by assuming a particular water model will be used (*e.g.* TIP3P water with the CHARMM and AMBER protein force fields). The refinement includes properties such as the relative energies of peptide conformations in the presence of water, and solvation free energies. The choice of the water model is therefore woven into the force field parameters (*e.g.* charges, Lennard-Jones radii and well depths, torsional barriers, *etc.*) for the biomolecules. Substituting a different water molecule will upset the balance between solute–solute, solute–solvent, and solvent–solvent interactions achieved in the force field refinement, with unpredictable and potentially adverse consequences. A natural question is: why don't the developers of biomolecular force fields reoptimize their parameters using a better water model? The force field developers are well aware of the deficiencies of the water models assumed in current generation biomolecular force field, but reoptimizing with a new water model is a task that would require

 This journal is © The Royal Society of Chemistry 2009

a substantial fraction of the enormous effort required to develop a new force field from scratch. Development of the next generation force fields (based on polarizable potentials) is under way, but it will take a lot more time and effort before they are fully developed and validated to the point where they will be ready to replace the ones currently in use.

Professor Vega responded: For the community studying pure water, I think it is clear that other water models provide a much better description of pure water than TIP3P. Concerning mixtures and biomolecules in water, this deserves some discussion. It is true that force fields are developed so that certain target experimental properties depending on the solute–solvent interactions are reproduced. Obviously if the water model is replaced there is no guarantee that the fitted properties are still reproduced. However, it should be pointed out that there is no guarantee that the original force field (with its water model) is going to reproduce any property beyond that used in the fit. The question is particularly relevant when other properties of the pure solvent are not accounted by the model. On the other hand, the consequences of replacing the original water model of the force field by another model have not been fully explored. The idea of replacing the original water model for which the force field was optimized by another water model should not be rejected as suggested by Nutt and Smith,[1] until further evidence (against or in favour) is obtained. Let me provide several examples to show where replacing the water model of a force field may be a good idea.

Firstly, antifreeze proteins (used by fish in the Antarctic sea) are able to prevent the formation of ice (even at temperatures below 0°C). Simulations on these antifreeze proteins have recently been performed. In these studies TIP3P (the original model of the force field) was replaced by TIP5P and the obtained results were quite reasonable.[2,3] The reason for this replacement is that TIP3P freezes at 146 K into ice Ih and at 220 K in ice II. Performing simulation studies of the antifreeze protein at −10°C using TIP3P would not be a very clever idea since the model has a much lower freezing temperature. Since TIP5P reproduces the experimental melting point of water, replacing TIP3P by TIP5P seems a reasonable idea to study the behaviour of this type of proteins. No dramatic change was observed in the behaviour or the protein when replacing TIP3P by TIP5P and the protein was stable during the time of the simulation. Obviously the behaviour of water changed dramatically.[2–4]

Secondly, there is a long debate about the hydrophobic effect and its origin. Probably the simplest molecule to study the hydrophobic effect is methane. The methane–water mixture seems a very simple model to study, and probably the feeling is that the behaviour of such a simple mixture should be reproduced by current force fields. However, current force fields fail in describing the variation with temperature of the solubility of methane in water. It has been proved by Paschek[5] and by our group[6] that in order to predict the solubility of methane in water at different temperatures two conditions should be met: (a) the water model should describe correctly the dependence of the density with T; and (b) the polarisability must be introduced or the water–methane cross interaction adjusted.[6,7] Since TIP3P fails in describing the variation with T of the density of pure water it fails in describing the solubility of methane in water. When using TIP4P/2005 it has been possible to reproduce the variation of the solubility in pure water for a wide range of temperatures.

I understand that developing good force fields is quite a difficult task and that the huge effort of a lot of people to provide good potentials for the rest of the scientific community should be acknowledged. However, I do not understand why we should continue using and developing force fields around poor water models. I think it would be quite more useful to explore in more detail the consequences of replacing the "standard" water models of the force fields by more satisfactory water models. In case this approach fails, it seems a good idea to develop new force fields around better water models as, for instance, TIP4P/2005. Having a good polarisable model for water would be even better, and sooner or later it will appear, but before accepting a polarisable model I would suggest that the "decathlon" test (*i.e.* the ten properties considered in this work) is passed so that the community agrees that the model represents a real improvement.

1 D. R. Nutt and J. C. Smith, *J. Chem. Theory Comput.*, 2007, **3**, 1550.
2 H. Nada and Y. Furukawa, *J. Phys. Chem. B*, 2008, **112**, 7111.
3 D. R. Nutt and J. C. Smith, *J. Am. Chem. Soc.*, 2008, **130**, 13066.
4 N. Smolin and V. Daggett, *J. Phys. Chem. B*, 2008, **112**, 6193.
5 D. Paschek, *J. Chem. Phys.*, 2004, **120**, 6674.
6 H. Docherty, A. Galindo, C. Vega and E. Sanz, *J. Chem. Phys.*, 2006, **125**, 074510.
7 P. J. Dyer, H. Docherty and P. T. Cummings, *J. Chem. Phys.*, 2008, **129**, 024508.

 This journal is © The Royal Society of Chemistry 2009

"Similarities" between confined and supercooled water

Maria Antonietta Ricci,* Fabio Bruni and Alessia Giuliani

Received 4th April 2008, Accepted 16th May 2008
First published as an Advance Article on the web 9th September 2008
DOI: 10.1039/b805706k

Neutron diffraction and deep inelastic neutron scattering experiments performed on bulk stable and supercooled water are compared with the same experiments performed on water confined in silica substrates. Similarities and differences between the two cases clearly show up, as far as both microscopic structure and single proton dynamics are concerned. In particular in both supercooled bulk water and water under confinement we observe a closer average distance between first neighboring oxygen sites and shortening of the H-bonds. In contrast the number of H-bonds per molecule and the number of interstitial water molecules are severely reduced under confinement, and the second peak of the oxygen-oxygen radial distribution function is shifted to shorter distances, compared to the bulk phase. Based on these results a possible scenario for understanding changes evidenced by deep inelastic neutron scattering when water is either confined or supercooled is proposed.

I. Introduction

If water is a fascinating liquid, its supercooled or confined phases represent an even more mysterious universe, yet to be fully explored. Understanding the properties of water in the metastable phase below $T_f = 273$ K is a challenge for chemical physics, because most of its anomalies become more pronounced. The trouble is that in this phase there is a gap between the temperature of homogeneous nucleation T_H and the crystallization temperature T_X, where theory and simulations predict that the solution of the water mysteries should be found,[1] but experiments are not feasible. At least three different scenarios, namely the retracing spinodal,[2,3] the second critical point,[1,4] and the singularity free scenario,[5,6] are compatible with the known water phenomenology and thermodynamics and await experimental validation. The observation that T_H decreases upon confinement of water in small volumes has consequently suggested that experiments on confined water could circumvent the difficulty of entering the not allowed thermodynamic region and stimulated an intense research activity. It goes without saying that confined water is of great interest by itself, as water is confined in many real situations of interest for instance in biology or geology. On the other side, materials in confined spaces may exhibit rich physical behaviors that are not observable in the bulk, such as for instance the emergence of new kinds of phase transitions and shifts of the common transitions within the thermodynamical space,[7] of relevance depending on the severity of confinement.[8] Water does not make exception in this respect.[9] Moreover it is well established that water in confinement is not a homogeneous system, as its density profile exhibit a maximum in the vicinity of the substrate, regardless of the hydrophilic or hydrophobic character of the interaction.[9–11] These observations

Dipartimento di Fisica E. Amaldi, Università degli Studi Roma Tre, via della Vasca Navale 84, 00147 Roma, Italy. E-mail: riccim@fis.uniroma3.it

raise the question whether and to what extent the behavior of bulk water can be extrapolated from experiments performed on confined water.

Here we report a comparison between results obtained by two experimental experimental techniques applied to bulk water and water confined in different substrates in order to determine the microscopic structure of the liquid and the dynamics of water protons at the attosecond timescale. Both experimental techniques are briefly described in section II; the microscopic structure is discussed in section III and section IV reports the comparison between the momentum distribution functions of protons in bulk and confined water.

II. Experimental techniques

All the experiments have been performed at the ISIS spallation neutron source,[12] by using the SANDALS diffractometer[13] and the VESUVIO inverse geometry spectrometer.[14]

A. Neutron diffraction with isotopic H/D substitution

This experimental technique follows the seminal work of Enderby and coworkers[15] and is based on the unique characteristic of neutrons of being sensitive to the isotopic state of the nuclei, through the scattering length b_α. Thus changing the isotopic composition of the liquid gives access to a different diffraction pattern: this is a key requirement to fully investigate the microscopic structure of a molecular liquid composed by x atomic species, which is described by a set of $\frac{x(x+1)}{2}$ radial distribution functions (RDF), $g_{\alpha\beta}(r)$. The case of water is particularly lucky, because the scattering lengths of hydrogen and deuterium ($b_H = -3.742$ fm, $b_D = 6.674$ fm)[16] cause a large contrast between the diffraction patterns measured at different H/D content and three experiments are fully exhaustive for the extraction of the O–O, O–H and H–H RDF, under the hypothesis that the isotopic effects on the liquid structure are negligible.

The output of the experiment, after correction of all systematic effects and reduction to an absolute scale[17] is the interference differential cross sections (IDCS), $F(Q)$, defined in eqn (1)–(3) and measured in barn/atom sr:

$$F(Q) = \Sigma_\alpha \Sigma_{\beta \geq \alpha} w_{\alpha\beta} S_{\alpha\beta}(Q) \tag{1}$$

where Q, the momentum transferred in the interaction between a neutron and a nucleus, is defined as a function of the neutron wavelength, λ, and scattering angle, 2θ by:

$$Q = \frac{4\pi}{\lambda} sin\theta \tag{2}$$

and the weighting factors

$$w_{\alpha\beta} = c_\alpha c_\beta b_\alpha b_\beta (2 - \delta_{\alpha\beta}) \tag{3}$$

depend on the neutron scattering lengths, b_α and b_β,[16] of the $\alpha\beta$ atom pairs, while the Kronecker $\delta_{\alpha\beta}$ avoids double counting of like terms in the summation. The $S_{\alpha\beta}(Q)$ are called partial structure factors (PSF) and are defined as Fourier transforms of the corresponding RDF of the $\alpha\beta$ pair:

$$\left(S_{\alpha\beta}(Q) - 1\right) = 4\pi\rho \int_0^\infty r^2 \left(g_{\alpha\beta}(r) - 1\right) \frac{\sin(Qr)}{Qr} \mathrm{d}r \tag{4}$$

Data transformation from Q to r space, has been in all cases presented here obtained by using a Monte Carlo code, developed at the end of the 90's by Soper,[18]

 This journal is © The Royal Society of Chemistry 2009

and known as empirical potential structure refinement (EPSR). This allows to record configurations of a three-dimensional ensemble of molecules compatible with the experimental data, giving access to RDFs, along with additional structural informations not available in the standard data treatment based on Fourier transform. This method avoids truncation errors in the RDF and allows checking for the absence of residual systematic errors, which could give unphysical features in the standard method. The routine requires as input the sample composition and thermodynamic parameters and a model for the site-site interactions, which is used as "reference" to start the simulation and produce an initial equilibrated ensemble. After equilibration, an empirical correction to the reference potential is evaluated iteratively within the routine, in order to achieve the best fit to the experimental $F(Q)$ functions. Such iterative procedure leads to the refinement of the so-called "empirical potential", which will then be used while accumulating molecular configurations, during the production run.

When water is confined in a substrate, such as for instance porous vycor glass, data analysis requires accounting for excluded volume effects and interaction between water and substrate. The latter is relevant, as the surface to volume ratio exceeds by orders of magnitude its value for standard bulk samples and consequently the contribution to the $F(Q)$ functions of substrate-water PSFs cannot be neglected. For instance in the case of vycor, presented here, the substrate is basically SiO_2, but the presence of silanol groups on the surface brings the need to distinguish between bridging (bO) and non bridging (nbO) oxygens, leading to a total of 21 PSFs in the measured $F(Q)$ function. Moreover the existence of regions where water molecules are not allowed spoils the significance of a direct comparison of the RDFs of confined water with those of the bulk. Indeed the local density of water molecules close to a given water molecule will be higher than the average water density by a factor of the order of V/V_p, where V and V_p are the total sample volume and the fraction occupied by water, respectively. This means that, in the case of confinement in vycor presented here, the water-water RDFs decay by more than a factor of 3 in going from short to longer range. A detailed treatment of this "excluded volume effects" can be found in ref. 19 and 20. In view of the above difficulties, data analysis through the EPSR routine, along the lines described in ref. 11 is mandatory. Shortly this requires building an atomistic model of the substrate and refinement of its structure against the experimental data recorded for the dry substrate; equilibration of the water sample within the substrate cavity and subsequent refinement of its structure against the experimental data. Moreover, in order to correctly normalize the RDFs, the uniform atom distribution has to be computer simulated within the same confined geometry as for the water sample. Further outcome of this analysis is that, after achieving a good fit of the experimental data, a shell inspection of the water structure can be performed moving from the middle of the cavity towards its walls, along the same lines suggested by computer simulation work.[21] Indeed it is expected, based on the non uniform density profile within the pore[9–11,21] that the microscopic structure of water evolves on going from the middle to periphery of the pore.

B. Deep inelastic neutron scattering

Deep inelastic neutron scattering (DINS) technique[22,23] is based on measurements at high energy, $\hbar\omega$, and high momentum, $\hbar Q$, transfers, thus providing a probe of both the short-time ($t \leq 10^{-15}$ s) dynamics and local ($r \leq 1$ Å) environment of the atoms in materials, and it represents a unique tool to directly measure single-particle physical quantities. The high energy and momentum transfers achieved allow to describe the scattering process within the framework of the Impulse Approximation (IA).[23–25] This assumes that a single particle of the system, struck by the scattering probe, recoils freely from the collision, with inter-particle interaction in the final state being negligible. The scattering cross section is then expressed in terms of the single particle momentum distribution, $n(p)$, whose variance is in turn related to the average kinetic energy $\langle E_k \rangle$ of the target nuclei. This means that the dynamical structure factor, $S_{IA}(\vec{Q},\omega)$, is related to the $n(p)$ through the relation:

$$S_{\mathrm{IA}}(\vec{Q},\omega) = \int n(\vec{p})\ \delta\left(\omega - \frac{\hbar Q^2}{2M} - \frac{\vec{p}\vec{Q}}{M}\right) \mathrm{d}\vec{p} \tag{5}$$

where $\frac{\hbar^2 Q^2}{2M}$ is the recoil energy of the struck atom of mass M. This dependence on the atomic mass makes the intense hydrogen signal well separated in neutron time of flight from those from the heavier nuclei over a wide scattering angle range.The dynamical structure factor is then expressed in terms of a neutron Compton profile (NCP):[23] $F(y) = \frac{\hbar Q}{M} S_{\mathrm{IA}}(\vec{Q},\omega)$, where $y = \frac{M}{\hbar Q}\left(\omega - \frac{\hbar Q^2}{2M}\right)$ is the West scaling variable.[23] The NCP lineshape is convolved with the instrumental resolution function, and represented as a series expansion in Hermite polynomials:

$$F(y) = \frac{e^{-\frac{y^2}{2\sigma^2}}}{\sqrt{2\pi}\sigma}\left[1 + \sum_n \frac{c_n}{2^{2n} n!} H_{2n}\left(\frac{y}{\sqrt{2}\sigma}\right)\right] \tag{6}$$

The coefficients c_n and σ, appearing in the series expansion, are determined by a least squares fitting procedure of the experimental data, along the lines described in ref. 23 and the momentum distribution along the $\vec{Q}$ direction, $n(p)$, is finally calculated as:

$$n(p) = -\frac{1}{2\pi\hbar^3 y}\frac{\mathrm{d}}{\mathrm{d}y}F(y) = \frac{e^{-\frac{p^2}{2\sigma^2}}}{(\sqrt{2\pi}\sigma)^3}\sum_n c_n(-1)^n L_n^{\frac{1}{2}}\left(\frac{p^2}{2\sigma^2}\right) \tag{7}$$

where $L_n^{1/2}$ are generalized Laguerre polynomials. Deviations from the IA give rise to the so-called final state effects:[24] the distortion to the lineshape of eqn (6) can be in this case accounted for by the addition of further terms proportional to $\frac{1}{q} H_3\left[\frac{y}{\sqrt{2}\sigma}\right]$.

When water is confined in a hydrophilic substrate, such as silica xerogel used in the experiments reported here, both water and silanol hydrogens will contribute to scattered intensity at the same time of flight. Consequently the measured experimental profile, $F_{hyd}^{exp}(y)$, is expressed as:[26]

$$F_{hyd}^{exp}(y) = xF_{dry}^{exp}(y) + (1 - x)F^{\mathrm{exp}}{}_{\mathrm{wat}}(y) \tag{8}$$

where $F_{dry}^{exp}(y)$ is the measured NCP of the dry substrate, $F_{wat}^{exp}(y)$ the water hydrogens contribution and x the silanol groups concentration. The σ parameter in eqn (6) satisfies in this case the relation: $\sigma_{hyd}^2 = x\sigma_{dry}^2 + (1 - x)\sigma_{wat}^2$ and once σ_{dry} is extracted from the dry sample measurements, x and σ_{wat} can be calculated by fitting the experimental NCP, to finally get the water hydrogens mean kinetic energy:

$$\left\langle E_{\mathrm{k}}^{\mathrm{wat}}\right\rangle = \frac{3\hbar^2}{2M}\int_{-\infty}^{\infty} y^2 F_{\mathrm{wat}}^{\mathrm{exp}}(y)dy = \frac{3\hbar^2}{2M}\sigma_{\mathrm{wat}}^2 \tag{9}$$

III. Microscopic structure

Pioneering studies of the structure of confined water have been performed by Steitler *et al.*[27,28] and Bellissent-Funel *et al.*.[29] Both groups have performed neutron diffraction experiments on D_2O within silica nanoporous materials, at different temperatures, and report the isochoric temperature derivative (ITD) of the total neutron weighted radial distribution function:

 This journal is © The Royal Society of Chemistry 2009

$$\frac{\Delta d(r, \Delta T)}{\Delta T} = \frac{2}{\pi \Delta T} \int_0^{\infty} Q[S(Q, T) - S(Q, T_0)] \sin(Qr) dQ \tag{10}$$

where $\Delta T = T - T_0$ and $\rho(T) = \rho(T_0)$. This quantity is often used to evidence the amplitude and r-range of the temperature evolution of the microscopic structure of water around its maximum density. The above quoted papers arrive at quite distinct conclusions about the comparison of the ITD measured in the bulk and under confinement. As a matter of fact ref. 28 reports differences with bulk water in the region of the hydrogen bond peak (at ~ 2 Å) only for the most severe confinement. On the contrary the amplitudes of the temperature derivatives reported in ref. 29 is lower than that of bulk water by a factor 4 in the whole r-range and suggest that the major temperature effects under confinement are visible on the second neighboring shell, contrarily to what happens in the bulk, where the hydrogen bond peak is the most sensitive to temperature changes.

MD simulations performed on simple point charge/extended (SPC/E) water[30] confined within an atomistic model of porous vycor[21] suggest instead that the structure of water close to the center of the pore is similar to that of bulk water at a temperature of 30 K higher. On the contrary the structure of water molecules within 5Å from the substrate walls is sensibly distorted with respect to the bulk and interestingly is almost unaffected by temperature changes, as if it were in a glassy state. The latter evidence may explain the results of ref. 28, and the observation that the nucleation temperature decreases with the hydration level.

The first experimental determination of the partial RDF of water in confinement has been published in 1998[20,31] and recently reanalyzed through the EPSR routine,[11] which allows also a layer analysis similar to what reported in ref. 21. Here we will compare the RDF's extracted in ref. 11 with data on bulk water, both supercooled[32] and at two temperature states in the stable phase. All data have been collected on SANDALS[13] and analyzed using the same routines, along the lines reported in section II. The density and temperature of the experiments are reported in Fig. 1, along with data for bulk ambient water. It is worth noting that the overall density change over the investigated bulk phase is of the order of 1%, while the density of water confined in vycor glass is ~10% lower than that of ambient water and that the density profile over the pore diameter, as calculated within the EPSR code is close to that reported in the atomistic MD simulation.[21]

The partial RDF's are compared in Fig. 2–4. In bulk water, the first peak of the $g_{OO}(r)$ (see Fig. 2) moves to shorter distances as the temperature decreases towards

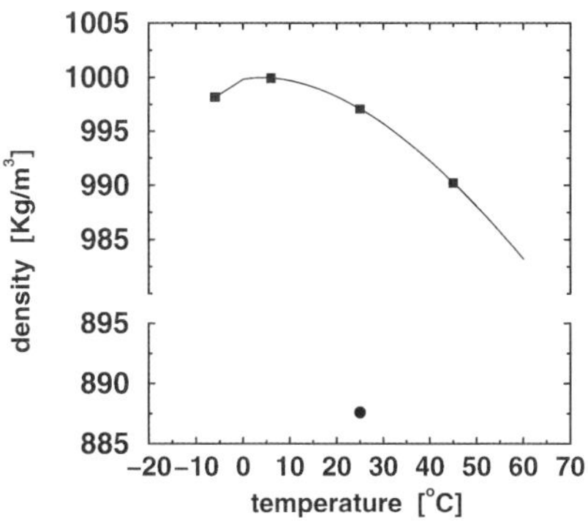

Fig. 1 The solid squares represent the density vs. temperature states considered in the bulk; the circle refers to water confined in vycor glass. Notice that the vertical axis is broken and the difference of density between bulk and confined ambient water is much greater than that observed in the whole temperature range from 46 and –6 °C.

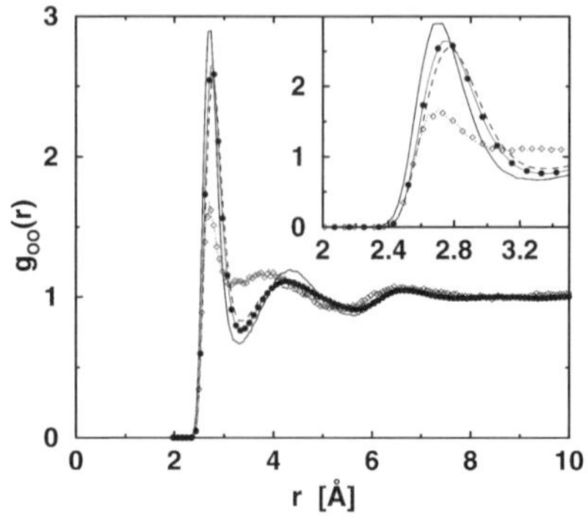

Fig. 2 Comparison of the oxygen-oxygen radial distribution functions at all measured state points, namely: bulk water at $T = -6$ °C (solid line); $T = 6$ °C (solid line with circles); $T = 46$ °C (dashed line); water confined in vycor glass (dashed line with diamonds). The inset reports the first peak on a larger scale: notice the shift towards shorter distances in both supercooled and confined water.

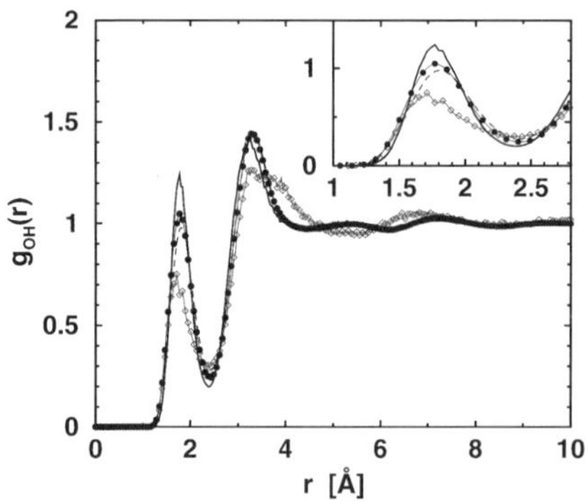

Fig. 3 Comparison of the oxygen-hydrogen radial distribution functions at all measured state points (same symbols as in Fig. 2). The inset reports the hydrogen bond peak on a larger scale: notice the reduced intensity in confined water.

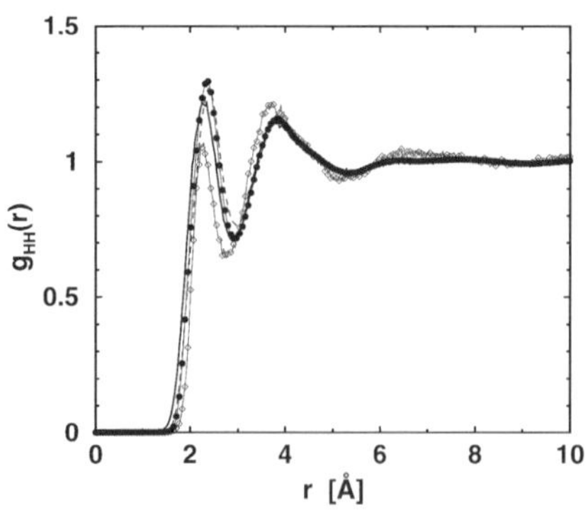

Fig. 4 Comparison of the hydrogen-hydrogen radial distribution functions at all measured state points (same symbols as in Fig. 2).

 This journal is © The Royal Society of Chemistry 2009

the metastable phase, and all structures become sharper. Under confinement at ambient conditions the first peak comes at ~2.7 Å, as in the bulk supercooled state, but its intensity is severely reduced, and the second peak moves to shorter distances. The latter finding is a signature of sensible distortion of the tetrahedral arrangement of neighboring molecules upon confinement. This is consistent with the reduced intensity of the first peak of the $g_{OH}(r)$ (see Fig. 3), which brings a reduced number of H-bonds (2.2 instead of 3.6 per molecule). Nevertheless, as shown in the inset of Fig. 3, these are on the average shorter (*i.e.* stronger) than in the bulk. The increased orientational disorder with respect to the bulk states is also demonstrated by changes of the second peak of the $g_{OH}(r)$ and overall changes of the $g_{HH}(r)$ (see Fig. 4). According to ref. 21, these changes should be mainly due to the thin (~5Å) layer of molecules close to the confining surface. We have therefore performed on the confined water data the same layer analysis suggested by Gallo *et al.*.[21] This confirms that the microscopic structure of a thin layer of water molecules close to the substrate is markedly different from that of molecules in the inner layers, while all the inner layers have the same structure, although strongly distorted compared to bulk water at all investigated temperatures (see Fig. 5). Thus our EPSR data refinement, contrarily to the MD simulations,[21] points out that the distortions observed on the microscopic structure of water under confinement propagate throughout the pore.

The relative orientation of neighboring molecules can also be investigated looking at the distribution function of the $O\hat{O}O$ angle (see Fig. 6): at $T = 46$ °C this function is characterized by an intense and broad peak centered at ~104°, due to the tetrahedral arrangement of water molecules within the H-bonded clusters. A sharper peak at ~56° is instead due to the presence in the first neighbor's shell of molecules not directly H-bonded to the central one, and usually referred to as "interstitial". When the temperature is lowered towards the supercooled state both these features become sharper. We notice that the distribution function obtained for confined water shows an even broader main peak, while the feature at sharp angles is almost completely missing: interestingly the reduced number of interstitial molecules may be the origin of the average reduced density of water confined in vycor glass.

IV. Momentum distribution and proton mean kinetic energy

The first DINS experiments performed on water at a spallation neutron source date back to the 2000.[36] Since then the experimental technique and data analysis

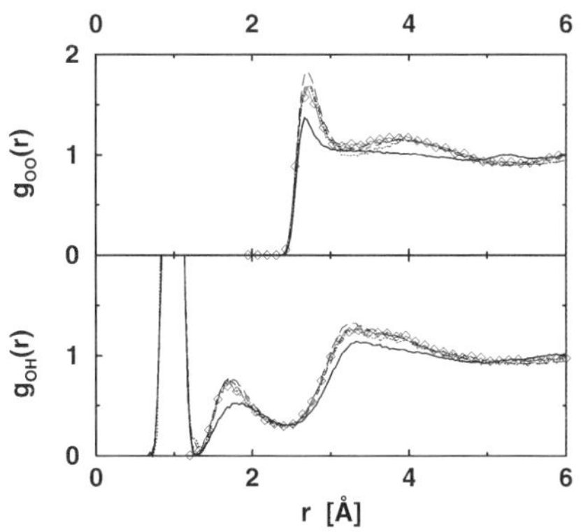

Fig. 5 Shell analysis of the *OO* (top) and *OH* (bottom) RDFs: the solid line with diamonds is the total $g(r)$ function; the dotted line represents the contribution of water molecules in the inner cylinder of radius 5Å; the dashed and long-dashed lines represent the contribution of water molecules in two annular shells between 5 and 10Å and between 10 and 15Å respectively; the solid line is the contribution of water molecules in the shell between 15Å and the vycor surface.

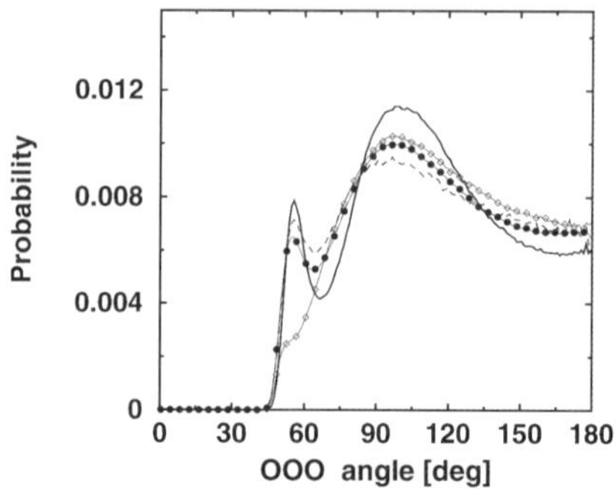

Fig. 6 Distribution function of the $O_w\hat{O}_wO_w$ angle (same symbols as in Fig. 2). Notice the reduced contribution from interstitial water molecules at $\sim$ 56° under confinement.

procedure have progressed: enhanced detector stability and resolution, along with new computer routine for correction of systematic effects and conversion of the experimental cross-section to the $n(p)$ function allow to determine the lineshape of the momentum distribution, not only its second momentum.[23] DINS experiments are sensitive to details of the structural environment of a water proton in a double well of a hydrogen bond, as in principle can distinguish between a proton wavefunction with amplitude in both wells and a mixture of states where the proton is localized in a well or in another. The state of a proton delocalized in a double well is usually referred to as coherent. The signature of proton coherence has to be found in the high-p tails of the momentum distribution, more evident if compared to a purely harmonic distribution, as for instance that reported in ref. 37. Consequently it is customary to report the radial momentum distribution: $4\pi p^2 n(p)$ vs. p, in order to enhance the tails of the $n(p)$ function.[23] This function is reported in Fig. 7A for water bulk phases[33–35] and in Fig. 7B for water confined in xerogel.[26]

The radial momentum distribution of both water stable phases (ambient liquid and ice)[33,34] are slightly asymmetric functions, with a maximum moving from $\sim$6.7 Å^{-1} to $\sim$5.8 Å^{-1} going from the liquid to the crystalline phase. On the contrary the signature of proton coherence on a double well potential shows up in the metastable supercooled phase,[35] as a secondary maximum between 17 and 18 Å^{-1}. Interestingly, the same phenomenology is found when water is confined in porous xerogel at ambient temperature (see Fig. 7B): in this case the more severe is the confinement, the more evident is the proton coherence. Similar confinement effects are visible at the surface of a protein.[38]

Such changes of the lineshape of the radial momentum distribution bring along also differences among the second momenta, that is the proton mean kinetic energy, $\langle E_k \rangle$, reported in Fig. 8. Data for bulk water stable phases (open circles and square) are in agreement with the theoretical predictions (dotted line) made by a semi-classical calculation of the translational and rotational contributions and a full quantum calculation for the vibrational modes, under the assumption of decoupling of motions.[39] Incidentally this theoretical prediction correctly extrapolates to the measured mean kinetic energy at sub- and supercritical states.[39] Both confined water data and supercooled ones show instead large deviations from the theoretical values. In the case of water confined in xerogel this finding has been rationalized as due to the interaction between water and the confining substrate, which may modify the potential landscape seen by the protons.[26] That under a quite moderate supercooling the mean kinetic energy of bulk water protons could change by $\sim$ 50% is instead surprising and not yet completely understood.[35] Nevertheless the comparison with water under confinement may suggest some speculative clue.

As a matter of fact DINS is a local probe that can be used to infer details of the effective potential seen by a proton in water. The $n(p)$ function of a statistical

 This journal is © The Royal Society of Chemistry 2009

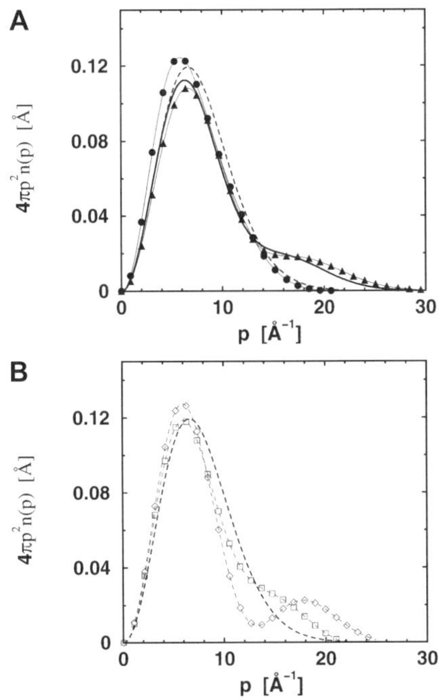

Fig. 7 Spherical averages of the proton momentum distribution for: A) bulk ambient water (dashed line); ice Ih (solid line with solid circles); supercooled water at T = 269 K (solid line) and at T = 271 K (solid line with solid triangles); B) water confined in xerogel with pore diameter of 82 Å (dashed line with open squares) and of 24 Å (dashed line with diamonds); data for bulk ambient water are reported again as dashed line for comparison.

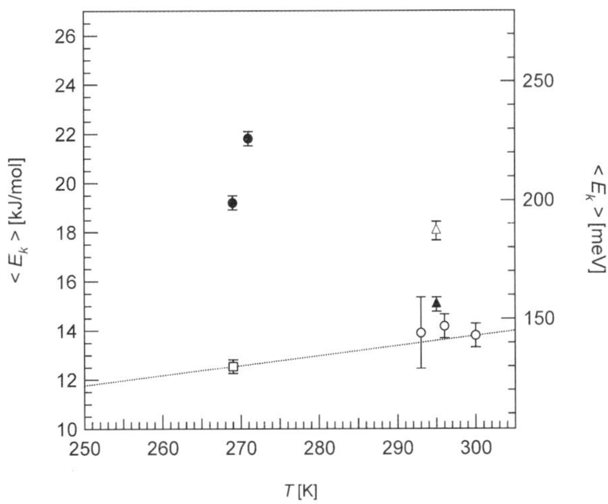

Fig. 8 Mean kinetic energy vs. temperature for water protons in: bulk water at near ambient conditions (open circles); ice (open square); supercooled water (solid circles); water confined in xerogel with pore diameter of 84 Å (solid triangle) and 24 Å (open triangle). Data are reported with their error bar. Linear temperature dependence of $\langle E_k \rangle$ (see text) is shown as a dotted line.

mixture of H-bonds in which hydrogens are localized in the vicinity of an oxygen or the other along a bond is indeed predicted to have a single maximum. On the contrary it has oscillations at wavevector values proportional to the inverse of the distance, d, between the centers of the bimodal distribution of hydrogens:

$$\Delta p = \frac{2\pi}{d} \tag{11}$$

when the ground state wavefunction is delocalized over two minima of the potential well.[24] Hence in both supercooled and confined water there is evidence of proton coherent delocalization along the hydrogen bonds, at variance with water in the bulk stable phase. This suggests that the role of the substrate in the case of confined water cannot be the primary direct origin of proton coherence. The potential well seen by an hydrogen is instead primarily determined by its two first neighbor oxygens and to a lesser extent by many body contributions coming from the entire surrounding medium. In other words, the similarities observed in the radial distribution functions of metastable and confined water may suggest a key for interpreting the present results. In both systems we have observed a shortening of the average distance between first neighbor oxygens, along with a shortening of the H-bond, most evident in the confined fluid. The first finding (*i.e.* shortening of the *O–O* distance between first neighbors) may be the origin of proton coherence along the H-bond, while the second observation (*i.e.* strengthening of the H-bond under confinement) is compatible with the larger separation between the first and second maximum of the $n(p)$ [see eqn (11)] function under confinement, that is with a coherence of the proton over a shorter d range. The details of the lineshape of the $n(p)$ functions are instead determined by the details of the effective potential well, such as height and steepness of the barrier between the two coherent proton states and may be influenced by the medium range order in the liquid. If we make the assumption that water protons may be considered as particles in an effective two-body potential well, then the higher kinetic energy of protons in supercooled water suggests that they see a deeper well, compared to the confined liquid case. This not negligible detail of the potential landscape is likely due to the different medium range order observed by neutron diffraction with isotopic substitution (NDIS) experiments, namely to the strong reduction of the number of interstitial molecules and to the collapse of the second shell of water molecules in the vicinity of the first one. Interactions with the substrate may additionally contribute to the above quoted differences: on the other hand their main effect is reflected in the structural modification.

V. Conclusions

We have compared structure and single proton dynamics in bulk water with the same quantities measured in the case of water confined in silica substrates, with the aim of understanding whether confinement of water in small volumes may or not be the right way to circumvent the difficulties inherent in deep supercooling of bulk water. The conclusion is that in both structure and dynamics there are similarities and clear differences as well.

In particular as far as the microscopic structure is concerned, in both supercooled bulk water and water under confinement in a silica substrate we observe a closer average distance between first neighboring oxygen sites and shortening of the H-bond. Nevertheless the overall structure of the liquid is quite different, since the number of H-bonds per molecule as well as the number of interstitial water molecules is severely reduced under confinement, and the medium range order of the

 This journal is © The Royal Society of Chemistry 2009

extended H-bond network of water is also distorted, as evidenced by changes of position and shape of the second neighbor shell and distribution function of the $O\hat{O}O$ angle. As far as the momentum distribution of the water hydrogens is concerned, DINS experiments in both supercooled bulk water and water confined in silica substrates evidence coherent delocalization of the water protons along the H-bonds, within a double well potential landscape. Nevertheless the proton mean kinetic energy and details of the potential landscape are different in the two cases. A parallel study of the DINS spectra and microscopic structure, as presented here, may be helpful in rationalizing these findings, along the lines reported at the end of Section IV. That is the shortening of the average distance between first neighbor oxygens may be the origin of proton coherence along the H-bond, while the strengthening of the H-bond under confinement may be the signature of a larger separation between the first and second maximum of the $n(p)$ function under confinement in comparison with the supercooled state. The different values of the mean kinetic energy may be due to the combined effect of the different arrangement of water molecules at longer distances and of the interaction with the confining substrate.

Obviously the present conclusions need validation through future experimental work, performed more systematically on different substrates, hydration levels and temperatures, nevertheless they draw attention on the need for a cautious comparison between metastable and confined water: these systems do have similarities, but cannot in any respect be considered the same liquid.

Acknowledgements

This work has been performed within the Agreement No. 01/9001 between STFC and CNR, concerning collaboration in scientific research at the spallation neutron source ISIS and with partial financial support of CNR.

References

1 O. Mishima and H. E. Stanley, *Nature*, 1998, **396**, 329.
2 R. J. Speedy, *J. Phys. Chem.*, 1982, **86**, 982; R. J. Speedy, *J. Phys. Chem.*, 1982, **86**, 3002.
3 M. C. D'Antonio and P. G. Debenedetti, *J. Chem. Phys.*, 1987, **86**, 2229.
4 P. H. Poole, F. Sciortino, U. Essmann and H. E. Stanley, *Nature*, 1992, **360**, 324.
5 P. H. Poole, F. Sciortino, T. Grande, H. E. Stanley and C. A. Angell, *Phys. Rev. Lett.*, 1994, **73**, 1632.
6 S. Sastry, P. G. Debenedetti, F. Sciortino and H. E. Stanley, *Phys. Rev. E*, 1996, **53**, 6144; L. P. N. Rebelo, P. G. Debenedetti and S. Sastry, *J. Chem. Phys.*, 1998, **109**, 626.
7 L. D. Gelb, K. E. Gubbins, R. Radhakrishnan and M. Sliwinska-Bartkowiak, *Rep. Prog. Phys.*, 1999, **62**, 1573.
8 C. Alba Simionesco, 2008, private communication.
9 T. M. Truskett, P. G. Debenedetti and S. Torquato, *J. Chem. Phys.*, 2001, **114**, 2401.
10 P. Gallo, M. Rovere and E. Spohr, *Phys. Rev. Lett.*, 2000, **85**, 4317.
11 H. Thompson, A. K. Soper, M. A. Ricci, F. Bruni and N. T. Skipper, *J. Phys. Chem.*, 2007, **B111**, 5610.
12 http://www.isis.rl.ac.uk.
13 A. K. Soper, in *Proceedings of the Conference on Advanced Neutron Sources 1988*, ed. D. K. Hyer, IOP Conf. Proc. London, 1989, no. 97, p. 353.
14 R. Senesi, C. Andreani, Z. Bowden, D. Colognesi, E. Degiorgi, A. L. Fielding, J. Mayers, M. Nardone, J. Norris, M. Praitano, N. J. Rhodes, W. J. Stirling, J. Tomkinson and C. Uden, *Physica*, 2000, **B276–278**, S189.
15 G. W. Neilson and J. E. Enderby, *Proc. R. Soc. London, Ser. A*, 1983, **390**, 353.
16 V. F. Sears, *Neutron News*, 1992, **3**, 26.
17 A. K. Soper, W. S. Howells and A. C. Hannon, ISIS Facility, Science and Technology Facilities Council, Didcot, UK, 2007, see http://www.isis.rl.ac.uk/disordered.
18 A. K. Soper, *Chem. Phys.*, 1996, **202**, 295; A. K. Soper, *Chem. Phys.*, 2000, **258**, 121; A. K. Soper, *Mol. Phys.*, 2001, **99**, 1503; D. T. Bowron, J. L. Finney and A. K. Soper, *J. Phys. Chem. B*, 1998, **102**, 3551.
19 A. K. Soper, *J. Phys.: Condens. Matter*, 1997, **9**, 2399.

20 A. K. Soper, F. Bruni and M. A. Ricci, *J. Chem. Phys.*, 1998, **109**, 1486.
21 P. Gallo, M. A. Ricci and M. Rovere, *J. Chem. Phys.*, 2002, **116**, 342.
22 G. F. Reiter, J. Mayers and J. J. Noreland, *Phys. Rev.*, 2002, **B65**, 104305.
23 C. Andreani, D. Colognesi, J. Meyers, G. F. Reiter and R. Senesi, *Adv. Physics*, 2005, **54**, 377.
24 G. F. Reiter and R. Silver, *Phys. Rev. Lett.*, 1985, **54**, 1047.
25 G. I. Watson, *J. Phys.: Condens. Matter*, 1996, **8**, 5955.
26 V. Garbuio, C. Andreani, S. Imberti, A. Pietropaolo, G. F. Reiter, R. Senesi and M. A. Ricci, *J. Chem. Phys.*, 2007, **127**, 154501.
27 D. C. Steitler, J. C. Dore and C. W. Wright, *Mol. Phys.*, 1983, **48**, 1031.
28 D. C. Steitler and J. C. Dore, *Mol. Phys.*, 1985, **56**, 1001.
29 M.-C. Bellissent-Funel, J. Lal and L. Bosio, *J. Chem. Phys.*, 1993, **98**, 4246.
30 H. J. C. Berendsen, J. R. Grigera and T. P. Straatsma, *J. Phys. Chem.*, 1987, **91**, 6269.
31 F. Bruni, M. A. Ricci and A. K. Soper, *J. Chem. Phys.*, 1998, **109**, 1478.
32 A. Botti, F. Bruni, A. Isopo, M. A. Ricci and A. K. Soper, *J. Chem. Phys.*, 2002, **117**, 9570.
33 A. Pietropaolo, C. Andreani, A. Filabozzi, R. Senesi, G. Gorini, E. Perelli Cippo, M. Tardocchi, N. J. Rhodes and E. M. Schooneveld, *JINST*, 2006, **1**, P04001.
34 G. F. Reiter, J. C. Li, J. Mayers, T. Abdul-Redah and P. Platzman, *Braz. J. Phys.*, 2004, **34**, 142.
35 P. Pietropaolo, R. Senesi, C. Andreani, A. Botti, M. A. Ricci and F. Bruni, *Phys. Rev. Lett.*, 2008, **100**, 127802.
36 C. H. Uffindell, A. I. Kolesnikov, J. C. Li and J. Mayers, *Phys. Rev. B*, 2000, **62**, 5492.
37 J. A. Morrone, V. Srinivasan, D. Sebastiani and R. Car, *J. Chem. Phys.*, 2007, **126**, 234504.
38 R. Senesi, A. Pietropaolo, A. Bocedi, S. E. Pagnotta and F. Bruni, *Phys. Rev. Lett.*, 2007, **98**, 138102.
39 C. Andreani, D. Colognesi, E. Degiorgi and M. A. Ricci, *J. Chem. Phys.*, 2001, **115**, 11243.

 This journal is © The Royal Society of Chemistry 2009

Structural and mechanical properties of glassy water in nanoscale confinement

Thomas G. Lombardo, Nicolás Giovambattista and Pablo G. Debenedetti*

Received 1st April 2008, Accepted 16th May 2008
First published as an Advance Article on the web 25th September 2008
DOI: 10.1039/b805361h

We investigate the structure and mechanical properties of glassy water confined between silica-based surfaces with continuously tunable hydrophobicity and hydrophilicity by computing and analyzing minimum energy, mechanically stable configurations (inherent structures). The structured silica substrate imposes long-range order on the first layer of water molecules under hydrophobic confinement at high density ($\rho \geq 1.0$ g cm^{-3}). This proximal layer is also structured in hydrophilic confinement at very low density ($\rho \sim 0.4$ g cm^{-3}). The ordering of water next to the hydrophobic surface greatly enhances the mechanical strength of thin films (0.8 nm). This leads to a substantial stress anisotropy; the transverse strength of the film exceeds the normal strength by 500 MPa. The large transverse strength results in a minimum in the equation of state of the energy landscape that does not correspond to a mechanical instability, but represents disruption of the ordered layer of water next to the wall. In addition, we find that the mode of mechanical failure is dependent on the type of confinement. Under large lateral strain, water confined by hydrophilic surfaces preferentially forms voids in the middle of the film and fails cohesively. In contrast, water under hydrophobic confinement tends to form voids near the walls and fails by loss of adhesion.

1 Introduction

Water is the most abundant substance on the Earth's surface, of which it covers approximately 70%.[1] There is hardly a human activity from agriculture to travel and from manufacture to artistic creation that is not dependent on, influenced or inspired by the properties of water. Life itself as we know it depends on water (see *e.g.* ref. 2–5). About 70% of human body weight is water, and 70% of this water is found in intracellular fluid.[6] Thus, most biochemical processes such as protein synthesis occur in aqueous solution. It has been proposed that water be considered as 'an integral and active component of biomolecular systems' rather than as an 'inert environment'.[7] To the physical scientist, water remains a source of continued interest and curiosity on account of its many anomalous properties. The peculiar behavior of many of water's properties, which render it qualitatively different from most other substances, becomes more pronounced at low temperature.[8–10] In the liquid phase, examples include the increase of the isothermal compressibility upon isobaric cooling[11] and the enhancement of diffusivity upon isothermal compression.[12] In the glass state, water exhibits polyamorphism, existing in at least two different forms,[13–16] which may be separated by a first-order transition.[8,17] Understanding the molecular basis of this complex behavior remains an active area of research.[17]

Department of Chemical Engineering, Princeton University, Princeton, New Jersey, USA. E-mail: pdebene@princeton.edu

In many situations, water is found in confined geometries or as a thin film in contact with surfaces. Examples are numerous, encompassing fields such as geology (*e.g.* in zeolites[18] and mineral inclusions[19]) as well as technological applications (*e.g.* in nanofluidics[20,21] and cryopreservation[22]). Understanding the behavior of confined and interfacial water is particularly important in biology. The cell interior is a crowded environment where proteins are separated on average by 2–3 water layers.[23] The dynamics of proteins in such an environment is tightly correlated with water dynamics.[24] Water's behavior under confinement is incompletely understood, depending sensitively on the properties of the confining surfaces such as surface hydrophilicity,[25–27] chemical heterogeneity[28] and topography.[29,30] Recent experiments that attempt to shed light on supercooled water's bulk properties from measurements performed in nanoscopic pores[31,32] highlight the importance of understanding water's properties in the presence of confining surfaces.[33–35]

The characterization of confined water properties at low temperature (either in the glass state, or close to the system's glass transition temperature, T_g) is particularly relevant in pharmaceutical applications.[36–39] This is primarily because one of the most common techniques to preserve pharmaceuticals, including proteins as well as small-molecular-weight drugs and peptides, is lyophilization.[37] Lyophilization is the process of freezing the substance to be stored into an amorphous state, followed by drying.[40] In addition to the substance to be preserved, lyophilized formulations consist of a glassy matrix[41] composed typically of sugars with up to 4% by weight of water, plus excipients such as fillers and anti-oxidants.[42,43] Cooling the system into the glassy state greatly reduces the rate of transport processes, thereby protecting proteins or other labile biochemicals against aggregation or chemical reactions which can otherwise damage the product.[37,44–46] The drying process consists of sublimation of water from the glassy matrix: water is a good plasticizer and unless its concentration is kept sufficiently low it will greatly facilitate molecular mobility.[47–51] Although lyophilization is commonly used in the pharmaceutical and food processing industries, it is incompletely understood at the molecular level. Lyophilization generates both freezing and drying stresses. These stresses, as well as aging in the glass state, can result in undesired molecular mobility.[36,37,52,53] Such residual stresses are intimately related to the way in which water behaves under confinement and at extended interfaces with large solutes, such as proteins, both in the glass state and as the system is cooled from the liquid phase across T_g.

Ice formation inside the cell is injurious and can be lethal for organisms living in subfreezing conditions.[54] One strategy for survival at these conditions involves depressing water's freezing point.[55] Alternatively, ice nucleation can be inhibited kinetically by antifreeze proteins, such as those commonly found in Antarctic fish,[56] and also observed in insects at low temperature.[57] The mechanisms by which biological surfaces can suppress ice nucleation are poorly understood.

Cold water close to biological surfaces is believed to play an important role in the low-temperature unfolding of proteins.[58] This process, which is the subject of much current interest (see *e.g.* ref. 59 and 60), is believed to have important implications for the possibility of life at extreme conditions.[61] Finally, we mention the fact that thin films of glassy water have played an important role in the ongoing quest to understand the thermodynamics of deeply supercooled bulk water.[62–65]

The selected examples above illustrate the importance of understanding the behavior of supercooled and glassy water in confinement. Here, we are specifically interested in confined glassy water, and we take a step in this direction by exploring the structure and mechanical properties of amorphous solid water confined between two parallel silica-based nanoscale surfaces separated by $d = 0.8$ and 1.6 nm. We consider a family of surfaces with tunable hydrophilicity and study glassy water configurations obtained by energy minimization (inherent structures[66]). We explore the effect of d and surface hydrophilicity on water stresses and on the amorphous films' mechanical failure modes. The picture that emerges is quite rich. Water responds to lateral strain, surface chemistry and confinement in ways that include

 This journal is © The Royal Society of Chemistry 2009

the formation of layered phases with unusually large mechanical strength, and the appearance of surface-chemistry-sensitive failure modes. It is therefore anticipated that future similar studies, but with the added complexities of biologically-relevant confining surfaces[67] and carbohydrates, will provide valuable insights into the mechanisms underlying the preservation of labile biomolecules in glassy matrices.

The present work was inspired by recent theoretical and computational studies of atomic glassy films using the potential-energy landscape formalism.[68–70] We also note the recent related study of inherent structures of a confined Lennard-Jones binary mixture by Attili *et al.*[71] This paper is organized as follows. Section 2 provides details on the molecular simulation and energy minimization techniques and specifies the various systems studied. Analysis and results of glassy configurations of confined water are presented and discussed in section 3. The major conclusions arrived at in this work and the open questions arising as a result of our study are listed in section 4.

2 Simulation methods

A powerful framework for examining liquids and glasses is the potential-energy landscape (PEL) approach. The basic concepts of the PEL originate in the work of Goldstein, more than 30 years ago.[72,73] Stillinger and Weber incorporated these ideas into a mathematical theory that placed the PEL interpretation on firm conceptual ground.[66,74] The PEL is the hypersurface defined by the potential energy (PE) of the system as a function of its degrees of freedom. For a given system composed of N particles in a fixed volume, there exist $(3 + n)N$ degrees of freedom. Each particle is specified by three center-of-mass (CM) coordinates and n orientational, vibrational and conformational degrees of freedom. The PEL is the potential-energy hypersurface in $[(3 + n)N + 1]$-dimensional space, and the system configuration is represented by a single point on the surface. This multidimensional surface cannot be visualized, but the local minima and potential-energy barriers between them are often characterized with three-dimensional descriptions such as peaks, basins and ridges. As the particles of the system move, the representative point of the system on the PEL also moves.[66,75,76] For the case of high-temperature liquids, kinetic energy allows the system to visit most of the basins. However, as the temperature is reduced, potential-energy barriers become relevant and the number of basins accessible to the system decreases. Upon further cooling, the system becomes trapped in a single basin, entering the glassy state. Here we use the term basin to denote the collection of configurations that are mapped onto the same local minimum (or its $N!$ permutationally identical versions obtained by particle interchanges) upon energy minimization.

Computationally, glassy configurations can be generated by two methods: rapid cooling and potential-energy minimization (which can be regarded as the infinite-cooling-rate limit of the former process). This study focuses on the latter approach, and in particular on inherent structures (IS), the local minima of the PEL. Inherent structures are calculated by minimizing the potential energy of an equilibrium configuration with respect to all of its degrees of freedom. The local minima of the PEL are mechanically stable packings that in principle lack long-range order, and are hence representative of glasses.

We analyze the effects of wall-to-wall separation, density and surface chemistry for glassy water confined between two parallel plates. Water is modeled with the extended simple-point-charge (SPC/E) model: a rigid three-site representation of water with a tetrahedral bond angle and partial charges at the oxygen and hydrogen sites, in addition to Lennard-Jones (LJ) dispersive interactions between oxygen atoms.[77] The properties of bulk SPC/E water have been studied extensively in the liquid and glassy states, and the model exhibits many of water's anomalies.[78–80] We confine water between two parallel atomically detailed silica-based walls with force fields developed by Lee and Rossky;[81] these walls have been used in several recent studies to examine the hydration and phase behavior of water under confinement.[27,28,82] The underlying structure of the confining walls consists of four layers of SiO_2 which reproduce the (111) octahedral face of cristobalite.

The unit cell of the silica-based walls is an idealized tetrahedron with O–O and Si–O distances of 0.247 nm and 0.151 nm respectively. All Si and O wall atoms interact with the O atom of water molecules *via* a LJ interaction. When only LJ interactions are present, the surface displays a hydrophobic character;[82] to increase the plate's affinity for water the surface of each plate is fully hydroxylated, producing charged Si–O–H groups with a bond angle of 109.27°. These charged Si, O and H atoms exist only at the surface of the wall and have Coulombic interactions with the three charged sites on each SPC/E water molecule; the addition of electrostatic interactions creates a hydrophilic surface.[27,28,82] The Si and O atoms are fixed, but the H atoms are allowed to move with fixed O–H length and Si–O–H bond angle; the motion of each H atom therefore traces a circle parallel to the wall. The LJ parameters for all Si and O wall atoms, and the charges for surface Si–O–H atoms are given in Table 1. All LJ interactions, including those between O atoms of SPC/E water molecules, are truncated at 0.79 nm with a switching function such that the potential and its first two derivatives are zero at the cutoff.[83] This is essential for the system to be amenable to energy minimization techniques used to generate glassy configurations (see below). More details about the walls can be found in ref. 81 and 82.

Most natural surfaces have interactions with water molecules that are between the hydrophobic and hydrophilic cases described above. To model these surfaces, we tune the polarity of the wall by rescaling the charges of the Si–O–H groups by a factor k, where $0 \le k \le 1$. When $k = 1$ the surface atoms retain their full charge listed in Table 1. When $k = 0$ the walls have no charge and the H atoms at the wall surface vanish from the potential because they lack any interaction with water molecules. A study of the contact angle of SPC/E water next to these walls by Giovambattista *et al.*[27] revealed that the walls are hydrophobic (contact angle greater than 90°) when $k \le 0.4$ and hydrophilic (contact angle less than 90°) for values of $k > 0.4$.

Water molecules are confined to a small distance d in the z direction by placing two walls parallel to the xy plane and equidistant from the $z = 0$ plane. The cubic simulation box is periodic in all directions, but is chosen such that the walls span the simulation box in the x and y directions (*i.e.* the x and y dimensions of the walls and all sides of the cubic simulation box are 6.93 nm in length). Thus the SiO_2 lattice of the walls is uninterrupted in the x and y directions, and water is confined to a small distance d in the z direction. The separation distance between the walls, d, is defined to be the distance between the planes that contain the H atoms on the two walls. For the case when $k = 0$ and the H atoms vanish, the plane where the H atoms would lie is used (*i.e.* the position of the Si and O wall atoms is unchanged for all simulations with the same d). The confined volume is then defined to be the product of the square of the length of the simulation box times d; accordingly, the average mass density of the confined water, ρ, is the ratio of the mass of water molecules to the confined volume.

Table 1 Potential parameters for plate–water interactions (taken from ref. 81)

Atom	ε/kJ mol^{-1} [a]	σ/nm [a]	Charge/e [b]
O	0.6487	0.3154	−0.71
Si	0.5336	0.3795	0.31
H			0.40

[a] Lennard-Jones parameters for plate–water O atom interactions. [b] The full charges on Si–O–H groups ($k = 1$) at the surface of the walls. These point charges interact *via* Coulombic interactions with all the charges present on SPC/E water.

 This journal is © The Royal Society of Chemistry 2009

Confinement of a fluid by solid surfaces parallel to the xy plane results in inhomogeneities in the z direction. Consequently, the stress tensor for such materials is anisotropic, and two relevant pressures exist: a transverse pressure measured in the x and y directions ($P^{\parallel}$), and a normal pressure calculated in the z direction ($P^{\perp}$).[84] The virial expression was used to calculate the pressure parallel to the walls $P^{\parallel}$, accounting for the fixed locations of the atoms in the wall.[85] The pressure normal to the walls, $P^{\perp}$, was calculated by summing the total z projection of the force between water molecules and wall atoms, and then dividing by the surface area of the walls. For both pressures we adopt the convention that lateral and normal compression corresponds to positive stresses, and expansion in these directions corresponds to negative stresses.

To investigate the effects of wall–wall separation (d), density (ρ) and surface polarity (k), we studied 108 different sets of conditions that correspond to all combinations of: $d = 0.8$ and 1.6 nm, $k = 0$ to $k = 1$ in increments of 0.2, and $\rho = 0.4$ to 1.2 g cm^{-3} in increments of 0.1 g cm^{-3}. Constant NVT molecular dynamics simulations were used to generate equilibrium configurations of water confined at the above-listed conditions and $T = 300$ K. The rigid-body equations of motion were integrated using a half-step leap-frog Verlet scheme[86] for translation and an iterative quaternion algorithm[87] for rotation with a step size of $\delta t = 1$ fs. Each simulation was equilibrated at $T = 300$ K for 100 ps; 25 equilibrium configurations were then saved for each run at 5 ps intervals for use in generating ISs. The temperature was fixed for the entire simulation (*i.e.* equilibration and production) with a Berendsen thermostat.[88] Electrostatic interactions are treated using the Ewald sum technique with a cutoff of distance of 0.79 nm and parameters $m_{\max} = 5^3$ (for the number of vectors in the reciprocal space sum) and $\alpha = 4$ nm^{-1} (for the width of the screening-charge Gaussian distribution).[89,90] The linked-cell-list and neighbor-list methods[86,91] are employed to facilitate fast calculation of pair interactions.

The inherent structures (IS) were generated by minimizing the PE of an equilibrium configuration with respect to all degrees of freedom. Each water molecule has six degrees of freedom: the CM is specified with three Cartesian coordinates and the orientation is defined by three Euler angles.[92] An additional degree of freedom exists for each H atom on the walls (only present when $k > 0$, see above) because these atoms are permitted to reorient in a circle.† The PE hypersurface is then a function of $6N$ or $6N + N_{\mathrm{H}}$ variables (N_{H} is the number of hydrogen atoms on the walls) when $k = 0$ and $k > 0$ respectively. The PE was minimized by performing a conjugate-gradient numerical approximation to steepest-descent energy minimization,[77,93] and the minimization was considered complete when the change in energy between two successive minimization steps was $\leq 10^{-15}$ kJ mol^{-1}. The three-dimensional analogue would correspond to the point on the PEL that represents the system moving downhill until it reaches a local minimum. The minima of the PEL satisfy

$$\nabla_{\Omega}\Phi = 0$$

where Φ is the potential-energy surface and $\boldsymbol{\Omega}$ is the vector whose components are the degrees of freedom of the system (*i.e.* the CM positions and orientations of all water molecules and the angular coordinate of each H atom on the wall). The ISs are therefore mechanically stable arrangements of the molecules since the components of $\nabla_{\Omega}\Phi$ are the net forces and torques on the molecules. This minimization was performed on all 25 equilibrium configurations produced at each set of conditions (*i.e.* set of k, ρ and d), allowing IS properties to be averaged over 25 different configurations.

† The hydrogen atoms move with fixed bond length and bond angle such that the position of each H atom can be specified with a single angle between 0 and 2π.

3 Results and discussion

Results are presented in two sections: section 3.1 addresses the mechanical properties of ISs; section 3.2 provides information on structural order present in ISs. All IS calculations are averages over 25 initial equilibrated configurations.

3.1 Mechanical properties

One approach to studying failure in glassy materials is to investigate mechanically stable configurations (*i.e.* ISs) under different states of strain. Such calculations provide insight into how glasses deform and fail. IS configurations can be collected at various macroscopic strain states by minimizing the potential energy of equilibrium configurations over a range of densities.‡ LaViolette and Stillinger first considered the relationship between the inherent structure pressure of bulk materials and density;[94] this relationship has since been termed the equation of state of an energy landscape (EOSEL).[95] In particular, a study by Sastry *et al.*[96] highlighted the fact that for atomic materials the pressure of ISs displays a minimum as a function of density. The density at which this minimum occurs is now referred to as the Sastry density, ρ_s. This corresponds to a state of maximum isotropic tensile stress, with the maximum isotropic tension given by $-P(\rho_s)$. IS configurations at ρ_s are at a state of maximum stress, and energy minimization at $\rho < \rho_s$ generates ISs that are in a state of strain beyond the point of mechanical failure. This causes the material to fissure and form voids that develop from "weak spots" present in the equilibrium configuration. Thus, ρ_s is the minimum density for which ISs are both mechanically stable and spatially homogeneous. Accordingly, ρ_s is an important material property, namely the lowest density at which a homogeneous glass can be formed. Subsequent studies have shown that these characteristics of the EOSEL are quite general, and apply also to molecular glasses.[97]

The investigation of Sastry *et al.* examined materials in bulk; using this energy minimization approach recent work has begun to explore, both theoretically[68,70] and computationally,[69] the properties of nanoscale films. One important difference with respect to bulk behavior is the anisotropy in the stress tensor caused by confinement (see section 2). Two relevant pressures exist, a normal pressure, $P^{\perp}$, and a transverse or parallel pressure, $P^{\parallel}$. Consequently there are two Sastry densities $\rho_s^{\parallel}$ and $\rho_s^{\perp}$. A numerical study by Shah and Truskett[69] examined LJ particles confined by structureless walls and found that these two densities and their associated maximum sustainable tensions [$-P^{\parallel}(\rho_s^{\parallel})$ and $-P^{\perp}(\rho_s^{\perp})$] correspond to different modes of mechanical failure and directional material strengths. The direction with a larger value of ρ_s (*i.e.* maximum stress is reached at a smaller state of transverse strain) is more susceptible to failure. Additionally, voids were found to emerge at the larger of the two Sastry densities (*i.e.* at the smaller state of strain that corresponds to maximum stress). In this section we examine the corresponding behavior for water films confined by silica-based walls.

Fig. 1 shows the EOSEL, for $P^{\parallel}$ and $P^{\perp}$, at the two values of wall separation studied here ($d = 0.8$ and 1.6 nm) and over the full range of wall polarities ($0 \leq k \leq 1$). Thin films exhibit complex behavior, especially under hydrophobic confinement ($k \leq 0.4$). In this case, the transverse stress (Fig. 1a) exhibits both a high-density extremum at 1.1 g cm^{-3} and an inflection point with almost vanishing slope at 0.8 g cm^{-3}. The normal stress (Fig. 1b) displays a single extremum across the entire polarity range, although inflections at high density become evident for $k \leq 0.4$. For $k \geq 0.8$, corresponding to strong water–wall interactions, the transverse and normal directions have single minima between $\rho = 0.8$ and 0.9 g cm^{-3}. Thick

‡ Decreasing density at fixed separation distance corresponds to increasing the transverse strain.

 This journal is © The Royal Society of Chemistry 2009

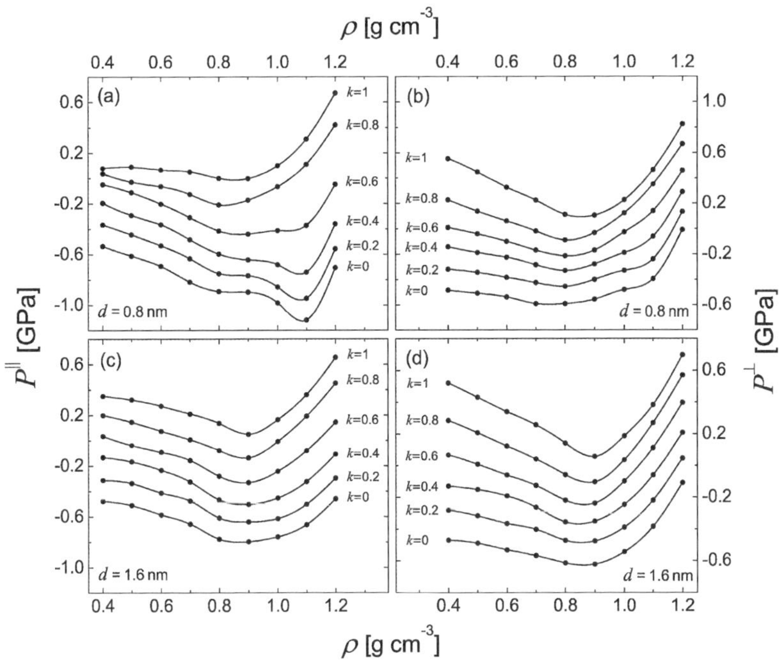

Fig. 1 Equation of state of the energy landscape (pressure as a function of density for inherent structures) of water under various confinement conditions: (a) $P^{\parallel}$ for $d = 0.8$ nm, (b) $P^{\perp}$ for $d = 0.8$ nm, (c) $P^{\parallel}$ for $d = 1.6$ nm, and (d) $P^{\perp}$ for $d = 1.6$ nm. Each plot shows all six studied values of k with the curves shifted vertically for clarity. The values of k for all plots from the bottom to top are $k = 0$ [−0.4 GPa], $k = 0.2$ [−0.2 GPa], $k = 0.4$, $k = 0.6$ [+0.2 GPa], $k = 0.8$ [+0.4 GPa], $k = 1$ [+0.6 GPa]. The standard deviation among the computed pressures, calculated across 25 inherent structures for each data point, is smaller than the size of the data symbols.

films ($d = 1.6$ nm) display a single minimum at $\rho = 0.9$ g cm^{-3} in the EOSEL in the transverse and normal directions.

In analogy with bulk behavior, the single minima shown in Fig. 1c and d for thick films correspond to mechanical stability limits, with progressively larger voids appearing as the limiting strain is exceeded (*i.e.* the Sastry density is at the minimum of the EOSEL). In contrast, the high-density extremum in the transverse stress for thin films at $\rho = 1.1$ g cm^{-3}, seen for films confined by non-polar and weakly polar walls ($k \leq 0.4$), corresponds to strain-induced disruption of a highly structured trilayer film whose two outer layers exhibit long-range crystalline order. Voids emerge in these films between $\rho = 0.8$ and 0.9 g cm^{-3}. Thus the high-density minimum is not associated with loss of mechanical stability, and the Sastry density for these films corresponds to the low-density inflection point in Fig. 1a for $k \leq 0.4$. This will be discussed in detail in section 3.2.1, where the structure of the films is analyzed. It then follows that for all films, $\rho_s^{\parallel}$ and $\rho_s^{\perp}$ lie between 0.8 and 0.9 g cm^{-3} for the transverse and normal directions.§ This is the density range where loss of mechanical stability occurs.

The effect of wall polarity (k) and film thickness (d) on the mechanical strength of thin water films can be determined by comparing the maximum tension for each value of k and d. Fig. 2 plots the tension ($-P^{\parallel}$ and $-P^{\perp}$) of IS configurations at the minimum of the EOSEL in Fig. 1. For thick films ($d = 1.6$ nm) the maximum

§ In this work, the strain resolution is 0.1 g cm^{-3}. Differences between $\rho_s^{\parallel}$ and $\rho_s^{\perp}$ cannot be determined more precisely; additional simulations would be required for this.

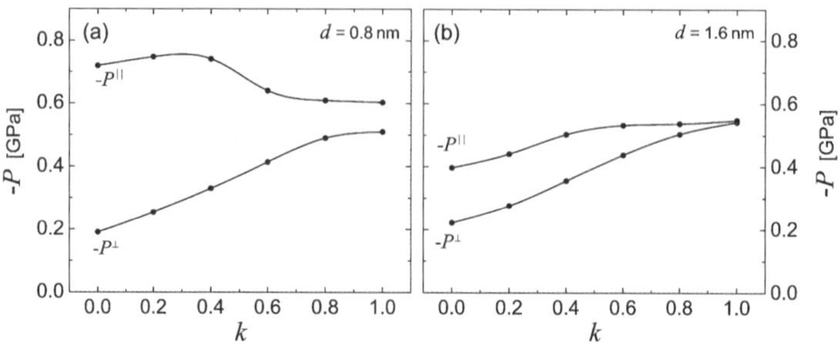

Fig. 2 Maximum sustainable tension for the transverse and normal directions as a function of wall polarity k for separations of (a) $d = 0.8$ nm and (b) $d = 1.6$ nm. These points correspond to the minimum pressure values of the curves in Fig. 1. The standard deviation among the computed pressures, calculated across 25 inherent structures for each data point, is smaller than the size of the data symbols.

tension increases with k in both the transverse and normal directions; the film can withstand more stress as the interaction strength between water and the wall is increased. This trend is also valid for thin films ($d = 0.8$ nm) in the normal direction, but not in the transverse direction. Fig. 2a shows that the maximum lateral stress increases with k in the hydrophobic regime ($k \leq 0.4$), but decreases with k in the hydrophilic region ($k \geq 0.6$). Understanding this behavior requires examination of the structure of confined water near hydrophobic surfaces. This will be addressed in section 3.2.1. A notable feature of Fig. 2 is the strong correlation between substrate hydrophobicity and stress anisotropy. Under hydrophobic confinement, water films are much stronger in the transverse than in the normal direction. However, for hydrophilic confinement $-P^{\parallel}$ and $-P^{\perp}$ are considerably closer.

Minimizing the potential energy of confined water at low density results in the formation of ISs that contain voids. We define the void space to be the volume comprising all points that lie a distance of more than 0.316 nm (the LJ σ value for SPC/E water) from any O atom belonging to a water molecule or the wall. The location of these voids provides information about the failure mechanism. Void probability profiles were determined by performing 10^6 trial point insertions in a slice of thickness $\delta z = 0.02$ nm. The ratio of successful insertions (point belongs to void space) to total insertions is the void probability within that slice. Fig. 3 shows the void probability profiles in thin ($d = 0.8$ nm) glassy films for the hydrophobic ($k = 0$) and the hydrophilic ($k = 1$) walls, and for densities $0.4 \leq \rho \leq 0.9$ g cm^{-3}.

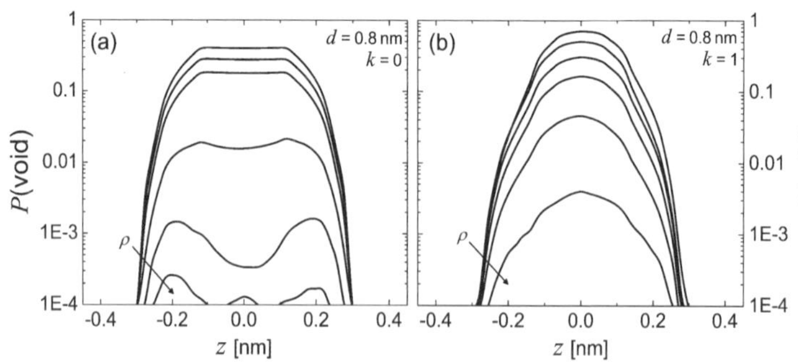

Fig. 3 Probability of finding a void in the water film for walls where (a) $k = 0$, and (b) $k = 1$. Both are for a separation of $d = 0.8$ nm with the middle of the film at $z = 0$ and the wall surfaces at $|z| = 0.4$ nm. Density increases from 0.4 g cm^{-3} to 0.9 g cm^{-3} in intervals of 0.1 g cm^{-3} as indicated by arrows. The profiles are similar for $d = 1.6$ nm.

 This journal is © The Royal Society of Chemistry 2009

At $\rho \geq 1.0$ g cm^{-3} the void probability is negligible. Therefore, the values of $\rho_S^{\parallel}$ and $\rho_S^{\perp}$ discussed above are consistent with the density at which voids emerge. In addition, the z position where voids are most likely to form is strongly dependent on the polarity of the confining surfaces. For $k = 0$, the void profile at $\rho = 0.7$, 0.8 and 0.9 g cm^{-3} develops maxima near the walls; when $k = 1$, the profile has a single peak in the middle of the film. These void profiles are characteristic of the other values of k [*i.e.* when $k \leq 0.4$ ($k \geq 0.6$) the void profiles have the same general shape as the $k = 0$ ($k = 1$) case].

Fig. 4 shows two examples of voids present in the ISs of water at $\rho < \rho_S$, when confined by hydrophobic (Fig. 4a) and hydrophilic (Fig. 4b) surfaces. These pictures are consistent with the void profiles seen in Fig. 3. The void that forms in water under hydrophobic confinement is near the wall, while hydrophilic confinement leads to mechanical failure in the middle of the film. The difference in void location can be understood by comparing the self-interaction of water with the strength of its interaction with the wall when $k = 0$ and $k = 1$. Fig. 5 displays the profile of the potential energy per water molecule under hydrophobic ($k = 0$) and hydrophilic ($k = 1$) confinement. In Fig. 5a and b the potential energy is split into water–water and water–wall contributions. Fig. 5c shows the total potential energy per water molecule for $k = 0$ and $k = 1$; this is the sum of the curves in Fig. 5a and b. Under hydrophobic confinement with $k = 0$, only dispersive interactions exist between water and the wall, and Fig. 5a shows that the water–water interaction energy is much stronger than the water–wall energy, even close to the confining surface. This is due to the strong electrostatic interactions between water molecules. This results in "weak spots" near the walls that, at $\rho < \rho_S$, develop into voids upon energy minimization. The detachment of water molecules from the wall is indicative of adhesive failure. Under hydrophilic confinement, Fig. 5b shows that the attractive energy for water–wall interactions is stronger than that for water–water interactions when $|z| \geq 0.7$ nm. Thus near the confining surface, the interaction between water and the wall is stronger than that between water molecules. This causes the formation of voids in the middle of the film and leads to cohesive failure at $\rho < \rho_S$.

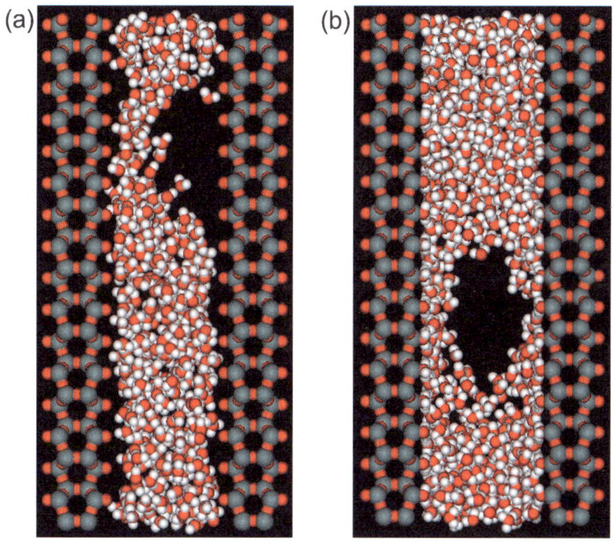

Fig. 4 Voids present in ISs of water confined between (a) hydrophobic surfaces with $k = 0$, and (b) hydrophilic walls with $k = 1$. Both are IS configurations at $\rho = 0.8$ g cm^{-3} with a separation of $d = 1.6$ nm. Water molecules have been removed for ease of visualization from the front and rear of the void (*i.e.* the voids do not pass through the entire simulation box). Si, O and H atoms are represented by gray, red and white spheres respectively.

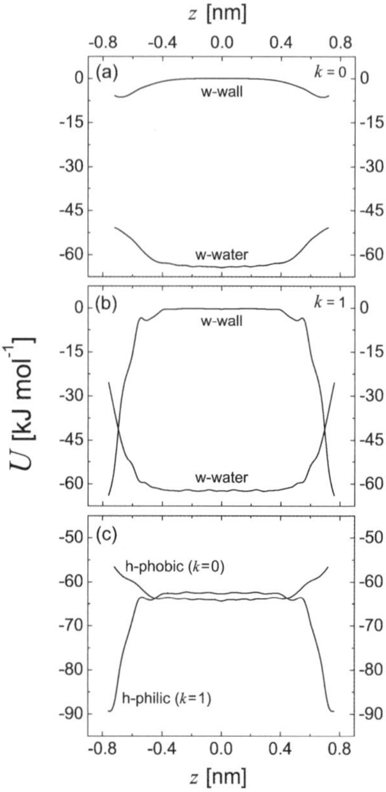

Fig. 5 Potential-energy profiles of water molecules under hydrophobic confinement with $k = 0$ (a) and hydrophilic confinement with $k = 1$ (b) at $\rho = 0.8$ g cm^{-3} and $d = 1.6$ nm (the same conditions as in Fig. 4). Energies shown are on a per water molecule basis, and the potential energy is split into water–water (w-water) and water–wall (w-wall) contributions. (c) The total potential energy per water molecule (water–water and water–wall contributions) for hydrophobic and hydrophilic confinement at the same conditions as in (a) and (b) respectively.

3.2 Structure

Inherent structures formed from equilibrium liquid configurations are amorphous and lack long-range order; hence these configurations are representative of glasses. However, the atomic structure of the silica walls and the interactions between water molecules and the walls impose two distinct types of order. One type is evident at high density near hydrophobic walls and the other is present at low density alongside hydrophilic walls. Accordingly, section 3.2.1 examines ISs for $k = 0$, $d = 0.8$ nm and $\rho = 1.2$ g cm^{-3}, and section 3.2.2 analyzes IS configurations with $k = 1$, $d = 0.8$ nm and $\rho = 0.4$ g cm^{-3}.

3.2.1 Hydrophobic Confinement. Confinement negates the structural isotropy characteristic of bulk liquids. One approach to studying the structure of confined materials is to examine the local density along the direction of confinement. A density profile along the z direction is calculated by assigning water molecules to 0.02 nm thick bins in the z direction, according to the position of the O atom. The density in each bin is computed from the total mass of water molecules in the slab, divided by its volume. The density profiles for $k = 0$ and $d = 0.8$ nm are shown in Fig. 6 for the entire range of studied mean densities. Fig. 6a shows a bilayer structure that is present at average densities up to $\rho = 0.8$ g cm^{-3}. At $\rho \geq 0.9$ g cm^{-3} a third

 This journal is © The Royal Society of Chemistry 2009

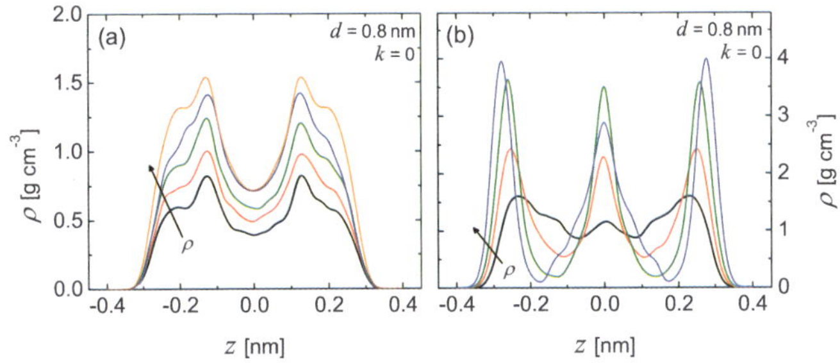

Fig. 6 Density profiles for water confined between hydrophobic walls with $k = 0$ and $d = 0.8$ nm. The wall surfaces are located at $|z| = 0.4$ nm. (a) The underlying structure is bilayer when $\rho \leq 0.8$ g cm^{-3}, and (b) trilayer when $\rho \geq 0.9$ g cm^{-3}. On plot (a) the average densities are $\rho = 0.4$ g cm^{-3} (black), $\rho = 0.5$ g cm^{-3} (red), $\rho = 0.6$ g cm^{-3} (green), $\rho = 0.7$ g cm^{-3} (blue) and $\rho = 0.8$ g cm^{-3} (orange). On plot (b) the average densities are $\rho = 0.9$ g cm^{-3} (black), $\rho = 1.0$ g cm^{-3} (red), $\rho = 1.1$ g cm^{-3} (green) and $\rho = 1.2$ g cm^{-3} (blue). Arrows indicate the direction of increasing density.

layer of water molecules develops as seen in Fig. 6b. The density range over which this transition occurs coincides with the inflection point (*i.e.* the Sastry density) in the EOSEL for thin films confined by hydrophobic walls (Fig. 1a). The bilayer-to-trilayer transition occurs for $d = 0.8$ nm when $k \leq 0.6$. For thicker films ($d = 1.6$ nm), six layers of water are present at high density for all values of k and there is no clear layering transition.

Examining configurations of water molecules that belong to a single peak in the density profile provides detailed structural information. Fig. 7 shows a characteristic arrangement of water molecules that constitute one of the two symmetric outer peaks (*i.e.* the O atom of the water molecule is located at $|z| > 0.18$ nm) of the density profile for $\rho = 1.2$ g cm^{-3}. This monolayer of water molecules is approximately 0.2 nm thick and self-assembles into a hexagonal lattice. The same structure is evident in the first layer of water molecules next to the wall when $\rho \geq 1.0$ g cm^{-3} and $k \leq 0.6$ for $d = 0.8$ and 1.6 nm. The hexagonal order becomes less pronounced as the density decreases and k increases.

In the perfect lattice, water molecules can adopt two configurations: (i) with a single H atom pointing away from the wall and (ii) with both H atoms in a plane nearly parallel to the surface. These configurations alternate, and water molecules form six-membered rings, giving rise to a hexagonal lattice. Thus each water molecule participates in four hydrogen bonds (HB): three of these point roughly parallel to the wall and correspond to HBs with water molecules in the same layer, and one

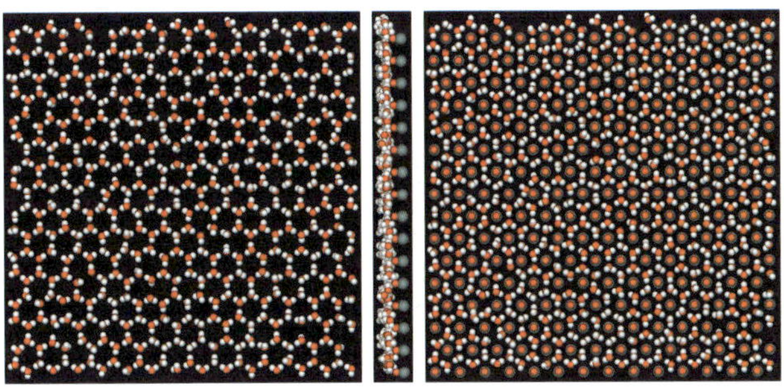

Fig. 7 Inherent structure configuration showing water molecules within 0.22 nm of a hydrophobic wall with $k = 0$, $\rho = 1.2$ g cm^{-3} and $d = 0.8$ nm (left); profile view of the same layer with surface wall atoms included (middle); with surface Si and O wall atoms included (right). Si, O and H atoms are represented by gray, red and white spheres respectively.

points normal to the substrate surface and corresponds to a HB with a water molecule in the adjacent layer. The structure in Fig. 7 has defects that can be attributed to the presence of a middle layer of water molecules that is disordered.

Closer inspection of the monolayer in Fig. 7 reveals that the center of each hexagonal ring of water molecules is occupied by an oxygen atom on the surface of the wall. This strict correlation between the substrate and water suggests that the lattice is templated by the wall. At high density, water molecules are pushed into the small spaces between the top layer of Si and O atoms on the substrate surface. These water molecules self-assemble into a hexagonal lattice in order to participate in four HBs. Thus it is the atomic structure of the wall that imposes the hexagonal lattice on the water. This conclusion is supported by the amorphous character of the middle peak (or peaks when $d = 1.6$ nm) of the density profile. Without the hexagonal template provided by the wall, no long-range order exists for water molecules in the middle of the film.

This structure has been previously observed[82] for SPC/E water confined by identical silica walls at a separation of $d = 0.6$ nm and $T = 300$ K. At this value of d only two layers of water are present, and in the bilayer ice these layers are in registry. It was found that the crystal structure was very sensitive to the value of d and this hexagonal lattice was not evident at $T = 300$ K in simulations with $d = 0.8$ or 1.6 nm. Those simulations were done at constant pressure and examined confined water that was in contact with a bulk water reservoir.[82] The highest densities studied were approximately 1.0 g cm^{-3}, which is the lowest density at which the hexagonal lattice is evident in the IS configurations.

It is now appropriate to resume the discussion of the EOSEL for $P^{\parallel}$ at $d = 0.8$ nm (see Fig. 1a and 2a) and examine why the minimum of the EOSEL does not correspond to the emergence of voids at $k \leq 0.4$ and $d = 0.8$ nm. Under hydrophobic confinement with $k = 0$ there is a sharp minimum in $P^{\parallel}$ at $\rho = 1.1$ g cm^{-3}; as k increases this minimum becomes less prominent. At $k = 0.6$, only a shoulder in the EOSEL is present at $\rho = 1.1$ g cm^{-3}, and at $k \geq 0.8$ the feature has completely receded. This behavior is caused by the hexagonal ordering of water molecules near the wall at these conditions. The ordered arrangement of water at both walls increases the strength of the thin film, resulting in a deep minimum in $P^{\parallel}$ at $\rho =$ 1.1 g cm^{-3}. Thus, for $k \leq 0.4$ at a separation $d = 0.8$ nm, the minimum of the EOSEL does not correspond to the emergence of voids in system, but represents the density below which the arrangement of water next to the wall begins to disorder. Accordingly, for these conditions $-P^{\parallel}$ at the minimum represents the maximum stress that the film can withstand before significant defects become present in the lattice near the wall.

This observation raises the question of why similar behavior is not evident for thick films ($d = 1.6$ nm). The answer lies in the limited ability of the wall to impose order beyond the first layer of water molecules. The hexagonal template of the walls results in the ordering of a single layer of water next to the surface. At $d = 0.8$ nm this means that the two outer layers are organized into hexagonal lattices and the one middle layer is amorphous. However, at $d = 1.6$ nm there are four amorphous inner layers and two ordered outer layers. The trilayer thin film ($d = 0.8$ nm) is thus much stronger than its six-layer thicker counterpart ($d = 1.6$ nm): the maximum sustainable tension for $d = 0.8$ nm is more than 300 MPa greater than for $d = 1.6$ nm.

3.2.2 Hydrophilic confinement. Structural ordering of water molecules is also evident near hydrophilic surfaces. Examination of the density profiles in Fig. 8a for $k = 1$, $d = 0.8$ nm and $0.4 \leq \rho \leq 0.8$ g cm^{-3} reveals two sharp outer peaks next to the confining surfaces. The first peak (located closest to the confining walls) corresponds to $0.36 < |z| < 0.4$ nm. The second peak is slightly closer to the middle of the film and corresponds to $0.28 < |z| < 0.36$ nm. At $\rho = 0.4$ g cm^{-3} the outer peak is significantly higher than the inner peak. However, as the average density increases from $\rho = 0.4$ to 0.8 g cm^{-3}, the outer peak recedes and the inner peaks grows.

 This journal is © The Royal Society of Chemistry 2009

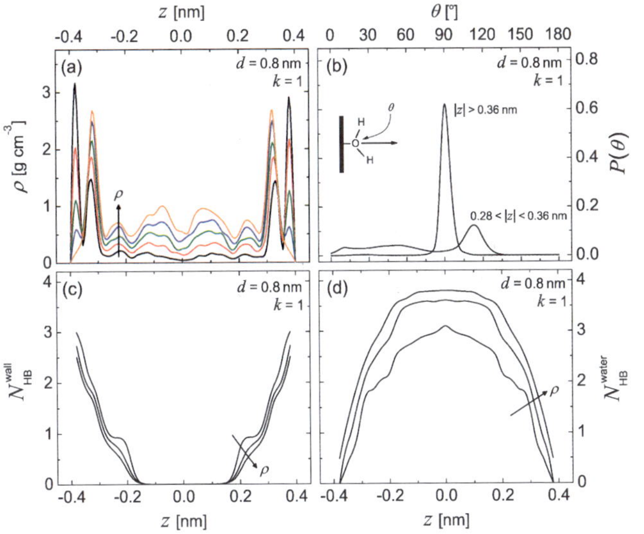

Fig. 8 (a) Density profiles for water confined between hydrophilic walls with $k = 1$ and $d = 0.8$ nm for low density, $\rho \leq 0.8$ g cm^{-3}. The wall surfaces are located at $|z| = 0.4$ nm. The average densities are $\rho = 0.4$ g cm^{-3} (black), $\rho = 0.5$ g cm^{-3} (red), $\rho = 0.6$ g cm^{-3} (green), $\rho = 0.7$ g cm^{-3} (blue) and $\rho = 0.8$ g cm^{-3} (orange). Arrow indicates the direction of increasing density. (b) Probability distribution of the angle formed between the O–H bond of a water molecule and a vector normal to the wall that points into the water film. The two curves are $P(\theta)$ for molecules in the outer peak of the density profile with $|z| > 0.36$ nm, and the inner peak with $0.28 < |z| < 0.36$ nm. (c) Average number of HBs per water molecule formed between water and wall atoms with confining surfaces located at $|z| = 0.4$ nm. (d) Average number of HBs per water molecule formed with other water molecules. In (c) and (d) the densities are $\rho = 0.4$, 0.8 and 1.2 g cm^{-3}. Arrows indicate the direction of increasing density.

Once $\rho \geq 0.8$ g cm^{-3}, the outer peak no longer exists and the inner peak corresponds to a local density over 2.5 g cm^{-3}. Since each peak corresponds to a layer of water molecules, this behavior indicates a significant change in the layering of water near the walls as a function of mean density. It is important to note that simulations at $T = 300$ K have shown the stability limit of liquid water confined by hydrophilic plates ($k = 1$) to be approximately 0.7 g cm^{-3} for all values of d.[82] Thus the IS configurations that result in the behavior evident in Fig. 8 are calculated from an inhomogeneous phase-separated system.

To probe the structure of water molecules near the confining surface we focus on their orientation, and in particular the angle θ formed by an O–H bond of a water molecule and a vector normal to the wall surface pointing into the film (see diagram in Fig. 8b). Fig. 8b shows the probability distribution of θ for water molecules located at $|z| > 0.36$ nm and $0.28 < |z| < 0.36$ nm (the outer and inner peaks of the density profile respectively) for $0.4 \leq \rho \leq 0.8$ g cm^{-3}. $P(\theta)$ shows a single maximum at $\theta = 90°$ for molecules that constitute the outer peak. This indicates that water molecules closest to the wall ($|z| > 0.36$ nm) are oriented parallel to the substrate surface and lie in the xy plane. $P(\theta)$ for molecules in the inner peak of the density profile ($0.28 < |z| < 0.36$ nm) has a small maximum near the tetrahedral angle ($\theta = 109°$); molecules in this region show comparatively less orientational preference.

Under hydrophilic confinement, water is capable of forming HBs with the charged Si–O–H groups on the substrate surface, as well as with other water molecules. Fig. 8c shows the average number of HBs (per water molecule) that are formed between a water molecule and the wall as a function of z; Fig. 8d shows the corresponding average number of water–water HBs per molecule. Hydrogen bonds are defined by standard geometric criteria that specify distance and angle

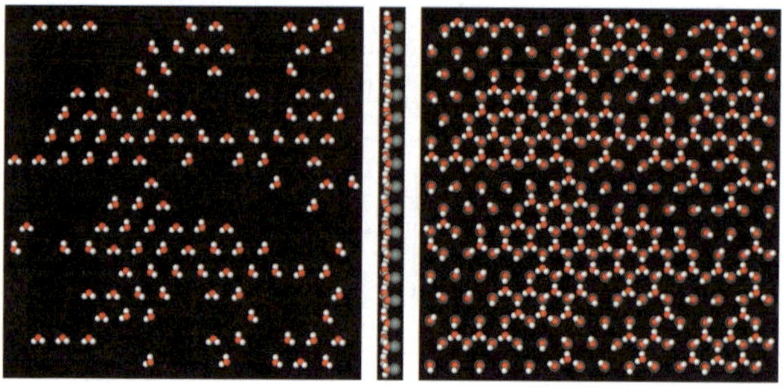

Fig. 9 Inherent structure configuration showing water molecules within 0.04 nm of a hydrophilic wall with $k = 0$, $\rho = 0.4$ g cm^{-3} and $d = 0.8$ nm (left); profile view of the same layer with surface wall atoms included (middle); with surface Si, O and H wall atoms included (right). Si, O and H atoms are represented by gray, red and white spheres respectively.

requirements [98] and we count the number of HBs that each water molecule participates in (*i.e.* coordination number). These plots indicate that water molecules close to the hydrophilic wall ($|z| > 0.36$ nm) form on average three HBs with Si–O–H groups at the surface, and less than one HB with other water molecules. Also, as the density increases, the average number of HBs between water and the wall decreases, and the average number of water–water HBs increases.

Fig. 9 shows a representative configuration of water molecules that constitute the outer peak of the density profile, $0.36 < |z| < 0.4$ nm for $\rho = 0.4$ g cm^{-3}. In this layer each water molecule is oriented parallel to the *xy* plane, in agreement with $P(\theta)$ in Fig. 8b. Furthermore, if this configuration is placed on top of the Si–O–H groups that compose the surface of each wall, it is clear that the water molecules and the charged H atoms on the wall surface are arranged such that three HBs are formed between each water molecule and the wall.¶ This is consistent with Fig. 8c which shows that when $\rho < 0.8$ g cm^{-3}, water molecules located at $0.36 > |z| > 0.4$ nm form three HBs with the wall. A similar inspection of the configuration of water molecules in the second peak ($0.28 > |z| > 0.36$ nm) reveals that this layer lacks long-range order at all studied densities. Fig. 8c and d show that water molecules in this layer form on average one less HB with the wall and one more HB with other water molecules. Thus, the loss of one HB with the wall is enough to prevent structural order near hydrophilic walls. Very similar behavior of the density profiles in Fig. 8a and the structure in Fig. 9 is evident when the separation is increased to $d = 1.6$ nm (Figures not shown).

The structure apparent in Fig. 9 results from the ability of water molecules to participate in four HBs. At low density ($\rho = 0.4$ g cm^{-3}), this propensity for hydrogen bonding is satisfied by forming in-plane bonds with the wall as explained above. Furthermore, under these conditions there is very little water in the region between the confining surfaces [*e.g.* $|z| < 0.3$ nm; see Fig. 8a], and hence hydrogen bonding between water molecules close to the surface and those between the confining surfaces is minimal. Thus, the arrangement of water molecules in Fig. 9 results from three features of the confining walls: (i) the strong electrostatic interactions of charged Si–O–H groups with water molecules, (ii) cavities on the structured silica surface that can accommodate one water molecule, and (iii) the ability of hydrogen atoms on the silica surface to reorient with fixed bond length and bond angle. This reasoning also explains why the outside peak of the density profile decreases with density. At higher densities, the abundance of water molecules creates a competition for HBs between neighboring water molecules and the surface. Hence,

¶ During energy minimization the position of each H atom on the wall surface is optimized in addition to the position and orientation of each water molecule (see section 2).

 This journal is © The Royal Society of Chemistry 2009

on average, water molecules form fewer HBs with the surface and more with other water molecules as density increases (see Fig. 8c and d). This prevents the arrangement seen in Fig. 9 from occurring.

4 Conclusions

In this work, we have used molecular simulations and the PEL paradigm to explore the structure and mechanical properties of thin glassy-water films confined between atomically detailed, silica-based surfaces with continuously tunable hydrophobicity and hydrophilicity. The mechanical properties of water films were explored by calculating how the state of strain affects mechanically stable configurations. Loss of mechanical stability occurs between 0.8 and 0.9 g cm^{-3} for all values of wall polarity (k) and separation (d). Below such densities, progressively larger fissures appear and mechanically stable configurations are no longer homogeneous. Thus the Sastry density (ρ_s) was determined to be in the 0.8–0.9 g cm^{-3} range for all polarities and separations investigated. In the bulk, ρ_s corresponds to the minimum in the EOSEL; our results in confinement are consistent with this except for thin films ($d = 0.8$ nm) bounded by hydrophobic ($k \leq 0.4$) walls. Under these conditions, ρ_s corresponds to a low-density inflection point in the EOSEL, and the high-density minimum in the stress–strain curve corresponds to the disruption of a hexagonal lattice of water molecules in the two outer layers of a trilayer film. This specific structure also results in a substantial stress anisotropy for confined water. The maximum transverse stress is much larger than the normal stress for water confined by hydrophobic surfaces. In contrast, water confined by hydrophilic surfaces shows closer values for the maximum normal and transverse stress. Additionally, the substrate–water interactions were observed to have a strong influence on the mode of mechanical failure. For values of $k \leq 0.4$ (hydrophobic) the film failed adhesively, with the development of voids next to the confining surface. For $k \geq 0.6$ (hydrophilic) voids form in the middle of the film, leading to cohesive failure.

A salient result of this study is the ability of the confining surfaces to impose long-range structural order on the first layer of water molecules next to the wall. A hexagonal lattice is evident at high density under hydrophobic confinement. At $\rho \geq 1.0$ g cm^{-3}, water molecules are forced close to the wall and into the small cavities between the substrate surface atoms. The architecture of these cavities is such that water molecules adopt one of two alternating orientations in an attempt to participate in four HBs. The result is a hexagonal lattice of water molecules in which the center of each hexagon is occupied by a surface oxygen atom. Under hydrophilic confinement and low density, water molecules position themselves to form three HBs with Si–O–H groups on the surface. It is the strong electrostatic interactions between water and the charged surface groups and the ability of H atoms on the surface to reorient that cause this behavior. In both of these cases the hydrophobic and hydrophilic walls template an ordered monolayer and submonolayer of water molecules respectively; the atomic details and different interactions permit the formation of these complex structures. The ability of ordered surfaces to template structured water configurations, and the profound influence that such configurations impart to the entire film suggests that smooth walls, though frequently used in simulations, may not be able to capture an important range of properties and phenomena associated with thin films of molecular glasses.

Glasses are nonequilibrium materials, hence their properties are history-dependent, and differences in the formation process lead to materials with distinct characteristics. This study examined glassy configurations formed *via* energy minimization, a process akin to quenching a liquid instantaneously. Experimentally, however, glasses are formed by cooling a liquid at a finite rate that is sufficiently fast to avoid crystallization. A complementary study in which confined liquid configurations are cooled at various finite rates into the glassy state is in progress and will be the subject of a future publication.

The large number of factors affecting confined materials leaves much room for future work. This study demonstrated the strong influence of confining surfaces on water; in particular, the architecture and interactions of the confining substrate are capable of inducing long-range order in the first layer of water close to the substrate. These structures were found to have important consequences for the mechanical properties of thin water films, especially for hydrophobic confinement. Future studies should address how features such as surface curvature and heterogeneity affect the properties and structure of confined glassy water. The surfaces used in this study can easily be modified to include various patterns or patches of hydrophobicity/hydrophilicity. Alternatively, an entirely different confining surface such as a protein,[67] can be studied to provide results that are relevant to biotechnology and the preservation of labile biomolecules in the glassy state. Improved knowledge of these topics would represent a significant advance in the quest for a deeper molecular-based understanding of nano-confined materials and their applications.

Acknowledgements

We thank F. H. Stillinger for fruitful discussions. P. G. D. gratefully acknowledges the financial support of the National Science Foundation Collaborative Research in Chemistry Grant No. CHE0404699, and of Unilever Research U. S.

References

1 D. Eisenberg and W. Kauzmann, *Structure and Properties of Water*, Oxford University Press, New York, USA, 1969.
2 F. Franks, *Water: a Matrix of Life*, The Royal Society of Chemistry, Cambridge, UK, 2nd edn, 2000.
3 P. Ball, *Life's Matrix: A Biography of Water*, University of California Press, California, 2001.
4 C. Tanford, *The Hydrophobic Effect: Formation of Micelles and Biological Membranes*, Wiley, New York, 2nd edn, 1980.
5 W. Kauzmann, *Adv. Protein Chem.*, 1959, **14**, 1.
6 J. M. Orten and O. W. Neuhaus, *Human Biochemistry*, Mosby, St. Louis, 10th edn, 1982.
7 Y. Levy and J. N. Onuchic, *Annu. Rev. Biophys. Biomol. Struct.*, 2006, **35**, 389.
8 P. G. Debenedetti and H. E. Stanley, *Phys. Today*, 2003, **56**, 40.
9 C. A. Angell, *Annu. Rev. Phys. Chem.*, 1983, **34**, 593.
10 P. G. Debenedetti, *J. Phys.: Condens. Matter*, 2003, **15**, R1669.
11 H. Kanno and C. A. Angell, *J. Chem. Phys.*, 1979, **70**, 4008.
12 K. T. Gillen, D. C. Douglass and J. R. Hoch, *J. Chem. Phys.*, 1972, **57**, 5117.
13 P. H. Poole, F. Sciortino, U. Essmann and H. E. Stanley, *Nature*, 1992, **360**, 324.
14 O. Mishima, L. D. Calvert and E. Whalley, *Nature*, 1985, **314**, 76.
15 T. Loerting, C. Salzmann, I. Kohl, E. Mayer and A. Hallbrucker, *Phys. Chem. Chem. Phys.*, 2001, **3**, 5355.
16 T. Loerting and N. Giovambattista, *J. Phys.: Condens. Matter*, 2006, **18**, R919.
17 C. A. Angell, *Science*, 2008, **319**, 582.
18 R. M. Barrer, *Zeolites and Clay Minerals As Sorbents and Molecular Sieves*, Academic Press, London, 1978.
19 E. Roedder, *Rev. Mineral.*, 1984, **12**, 1.
20 Special issue 'Lab on a Chip': Nature, 2006, **442**, 367.
21 Y. Gogotsi, N. Naguib and J. A. Libera, *Chem. Phys. Lett.*, 2002, **365**, 354.
22 B. J. Fuller, N. Lane, and E. E. Benson, *Life in the Frozen State*, CRC Press, Boca Raton, Florida, 2004.
23 P. Mentre, *Cell. Mol. Biol.*, 2001, **47**, 709.
24 V. Helms, *ChemPhysChem*, 2007, **8**, 23.
25 J. Dzubiella, J. M. J. Swanson and J. A. McCammon, *Phys. Rev. Lett.*, 2006, **96**, 087802.
26 F. Bresme and A. Wynveen, *J. Chem. Phys.*, 2007, **126**, 044501.
27 N. Giovambattista, P. G. Debenedetti and P. J. Rossky, *J. Phys. Chem. B*, 2007, **111**, 9581.
28 N. Giovambattista, P. G. Debenedetti and P. J. Rossky, *J. Phys. Chem. C*, 2007, **111**, 1323.
29 L. Hua, X. H. Huang, P. Liu, R. H. Zhou and B. J. Berne, *J. Phys. Chem. B*, 2007, **111**, 9069.
30 Y. K. Cheng and P. J. Rossky, *Nature*, 1998, **392**, 696.

 This journal is © The Royal Society of Chemistry 2009

31 F. Mallamace, M. Broccio, C. Corsaro, A. Faraone, D. Majolino, V. Venuti, L. Liu, C. Y. Mou and S. H. Chen, *Proc. Natl. Acad. Sci. U. S. A.*, 2007, **104**, 424.
32 L. Liu, S. H. Chen, A. Faraone, C. W. Yen and C. Y. Mou, *Phys. Rev. Lett.*, 2005, **95**, 117802.
33 J. Swenson, *Phys. Rev. Lett.*, 2006, **97**, 189801.
34 S. Cerveny, J. Colmenero and A. Alegría, *Phys. Rev. Lett.*, 2006, **97**, 189802.
35 S. H. Chen, L. Liu and A. Faraone, *Phys. Rev. Lett.*, 2006, **97**, 189803.
36 S. Yoshioka and Y. Aso, *J. Pharm. Sci.*, 2007, **96**, 960.
37 W. Wang, *Int. J. Pharm.*, 2000, **203**, 1.
38 J. Lee and S. Timasheff, *Biochemistry*, 1974, **13**, 257.
39 R. Hatley, F. Franks and S. Mathias, *Process Biochem.*, 1987, **22**, 169.
40 T. A. Jennings, *Lyophilization: Introduction and Basic Principles*, Interpharm Press, Englewood, CO, 1999.
41 C. J. Roberts and P. G. Debenedetti, *AIChE J.*, 2002, **48**, 1140.
42 S. D. Allison, M. C. Manning, T. W. Randolph, K. Middleton, A. Davis and J. F. Carpenter, *J. Pharm. Sci.*, 2000, **89**, 199.
43 W. Garzon-Rodriguez, R. L. Koval, S. Chongprasert, S. Krishnan, T. W. Randolph, N. W. Warne and J. F. Carpenter, *J. Pharm. Sci.*, 2004, **93**, 684.
44 D. Q. M. Craig, P. G. Royall, V. L. Kett and M. L. Hopton, *Int. J. Pharm.*, 1999, **179**, 179.
45 M. Pikal, Freeze-Drying/Lyophilization of Pharmaceutical and Biological Products, in *Drugs and the Pharmaceutical Sciences*, ed. L. Rey and J. May, Informa Health Care, U. S. A., 2004, p. 63.
46 S. Shamblin, in *Lyophilization of Biopharmaceuticals (Biotechnology: Pharmaceutical Aspects)*, ed. H. Constantino and M. Pikal, American Association of Pharmaceuticals, Arlington, VA, 2004, p. 229.
47 C. A. Oksanen and G. Zografi, *Pharm. Res.*, 1990, **7**, 654.
48 C. Ahlneck and G. Zografi, *Int. J. Pharm.*, 1990, **62**, 87.
49 B. C. Hancock and G. Zografi, *Pharm. Res.*, 1994, **11**, 471.
50 Y. H. Lam, R. Bustami, T. Phan, H. K. Chan and F. Separovic, *J. Pharm. Sci.*, 2002, **91**, 943.
51 V. Molinero and W. A. Goddard, *Phys. Rev. Lett.*, 2005, **95**, 045701.
52 J. Carpenter, B. Chang, and T. Randolph, in *Lyophilization of Biopharmaceuticals (Biotechnology: Pharmaceutical Aspects)*, ed. H. Constantino and M. Pikal, American Association of Pharmaceuticals, Arlington, VA, 2004, p. 423.
53 S. Yoshioka, Y. Aso and S. Kojima, *Pharm. Res.*, 2001, **18**, 256.
54 F. Franks, *Biophysics and Biochemistry at Low Temperatures*, Cambridge University Press, Cambridge, UK, 1985.
55 K. Storey and J. Storey, in *Insects at Low Temperature*, ed. R. E. Lee and D. L. Denlinger, Chapman and Hall, New York, 1991.
56 A. L. Devries, *Annu. Rev. Physiol.*, 1983, **45**, 245.
57 J. Duman, L. Xu, L. Neven, D. Tursman, and D. Wu, in *Insects at Low Temperature*, ed. R. E. Lee and D. L. Denlinger, Chapman and Hall, New York, 1991.
58 P. L. Privalov, *Crit. Rev. Biochem. Mol. Biol.*, 1990, **25**, 281.
59 M. S. Pometun, R. W. Peterson, C. R. Babu and A. J. Wand, *J. Am. Chem. Soc.*, 2006, **128**, 10652.
60 D. Paschek, S. Nonn and A. Geiger, *Phys. Chem. Chem. Phys.*, 2005, **7**, 2780.
61 P. G. Debenedetti, *Metastable Liquids: Concepts and Principles*, Princeton University Press, Princeton, NJ, 1996.
62 R. S. Smith, T. Zubkov and B. D. Kay, *J. Chem. Phys.*, 2006, **124**, 114710.
63 R. S. Smith and B. D. Kay, *Nature*, 1999, **398**, 788.
64 S. M. McClure, D. J. Safarik, T. M. Truskett and C. B. Mullins, *J. Phys. Chem. B*, 2006, **110**, 11033.
65 S. M. McClure, E. T. Barlow, M. C. Akin, D. J. Safarik, T. M. Truskett and C. B. Mullins, *J. Phys. Chem. B*, 2006, **110**, 17987.
66 F. H. Stillinger, *Science*, 1995, **267**, 1935.
67 N. Giovambattista, C. F. Lopez, P. J. Rossky and P. G. Debenedetti, *Proc. Natl. Acad. Sci. U. S. A.*, 2008, **105**, 2274.
68 J. Mittal, P. Shah and T. M. Truskett, *J. Phys. Chem. B*, 2004, **108**, 19769.
69 P. Shah and T. M. Truskett, *Mech. Mater.*, 2006, **38**, 924.
70 T. M. Truskett and V. Ganesan, *J. Chem. Phys.*, 2003, **119**, 1897.
71 A. Attili, P. Gallo and M. Rovere, *Phys. Rev. E*, 2005, **71**, 031204.
72 M. Goldstein, *J. Chem. Phys.*, 1969, **51**, 3728.
73 M. Goldstein, *J. Chem. Phys.*, 1977, **67**, 2246.
74 F. H. Stillinger and T. A. Weber, *Phys. Rev. A*, 1982, **25**, 978.
75 F. H. Stillinger and J. A. Hodgdon, *Phys. Rev. E*, 1994, **50**, 2064.

76 P. G. Debenedetti and F. H. Stillinger, *Nature*, 2001, **410**, 259.
77 W. H. Press, S. Teukolsky, W. Vetterling, and B. Flannery, *Numerical Recipes: The Art of Scientific Computing*, Cambridge University Press, New York, 3rd edn, 2007.
78 J. R. Errington and P. G. Debenedetti, *Nature*, 2001, **409**, 318.
79 L. A. Baez and P. Clancy, *J. Chem. Phys.*, 1994, **101**, 9837.
80 F. W. Starr, F. Sciortino and H. E. Stanley, *Phys. Rev. E*, 1999, **60**, 6757.
81 S. H. Lee and P. J. Rossky, *J. Chem. Phys.*, 1994, **100**, 3334.
82 N. Giovambattista, P. J. Rossky and P. G. Debenedetti, *Phys. Rev. E*, 2006, **73**, 041604.
83 A. R. Leach, *Molecular Modelling: Principles and Applications*, Prentice Hall, Harlow, England, 2nd edn, 2001.
84 M. Schoen and S. Klapp, Nanoconfined Fluids: Soft Matter Between Two and Three Dimensions, in *Reviews of Computational Chemistry*, ed. K. B. Lipowitz and T. Cundari, John Wiley and Sons, New York, 2007, vol. 24.
85 W. Smith and P. Rodger, The Pressure in Systems with Frozen Atoms, in *Collaborative Computational Projects 5 (CCP5)*, 2002, see http://www.ccp5.ac.uk/infoweb/wsmith22/wsmith22.pdf.
86 M. P. Allen and D. J. Tildesley, *Computer Simulation of Liquids*, Oxford University Press, New York, NY, 1987.
87 I. P. Omelyan, *Phys. Rev. E*, 1998, **58**, 1169.
88 H. J. C. Berendsen, J. P. M. Postma, W. F. Vangunsteren, A. Dinola and J. R. Haak, *J. Chem. Phys.*, 1984, **81**, 3684.
89 T. M. Nymand and P. Linse, *J. Chem. Phys.*, 2000, **112**, 6152.
90 A. Y. Toukmaji and J. A. Board, *Comput. Phys. Commun.*, 1996, **95**, 73.
91 D. Frenkel and B. Smit, *Understanding Molecular Simulation: From Algorithms to Applications*, Academic Press, San Diego, CA, 2nd edn, 2002.
92 H. Goldstein, *Classical Mechanics*, Addison-Wesley Pub. Co., Reading, MA, 2nd edn, 1980.
93 F. H. Stillinger and T. A. Weber, *J. Phys. Chem.*, 1983, **87**, 2833.
94 R. A. LaViolette and F. H. Stillinger, *J. Chem. Phys.*, 1985, **82**, 3335.
95 P. G. Debenedetti, F. H. Stillinger, T. M. Truskett and C. J. Roberts, *J. Phys. Chem. B*, 1999, **103**, 7390.
96 S. Sastry, P. G. Debenedetti and F. H. Stillinger, *Phys. Rev. E*, 1997, **56**, 5533.
97 V. K. Shen, P. G. Debenedetti and F. H. Stillinger, *J. Phys. Chem. B*, 2002, **106**, 10447.
98 R. Kumar, J. R. Schmidt and J. L. Skinner, *J. Chem. Phys.*, 2007, **126**, 204107.

 This journal is © The Royal Society of Chemistry 2009

Water nanodroplets confined in zeolite pores†

François-Xavier Coudert,‡[a] Fabien Cailliez,[b] Rodolphe Vuilleumier,§[c] Alain H. Fuchs[b] and Anne Boutin*[a]

Received 28th March 2008, Accepted 2nd June 2008
First published as an Advance Article on the web 19th September 2008
DOI: 10.1039/b804992k

We provide a comprehensive depiction of the behaviour of a nanodroplet of ≃20 water molecules confined in the pores of a series of 3D-connected isostructural zeolites with varying acidity, by means of molecular simulations. Both grand canonical Monte Carlo simulations using classical interatomic forcefields and first-principles Car–Parrinello molecular dynamics were used in order to characterise the behaviour of confined water by computing a range of properties, from thermodynamic quantities to electronic properties such as dipole moment, including structural and dynamical information. From the thermodynamic point of view, we have identified the all-silica zeolite as hydrophobic, and the cationic zeolites as hydrophilic; the condensation transition in the first case was demonstrated to be of first order. Furthermore, in-depth analysis of the dynamical and electronic properties of water showed that water in the hydrophobic zeolite behaves as a nanodroplet trying to close its hydrogen-bond network onto itself, with a few short-lived dangling OH groups, while water in hydrophilic zeolites "opens up" to form weak hydrogen bonds with the zeolite oxygen atoms. Finally, the dipole moment of confined water is studied and the contributions of water self-polarisation and the zeolite electric field are discussed.

1 Introduction

Water in zeolites represents a model system for a wide range of experimental as well as theoretical investigations, aimed at understanding the effect of confinement on the structure,[1] dynamics[2,3] and thermodynamics[4–8] of molecular fluids. Understanding the behaviour of nano-confined water is also relevant to such important problems as protein stability,[9–13] transport in ion channels,[14–16] nano-fluidic devices[17] and the thermodynamics of colloidal assemblies.[18,19] From a practical point of view, water plays a key role in many applications of zeolites such as ion-exchange and liquid phase separation.

Several computer simulation studies of water in zeolites have been reported in the past few years, using a variety of techniques including energy minimisation (with

[a]*Laboratoire de Chimie Physique, Univ Paris-Sud, F-91405 Orsay, France. E-mail: anne.boutin@lcp.u-psud.fr*
[b]*École Nationale Supérieure de Chimie de Paris (Chimie ParisTech) and Université Pierre et Marie Curie, Paris 6, F-75005 Paris, France*
[c]*Laboratoire de Physique Théorique de la Matière Condensée, Université Pierre et Marie Curie, Paris 6, F-75005 Paris, France*

† Electronic supplementary information (ESI) available: The two possible structures of $[Al_9Si_{15}O_{48}]^{9-}$ satisfying Loewenstein's rule. For ESI see DOI: 10.1039/b804992k

‡ Present address: University College London, Gower Street, London, United Kingdom WC1E 6BT
§ Present address: Département de Chimie, École Normale Supérieure, 75005 Paris, France

effective shell model-ionic potentials),[20–24] Car–Parrinello first-principles molecular dynamics (MD),[25–29] classical MD or Monte Carlo (MC) simulations using conventional two-body or polarisable forcefields.[7,8,30–39]

Here we report for the first time, to our knowledge, a comprehensive description of the behaviour of 20 water molecules confined in the pores of a series of 3D-connected isostructural zeolites with varying acidity, by means of both classical grand canonical MC simulations and first-principles Car–Parrinello MD simulations. We characterise the behaviour of confined water by a range of properties, from thermodynamic quantities to electronic properties such as dipole moment, including structural and dynamical information.

2 Methods and systems

2.1 Systems studied

To understand the behaviour of water confined in the nanopores of a zeolite and the influence of the cationic content (or acidity) of the zeolite on the properties of water, we have studied three zeolites with the same structure and different Si : Al ratios. The chosen system was the LTA structure type, which displays supercages (also called α cages) of diameter $\simeq$13 Å, connected to one another in a cubic symmetry by 8-ring windows of diameter $\simeq$6 Å, and smaller sodalite cages of diameter $\simeq$7 Å (see Fig. 1). As opposed to previous simulations of 1D water chains confined in zeolitic materials such as Li-ABW, bikitaite,[25,26] and VPI-5,[27] the medium-size, 3D-connected pores of LTA make it a good candidate for studying the effect of confinement on liquid-water condensation in open framework nanoporous materials. Furthermore, the small size of the LTA unit cell (approximately 12 Å) makes it a reasonable system for first-principles molecular dynamics simulations.

Synthesis of all-silica zeolites remains an experimental challenge. Although a synthesis of the reportedly all-silica LTA zeolite was published in the 1980s,[40] the proposed procedure (thermal dealumination) is known to result in materials exhibiting non-negligible concentrations of defects[41] and no direct structural characterisation was reported in this study. To our knowledge, this synthesis was never reproduced. More recently, the all-silica LTA zeolite was synthesised and characterised by the groups of Corma[42] and Patarin.[43,44] Because the work reported in the present communication was started before these experimental results were published, the structure we used is an hypothetical structure taken from ref. 45. Its unit cell structure ($Si_{24}O_{48}$) is cubic ($Pm\bar{3}m$ symmetry and a = 11.9 Å) and contains one supercage and one sodalite cage. It is very similar in both unit cell

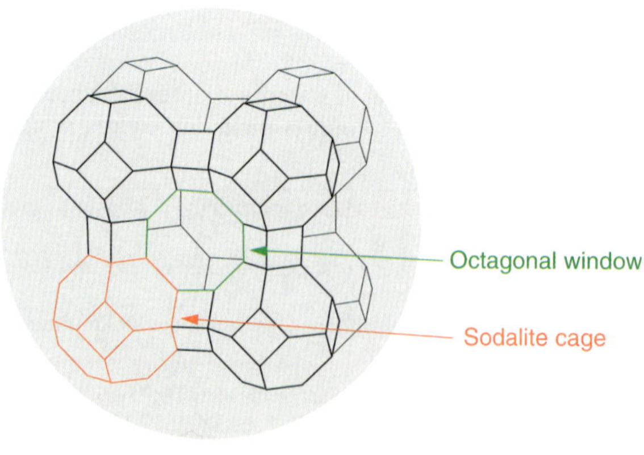

Fig. 1 Schematic view of the LTA structure showing an α cage (center) with its eight surrounding sodalite cages. α cages are linked to one another by octagonal windows.

 This journal is © The Royal Society of Chemistry 2009

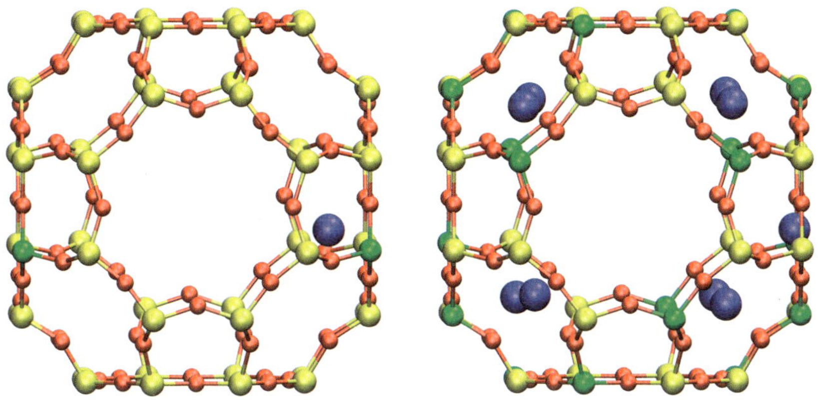

Fig. 2 Left: structure of the Na_1-LTA zeolite, containing one aluminium atom (in green) and one extra-framework Na^+ cation (in blue). The cation is represented in its most stable crystallographic site, known as Na(3).[47] Right: structure of Na_9-LTA zeolite with 9 aluminium atoms and 9 Na^+ cations. Eight cations are in sites Na(1), and one is in site Na(3).

parameter and atomic positions to the experimental structures reported by Patarin and Corma.[42–44]

When it comes to cationic zeolites with the LTA structure, a large number of materials have been synthesised. Most natural and synthetic zeolites feature sodium as an extra-framework cation. We have thus studied two different cationic LTA zeolites, containing 1 and 9 sodium cations respectively. The unit cell formulae are $NaSi_{23}O_{48}Al$ and $Na_9Si_{15}O_{48}Al_9$, corresponding to Si : Al ratios of 23 and 1.67 respectively. To make our discussion easier to read, these two materials will hereafter be named Na_1-LTA and Na_9-LTA. Na_1-LTA corresponds to a single aluminium atom and a single Na^+ cation per unit cell. As all T atoms¶ in the LTA structure are symmetrically equivalent, there is only one possible framework structure for this material, which is represented in Fig. 2. As there has been no synthesis of a LTA zeolite with this particular Si : Al ratio, we have used the same structure and unit cell parameter as the all-silica LTA zeolite mentioned above.

The choice of Na_9-LTA for the present study was dictated by our intention to model a zeolite with a high cationic content (low Si : Al ratio) and the constraints of the first-principles approach. This latter arises from the Loewenstein rule,[46] which states that no two aluminium atoms of a zeolite structure should be linked directly by an oxygen atom. This results in highly cationic zeolites having unit cells many times larger than the unit cell of the LTA structure, which are then too costly to perform first-principles molecular dynamics simulations. For example, the unit cell parameter of the well-known NaA zeolite ($Na_{12}Si_{12}Al_{12}O_{48}$, Si : Al = 1) is twice that of the pure-silica LTA structure. We thus performed a systematic exploration of all possible distributions of n aluminium atoms among the 24 T atoms of a single LTA unit cell that conform to the Loewenstein rule. These calculations proved that the only values of n compatible with both constraints are in the range $1 \leq n \leq 9$. They furthermore showed that for $n = 9$, corresponding to the lowest Si : Al ratio, there are only two possible distributions. These two structures, though not symmetrically equivalent, are very similar (see Fig. S1†). We have thus taken the first one as a starting point for our simulations, along with a unit cell parameter taken from the experimental structure of zeolite NaA ($a = 11.26$ Å).[47]

¶ In a zeolite framework, the tetrahedrally bonded Si and Al atoms are referred to as the T atoms.

2.2 Monte Carlo simulations

2.2.1 Models and forcefields. The development of models and forcefields to describe the interaction of zeolites with various guest molecules is a very active field of research.[48] In this study, the focus is put on the properties of confined water and their evolution with the nature of the zeolite. No chemical change occurs during water condensation in any of the LTA zeolites studied and Monte Carlo simulations were thus performed in the classical limit. We used existing forcefields, previously designed in the group of Fuchs and known to yield good transferability for both all-silica[7,49] and cationic zeolites.[50] For Na_1-LTA and Na_9-LTA systems, an average tetrahedrally bonded (T) atom was used, instead of distinguishing aluminium and silicon atoms.[50] It has been shown that this yields a reasonable description of sodium cation distributions in faujasites.[50,51]

The zeolite framework was considered fixed. Its interaction with extra-framework cations and guest molecules included a Coulombic term and a repulsion–dispersion term; the latter was modeled by a Buckingham (or exp 6) potential for the cations[52] and a Lennard-Jones potential[50] for water molecules. A summary of the forcefield parameters (partial charges, Lennard-Jones and Buckingham parameters) can be found in Tables 1 and 2). The rigid, non-polarisable TIP4P model[53] was used for water. Lorentz–Berthelot rules have been used to obtain cross-potential parameters.

2.2.2 Grand canonical Monte Carlo simulations. Adsorption isotherms were obtained using grand canonical Monte Carlo simulations[54,55] to compute the average number of adsorbed water molecules for several values of the chemical potential. The chemical potential of water, μ, was related to the thermodynamic pressure, P, by using the ideal gas law in the vapour phase, and by computing the $\mu(P)$ relation

Table 1 Forcefield parameters used for modeling repulsion and dispersion interactions in Monte Carlo simulations

Interaction	Parameter	Value
Na–Na	σ	2.584 Å
	ε	50.27 K
O_{water}–O_{water}	σ	3.154 Å
	ε	78.03 K
$O_{zeolite}$–$O_{zeolite}$	σ	3.000 Å
	ε	93.53 K
Na–$O_{zeolite}$	α	61.1×10^6 K
	β	4.05 Å^{-1}
	γ	76.52×10^4 K Å^6

Table 2 Partial charges borne by atoms in the Monte Carlo simulations

Molecule	Charges/*e*
TIP4P water	$q_O = -1.04$
	$q_H = +0.52$
All-silica LTA framework	$q_O = -0.7$
	$q_{Si} = +1.4$
Na_1-LTA framework	$q_O = -0.7125$
	$q_T = +1.3833$
Na_9-LTA framework	$q_O = -0.8125$
	$q_T = +1.25$
Na^+	$q_{Na} = +1$

 This journal is © The Royal Society of Chemistry 2009

in liquid water as described in ref. 7. Preinsertion,[56] orientation[57,58] and jump[50] bias moves were used to accelerate the convergence of the Monte Carlo simulations.

All Monte Carlo runs were performed at 300 K on a 2 × 2 × 2 supercell to minimise the influence of periodic boundary conditions. Typical runs included 10 to 200 million steps to equilibrate the system followed by a production run of 30 to 100 million steps. Long-range electrostatic interactions were taken into account using the Ewald summation technique. In order to improve the efficiency of the calculations, the electrostatic and repulsion–dispersion interaction energies between the rigid zeolite framework and both water molecules and extra-framework cations were precomputed on a grid (with a grid mesh of 0.2 Å) and stored for use during the simulation.

2.2.3 Grand potential computation. The question of the nature of the transition in adsorption/desorption experiments is often subject to debate. Is this transition a real phase transition, between a gas-like and a liquid-like fluid? Likewise, when hysteresis occurs, its origin also raises several questions. Is it due to a bad sampling of the system or can it be related to the existence of multiple metastable states? In order to bring an answer to these issues, knowledge of the thermodynamic potential of the system is necessary. Such a quantity is not straightforward to obtain and requires the use of sophisticated techniques. Recently, Wang and Landau developed such a technique that is based on the computation of the density of states (DOS) on-the-fly, using biased Monte Carlo simulations.[59,60] (Once the DOS is known, the calculation of the thermodynamic potential is straightforward.) This method has been used successfully to study a wide range of systems, including Ising models,[59] Lennard-Jones fluids[61] and clusters,[62] or proteins.[62,63] It has also been revisited in the context of expanded ensembles.[64–66] In expanded ensembles, substates of the system are defined by the introduction of a new variable often called the reaction coordinate ξ. The aim of those EXEDOS (expanded ensemble density of states) simulations is not to determine the density of states as a function of the energy, like in the original Wang–Landau algorithm, but as a function of the reaction coordinate, in order to compute variations of thermodynamic quantities along ξ. When dealing with adsorption, the thermodynamic potential is the grand potential Ω, since simulations are conducted in the Grand Canonical ensemble. We have applied the EXEDOS methodology to compute Ω as a function of the number of adsorbed water molecules in the zeolite framework. To this end, we have computed the density of states of the system as a function of N, noted $Q(N)$.

The probability to generate randomly a configuration with N water molecules is proportional to $Q(N)$. A flat distribution is thus obtained if each state is visited with probability proportional to $1/Q(N)$. The acceptance probabilities of insertion and destruction moves have been modified accordingly (rotations, translations, and jump moves are still accepted according to the conventional Metropolis criteria):

$$p(N \rightarrow N+1) = \min\left(1, \frac{V}{(N+1)\Lambda^3} \frac{Q(N)}{Q(N+1)} e^{-\beta \Delta E}\right) \quad (1)$$

$$p(N \rightarrow N-1) = \min\left(1, \frac{N\Lambda^3}{V} \frac{Q(N)}{Q(N-1)} e^{-\beta \Delta E}\right) \quad (2)$$

Λ is the de Broglie wavelength and ΔE is the energy change between the new and the old configuration. $Q(N)$ is determined iteratively during the simulation. Initially, Q is uniformly set to 1 for the entire range of values accepted for N. Q is then updated at each insertion or destruction trial move between two configurations containing respectively N_1 and N_2 water molecules. If the move is accepted, $Q(N_2)$ is multiplied by a factor f (with $f > 1$); if the move is rejected, $Q(N_1)$ is multiplied by f. At the same time, a N-histogram $H(N)$ is accumulated. The simulation is stopped when each bin has been visited at least $100/\sqrt{[\ln f]}$ times, following Zhou and Bhatt.[67] At this time, the multiplicative factor is decreased, so that $f \rightarrow \sqrt{f}$. The histogram is reset to zero and a new cycle is started, until the histogram is flat again. This process is repeated until f reaches a small enough value. f is set to e^1 for the initial cycle, and 20 cycles are made, so that the last value of f is approximately equal to $e^{10^{-6}}$. For small values of f (typically $f < e^{10^{-4}}$), the "minimum visit per bin" criterion may be too restrictive and the simulation is stopped if the total number of MC moves is superior to $200\,000/\sqrt{[\ln f]}$, as suggested by Poulain *et al.*[62]

Once $Q(N)$ has been determined, one is able to calculate Ω for the system, using:

$$\Omega(\mu, V, T) = -k_B T \ln\left(\sum_N Q(N) e^{\beta\mu N}\right) \tag{3}$$

For a given value of the chemical potential μ, the Landau free energy between two states containing respectively N_1 and N_2 molecules is:

$$\Omega_L(\mu, N_2) - \Omega_L(\mu, N_1) = -k_B T \ln\left(\frac{Q(N_2)}{Q(N_1)} e^{\beta\mu(N_2 - N_1)}\right) \tag{4}$$

This allows one to determine Ω_L to within an additive constant. However, in our particular case, we can determine this constant, since we know that $N = 0$ only corresponds to one configuration for any value of μ, so that $\Omega_L(\mu, 0) = 0$ kJ mol^{-1}. Using the $\mu(P)$ relationship for bulk water, we can also present variations of Ω (or Ω_L) as a function of the external pressure of the reservoir.

2.3 First-principles simulations

2.3.1 Molecular dynamics. The molecular dynamics of anhydrous and hydrated zeolite were performed with the Car–Parrinello approach[68] as implemented in the CPMD software package,[69] using the Kohn–Sham formulation of DFT. We employed the gradient-corrected BLYP functional.[70,71] A plane-wave basis set with an energy cutoff of 70 Ry was used and only the valence electrons were explicitly considered, with semilocal norm-conserving Trouillier–Martins pseudopotentials taking into account the interactions between core and valence electrons. All molecular simulations were performed using periodic boundary conditions on a single cubic unit cell of the zeolite. Hydrogen atoms were deuterated in order to improve decoupling of the dynamics of the ionic and electronic subsystems and allow for a larger time step in the integration of the equations of motion. A timestep of 7 a.u. (0.17 fs) and a fictitious electron mass of 1000 a.u. were used.

Molecular dynamics simulations of the anhydrous zeolite were started from the hypothetical structure for the purely siliceous LTA zeolite available in the literature.[45] For the hydrated zeolites, starting configurations were extracted from the previously performed grand canonical Monte Carlo simulations. A first equilibration run of at least 1.2 ps was performed on each system, during which the temperature was controlled to 300 K by a Nosé–Hoover thermostat[72,73] with a frequency of 3800 cm^{-1}. Production runs of 5 to 6 ps in the microcanonical NVE ensemble were then performed, from which were extracted all the quantities reported in this article.

2.3.2 Analysis of electronic properties. The total polarisation $M(t)$ of the periodic unit cell was defined using the Berry phase approach of Resta.[74,75] The IR absorption coefficient of the system, $\alpha(\omega)$, was then computed by calculating the autocorrelation function of the polarisation, within the framework of linear response theory:

$$\alpha(\omega) \propto \frac{\omega \tanh(\beta\hbar\omega/2)}{n(\omega)V} \int_{-\infty}^{\infty} \mathrm{d}t\, \mathrm{e}^{-\mathrm{i}\omega t} \left\langle \sum_{ij} \mu_i(t)\mu_j(0) \right\rangle \tag{5}$$

where $n(\omega)$ is the refractive index. A harmonic quantum-classical correction was then applied to the classical correlation function to account for the quantum aspects of the hydrogen motion, as was shown to be adequate in previous works.[76–78] The resulting infrared spectrum was then smoothed by a linear filter as in ref. 77 and 78.

Dipole moments of individual water molecules were also computed: in order to do so, the electronic wavefunctions are replaced by the maximally localised Wannier functions[79] and the Wannier centers are assigned to each water molecule.[80,81,77] The calculation of both the macroscopic polarisability and the Wannier functions was performed every 1.7 fs of the dynamics, for each system.

To characterise the presence of hydrogen bonds and their strength from an electronic point of view, we propose to study atomic polar tensors on hydrogen atoms. These tensors and particularly their strong anisotropy can be shown to be a signature of hydrogen bonding in liquid water.[82] The atomic polar tensor of a given atom n is defined as

$$\alpha_{ij}^{(n)} = \frac{\partial \boldsymbol{M}_i}{\partial \boldsymbol{R}_{n,j}}, \text{ where } i \text{ and } j \text{ can be } x,\ y \text{ or } z \tag{6}$$

where M_i is the component of the total polarisation of the system along axis i and R_n is the position of atom n.

In this study, the atomic polar tensors were calculated in the density functional linear response theory,[83] using Putrino *et al.*[84] implementation in the CPMD code[69] of the linear response to an external electric field. The atomic polar tensors (APT) of all atoms in a given configuration of the system were computed by three calculations of the linear response of the system to an external electric field, according to the following formula:

$$\alpha_{ij}^{(n)} = \frac{\partial \boldsymbol{F}_j^{(n)}}{\partial \boldsymbol{E}_i} \tag{7}$$

where $F^{(n)}$ is the force acting on atom n and E is the external electric field. For each system studied, this analysis was performed on 200 configurations extracted at regular intervals from the NVE dynamics.

From the APT tensor we derived an atomic charge, namely the Born charge, as the trace of the APT tensor. To better characterise the APT, which is a non-symmetric 3 × 3 tensor, we also determined the singular values of the APT from singular value decomposition. The three singular values (ξ_1, ξ_2 and ξ_3) have the dimension of a charge and their differences characterise the anisotropy of the atomic polar tensor.

3 Results and discussion

3.1 Thermodynamics of water in LTA zeolites

3.1.1 Water adsorption isotherms. Water adsorption isotherms for the three different zeolite models are shown in Fig. 3. For Na_9-LTA, the isotherm is a typical type I isotherm, according to the IUPAC classification scheme,[85] commonly found in the case of hydrophilic aluminosilicate zeolites. It is characterised by a strong water uptake at low pressure (high value of Henry's constant). In contrast, the isotherm for the all-silica zeolite is typically that of water intrusion into an hydrophobic nanoporous solid:[49,86] a negligible water uptake is observed up to a pressure of $\simeq$100 MPa (*i.e.* much higher than the saturated vapour pressure of 3500 Pa), at which an abrupt

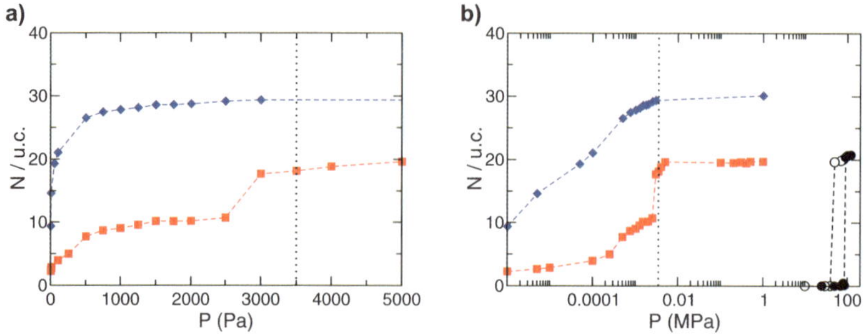

Fig. 3 (a) Computed adsorption isotherms of water in Na_1-LTA (red squares) and Na_9-LTA (blue diamonds) at 300 K. (b) Intrusion (filled circles) and extrusion (open circles) isotherm of water in all-silica LTA (in black). Adsorption curves for Na_1-LTA and Na_9-LTA are also reported for comparison. A vertical dotted line indicates P_0, the water saturation vapour pressure at 300 K.

condensation transition takes place. Moreover, in that case, a hysteresis is observed: water extrusion takes place at $\simeq$45 MPa.

In the case of Na_1-LTA, three different stages can be distinguished during water adsorption. At low pressure, 2 to 5 water molecules per unit cell are involved in small clusters around sodium cations, equally distributed in each of the eight unit cells of the simulation box (see Fig. 4a). Around 250 Pa, a first transition occurs that corresponds to the filling of half the unit cells of the simulation box (Fig. 4b). This is accompanied by a redistribution of the cations, so that 4 unit cells are filled with 2 cations and water, whereas the other 4 are almost empty. This is the stable situation until 2500–3000 Pa where a second transition occurs leading to the total filling of all the supercages. This transition is accompanied once again by the displacement of sodium cations so that, in the end, the system is homogeneous with each unit cell filled with water and one cation (Fig. 4c). It is interesting to point out the importance of simulating eight unit cells instead of one, because it would otherwise not have been possible to observe such cationic redistribution and inhomogeneous pore filling.

The change in shape of the adsorption isotherms from all-silica to Na_9-LTA can be regarded as an evolution from a purely hydrophobic nanoporous solid to a hydrophilic one, the intermediate Na_1-LTA case being that of a hydrophobic medium containing strong hydrophilic defects. If we ignore the two-step filling mechanism (which is a consequence of the mobility of sodium cations), the adsorption isotherm in Na_1-LTA is of type IV: adsorption at low pressures followed by a plateau, and

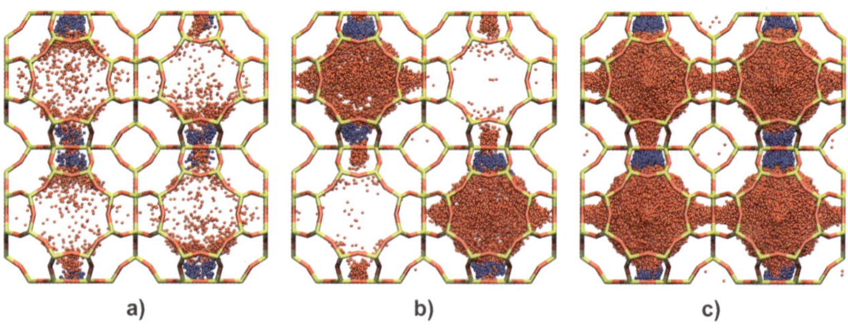

Fig. 4 Water filling of Na_1-LTA zeolite from GCMC simulations. Each dot represents the position of molecules (water in red and sodium in blue) accumulated over the simulation run (after convergence has been reached) at 250 Pa (a), 1000 Pa (b), and 3000 Pa (c). Zeolite framework is shown in sticks (oxygen atoms in red and silicon atoms in yellow). For the sake of clarity, only 4 unit cells are shown.

 This journal is © The Royal Society of Chemistry 2009

Table 3 Filling of LTA at saturation, porous volume and computed density of water. The porous volumes and densities shown in brackets take into account the excluded volume due to sodium cations

System	N per u.c.	V_P per u.c./Å^3	ρ_{conf}/g cm^{-3}
LTA all-silica	21	754	0.83
Na_1-LTA	20	754 (745)	0.79 (0.80)
Na_9-LTA	30	983 (901)	0.91 (0.99)

a condensation transition occurring below the saturated vapour pressure P_0.[85] To our knowledge, this is the first time that such a "pure" type-IV isotherm is predicted in a nanoporous solid. In previous works,[39,87] we observed that the inclusion of a silanol-type defect could result in type-IV isotherm, but the condensation transition occurred above P_0.

We now discuss the effect of the cations on water density. Values for the three systems are shown in Table 3. Porous volumes have been computed using an approach similar to that of the calculation of Connolly surfaces, and is described in detail elsewhere.[7,88] In the case of all-silica and Na_1-LTA systems, a negligible number of water molecules is found in sodalite cages. Thus, the porous volume only takes into account supercages. In the case of Na_1-LTA and Na_9-LTA, we also compute a porous volume and a density that takes into account the presence of sodium cations (considered as a sphere with a radius being half the σ value of sodium, *i.e.* 1.29 Å). Water in all-silica LTA is strongly depleted with respect to bulk water, as observed earlier in other siliceous zeolites.[7,88] The presence of only one sodium cation also lowers water density. This is because the presence of one cation is not sufficient to modify strongly water–water interactions (see Section 3.2.1), and because of the volume occupied by the sodium cation. On the other hand, the presence of the nine cations results in a large increase of water density, up to 0.99 g cm^{-3}, which is similar to bulk water density. Cations modify water structure in the LTA framework, so that a larger amount of water can be introduced. However, as will be shown in Section 3.2, water structure in Na_9-LTA is different from the bulk water.

3.1.2 On the nature of the condensation transition. In order to determine the nature of the condensation transition in all-silica LTA, we have computed the evolution of the grand potential Ω with pressure. Fig. 5a shows the profile of Ω_L as a function of the water content for various pressures. Below roughly 20 MPa, only one stable state exists, that corresponds to an empty LTA zeolite. When pressure is increased, a filled metastable state appears, that is however less stable than the empty

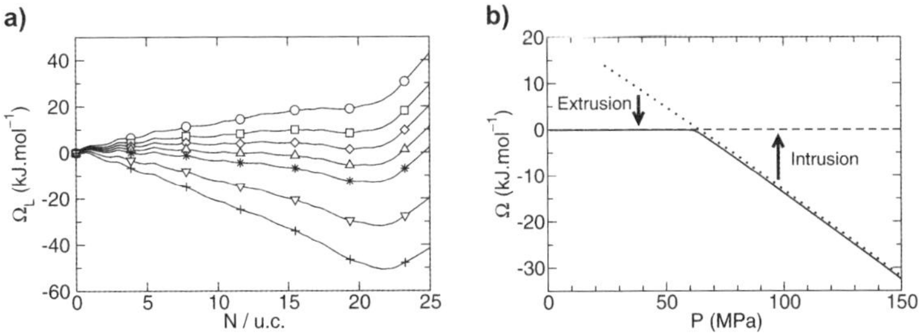

Fig. 5 (a) Landau free-energy profile as a function of water content of all-silica LTA at 10 MPa (circles), 40 MPa (squares), 60 MPa (diamonds), 80 MPa (up triangles), 100 MPa (stars), 150 MPa (down triangles), and 200 MPa (plus symbols). (b) Variation of the grand potential Ω at equilibrium with pressure (thin plain line). The dashed and dotted lines represent the value of Ω for the metastable empty and filled states respectively.

state. Around 70 MPa, the filled configuration becomes the most stable state. The existence of two stable states is characteristic of a first-order transition between gas-like and liquid-like phases. The conversion between those two states is possible over a wide range of pressure (20 → 190 MPa), as seen in Fig. 5, but the system has to overcome a macroscopic energy barrier. Monte Carlo simulations proceed *via* microscopic moves. In a finite time MC run, the most stable state may not be reached, and this is why hysteresis can be observed. In the present case, although the equilibrium pressure is approximately 70 MPa, intrusion and extrusion occur at 85 and 40 MPa respectively in the simulations.

This is, to our knowledge, the first time that the Wang–Landau algorithm has been used in the context of fluid intrusion in porous solids, and that the existence of a first-order transition is really demonstrated. We believe that this represents important progress for the study of adsorption (intrusion) processes, since we are now able to derive thermodynamic quantities that were hardly accessible until now.

3.2 On the hydrogen bonding of water and the zeolite

3.2.1 Structure and orientation dynamics of water. To study the interaction of water with the zeolite and add a microscopic insight to the thermodynamic vision of water developed in the previous section, we have first studied the structure and orientation of water confined in the all-silica, Na_1-LTA and Na_9-LTA zeolites. The arrangement of water molecules in the α cages of the zeolites is irregular and reveals none of the features exhibited by gas-phase water clusters.[89] In order to probe the nature of the hydrogen-bond network of confined water, we have studied the orientation of the molecules around neighbouring water molecules and zeolite oxygen atoms. The key quantities for that task are the OO′ distances and HOO′ angles (respectively noted d and α in the following, see Fig. 6), where O and H are atoms from a same water molecule, and O′ is a neighbouring oxygen atom. Following the common practice, we used a geometric criterion[90] to characterise hydrogen bonds and quantitatively study the evolution of water structure between the different zeolites: we consider the hydroxyl group OH of a water molecule hydrogen bonded to an acceptor atom O′ if $d \leq 3.5$ Å and $\alpha \leq 30°$. The limits of the criterion are based on the position of the minimum of the 2D distribution function of liquid water orientation. This criterion is widely used in the literature, as well as a few stricter variants imposing an additional restriction on the OH distance. Using this criterion, we have studied the evolution of the number of both water–water and water–zeolite hydrogen bonds when the Si : Al ratio of the zeolite is lowered (see Table 4a). First, it can be seen that the number of water–zeolite hydrogen bonds in the all-silica zeolite is fairly small and that the number of water–water hydrogen bonds is still large, while smaller than for same-density bulk water (around 1.6 per water molecule). The confined water in the hydrophobic, all-silica zeolite can thus be seen as a nanodroplet of liquid water with its hydrogen-bond network largely closed onto itself, and a few OH groups oriented towards the zeolite. When we move to cationic zeolites, the number of water–zeolite bonds increases while water–water hydrogen bonds diminish; the net result in the total number of hydrogen bonds per

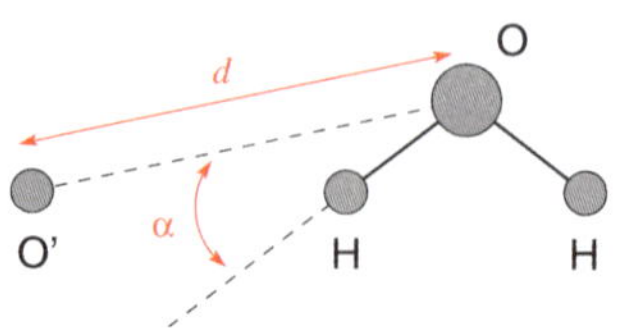

Fig. 6 Definition of the parameters d and α used to characterise the water orientation and on which the geometric criterion for hydrogen bonds is based.

 This journal is © The Royal Society of Chemistry 2009

Table 4 Number of hydrogen bonds among the water molecules confined inside the α cage of LTA zeolites with different numbers of cations, using a geometric criterion (see text), from Monte Carlo simulations (upper table) and first-principles molecular dynamics (lower table). For each zeolite, the number of water–water hydrogen bonds and water–zeolite hydrogen bonds is indicated as well as the ratios per water molecule

(a) From Monte Carlo simulations					
Zeolite	Water–water H bonds		Water–zeolite H bonds		All H bonds
	Total	Per H_2O	Total	Per H_2O	Per H_2O
All-silica	25.6	1.35	3.9	0.20	1.55
One cation	23.2	1.22	6.0	0.32	1.54
Nine cations	17.6	0.93	13.0	0.68	1.61
(b) From first-principles molecular dynamics					
Zeolite	Water–water H bonds		Water–zeolite H bonds		All H bonds
	Total	Per H_2O	Total	Per H_2O	Per H_2O
All-silica	30.8	1.54	4.1	0.21	1.75
One cation	27.2	1.43	7.1	0.37	1.80
Nine cations	22.9	1.21	14.0	0.74	1.95

water molecule is a slight increase with acidity. As the acidity of the zeolite framework and the number of cations grow, the droplet of confined water "opens up" and forms more hydrogen bonds with the zeolite, at the expense of the water–water hydrogen-bond framework.

To understand the interaction between the zeolite and the confined water further, the classical Monte Carlo simulations were complemented with first-principles molecular dynamics simulations in order to provide valuable information on the intermolecular and vibrational dynamics of the system as well as to probe its electronic properties (the details of these Car–Parrinello DFT simulations can be found in Section 2.3). Table 4b shows the number of water–water and water–zeolite hydrogen bonds counted by the same geometric criterion, for all three zeolites studied. Comparison to Table 4a shows a similar evolutions for both classical models and DFT-based simulations. The absolute number of hydrogen bonds counted is, however, larger in the latter case. This is in keeping with observations that Kohn–Sham DFT liquid water described with the BLYP exchange–correlation function is overstructured at ambient conditions. But the use of molecular dynamics simulations also allows us to get a picture of the dynamics of these hydrogen bonds, in the form of the autocorrelation function of the existence of hydrogen bonds. Fig. 7 shows this autocorrelation function for water–zeolite hydrogen bonds in all three zeolites. It can be seen that the OHO′ alignments, counted as hydrogen bonds by the geometric criterion, decay very rapidly in the case of all-silica zeolite and Na_1-LTA, while the alignments in Na_9-LTA take longer to subside. The hydrogen-bond lifetime for each zeolite was obtained by an exponential fit of the 0.05 to 1.5 ps region of the autocorrelation function, where they display monoexponential behaviour. The lifetimes of water–zeolite hydrogen bonds, reported on Fig. 7, range from 125 to 520 fs and are smaller than the lifetime of water–water hydrogen bonds both in bulk water (1.5 ps, following the same procedure described above) and in confined water (between 1.3 and 1.6 ps depending on the number of cations). The very short lifetime of the "geometric H bonds" in both the all-silica zeolite and Na_1-LTA is incompatible with the chemist's view of a hydrogen bond. It is thus plausible that what the geometric criterion counts as hydrogen bonds in these cases are occasional, short-lived $OH\cdots O_{zeolite}$ alignments, corresponding to dangling OH groups of the confined water. On the other hand, the lifetime of 520 fs observed in Na_9-LTA is compatible with weak water–zeolite hydrogen bonds. To characterise these two different situations further, we gain insight from the first-principles dynamics and turn to a electronic characterisation of hydrogen bonds.

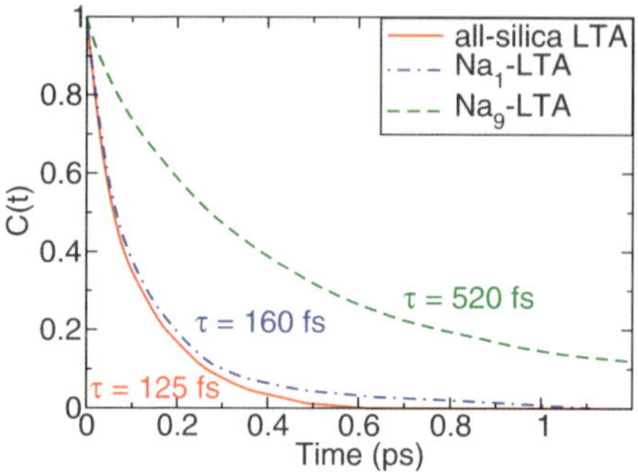

Fig. 7 Autocorrelation function $C(t)$ of the existence of water–zeolite hydrogen bonds, as defined per the geometric criterion described in the text, for water confined in different zeolites of identical framework type: all-silica (red), one cation (blue) and nine cations (red). The lifetime of the bond, indicated on each curve, is obtained by an exponential fit of the 0.05–1.5 ps region of the autocorrelation function.

3.2.2 Electronic characterisation of hydrogen bonds: the atomic polar tensor. By using first-principles molecular dynamics, we can clearly depict the electronic properties of the confined water and thus help understanding its interaction with the confining zeolites. This is needed to characterise hydrogen bonds in a way that will go past the limits of simple geometric criteria. To probe the electronic nature of the hydrogen bonds, we studied the anisotropy of the atomic polar tensor of the hydrogen atoms. In the presence of a hydrogen bond, the hydrogen atom involved exhibits a fairly anisotropic atomic polar tensor. This anisotropy can be characterised by the three singular values of the atomic polar tensor (see Section 2.3). For example, in the case of the water dimer shown in Fig. 8, it can be seen that the ellipsoid defined from the singular value decomposition of the APT tensor of the hydrogen atom involved in the hydrogen bonding exhibits a clearly oblong shape, with one singular value much larger than the others and the corresponding singular vector directed along the axis of the hydrogen bond. Physically, the ellipsoid shown in Fig. 8 shows the magnitude of dipole moment change upon a small displacement of the hydrogen atoms in the direction of the principal axis. For an hydrogen involved in an H bond, a displacement parallel to the H bond can be seen to lead to a much larger change of dipole moment than for a displacement perpendicular to it. This arises from noticeable charge transfer across the H bond when the hydrogen is moved along the H bond.

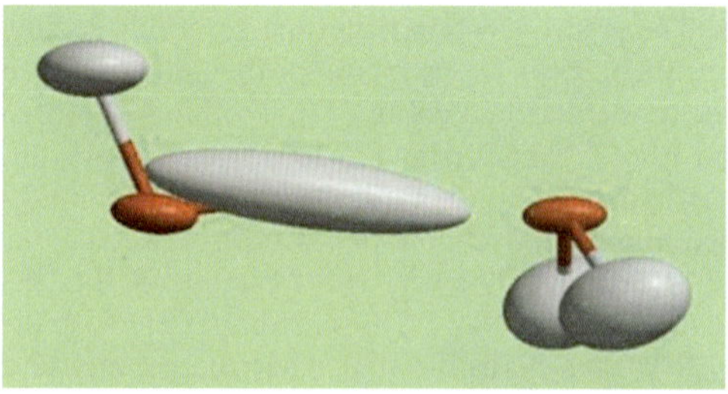

Fig. 8 Graphical representation of the atomic polar tensor (APT) of atoms in a water dimer. The ellipsoids are drawn from the singular value decompositions of the APT tensors. They describe the magnitude of dipole moment changes upon displacement of the atoms in the principal directions of the ellipsoids. The lengths of the principal axis of these ellipsoids are thus the singular values of the APT tensors. It can be noted that the hydrogen involved in the hydrogen bond exhibits a significant anisotropy of the APT tensor.

 This journal is © The Royal Society of Chemistry 2009

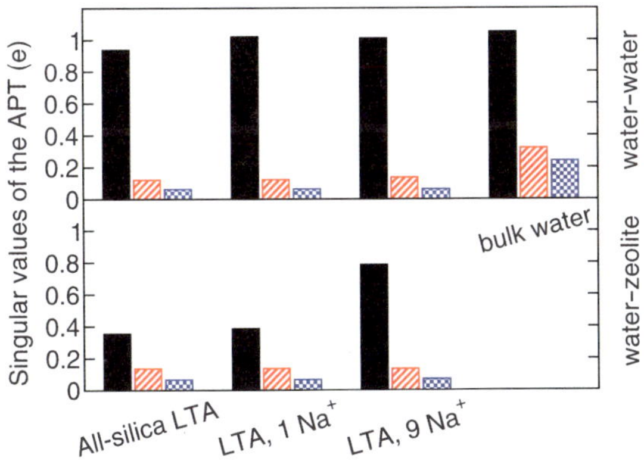

Fig. 9 Average singular values of the symmetric of the atomic polar tensor (see Section 2.3) for hydrogen atoms of water confined in the three zeolites (all-silica, one cation and nine cations): ξ_1 in black, ξ_2 in red and ξ_3 is blue. The unit of the singular values is the atomic unit of charge, e. The upper panel features the singular values for hydrogen atoms involved in water–water hydrogen bonds; the values for bulk water are indicated for comparison. The lower panel features singular values for hydrogen atoms oriented towards the zeolite.

We have thus applied this method to study the strength of both water–water and water–zeolite bonds identified by the geometric criterion; Fig. 9 shows the average eigenvalues in each case, as well as for bulk water. It can be seen that the strength for water–water hydrogen bonds in the confined water is fairly similar to that of bulk water: the largest singular value, ξ_1, is close to 1 e. Singular values ξ_2 and ξ_3, however, are smaller than in the bulk.

The water–zeolite hydrogen bonds, on the other hand, can be seen to be significantly weaker than the water–water bonds. Moreover, we see a gradual increase of ξ_1 when we go from the all-silica LTA ($\langle\xi_1\rangle \simeq 0.36$) to Na_1-LTA ($\langle\xi_1\rangle \simeq 0.40$) and then Na_9-LTA ($\langle\xi_1\rangle \simeq 0.77$). Both findings are consistent with the lifetimes reported above, but quantify the nature and strength of these hydrogen bonds in a different perspective: the difference in lifetime, for example, between water–zeolite bonds in Na_9-LTA and the bulk water–water hydrogen-bond lifetime is markedly more important than the difference in the anisotropy of the atomic polar tensor. The picture they draw for the evolution of water–zeolite interaction with increasing cationic content of the zeolite, however, is the same: while the water confined in the all-silica zeolite can be considered as a droplet forming no hydrogen bonds with the zeolite, but exhibiting a few dangling OH groups, water confined in the same-structure zeolite with a Si : Al ratio of 1.67 forms relatively weak hydrogen bonds with the oxygen atom of the zeolite framework. This difference in the water–zeolite interaction strengthens the explanation of the thermodynamic properties of water condensation in both all-silica and cationic zeolites, as presented in Section 3.1: the hydrophobicity of the internal surface of all-silica zeolites, due to the extreme weakness of the water–zeolite attraction compared to the water–water dispersion energy, leads to intrusion in the liquid phase, while the larger water–zeolite interaction of acidic zeolites leads to adsorption in the gas phase.

The Na_1-LTA, on account of the thermodynamic, dynamic and electronic results shown here, can be seen as a situation midway between the two cases just discussed. Because of the strong electric field created by the presence of an aluminium atom and a cation in the zeolite, water molecules are strongly attracted in the vicinity of the cation, leading to a first water uptake in the gas phase (less than 500 Pa). However, in spite of this local attraction, the internal surface is still only weakly attractive for water, as confirmed by the study of the nature of the water–zeolite interaction in Na_1-LTA. This leads to the condensation of water at a pressure close to the saturation vapour pressure.

3.2.3 Infrared spectrum and vibrational spectroscopy of the all-silica zeolite. To link our simulations to experimentally observable properties, we have also studied the vibrational properties of the all-silica zeolite and the water confined inside its pores. No experimental data is available today on this system, but vibrational spectroscopies have been widely used to investigate the behaviour of water at interfaces.[91,92] We have calculated the infrared spectrum of both the anhydrous zeolite and the hydrated system. Through the use of the Berry phase approach and maximally localised Wannier functions described in Section 2.3, it was also possible in that latter case to separate the contributions of water and zeolite to the total spectrum. We have first shown that the calculated infrared spectrum of the anhydrous all-silica LTA (Fig. 10, lower panel) is in reasonable agreement with the experimental data[93] (Fig. 10, upper panel) of the material synthesised by Fyfe *et al.*[40] The general shape of the spectrum is reproduced; the positions of the absorption bands present relatively large shifts (up to 200 cm^{-1}). These are attributed to the fact that the simulation was performed at constant volume with a unit cell parameter that was not optimised, leading to constraints on the vibrations modes of the zeolite, as well as the dealumination procedure used in ref. 40, which is known to yield materials with non-negligible concentrations of defects.[41]

For the hydrated zeolite, we have shown that the presence of water has no significant effect on the infrared spectrum of the zeolite. Moreover, the vibration modes of the zeolite and the confined water are not correlated and the total spectrum of the system matches the sum of the two separate spectra for water and the zeolite, at frequencies higher than 300 cm^{-1}.[29] We have then focused our study on the OH-stretch region of the confined water IR spectrum, shown in Fig. 11. It can be seen that, while the ν_{OH} peak of bulk water is large and structureless, that of water confined in all-silica LTA exhibits a narrower high-frequency peak on top of a larger band. This peak is attributed to dangling OH groups, which are not part of the water hydrogen-bond network and are oriented towards the zeolite. Such a peak has been reported for water near a variety of hydrophobic surfaces[94,95] and can thus be used as a signature of hydrophobicity. As an example of systems where it has been observed, the lower panel of Fig. 11 shows the IR SFG (sum-frequency generation) spectrum of water near an H_2O/CCl_4 interface, from Scatena *et al.*[95] In the present case of water confined in an all-silica zeolite, we have further confirmed the attribution of this high-frequency ν_{OH} peak to dangling OH groups by studying the correlation between the average stretching frequency of water OH groups and the percentage of simulation time during which they are involved in water–water hydrogen bonds (according to the geometric criterion). Fig. 12 shows that there is a clear correlation

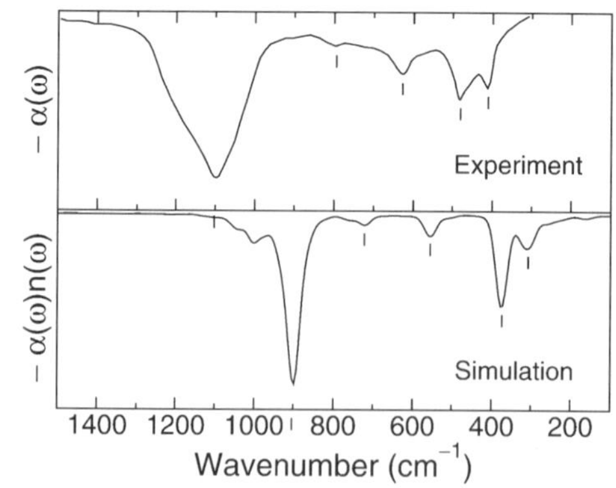

Fig. 10 Infrared spectrum of the dehydrated, LTA zeolite: computed from first-principles molecular dynamics of the all-silica zeolite (lower panel); compared to experimental data from a reportedly all-silica zeolite[40] (upper panel).

 This journal is © The Royal Society of Chemistry 2009

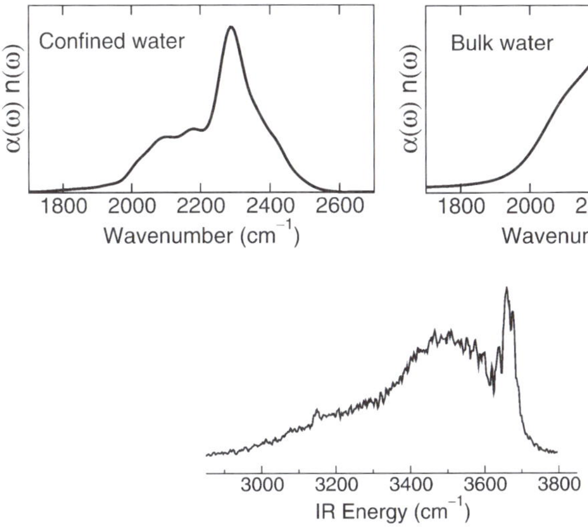

Fig. 11 OH stretch band of the infrared spectrum of 20 water molecules confined in all-silica LTA zeolite (left panel) and 32 bulk water molecules (right panel). The lower panel shows the Vis-IR SFG spectrum of water near an H_2O/CCl_4 interface, from Scatena *et al.*,[95] exhibiting a high-frequency peak considered as a signature of hydrophobicity. The simulated systems (top panels) are deuterated, hence the difference in vibration frequency.

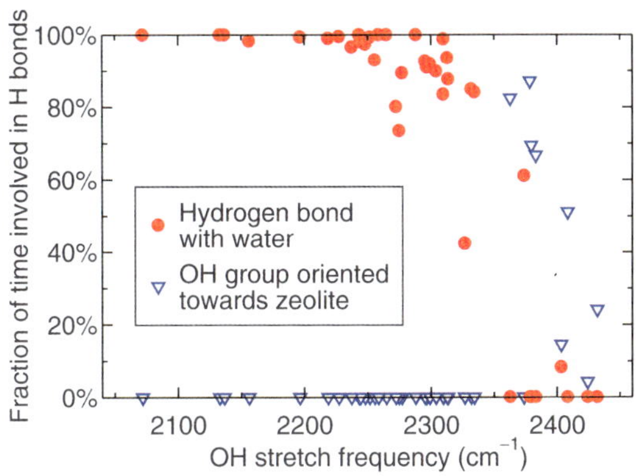

Fig. 12 Correlation between the average stretching frequency (horizontal axis) of a water OH group confined in an all-silica LTA and the percentage of simulation time during which this OH group is involved in water–water hydrogen bonds (vertical axis, red circles) and is oriented towards the zeolite (vertical axis, blue triangles).

between these two quantities: OH groups with $\nu_{OH} \geq 2250$ cm^{-1}, which compose the high-frequency peak, are markedly less involved in the water–water hydrogen-bond network than the other groups.

3.3 Dipole of the confined water

The dipole moment of the water molecules is an important quantity in order to understand the behaviour of confined water and to help assess the quality of the simulation models used. The histogram of the dipole moments of water molecules confined in the α cage of the three zeolites under study are reported in Fig. 13, as well as the distribution obtained for a bulk water simulation including 32 water molecules in a periodic unit cell. It can be seen that the distributions for confined

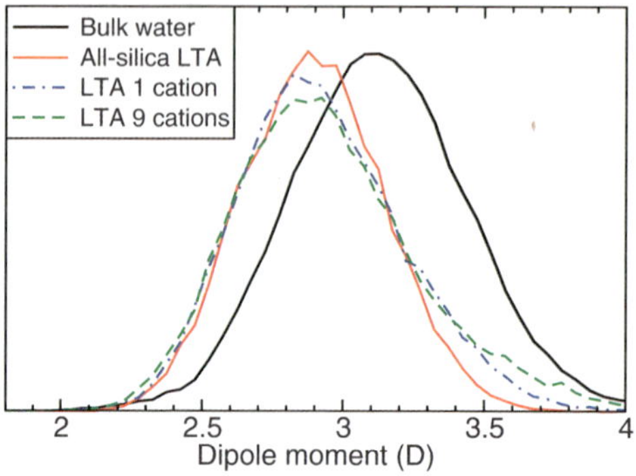

Fig. 13 Distribution of the dipole moment of water molecules in confined in the all-silica LTA (red), Na_1-LTA (blue) and Na_9-LTA (green). The distribution of dipole moments in bulk water is drawn in black.

water peak around 2.9 D, which is close to the peak value of bulk water at 3.1 D. The self-polarisation of confined water is thus already quite large in these relatively small nanopores; from the point of view of the dipole moment, the confined water is liquid-like. Moreover, Fig. 13 and Table 5 reveal that the dependence of the dipole moment on the acidity of the zeolite is very small. The histograms for the all-silica, Na_1-LTA and Na_9-LTA zeolites have the same peak value, and the only statistically significant difference is that zeolites with more cations present a less symmetric distribution, with a more important high-dipole wing, attributed to water molecules in close contact with the cations.

The relatively small influence of the cationic content of the zeolite is somewhat counterintuitive, as the presence of ionic species in close proximity to the water molecules might be expected to polarise the water molecules significantly. Indeed, calculations performed on a system comprised of a single water molecule close to the Na^+ cation in Na_1-LTA have yielded a water dipole moment of $\simeq$2.4 D, which represents a large increase (0.6 D) from the gas-phase value. These facts lead to the logical conclusion that the dipole moment of the confined water is almost entirely due to the self-polarisation of water, while the influence of both the zeolite and the extra-framework cations is faint. Fig. 14 presents the histogram of the dipole moments of 15 and 20 water molecules confined in all-silica LTA, which have their maximum at 2.7 and 2.9 D respectively. This shows that the dependence of the dipole on the water content of the zeolite is more important that the dependence on the

Table 5 Dipole moment and quadrupole eigenvalues of water confined in different LTA zeolites. The values for bulk water, ice and an isolated water molecule are also shown for comparison purposes

System	Dipole/D	Quadrupole eigenvalues/10^{-26} esu cm^{2} [a]		
15 H_2O in all-silica LTA	2.79	−3.04	−0.17	3.21
20 H_2O in all-silica LTA	2.83	−3.18	−0.14	3.32
20 H_2O in Na_1-LTA	2.83	−3.14	−0.12	3.27
20 H_2O in Na_9-LTA	2.88	−3.06	−0.12	3.18
1 H_2O in Na_1-LTA	2.38	−2.57	−0.26	2.82
Isolated water molecule	1.84	−2.47	−0.10	2.57
Bulk liquid water	3.0	−3.19	−0.19	3.38
Ice	3.44	−3.69	−0.12	3.81

[a] esu = electrostatic unit of charge

 This journal is © The Royal Society of Chemistry 2009

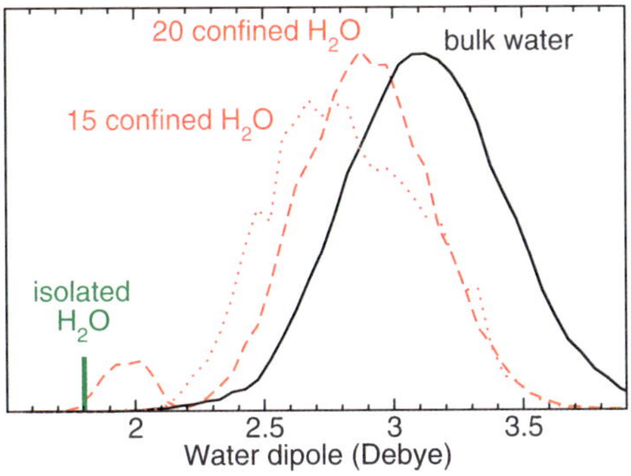

Fig. 14 Distribution of the dipole moment of water molecules in bulk water (black, full line), of 20 water molecules confined in an all-silica LTA zeolite (red, dashed line) and of 15 water molecules inside the all-silica zeolite (red, dotted line). The dipole moment of a water molecule in the gas phase is also presented (green line).

number of cations, and that the self-polarisation of water is the main factor determining the dipole in the systems we study here. This conclusion was further strengthened by the analysis of the quadrupole moments of the confined water molecules, presented in Table 5. The quadrupoles are only weakly affected by the changes in the cationic content of the zeolite while they have a more pronounced dependence on the number of water molecules.

3.4 Partial charges from population analysis and their use as forcefield parameters

There are many different ways to calculate charges from electronic structure. We have here investigated the atomic charges of zeolite atoms from four different definitions. One of our goals was to compare the obtained values from these definitions with charges used in the empirical potentials that were shown to reproduce correct thermodynamics. Two very common methods are the Mulliken population analysis and the Hirshfeld population analysis. However, of possible definitions of atomic charges, those fitted to reproduce the electrostatic potential (called ESP charges) seem the best suited for extracting charges for building an empirical potential. On top of these common atomic charge definitions we have also computed the Born charges obtained as the trace of the APT tensor discussed above. Table 6 shows the resulting charges for atoms of the all-silica zeolite. All these charges give a large spectrum of values. None seem, however, to agree with the range of empirical charges (-0.6 to -0.8 e for O) that can correctly describe the hydrophobicity of silicalite-1.[7]

We have also checked the dependence of Mulliken charges on the electronic method used. To this end we selected a few snapshots from the first-principles molecular dynamics of empty all-silica zeolite and computed Mulliken charges with a Gaussian basis set, using the CP2K program[96]|| instead of a plane-wave basis set. The resulting charges are -0.6 e for oxygen atoms and 1.2 e for silicon atoms, yet again different from other approaches.

Nevertheless, if Mulliken charges are computed for the hydrated all-silica zeolite as well, no variation is observed with respect to the empty zeolite. This is consistent with observations made previously in this paper (from the vibrational dynamics analysis for instance) that the coupling between water and the all-silica zeolite is very small. This is also in line with the charges on zeolite atoms independent of the filling commonly assumed in empirical forcefields.

|| DFT calculations were performed with the BLYP functional and Goedecker–Tetter–Hutter pseudopotentials in the GPW approach, with the DZVP basis set and a cutoff of 300 Ry for the description of the electronic density in plane waves.

Table 6 Average partial charges on the oxygen and silicon atoms of the all-silica zeolite, depending on the analysis method used: Mulliken, Hirshfeld, ESP and Born (see Section 3.4)

Method	Oxygen	Silicon
Mulliken	−1.007	2.175
Hirshfeld	−0.293	0.586
ESP	−1.901	3.801
Born	−1.557	3.114

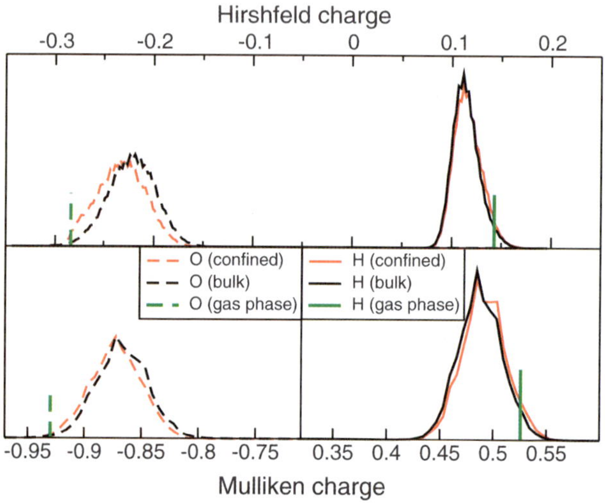

Fig. 15 Distribution of partial charges of the water oxygen and hydrogen atoms according to the Mulliken and Hirshfeld population analysis methods. The distributions were computed for water confined in an all-silica LTA (red), for bulk water (black) and for a single gas-phase water molecule (green).

Turning now to water, the distribution of Mulliken and Hirshfeld charges are shown in Fig. 15. For these two definitions, a decrease of atomic charges is found from an isolated water molecule to bulk liquid water or confined water. Both Mulliken and Hirshfeld population analyses thus give rise to atomic charges that are in contradiction with the increase of water dipole moment. This may be understood by noting that H bond leads to an increase of electron density in the intermolecular region between O and H. Instead of an increase of molecular polarisation, the Mulliken and Hirshfeld population analyses attribute some of that density to the partner hydrogen of a donating oxygen, leading to a decrease of absolute values of charges arising from this apparent intermolecular charge transfer.

On the contrary, the Born charges (see Fig. 16) show an increase of atomic charges upon going from an isolated molecule to a confined water molecule and to a bulk water molecule. This increase of Born atomic charges is, furthermore, in near-quantitative agreement with the dipole moment variation: assigning these charges to hydrogen and oxygen atoms, we find dipole moments of 1.4, 2.77 and 3.2 D for an isolated molecule, a confined molecule and a bulk water molecule respectively. These charges also appear to be in the range of partial charges found in empirical forcefields for liquid water.

From considerations of infrared intensities,[82] it is possible to define an alternative to the Born charge from the APT tensor by

 This journal is © The Royal Society of Chemistry 2009

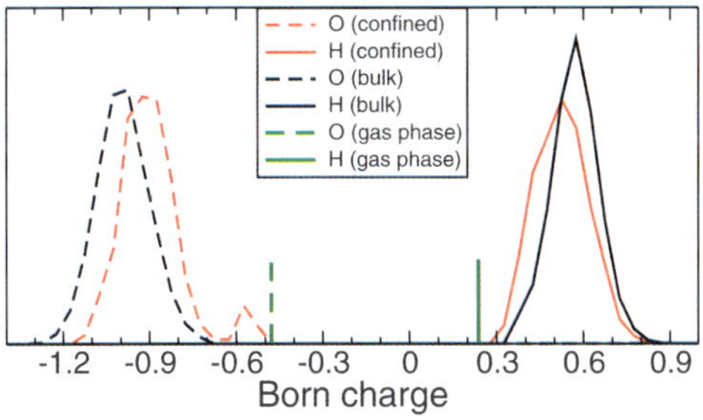

Fig. 16 Distribution of partial charges of the water oxygen and hydrogen atoms according to the Born population analysis method. The distributions were computed for water confined in an all-silica LTA (red), for bulk water (black) and for a single gas-phase water molecule (green).

$$\xi^2 = \frac{\xi_1^2 + \xi_2^2 + \xi_3^2}{3}, \tag{8}$$

where ξ_1, ξ_2 and ξ_3 are the singular values of the APT tensor as discussed above and ξ has thus the dimension of a charge. These charges are found to be slightly lower that the Born charges ($\xi_H = 0.39\ e$ for confined water and $0.54\ e$ for bulk liquid) but show the same trend, with an increase of molecular polarisation from confined to bulk water.

APT-derived charges, and in particular Born charges, appear as interesting candidates to define meaningful atomic partial charges, at least for molecular systems. As mentioned before, the Born charges of the zeolite atoms appear, however, too large with respect to the range of charges used in classical forcefields.

4 Concluding remarks

The use of both classical GCMC and first-principles CPMD simulations has allowed us to describe in some detail the behaviour of a nanodroplet of water confined in the supercage of LTA zeolites. Each method has its (known) strengths and drawbacks, and provides information in a physical chemistry domain where the other does not. Whenever the same kind of information can be obtained by both techniques, the agreement was found satisfactory. This was the case, for instance, for the number of hydrogen bonds between water molecules and the framework atoms as well as between water molecules themselves. Work is in progress to get a better insight into the hydrogen-bond reorientation dynamics of water in such confined systems.[97]

The all-silica version of LTA as well as two model systems with 1 and 9 sodium cations were chosen for this study. From the thermodynamic point of view, we have identified the all-silica LTA zeolite as hydrophobic, and the cationic Na_9-LTA zeolites as hydrophilic. The condensation transition in the first case was demonstrated to be of first order. The adsorption isotherm in the case Na_1-LTA and Na_9-LTA was of types IV and I respectively. Furthermore, the in-depth analysis of the dynamic and electronic properties of water showed that water in the hydrophobic zeolite behaves as a nanodroplet trying to close its hydrogen bonds onto itself, with a few short-lived dangling OH groups, while water in the hydrophilic zeolites opens up to form weak hydrogen bonds with the oxygen atoms of the framework. We conclude that a thoughtful combination of methods, in which each technique is used for what it was designed for, can provide a nice insight into the physics and chemistry of such a complex interfacial system as water confined in zeolite nanopores.

Acknowledgements

This work was supported by the French Agence Nationale de la Recherche, under contracts # JC05_51932 and BLANC06-3_144027. We acknowledge the Institut du Développement et des Ressources en Informatique Scientifique (IDRIS) and ENSCP for computer time allocation. We thank Isabelle Demachy, Florent Calvo and Nicolas Desbiens for fruitful discussions.

References

1 T. Ohba, H. Kanoh and K. Kaneko, *J. Am. Chem. Soc.*, 2004, **126**, 1560–1562.
2 S. Devautour, S. A. Abdoulaye, J.-C. Giuntini and F. Henn, *J. Phys. Chem. B*, 2001, **105**, 9297–9301.
3 P. Demontis and G. B. Suffritti, *Chem. Rev.*, 1997, **97**, 2845–2878.
4 B. Boddenberg, G. U. Rakhmatkariev, S. Hufnagel and Z. Salimov, *Phys. Chem. Chem. Phys.*, 2002, **4**, 4172–4180.
5 J.-C. Moise, J.-P. Bellat and A. Méthivier, *Microporous Mesoporous Mater.*, 2001, **43**, 91–101.
6 N. Giovambattista, P. J. Rossky and P. G. Debenedetti, *Phys. Rev. E*, 2006, **73**, 041604.
7 N. Desbiens, A. Boutin and I. Demachy, *J. Phys. Chem. B*, 2005, **109**, 24071–24076.
8 C. E. Ramachandran, S. Chempath, L. J. Broadbelt and R. Q. Snurr, *Microporous Mesoporous Mater.*, 2006, **90**, 293–298.
9 W. Kauzmann, *Adv. Protein Chem.*, 1959, **14**, 1–63.
10 K. A. Dill, *Biochemistry*, 1990, **29**, 7133–7155.
11 D. Chandler, *Nature*, 2002, **417**, 491–491.
12 C. L. Brooks, J. N. Onuchic and D. J. Wales, *Science*, 2001, **293**, 612–613.
13 C. M. Dobson, A. Sali and M. Karplus, *Angew. Chem., Int. Ed.*, 1998, **37**, 868–893.
14 O. Beckstein and M. S. P. Sansom, *Proc. Natl. Acad. Sci. U. S. A.*, 2003, **100**, 7063–7068.
15 G. Hummer, J.-C. Rasaiah and J. P. Noworyta, *Nature*, 2001, **414**, 188–190.
16 J. Dzubiella and J.-P. Hansen, *J. Chem. Phys.*, 2005, **122**, 234706.
17 A. Otten and S. Herminghaus, *Langmuir*, 2004, **20**, 2405–2408.
18 J. W. G. Tyrell and P. Attard, *Phys. Rev. Lett.*, 2001, **87**, 176104.
19 R. Steitz, T. Gutberlet, T. Hauss, B. Klösgen, R. Krastev, S. Schemmel, A. C. Simonsen and G. H. Findenegg, *Langmuir*, 2003, **19**, 2409–2418.
20 R. A. Jackson and C. R. A. Catlow, *Mol. Simul.*, 1988, **1**, 207–224.
21 M. J. Sanders, M. Leslie and C. R. A. Catlow, *J. Chem. Soc., Chem. Commun.*, 1984, 1271–1273.
22 N. H. de Leeuw and S. C. Parker, *J. Am. Ceram. Soc.*, 1999, **82**, 3209–3216.
23 F. Manon-Higgins, N. H. de Leeuw and S. C. Parker, *J. Mater. Chem.*, 2002, **12**, 124–131.
24 R. Ruiz-Salvador, D. W. Lewis, J. Rubayo-Soneira, G. Rodriguez-Fuentes, L. R. Sierra and C. R. A. Catlow, *J. Phys. Chem. B*, 1998, **100**, 6722–6730.
25 E. Fois, A. Gamba, G. Tabacchi, S. Quartieri and G. Vezzalini, *Phys. Chem. Chem. Phys.*, 2001, **3**, 4158–4163.
26 E. Fois, A. Gamba, G. Tabacchi, S. Quartieri and G. Vezzalini, *J. Phys. Chem. B*, 2001, **105**, 3012–3016.
27 E. Fois, A. Gamba and A. Tilocca, *J. Phys. Chem. B*, 2002, **106**, 4806–4812.
28 C. Ceriani, A. Gamba, G. Tabacchi, O. Ferro, S. Quartieri and G. Vezzalini, *Am. Mineral.*, 2004, **89**, 102–109.
29 F.-X. Coudert, R. Vuilleumier and A. Boutin, *ChemPhysChem*, 2006, **7**, 2464–2467.
30 C. Bussai, S. Fritzsche, R. Haberlandt and S. Hannongbua, *J. Phys. Chem. B*, 2003, **107**, 12444–12450.
31 K. S. Smirnov and D. Bougeard, *Chem. Phys.*, 2003, **292**, 53–70.
32 C. Beauvais, A. Boutin and A. H. Fuchs, *ChemPhysChem*, 2004, **5**, 1791–1793.
33 G. Maurin, R. G. Bell, S. Devautour, F. Henn and J.-C. Giuntini, *J. Phys. Chem. B*, 2004, **108**, 3739–3745.
34 K. Shirono, A. Endo and H. Daiguji, *J. Phys. Chem. B*, 2005, **109**, 3446–3453.
35 N. A. Ramsahye and R. G. Bell, *J. Phys. Chem. B*, 2005, **109**, 4728–4747.
36 K. Shirono and H. Daiguji, *Chem. Phys. Lett.*, 2006, **417**, 251–255.
37 N. Desbiens, I. Demachy, A. H. Fuchs, H. Kirsch-Rodeschini, M. Soulard and J. Patarin, *Angew. Chem., Int. Ed.*, 2005, **44**, 5310–5313.
38 D. Bougeard and K. S. Smirnov, *Phys. Chem. Chem. Phys.*, 2007, **9**, 226.
39 M. Trzpit, M. Soulard, J. Patarin, N. Desbiens, F. Cailliez, A. Boutin, I. Demachy and A. H. Fuchs, *Langmuir*, 2007, **23**, 10131–10139.

 This journal is © The Royal Society of Chemistry 2009

40 C. A. Fyfe, G. J. Kennedy, G. T. Kokotailo and C. T. DeSchutter, *J. Chem. Soc., Chem. Commun.*, 1984, 1093–1094.
41 D. Siantar, W. Millman and J. Fripiat, *Zeolites*, 1995, **15**, 556–560.
42 A. Corma, F. Rey, J. Rius, M. J. Sabater and S. Valencia, *Nature*, 2004, **431**, 287–290.
43 P. Caullet, J. L. Paillaud, A. Simon-Masseron, M. Soulard and J. Patarin, *C. R. Chim.*, 2005, **8**, 245–266.
44 J. Patarin, private communication, 2006.
45 C. Baerlocher, W. M. Meier and D. H. Olson, *Atlas of zeolite framework types*, Elsevier, Amsterdam, 2001.
46 W. Loewenstein, *Am. Mineral.*, 1954, **39**, 92–96.
47 R. Y. Yanagida, A. A. Amaro and K. Seff, *J. Phys. Chem.*, 1973, **77**, 805–809.
48 H. Fuchs and A. K. Cheetham, *J. Phys. Chem. B*, 2001, **105**, 7375–7383.
49 N. Desbiens, I. Demachy, A. H. Fuchs, H. Kirsch-Rodeschini, M. Soulard and J. Patarin, *Angew. Chem., Int. Ed.*, 2005, **44**, 5310–5313.
50 A. Di Lella, N. Desbiens, A. Boutin, I. Demachy, P. Ungerer, J.-P. Bellat and A. H. Fuchs, *Phys. Chem. Chem. Phys.*, 2006, **8**, 5396–5406.
51 C. Beauvais, X. Guerrault, F.-X. Coudert, A. Boutin and A. H. Fuchs, *J. Phys. Chem. B*, 2004, **108**, 399–404.
52 E. Jaramillo and S. M. Auerbach, *J. Phys. Chem. B*, 1999, **103**, 9589–9594.
53 W. L. Jorgensen and J. D. Madura, *Mol. Phys.*, 1985, **56**, 1381–1392.
54 D. Nicholson and N. G. Parsonage, *Computer Simulation and the Statistical Mechanics of Adsorption*, Academic Press, New York, 1982.
55 D. Frenkel and B. Smit, *Understanding molecular simulation: from algorithms to applications*, Academic Press, London, 2002.
56 A. D. Mackie, B. Tavitian, A. Boutin and A. H. Fuchs, *Mol. Simul.*, 1997, **19**, 1–15.
57 R. F. Cracknell, D. Nicholson, N. G. Parsonage and H. Evans, *Mol. Phys.*, 1990, **71**, 931–943.
58 V. Lachet, A. Boutin, B. Tavitian and A. H. Fuchs, *J. Phys. Chem. B*, 1998, **102**, 9224–9233.
59 F. Wang and D. Landau, *Phys. Rev. E*, 2001, **64**, 056101.
60 F. Wang and D. Landau, *Phys. Rev. Lett.*, 2001, **86**, 2050–2053.
61 M. Shell, P. Debenedetti and A. Panagiotopoulos, *Phys. Rev. E*, 2002, **66**, 056703.
62 P. Poulain, F. Calvo, R. Antoine, M. Broyer and P. Dugourd, *Phys. Rev. E*, 2006, **73**, 056704.
63 N. Rathore and J. de Pablo, *Abstr. Am. Chem. Soc.*, 2002, **224**, U485–U485.
64 E. Kim, R. Faller, Q. Yan, N. Abbott and J. de Pablo, *J. Chem. Phys.*, 2002, **117**, 7781–7787.
65 N. Rathore, Q. Yan and J. de Pablo, *J. Chem. Phys.*, 2004, **120**, 5781–5788.
66 M. Chopra, M. Müller and J. de Pablo, *J. Chem. Phys.*, 2006, **124**, 134102.
67 C. Zhou and R. Bhatt, *Phys. Rev. E*, 2005, **72**, 025701.
68 R. Car and M. Parrinello, *Phys. Rev. Lett.*, 1985, **55**, 2471.
69 *CPMD*, copyright IBM Corp. 1990–2006, copyright MPI für Festkörperforschung Stuttgart 1997–2001.
70 A. D. Becke, *Phys. Rev. A*, 1988, **38**, 3098.
71 C. Lee, W. Yang and R. G. Parr, *Phys. Rev. B*, 1988, **37**, 785.
72 S. Nosé, *Mol. Phys.*, 1984, **52**, 255–268.
73 W. G. Hoover, *Phys. Rev. A*, 1985, **31**, 1695–1697.
74 R. Resta, *Phys. Rev. Lett.*, 1998, **80**, 1800.
75 P. Silvestrelli and M. Parrinello, *J. Chem. Phys.*, 1999, **111**, 3572.
76 R. Ramrez, T. Lopez-Ciudad, P. Kumar and D. Marx, *J. Chem. Phys.*, 2004, **121**, 3973–3983.
77 M.-P. Gaigeot and M. Sprik, *J. Phys. Chem. B*, 2003, **107**, 10344.
78 M. Gaigeot, R. Vuilleumier, M. Sprik and D. Borgis, *J. Chem. Theory Comput.*, 2005, **1**, 772.
79 N. Marzari and D. Vanderbilt, *Phys. Rev. B*, 1997, **56**, 12847.
80 M. Sharma, R. Resta and R. Car, *Phys. Rev. Lett.*, 2005, **95**, 187401.
81 R. Iftimie and M. E. Tuckerman, *J. Chem. Phys.*, 2005, **122**(21), 214508.
82 R. Vuilleumier, Liquid water dynamical partial charges and infrared intensities, http://hal.archives-ouvertes.fr/hal-00267645/en/, to be submitted.
83 X. Gonze, *Phys. Rev. B*, 1997, **55**, 10337–10354.
84 A. Putrino, D. Sebastiani and M. Parrinello, *J. Chem. Phys.*, 2000, **113**, 7102–7109.
85 F. Rouquerol, J. Rouquerol, and K. Sing, *Adsorption by Powders and Porous Solids*, Academic Press, New York, 1999.
86 V. Eroshenko, R.-C. Regis, M. Soulard and J. Patarin, *J. Am. Chem. Soc.*, 2001, **123**, 8129–8130.

87 F. Cailliez, G. Stirnemann, A. Boutin, I. Demachy and A. H. Fuchs, *J. Phys. Chem. C*, 2008, **112**, 10435–10445.
88 F. Cailliez, N. Desbiens, I. Demachy, M. Trzpit, M. Soulard, J. Patarin and A. H. Fuchs, *Stud. Surf. Sci. Catal.*, 2008, **174**, 683–688.
89 D. J. Wales, J. P. K. Doye, A. Dullweber, M. P. Hodges, F. Y. Naumkin, F. Calvo, J. Hernández-Rojas and T. F. Middleton, Cambridge Cluster Database, http://www-wales.ch.cam.ac.uk/CCD.html.
90 A. Luzar and D. Chandler, *Nature*, 1996, **379**, 55–57.
91 C. Bain, *J. Chem. Soc., Faraday Trans.*, 1995, **91**, 1281–1296.
92 Y. R. Shen and V. Ostroverkhov, *Chem. Rev.*, 2006, **106**, 1140–1154.
93 Y. Huang and Z. Jiang, *Microporous Mater.*, 1997, **12**, 341–345.
94 P. B. Miranda and Y. R. Shen, *J. Phys. Chem. B*, 1999, **103**, 3292–3307.
95 L. Scatena, M. Brown and G. Richmond, *Science*, 2001, **292**, 908–912.
96 The CP2K developers' group, http://cp2k.berlios.de/, 2007.
97 D. Laage, F.-X. Coudert and A. Boutin, to be published.

 This journal is © The Royal Society of Chemistry 2009

Dynamic properties of confined hydration layers

Susan Perkin,[*ab] **Ronit Goldberg,**[c] **Liraz Chai,**[c] **Nir Kampf**[c] **and Jacob Klein**[*ac]

Received 28th March 2008, Accepted 16th May 2008
First published as an Advance Article on the web 8th October 2008
DOI: 10.1039/b805244a

Prompted by the recent discovery that water and aqueous monovalent Na^+ solutions remain fluid-like when confined to films of a few molecular layers between mica surfaces,[Raviv *et al.*, *Nature*, 2001, **413**, 51–54; and Raviv and Klein, *Science*, 2002, **297**, 1540–1543] we now extend the previous study by comparing the shear- and normal-force properties of 0.1 M Na^+, Cs^+ and Ni^{2+} salt solutions which demonstrate a diverse range of behaviours under confinement. In the case of hydrated Na^+ we extend the previous study to higher pressures, up to ~10 atmospheres, and record similar fluidity of the hydration layers at these elevated pressures. Aqueous Cs^+ films under confinement between mica sheets have been found to be unable to support an applied load—that is to say they do not demonstrate any hydration repulsion regime—as a result of their lower hydration energy [see Goldberg *et al.*, *Phys. Chem. Chem. Phys.*, 2008, **10**, 4939–4945] which contrasts with the properties of Na^+. We show that 0.1 M Ni^{2+} solution remains close to its bulk viscosity down to nanometre thin films, but does not demonstrate a hydration repulsion. By comparing the properties of this range of cations, with differing valency and hydration, we aim to examine the conditions under which ions serve as effective lubricants and what we call the 'hydration lubrication' mechanism.

1. Introduction

The forces that act between surfaces across liquid media have been studied extensively over the past century or so, not least because of their relevance to understanding and controlling colloidal and bio-colloidal stability.[1,2] Direct force *vs.* surface-separation measurements between well-characterised, molecularly smooth surfaces across a variety of different liquids became possible with the development of the mica surface force balance (SFB): some 30 years of experiments using various forms of the SFB, as well as other methods, have achieved an extensive understanding of many aspects of equilibrium normal surface forces in both non-aqueous and particularly aqueous systems.[3–8] The Derjaguin–Landau–Verwey–Overbeek (DLVO) theory of colloidal stability[1,9]—still the benchmark for interpreting surface interactions—has been demonstrated to apply well to certain aqueous systems at low ionic strengths. In addition, the detailed direct studies of normal forces have revealed, beyond the mean field theory, the importance of structural and hydration forces,[10–12] especially at small surface separations, as well as of hydrophobic interactions.[13,14]

In contrast to normal forces between surfaces across thin aqueous films, which are generally well understood, the issue of *lateral* forces across such films has been far less studied.[15–17] These reflect the dynamic rather than the equilibrium properties

[a]*Department of Physical and Theoretical Chemistry, University of Oxford, Oxford, UK*
[b]*Department of Chemistry, University College London, UK*
[c]*Department of Materials and Interfaces, Weizmann Institute of Science, Rehovot, Israel*

of the confined films, with clear relevance to tribology, slip[18] and lubrication issues. In the past few years studies on the shear of thin water films have indicated that the properties of these films differ significantly from those of non-aqueous solvents.[19,20] In particular, water appears to retain its bulk fluidity even when confined by solid surfaces to films of sub-nanometre thickness.[19,21] This is in marked contrast to organic solvents whose viscosity diverges for films thinner than a few nanometres, and which tend to become solid-like under such confinements.[22–24] The difference has been attributed to the densification of liquid water under the van der Waals attraction of the confining walls which suppresses its tendency to solidify, an effect consistent with detailed molecular dynamics studies.[20,25]

A related but even more remarkable observation concerns recent experiments demonstrating an extreme lubricating effect of hydrated ions[26,27] and other hydrated charged species[28,29] when confined to nanometrically-thin films between charged surfaces. Such films can support a normal load due to the presence of the trapped hydrated ions and the resulting hydration repulsion. The central hypothesis put forward to explain these first results is based on the thermodynamic stability of the hydration sheath around the confined ions—which supports the normal load—concurrent with the kinetic lability of the primary hydration sheath about such ions, which facilitates rapid exchange of water molecules and thus results in fluid-like response under shear.[26,30] This effect has recently been invoked to explain the extreme lubricity of charged, highly compressed polymer brushes.[28] Very recently, a massive reduction in friction by boundary lubricants under water has also been attributed to the hydration of their polar head-groups, with a resulting locally-fluid (and thus easily sheared) layer at the head-group/substrate interface.[29]

In the present work we inspect in more detail the normal and shear characteristics of three contrasting hydrated metal cations between negatively charged (mica) surfaces, extending the earlier studies[21,26] and drawing conclusions through comparison of ions with differing hydration and dynamic characteristics.[31] By increasing the applied load we demonstrate the lubricating behaviour of aqueous Na^+ up to pressures relevant in biological and mechanical shearing environments. The overall aim of the present study is to demonstrate the contrast between Cs^+, Na^+ and Ni^{2+} ions, and especially their lubrication behaviour, resulting from their differing charge and hydration status at interfaces.

2. Experimental procedure

2.1 Force measurements

The surface force balance (SFB) used in this study to record normal and shear forces between mica surfaces across electrolyte solutions has been described previously.[23] To summarise: large (~cm^2) single-crystal facet mica sheets, of 1–2.5 μm thickness, are back-silvered and mounted onto hemi-cylindrical lenses such that the mica sheets face one another in a crossed-cylinder configuration. One of these lenses is then translated relative to the other in normal and lateral directions by applying varying voltages to the sectored piezoelectric tube upon which it is mounted. The distance between the crossed cylinders at their point of closest separation, D, is measured directly to ±3 Å *via* fringes of equal chromatic order (FECO) observed in a grating spectrometer.[32] The shape of the FECO also reveals the interaction geometry. The normal and lateral (shear) forces, F_N and F_S respectively, resulting from the normal and lateral displacements are detected as the bending of springs either in the FECO (normal forces) or by an air-gap capacitance probe (shear forces). Normal forces are presented as normalised by the mean radius of curvature of the crossed cylinders, R, *i.e.* as F_N/R; this value is directly proportional to the interaction energy per unit area between parallel plates.[33] The resolution of the shear spring deflection, and therefore shear force, depends on the isolation of the system from ambient vibrations: before signal processing this is ~4–20 nm (corresponding to 1–5 μN). This shear force data

 This journal is © The Royal Society of Chemistry 2009

collected as a capacitance signal is then Fourier transformed to give a frequency spectrum. The amplitude of response at the frequency of applied shear motion, ω_0, can then be extracted and normalised if necessary, as described earlier.[21] Comparison is then made between the response signals as a function of surface separation, including at large separations where there is no direct shear coupling although some response is observed due to indirect coupling through the apparatus. Within each experiment the scatter is in this way reduced to $\delta F_S \approx$ 20–50 nN, with variation from experiment to experiment due to differences in experimental setup leading to differences in the through-apparatus indirect coupling of the shear motion.

Before each experiment stringent cleaning procedures for all glassware, tools and apparatus parts were followed, involving strong oxidising (piranha) solutions followed by water and organic solvents, as described earlier.[21] All preparation was carried out in a laminar flow cabinet (Bassaire Ltd) to avoid particulate contamination. Each experiment involved calibration of the mica contacting geometry and mica thickness in air, followed by similar calibration in pure 'conductivity' water. Only if the normal forces were in agreement with DLVO theory and the shear response showed no coupling in the last few nm of the water film before mica–mica contact[21] was the experiment deemed to be clean enough to proceed to the next stage.

Calibration of the mica–mica surface separation in water yields a value of approx. -5 ± 5 Å relative to the air contact value, known to be due to dissolution in water of the air-adsorbed carbonaceous and water layer.[34] All measurements of surface separation (D) are relative to this point of mica–mica contact under water, $D_0 = 0.0$ nm. Therefore a layer of solid-like bound water on the mica surface (as suggested in X-ray reflectivity (XRR) measurements of the mica/water interface[35]), leading to the possibility of a bilayer of water between the mica surfaces when in contact, is included within this calibration of D_0 and subsequent D values measured relative to this (possible) water layer. For this reason we cannot detect the presence or absence of this single water layer adsorbed onto the mica, which if present behaves in a solid-like manner and can be considered as part of the solid crystal-lattice structure, in contrast to the adjacent liquid water.[21,34,35]

2.2 Materials

The pure 'conductivity' water used for these experiments was obtained by passing tap water through a Millipore water purification apparatus, consisting of a RiOs5 and Milli-Q Gradient A10 system. The water has resistivity greater than 18.2 MΩ cm, and a total organic content of 2–4 ppb. Mica was ruby muscovite, grade 1 (S&J Trading Inc., New York). The mica was cleaved for each experiment using one of two possible methods: downstream melt-cutting and the torn method. These methods were devised recently to avoid any possibility of the effect of Pt nanoparticles.[34] Salts were as follows: NaCl 99.999%, Aldrich, crystals treated with UV for 30–40 min before making up solutions; $CsNO_3$ 99.999%, Aldrich; $Ni(NO_3)_2$ 99.999%, Aldrich.

3. Results

3.1 Normal forces in Na^+ and Cs^+ salt solutions

Experiments were conducted to measure the normal forces between mica surfaces across aqueous Na^+ and Cs^+ salt solutions, each at $\sim 1 \times 10^{-1}$ M concentration, as a function of surface separation, and the results are presented in Fig. 1. Looking first to the normal interaction force between the mica sheets across Na^+ solution, we see that the interaction is short-ranged—of the order of a few nm—and strongly repulsive in agreement with previous studies. The confined hydrated Na^+ ions resist compression or squeeze-out from the film at all values of applied normal loads in this investigation, which are an order of magnitude higher than those applied in previous studies.[26,27,36] The mica surfaces do not adhere and the forces are reversible. This is in

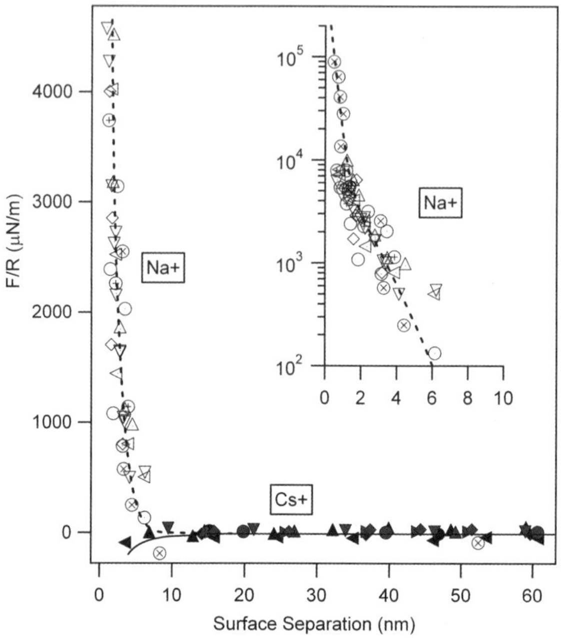

Fig. 1 Normal interaction forces between mica sheets across $\sim 1.0 \times 10^{-1}$ M aqueous solutions of NaCl (all open symbols) and $CsNO_3$ (all filled symbols). The main Figure compares the normalised F/R values as a function of surface separation in these two solutions on a linear scale. The data for $CsNO_3$, collected in two different laboratories, shows no interaction until the surfaces are close to $D = 1.0$–2.0 nm, at which point they are attracted into adhesive contact at a separation of $D = +0.6$ nm relative to the closest separation in pure water (before addition of salt). The solid line in the main Figure represents a DLVO expression using the algorithm of Chan *et al.*[55] to solve the non-linear Poisson–Boltzmann equation for the electrostatic repulsion, fitted to the Cs^+ data with $\Psi_0 = 10$ mV and $c = 1.0 \times 10^{-1}$ M monovalent salt, added to the van der Waals interaction with A_H(mica–water–mica) $= 2 \times 10^{-20}$ J. Due to the experimental scatter, the data supports values of $\Psi_0 < 25$ mV. The data for NaCl show a short-ranged repulsion with no adhesion at small separations, down to 0.8 ± 0.3 nm. This data is shown on an extended and logarithmic scale in the inset. The dotted lines in both the main Figure and the inset show a fit to the Na^+ data using the DLVO expression as above, with $\Psi_0 = 60$ mV and $c = 0.6 \times 10^{-1}$ M, and added to this an exponential term for the hydration repulsion of $E_h\exp(-D/D_h)$ with $E_h = 0.12$ J m^{-2}, $D_h = 0.25$ nm.

contrast to the mica–mica interaction in pure water or solutions of Na^+ salts at concentrations below the critical hydration concentration, where the surfaces jump together from 2–7 nm into primary minimum contact under the influence of the van der Waals interaction.[3,33,37] This short-range repulsive interaction, observed only above a critical value of the bulk salt concentration and not accounted for within the DLVO theory, has been called the 'hydration repulsion'[36,38] to reflect the suggestion that it is due to the tenaciously bound water surrounding the Na^+ ions, preventing dehydration and condensation of the ions into the mica lattice. Here we see that this hydration repulsion extends up to elevated normal force values of $F_N/R \approx 10^5$ μN m^{-1}. Using the theory of Hertz for a non-adhesive deformable solid sphere in contact with a flat plate, we calculate the contacting area, A:[33]

$$A \approx \pi(RF_N/K)^{2/3}$$

where R is the radius of the sphere, corresponding to the mean radius of our crossed cylinders at the contact region determined from the FECO, and K is related to the effective modulus of the surfaces and has a value of $K \sim 1 \times 10^9$ Nm^{-2}. This method

 This journal is © The Royal Society of Chemistry 2009

yields a value of $A = 1 \times 10^{-9}$ m^2 at this maximal applied F_N, which is in accordance with the value estimated from the flattening of the fringes viewed in the FECO and corresponds to a mean pressure over the contacting area of 1.1 × 10^6 Pa. By assuming an exponential dependence of the hydration force on the separation between the mica plates, and that this additional force is additive with the mean-field DLVO contribution, we deduce a value of $\Psi_0 = -60$ mV for the surface potential of the mica and 0.25 nm for the decay length of the hydration repulsion (see caption to Fig. 1). These values are close to those observed in the past,[26] and we show here that this expression applies up to enhanced normal force values.

In contrast to this hydration-repulsive behaviour, the normal interaction force between mica sheets across a 0.1 M Cs^+ salt solution is quite different.[27,31] We observe no osmotic repulsion between the mica sheets within our force resolution until at 1–2 nm a van der Waals attraction causes a small jump-in to contact at $D = +0.6$ nm relative to mica–mica contact in salt-free water.[39] At this point the surfaces are observed to have a small flattened contact, indicating adhesion, and indeed a finite pull-off force is required to separate the surfaces from this point: from the pull-off force, $F_{\text{pull-off}}$, we calculate the solid–liquid surface tension, $\gamma = F_{\text{pull-off}}/3\pi R = -0.72 \pm 0.2$ mN m^{-1}.[31]

The absence of osmotic repulsion between the surfaces in the case of Cs^+, beyond the scatter of the experimental data, implies that there is significantly reduced excess concentration of ions in the electrical double layer at the mica surface. This in turn implies a very low surface potential (and surface charge), relative to mica surfaces in pure water, and is likely to be due to neutralisation of the mica (negative) surface charge by adsorbed Cs^+ ions. We can calculate an upper limit for (the modulus of) the residual negative surface charge on the mica, $\sigma_0^{\max}$, due to the extensive neutralisation by Cs^+ ions, using the experimental resolution-limited upper limit of surface potential measured in our experiments, Ψ_0, and using the Grahame equation:[33]

$$\sigma_0^{\max} = (8k_B T \varepsilon_0 \varepsilon_r c_\infty)^{1/2} \sinh(e\Psi_0^{\max}/2k_B T)$$

where k_B is the Boltzmann constant, T the temperature, c_∞ is the concentration of bulk 1 : 1 salt solution, e is the charge of an electron and ε_0 and ε_r are the permittivity of free space and relative permittivity of the medium respectively. From the upper limit value of $\Psi_0^{\max} = -25$ mV (caption to Fig. 1) we find $\sigma_0 < 0.02$ Cm^{-2} in 0.1 M Cs^+ solution (with Debye length $\kappa^{-1} = 0.97$ nm);[31] compared to ~0.06 Cm^{-2} in equivalent Na^+ solution and ~0.002 Cm^{-2} in pure water.

3.2 Shear forces in Na^+ salt solutions

In order to investigate the dynamic properties of the film of confined hydrated Na^+ ions within the regime of hydration repulsion described above, we have applied a lateral shearing motion to one of the confining mica surfaces and recorded the response of the other surface to this motion. The results of these measurements are shown in Fig. 2. In all cases we observe no shear force response above the noise level of the measurement, indicating high fluidity of the confined hydrated film between the surfaces: At all of the shear rates and applied pressures investigated the aqueous hydration film provides extremely efficient lubrication. From this observation we can extract an upper limit for the effective viscosity of the confined liquid film, η_{eff}, between the surfaces, since:

$$\sigma_s = \eta_{\text{eff}} \gamma$$

where $\sigma_s = (Fs/A)$ is the transmitted shear stress in response to the applied shear rate $\gamma = (v_s/D)$ where v_s is the shear velocity. The shear force resolution of our measurement, $\delta F_S = 30$ nN (as outlined in the Figure caption), determines the upper limit for the shear stress value, σ_s^{upper}, through $\sigma_s^{\text{upper}} = \delta F_S/A$. Using the values from trace

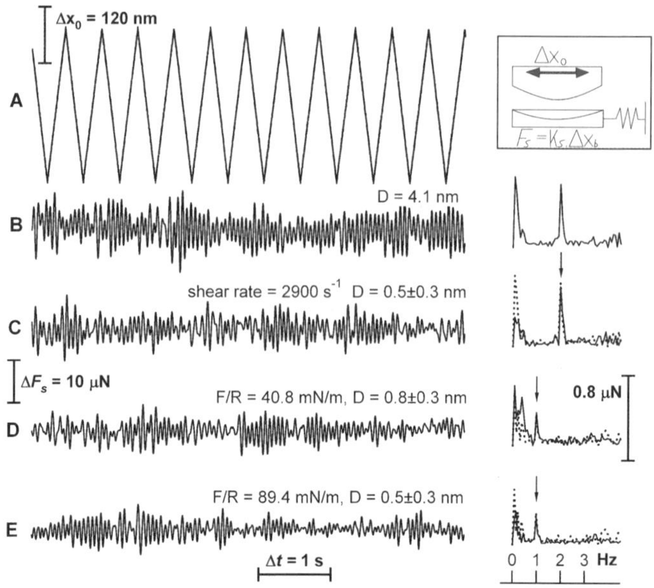

Fig. 2 Shear forces between mica sheets across films of aqueous Na^+ solution at varying film thicknesses. A: example of applied triangle-wave shearing motion, at 2 Hz and 330 nm peak-to-peak shearing amplitude. The cartoon inset shows how one surface is sheared laterally relative to the other (with lateral position Δx_0), and any coupling of the surfaces at that point results in the bending of a spring, by Δx_b, which when multiplied by the spring constant, K_s, gives the shear force, F_S. B: shear force response to the applied shearing, with surfaces separated by 4.1 nm of aqueous solution, recorded using a capacitance probe to detect shear spring deflection and taken directly from the oscilloscope reading. To the right of the oscilloscope reading is the Fourier transformation (FT) of this data showing a peak at the applied frequency of 2 Hz (applied frequencies are indicated with arrows) due to indirect coupling of the surface through the apparatus. This signal is identical to traces measured at all larger separations, within the scatter of ~30 nN. C: shear force response to the same applied shearing motion with the surfaces separated by 0.5 ± 0.3 nm of hydrated Na^+ ions, representing a high shear rate of 2.9×10^3 s^{-1}. The FT of the oscilloscope reading, to the right-hand side, is shown with a solid line and the FT of the 'large D' signal is superimposed with the dashed line. The FT traces are identical within the 30 nN scatter. D and E: shear force response traces for mica surfaces separated by 0.5 and 0.8 nm (±0.3 nm) respectively, with large applied normal force as shown on the traces corresponding to the hydration repulsion regime. In each case the FT trace has the frequency of applied shear marked with an arrow; the solid line is the FT of the response data; and the dotted traces overlaid onto these are FTs of the shear force response to an identical applied shear amplitude and frequency when the surfaces are far apart.

C—corresponding to the highest shear rate applied in this study—we therefore find the upper limit of viscosity: $\eta_{\rm eff}^{\rm upper} = 0.03$ Pa s. Although this value is 30 times higher than the bulk viscosity of water of ~0.001 Pa s, we emphasise that this is an upper limiting value. In the lower two traces of Fig. 2 we show that by applying a normal load to the confined hydrated film within the hydration repulsion regime as described above, we were able to verify this fluidity of the film up to mean pressures of 1×10^6 Pa. These data, measured in 4 separate experiments (different mica sheets) and with different contact points within each experiment, corroborate the previous report of fluidity of confined hydration layers[26,34] and extend towards higher shear rates and applied normal pressures within the hydration repulsion region.

3.3 Shear forces in Cs^+ salt solutions

The shear forces between mica sheets approaching contact in Cs^+ solutions and in pure water are shown in Fig. 3. These very recent results[31] demonstrate a significant

 This journal is © The Royal Society of Chemistry 2009

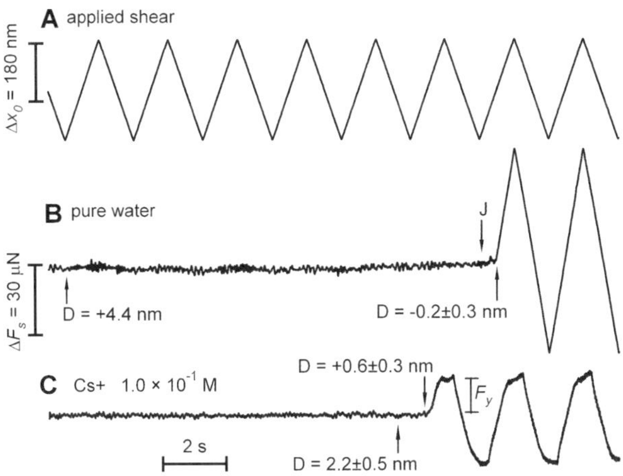

Fig. 3 Shear forces between mica sheets across films of pure water and aqueous Cs^+ solution. A: applied triangle-wave shear motion of the upper surface. B: shear force response between mica sheets across pure water during approach of the surfaces. The shear force response is below the noise-limited resolution from large separations (shown here are separations below 4.4 nm), beyond the point at which the surfaces jump into contact (shown with 'J'), down to mica–mica contact at $D = -0.2 \pm 0.3$ nm, as shown. At this point the surfaces become instantly rigidly coupled, and the lower surface moves in parallel with the upper one. C: shear force response (to applied shear similar to that in A) between mica sheets across 1.0×10^{-1} M $CsNO_3$ during approach of the surfaces. As in pure water, there is no shear force response measurable at all separations down to closest approach—this time at $+0.6 \pm 0.3$ nm—at which point the surfaces become coupled. However, in contrast to the pure water case, the yield force is lower and once the yield force is reached the surfaces slide across each other.

difference between the shear properties of mica sheets contacting across pure salt-free water and across an aqueous 1.0×10^{-1} M solution of Cs^+ ions. Additionally, in parallel with the normal force behaviour, we also see a significant contrast with the results of similar shear measurements in 1.0×10^{-1} M Na^+ solutions (Fig. 2).

Fig. 3A shows the applied shear motion of the upper surface and Fig. 3B shows the shear force response between mica sheets across pure water of film thicknesses between 4.4 and ~0.0 nm. This result, including in particular the jump of the surfaces into adhesive contact at the point J on the trace, has been analysed in detail previously.[21] From these data, in conjunction with measurements of normal force interactions in pure water,[21] it has been shown that the viscosity of the water remains close to the bulk value right down to $D = D_0$. At this point the surfaces spontaneously adhere under the influence of the van der Waals forces, and the shear response demonstrates rigid coupling between the surfaces. Fig. 3C shows the shear behaviour of mica surfaces approaching contact across 1.0×10^{-1} M solutions of $CsNO_3$. The shear response is within the very low noise level of the measurement at all separations, including during the small jump-in, to $D = +0.6 \pm 0.3$ nm, corresponding to the point at which the surfaces spontaneously flatten and the interaction becomes adhesive. At this point the shear response changes spontaneously to a coupled motion until the yield force is reached: the surfaces move together as one is translated by Δx_0, with increasing shear force, $F_S = K_s \Delta x_0$, until the shear force reaches a 'yield point', $F_S = F_y$, at which point the surfaces slide past each other as seen by the flat-topped part of the shear trace. This process then repeats itself as the translated surface moves back in the other direction. F_y values were measured in the range $15\ \mu N < F_y < 21\ \mu N$. We note that these normal and shear measurements across $CsNO_3$ solutions involved over 10 independent experiments, with different mica sheets, in two different laboratories.

3.4 Normal and shear forces in Ni^{2+} salt solutions

Fig. 4 presents the normal and shear interactions between mica sheets across solutions of the divalent transition metal salt $Ni(NO_3)_2$. Fig. 4C shows the normal interaction forces across pure water and across 1.0×10^{-1} M $Ni(NO_3)_2$ solution. In both cases the surface forces are repulsive, due to osmotic repulsion between the diffuse double layers: in pure water the interaction is long-ranged, due to the low screening and long Debye length, whereas in the 1.0×10^{-1} M $Ni(NO_3)_2$ solution the screening of the surface charge results in the expected shorter-range repulsive interaction. At small separations an attractive van der Waals force causes the surfaces to jump together into contact to $D = D_0$, not only in pure water but also in the Ni^{2+} solution. That is to say there is no hydration-repulsion regime in the Ni^{2+} solution at a bulk concentration of 1.0×10^{-1} M. After attraction into contact the adhesion force between the surfaces is measured by pulling the surfaces apart and recording the jump-out distance. For mica sheets in 1.0×10^{-1} M $Ni(NO_3)_2$ solution we found $F_{\text{pull-off}}/R = 25$ mN m^{-1} (a single measurement), from which we obtain the solid–liquid surface tension, $\gamma = 2.7$ mN m^{-1}. This is within the range of values observed for mica sheets adhered in pure water as measured in these and similar experiments.[21]

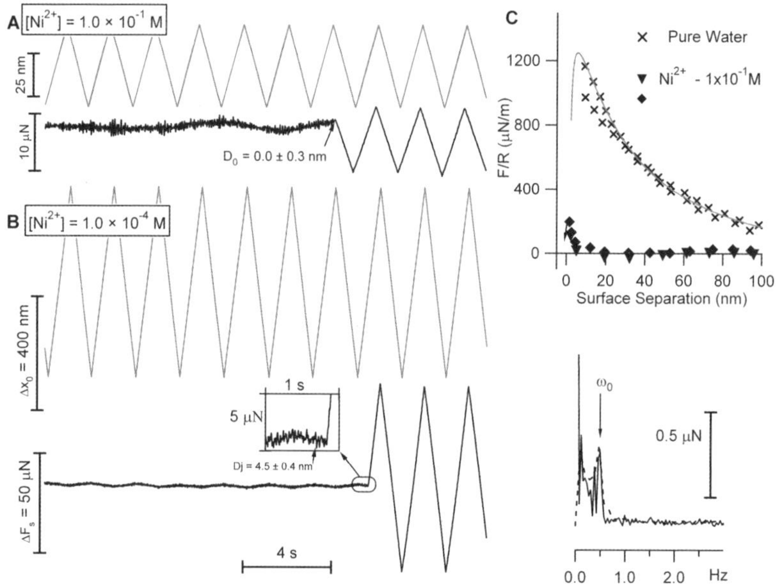

Fig. 4 A: applied triangle-wave shear motion (upper trace) and shear force response (lower trace) between mica sheets across 1.0×10^{-1} M $Ni(NO_3)_2$ solution during approach of the surfaces. At the point when the surfaces reach contact (after jump towards each other due to van der Waals attraction) they become coupled and both surfaces move with identical triangle-wave motion. Before that point there is no shear force response resolved above the noise level. B: as for A, with solution concentration 1.0×10^{-4} M and high shear amplitude. The inset within B shows a zoom-in of the response in the region before and immediately after contact between the mica surfaces. The jump distance $D_j = 4.5 \pm 0.4$ nm. To the right of the oscilloscope shear response trace is an FT of the response signal up until the jump—shown by the solid line—and, for comparison, the FT of a response signal when the surfaces are far apart. The component at the applied frequency, ω_0 (indicated with an arrow), is identical for these two traces within the scatter, leading to an upper limit for the shear force response before the jump of $\delta F_S = 40$ nN. C: normal forces between mica sheets in pure water (crosses) and in 1.0×10^{-1} M $Ni(NO_3)_2$ solution. The grey line shows a DLVO fit to the water data using the algorithm of Chan *et al.*[55] to solve the non-linear Poisson–Boltzmann equation. The fitting parameters are the surface potential, $\Psi_0 = 115$ mV, and the bulk concentration of monovalent ion, $c = 3.0 \times 10^{-5}$ M.

 This journal is © The Royal Society of Chemistry 2009

In Fig. 4A and B we present the shear force response between mica sheets across solutions of Ni^{2+} salt solutions, at concentrations of 1.0×10^{-1} and 1.0×10^{-4} M respectively, during the approach of the mica sheets from ~6–8 nm separations down to contact. The shear force response, F_S, remains within the noise level of the measurement, δF_S, before and during the jump into contact of the mica surfaces, at which point they become rigidly coupled with yield force greater than the maximum applied shear force. The jump time is rapid—estimated to be within 0.1–0.3 s in each case—and the jump distances vary according to the concentration: $D_j = 1.7 \pm 0.3$ nm at $[Ni^{2+}]_{bulk} = 1.0 \times 10^{-1}$ M, and $D_j = 4.5 \pm 0.4$ nm at $[Ni^{2+}]_{bulk} = 1.0 \times 10^{-4}$ M. By performing a Fourier-transform analysis to extract the amplitude of F_S at the applied frequency of the shear, we find that there is no difference between the response immediately before the jump into contact and the response at large surface separations between the mica plates. That is to say, there is no shear response within this improved sensitivity of $\delta F_S = 40$ nN for films of thicknesses down to the point of the jump-in. In addition during the jump itself, while the film thins from 4.5 nm to zero, F_S remains within the noise level until the spontaneous rigid coupling of the surfaces at $D = D_0$.

4. Discussion

4.1 Normal forces in Na^+ and Cs^+ salt solutions

We begin with a comparison of the normal force between mica sheets across aqueous Na^+ and Cs^+ solutions, which demonstrates two contrasting features. Whereas the mica sheets experience an osmotic repulsion in the case of Na^+, there is no measurable repulsion between the mica surfaces across Cs^+ solutions. And whilst the mica plates are held apart by a film of hydrated Na^+ ions at separations below 1.0 nm, there is no hydration-repulsion regime for aqueous Cs^+ confined between mica surfaces.[31]

The absence of hydration repulsion between mica surfaces across Cs^+ solutions, as well as the very low value of the surface charge, implies that Cs^+ ions are not bound to the mica surface in a fully hydrated state in the first solution layer adjacent to the negative lattice site. Instead, the Cs^+ ions appear to energetically favour either expulsion from the film or condensation into the mica lattice. In contrast, Na^+ is observed to form a molecular film of strongly bound and hydrated ions at the mica surface. Expulsion of Cs^+ from the film, by exchanging with a hydrated proton (H_3O^+) to maintain charge neutrality, has until recently been the primary interpretation of the fate of ions when mica surfaces contact across aqueous solution.[40] However, we suggest that a more likely explanation in the case of Cs^+ is the condensation of Cs^+ into the mica-lattice negative sites in a partially de-hydrated or fully-dehydrated state.[31] The evidence for this is now discussed.

The Cs^+ ion is large in size ($r = 0.169$ nm)[41] with single unit positive charge, combining to give a small charge to radius ratio. This results in a relatively low hydration energy, ΔG^{hyd},[42] which can be interpreted in terms of the energy per coordination water: $\Delta G^{hyd}/6\text{–}12 = -(24\text{–}47)$ kJ mol^{-1};[27,43] equivalent to 9–18 k_BT at room temperature. Therefore we expect the Coulombic interaction between the ion and the surrounding dipolar water molecules to have significantly less effect on the ordering of water molecules around a Cs^+ ion compared to Na^+ ($\Delta G^{hyd} = 24$–$42\ k_BT$ per water molecule) or particularly in Ni^{2+} ($\Delta G^{hyd} = 139\ k_BT$ per water molecule). This simplistic argument is borne out in detailed simulations and studies.[43–46] In relation to the present work, this explains the increased surface-bound concentration of Cs^+ on mica relative to Na^+: Cs^+ is more favourably bound to the surface (as a de-hydrated ion) due to its less thermodynamically favourable solvation. Indeed, very recent studies using XRR suggest that Cs^+ ions adsorb to mica (not under confinement) with no intercalated water layer between the ion and the muscovite negative surface site;[47] and X-ray photoelectron spectroscopy (XPS) studies confirm the ion-exchange of Cs^+ on mica.[31,48]

This favourable dehydration and binding at the mica surface for Cs^+ explains why we do not observe the hydration repulsion regime in this case: the Cs^+, sitting inside the mica ditrigonal siloxane cavity,[47] is not holding a sheath of water molecules tightly around it and thereby allows the mica sheets to reach close contact without repulsion. In addition, we note that this surface-binding of partially dehydrated Cs^+ is likely to occur on the mica irrespective of confinement—in the XRR study there was no confinement—and so the mica surface charge is neutralised by the Cs^+ without the need for charge-regulation (condensation due to counterion osmotic pressure) arguments. Therefore this also accounts for the lack of double-layer repulsion between the surfaces in $CsNO_3$ solution at all surface separations as well as the low surface charge density, σ_0.

We observe a significantly lower adhesion value for mica sheets reaching contact across Cs^+ solution compared to pure water, and in addition the closest separation (at which the adhesion takes place) is some 0.6 nm further apart in Cs^+ solution compared to pure water. We again attribute these observations to the presence of dehydrated Cs^+ ions at the mica surfaces when the surfaces reach contact, as follows:[31] the large Cs^+ cation ($r_{Cs+} = 0.169$ nm) occupying the ditrigonal mica sites would lead to (i) distortion of the mica lattice structure relative to its bond lengths and angles with the natural K^+ counterion ($r_{K+} = 0.133$ nm); and (ii) protrusion of the Cs^+ outside of the mica lattice. These two factors combine to result in the observed reduced adhesion, at a closest separation of 0.6 nm, and reduced sliding friction (discussed below) between the mica sheets, since the mica sheets are not able to reach the same intimate contact as in pure water.

4.2 Shear forces in Na^+ salt solutions

We have shown that molecularly thin films or 'hydration layers' containing Na^+ ions and water confined between mica surfaces $\sim 0.8 \pm 0.3$ nm apart and under pressures up to 10 times atmospheric pressure remain fluid-like with an upper limit of viscosity η^{upper} no larger than 30 times that of bulk water.

This highly fluid behaviour under extreme confinement and shear is in direct contrast to the performance of non-associating simple liquids, such as octamethylcyclotetrasiloxane (OMCTS), cyclohexane or toluene, under similar conditions: these non-associating liquids undergo liquid-to-solid phase transition, resulting in a finite yield stress once confined to 6–9 molecular layers.[22,23] We suggest that the lubricity of hydrated films of Na^+ relies upon two factors: the capacity of water to retain its bulk fluidity under confinement and in hydration layers around charged species;[21,26] and the strong binding of water molecules within the hydration shell around Na^+ which supports an applied load, therefore preventing primary minimum contact (which is adhesive and leads to high shear forces)[31] between the mica surfaces. The shear lubricity of these load-bearing hydrated films have been rationalised in terms of the rapid kinetics of exchange of water molecules within hydration spheres with adjacent water molecules, as well as the rapid rotational dynamics and diffusivity of the water molecules within the thin film. The bulk water exchange rate, k_{ex}, for a water ligand in the primary hydration sphere of Na^+ is $\sim 10^9$ s^{-1}, a factor of at least 10^4 faster than the fastest shear rates accessible in this experiment. In addition, the rotational relaxation time of water molecules, $\sim 10^{-11}$ s in bulk water, is thought to be a factor in the persistent fluidity of the confined hydration layers.[30]

Recent experiments of Sakuma *et al.*[49] and simulations of Leng and Cummings[25,30] have corroborated these results for Na^+ hydration layers of ≥ 0.92 nm thickness. This is in good agreement with the surface separation at which we find the highly fluid hydration layer between mica surfaces: D values measured are 0.5–1.0 nm in Na^+ salt within the hydration-repulsion region of the interaction profile, at F/R values up to 10^5 μN m^{-1}. We note that there may additionally be one water layer at each surface not accounted for within the D values quoted in the SFB

 This journal is © The Royal Society of Chemistry 2009

measurement if bound water remains at the surfaces during calibration of D_0 of mica contact in water (as discussed in the methods section).

A less fluid layer is described in the simulations of thinner films; rotational relaxation times of water in films of 'bilayer ice' ($D \leq 0.6$ nm) was found to be up to 10^{-5} s and viscosity is significantly increased at this separation.[25] It is interesting that this seemingly 'critical separation' for fluidity occurs at the same separation value as the steep repulsive wall observed in normal force measurements. This corroborates our conclusion that the presence of hydrated ions—cations coordinated to the negative mica sites but surrounded by a sheath of water molecules—constitutes the lubricating layer. At closer separations, not accessible in the present experiments due to the steep repulsive wall of hydration repulsion, the cations are presumably stripped of their hydration layers and the lubricating fluidity may not occur.

4.3 Shear forces in Cs^+ salt solutions

In contrast to the shear forces across Na^+ solutions at small separations discussed above, mica sheets shearing across Cs^+ solutions exhibit discontinuous behaviour: The shear force remains below our detection limit for $D \geq 0.6 \pm 0.3$ nm, while at $D = 0.6 \pm 0.3$ nm—the closest separation of the surfaces—a shear force is transmitted between the surfaces with yield force in the range $15.0\ \mu N < F_y < 21.5\ \mu N$.

We can also compare this behaviour in Cs^+ solution with the shear forces in pure water (Fig. 3B), to find that (i) the separation of the mica sheets at which the shear stress increases above the experimental resolution is larger in Cs^+ than in pure water (by ~0.6 nm), and (ii) in Cs^+ solution the yield stress at the closest separation is significantly decreased compared to micas reaching contact from pure water.

In the discussion of normal forces above we arrived at the conclusion that Cs^+ ions reside at the mica surfaces in an un-hydrated form, particularly when the mica sheets are brought to sub-nanometre separations. This conclusion is strongly supported by these shear force measurements, since the nature of the contacting mica sheets is clearly different when contact occurs from Cs^+ solution. The bound Cs^+ ions must be un-hydrated for geometrical reasons: hydrated Cs^+ would not fit inside the mica sites nor fit between the mica sheets at this closest separation. Therefore it appears that unhydrated Cs^+ ions, incorporated into the mica lattice but extending outwards from it, are leading to a reduction in the shear stress sustainable between the Cs-mica sheets. We note that variations in the value of F_y could result from variation in the mica–mica twist angle, which was not finely controlled in these experiments.[50]

4.4 Normal and shear forces in Ni^{2+} salt solutions

The short-ranged (2–5 nm) small repulsive interaction between mica sheets across 1.0×10^{-1} M $Ni(NO_3)_2$ solution is in broad agreement with previous studies of divalent ions.[12,51] At a surface separation of 1.7 ± 0.3 nm the interaction becomes attractive, and the surfaces are pulled into contact at $D = D_0$. The interaction cannot be well explained using the mean-field DLVO theory; the attractive interaction appears to be accounted for only by invoking a correlation interaction due to the divalent nature of the ions,[12,52,53] although we do not discuss this further here. Our main focus is the absence of hydration repulsion at 1.0×10^{-1} M solution and the resulting shear forces as the surfaces approach contact.

We have argued above that the hydration repulsion is observed in Na^+ solution but not in Cs^+ as a result of the repulsion between strongly bound hydration sheaths of water molecules around the Na^+ ions. However, we now look to the case of Ni^{2+}, which has a hydration energy value 5 times higher than Na^+ ($\Delta G^{hyd}(Ni^{2+}) = -2068$ kJ mol^{-1}, due to the double charge and small ionic radius of 0.07 nm,[42] compared to $\Delta G^{hyd}(Na^+) = -411$ kJ mol^{-1}) and see no hydration-repulsion regime at this bulk concentration. We suggest that this may be due to the less strong coordination of the highly hydrated Ni^{2+} ions with the mica sites—due to the less (enthalpically) favourable partial-dehydration required for closest approach of the ion to the surface—resulting in attraction of the surfaces to $D = D_0$ and balance of the mica charge by protons at D_0 as the hydrated Ni^{2+} ions are expelled from the confined film. This conclusion is additionally backed up by the observations that the adhesion force after jump-in to contact and the surface separation at this point of closest approach are within the boundaries of values measured in pure water, confirming that the Ni^{2+} ions are unlikely to be held between the mica surfaces at this point in contrast to what we have seen in the case of Cs^+.

We now calculate the effective viscosity of the Ni^{2+} solution confined to a film between the surfaces before the jump-in to contact. It has been demonstrated recently how the effective viscosity of the film, η_{eff}, can be calculated by analysis of the dynamics of the escaping hydration film during the jump together of the surfaces into mica–mica contact under influence of van der Waals interactions.[21] The 'jump-time', τ_j, is related to the viscosity of the film according to the following expression:

$$\tau_j = (18\pi R\eta_{eff}/A)[D_j^2 - D_0^2]$$

where A is the Hamaker constant determining the van der Waals interaction. Using the experimentally measured value of the jump distance of 1.7 nm in 1×10^{-1} M

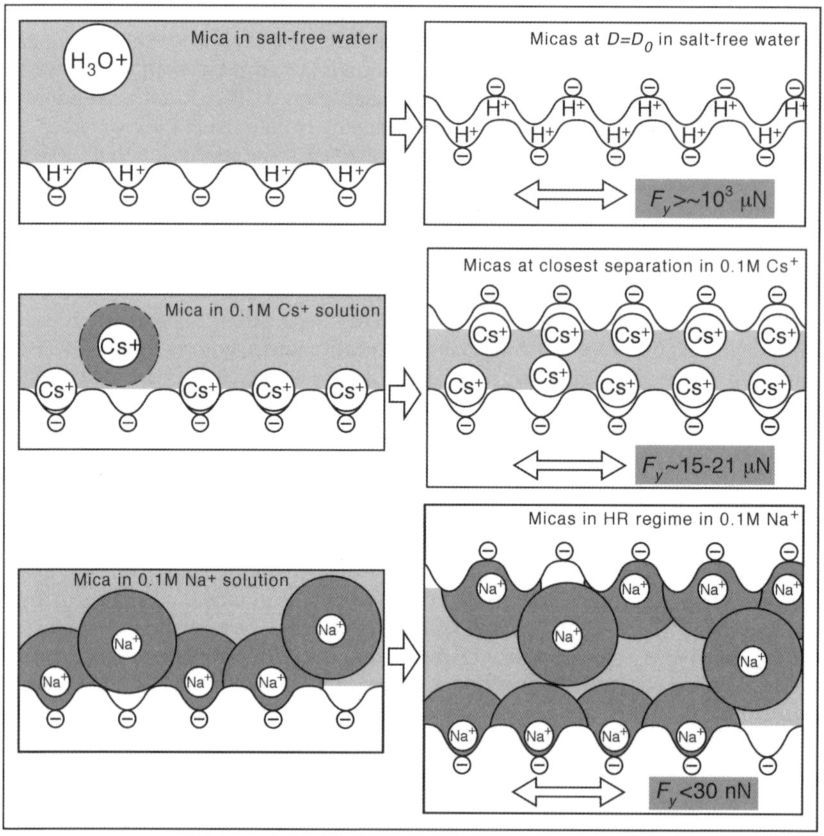

Fig. 5 Cartoon, approximately to scale, indicating the contrasting behaviour of ions at mica surfaces. On the left-hand side the negatively charged mica surface is neutralised by counterions in pure water (top), 1×10^{-1} M Cs^+ ions (middle), and 1×10^{-1} M Na^+ ions (bottom). On the right-hand side we show two mica surfaces in contact, with different counterions between the surfaces, and their differing degrees of hydration, as discussed in the text.

 This journal is © The Royal Society of Chemistry 2009

$Ni(NO_3)_2$ solution and the experimentally estimated jump times of 0.3 ± 0.2 s, we determine the effective viscosity of the film in these two cases to be $1.2 < \eta_{eff} < 6.1$ mPa s.[54] These results suggest that the film of hydrated Ni^{2+} ions squeezing out from between the mica surfaces has viscosity close to that of bulk water, and to that expected for bulk 1×10^{-1} M $Ni(NO_3)_2$ solution.[42]

This study of $Ni(NO_3)_2$ solution demonstrates that although 1×10^{-1} M $Ni(NO_3)_2$ solution remains highly fluid in thin films between mica sheets, this very strongly hydrated ion does not provide 'hydration lubrication' at this bulk concentration due to its apparent weaker interaction with the mica surface compared to Na^+ and Cs^+ ions.

5. Conclusion

The three ions investigated in this study, each at 1×10^{-1} M concentration, demonstrate three very different behaviours with respect to confinement in thin films and the resulting lubrication of the shearing surfaces. We tentatively attribute these contrasting properties as demonstrated with a schematic cartoon in Fig. 5. Na^+ appears to remain bound to the mica surfaces, in hydrated form, on confinement to a film of thickness 0.8 ± 0.3 nm and under external applied loads of up to 10 atmospheres pressure. Under this confinement the hydrated film remains fluid in the shear plane, with viscosity close to that of bulk solution, and thereby provides effective lubrication of the shearing surfaces by relieving the applied shear stress with little frictional dissipation. Cs^+ also remains bound to the mica surfaces under molecular level confinement—a conclusion derived from the shear behaviour of the surfaces at closest approach, the distance of closest approach and the adhesion value—but in contrast to Na^+ films, the Cs^+ ion does not hold a bound hydration layer. As the mica surfaces approach, the van der Waals attraction between them is sufficient to expel all water from the film, whereupon a finite yield stress is recorded between the shearing surfaces. Ni^{2+} solutions at similar 1×10^{-1} M concentrations demonstrate an example of hydrated ions which are not bound to the mica surfaces on close approach of the surfaces: the hydrated ions are expelled from the film as the mica surfaces come into lattice-lattice contact. Although the aqueous Ni^{2+} film is demonstrated to have low viscosity until the point of jump into contact of the surfaces, there is no lubrication between the surfaces at their closest separation (at this bulk concentration) since no fluid hydration layer is present. The adhesion, shear force and surface separation of the mica sheets at this point of adhesive contact all corroborate the suggestion that Ni^{2+} does not remain between the surfaces—in contrast to Cs^+—and the negative mica sites are neutralised instead by protons when the mica sheets come into contact. This study has allowed us to formulate two conditions which must be satisfied in order for 'hydration lubrication' to take place: (i) ions must remain bound to the shearing surfaces under confinement; and (ii) the surface-bound ions must retain their hydration sheath under confinement and applied load. When these conditions are fulfilled, hydration lubrication is mediated by the (thermodynamically) bound and (kinetically) fluid hydration layers attached to interfacial charged species. This mechanism is likely to occur also at more complex shearing, strongly hydrated interfaces, such as occur in synovial joints and mucosal surfaces, and sliding over wet sand at high pressures.

Acknowledgements

We thank Merton College, Oxford (for SP Fellowship), the EPSRC (UK), the Israel Science Foundation and the Minerva Foundation at the Weizmann Institute for support of this work and donors of the American Chemical Society Petroleum Research Fund for support under grant 45694-AC 7. We thank Ben Ocko and Jia Wang for useful discussions and suggestions concerning caesium hydration.

References

1 E. J. W. Verwey and J. T. G. Overbeek, *Theory of the Stability of Lyophobic Colloids*, Elsevier, Amsterdam, 1948.

2 D. F. Evans and H. Wennerström, *The Colloidal Domain*, Wiley, New York, 2nd edn, 1999.

3 J. Israelachvili and G. Adams, Measurement of forces between two mica surfaces in aqueous electrolyte solutions in the range 0–100 nm, *J. Chem. Soc., Faraday Trans. 1*, 1978, **74**, 975–1001.

4 H. Christenson, R. Horn and J. Israelachvili, Measurement of forces due to structure in hydrocarbon liquids, *J. Colloid Interface Sci.*, 1982, **88**(1), 79–88.

5 J. Klein, Forces between mica surface bearing adsorbed macromolecules in liquid media, *J. Chem. Soc., Faraday Trans. 1*, 1983, **79**, 99–118.

6 W. Ducker, T. Senden and R. Pashley, Direct measurement of colloidal forces using an atomic force microscope, *Nature*, 1991, **353**, 239–241.

7 V. A. Parsegian, R. P. Rand and N. L. Fuller, Direct osmotic stress measurements of hydration and electrostatic double-layer forces between bilayers of double-chained ammonium acetate surfactants, *J. Phys. Chem.*, 1991, **95**, 4777–4782.

8 W. Briscoe and R. Horn, Direct measurement of surface forces due to charging of solids immersed in a non-polar liquid, *Langmuir*, 2002, **18**, 3945–3956.

9 B. V. Derjaguin and L. Landau, Theory of the stability of strongly charged lyophobic sols and the adhesion of strongly charged particles in solutions of electrolytes, *Acta Physicochim. URSS*, 1941, **14**, 633–662.

10 R. Pashley, Hydration forces between mica surfaces in aqueous electrolyte solutions, *J. Colloid Interface Sci.*, 1981, **80**(1), 153–162.

11 J. Israelachvili and R. Pashley, Molecular layering of water at surfaces and origin of repulsive hydration forces, *Nature*, 1983, **306**, 249–250.

12 R. Kjellander, S. Marčelja, R. Pashley and J. P. Quirk, A theoretical and experimental study of forces between charged mica surfaces in aqueous $CaCl_2$ solutions, *J. Chem. Phys.*, 1990, **92**(7), 4399–4407.

13 J. Israelachvili and R. Pashley, The hydrophobic interaction is long range, decaying exponentially with distance, *Nature*, 1982, **300**, 341–342.

14 H. Christenson and P. M. Claesson, Direct measurements of the forces between hydrophobic surfaces in water, *Adv. Colloid Interface Sci.*, 2001, **91**, 391–436.

15 R. Horn, D. T. Smith and W. Haller, Surface forces and viscosity of water measured between silica sheets, *Chem. Phys. Lett.*, 1989, **162**(45), 404–408.

16 M. Antognozzi, A. D. Humphris and M. J. Miles, Observation of molecular layering in a confined water film and study of the layers' viscoelastic properties, *Appl. Phys. Lett.*, 2001, **78**, 300–302.

17 J. P. Cleveland, T. E. Schaffer and P. K. Hansma, Probing oscillatory hydration potentials using thermal-mechanical noise in an atomic force microscope, *Phys. Rev. B*, 1995, **12**, R8692–R8695.

18 E. Lauga, M. P. Brenner and H. A. Stone, Microfluidics: the no-slip boundary condition, in *Handbook of Experimental Fluid Dynamics*, ed. J. Foss, C. Tropea and A. Yarin, Springer, New York, 2005, ch. 15.

19 U. Raviv, P. Laurat and J. Klein, Fluidity of water confined to subnanometre films, *Nature*, 2001, **413**, 51–54.

20 E. A. Jagla, Boundary lubrication properties of materials with expansive freezing, *Phys. Rev. Lett.*, 2002, **88**, 245504.

21 U. Raviv, S. Perkin, P. Laurat and J. Klein, Fluidity of water confined down to subnanometer films, *Langmuir*, 2004, **20**(13), 5322–5332.

22 S. T. Cui, P. T. Cummings and H. D. Cochran, Molecular simulation of the trasition from liquidlike to solidlike behavior in complex fluids confined to nanoscale gaps, *J. Chem. Phys.*, 2001, **114**(16), 7189–7195.

23 J. Klein and E. Kumacheva, Simple liquids confined tomolecularly thin layers. I. Confinement-induced liquid-to-solid phase transitions, *J. Chem. Phys.*, 1998, **108**(16), 6996–7008.

24 M. L. Gee, P. M. McGuiggan and J. N. Israelachvili, Liquid to solid-like transitions of molecularly thin films under shear, *J. Chem. Phys.*, 1990, **93**, 1895.

25 Y. Leng and P. T. Cummings, Hydration structure of water confined between mica surfaces, *J. Chem. Phys.*, 2006, **124**, 074711.

26 U. Raviv and J. Klein, Fluidity of bound hydration layers, *Science*, 2002, **297**, 1540–1543.

27 L. Chai, R. Goldberg, N. Kampf and J. Klein, Selective adsorption of poly(ethylene oxide) onto a charged surface mediated by alkali metal ions, *Langmuir*, 2008, **24**, 1570–1576.

28 U. Raviv, S. Giasson, N. Kampf, J.-F. Gohy, R. Jerome and J. Klein, Lubrication by charged polymers, *Nature*, 2003, **425**, 163–165.

 This journal is © The Royal Society of Chemistry 2009

29 W. H. Briscoe, S. Titmuss, F. Tiberg, R. K. Thomas, D. McGillivray and J. Klein, Boundary lubrication under water, *Nature*, 2006, **444**(7116), 191.
30 Y. Leng and P. T. Cummings, Fluidity of hydration layers nanoconfined between mica surfaces, *Phys. Rev. Lett.*, 2005, **94**, 026101.
31 R. Goldberg, L. Chai, S. Perkin, N. Kampf and J. Klein, Breakdown of hydration repulsion between charged surfaces in aqueous Cs^+ solutions, *Phys. Chem. Chem. Phys.*, 2008, **10**, 4939–4945.
32 J. Israelachvili, Thin film studies using multiple-beam interferometry, *J. Colloid Interface Sci.*, 1973, **44**(2), 259–272.
33 J. Israelachvili, *Intermolecular and Surface Forces*, Academic Press, London, 2nd edn, 1991.
34 S. Perkin, L. Chai, N. Kampf, U. Raviv, W. H. Bruge, I. Dunlop, S. Titmuss, M. Seo, E. Kumacheva and J. Klein, Forces between mica surfaces, prepared in different ways, cross aqueous and non-aqueous liquids confined to molecularly-thin films, *Langmuir*, 2006, **22**(14), 6142–6152.
35 L. Cheng, P. Fenter, K. L. Nagy, M. L. Schlegel and N. C. Sturchio, Molecular-scale density oscillations in water adjacent to a mica surface, *Phys. Rev. Lett.*, 2001, **87**(15), 156103.
36 R. Pashley, DLVO and hydration forces between mica surfaces in Li^+, Na^+, K^+ and Cs^+ electrolyte solutions: a correlation of double-layer and hydration forces with surface cation exchange properties, *J. Colloid Interface Sci.*, 1981, **83**(2), 531–546.
37 U. Raviv, P. Laurat and J. Klein, Time dependence of forces between mica surfaces in water and its relation to the release of surface ions, *J. Chem. Phys.*, 2002, **116**(12), 5167–5172.
38 D. M. Le Neveu and R. P. Rand, *Biophys. J.*, 1977, **18**, 209–230.
39 In early SFB experiments studying a range of alkali metal ions, Pashley reported[36] hydration repulsion in CsCl solutions already at 1 mM, in contrast to the present results where a jump into adhesive contact is seen. In other experiments from our group (not shown) we have also occasionally observed short-range repulsions in Cs^+ solution but we attribute these to unidentified species adsorbed on the mica.
40 R. Pashley, Hydration forces between mica surfaces in electrolyte solutions, *Adv. Colloid Interface Sci.*, 1982, **16**, 57–62.
41 E. R. Nightingale, Phenomenological theory of ion solvation. Effective radii of hydrated ions, *J. Phys. Chem.*, 1959, **63**, 1381–1387.
42 J. Burgess, *Metal ions in solution*, Ellis Horwood Ltd, Chichester, 1978.
43 B. Hribar, N. T. Southall, V. Vlachy and K. A. Dill, How ions affect the structure of water, *J. Am. Chem. Soc.*, 2002, **124**, 12302–12311.
44 R. A. Robinson and R. H. Stokes, *Electrolyte Solutions*, Dover Publications, Mineola, NY, 2nd edn, 1959.
45 H. S. Frank and M. W. Evans, Free volume and entropy in condensed systems III. Entropy in binary liquid mixtures; partial molal entropy in dilute solutions; structure and thermodynamics in aqueous electrolytes, *J. Chem. Phys.*, 1945, **13**, 507.
46 F. Hofmeister, *Arch. Exp. Pathol. Pharmakol.*, 1888, **24**(29), 247–260.
47 M. L. Schlegel, K. L. Nagy, P. Fenter, L. Cheng, N. C. Sturchio and S. D. Jacobsen, Cation sorption on the muscovite (001) surface in chloride solutions using high-resolution X-ray reflectivity, *Geochim. Cosmochim. Acta*, 2006, **70**, 3549–3565.
48 P. M. Claesson, P. Herder, P. Stenius, J. C. Eriksson and R. Pashley, An ESCA and AES study of ion exchange on the basal plane of mica, *J. Colloid Interface Sci.*, 1986, **109**(1), 31–39.
49 H. Sakuma, K. Otsuki and K. Kurihara, Viscosity and lubricity of aqueous NaCl solution confined between mica surfaces studied by shear resonance measurement, *Phys. Rev. Lett.*, 2006, **96**, 046104.
50 P. M. McGuiggan and J. Israelachvili, Adhesion and short-range forces between surfaces. Part II: Effects of surface lattice mismatch, *J. Mater. Res.*, 1990, **5**(10), 2232–2243.
51 R. Pashley and J. Israelachvili, DLVO and hydration forces between mica surfaces in Mg^{2+}, Ca^{2+}, Sr^{2+} and Ba^{2+} chloride solutions, *J. Colloid Interface Sci.*, 1984, **97**(2), 446–455.
52 R. Kjellander, S. Marčelja, R. Pashley and J. P. Quirk, Double-layer ion correlation forces restrict calcium-clay swelling, *J. Phys. Chem.*, 1988, **92**, 6489–6492.
53 P. Petrov, S. J. Miklavic and T. Nylander, Forces between charged macroscopic surfaces in polyamide solutions, *J. Phys. Chem.*, 1994, **98**, 2602–2607.
54 This analysis does not take into account the effect of any correlation forces on the jump time.
55 D. Chan, R. Pashley and L. R. White, A simple algorithm for the calculation of the electrostatic repulsion between identical charged surfaces in electrolyte, *J. Colloid Interface Sci.*, 1980, **77**(1), 283–285.

Study of a nanoscale water cluster by atomic force microscopy

Manhee Lee,[a] Baekman Sung,[a] N. Hashemi[b] and Wonho Jhe*[a]

Received 6th May 2008, Accepted 12th June 2008
First published as an Advance Article on the web 9th September 2008
DOI: 10.1039/b807740c

We present a novel method for investigating a nanometric cluster of water molecules, which includes the formation and manipulation of nanometric water, and the measurement of its mechanical properties. The atomic force microscope based on the quartz tuning-fork sensor is employed to form and manipulate the nanometric water, and the theoretical tool of amplitude-modulation atomic force microscopy is used to obtain the elasticity, viscosity and dissipation energy of it. With high vertical resolution less than ~0.1 nm and high force sensitivity of ~0.01 N m^{-1}, this tool facilitates the stable formation and manipulation of a nano-water cluster (~10^4 molecules) in air without 'jump-to-contact' instability, as well as quantitative measurements of its physico-chemical properties.
PACS® numbers: 47.55.nk, 62.10.+s, 68.37.Ps

Understanding nanometric water is essential to solve a wide range of common problems in nanoscience and technology such as tribology, scanning probe microscopy (SPM), micro-electro-mechanical systems (MEMS), and biology. Frictions experienced in everyday life are associated with the mechanical properties of nanometric water formed between two surfaces,[1,2] and a water layer of a few nanometers' thickness existing on every surface affects the performance of SPM and MEMS[3] due to the strong adhesion force.[4] This nanoscale water is also an important environment in biological processes. For example, protein folding and enzyme activation cannot be made without nanometric water both surrounding and confined within the proteins,[5,6] and the ions are transported through a nanometre-sized ion channel which is filled with nanometric water.[7] Due to this ubiquity and importance in nanoscience and nanotechnology, the study on nanoscale water (in general, any liquid) becomes more and more important.

One of the methods to investigate the nanoscale water is 'touching' it directly using probe-based techniques such as surface force apparatus (SFA) and atomic force microscopy (AFM). By using SFA,[4] various aspects of nanoconfined liquids were observed; oscillatory solvation force, adhesion force and confinement-induced phase transition of some liquids.[8] However, the detailed behavior of a liquid confined within a tip–sample separation of less than ~1 nm could not be measured because of the 'jump-to-contact' instability in the region. Moreover, only a 2D liquid film has been considered due to the relatively large probe size (~1 μm). Although AFM is capable of investigating a nanoscale-water bridge formed between the tip and sample surface,[9] the conventional cantilever-based AFM also suffers from the 'jump-to-contact' instability, so that one can not arbitrarily form and manipulate the nanoscale water in air.

[a]*Center for Nano-Liquid, Department of Physics and Astronomy, Seoul National University, Seoul, 151-747, Korea. E-mail: whjhe@snu.ac.kr*
[b]*Department of Mechanical Engineering, Virginia Polytechnic Institute and State University, Blacksburg, Virginia 24061, USA*

In this paper, we present a novel method of investigating a nanometric water cluster by using the quartz tuning-fork-based amplitude-modulation (AM) AFM. Combing this technique with related theoretical tools recently developed,[10] we achieve stable formation and manipulation of a nano-cluster of water molecules in air, and perform a quantitative measurement of the elasticity, viscosity and dissipation energy of the nanoscale water.

For the study of a nanometric water cluster, it is essential to realize a stable formation of the nanometric water and to perform a sensitive measurement of its properties. This can be achieved by using our novel method; use of the quartz tuning-fork as a force sensor, which exhibits a very strong rigidity of $\sim$1000 N m^{-1} as well as a very high quality factor of order $\sim$1000. The high rigidity of the tuning-fork allows stable formation of nanometric water bridge in air by capillary condensation[11] without any 'jump-to-contact' instability. Moreover, it also allows one to perform very sensitive measurements since the high quality factor compensates the decrease of sensitivity due to the rigid spring constant. For a local detection, the tuning-fork actuator is combined with a conventional cantilever tip (100 nm radius, n$^+$-silicon, resistivity of 0.01–0.02 Ω cm, Nanosensors), by gluing the cantilever tip on the tuning-fork and then cutting away the leg of the whole body of cantilever except the tip itself. The tuning-fork probe and the mica substrate, prepared by careful chemical cleansing with dilute acid after *in situ* cleavage,[11] are placed in a humidity control chamber, as shown in Fig. 1.

For precise and quantitative measurement, an accurate calibration of the probe properties as well as an unambiguous determination of the absolute zero distance between tip and substrate is required. The probe calibration can be made accurately by fitting the resonance curve to the tuning-fork's resonance function described in ref. 12 (Fig. 2a). On the other hand, the absolute zero point is determined by the full-approach curves (Fig. 2b), where the amplitude and the phase signals exhibit saturation characteristics after about 2 nm movement toward the substrate from the point of capillary condensation, marked as $z = 0$. Thereby, we assume the capillary condensation, where the abrupt change of the amplitude and phase signals are observed, occurs at the tip–sample separation of $\sim$2 nm in our environmental conditions (relative humidity (RH) = 23%, 18 °C).

Such a calibrated tuning-fork-based AM-AFM is now used for a stable formation of a nano-cluster of water molecules in the following way. The laterally oscillating AFM tip approaches the substrate within a few nanometre distance until capillary condensation occurs, which produces a few-nanometre-sized bridge of water molecules between the tip and the substrate. Once the capillary condensation has occurred, it is stretched vertically upward by subsequent retraction of the tip to avoid the mechanical hard contact with the substrate. During these procedures, the amplitude and the phase changes are recorded, from which true mechanical

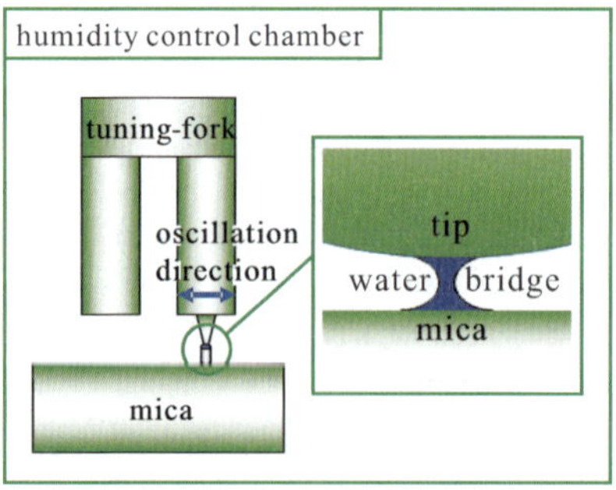

Fig. 1 Schematic diagram of tuning-fork-based atomic force microscope in a humidity chamber.

 This journal is © The Royal Society of Chemistry 2009

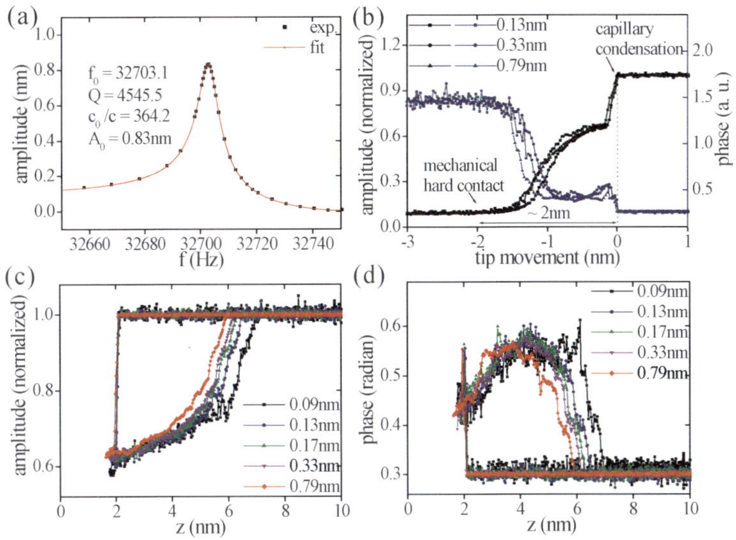

Fig. 2 (a) Free resonance curve of the tuning-fork probe and its fit to the model described in ref. 12. (b) Full approach curves for different free oscillation amplitudes; $A_f = 0.13$, 0.33 and 0.79 nm. (c) and (d) The amplitude and the phase changes during the tip approach and subsequent retraction ($v \simeq 1$ nm s^{-1}) for various free oscillation amplitudes $A_f = 0.09$, 0.13, 0.17, 0.33 and 0.79 nm on a fixed operating frequency, $f = 32$ 702 Hz.

signals are obtained by using the formulae described in ref. 12 to remove the artifact signals which occur in quartz resonators.

Fig. 2c and 2d show the mechanical motion of the probe, the amplitude and the phase, where we set the capillary condensation point to $z = 2$ nm, as noted above (Fig. 2b). The probe is operated at the frequency $f = 32$ 702 Hz and approaches or retracts with velocity of $v \simeq 1$ nm s^{-1}, which is far faster than the thermal drift of the system $v_{\mathrm{drift}} \approx 0.03$ nm s^{-1}. When the capillary condensation has occurred, the amplitude and the phase signals are abruptly changed and then gradually recovered as the tip is retracted to the rupture points, exhibiting hysteresis. These are the characteristic features of the capillary interaction because any other force between the tip and the substrate, such as the van der Waals (vdW) or electrostatic interactions, are exerted on the probe rather gradually, not abruptly, and also do not exhibit the hysteresis.[13] Interestingly, the rupture distance is increased little by little with the decrease of free oscillation amplitude A_f, which indicates that a more stable formation of the nanometric water bridge is made when a smaller oscillation amplitude is used.

The nanometric water bridge formed between the tip and the substrate can be modeled by sphere-flat geometry,[14] as shown in Fig. 3, where R is the tip radius, x

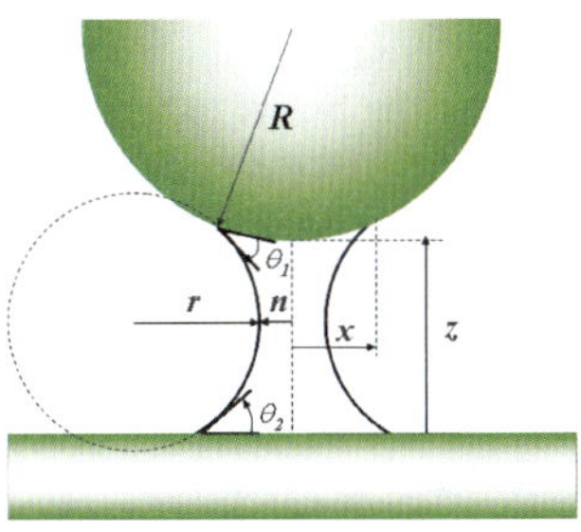

Fig. 3 Schematic of a sphere on a flat surface with a liquid meniscus between them.

the contact point between the tip and the liquid, z the tip–sample distance, and n and r the two principal radii. The contact angles, θ_1 and θ_2, are usually assumed to be zero; $\theta_1 = \theta_2 = 0$, for the hydrophilic Si tip and for the mica sample. For a given z, two parameters, n and r, can be uniquely determined once the meniscus volume is determined, which can be done by using the relation Willett *et al.*[15] have found on the basis of numerical solutions of the Young–Laplace equations,

$$z_r = (1 + \theta/4)V^{1/3} - \frac{2}{5R}(1 + \theta/4)V^{2/3}, \tag{1}$$

where z_r is the rupture distance, V the volume of water bridge, R the tip radius, and θ the contact angle in radians ($\theta = \theta_1 = \theta_2$). For our rupture distance of ~6.5 nm (Fig. 2c and 2d), a liquid volume of 297.9 nm^3 is obtained (about 36 000 molecules of water).

We note that the rupture distance of 6.5 nm is not satisfied with the Kelvin equation,[16,17] which requires a Kelvin radius of curvature $r_k = -0.37$ nm to remain constant value under the given temperature and relative humidity in our experiment, and from which a rupture distance of $z_r = 0.32$ nm is predicted for the zero contact angles. Thus, the nanometric water bridge formed in our experiments does not satisfy the boundary condition of constant curvature expected by the Kelvin equation, which indicates that the liquid phase does not remain in equilibrium with the vapor, but liquid–vapor exchanges are negligible during the subsequent retraction of the tip after the capillary condensation is occurred. For the latter case, the constant volume boundary condition is available, which is now employed to analyze the thermodynamics of our nanometric water bridge with the following equations:[16,17]

$$(R + r)^2 = (n + r)^2 + (R + z - r)^2, \tag{2}$$

$$p = \gamma\left(\frac{1}{n} - \frac{1}{r}\right), \tag{3}$$

$$F_c = p(n + r)^2\pi, \tag{4}$$

$$F_t = 2\pi\gamma\frac{R(n + r)}{R + r}, \tag{5}$$

where eqn (2) is the constraint of our system given by the zero contact angle, p the Laplace pressure, F_c the capillary force induced by the Laplace pressure, γ (= 72.8 mJ m^{-2}) the surface tension for water at 20 °C, and F_t the surface tension force induced by surface tension.

The thermodynamic parameters calculated from eqn (2), (3), (4) and (5) are shown in Table 1, where the nominal values for the capillary condensation and the rupture distances were chosen as 2.00 nm and 6.50 nm, respectively.[18] The negative sign of p represents the concave shape of the bridge as well as the attractive force between two surfaces. While the Laplace pressure p, or equivalently F_c, has the negative value at the capillary condensation point, it is changed to be a positive one at the rupture distance. This means that when the nano-bridge is formed, it exerts high attractive force (approx. −8.9 nN) between the two surfaces, and then it becomes thin and long reducing the attractive force during its stretch. But near $z \simeq 6.5$ nm the attractive force is changed to the repulsive one (~0.3 nN), whereby the nano-bridge becomes highly unstable so that it is ruptured.

Interestingly, even at the rupture point, the neck radius n has a small but finite value of ~3 nm. Although this critical value of rupture width is obtained by classical

 This journal is © The Royal Society of Chemistry 2009

Table 1 Thermodynamic parameters of the nanometric bridge of ~36 400 water molecules formed between the tip and the substrate. The condensation and rupture distances z are set at 2 nm and 6.50 nm respectively, from which r, n, p, F_c and F_t are calculated by using eqn (2), (3), (4) and (5) with the boundary conditions of constant volume and zero contact angle

	Condensation	Rupture
z/nm	2.00	6.50
r/nm	1.13	3.35
n/nm	6.22	2.98
x/nm	7.27	6.12
p/MPa	−52.5	2.7
F_c/nN	−8.9	0.3
F_t/nN[a]	3.3	2.8
F_c/F_t	−2.7	0.1

[a] Surface tension force F_t is exerted on the tip surface in the radial direction, which results in a vertical attractive force.

thermodynamic theory, it is also in agreement with the theoretical results based on the lattice gas Monte Carlo simulations.[20] We also note that about 1 nm slip (x = 7.27 → 6.12 nm) of contact line between the tip surface and the liquid phase is occurred during the stretch up to rupture distance, which is often revealed by the 'stick–slip' behavior of mechanical friction.[11]

Mechanical properties of the small cluster of water molecules can be obtained with the theory of AM-AFM,[10] by substituting the amplitude and the phase information in Fig. 2c and 2d into the following set of equations for the elasticity k_{int}, viscosity b_{int}, elastic force F_k, viscous force F_b and dissipation energy E_{dis}:

$$\frac{k_{\text{int}}}{k} = \frac{A_0}{QA}\sin\theta + \left(\frac{w}{w_0}\right)^2 - 1, \tag{6}$$

$$\frac{b_{\text{int}}}{b} = \frac{w_0 A_0}{wA}\cos\theta - 1, \tag{7}$$

$$F_k = k_{\text{int}} A, \tag{8}$$

$$F_b = b_{\text{int}} w A, \tag{9}$$

$$\frac{E_{\text{dis}}}{kA_0^2} = -\frac{\pi}{Q}\left[\frac{A}{A_0}\cos\theta - \frac{w}{w_0}\left(\frac{A}{A_0}\right)^2\right], \tag{10}$$

where A and θ are the amplitude and the phase of tip motion respectively, A_0 is the peak amplitude at the resonance frequency w_0, w the operating frequency, and Q the quality factor of the probe. The rigidity of the probe k (= 1478.2 N m^{-1}) can be obtained either by theoretical calculation[19] or thermal spectrum, and the damping coefficient b of the free space is given by $b = k/(Qw_0)$. We note that eqn (6) and (7) are corresponded to the first-order solution of the tapping-mode configuration of AM-AFM,[10] which are still available and generally work in this shear-mode configuration as shown in ref. 12.

Fig. 4 presents k_{int}, b_{int}, F_k, F_b and E_{dis} of the nanometric water column. It is clearly shown that the measured $k_{\text{int}} \simeq 0.1$ N m^{-1} and $b_{\text{int}} \simeq 0.6$ μN s m^{-1} are

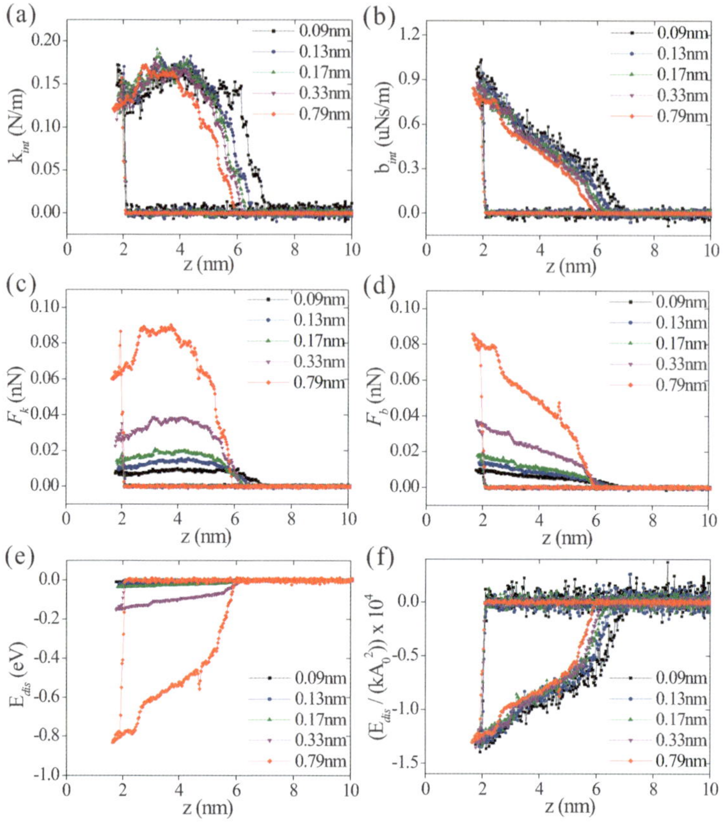

Fig. 4 Mechanical properties of nanometric cluster of water molecules: elasticity (a), viscosity (b), elastic force (c), viscous force (d), dissipation energy (e), and dissipation energy normalized by kA_0^2 (f) obtained from the amplitude and the phase curves in Fig. 2c and 2d.

not much dependent on the measured amplitudes in the range from 0.1 nm to 1 nm, which manifests the linear characteristics of F_k and F_b that are linearly increased with increasing the oscillation amplitude. Although frictions which are experienced by an object sliding on a surface have been macroscopically believed to be proportional to the velocity of the object motion, the probe shearing the mica surface experiences the velocity-dependent force F_b (Fig. 4d) as well as displacement-dependent force F_k (Fig. 4c). Moreover, almost same amount of forces are contributed by F_k and F_b to the friction. This is due to the viscoelasticity, k_{int} and b_{int}, of the nanometric water cluster between two surfaces.

The measured damping $b_{\text{int}} \simeq 0.6\ \mu\text{N s m}^{-1}$ at $z = 3$ nm is 10 times greater than the hydrodynamic damping acting on the tip moving on a bulk fluid with the bulk constant viscosity ($\eta_{\text{water}} \cong 10^{-3}$ Pa s), which leads to a damping of $b_{\text{water}} \simeq 0.06\ \mu\text{N s m}^{-1}$ at $z = 3$ nm.[21,22] The b_{int}, or F_b, causes a dissipation energy E_{dis} of order 0.1 to 1 eV, of which the origin is different from the one of the dissipation of order ~10 eV induced by the condensation–rupture process during each oscillation cycle in large amplitude operation, suggested in ref. 23 On the other hand, b_{int} as well as k_{int} of the nanometric cluster of water observed in our experiment seem to be intrinsic quantities associated with boundary surface tension and internal structure of the hydrogen bonding network.

In conclusion, the presented method utilizing quartz tuning-fork-based AM-AFM and the developed theoretical tools available allow stable formation and

 This journal is © The Royal Society of Chemistry 2009

manipulation of the nanometric water cluster as well as quantitative measurement of its thermodynamic–mechanical properties. Due to the reliable small amplitude operation of our experimental system, this technique enables one to perform a non-perturbative measurement, providing very intrinsic properties ($k_{\rm int}$ and $b_{\rm int}$) associated with the hydrogen bonding network of a nanoscale water cluster rather than the tip-motion-induced ones, such as the large tip-motion-induced dissipation energy.[23] These novel measurements may provide a useful means of extending our understanding of hydrogen bonds in the nanoscale and their cooperative effects in a small water nanocluster,[24] with which numerous anomalies including the large heat capacity, high thermal conductivity and high melting, boiling and critical points are closely associated.

Acknowledgements

We are grateful to S. Jang and H. E. Noh for helpful discussions. This work was supported by the Acceleration Research Program Program of Korean Ministry of Science and Technology. WJ also appreciates support from the LG Yonam Foundation.

References

1 J. Chen, I. Ratera, J. Y. Park and M. Salmeron, *Phys. Rev. Lett.*, 2006, **96**, 236102.
2 K. B. Jinesh and J. W. M. Frenken, *Phys. Rev. Lett.*, 2006, **96**, 166103.
3 C. H. Mastrangelo, *Tribol. Lett.*, 1997, **3**, 223.
4 J. Israelachvili, *Intermolecular and Surface Forces*, Academic Press, London, 2003.
5 M. S. Cheung, A. E. Garcia and J. N. Onuchic, *Proc. Natl. Acad. Sci. U. S. A.*, 2002, **99**, 685.
6 G. A. Papoian, J. Ulander, M. P. Eastwood, Z. Luthey-Schulten and P. G. Wolynes, *Proc. Natl. Acad. Sci. U. S. A.*, 2004, **101**, 3352.
7 Y. J. Jeon, H. Kim, S. Jon, N. Selvapalam, D. H. Oh, I. Seo, C. S. Park, S. R. Jung, D. S. Koh and K. Kim, *J. Am. Chem. Soc.*, 2004, **126**, 15944.
8 J. Klein and E. Kumacheva, *Science*, 1995, **269**, 816.
9 R. Szoszkiewicz and E. Riedo, *Phys. Rev. Lett.*, 2005, **95**, 135502.
10 Mahee Lee and Wonho Jhe, *Phys. Rev. Lett.*, 2006, **97**, 036104.
11 H. Choe, M.-H. Hong, Y. Seo, K. Lee, G. Kim, Y. Cho, J. Ihm and W. Jhe, *Phys. Rev. Lett.*, 2005, **95**, 187801.
12 M. Lee, J. Jahng, K. Kim and W. Jhe, *Appl. Phys. Lett.*, 2007, **91**, 023117.
13 N. Hashemi, H. Dankowicz and M. R. Paul, *J. Appl. Phys.*, 2008, **103**, 093512.
14 T. Stifter, O. Marti and B. Bhushan, *Phys. Rev. B*, 2000, **62**, 13667.
15 C. D. Willett, M. J. Adams, S. A. Johnson and J. P. K. Seville, *Langmuir*, 2000, **16**, 9396.
16 O. H. Pakarinen, A. S. Foster, M. Paajanen, T. Kalinainen, J. Katainen, I. Makkonen, J. Lahtinen and R. M. Nieminen, *Model. Simul. Mater. Sci. Eng.*, 2005, **13**, 1175.
17 A. Chau, S. Rignier, A. Delchambre and P. Lambert, *Model. Simul. Mater. Sci. Eng.*, 2007, **15**, 305.
18 Although the tip–sample distances for the capillary condensation and the rupture are estimated roughly, it is found that a small variation of the values does not affect the results presented here including finite rupture width, positive Laplace pressure at the rupture point, and slip behavior on the tip surface.
19 F. J. Giessibl, *Appl. Phys. Lett.*, 1998, **73**, 3956.
20 J. Jang, G. C. Schatz and M. A. Ratner, *Phys. Rev. Lett.*, 2004, **92**, 085504.
21 Olga I. Vinogradova, *Langmuir*, 1995, **11**, 2213.
22 Here we used the approximated solution for hydrodynamic damping, $b = 6\pi\eta R^2/z$, where b is the damping coefficient, η the bulk viscosity, R the tip radius, and z the tip–sample distance, as described in ref. 21.
23 E. Sahagun, P. Garcia-Mochales, G. M. Sacha and J. J. Saenz, *Phys. Rev. Lett.*, 2007, **98**, 176106.
24 F. N. Keutsch and R. J. Saykally, *Proc. Natl. Acad. Sci. U. S. A.*, 2001, **98**, 10533.

Water at an electrochemical interface—a simulation study

Adam P. Willard,[a] Stewart K. Reed,[b] Paul A. Madden†*[b] and David Chandler[a]

Received 2nd April 2008, Accepted 2nd June 2008
First published as an Advance Article on the web 29th September 2008
DOI: 10.1039/b805544k

The results of molecular dynamics simulations of the properties of water in an aqueous ionic solution close to an interface with a model metallic electrode are described. In the simulations the electrode behaves as an ideally polarizable hydrophilic metal, supporting image-charge interactions with charged species, and it is maintained at a constant electrical potential with respect to the solution so that the model is a textbook representation of an electrochemical interface through which no current is passing. We show how water is strongly attracted to and ordered at the electrode surface. This ordering is different to the structure that might be imagined from continuum models of electrode interfaces. Further, this ordering significantly affects the probability of ions reaching the surface. We describe the concomitant motion and configurations of the water and ions as functions of the electrode potential, and we analyze the length scales over which ionic atmospheres fluctuate. The statistics of these fluctuations depend upon surface structure and ionic strength. The fluctuations are large—sufficiently so that the mean ionic atmosphere is a poor descriptor of the aqueous environment near a metal surface. The importance of this finding for a description of electrochemical reactions is examined by calculating, directly from the simulation, Marcus free-energy profiles for transfer of charge between the electrode and a redox species in the solution and comparing the results with the predictions of continuum theories. Significant departures from the electrochemical textbook descriptions of the phenomenon are found and their physical origins are characterized from the atomistic perspective of the simulations.

1 Introduction

The altered solvation, dielectric and dynamical properties of water molecules close to electrode surfaces have an important influence on electrochemical reactions. There have been numerous simulation studies of aqueous solutions close to charged solid surfaces which have cast light on the ordering of water molecules by the solid surface and begun to make the connection between the layers with altered dielectric characteristics invoked in continuum models of the electrode capacitance and the ordered molecular films which are known from surface science. Ref. 1 contains an excellent summary of work to date, and ref. 2 and 3 a more longstanding review. To complete the link between molecular behaviour and electrochemical observations

[a]*Department of Chemistry, University of California, Berkeley, California 94720, USA*
[b]*School of Chemistry, University of Edinburgh, Edinburgh, UK EH9 3JJ. E-mail: paul.madden@queens.ox.ac.uk*

† Present address: The Queen's College, University of Oxford, Oxford, UK OX1 4AW

This journal is © The Royal Society of Chemistry 2009

we need a realistic representation of the electrochemical interface and a direct way of calculating the electrochemical observable, namely the dependence of the rate of the electrochemical electron transfer on the potential applied to the electrode.[4]

What is required to achieve the second of these objectives is suggested by the Marcus theory of electron transfer.[5,6] In a Marcus description of the oxidation of some solution species R to an oxidized species O by transfer of an electron to an electrode maintained at some potential V with respect to the solution, we construct curves (as illustrated schematically in Fig. 1) which describe how the free energies of O and R depend upon some reaction coordinate, which is envisaged as reflecting the influence of fluctuating solvent degrees of freedom. The Marcus expression for the rate of electron transfer can be calculated from the probability that the system will access the configuration where the two curves cross. Note that the free-energy curve for O includes the potential energy of the electron on the electrode (eV), so that the position and height of the crossing point depend on the electrode potential. The two curves are coupled by a term (γ) which reflects the tunneling of the electron between the redox centre and the electrode, and this is expected to depend exponentially on its distance from the electrode surface. We should, therefore, be thinking about the dependence of the Marcus curves on the proximity to the electrode surface, to which two factors contribute. Firstly, the *difference* between the *direct* interactions of the O and R species with the charged surface itself will produce a differential shift on them, and therefore affect the crossing point. Secondly, if the redox species is close to the electrode, the competing interactions of the water molecules in its coordination shell with the surface and the solute itself may result in a change in the character of the fluctuations of the reaction coordinate. Both of these factors may be affected by the potential applied to the electrode. The Marcus curves contain the information which is required to calculate the electron transfer rate for a redox ion at a given distance from the electrode surface, but to complete the calculation of the rate we also need to know the probability of the reactant reaching this position, and this too will be affected by the potential exerted by the electrode and its influence on the solvating properties of the water molecules.

Blumberger, Sprik and co-workers[7,8] have demonstrated how the Marcus curves can be calculated for a homogeneous electron transfer reaction within an *ab initio* molecular dynamics scheme. Following Warshel[9] they emphasize the advantages for computation of choosing the "vertical energy gap" as the reaction coordinate. The vertical energy gap for oxidation is calculated for a single configuration in a simulation by switching the identity (and all associated interaction parameters) of a redox species initially in its reduced form R to its oxidized form without allowing

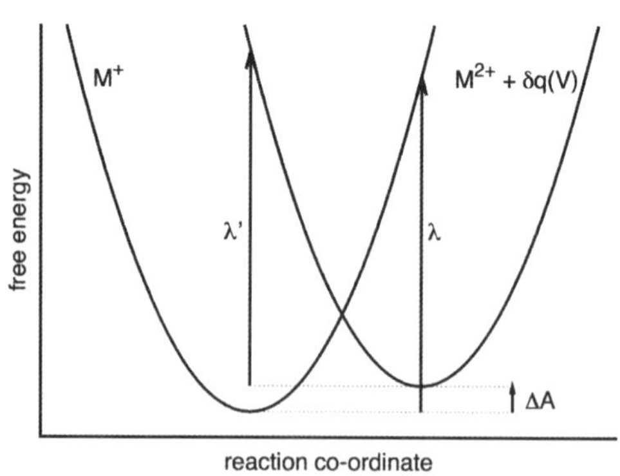

Fig. 1 Schematic illustration of the Marcus construction for understanding the electrochemical oxidation of some species R = M^+ to O = M^{2+} with transfer of the charge (δq) to an electrode maintained at a potential V. The quantities ΔA, λ and λ' are the reaction free-energy and reorganization energies for oxidation and reduction respectively.

 This journal is © The Royal Society of Chemistry 2009

any changes in the nuclear coordinates (hence a "vertical" transition in the Franck–Condon sense) and evaluating the energy difference between the final and initial states. The Marcus curves for the oxidized and reduced species may then be estimated from the probability distributions of the vertical energy gaps obtained by repeatedly sampling through the course of an molecular dynamics (MD) simulation and assuming this distribution is Gaussian. The Gaussian assumption is required by this method because regions of the distribution pertinent to the charge-transfer reaction are not generally accessed in a straightforward simulation. In the calculations reported here, the Gaussian approximation is tested and shown to be accurate. More generally, the Gaussian approximation has been tested and found to be accurate[10,11] provided proper account is taken of molecular boundary conditions[12]

Use of this scheme to study the electrochemical electron transfer process in a simulation is illustrated in Fig. 2. An aqueous solution is contained between two crystalline arrays of atoms which comprise the (metallic) electrodes, these are maintained at a definite electrical potential. The solution contains the redox species in its reduced and oxidized forms and, periodically during the simulation, an ion (the "redox target") is selected and its redox state is switched and the energy difference between final and initial states is evaluated. By selecting ions at different distances from the electrodes and by examining how the vertical energy gaps depend on the applied potential, we can build up the necessary information to study how the nature of the water at the electrode surface affects the electrochemical electron transfer rate *via* the Marcus construction.

In order to make contact with experimental studies the calculation needs to be done with as realistic a representation of the constant-potential electrode and interfacial water as can be managed. Ideally, it would be done within an *ab initio* MD scheme, as this would enable the difficult-to-characterize interactions between the solution and the electrode to be modeled without the introduction of interaction potentials.[1,13,14] However, the time and length scales involved in the relaxation of the solution in the vicinity of the electrodes (which we will characterize below) are far too long to allow a full self-consistent description of the screening of the electrode potential within an *ab initio* scheme.[15] Recently we introduced a way of incorporating some of the essential physical effects necessary for a realistic description of the interfacial charge-transfer process into a simulation which uses interaction potentials and therefore enables simulations of much larger time and length scales than is possible *ab initio*.[16] In particular, model *metallic* electrodes maintained at

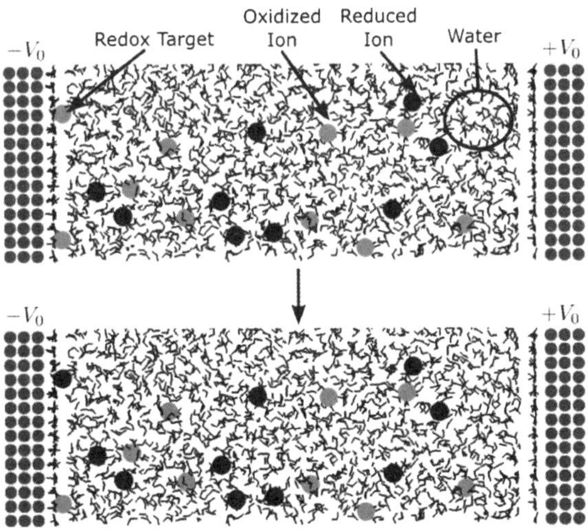

Fig. 2 Schematic depiction of electron transfer.

a *constant electrical potential* may be introduced into such a simulation, following a technique introduced by Siepmann and Sprik.[17] Because the electrodes behave as ideally polarizable metals they support image-charge interactions between charged species and the electrode; these, as we shall see, have an important influence on the electron-transfer process. Because the electrodes are maintained at a constant potential, when the charge of the redox species is changed to sample the vertical energy gap that charge is transferred in full to the electrodes, so that the source of the dependence of the electron-transfer rate on the electrode potential is included in the calculation. The electrode potential and the potential felt by the molecules and ions in the solution region are calculated self-consistently. Calculations using these methods have already been performed to examine the Marcus curves in simulations of redox-active molten salts.[18]

We begin with a brief description of the methods and interaction potentials used to simulate pure water and aqueous solutions of LiCl and the Ru^{2+}/Ru^{3+} couple confined between model platinum electrodes. We then examine the structure and dynamical properties of the electrode-adsorbed water and the way they are affected by the application of a potential to the electrode. In sections 3 and 4 we consider the consequences of this adsorbed water for the approach of ions to the electrode surface and the effect of the adsorbed water and the ionic atmosphere on the electrical potential in the vicinity of the electrode. In classical models of electrochemical charge transfer this potential is invoked to represent the dependence of the energies of the oxidized and reduced species on the proximity to the electrode. We then present preliminary results for the Marcus curves for the Ru^{2+}/Ru^{3+} system and discuss the physical factors which determine their dependence on the applied potential and the proximity to the electrode.

2 Scope of the model

The electrochemical interface is affected by many phenomena and a comprehensive representation of all of them within a single simulation is beyond current capabilities. Our focus here is on the solution side of the interface, on the properties of the water molecules at the interface and their influence on the electrical potential. As such we will present a significantly simplified model of the electrode itself in which we ignore the motion of the electrode atoms and therefore neglect effects like the restructuring of the electrode surface under chemical or electrical influences.[19] Furthermore the representation of the electrode as a metal is a simplified one, designed to capture the correct macroscopic response to an electrical potential appropriate to a metal rather than to deal with a correct microscopic description of surface electronic states *etc.*

As illustrated in Fig. 2 and 3, the electrodes each consist of three layers of atoms arranged in a face-centered cubic (fcc) lattice with the 100 face exposed to the solution; the lattice parameter is appropriate to Pt. Following Siepmann and Sprik[17] each electrode atom i carries a Gaussian charge distribution of fixed width but variable amplitude (q_i). These charges are Coulombically coupled to all other charges in the system. They are treated as additional dynamical degrees of freedom whose values are adjusted at each timestep in the molecular dynamic procedure in order to variationally minimize an appropriate energy functional. The energy functional is chosen[17] so that at its minimum the electrical potential on every electrode atom is the same (as appropriate to a metal) and equal to some pre-set value V_0. The use of a variational principle allows forces and response behaviour to be calculated straightforwardly *via* an application of the Hellmann–Feynman Theorem, as used to good effect in *ab initio* MD simulations. The methodology for the simulation of the electrodes is described in great detail in ref. 16 where it is shown that the electrodes become polarized in the presence of a charge in the solution region in a way which corresponds to the classical image-charge response. We illustrate this response in Fig. 3 where the charge induced on the electrode atoms by the instantaneous

 This journal is © The Royal Society of Chemistry 2009

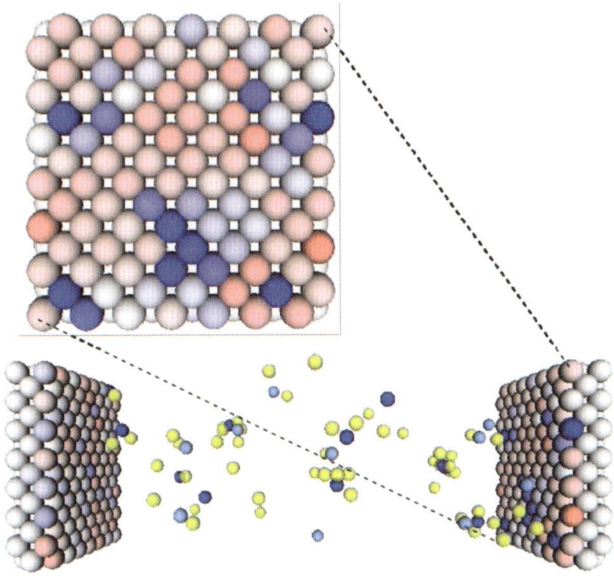

Fig. 3 Snapshot of the distribution of charge on the polarizable model electrode. Electrode atoms are shaded according to their partial charge, with blue shading corresponding to positive charge and red shading corresponding to negative charge. In this rendering, the darkest shade of blue corresponds to a partial charge of 0.1 *e*, and the darkest shade of red to a partial charge of −0.1 *e*. In the solvent region the yellow spheres represent Cl^- ions, and the light-blue and dark-blue spheres represent Ru^{2+} and Ru^{3+} respectively. Water molecules have been omitted for clarity.

configuration of the charges in solution is shown by colour-coding the electrode atoms. The bright-blue region, for example, is caused by the presence of an anion within the first molecular layer of the solution close to this position. Furthermore, if the charge on one of the ions in solution is changed, the charge difference is fully transferred to the electrodes to maintain charge neutrality with a constant electrode potential:[18] it is this feature which enables us to calculate Marcus curves for electrochemical charge transfer.

To describe the interactions between water molecules we use the SPC/E potential,[20] which is known to give a dielectric constant for water close to the experimental value. The interactions of water molecules with a metallic electrode are complex and cannot be modeled accurately with a simple two-body potential; since the behaviour of water at the interface is the central purpose of our study we were concerned to represent this interaction as carefully as is possible through the introduction of a potential. Experiments[21] and ab-initio studies[22] have shown that water molecules interact with a crystalline platinum surface by adsorbing on top sites and orienting their dipole along the plane of the electrode. In their study of water at an STM (scanning tunnelling microscope) tip, Siepmann and Sprik[17] parameterized a two- and three-body potential to describe the adsorption of water molecules on a platinum surface; their potentials are particularly appropriate for our study since they do not include the consequences of image interactions, which are dealt with through the polarizable electrode model as in our simulations. We have used these potentials exactly as described in their paper.

We have not attempted the same level of realism with the interactions between the ions and the electrode surface. Experimental studies show that anions interact quite strongly with transition metal surfaces to the extent that complete surface coverage of ordered layers of Cl^- is observed on positively charged single-crystal electrodes above about 0.5 V from molar solutions.[23] Guymon *et al* have shown how suitable potentials to describe these strong interactions could be obtained from *ab initio* calculations.[24] However, adsorption of this strength would present a significant problem for our simulations since it would mean that the solution region would be strongly depleted in Cl^- ions. To represent the interface under these conditions we would need to equilibrate our system in the presence of a reservoir of electrolyte; furthermore equilibrating this system would be *very* slow, as we shall see. We have, therefore, for the present study introduced only weakly attractive interaction potentials between the ions and the electrode surface—we use exponential-6 potentials to represent the short-range interactions with the parameters chosen as if the atoms of the metallic walls were themselves Cl^- ions. These potentials are too weakly attractive, compared to the water–electrode interactions, to lead to the kind of anion adsorption phenomena seen in the experimental studies of Cl^--containing electrolytes. We note that fluoride ions are not thought to form adsorbed layers[23] and so the picture of the interface we present may be more representative of a fluoride- than a chloride-containing solution.

The interactions between the other species present in solution were modeled with pair potentials; this too reflects a compromise in the realism of the calculations as it is known that polarization effects have a significant influence on the way that ions interact and coordinate water.[25] The water–ion interactions were modeled with a Lennard-Jones potential acting between the ion and the oxygen center of the water molecule. The parameters used in these interactions were adapted from Lynden-Bell *et al.*[26] The ruthenium ion–water interactions were parameterized with a purely repulsive potential:

$$U_{\mathrm{Ru,O}} = A/|\boldsymbol{r}_{\mathrm{Ru}} - \boldsymbol{r}_{\mathrm{O}}|^9 \tag{1}$$

with the parameter (A = 49 977.9 kJ mol^{-1} Å^9 for both Ru^{2+} and Ru^{3+}) chosen so that the first peaks of the ion–water radial distribution agreed with those obtained in an *ab initio* MD study.[27] The ions interact with each other with exponential-6 potentials with the RuCl potentials taken from lanthanides of corresponding ionic size.[28,29]

3 Pure water results

We begin by showing, in Fig. 4, results obtained for the profiles across the cell of the mean electrical (or "Poisson") potential, Ψ, in pure water. This is obtained by integrating Poisson's equation,

$$\nabla^2\Psi = -\frac{\rho}{\varepsilon_0}, \tag{2}$$

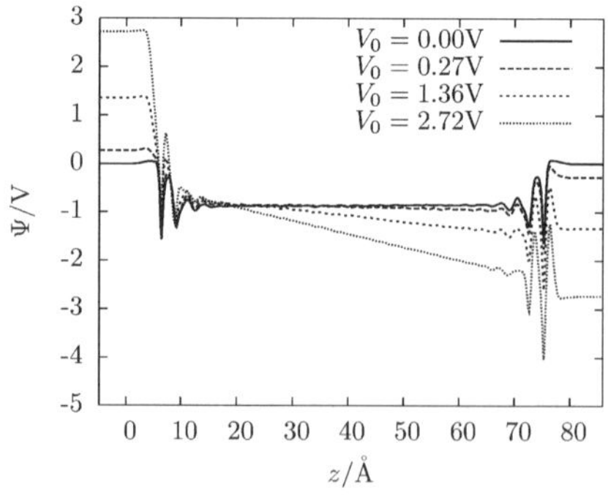

Fig. 4 The Poisson potential across the model electrochemical cell for pure water. The oscillations near the electrodes result from ordering of the solvent in the vicinity of the electrode.

 This journal is © The Royal Society of Chemistry 2009

with the mean charge density, ρ, calculated from the simulation for different values of the applied electrode potential, V_0, as the source term (ε_0 is the permittivity of free space). The Poisson potential is the potential used in describing the potential at the electrochemical interface in classical theories and is therefore an important point of contact between our calculations and textbook descriptions of the electrochemical interface.[4,30] The potential is constant on the interior of the electrodes and equal to the applied potential. It then drops rapidly and oscillates across an interfacial region about 12 Å wide, for reasons we will discuss in detail below, before settling down to acquire the constant slope appropriate to the behaviour of the potential in a bulk dielectric subject to an external potential. Notice that, other than at $V_0 = 0$, the potential drops across the two interfaces are not symmetrical because of the different microscopic arrangements of the water molecules at positively and negatively charged surfaces.

We can calculate a value for the dielectric constant of water from the behaviour of the potential across the bulk region. By integrating the mean charge in the region of the cell to the left of 20 Å we can obtain a value for the charge, Q, on one plate of a virtual parallel-plate capacitor placed at this position; the region to the right of 61.5 Å has an equal and opposite charge and can be regarded as the other plate. The potential drop between the two plates is $\Delta\Psi$ and we can obtain values for the capacitance C from $Q = C\Delta\Psi$ for the different applied potentials. This calculated capacitance C can be compared with the theoretical expression for capacitance of parallel plates filled with a medium of dielectric constant ε:

$$C = \varepsilon\varepsilon_0 A/d \tag{3}$$

where A is the cross-sectional area of the cell and d the distance between the plates. The calculation shows that the dielectric constant for SPC/E water depends on the applied electrode potential. Specifically, when the electrode potential has the values $V_0 = 0.27$, 1.36 and 2.72 V, the dielectric constant for the bulk water is $\varepsilon = 75.07$, 61.50 and 57.30 respectively. The low-voltage result is in reasonable agreement with direct simulation studies (68 ± 5.8),[31] which is good confirmation that the potential and the response of the water molecules to it are correct (see also ref. 16). The reduction in the apparent dielectric constant at higher voltages is consistent with a saturation effect,[32] note that the potential difference of $\Delta\Psi \simeq 1.1$ V, obtained with $V_0 = 2.72$ V, across our virtual capacitor of width 41.5 Å is equivalent to an electric field of $\simeq 2.6 \times 10^8$ V m^{-1}.

The rapid oscillation of the potential close to the interfaces is due to the strong adsorption of a layer of water molecules at the electrode surfaces. This is illustrated in Fig. 5. The oxygen atoms of the water molecules form a commensurate layer on the top sites of the underlying 100 fcc surface. With zero applied potential the water molecules lie with at least one O–H bond in the plane of the interface, with the H-atom pointing towards a neighbouring oxygen, and form ordered domains which reorient on a long time scale. Although the predominant orientation is in-plane, at zero applied potential there is a net orientation of the negative ends of the molecular dipoles towards the surface and this induces a small positive charge on the electrode atoms.[32]

When the potential is applied to the cell, the water molecules in the first layer partially reorient in the interfacial field. This is illustrated in the right-most panel of Fig. 5 where one of the OH bonds of the water molecules at the negatively charged electrode may point towards the electrode surface (note that this potential is very large for aqueous electrochemistry). The consequences of this for the Poisson potential can be seen by reference to the right-hand interface in Fig. 4. Whereas at $V_0 = 0$ the potential initially drops on moving into the electrolyte, consistent with the negatively charged oxide ions being closest to the surface, at the applied potential of $V_0 = -2.72$ V, the potential now rises showing an excess of positive charge lying close to the surface.

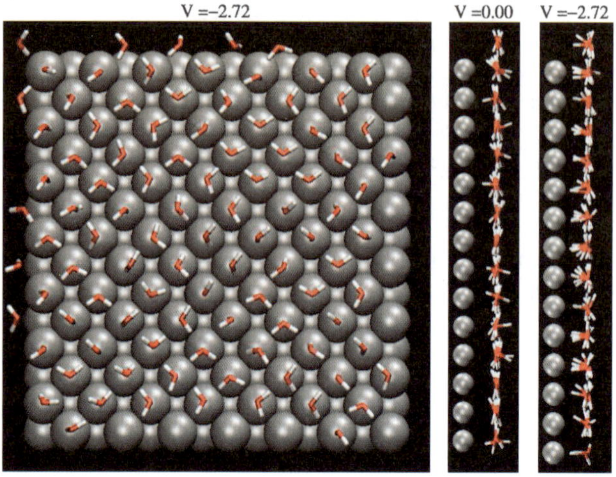

Fig. 5 A snapshot of the adsorbed water layer, and their profiles at different values of the applied potential V_0. The orientation of adsorbed water molecules is noticeably altered by the change in electrode potential.

We can track these changes by showing the probability distributions of the orientation of O–H bonds in the first and second layers of water molecules as the applied potential is changed. We compute the probability distribution function $P(\cos\,\theta)$, where θ is the angle between an OH bond vector (the vector extending from the oxygen centre of a water molecule to the centre of one of the associated hydrogen atoms) and the outward normal vector to the electrode surface. Fig. 6 shows the distribution $P(\cos\,\theta)$ for different values of applied potential for the two electrodes. The left-hand panel of Fig. 6 shows the results at the positively charged electrode and the right-hand panel at the negatively charged one. At all values of the potential, the probability distribution is peaked around $\cos\,\theta = 0$, corresponding to configurations for which the OH vector of a water molecule is aligned with the plane of the electrode surface. At zero applied potential the distribution $P(\cos\,\theta)$ for the adsorbed molecules is the same for the two electrodes, but even at $V_0 = 0$ there is an excess of outward (negative $\cos\,\theta$) over inward-pointing OH bonds, which give rise to the potential drop between the electrode and solution noted above. At different values

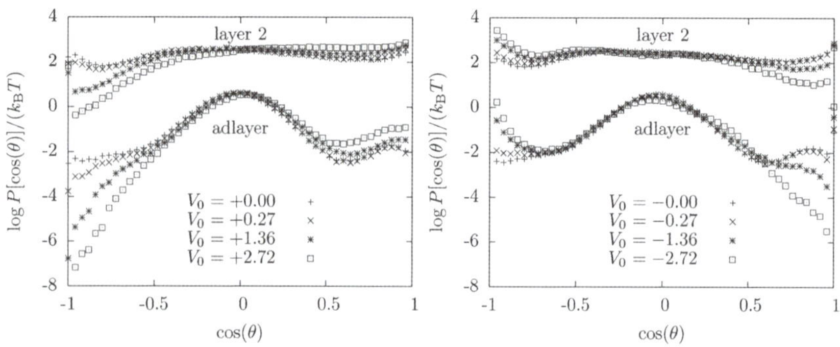

Fig. 6 The distribution of $\cos\,\theta$ where θ is the angle between a water's O–H bond vector and the vector normal to the electrode surface (pointing into the bulk). The left-hand panel is the distribution taken over the first two solvent layers adjacent to the positive electrode, and the right-hand one the distribution taken over the first two solvent layers adjacent to the negative electrode.

 This journal is © The Royal Society of Chemistry 2009

of applied potential the distribution of OH vectors is changed significantly. At the positive electrode (left panel of Fig. 6) the main effect of the electrode potential is to deplete the population of OH bonds pointing into the electrode ($\cos\theta < 0$). At the negative electrode (right panel of Fig. 6), however, at increased electrode potential there emerges a large population of OH vectors which point into the electrode. This change in orientational structure at the negative electrode is yet another demonstration of the asymmetry between the solvent structure at the positive and negative electrodes.

The orientation of the adsorbed water molecules influences the charge induced on the electrode atoms on which they sit. Fig. 7 shows the distributions of the charges induced on the atoms which make up the outermost layer of atoms on the negatively charged electrode. At zero applied potential the average charge is small and positive, for the reasons discussed above, but there is a significant number of negatively charged atoms which are associated with an adsorbed water molecule with an inward-pointing O–H bond. As the electrode potential becomes increasingly negative, so does the mean charge, but the bimodal character of the distribution becomes even more pronounced as more water molecules flip an O–H bond towards the electrode.

One question which arises from these results is the extent to which they are influenced by the inclusion of image-charge interactions in the potential model. A way of addressing this question is to carry out constant charge simulations in which the values of the charges on the electrode atoms are fixed at the values of the average charges on the first, second and third layers of the electrode atoms obtained in constant potential runs with an electrode potential V_0. These static charge distributions generate similar electrode potentials to the V_0 values used in the constant potential runs used to generate them. It can be seen, from the right-hand panel of Fig. 7, that the dynamical nature of the charges in the constant potential simulations has only a small effect on the mean orientational distributions of the water molecules in the adsorbed layer and none on the second layer. However, as illustrated in Fig. 7 left-hand panel, there is a local response of the electrode in the constant potential simulations and this is responsible for the emergence of a significant population of electrode-pointing OH bonds in the directly adsorbed molecules; it is not observed in the simulations run at constant charge. The relatively small effect of the image charges on *average* interfacial structure parallels the findings in the molten salt simulations.[16]

The layering of solvent and the molecular orientations within the layers adjacent to the electrode affect the capacitance of the electrode, which can be measured

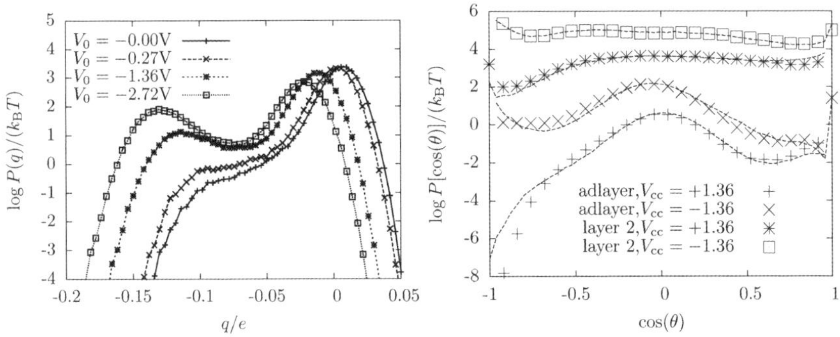

Fig. 7 The left-hand panel shows the distribution of the charges on the atoms which make up the innermost layer of the negative electrode. The changes in the orientations of the water molecules are accompanied by the development of a bimodal distribution of wall charges. The right-hand panel contrasts the O–H bond cos θ distributions obtained in constant potential (Fig. 6) and constant charge simulations; the former are shown by lines and the latter by symbols.

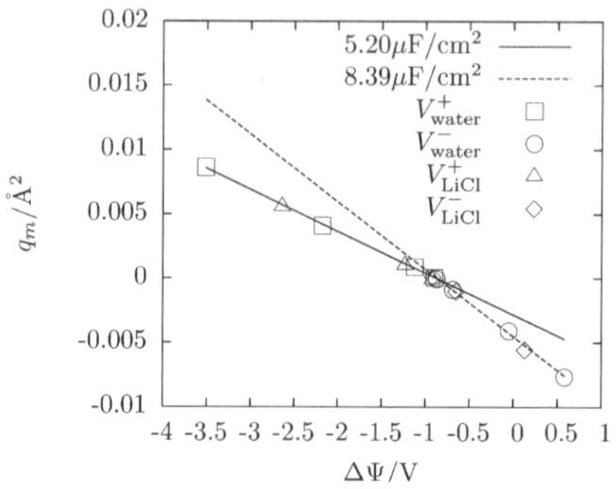

Fig. 8 The variation of the potential drop across the double layer, $\Delta\Psi$, with electrode charge density, q_m, for water and LiCl(aq) for the positive and negative electrodes. The lines, whose slopes represent the differential capacitance of the positive and negative double layers, are least-squares fits to the water data.

experimentally. The capacitance of the first two layers of solvent can be calculated through the differential capacitance, $C = \partial q_m/\partial\Delta\Psi$, where q_m is the charge density on the electrode and $\Delta\Psi$ is the potential drop across the first two layers of water. Fig. 8 shows the dependence of $\Delta\Psi$ on q_m for several values of the applied potential. The plot reveals that the potential of zero charge (pzc) for our simulated system is at −0.8 V. Comparing this quantity with experiment is not straightforward, since in the experimental measurements the potential is quoted with respect to a reference electrode whereas we can access directly the potential difference between the interior of the electrode and the solution. The experimental value with respect to a standard hydrogen electrode is +0.41 V.[4] The capacitance for electrode potentials which are on the positive side of the potential of zero charge ($C = 8.39$ μF cm^{-2}) is larger than for negative potentials, $C = 5.20$ μF cm^{-2}. Both results are considerably lower than the double-layer capacitance values deduced from experimental measurements which are in the range of $C = 20$–50 μF cm^{-2} [19,33] close to the pzc. The substantial difference between our calculated value and experiment is not surprising as our model does not include a realistic description of the electron density at the metallic surface; the surface dipole potential arises from the extension of the metal electrons into the interface beyond the nuclei and makes a large contribution to the capacitance.[33] This effect can be included in a jellium model for the metal, as included in the theory of Schmickler and Henderson.[34]

4 Results for electrolyte solutions

In Fig. 9 we show the Poisson potential for an approximately one molar solution of LiCl. In contrast to the pure water case (Fig. 4), in the bulk region away from the interfaces the Poisson potential is now constant as a consequence of the screening by the ions present in the solution. The oscillations in the potential across the interfacial region closely resemble those in pure water at the same values of the applied potential.

That the Poisson potential is exhibiting perfect screening is quite surprising as the profiles of the ion density obtained by averaging over the whole simulation runs are manifestly not well equilibrated (see Fig. 10). Even at $V_0 = 0$ we see an excess of ions at the left-hand side of the cell, whereas the equilibrium ion density profile should be constant, except close to the electrodes. It would appear that efficient screening can be caused by an appropriate *local* arrangement of cations and anions, relaxation of the whole ion density profile is not necessary. The failure to reach a fully equilibrated ion density profile arises primarily because of the slow rate of

 This journal is © The Royal Society of Chemistry 2009

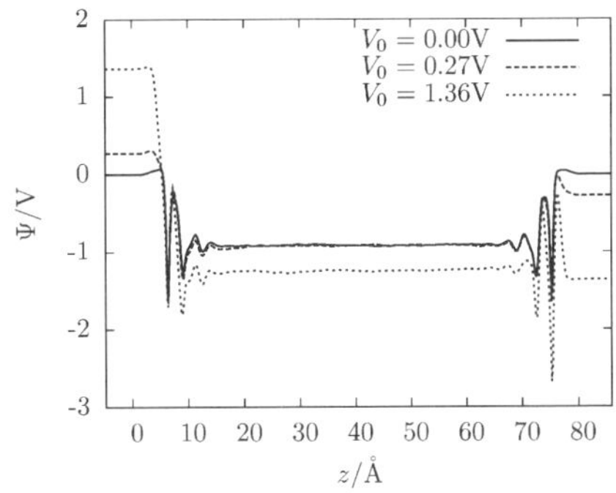

Fig. 9 The Poisson potential for a solution of LiCl(aq). In contrast to the Poisson profile for pure water (shown in Fig. 4), the LiCl solution screens the electrodes perfectly beyond roughly 15 Å from the electrode surface.

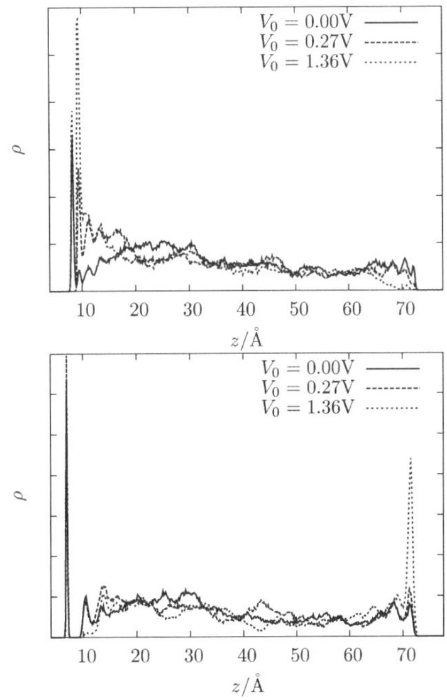

Fig. 10 The density profiles of the Li^+ and Cl^- ions across the simulation cells for various values of the applied potential. The results are the average of runs of 2.5 ns; despite the lengths of these runs we see that the average density profiles do not accord with those expected at full equilibrium. Nevertheless, the Poisson potential profiles calculated in these runs do show perfect screening (see Fig. 9).

relaxation of the concentration by diffusion (and, perhaps, an inappropriate initialization of the ion positions in the simulations). The rate should depend on the diffusion coefficient divided by the square of the distance between the electrodes, and because we have used a large cell in the hope of seeing the interfaces well separated by bulk, the relaxation times have become extremely long. There is a second slow relaxation process, however. Examination of the Li^+ profile close to the left-hand (positively charged) electrode shows a sharp peak in the region associated with the adsorbed layer of water. This peak arises from the presence of a single cation ion in this layer throughout the $V_0 = 0$ and 0.27 V runs; it was placed there in the initial configuration and remained until the electrode potential was increased to 1.36 V. That this process is so slow is because exchange of an ion between the strongly adsorbed layer of water and the bulk is very slow, effectively because the ion cannot carry its coordinating water molecules between the two regions.

We can examine the barrier which arises to prevent exchange between the adsorbed layer and the bulk by umbrella sampling techniques. We compute the mean force, $F(z)$, in the z direction perpendicular to the electrode on an atom i in the simulation by constraining the atom i at some position z_0 in a harmonic potential $U_{z_0}(z) = \frac{k}{2}(z - z_0)^2$, where k is the force constant of the harmonic well, taken to be 100 k_BT Å^{-2}. The mean force on atom i at position z_0 can be estimated as $F(z_0) = -k(\bar{z} - z_0)$ where $\bar{z}$ is the average value of z for species i constrained to $U_{z_0}(z)$ during a simulation. We performed the mean force calculations on a Li^+ ion and an oxygen centre of a water molecule in a LiCl(aq) solvent at zero applied potential ($V_0 = 0.00$). In addition we computed $F(z)$ for the oxygen center in a pure water with electrode potentials $V_0 = \pm 2.72$ V. This method for generating the mean force (and subsequently the potential of mean force (PMF) by integration) can be sensitive to the set of initial conditions.[35] One set of initial conditions, which corresponds to electrode desorption, was initiated by choosing an already adsorbed species setting z_0 at the adsorption distance and equilibrating with $U_{z_0}(z)$ for 1 ps. The next member of this set of initial conditions was created by setting $z_0 \rightarrow z_0 = +0.26$ Å and again equilibrating with $U_{z_0}(z)$ for 1 ps. This process is continued, in increments of 0.26 Å for approximately 5 Å. Another set of initial conditions, corresponding to electrode adsorption were generated in an analogous fashion by selecting an atom in the bulk and moving z_0 towards the electrode in 0.26 Å steps. The mean force was computed by averaging $\bar{z}$ over a 20 ps trajectory.

The potentials of mean force obtained by integration over $F(z)$ show a large degree of hysteresis (Fig. 11), which often arises when there is a free-energy barrier in the chosen coordinate (z) frustrating equilibration on short time scales. In other words, the reaction mechanism for electrode adsorption is not correctly characterized simply by a species distance from the electrode. For the atom to move into the adlayer it is necessary for an already adsorbed water molecule to vacate an

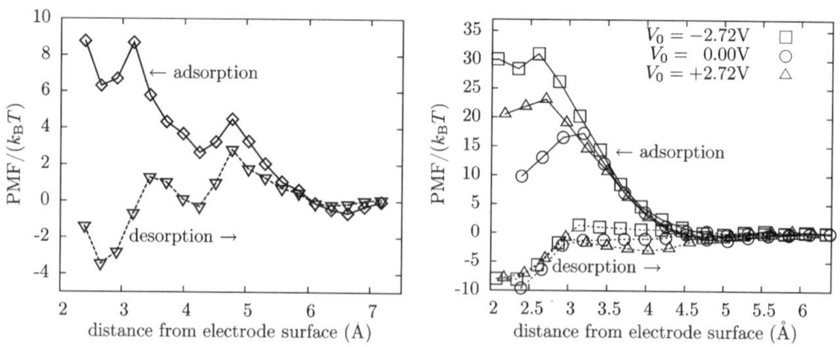

Fig. 11 The potentials of mean force obtained for an Li^+ ion (left) and for the oxygen atom (right) of a water molecule calculated with initial conditions appropriate to electrode desorption and electrode adsorption. The curves with dashed lines represent electrode desorption and the curves with solid lines represent electrode adsorption. The results show a large degree of hysteresis, consistent with fact that as a species is brought into the adlayer, a resident water molecule must be displaced.

 This journal is © The Royal Society of Chemistry 2009

adsorption site, thus we might expect that a more suitable reaction coordinate would describe the collective rearrangement which this entails. Nonetheless, the calculated potential of mean force (PMF) curves are informative.

If we focus firstly on the adsorption PMF for Li^+, we see that there is a substantial barrier at about 5 Å for the movement of the ion from the bulk into a relatively stable position where the cation sits between the first and second adlayers located about 4.4 Å from the electrode surface; this position is illustrated at the left-hand electrode in Fig. 2. This barrier arises from the reorganization of the solvation shell of the Li^+ ion which is necessary for it to be accommodated in this layer. Note that a similar barrier appears in the desorption pathway. There is then a second barrier before the ion is adsorbed at the electrode surface. In this region the hysteresis in the two curves is pronounced. On the adsorption pathway the ion must force an already adsorbed water molecule out of the way, so the energy increases steeply. On desorption from the adlayer there is also a large energy increase as the process leaves an empty adsorption site on the electrode surface.

The PMF for water shows no barrier for exchange of water molecules between the bulk and the second adlayer. A large degree of hysteresis then sets in, associated with the replacement of an already adsorbed water molecule by a molecule from the bulk. The barrier to desorption from the first adlayer suggested by these data is of the order of 10 k_BT, sufficient to lead to very slow exchange of the adsorbed water and the bulk.

5 Calculation of the Marcus curves for electron transfer

In order to examine how the electrical potential and the water structure in the interfacial region affect the rate constants for electrochemical charge transfer we have followed the scheme illustrated in Fig. 2 for the aqueous Ru^{2+}/Ru^{3+} couple close to the model metallic electrode. Similar calculations have been reported recently for a redox-active molten-salt system,[18] where the problems caused by the very slow equilibration of the concentration profiles we have noted above are not so marked and where the statistical precision necessary to validate the calculations was relatively easily obtained. We refer the reader to that paper[18] for full details of the calculation and merely recapitulate some essential details here.

We calculate the probability distribution functions for the vertical transition energy between the two redox states, $\delta E_{Ru^{2+}} \rightarrow Ru^{3+}$ for oxidation and $\delta E_{Ru^{3+}} \rightarrow Ru^{2+}$ for reduction. The vertical transition consists of changing the identity (*i.e.* charge and interaction potentials) of a single ion at some configuration along an MD trajectory with the electrode potentials set at some value V_0 and, without changing the atomic positions (as befits the vertical or diabatic nature of the Marcus curves), relaxing the electrode charges. The vertical transition energy is the difference in the total interaction energy between the initial and final states. There is a constant term in this energy gap that depends upon the metal of which the electrode is made (through its work function) and the reacting ion (through the gas-phase ionization energy: $Ru^{2+} \rightarrow Ru^{3+} + e^-$), but independent of the electrolyte solution. We have arbitrarily set the value of this constant to make the mean energy gap of Ru^{2+} to be the negative of that for Ru^{3+} when $V_0 = 0$; the consequences of this will be illustrated below.

In both oxidation and reduction, as discussed in detail in ref. 18, a balancing charge is transferred to the electrodes and the energy of this, which depends on the potential applied to the electrodes, is included in the vertical transition energy.[36] We then return the identity of the ion to its initial value and continue the MD trajectory. By repeatedly sampling these transition processes for redox species found at a given distance from the electrodes we may build up probability distributions for the energy gaps, $P_{Ru^{2+}}(\delta E_{Ru^{2+}} \rightarrow Ru^{3+})$ and $P_{Ru^{3+}}(\delta E_{Ru^{3+}} \rightarrow Ru^{2+})$, at different positions in the cell. Examples of the probability distributions are shown in Fig. 12. They are calculated for a sample of ions located in the middle of the simulation cell, and compared to those of ions adjacent to one of the electrodes.

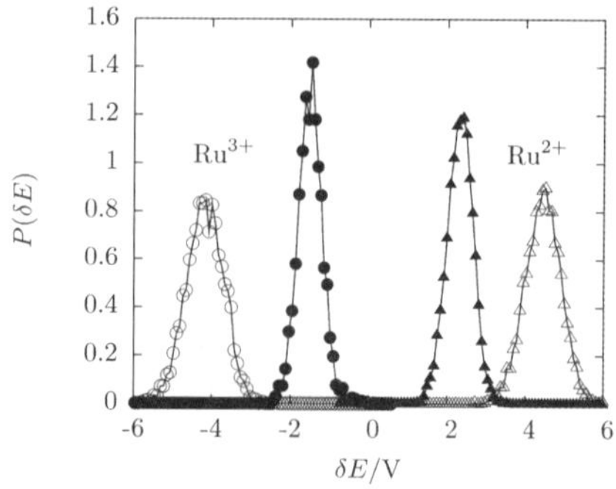

Fig. 12 The probability distributions $P_{\mathrm{Ru}^{2+}}(\delta E_{\mathrm{Ru}^{2+}} \rightarrow \mathrm{Ru}^{3+})$ (triangles) and $P_{\mathrm{Ru}^{3+}}(\delta E_{\mathrm{Ru}^{3+}} \rightarrow \mathrm{Ru}^{2+})$ (circles) at different positions in the cell. The distributions plotted with open symbols are taken in the centre of the cell, *i.e.* $37 \leq z \leq 39$ Å, and the distributions plotted with filled symbols are taken adjacent to the right-hand electrode, *i.e.* $71 \leq z \leq 75$ Å.

The mean positions of the distributions and their widths are found to depend quite strongly on the distance of the redox ion from the electrodes, as we will discuss below. The distributions are found to be rather accurately Gaussian, which is the expectation from Marcus theory if the surrounding medium responds linearly to the change in the identity of the redox species. Our potentials describing the interaction of the Ru^{2+} and Ru^{3+} with water were chosen so that both cations had similar coordination shells and, as previous studies of redox processes in the bulk have shown,[37,8,27] under these conditions it is likely that the linear response limit is applicable. Our data seems to be consistent with a linear response even when we consider the redox process for ions close to the electrode surface, despite the strength of the interactions and the restricted nature of the water molecules in the first adsorbed layer.

Following Sprik and co-workers,[7,8] and making use of the special properties of the mean vertical gap for oxidation ($\Delta E = \delta E_{\mathrm{Ru}^{2+}} \rightarrow \mathrm{Ru}^{3+} = -\delta E_{\mathrm{Ru}^{3+}} \rightarrow \mathrm{Ru}^{2+}$) as a reaction coordinate,[6,9] we may evaluate the free energies of the Ru^{2+} and Ru^{3+} ions along this reaction coordinate from the probability distributions

$$A_{\mathrm{Ru}^{2+}}(\Delta E) = -k_{\mathrm{B}}T \ln P_{\mathrm{Ru}^{2+}}(\Delta E = \delta E_{\mathrm{Ru}^{2+} \rightarrow \mathrm{Ru}^{3+}}) + \bar{A}_{\mathrm{Ru}^{2+}} \tag{4}$$

and

$$A_{\mathrm{Ru}^{3+}}(\Delta E) = -k_{\mathrm{B}}T \ln P_{\mathrm{Ru}^{3+}}(\Delta E = -\delta E_{\mathrm{Ru}^{3+} \rightarrow \mathrm{Ru}^{2+}}) + \bar{A}_{\mathrm{Ru}^{3+}}, \tag{5}$$

where $\bar{A}_{\mathrm{Ru}^{2+}}$ corresponds to the free-energy at the minimum of the Ru^{2+} curve. Furthermore, when the vertical energy gap is taken as the reaction coordinate the two free-energy curves are linearly dependent, *i.e.*:[8,38]

$$A_{\mathrm{Ru}^{3+}}(\Delta E) - A_{\mathrm{Ru}^{2+}}(\Delta E) = \Delta E \tag{6}$$

.

This apparently simple relationship is remarkably powerful; it means that we can establish a relationship between the origins of the two curves ($\bar{A}_{\mathrm{Ru}^{3+}}$ and $\bar{A}_{\mathrm{Ru}^{2+}}$) and also sample the free-energy surfaces for values of the reaction coordinate which are well away from the most stable configurations simply by calculating the energy gap in the free-running simulation. The ability to sample the curves away from their minima means that we can obtain information on the Marcus curves in the vicinity of their crossing point, which is the region which determines the kinetics of the electron transfer event.

The data points obtained from (4)–(6) are plotted in Fig. 13 for $V_0 = 0$. Note that our choice of the arbitrary energy added to the gap to represent the work function

 This journal is © The Royal Society of Chemistry 2009

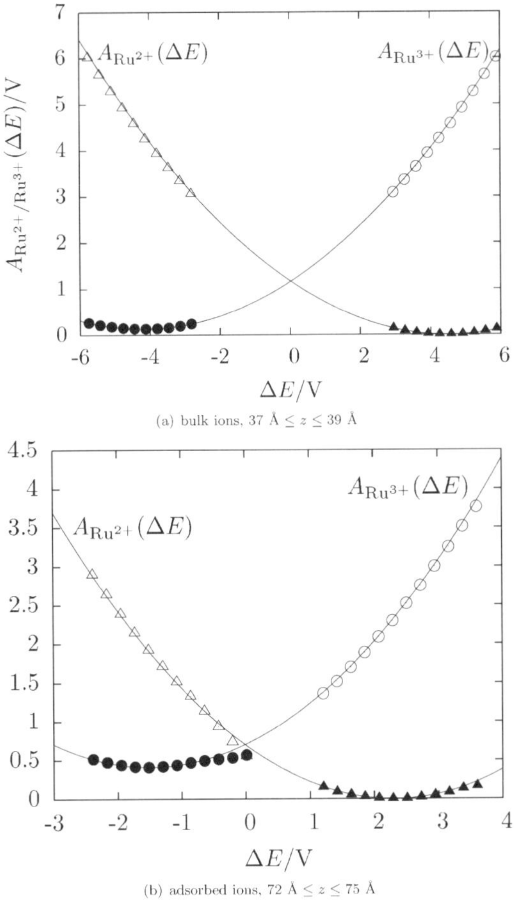

Fig. 13 The free energies of Ru^{2+} (triangles), and Ru^{3+} (circles) along the reaction coordinate ΔE. The filled symbols are the distributions as computed from simulation data and the hollow symbols are constructed using the relationship in eqn (6). Solid lines are the linear response (Marcus) prediction, computed from the mean values of the energy gaps for oxidation and reduction, $\langle \Delta E_1 \rangle$ and $\langle \Delta E_2 \rangle$, along with eqn (7)–(10). Panel (a) is a representative distribution representative of bulk ions and panel (b) is the distribution of ions adsorbed at the right-hand electrode.

and ionization energy has resulted in only a small difference between the mean free energies of the oxidized and reduced forms for the mid-cell position: experimentally, the reduction potential for this couple is 0.249 V with respect to the standard hydrogen electrode, so the relative positions of the minima in the curves should be similar to reality and the electron transfer in the "normal" Marcus regime at $V_0 = 0$.

If the probability distributions really are Gaussian, eqn (4) and (5) show that the Marcus free-energy curves will be harmonic about the mean values of the reaction coordinate for the oxidation and reduction processes, *i.e.* the peak positions of the respective probability distributions which we denote $\langle \Delta E_1 \rangle$ and $\langle \Delta E_2 \rangle$ respectively. It was shown by Tachiya[38] that under this Gaussian assumption all properties of the Marcus curves can be *predicted* simply from a knowledge of $\langle \Delta E_1 \rangle$ and $\langle \Delta E_2 \rangle$; the necessary manipulations are described in the previous paper.[8,18] These predicted curves are shown by solid lines in Fig. 13 and are seen to provide an accurate representation of the data. This applies both for the data obtained for redox ions close to the centre of the cell and also close to the electrode surfaces, despite the fact that the values of $\langle \Delta E_1 \rangle$ and $\langle \Delta E_2 \rangle$ themselves depend quite strongly upon the distance from the electrode. The position-dependence of the widths of the probability distributions which we noted in discussing Fig. 12 is therefore seen to be contained within the Gaussian description of the fluctuations in the reaction coordinate and related to the position-dependence of $\langle \Delta E_1 \rangle$ and $\langle \Delta E_2 \rangle$.

The parameters which are normally used to describe the shapes of the Marcus curves are ΔA and the reorganization energy λ (see Fig. 1). In the Gaussian/linear response regime, both may be written in terms of the mean energy gaps:[8,38]

$$\Delta A = \frac{1}{2}(\langle \Delta E_1 \rangle + \langle \Delta E_2 \rangle) \tag{7}$$

and

$$\lambda = \frac{1}{2}(\langle \Delta E_1 \rangle - \langle \Delta E_2 \rangle), \tag{8}$$

with $\lambda = \lambda'$ in Fig. 1. In this regime the activation free energy for electron transfer is given by the famous expression:[5]

$$\Delta A^{\ddagger} = \frac{(\Delta A + \lambda)^2}{4\lambda}. \tag{9}$$

The dependence of ΔA and λ on the position of the redox ion in the cell and on the applied potential is illustrated in Fig. 14. The behaviour of these parameters parallels that seen in the molten-salt simulations[18] and we refer to that paper to fully vindicate the interpretations of the data which we offer below.

The reorganization energy λ is seen to be virtually independent of the applied potential, but strongly dependent on the position of the ion in the cell. The latter is associated with the way in which the polarization of the electrodes (image-charge effect) contributes to the vertical energy gap. In passing from the initial state, say Ru^{2+}, to the final state Ru^{3+} we create a unit positive charge at the location of the redox ion. When we allow the relaxation of the electrode charges to re-establish the constant potential condition we not only allow the transfer of one unit of negative charge to the electrodes, we also allow the electrode to be polarized by the newly-created positive charge. The interaction between the newly created image charge and the change in the charge of the redox ion *is not screened* because the

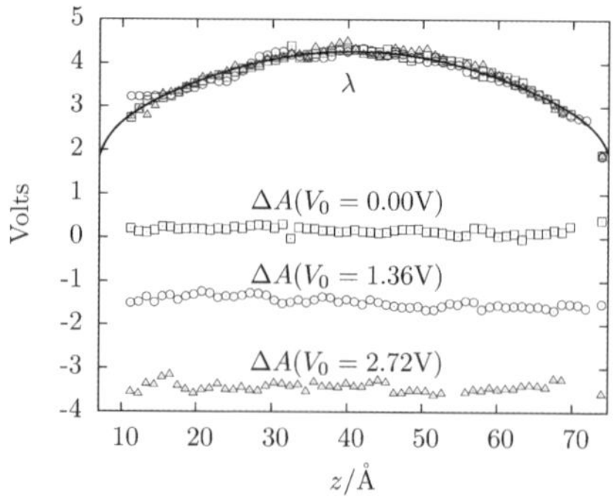

Fig. 14 Dependence of the reaction free energy ΔA (lower three curves) and the reorganization energy λ (upper three curves) on cell position at different values of electrode potential V_0. The solid lines show the variation of the energy (shifted by an additive constant) of a test charge as it is moved across an otherwise empty cell, which reflects the energy associated with the image-charge effects in our simulations.

positions of the electrolyte atoms do not relax after the excitation event in a diabatic description of the charge-transfer process. Since the image effect contributes to $\langle \Delta E_1 \rangle$ and $\langle \Delta E_2 \rangle$ with equal magnitude but opposite sign, it does not affect the value of ΔA, which is seen to be z independent. The reorganization energy is, however, strongly affected by the image effect. Marcus[39] obtained an expression for the reorganization energy appropriate to an ion in a dielectric fluid at a distance d from a single metallic surface:,

$$\lambda_{\text{Marcus}}(d) = \delta q^2 \left[\frac{1}{\varepsilon_\infty} - \frac{1}{\varepsilon_\text{s}} \right] \left(\frac{1}{2a} - \frac{1}{2d} \right), \tag{10}$$

where δq is the charge difference between the reduced and oxidized species. When d is large, this expression gives the reorganization energy for a redox process in the bulk fluid: it contains the contribution the non-electronic part of the dielectric response of the fluid (*i.e.* that caused by reorganization of the nuclear positions) to the change in the charge of a redox species with radius a; as such it involves (in the first bracket) the difference between the static ($\varepsilon_\text{s}^{-1}$) and infinite frequency (ε_∞^{-1}) longitudinal dielectric susceptibilities. We cannot compare directly with this expression because our sample geometry has *two* metallic surfaces and is periodic in the transverse direction. However, we can compare directly with the position-dependent energy of a charge introduced into an empty simulation cell; this is the effective image interaction energy in our periodic system[18] when the newly created charge is in a vacuum. Away from the interfaces, any difference between this quantity and the reorganization energy should reflect the effective dielectric screening function of our simulated electrolyte [*i.e.* the factor analogous to the square-bracketed term in eqn (10)]. In fact, we see that the two curves coincide well showing that the factor is indistinguishable from 1. In our simulated system the water molecules and ions are not polarizable, so ε_∞ is just unity and since for SPC/E water ε_s is about 70 we can only conclude that our data is consistent with the Marcus expression. Close to the electrodes, the reorganization energy does appear to depart from the modified Marcus expression, and this could be associated with the effect of the proximity of the electrode on the solvation characteristics of the water molecules there. However, the statistics in this domain are not good, as the ruthenium ions are even more reluctant to reorganize their solvation shells and approach the electrode than were the Li^+ ones. Better sampling methods for the vertical energy gaps are required before firm conclusions may be drawn.

In the central part of the simulation cell, the reaction free energy ΔA varies linearly with the potential applied to the electrode to which the electron is transferred. This reflects the change in the energy of the electron which is transferred to the electrode which, as we have emphasized, contributes to the free energy of the oxidized state. As the redox species approaches the electrode surface, the reaction free energy seems to be remarkably constant. It might have been expected to show the kind of fluctuating behaviour evident in the Poisson potential, since conventional electrostatic considerations would suggest that this potential should influence the relative energies of the doubly and triply charged ions. However, as discussed in the molten-salt context,[18] this is *not* the potential which should be used to discuss the changes in the energy levels of an ion. Rather, we should be considering the potential at the ion's centre due only to *the other* charges present in the system: this might be better called a Madelung potential. The difference between the two potentials is surprisingly large, as illustrated in Fig. 15 where we show the z dependence of the mean Madelung potentials experienced by the Ru^{2+} and Ru^{3+} ions compared with the Poisson potential.

Not only does the Madelung potential depend on the identity of the species on which the potential is evaluated, it is seen to be constant across the simulation cell

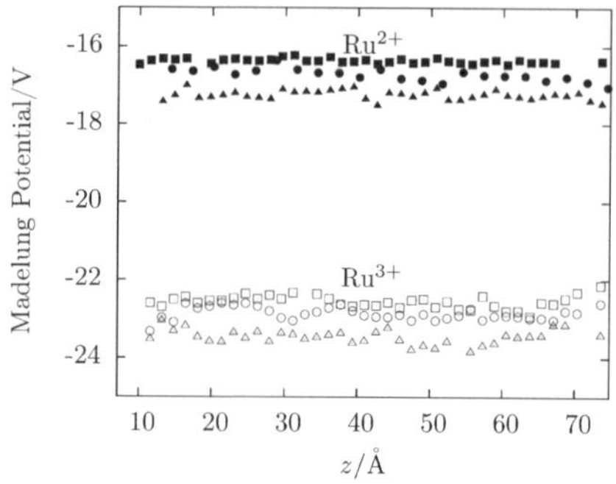

Fig. 15 Variation of the Madelung potential across the simulation cell for the Ru^{2+} (filled symbols) and Ru^{3+} (hollow symbols) ions at different values of the electrode potential, $V_0 =$ 0.00 V (squares), $V_0 = 1.36$ V (circles), and $V_0 = 2.72$ V (triangles). The Madelung potential, which is the potential felt by a species, does not exhibit the kinds of oscillations seen in the Poisson potential (Fig. 9) in the vicinity of the electrode.

except in the immediate vicinity of the electrode surfaces—its behaviour illustrates perfect screening much closer to the electrode surface. The z independence of the Madelung potential therefore provides a much better explanation of the insensitivity of ΔA to z than does the Poisson potential.

6 Summary and conclusion

The methods described have allowed a full, self-consistent calculation of the liquid structure and electrical potentials for an aqueous ionic solution close to a model metallic wall maintained at a constant electrical potential. The simulation is a direct realization of a model electrochemical interface, as it appears in textbooks. Using a realistic potential for water–platinum interactions, we find a strongly absorbed layer of water molecules on the electrode with the molecules oriented in the plane of the interface at zero potential, in common with earlier studies.[32] Despite the strength of the absorption, the water molecules do reorient as the electrode potential is changed and this affects the behaviour of the electrical potential across the interface and the differential capacitance of the electrode. The absorption of cations at the electrode is strongly inhibited by the requirement for them to reorganize their hydration shells to approach the electrode surface.

We have begun to characterize how the interfacial water affects the rate constant for electrochemical charge transfer by directly calculating the Marcus free-energy curves for the oxidized and reduced species at different positions in the cell with a particular choice of reaction coordinate. The fluctuations in the solvation structure which influence these curves were shown to be accurately Gaussian for the modeled Ru^{2+}/Ru^{3+} couple, consistent with linear response of the solvent to the charge state of the redox ion. The reorganization energy was strongly dependent on the distance of the redox species from the electrode surface and independent of the electrode potential. The effect was traced to image-charge interactions with the metal surface. With the statistics available to us at present, we could not detect an effect of the altered dynamical characteristics of the absorbed water on the solvation fluctuations when the redox species was close to the electrode surface. The reaction free energy ΔA measures the difference in the free energies of the oxidized and reduced states with the redox ion at a given distance from the electrode surface. Contrary to textbook expectations, its position dependence does not resemble the mean electrical potential. It only deviates from the bulk value in the immediate vicinity of the interface where the competition between solvating the electrode and solvating the redox species becomes a factor.

 This journal is © The Royal Society of Chemistry 2009

Acknowledgements

This work was supported by EPSRC, through Grant GR/T23268/01, as well as by the Director, Office of Science, Office of Basic Energy Sciences, Chemical Sciences, Geosciences, and Biosciences Division, U. S. Department of Energy under Contract No. DE-AC02-05CH11231.

References

1 C. D. Taylor, S. A. Wasileski, J. S. Filhol and M. Neurock, *Phys. Rev. B*, 2006, **73**, 165402.
2 C. D. Taylor, S. A. Wasileski and M. Neurock, *Curr. Opin. Solid State Mater. Sci.*, 2005, **9**, 49.
3 E. Spohr, *Electrochim. Acta*, 2003, **49**, 23.
4 J. O'M. Bockris, A. K. N. Reddy and M. Gamboa-Aldeco, *Modern Electrochemistry 2A*, Kluwer, New York, 2000.
5 R. A. Marcus, *J. Chem. Phys.*, 1956, **24**, 966; R. A. Marcus, *Rev. Mod. Phys.*, 1993, **65**, 599.
6 R. A. Marcus, *Discuss. Faraday Soc.*, 1960, **29**, 21.
7 J. Blumberger, I. Tavernelli, M. L. Klein and M. Sprik, *J. Chem. Phys.*, 2006, **124**; J. Blumberger, Y. Tateyama and M. Sprik, *Comput. Phys. Commun.*, 2005, **169**, 256.
8 J. Blumberger and M. Sprik, Redox free-energies from vertical energy gaps: ab initio MD implementation, in *Computer Simulations in Condensed Matter: from Materials to Chemical Biology, vol. 2*, ed. M. Ferrario, G. Ciccotti and K. Binder, Lecture Notes in Physics vol. 704, Springer, Berlin/Heidelberg, 2006, p. 481.
9 A. Warshel, *J. Phys. Chem.*, 1982, **86**, 2218.
10 R. A. Kuharski, J. S. Bader, D. Chandler, M. Sprik, M. L. Klein and R. W. Impey, *J. Chem. Phys.*, 1988, **89**, 3248–3257.
11 J. S. Bader and D. Chandler, *Chem. Phys. Lett.*, 1989, **157**, 501–504.
12 P. L. Geissler and D. Chandler, *J. Chem. Phys.*, 2000, **113**, 8154–8160.
13 D. L. Price and J. W. Halley, *J. Chem. Phys.*, 1995, **102**, 6603.
14 S. Izvekov, A. Mazzolo, K. VanOpdorp and G. A. Voth, *J. Chem. Phys.*, 2001, **114**, 3248.
15 S. Walbran and J. W. Halley in *Solid-Liquid Interface Theory—ACS Symposium 789*, ed. J. W. Halley, American Chemical Society, Washington D.C., 2001.
16 S. K. Reed, O. J. Lanning and P. A. Madden, *J. Chem. Phys.*, 2007, **126**, 084704.
17 J. I. Siepmann and M. Sprik, *J. Chem. Phys.*, 1995, **102**, 511.
18 S. K. Reed, P. A. Madden and A. Papadopoulos, *J. Chem. Phys.*, 2008, **128**, 124701.
19 D. M. Kolb, *Surf. Sci.*, 2002, **500**, 722.
20 H. J. C. Berendsen, J. R. Grigera and T. p. Straatsma, *J. Phys. Chem.*, 1987, **91**.
21 P. A. Thiel and T. E. Madey, *Surf. Sci. Rep.*, 1987, **7**, 385.
22 S. Holloway and K. H. Bennemann, *Surf. Sci.*, 1980, **101**, 327.
23 O. M. Magnussen, *Chem. Rev.*, 2002, **102**, 679.
24 C. G. Guymon, R. L. Rowley, J. N. Harb and D. R. Wheeler, *Condens. Matter Phys.*, 2005, **8**, 335.
25 D. Spangberg and K. Hermansson, *J. Chem. Phys.*, 2004, **120**, 4829.
26 S. Koneshan, J. C. Rasaiah, R. M. Lynden-Bell and S. H. Lee, *J. Phys. Chem. B*, 1998, **102**.
27 J. Blumberger and M. Sprik, *J. Phys. Chem. B*, 2005, **109**, 6793.
28 B. Morgan and P. A. Madden, *J. Chem. Phys.*, 2004, **120**, 1402.
29 F. Hutchinson, M. Wilson and P. A. Madden, *Mol. Phys.*, 2001, **99**, 811.
30 A. A. Kornyshev, *J. Phys. Chem. B*, 2007, **111**, 5545.
31 P. Höchtl, S. Boresch, W. Bitomsky and O. Steinhauser, *J. Chem. Phys.*, 1998, **109**, 4927.
32 I. C. Yeh and M. L. Berkowitz, *J. Electroanal. Chem.*, 1998, **450**, 313.
33 R. Parsons, *Solid State Ionics*, 1997, **94**, 91.
34 W. Schmickler and D. Henderson, *J. Chem. Phys.*, 1986, **85**, 1650.
35 M. Sprik and G. Ciccotti, *J. Chem. Phys.*, 1998, **109**, 7737.
36 As discussed at length in ref. 18, because of the adiabatic nature of our calculations with respect to the electrode charges, a fraction of the charge is transferred to each electrode, whereas in the normal description of electrochemical charge transfer we envisage a whole electron being transferred to a designated electrode. The value of the fraction depends systematically on the position of the ion in the cell and we may therefore correct our vertical transition energies to correspond to electron transfer to the anode, as described in detail.[18] We therefore quote our vertical gap energies as if the electron transfer event is occurring at the anode.
37 D. A. Rose and I. Benjamin, *J. Chem. Phys.*, 1994, **100**, 3545.
38 M. Tachiya, *J. Phys. Chem.*, 1989, **93**, 7050; M. Tachiya, *J. Phys. Chem.*, 2003, **97**, 5911.
39 R. A. Marcus, *J. Chem. Phys.*, 1965, **43**, 679; R. A. Marcus, *J. Chem. Phys.*, 1990, **94**, 1050.

General discussion

Dr Bakó opened the discussion of the paper by Dr Ricci: I enjoyed very much your lecture where you compared the apple with the peach. Both of them are fruits. If we look at Fig. 2 in your paper we can see that, in the case of bulk water (O···O prdf), we got a well-defined (very tetrahedral) structure. In the case of confined geometry you got a structure in which the tetrahedral order is totally disappear. This prdf resembles a prdf (it seems to me) that you can get if the system has a lot of OH^- anions. Can you see any overlapping effect between the OH intra- and inter-molecular peaks, which would prove your conclusion about the significant change on the potential energy curve? How large is the error caused by the excluded volume effect correction? Is it in the same size range as the error in the supercooled water case (*vs.* space and peak height)?

Dr Ricci replied: Yes, I can see an overlap of the OH intramolecular peak with the intermolecular H-bond peak, in the range 1.1–1.4 Å (data not shown in this paper). This overlap is, however, meaningless in the framework of your question.

A far as the distortion of the radial distribution functions is concerned, it is well known that a reduction of tetrahedral order (never a total breakdown) can also be achieved by solvation of ions in water. Here the silica surface and the silanols which decorate the pore surface play a similar role; nevertheless, we cannot decouple the interatomic interactions and the topological effects. A systematic study on different substrates, by changing the geometrical dimensions could perhaps be helpful. The corrections for the excluded volume effects are implemented within the EPSR routine, and the radial distribution functions are calculated with respect to the distribution of a uniform fluid confined within the same volume as the sample (see ref. 11 in our paper). As a consequence this correction does not contribute to increasing the uncertainty of the radial distribution functions of the confined fluid with respect to the bulk case, where the normalization is performed with respect to a uniform fluid within a cubic box.

Dr Bakó opened the discussion of the paper by Professor Debenedetti: When you have "failure" in the film and obtain voids, do you have any vapour inside or are they purely "vacuum voids"? You can get "vapour" inside not only by thermal motion but also by fragmentation. Additionally, do you know the size distribution and lifetime of these voids? So, to make it short: are they tiny bubbles with a well-developed vapour/liquid interface or just vacuum voids?

Professor Debenedetti replied: We do not see any molecules in the voids. It should be emphasized, though, that the calculations reported in the paper are energy minimizations, hence they provide no information on dynamics. We are currently performing calculations in which the glass is formed by quenching, rather than by energy minimization. One of the things we are interested in investigating is precisely the dynamics of void formation and the time-dependent mechanisms of failure. Because of the low temperatures involved, and small size of the voids, I would be surprised to find vapor molecules within the voids. We have calculated the spatial distribution of the probability of finding voids, but we have not computed the size distribution of the voids.

Professor Rovere commented: Firstly, in your Fig. 1 in the case of hydrophobic walls and thin film there is in the transverse P a minimum at high density, why there is not any signature of the instability in the corresponding $P^{\parallel}$?

Secondly, are you planning to calculate the Kauzmann temperatures for your systems ?

Professor Debenedetti answered: In reply to your first question, the high-density minimum in the transverse pressure as a function of density for the thin film case is the result of "melting" the highly-ordered structure that forms near the walls. The formation of such structures has a dramatic effect on the transverse stresses, but not on the normal stress. Since the long-range order is lateral, it makes sense intuitively that the layer close to the wall is able to support a large transverse stress, while being no different mechanically than a disordered (glass-like) layer with respect to normal stresses.

As for your second question, yes, how confinement affects the Kauzmann locus is a very interesting question, and we intend to pursue it.

Professor Havenith opened the discussion of the paper by Dr Coudert: Firstly, another group of systems which you want to study with the same methodology is MOFs (metal–organic frameworks). We have studied interfacial water in MOFs in the THz region where we found a coupling between phonon and water modes. Have you also calculated the spectrum of water nanodroplets in zeolites below 200 cm^{-1}. Do you find a coupling in this range or can you decompose it like in the IR?

Also, would you agree that the THz region is more sensitive to coupling due to the collective nature of these modes in contrast to the local modes in the IR?

Dr Coudert responded: Firstly, we do indeed find a coupling between water and zeolite dynamics for frequencies lower than 300 cm^{-1}, which manifests itself in the IR spectra because the total spectrum of the system is not the sum of the water and zeolite "partial" spectra. However, obtaining well-converged IR spectra for the low-frequency region requires simulation runs far longer than those we performed here, and we can thus make no further analysis or comment on this coupling.

I'd like to point out, however, that a very good tool for analysis of the dynamical coupling of such subsystems was recently published.[1] This method allows us to analyse the vibrational motions in a molecular dynamics trajectories in terms of effective (or "average") normal modes, where contributions of different atoms and subsystems are easily quantified.

Secondly, I do agree that terahertz is a nice probe for the collective modes of water and the coupling between water and zeolite, and a complement to IR spectra. A downside from the simulation point of view, as mentioned above, is that calculation of low-frequency modes is expensive, independently of the exact experimental properties we try to compute.

1 M. Martinez, M.-P. Gaigeot, D. Borgis and R. Vuilleumier, *J. Chem. Phys.*, 2006, **125**, 144106.

Professor Luzar remarked: Fig. 13 and 14 in your paper are showing confinement-induced dipole moments only for 20 waters. Would you agree that near an extended surface the effect would be much smaller?

Dr Coudert responded: The conclusion from the Figures is that the confinement effect on the dipole moments of liquid water is already quite small: for 15 water molecules in an LTA supercage, the average dipole moment is 2.7 D, and is already quite close to the bulk value for 20 molecules (2.9 D *vs.* 3.1 D), even though all molecules are very close to the internal surface of the zeolite. In the case of an extended surface, I would expect an even smaller effect on dipole moment of water molecules.

 This journal is © The Royal Society of Chemistry 2009

Professor Klein remarked: Were you able to calculate the viscosity or the diffusion coefficient of the water confined in your pores? Other studies (*e.g.* Raviv *et al.*[1,2]) have shown that water confined between two hydrophilic or hydrophobic walls retains a viscosity within a factor of 5 or so of its bulk viscosity value, down to confinements of *ca.* a monolayer or so, *i.e.* 3 Å.

1 U. Raviv, P. Laurat and J. Klein, *Nature*, 2001, **413**, 51.
2 U. Raviv, S. Giasson, J. Frey and J. Klein, *J. Phys.: Condens. Matter*, 2002, **14**, 9275.

Dr Coudert replied: In the present paper, we do not present long time dynamic properties of the confined water, such as viscosity or diffusion, but these are of course topics of interest to us, in order to obtain as complete as possible a picture of confined water. We have previously studied the effect of confinement on diffusion of water and solutes such as the hydrated electron.[1] We know that although confinement at such small scales slows down the dynamics of water, it keeps bulk-like properties; it would be interesting to see what can be said from our simulations about viscosity. On a related note, we're currently studying the effect on confinement of the orientational dynamics of water to check that the mechanism recently put forward for water reorientation in bulk water[2] and in the solvation shell of anions.[3]

1 F.-X. Coudert and A. Boutin, *Chem. Phys. Lett.*, 2006, **428**, 68.
2 D. Laage and J. T. Hynes, *Science*, 2006, **311**, 832.
3 D. Laage and J. T. Hynes, *Proc. Natl. Acad. Sci. U. S. A.*, 2007, **104**, 11167.

Professor Luzar continued the discussion of Dr Ricci's paper: Are the "similarities" between confined and supercooled water you describe in your paper related to findings of Teixeira *et al.*?[1,2] Those neutron scattering experiments show that dynamic properties of confined water are identical to those of bulk water at a temperature typically 30 °C below the real temperature. Specifically, water inside vycor pores at room temperature behaves like bulk water supercooled at −10 °C.

1 J. Teixeira, J. M. Zanotti, M.-C. Bellissent-Funel and S. H. Chen, *Phys. B*, 1997, **370**, 234.
2 For a review see, for example, J. Teixeira and A. Luzar, in *Hydration Processes in Biology: Theoretical and Experimental Approaches*, ed. M.-C. Bellissent-Funel, IOS Press, The Netherlands, 1999.

Dr Ricci answered: In the Teixeira *et al.* paper a quantitative relation between water residence time in bulk and confinement was established, based on neutron quasi-elastic scattering measurements as a function of temperature.

What I am trying to evidence in my paper is more on the qualitative side, also because the confining matrices used in neutron diffraction and DINS are not the same. As a matter of fact similarities in the microscopic structure and in the momentum distribution are limited to the shift of the first neighbor oxygen distance (all other structural features were indeed different) and to the evidence of proton coherence in both supercooled bulk and confined water. Moreover, the paper by Teixeira *et al.* was dealing with diffusive motions of the water molecule as a whole, while DINS technique is mostly sensible to the stretching mode, that is to the motion of a proton along the H bond. This is also where quantum effects may be better visible. In conclusion the take-home message could be that quantum coherence shows up when water oxygens come closer; this is quite different and new compared to the message from Teixeira *et al.*

Professor Jungwirth continued the discussion of the paper by Professor Debenedetti: Is the ice-like structure observed on the hydrophobic surface a general phenomenon or is it rather due to matching lattice constants for the particular system presented?

Professor Debenedetti replied: In simulations, ice-like structures have been found on a variety of hydrophobic confining surfaces, for example structureless hydrophobic slit pores;[1–3] carbon nanotubes;[4] atomically-detailed slit pores;[5,6] and atomically-detailed finite plates.[7] The detailed geometry of the various ice phases, however, depends on the nature of the confining surface.

1 K. Koga, X. C. Zeng and H. Tanaka, *Phys. Rev. Lett.*, 1997, **79**, 5262.
2 K. Koga, X. C. Zeng and H. Tanaka, *Chem. Phys. Lett.*, 1998, **285**, 278.
3 K. Koga and H. Tanaka, *J. Chem. Phys.*, 2005, **122**, 104711.
4 J. Bai, J. Wang and X. C. Zeng, *Proc. Natl. Acad. Sci. U. S. A.*, 2006, **103**, 19664.
5 R. Zangi and A. E. Mark, *Phys. Rev. Lett.*, 2003, **91**, 025502.
6 R. Zangi and A. E. Mark, *J. Chem. Phys.*, 2003, **119**, 1694.
7 N. Giovambattista, P. J. Rossky and P. G. Debenedetti, *Phys. Rev. E*, 2006, **73**, 041604.

Professor Yamaguchi commented: It has often been said that the internal pressure under confinement is several thousand bar. From your calculations, do you estimate the internal pressure under confinement with hydrophilic and hydrophobic surface?

Professor Debenedetti answered: The lateral and normal stresses in our confined glasses are given in Fig. 1 of our paper. In our calculations, the energy minimizations are done at constant volume (isochorically). This is analogous to isochoric quenches. Accordingly, many of the resulting glasses are under tension (negative pressure). Fig. 1 can be used to compute the density needed to generate confined glasses under ambient stresses. For example, from Fig. 1a and 1b, we can see that in hydrophilic confinement ($k = 1$), and a gap width of 0.8 nm, glasses with densities of *ca.* 1.2 and 1.1 g cm^3 will have lateral and normal stresses close to ambient pressure.

Dr Lee asked: The confined water between two hydrophobic walls is ice-like while the water confined between two hydrophilic walls is not. Does that mean the friction (both lateral and vertical forces) between hydrophobic plates is stronger than the one between hydrophilic plates?

Professor Debenedetti answered: In our calculations the walls are static. We have not studied friction, so I cannot comment on this point.

Professor Yarwood continued the discussion of the paper by Dr Coudert: First of all, congratulations on taking the trouble (and expense) of using a model which allows the IR spectra to be computed. At the moment, however, they are not in very good agreement with the experimental spectra. So, what needs to be done to improve the computed spectra and how will they help to elucidate the water–zeolite interactions?

Dr Coudert answered: Firstly, it is first to be noted that the experimental IR spectrum of the zeolite presented in our Fig. 10 and obtained by Fyfe *et al.*[1] in the 80s was measured on a low-silica zeolite synthesised by thermal dealumination. This procedure is known to result in materials exhibiting non-negligible concentrations of defects (aluminium atoms in the framework), although the purity of the zeolite wasn't reported (nor was any direct structural characterisation).

Secondly, even with the issue of the quality of the experimental sample aside, there is a general shift of the computer spectrum to lower frequencies than the experimental spectrum. A part of this shift can be attributed to the Car–Parrinello dynamics and would be lessened by a choice of smaller timestep and fictitious electron mass, at the expense of a largely increased CPU time. Another part of the shift can probably be attributed to the BLYP DFT functional used.

Finally, regarding the computation the IR spectra and how it can help elucidate water–zeolite interactions, I have nothing to add to what I said in my answer to Professor Havenith.

 This journal is © The Royal Society of Chemistry 2009

1 C. A. Fyfe, G. J. Kennedy, G. T. Kokotailo and C. T. DeSchutter, *J. Chem. Soc., Chem. Commun.*, 1984, 1093.

Dr Michaelides continued the discussion of the paper by Professor Debenedetti: Yesterday Dr Salmeron argued that "free OH" groups were important for and beneficial to wetting. Today you have discussed dangling OH bonds in your MD simulations as a characteristic of a hydrophobic interface. Can you comment please?

Dr Salmeron answered: It is a characteristic (as shown also by SFG experiments by Professor Shen's group) of the interface between water and hydrophobic surfaces that dangling OH bonds point towards the interface. But this is not in contradiction with the wetting nature of the free surface of water. Water would indeed grow (wet) its own surface that exposed OH bonds.

Professor Debenedetti replied: Our hydrophilic walls ($k = 1$) are fully hydroxylated. Our hydrophobic walls ($k = 0$) contain no OH groups. So I do not agree that OH bonds are a characteristic of a hydrophobic interface.

Dr Coudert responded: I think the answer to this difference is that we use the same languages for two very different types of OH groups. In the case of Dr Salmeron, the OH groups that are beneficial to wetting are "free, dangling H-bonds at the surface", oriented away from the surface and that can bind water molecules. In our case, we're talking of a nanoscale droplet of water and what we call "dangling OH groups" are non-hydrogen-bonded water OH groups that are oriented towards the surface. The properties (vibration frequency, length, polarisability, *etc.*) are widely different from OH groups involved in the hydrogen bond network of liquid water, and their presence (seen for example by IR spectroscopy) is a signature of the hydrophobic nature of the surface.

Dr Michaelides said: Can you tell me what you think are the characteristics of the ideal material for nucleating ice? Precisely how important are issues such as the lattice match or the strength of the interaction with the surface, *etc.*?

Professor Debenedetti replied: What makes it easy to nucleate ice in our simulations is the geometric match with the underlying surface. This is more important than surface electrostatic charges, which were absent in every case where ice formation occurred.

Dr Feibelman addressed Dr Ricci and Professor Debenedetti: Why can you supercool water? Does the percentage of broken H bonds correlate with the homogeneous nucleation temperature?

Dr Ricci replied: Water is not the unique liquid which can exist in a metastable supercooled state. Experimentally we have been able to supercool a quite large sample ($4 \times 4 \times 0.1$ cm^3) in a vanadium sample container coated with PTFE to avoid contact with the metal surface. In supercooled water the density decreases as the temperature decreases: this suggests that water gradually transforms into a more ordered phase, and the number of intact H bonds is not expected to decrease.

Professor Debenedetti replied: Water, like any liquid, can be supercooled because it is necessary to overcome a free energy barrier in order to form an ice nucleus large enough to grow spontaneously. There is a competition between the favorable thermodynamic driving force associated with forming the stable phase, and the energetic cost of forming an interface between the liquid and the ice nuclei. I don't know if there is a correlation between broken hydrogen bonds and the homogeneous nucleation temperature.

Professor Klein remarked: Professor Debenedetti commented that when confined to a 6 Å film between hydrophobic surfaces, water forms ice; perhaps this is dependent on the surface roughness. We found that water remains fluid between smooth hydrophobized surfaces as they jump together (see *e.g.* Raviv *et al.*[1]). However, I should note that we were not able in that study to control for registry between the surfaces, indeed they were certainly not in register as the hydrophobic layer was a surfactant monolayer and not a perfect crystal.

1 U. Raviv, S. Giasson, J. Frey and J. Klein, *J. Phys.: Condens. Matter*, 2002, **14**, 9275.

Professor Rovere continued the discussion of Dr Ricci's paper: Concerning the changes in the dynamical properties of supercooled confined water, I want to recall that some years ago we performed a molecular dynamics study of SPC/E water confined in a silica pore modelled to mimic vycor. We found that there is a layer of water close to the hydrophilic surface of the order of 4–5 Å where the molecules are in a sort of glassy state; instead the water in the center of the pore is similar to bulk water. In particular, for this free water we found a mode coupling transition similar to bulk water.[1]

1 P. Gallo, M. Rovere and E. Spohr, *Phys. Rev. Lett.*, 2000, **85**, 4317.

Dr Ricci replied: The present neutron diffraction data, after a layer analysis similar to that performed in both your *Phys. Rev. Lett.* and our *J. Phys. Chem.* papers,[1] reveal that the structure of the layer nearby the surface is strongly distorted compared to bulk water, and a sensible distortion is visible also in the layers close to the center of the pore, contrarily to what is found by molecular dynamics simulation. In other words these data do not seem to support the idea that water in the middle of the pore is similar to bulk water at a higher temperature.

1 P. Gallo, M. A. Ricci and M. Rovere, *J. Phys. Chem.*, 2002, **116**, 342.

Professor Vega remarked: If the radial distribution of a bulk fluid is computed considering only molecules being between say $Z = 0$ and $Z = 6$ Å then the result will be different from that obtained for the entire bulk fluid. Therefore "geometric confinement" provokes a change in the radial distribution even in a bulk fluid. Were you taking this effect into account when comparing the radial distribution functions of confined and bulk water?

Dr Ricci answered: We are perfectly aware of the excluded volume effect on the radial distribution functions and implemented the correction for this effect within the EPSR routine, as discussed in detail in ref. 11 and in my answer to Dr Bakó's question at the start of this discussion. Are you suggesting perhaps that we apply an excluded volume distortion to the bulk data instead of correcting those of water in vycor? Are you suggesting that this may be useful for decoupling the topological and chemical effects?

Professor Vega continued the discussion of Professor Debenedetti's paper: Firstly, what is the advantage of using potential energy minimization (*i.e.* looking at inherent structures) with respect to quenching (rapid cooling) to study the system you consider in your paper?

Secondly, do you have any explanation about why ice bilayer structures are formed under severe confinement (0.6 nm) at room temperature when the SPC/E model of water is used which has a much lower melting point (215 K) in the bulk? What is the effect of confinement on the melting point of water?

Professor Debenedetti responded: The advantage of using inherent structures and the energy landscape formalism is the associated theoretical superstructure. For

 This journal is © The Royal Society of Chemistry 2009

example, the notion of a Sastry point as a well-defined limit of stability for the glassy state is a direct consequence of the energy landscape approach. Knowledge of the depth distribution of potential energy minima, furthermore, allows the calculation of the configurational entropy. In ongoing calculations we are investigating quenches and comparing the results with energy minimization (inherent structure calculation). So far, we have not seen qualitative differences between these two approaches.

With regard to your second question, 0.6 nm corresponds to a water bilayer between the plates. The underlying geometry of the wall structure favors the formation of hexagonal monolayers. Hydrogen bonds between the two monolayers also contribute to the stability of the ice phase. This combination of geometric and energetic factors provides strong energetic stabilization, such that in our open-system simulations, in which two finite plates are immersed in bulk water, bilayer ice formation in the confined but open space between the plates occurs spontaneously at 300 K.

Dr Patel remarked: Empirical force fields perform a balancing act between the repulsions arising from the Lennard-Jones core and attractions due to the charges on the particles. By scaling the charges on the wall particles (k) to tune the degree of hydrophilicity, this balance is disrupted. As a result, the effective size of the wall particles (manifest in the position of the peak of the mean water density away from the wall) decreases as k is increased. Thus, the water molecules experience a slightly different strain for different values of k. Do you expect this effect to have a significant impact on the findings reported in this paper?

Professor Debenedetti answered: The particular atomistically-detailed force field that we use for the confining surfaces was developed as a model for a hydrophilic surface,[1] corresponding to $k = 1$. We have already shown that scaling the charges leads to physically realistic behavior for evaporation transitions ($k = 0$, ref. 2), and for wetting behavior as characterized by the contact angle ($0 \leq k \leq 1$, ref. 3). The one phenomenon where I believe results to be sensitive to the detailed geometry of the confining surfaces is the formation of highly-ordered layers, such as I describe in the paper for the case of high densities and hydrophobic confinement. Varying the size of surface atoms would have a dramatic influence on the formation of ordered layers.

1 S. H. Lee and P. J. Rossky, *J. Chem. Phys.*, 1994, **100**, 3334.
2 N. Giovambattista, P. J. Rossky and P. G. Debenedetti, *Phys. Rev. E*, 2006, **73**, 041604.
3 N. Giovambattista, P. G. Debenedetti and P. J. Rossky, *J. Phys. Chem. B*, 2007, **111**, 9581.

Professor Hodgson commented: It is perhaps worth pointing out that there is a clear analogy between the behaviour described in this model and the role of lateral strain in the growth of water films adsorbed next to hydrophilic metal surfaces. Wetting surfaces, such as Pt(111), Ru(0001) and Ni(111) optimise the density of water next to the metal surface, compressing the monolayer laterally compared to the density of bulk ice. This strain has direct implications for the wetting of this layer by further water, either the first layer must relax in response to multilayer adsorption (as happens on Ni(111)[1] and for thicker ice layers on Pt(111)[2–4]), allowing the interface layer to match the structure of bulk ice, or else the water multilayer does not wet, forming clusters with limited interaction with the first layer water, as is found on Pd(111)[2] and Ru(0001).[5] This situation is similar to that seen in the simulations of the Pd(100) interface discussed by Professor Madden and coworkers in his paper.[6]

1 M. Gallagher, A. Omer, S. Haq and A. Hodgson, *Surf. Sci.*, 2007, **601**, 268.
2 G. A. Kimmel, N. G. Petrik, Z. Dohnalek, and B. D. Kay, *J. Chem. Phys.*, 2007, **126**, 114702.
3 G. Zimbitas, S. Haq and A. Hodgson, *J. Chem. Phys.*, 2005, **123**, 174701.
4 K. Thürmer and N. C. Bartelt, *Phys. Rev. Lett.*, 2008, **100**, 186101.

5 S. Haq and A. Hodgson, *J. Phys. Chem. C*, 2007, **111**, 5946.
6 A. P. Willard, S. K. Reed, P. A. Madden and D. Chandler, *Faraday Discuss.*, 2009, DOI: 10.1039/b805544k.

Professor Finney asked: Why did you choose separations of 8 and 16 Å? 8 Å might naturally be expected to give a 3-layer structure for a water-like density and so might be thought to be "commensurate" with a structure with an integral number of layers. Would you expect qualitatively different behaviour for other separations that might be "*in*commensurate", *e.g.* 9 Å?

Professor Debenedetti replied: In simulations of two finite hydrophobic plates immersed in bulk water[1] we observed ice formation at ambient temperature only at a plate–plate separation of 0.6 nm, that is to say, when two water layers were present between the plates. Building on that previous work, in the present calculations we chose the small-gap case, 0.8 nm, to be considerably larger than the gap leading to bilayer ice (0.6 nm), yet small enough for interactions with the confining walls to play a dominant role. The large-gap case, 1.6 nm, was also chosen based on our previous calculations: at this distance we never observed ice, nor did we observe evaporation. This seemed to us to be a good limiting case to consider for energy minimizations. In hydrophobic confinement and at a separation of 0.8 nm, we see either trilayer or bilayer glasses, depending on the value of the mean density (see our Fig. 6). The same thing happens with the equilibrated liquid at 300 K (our unpublished work).

1 N. Giovambattista, P. J. Rossky and P. G. Debenedetti, *Phys. Rev. E*, 2006, **73**, 041604.

Professor Finney continued the discussion of the paper by Dr Ricci: I understand from your paper that the confined water results are based on data taken sometime ago and discussed in earlier publications. Does your current interpretation of those data differ in any way from that presented earlier, and if so how and why?

Dr Ricci answered: Yes, the data for bulk water in the stable phase have never been published so far, while those for water in vycor glass have already been published twice,[1,2] after being analyzed with different procedures. At the time of the first publication the EPSR code was not yet available and consequently it was not possible to discriminate between the water–water contributions and the water–substrate ones, which gave small but not negligible contributions to the three experimentally accessible partial structure factors. The information obtained at the time of the second publication is consequently richer, as the EPSR allows to extract all the individual radial distribution functions of relevance.

I want to take this opportunity to remind everyone that science proceeds in two ways: by progress in the experimental techniques and in the power and effectiveness of the data analysis procedure. In the case of neutron diffraction, for instance, the radial distribution functions can be obtained from the differential cross-section by simply performing a Fourier transform, but this procedure gives in practice spurious ripples in the *r*-space functions, which may alter the intensity of the genuine peaks, or introduce new unphysical ones. In order to avoid this problem the maximum entropy technique can be used, as in the *J. Chem. Phys.* paper:[1] this gives the less structured radial distribution function compatible with the experimental data. The risk here is that this function is too smooth and perhaps the peak intensities too low. The EPSR routine, available since 1996 (see ref. 18 in our paper), is nowadays the best technique as it constrains a Monte Carlo simulation, by using all the available experimental data at the same time, and gives access to all the radial distribution functions of the system under study. More than 60 publications based on this routine have already appeared in the best journals in this field. When the amount of experimental data available is less than the number of radial distribution functions of

 This journal is © The Royal Society of Chemistry 2009

interest, this routine is a unique tool for data analysis, although one needs to avoid over-interpretation of the data. It is remarkable that this code produces only radial distribution functions which are compatible with the physical configurations of molecules, thus evidencing the presence of residual systematic errors in the data, if any.

In conclusion, neutron diffraction data do not change with time. Nevertheless over the years we learn more about data analysis and sometimes reanalyzing old data, we can obtain slightly different radial distribution functions. In my experience, however, there are solid results, such as the peak positions and coordination numbers, which never change by changing the data analysis procedure.

1 A. K. Soper, F. Bruni and M. A. Ricci, *J. Chem. Phys.*, 1998, **109**, 1486.
2 H. Thompson, A. K. Soper, M. A. Ricci, F. Bruni and N. T. Skipper, *J. Phys. Chem.*, 2007, **B111**, 5610.

Dr Bakó commented: I totally agreed with the comment of Professor Vega. If you calculated the RDF only in the first shell you destroy the symmetry so the change in RDF comes from only technical reasons.

Dr Ricci responded: I totally disagree with you! Also during our layer analysis the confinement effects are properly taken into account. Moreover it is clear from our Fig. 5 that the first layer is different from the others layers: here we are not comparing with the bulk.

Professor Hodgson opened the discussion of the paper by Professor Madden: In your presentation you suggested that the tight binding of first layer water to the metal, and lack of interaction to the water multilayer, was an indication that the potential chosen to describe the metal–water interaction was over-binding compared to the water–water interaction. Actually I think this is unduly pessimistic, and not necessarily a direct indication that the potential used in the simulation is wrong. The existing surface science literature does not give much direct information about the structure of water at the Pt(100) interface, or any other fcc(100) surface for that matter, nor whether the first layer of water wets or re-structures as the multilayer covers it. We find the first layer of water adsorbed at a metal surface often does not wet, water preferring to form the exposed monolayer and large 3D multilayer ice clusters on Ru(0001).[1] In these cases it appears the first water layer is too tightly bound to undergo sufficient relaxation to stabilise a continuous multilayer film, similar to the behaviour in your simulation. On the other hand relaxation of the wetting layer is facile on Ni(111),[2] adsorption of a second water layer being sufficient to relax the lateral density of the wetting layer to form an incommensurate ice film. On Pt(111) the wetting layer does relax to form an incommensurate ice film,[3,4] but requires a critical ice thickness, indicating there is a significant energy cost in restructuring the water/Pt(111) interface.[5] If the interface is pinned into strict registry with the metal, *e.g.* by OH chemisorption, this lateral relaxation becomes impossible and the surface does not wet.[6] Along with a suitable lateral period for the wetting layer, calculations also indicate that the presence of dangling OH groups is critical in order to stabilise the multilayer and may be formed from an H-down layer in response to multilayer adsorption.[6] So while I agree that the water–metal potential is critical for describing the interface between first layer water and the liquid, do you really think there is any clear evidence that your choice is necessarily wrong? Surely the best one can do is to vary this interaction and see the range of behaviour observed?

1 S. Haq and A. Hodgson, *J. Phys. Chem. C*, 2007, **111**, 5946.
2 M. Gallagher, A. Omer, S. Haq and A. Hodgson, *Surf. Sci.*, 2007, **601**, 268.
3 G. Zimbitas, S. Haq and A. Hodgson, *J. Chem. Phys.*, 2005, **123**, 174701.
4 G. A. Kimmel, N. G. Petrik, Z. Dohnlek and B. D. Kay, *Phys. Rev. Lett.*, 2005, **95**, 166102.
5 K. Thürmer and N. C. Bartelt, *Phys. Rev. Lett.*, 2008, **100**, 186101,

6 G. Zimbitas, M. E. Gallagher, G. R. Darling and A. Hodgson, *J. Chem. Phys.*, 2008, **128**, 074701.

Professor Madden replied: Thank you for the comment and this valuable information. As I said, the source of our Pt–water potential was a study of absorption of a single molecule (ref. 17 of our paper) whereas the water–water interactions are described by the bulk water SPC/E model (ref. 20). Since the sources are so different, there is definitely the potential for them to be unbalanced. Your comment suggests that systematically varying these potentials or attempting a more uniform parameterisation based on condensed phase *ab initio* calculations would be a useful research direction, especially given the interesting experimental work which you have indicated. There would be the prospect of linking the behaviour you see from what might be called a surface chemistry perspective to quantities measured electrochemically, like the potential of zero charge.

Dr Michaelides remarked: Assuming that (in reality) water does not dissociate when the field is applied, I think the essential characteristic of the water density profile you raise concerns over, *i.e.* a strong contact layer which connects weakly to the water bulk, may not be unreasonable. There are several recent experiments from the surface science community (for example, from people like Professor Hodgson and Dr Salmeron) which indicate similar behaviour. Also could you comment on how reliable your potentials are for the strength of the water–Pt and water–water interactions?

Professor Madden replied: Thank you for the comment. I believe I dealt with your question in my reply to Professor Hodgson and have nothing more to add on that topic.

Professor Uosaki remarked: I must point out the wrong definition of your "electrode potential". Electrode potential should be defined for each electrode to represent the energy level of each electrode,[1,2] but your electrode potential is the bias between the two electrodes, *i.e.* the cell voltage. A bias of zero does not mean an electrode potential of zero. Imagine two metals connected by a conducting wire are placed in a solution without applying any external bias. According to your definition, the potentials of two electrodes are zero in this situation. However, they are equal but neither zero nor at the potential of zero charge. The two electrodes are at the open circuit potential which depends on the solution composition, *etc.* There exists a potential drop at the electrode/solution interface at this potential.

Also, it seems you described the system of polarizable electrodes but Pt, which you chose as an electrode, is the most well-known non-polarizable electrode. When the bias is applied various electrochemical reactions take place. One side (the cathode) becomes Pt–H and hydrogen evolution takes place and the other side (anode) becomes Pt oxide and oxygen evolution takes place.[3]

Furthermore, the choice of electrolyte is not ideal. Cl^- is known to adsorb on the Pt surface.[4] Your calculation does not seem to show the adsorption of Cl^-.

Finally, there are huge numbers of references on electrode/electrolyte interfaces in both theory and experiment. When one proposes a new theory, one needs to carefully examine them. I must mention that the Faraday Discussion in 1947 was on electrode processes and the first talk was entitled "The contact between a solid and a liquid electrolyte".[5]

1 S. Trasatti, *Pure Appl. Chem.*, 1986, **58**(7), 955.
2 K. Uosaki and H. Kita, in *Modern Aspects of Electrochemistry*, ed. R. E. White, J. O. M. Bockris and B. E. Conway, Plenum, NY, 1986, vol. 18, pp. 1–60.
3 J. O'M. Bockris and A. K. N. Reddy, Modern Electrochemistry 2, Plenum, New York, 1970.
4 N. Li, J. Lipkowski, J. Electroanal. Chem., 491(1-2), 95 (2000).
5 T. B. Grimley and N. F. Mott, *Discuss. Faraday Soc.*, 1947, **1**, 3.

 This journal is © The Royal Society of Chemistry 2009

Professor Madden answered: Thank you for your comment on the definition of electrode potentials, which will enable me to explain some of the relationships between our observations and practical electrochemistry. In the simulations we have control of the absolute value of the electrical potential of the electrodes; that is, when the applied potential is V_0 the energy change when a charge q is transferred from the vacuum to the electrode is qV_0. (In addition, a work function may be introduced to account for any non-V_0-dependent parts of this energy.) The potential behaves as the applied potential in accord with macroscopic electrostatics in other ways (see ref. 16 in our paper). In Fig. 9 we show the profile of the mean electrical potential across the cell containing an LiCl solution when different potentials are applied to the electrodes. As you can see at $V_0 = 0$ (zero bias) the potentials on both electrodes are indeed constant and zero and that there is a strong drop of about -1 V on passing into the electrolyte region; the potential there is also constant as befits an ionically conducting medium. This is exactly what you are referring to by the potential drop at the electrode/solution interface. Note that at zero applied potential the electrode does carry a charge (our Fig. 8). When a finite potential is applied to the electrodes the value of the potential drop changes. In the experimental situation, you cannot measure absolute potentials, only potential differences against a reference electrode. So, contrary to your assertion, I would say that the behaviour of the potential in the simulation is reproducing the electrochemical situation perfectly, what is different is that we really do know about potentials and not just potential differences.

With respect to polarizable electrodes, yes, I apologise. I believe you and I are using the term "polarizable" to describe different phenomena, and since I am discussing electrochemistry I should have been more careful. By ideally polarizable I mean that in the presence of an electric field the electrode material "polarizes" like a perfect metal; *i.e.* the field is perfectly screened inside the electrode so that the electrical potential is constant. In the electrochemical literature the term "ideally polarizable electrode" means that no chemical reactions occur at the electrode and the potential drop across the electrode/solution interface is achieved without any charge transfer (simply by the arrangement of charges and dipoles on the solution side). In my simulations electrolysis of water is not possible as I have not permitted OH bonds to break.

The choice of electrolyte was discussed in section 2 of our paper: there are good technical reasons why we did not want to have a strong attractive interaction between the electrodes and the chloride ions which would have resulted in strong absorption of chloride in competition with water, but this certainly reduces the ability to reproduce the real behaviour of a chloride solution. As far as the electrode–anion interactions are concerned our anions behave more like fluoride ions than chlorides.

As for your final point, I do not think we are proposing a new theory; all we have done so far is to produce an atomistic simulation model in which the electrical potentials, the Marcus-calculated rates *etc.* behave in just the same way as in conventional accounts of electrochemical behaviour. The simulation will allow microscopic theories of electrochemical behaviour to be examined (for example, how the solvent and ions contribute to the interfacial potential, or how the rate of charge transfer depends upon the applied potential) and, hopefully, might ultimately allow us to explain the overpotentials *etc.* in particular systems.

Mr Chen asked: This work is interesting, but I am concerned that some phenomena are not entirely accounted for in this study. As important as an accurate description of the electrode/water interface is for understanding the structure of water near the electrode, it seems clear that certain polarization and charge transfer effects are two of the most important physical effects that are missing from this study.

A polarizable water model is essential to study water in varying chemical environments. The SPC/E water model used in this study is parameterized to the bulk liquid phase, but is not particularly well suited for describing water molecules at an interface. Perhaps you could comment on whether your interaction potential that describes the surface–water interactions could account for surface waters exhibiting any deviation from bulk water properties?

In addition, charge transfer from solution to the electrode surfaces is not accounted for in the simulation. This would be a particularly important physical phenomenon when one is interested in what happens in the presence of strong electrostatic fields such as those near the electrode surfaces, which is unfortunately missing in this current work. Also, in response to the previous comment, in which a concern was raised that the adsorption of chloride on platinum as observed experimentally may make the simulation results questionable, I would like to mention the important work of Stuart and Berne[1] in which they studied the hydration behavior of chloride ion in small water clusters. What they have found is that the non-polarizable TIP4P water model gives rise to a hydrated chloride buried inside the water cluster, as would be expected by the Onsager theory of solvation. In contrast, the TIP4P-FQ polarizable water model[1] predicts that the chloride ion prefers to reside on the surface of the water droplet, which is in accordance with the available experimental evidence. It seems then that the previous question's concerns could be ameliorated due to a fortuitous cancellation of error, as the use of non-polarizable water model would bias the chloride ions away from the surface.

1 S. J. Stuart and B. J. Berne, *J. Phys. Chem.*, 1996, **100**, 11934.

Professor Madden responded: You are certainly correct that our model is not realistic in several ways, including the ones you mention. The purpose of the work was to explore the relationship between what is measured in electrochemistry and what may be calculated by simulation with as realistic a model as we could afford—as you can see from the paper, there are substantial statistical problems associated with such calculations necessitating very long runs and special computational methods, like umbrella sampling. So the strategy has been to start with the most essential ingredients in the interaction model, with the intention of moving to more realistic (and expensive) descriptions later, after we have understood the elements of the electrochemistry–simulation relationship better. Polarization effects on the water–water and water–ion interactions may well be important especially given the reduced symmetry at the interface and, as you say, such effects are known to influence the coordination of water by the ions. Polarization of the electrode by the molecules and ions is included in the present model. By charge transfer I presume that you mean the consequences of overlap of the electron clouds of the ions and the electrode surface and their effect on the interactions (as opposed to the actual electron transfer steps, which we do study through the Marcus construction). In parallel with this work we are undertaking similar simulations[1] but with a molten salt as electrolyte where the ions are all polarizable, and where the interactions between the ions and the electrodes are parameterized so as to reproduce the results of *ab initio* calculations. This model includes all the effects you mention, generalising it in an efficient way to aqueous systems is somewhat complicated by the shape of the water molecules. This project is near completion and our longer term intention is to carry over the same methods to the aqueous case.

1 M. Pounds, S. K. Reed and P. A. Madden, in preparation.

Dr Feibelman said: Pt(100) is unstable in UHV against reconstructing to a hexagonal surface layer with, I think, a 5×20 unit cell.

Professor Madden answered: Of course, in our simulations the Pt atoms were held fixed, so no reconstruction was possible. As to whether a study of electrochemistry at a Pt(100) surface is unrealistic, perhaps someone else will offer further information during this discussion.

 This journal is © The Royal Society of Chemistry 2009

Dr Feibelman opened the discussion of the paper by Professor Jhe: An interfacial force microscope was invented by Houston[1] which solves the snap-to-contact problem and allows measurements of adhesive and friction forces with a meniscus between tip and sample or an immersed tip.

1 S. A. Joyce and J. E. Houston, *Rev. Sci. Instrum.*, 1991, **62**, 710.

Professor Jhe answered: Although we already read and studied the interfacial force microscope and its excellent results, it may be helpful for us to check the recent progress of it. Nonetheless our work provides unique experimental results otherwise unrealizable or unavailable.

Professor Debenedetti opened the discussion of the paper by Dr Perkin: Can you tell us something about ion concentration effects in your experiments?

Also, you have invoked water exchange as an explanation for lubrication of the shearing surfaces. Is there any microscopic evidence (*e.g.* from simulations) in support of this explanation? Could the compressibility of the hydration layer play a role?

Dr Perkin answered: The concentration of ions in the bulk is important. There is a critical concentration, different for each ion, below which hydration repulsion does not occur (instead the mica sheets are attracted into primary minimum contact due to van der Waals forces and no hydrated ions remain between them). And with no hydration repulsion, there cannot be hydration lubrication with applied load. So the bulk concentration must be above this 'critical hydration concentration' (CHC) in order for the lubrication to occur. The CHC has been discussed in detail by Pashley and others.[1] However, as long as the bulk concentration is above this CHC, we record similar behaviour at different bulk concentrations (within the range of our parameters, *i.e.* up to the order of 0.1 M salt), perhaps suggesting a saturation of the surface with hydrated cations.

Yes, there has been direct simulation evidence for the 'hydration lubrication' due to water relaxation, by Leng and Cummings[2–4] who directly simulate the situation in a surface force balance.

1 R. Pashley, *Adv. Colloid Interface Sci.*, 1982, **16**, 57.
2 Y. Leng and P. T. Cummings, *Phys. Rev. Lett.*, 2005, **94**, 026101.
3 Y. Leng and P. T. Cummings, *J. Chem. Phys.*, 2006, **124**, 074711.
4 Y. Leng and P. T. Cummings, *J. Chem. Phys.*, 2006, **125**, 104701.

Dr Perkin discussed Dr Feibelman's concluding remarks: I would like to comment on the result which Dr Feibelman mentions, regarding the fluidity of water in thin films confined between solid surfaces. The viscosity of water confined between mica surfaces has been directly measured by Raviv *et al.*[1,2] using a surface force balance (SFB). It was shown, through both direct measurement of transmitted shear force and through hydrodynamic analysis of the escaping water during approach of the surfaces, that no shear stress is sustained by the water film and that the viscosity of the confined water is within a factor of 3 of bulk water at every film thickness down to $D = 0.0 \pm 0.3$ nm (this upper limit being resolution-limited).

Of relevance to any comparison between these results and measurements conducted using probe-techniques (such as AFM), it is important to note the following: (a) in the SFB the two confining mica surfaces are of identical composition, atomically smooth, and can be prepared cleanly (see Perkin *et al.*[3] for a detailed discussion); and (b) in the SFB the separation between the confining surfaces (*i.e.* the thickness of the water film) is determined absolutely, through the multiple beam interference spectrum. These features in combination have allowed the unambiguous determination of the fluidity of water down to sub-nanometre films, which could otherwise have been clouded by the possibility of contamination or roughness effects.

Indeed, the impossibility of actually measuring absolute layer thicknesses using AFM implies that one cannot carry out controls for the purity of the confined water layers or contamination of the confining surfaces when using scanning probe approaches. Such controls are an absolutely essential precursor to the SFB experiments, and experiments where purity of the surfaces or confined water does not pass these controls are always aborted.

1 U. Raviv, P. Laurat and J. Klein, *Nature*, 2002, **413**, 51.
2 U. Raviv, S. Perkin, P. Laurat and J. Klein, *Langmuir*, 2004, **20**, 5322.
3 S. Perkin, L. Chai, N. Kampf, U. Raviv, W. Briscoe, I. Dunlop, S. Titmuss, M. Seo, E. Kumacheva and J. Klein, *Langmuir*, 2006, **22**, 6142.

Professor Halle continued the discussion of the paper by Dr Perkin: It seems that your viscosity estimate assumes a macroscopic stick boundary condition at the mica surface. But did you (or anyone else) examine the accuracy of continuum hydrodynamics for a water slab of subnanometre thickness? In other words, might the high apparent fluidity be partly attributed to fluid slip at the mica surface?

Also, water exchange from the primary hydration shell of an Na^+ ion is probably about two orders of magnitude faster[1,2] than what you assume (which makes your argument even stronger). But do you actually know that the Na^+ ions at the mica surface are fully hydrated?

1 R. W. Impey, P. A. Madden and I. R. McDonald, *J. Phys. Chem.*, 1983, **87**, 5071.
2 T. Ikeda, M. Boero and K. Terakura, *J. Chem. Phys.*, 2007, **126**, 034501.

Dr Perkin replied: This question of slip at the mica surface and its implication for the viscosity calculation is an astute point, which has been addressed in the past in detail by Raviv *et al.*[1,2] The consensus is that there is little slip of water next to a hydrophilic surface such as mica,[3] and this has been demonstrated directly in experiments of Jacob Israelachvili[4] as well as more recently by Silbert and Klein where the slip length is found to be 0 nm within the scatter of the data which is of the order of 1nm (data to be published). It has been shown[5] that at sufficiently high shear rates there may be slip next to hydrophilic surfaces with a slip length of a few nanometres. However in the *J. Phys.: Condens. Matter* paper[2] mentioned above it was demonstrated that the effect on the deduced viscosity of this possible slip length of approximately a few nm is small. Thus we are confident in the low viscosity values we determine for our confined water films.

Our rate constants for water exchange were taken from sources such as Eigen[6] and Burgess[7] and as tabulated in standard texts such as Cotton and Wilkinson.[8] Our suggestion that the metal cations are hydrated while highly confined and demonstrating 'hydration lubrication' is primarily based on the closest separation distance of the mica surfaces in the hydration repulsion regime—between 0.5–0.8 nm for Na^+ for example—a separation much greater than the unhydrated ion radius but in line with the diameter, σ, of hydrated ions. The details of the hydration geometry within the hydration repulsion have been explored by Pashley in the 1980s.[9] It may be that ions are partially de-hydrated, perhaps in order to approach closer to the negative charge on the mica surface at the highest applied pressures, although this precise geometrical information cannot be determined from these experiments. What is clear, however, is that for the limit where the surface separation D decreases so that $D < \sigma$, some water of hydration must be 'squeezed away' from the confined ions.

1 U. Raviv, S. Perkin, P. Laurat and J. Klein, *Langmuir*, 2004, **20**, 5322.
2 U. Raviv, S. Giasson, J. Frey and J. Klein, *J. Phys.: Condens. Matter*, 2002, **14**, 9275.
3 Vinogradova, *Langmuir*, 1995, **11**, 2213.
4 J. Israelachvili, *J. Colloid Interface Sci.*, 1986, **110**, 263.

 This journal is © The Royal Society of Chemistry 2009

5 *e.g.* E. Bonaccurso, H.-J. Butt and V. S. J. Craig, *Phys. Rev. Lett.*, 2003, **90**, 144501.
6 M. Eigen, *Pure Appl. Chem.*, 1963, **6**, 105.
7 J. Burgess, *Metal Ions in Solution* (Ellis Horwood Series in Chemical Science), Ellis Horwood Ltd, Chichester, 1978.
8 Cotton and Wilkinson, *Advanced Inorganic Chemistry*, John Wiley and Sons Inc., 1988.
9 *e.g.* Pashley, *Adv. Colloid Interface Sci.*, 1982, **16**, 57.

Professor Kim commented: It is very interesting to see the superlubricity is observed only for the Na^+ ion, but not for Cs^+ and Ni^{2+} ions. Your justification is the difference in the hydration energy. If that is the case, do you see the superlubricity for Li^+ ions which is also expected to have a high hydration energy (even higher than the Na^+ ion).

Dr Perkin answered: The lubricity is indeed observed for hydrated Li^+ ions, and indeed for K^+ ions, confined to hydration layers between mica surfaces and has been reported recently in Chai *et al.*[1] (see Fig. 4 and 5); however, this has not yet been tested up to the ~MPa pressures reported in this paper for Na^+. Although the hydration energy is higher for Li^+ than Na^+—and so according to our argument we might expect the highest lubricity—the small size of the Li^+ ion means that the hydration number is thought to be lower (around 4 for Li^+, compared to up to 7 for Na^+) which may be also relevant.

1 L. Chai, R. Goldberg, N. Kampf and J. Klein, *Langmuir*, 2008, **24**, 1570.

Professor Bain communicated: As someone interested in the tribology of real surfaces, I'd like to ask whether you can extend your conclusions to practical lubrication of hard surfaces, which are not atomically smooth and which are normally subjected to contact pressures far in excess of those achievable in a surface force balance. For example, the much-vaunted layering transitions in hydrocarbon liquids between mica surfaces in an SFB disappear if the surface is roughened[1] and the liquid is squeezed out of the contact entirely at pressures of a few tens of MPa.[2] I sense that your work on hydrated ions may be less sensitive to atomic scale roughness, but what about the behaviour of interfacial ions at higher pressures?

1 *e.g.* J. Gao, W. D. Luedtke and U. Landman, *Tribol. Lett.*, 2000, **9**, 3.
2 D. A. Beattie, S. A. Winget and C. D. Bain, *Tribol. Lett.*, 2007, **27**, 159.

Dr Perkin communicated in reply: The effect of roughness is (as you say) twofold: the real contact area may be dramatically reduced which concomitantly increases the contact pressure at those points; and layering effects (such as solvation forces) are obviated due to the reduced uniformity and symmetry of the layers of liquid adjacent to the surface. In our paper we have attempted to address the question of the viscosity of an aqueous film at high pressure in the case of $Na^+(aq)$, by applying pressures up to ~MPa, and we observe that the viscosity of the film remains below 0.03 Pa s at this elevated pressure. We have on one occasion applied significantly higher pressures than this and observed a finite shear stress: this is currently under investigation and we hope to present the results for this in due course. There may also be ion-specific effects in the shear stress at high pressures. In the case of hydrocarbon liquids which show these oscillatory force-laws when confined between atomically smooth mica sheets, there is a liquid-to-solid phase transition on confinement.[1] All the evidence is that this liquid-like to solid-like transition under confinement between smooth solid surfaces is quite intrinsic to the purified organic liquid system that was studied (octamethyltetracyclosiloxane), though it may not be observable in a surface force balance if the surfaces are rough (or indeed if they are contaminated—see Perkin *et al.*[2] for a recent discussion of this). However, in the case of water we find no liquid-to-solid phase transition in our experiments with smooth surfaces—water remains close to bulk viscosity at all film thicknesses—and so I can see no intrinsic reason why the mechanism we propose, based on the rapid exchange and rotational relaxation of water in the film, should not also apply for rougher or non-uniform surfaces.

However, there is another important point to make regarding the extrapolation of the results we show here towards 'real' systems: the effect of the surface chemistry, and in particular the surface charge. We have concluded in this paper that in order for hydration lubrication to occur, hydrated ions must be 'retained' between the shearing surfaces. This can occur in our experiments using mica surfaces because they are highly (negatively) charged, and so may trap hydrated cations between them even under high applied load (as is the case of Na^+ for example). The same mechanism has been implicated to explain the lubrication of polyelectrolyte coated surfaces[3] and some hydrated surfactant coated surfaces;[4] in these cases the hydration is bound to the charges on the polymer- or surfactant-modified surfaces. It is interesting to note that many biological shearing interfaces hold such a high surface charge. So we might expect that for uncharged surfaces the same hydration lubrication mechanism cannot act to lubricate the surfaces. Early results (unpublished) of our investigation into lubrication at uncharged surfaces confirm this may indeed be the case.

1 J. Klein and E. Kumacheva, *J. Chem. Phys.*, 1998, **108**, 6996.
2 S. Perkin, L. Chai, N. Kampf, U. Raviv, W. Briscoe, I. Dunlop, S. Titmuss, M. Seo, E. Kumacheva and J. Klein, *Langmuir*, 2006, **22**, 6142
3 U. Raviv, S. Giasson, N. Kampf, J.-F. Gohy, R. Gerome and J. Klein, *Nature*, 2003, **425**, 163.
4 W. Briscoe, S. Titmuss, F. Tiberg, R. K. Thomas, D. McGillivray and J. Klein, *Nature*, 2006, **444**, 191.

Dr Christenson communicated: Your paper states that water films behave differently to films of non-polar liquids, which "tend to become solid-like under... confinement". There is an article by Patil *et al.*[1] which suggests that confinement-induced solidification inferred from dynamic AFM and SFA experiments is actually "jamming" due to slow relaxation times. Do you have any comments on this in relation to your findings?

1 S. Patil, G. Matei, A. Oral and P. M. Hoffman, *Langmuir*, 2006, **22**, 6485.

Dr Perkin communicated in response: The behaviour of organic liquids under confinement is very different from that of water or hydrated ions, which is the topic of our paper, as has been amply discussed (see *e.g.* Klein *et al.*[1]). The solid-like behaviour of (organic) confined liquids as determined in surface force measurements is deduced from the ability of such confined layers to sustain a shear stress (with a finite yield shear stress), which is one of the basic definitions of a solid. I do not think that the suggestion of jamming as a mechanism—as opposed say to glassy behaviour or crystallization—for such solidification in any way contradicts the conclusion of solid-like behaviour. Indeed, since there is evidence (*e.g.* Kumacheva and Klein[2] and papers by several other groups) that the yield shear stress for such confined liquids is considerably smaller than for the bulk crystal of the same liquid, it is very worthwhile to pursue the precise mechanism leading to solid-like behaviour.

1 J. Klein, U. Raviv, S. Perkin, N. Kampf, L. Chai and S. Giasson, *J. Phys.: Condens. Matter*, 2004, **16**, S5437.
2 E. Kumacheva and J. Klein, *J. Chem. Phys.*, 1998, **108**, 7010.

Dr Rao communicated: For making a comparison of the series, you chose to use Cl^- for Na but nitrate salts for Li and Ba. Is there any preference of nitrates over chlorides? It would be preferable to use the same anion while making a comparison of the cations. Any comments?

 This journal is © The Royal Society of Chemistry 2009

Dr Perkin communicated in reply: Our choice of anion was merely practical; the nitrates are usually preferable since chlorides degrade the silver mirrors we use for the interferometry, and you are right that it is generally preferable to use the same anion for comparison of cation behaviour. Indeed in a recent paper,[1] shear behaviour for the series $LiNO_3$, $NaNO_3$, KNO_3 and $CsNO_3$ was compared (though at lower pressures than the highest values we report here). However, in a few cases, including NaCl, the chloride was occasionally used. We have not so far observed any anion-specific effects, which we ascribe to the fact that the concentration of anions adjacent to the mica surface is much smaller than the concentration of cations, due to the high negative surface charge and surface potential of the mica. We could quantify this using a simple Boltzmann distribution for ion concentration, $[i]_x$, a distance x from the surface (so that at the surface $x = 0$ and the surface potential is Ψ_0): $[i]_0 = [i]_\infty \exp(-z_i e\Psi_0/kT)$, where $[i]_\infty$ is the bulk ion concentration, e is the electronic charge and z_i is the charge of the ion. For example in the case of Na^+ we recorded a surface potential value of $\Psi_0 = -60$ mV, from which we can calculate >2 orders of magnitude difference in the concentration of cations and anions at the surface.

1 L. Chai, R. Goldberg, N. Kampf and J. Klein, *Langmuir*, 2008, **24**, 1570.

Professor Kim continued the discussion of the paper by Professor Jhe: One comment and two questions for Professor Jhe. The comment is that, to me, the water bridge between the tip and the substrate looks more like a nano-disc (or at best a nano-cylinder) rather than a nano-wire with the aspect ratio of the diameter and length of the capillary bridge. It is very interesting to see how powerful the tuning fork method is in measuring the forces involved during tip approach and retraction through the condensed water.

Two questions that I have are as follows: in the first part of your paper, you assumed that there is no molecular exchange between the gaseous water and the water nano-wire once the condensation at the tip–substrate space is occurred. Based on this assumption, you calculated the principal radii of the capillary meniscus curvatures which are in turn used in calculation of the Laplace pressure. But the calculated Laplace pressures during the approach and retraction are drastically different, which is hard to understand. When we compare the gas collision rate at ambient conditions and the tip approach/retraction rate, it is obvious that the former is much faster than the latter. In that situation, I expect the capillary is always in equilibrium with the gas phase. In other words, the meniscus pressure (or disjoining pressure of the condensed bridge) should be determined the partial pressure and temperature of the gas and the molar volume of the condensed liquid. How can you explain this discrepancy between the equilibrium and your assumption? I think your assumption is not valid in the equilibrium condition.

In the second part of your paper, the elasticity and viscosity of the condensed water at the tip–substrate space were calculated from the measured force and phase signals. The qualitative trend of the viscosity (increasing as the tip–surface distance decreases) is consistent with the previously reported behavior although the magnitude that you calculated is many orders of magnitude lower. In the elasticity part, I do not see any strong and consistent dependence on the tip–surface separation distance until the tip is very close to the separation point. Can this behavior be justified? I wonder if there is any theoretical reason for this observed trend. Or may this be due to an insensitivity of the technique?

Professor Jhe answered: With regard to your first question, the formation of a capillary bridge consists of two steps: nucleation and growth. The time scale for the nucleation is found to be in the order of 1 ms in ambient conditions,[1] which is quite fast compared to the experimental time scale of seconds. However, it is not obvious that the growth time is also so fast that thermodynamic equilibrium can be always reached during the meniscus manipulation. If the manipulated meniscus is in thermodynamic equilibrium with the water vapor, the meniscus radius of curvature should be equal to the Kelvin radius (~1 nm), which is fixed to a constant value during the manipulation for a given relative humidity and temperature, so that the meniscus height is limited by the Kelvin radius. From careful calculation one can obtain the maximum height of the meniscus, or the rupture distance, is about twice the Kelvin radius. However, experimentally there were reported of rupture distances even of about 100 nm in several papers. This indicates that during the manipulation of the water meniscus, the meniscus is considered to be in mechanical equilibrium with the external pull-off force but not in thermodynamic equilibrium with the external water vapor.[2] While the thermodynamic equilibrium requires the constant Kelvin radius, the thermodynamic nonequilibrium does constant volume of water meniscus. Thus, we chose the constant volume boundary condition.

So far we do not have a concrete idea about your second question. We are now studying the viscoelasticity of the water bridge by changing the relative humidity, where we found that the overall shape of the elasticity curve is changed as the humidity is changed.

1 R. Szoszkiewicz and E. Riedo, *Phys. Rev. Lett.*, 2005, **95**, 135502.
2 L. Sirghi, R. Szoszkiewicz and E. Riedo, *Langmuir*, 2006, **22**, 1093.

Dr Held continued the discussion of the paper by Professor Madden: Firstly, a comment: although Dr Feibelman's comment is correct that the clean Pt(100) surface undergoes a hexagonal reconstruction it should also be noted that adsorbates tend to lift this reconstruction (this has been characterised in great detail for CO adsorption by the groups of Ertl[1] and of King[2]). It is therefore likely that the electrochemical interface of Pt(100) is un-reconstructed or has a periodicity not bigger than a few (1×1) unit cells.

Secondly, over the past few years it has been shown by a large number of studies that atomic co-adsorbates can induce dissociation of water on Pt-group surfaces (see *e.g.* Henderson[3] or Shavorskiy *et al.*[4]). This effect must play an important role at the electrochemical (electrolyte/Pt) interface. Therefore, a realistic model should include the possibility of water dissociation at the surface, which the present model does not seem to do. Are there ways of incorporating such effects in molecular dynamic simulations?

1 G. Ertl, *Surf. Sci.*, 1994, **299/300**, 742.
2 C. E. Wartnaby, A. Stuck, Y. Y. Yeo and D. A. King, *J. Chem. Phys.*, 1995, **102**, 1855, and references therein.
3 M. A. Henderson, *Surf. Sci. Rep.*, 2002, **46**, 1.
4 A. Shavorskiy, M. J. Gladys and G. Held, *Phys. Chem. Chem. Phys.*, 2008, **10**, 6150.

Professor Madden answered: Thank you for these valuable references. As I have said in response to other questions (and also in the paper) the present model is only a partial representation of the full chemistry of the Pt/water interface; in particular, bond-breaking in the water molecule is not allowed. A simulation which would allow for dissociation induced by the surface would be too time consuming to allow for other aspects of the interfacial structure to be examined. What your question points to is the need to break the whole problem down into discrete parts and to examine each aspect in a separate, specially constructed calculation.

Professor Luzar returned to the discussion of the paper by Dr Perkin: Would it be possible to study with your method a series of ions and maybe get an insight into how this dynamics contributes to Hofmeister effect?

Dr Perkin replied: This is a good suggestion. I think the Hofmeister effect is less clear for cations than for anions, and here we investigate only cation hydration lubrication. We have not yet investigated enough ions within the series to be able to draw

 This journal is © The Royal Society of Chemistry 2009

any trends. However, I anticipate that the driving forces acting here are more complex than simply polarisability or 'softness' since our results appear to depend on a delicate balance between thermodynamic (*e.g.* hydration energies) and kinetic (*e.g.* exchange or relaxation rates of the hydration layers) properties. However, a more complete investigation of the periodic series will clearly be interesting and we are presently working to carry out these experiments.

Professor Lyashchenko continued the discussion of the paper by Professor Madden: My question and a small comment are as follows: in your paper you use in the analysis the dielectric properties of a solution. What is the real values of dielectric constant and infinite dielectric permittivity that you use? The real dielectric constant of 1M LiCl solution (from high frequency dielectric data) is not the same as in case of pure water. It is about 70 or 65, but not 80 as in pure water. And in a double layer it will be less. The value of infinite dielectric permittivity is also not 1. This value may be 2 or 5. It is all connected with the electrolyte concentration. The real experimental data can be obtained from high frequency dielectric measurements of solutions in a wide concentration range.[1,2]

1 A. Lyashchenko and A. Zasetsky, *J. Mol. Liq.*, 1998, **77**, 61.
2 A. Lileev, Z. Filimonova and A. Lyashchenko, *J. Mol. Liq.*, 2003, **103–104**, 299.

Professor Madden replied: I think one must be careful how one uses the idea of a dielectric constant to discuss the behaviour of conducting solutions. In our simulations of pure water (an insulator) with an applied bias across the cell there is a uniform electric field (potential gradient) in the middle of the cell. From the potential drop and the stored charge we can calculate a dielectric constant and it agrees quite well with the dielectric constant of bulk water with this interaction potential. Across the electrode/solution interface in pure water the electric field is highly non-uniform, so one should not be using a dielectric constant to describe this behaviour: nevertheless, I can still measure a potential drop and a stored charge and by modelling the interface as a parallel plate condenser I could still deduce a "dielectric constant" for the water in this region. In the ionic solution (our Fig. 9) there is no electric field in the electrolyte region, as a consequence of perfect screening by the ionic charges. This is not dielectric behaviour. One could extract a dielectric permittivity from frequency-dependent conductivity measurements on such a solution by removing a term proportional to inverse frequency. However, it seems to me that where one wants a dielectric permittivity value in the chemistry of ionic solutions is to rationalise the solvation energies of ions; how this is related to the permittivity obtained from the conductivity measurements seems to me an interesting question. One of the motivations for our simulations was to have atomistic models with self-consistent electrical potentials so that we could examine such questions.

Professor Dr von Klitzing continued the discussion of the paper by Dr Perkin: What would happen at increased temperature? Could Ni^{2+} become a good lubricant and Na^+ a bad one due to lost ion hydration water molecules?

Dr Perkin responded: This is an interesting idea and worth following up. At the moment we are able to conduct measurements at temperatures within a relatively small range (*ca.* 15–35 °C), and we have not yet systematically varied the temperature within this range in these water measurements. It should also be possible—with modification of our set-up—to access a wider temperature range.

Professor Dr von Klitzing then continued the discussion of the paper by Professor Jhe: Regarding the step-wise force curve in the meniscus: what feature of the curve belongs to the pure water confinement effect and what feature reflects the surface tension between the water wire and the outer surfaces?

Professor Jhe answered: The interpretation of the stepwise force gradient was a very difficult problem for us. Until now, we thought that the stepwise characteristic comes from the atomic scale stick–slip motion between the tip surface and the water, but the nanoconfinement and the nanoscale surface tension may be also attributed to the stepwise gradient. That would be an interesting and challenging future research topic.

Miss Van Der Niet returned to the discussion of the paper by Professor Madden: I'd like to make an additional comment to the ones by Dr Feibelman and Dr Held about the reconstruction of the Pt(100) surface. The Pt(100) surface is known to reconstruct in UHV, but this reconstruction can be lifted.

However, the surface does not necessarily reconstruct under electrochemical conditions, where it has a stable potential window. The surface is more difficult to handle than Pt(111), but people are actually working with the unreconstructed Pt(100) surface. One should just take very good care of the electrode preparation conditions: cooling the electrode down slowly in an atmosphere containing a sufficient amount of hydrogen. The blank voltammetry is discussed in the work of Clavilier.[1–3] Work with a reactant added to the solution is done by the Feliu group in Spain.[4–7] Since an electrochemical system is studied in this paper, the choice of surface structure might not be as unfortunate as said before.

1 J. Clavilier, R. Durand, G. Guinet and R. Faure, *J. Electroanal. Chem.*, 1981, **127**(1–3), 281.
2 J. Clavilier and D. Armand, *J. Electroanal. Chem.*, 1986, **199**(1), 187.
3 A. Rodes, M. A. Zamakhchari, K. El Achi and J. Clavilier, *J. Electroanal. Chem.*, 1991, **305**(1), 115.
4 J. M. Orts, R. Gomez, J. M. Feliu, A. Aldaz and J. Clavilier, *Langmuir*, 1997, **13**(11), 3016.
5 V. Climent, A. Rodes, J. M. Orts, J. M. Feliu, J. M. Perez and A. Aldaz, *Langmuir*, 1997, **13**(8), 2380.
6 A. Rodes, R. Gomez, J. M. Orts, J. M. Feliu and A. Aldaz, *J. Electroanal. Chem.*, 1993, **359**(1–2), 311.
7 J. M. Orts, J. M. Feliu, A. Aldaz, J. Clavilier and A. Rodes, *J. Electroanal. Chem.*, 1990, **281**(1–2), 199.

Professor Madden replied: Thank you for this helpful remark and reference.

Professor Klein asked: In recent experiments[1] we observe directly the negative charging of a smooth gold surface in water with no added salt, which we attribute to the adsorption of OH^- ions onto the metal surface (and of course, there is a large literature on the charging of metals under water). Do your experiments indicate such an adsorption of OH^- ions onto the electrode surface?

1 L. Chai and J. Klein, *Langmuir*, 2007, **23**, 7777.

Professor Madden replied: One of the limitations of our model, as a representation of a the real physical system, is that it does not allow for the autoionization of water. To investigate the question you raise about the surface-induced dissociation of water and the absorption of hydroxide would require a very substantial *ab initio* investigation. Note, however, that a metal surface can acquire a charge without the dissociation of the water, even when maintained at zero potential. Our Fig. 8 shows an example where the surface has acquired a negative charge at $V = 0$. Whether this charge is the one you measure in your experiment is another matter—as we have seen in the discussion of the first two papers in this meeting, what the experiment sees is not always straightforwardly interpreted with a structural model.

Mr Hooton asked: Does DLVO theory come into this work: water and inorganic ions at electrochemical interfaces?

 This journal is © The Royal Society of Chemistry 2009

Professor Madden responded: DLVO theory includes an account of the screening of charged surfaces by dilute ionic solutions with the solvent modelled as a dielectric continuum and the ions having no excluded volume. Such ideas are derived from Poisson–Boltzmann theory. Our calculations address the behaviour of the same electrical potentials as are the targets of Poisson–Boltzmann theory but of course our solvent is explicitly molecular and it would appear from our results that the behaviour of the potential in the vicinity of the electrode is dominated by the molecular nature of the solvent. On this length scale it would appear that the continuum solvent, and point charge simplifications are missing a great deal.

Mr Hooton continued the discussion of the paper by Dr Perkin: I have a general question: are exclusion volume effects involved in this system? Perhaps the way the ions interact, or move relative to one another?

Dr Perkin answered: In one sense, we could say that the hydration repulsion itself is an 'excluded volume' or steric effect. It is due to the finite size of the hydrated ions between the surfaces, resulting in a deviation from mean-field DLVO theory which predicts an attraction into primary minimum mica–mica contact at sub-nanometre separations. In this paper we show how the hydration repulsion occurs for Na^+ but not for Cs^+ or Ni^{2+} at 0.1 M bulk concentration.

Dr Lee continued the discussion of the paper by Professor Madden: In some papers, electric-field-induced phase transitions of water were reported, where they said confinement of water is not enough for the phase transition but the proper electric field can cause phase transition of water into ice in ambient condition. Does the study in your paper reflect the electric-field-induced phase transition of water?

Professor Madden responded: We did not study fields of the requisite strength, so that in all cases the water is fluid and the molecules have very similar diffusion coefficients (*inter alia*) to those in the absence of a field.

Professor Jungwirth continued the discussion of the paper by Dr Perkin: Both hydration and interaction with the $Si–O^-$ group is cation-specific. The relative strength of cation–anion pairing in water for K^+ *vs.* Cs^+ is then a result of a subtle balance between these two effects which is, therefore, hard to establish *a priori*. Calculations shall be helpful in resolving this issue.

Dr Perkin replied: This is absolutely right, and we would welcome calculations to help us understand the preferences of different cations for the mica negative sites and therefore the relative tendencies to bind with and without hydration. In particular it will be interesting to see whether water molecules are expected to sit between the mica site and the cation: recent interpretations of XRR experiments by Park *et al.*[1] and Schlegel *et al.*[2] have concluded that some cations bind to mica with and some without this bridging water molecule.

1 C. Park, P. Fenter, K. L. Nagy and N. C. Sturchio, *Phys. Rev. Lett.*, 2006, **97**, 016101.
2 M. L. Schlegel, K. L. Nagy, P. Fenter, L. Cheng and N. C. Sturchio, *Geochim. Cosmochim. Acta*, 2006, **70**, 3549.

Dr Marques continued the discussion of the paper by Professor Jhe: You claim that "jumps" found in the gradient of the force are possibly due to the atomic roughness of the tip. Did you compare results with different tips? Could "jumps" be due to an internal restructuration of the water nano-neck (similar to the restructuration found in metallic nanowires)?

What is the physical difference between the energy dissipation found in the paper and the one described in the paper by Sahagún *et al.*?[1]

1 E. Sahagún, P. García-Mochales, G. M. Sacha and J. J. Sáenz, *Phys. Rev. Lett.*, 2007, **98**, 176106.

Professor Jhe responded: The first question is similar to Professor Dr von Klitzing's earlier question, which I addressed then.

As for your second question, the energy dissipation described in the paper by Sahagun *et al.* is caused by large tip motion (~50 nm), from which a nanoscale water bridge is created and destroyed within one oscillation cycle of the tip motion, which results in hysteresis in the energy curve and thereby energy dissipation. However, the energy dissipation described in our paper is caused by very small oscillation of the tip (~0.1 nm), so that the water bridge maintains a stable geometrical configuration and only the hydrogen bonding structure within the water bridge is modulated by the small oscillation. Thus we conclude that the energy dissipation is associated with the hydrogen bonding network of the water bridge.

Professor Wu continued the discussion of the paper by Professor Madden: How do you assign the potential of zero charge for your system? Why do you get such a negative potential of zero charge?

Professor Madden responded: As I explained in my answer to Professor Uosaki's question, we apply the potential and measure the charge which appears on the electrode in order to maintain this condition. As to why we get such a negative value at zero applied potential, firstly I should remind you that I am talking about the absolute value of the potential and not the potential difference with respect to some reference electrode. Notwithstanding this though, I would be surprised if our charge/potential values were the same as the real physical system. To get the capacitance across the electrode solution interface correct it would be necessary for our simulated system to reproduce the way that charge is ordered by the electrode in a real system. As we have shown, the behaviour of this is strongly affected by the orientation of water molecules and hence by the electrode–water interactions and it would be surprising if the potentials we are using captured this well. Even more problematic is that our model of the electrode has all the surface charge appearing on the atomic centres, *i.e.* a sharp layer of charge. At a real metal surface, in fact the conduction electrons will spill over into the interfacial region to some extent and partially screen the electric fields which develop there. As mentioned in the paper this effect was documented by Halley *et al.* (ref. 13 and *loc. cit.*)—please see also ref. 34.

Professor Bain communicated: Professor Jhe, I have two questions on your interesting paper. First, your model for the geometry of the water wire does not take into account the line tension at the three-phase contact line on the tip or the substrate. While line tensions are a controversial subject with little agreement on the magnitude (or even the sign!), one would expect line tensions to be significant when the characteristic length scale of the fluid body approaches the width of the interface, as it does in your experiments. How sensitive is your modelling to line tension?

Second, if I understand your paper correctly, you are not measuring the geometry of the water wire directly, but you are calculating the geometry based on the equilibrium, bulk properties of water (specifically the surface tension and contact angle). But if properties such as elasticity or viscosity deviate by an order of magnitude from their bulk values, how can you assume that other properties have not been similarly affected?

Professor Jhe communicated in reply: Yes, line tension is an important factor when we consider the mechanical forces and its balance in a fluidic system. However, in our case, we used a tip with 100 nm radius, which serves as an infinite upper plate, considering a few nanometre scale of the water meniscus. So, the water bridge that we made can be regarded as one between two infinite plates. In that case, the line

 This journal is © The Royal Society of Chemistry 2009

tension force is exerted on the tip surface in lateral direction isotropically, which results in zero lateral force as well as almost zero vertical attractive force ($= F_t \sin \alpha$, where α (~contact angle) is almost zero). Therefore, the force caused by negative pressure is simply considered in our case.

In the paper, we simply estimated the geometry of the water meniscus by using the thermodynamic equations, eqn (2)–(5), for bulk water meniscus. Although the equations are for a bulk system, we still believe that they are available for a simple estimation of the geometry and the associated thermodynamic parameters. For the accurate calculation, a theoretical method relevant to the system may be employed.

Dr Christenson communicated: From your paper it appears that the presence of an adsorbed film on the surfaces has not been taken into account. Since you assume a near-zero contact angle of the water on the surface it is inconceivable that there would be no adsorption. Such a film could easily be of nanometre thickness. Would not this change significantly the fitting of the results to your model?

Furthermore, I would like to know what the reproducibility of the measurements were for the same amplitude?

Professor Jhe communicated in response: There may exist an adsorbed thin film on the mica. Although the thin film can serve as another substrate on the original substrate mica, the water bridge formed on it may be contaminated and thereby the properties of water may be changed. Thus, a contamination-free experiment is essential for the study of the nanoscale water meniscus. And the reproducibility should be checked at the same experimental conditions such as amplitude, temperature, humidity, approach/retraction velocity, tip-state, and so on. At least it is found that a subsequent measurement shows the same result as the one obtained just before.

Dr Johnson communicated: You correctly state in the Table 1 footnote that the surface tension component of the capillary force is attractive between the tip and the mica. The corresponding two values in the table have positive signs. The same should be true for the Laplace component; *i.e.* attraction should be associated with positive force. Attraction occurs if the Laplace pressure is negative. Hence considering the condensation results tabulated in Table 1 the Laplace component F_c should be positive. In addition, the authors appear to be calculating the capillary force at one of the solid boundaries. It would be interesting to compare these values with those calculated within the liquid at the neck.

Professor Jhe communicated in reply: The first question is similar to Professor Bain's first question, which I addressed in detail. Your second comment is very interesting for us when we consider the small but finite radius of the neck at the ruptured distance. Thank you very much.

Concluding remarks

Peter J. Feibelman

Received 2nd October 2008, Accepted 3rd October 2008
First published as an Advance Article on the web 5th November 2008
DOI: 10.1039/b817311g

The organizers of "Water – from Interfaces to the Bulk" optimistically drew two disparate realms of water-related research, at very different stages of development, into a joint discussion. One area was biophysical—with its focus on how water behaves in the cellular environment. The other was the surface science of water, including the nature of the solid/water interface, and of that between liquid and vapor. This breadth of interests often made participation in the discussion challenging.

It is hard to disagree with the often expressed, by now virtually "boiler-plate", notions that water is among the most "complex" of materials, and that because it is everywhere it is also among the most important. The joint discussion made clear that deciding how to deal with the complexity, *i.e.* which water research problems it makes sense to address in the here and now, is worthy of serious consideration. In particular, a frontal attack on the "complexity" may not be the most productive approach.

The 40^+-year history of surface science,[1] my area of expertise, provides a good example of patient development of technique before any serious attempt to fulfil more or less extravagant early promises. It speaks well both of careful problem selection by surface scientists and of the tolerant understanding of their funding agents.

In its first two decades, 1964–1984, surface science was largely about technique development. The primary goal was the ability to say with confidence that experiments done in two different apparatuses, or two different laboratories, referred to the same surface structure and composition. The main tools for this purpose were low-energy electron diffraction (LEED) and Auger electron spectroscopy (AES) and, far from attempting to apply them to "real-world" problems, surface scientists in this era were eager to study systems in which impurity levels were below detection limits and as many atoms as possible were in identical environments. In such "ideal" surface systems, experimental information content was understood to be minimal, a vital benefit when computer and algorithmic power were much less than today's. Interpreting data from, or making predictions concerning systems of numerous inequivalent atoms was simply not possible.[2]

Even for ideal systems data reduction was a serious challenge, with the major obstacle that the experiments used to measure surface structure and composition derived their sensitivity from the strong interaction of probe and sample electrons. In the case of LEED, *e.g.* unlike X-ray structural analysis of bulk crystals, the first Born approximation cannot be used simply to invert diffraction data and arrive at atomic positions. A lengthy period was therefore unavoidable, during which multiple-scattering analysis needed to be developed and proven reliable.[3] Fulfilling promises to speed up the invention of new catalysts was therefore not how surface scientists spent their days, even while they may have worried about what they would say to grant managers at renewal time.

Sandia National Laboratories Albuquerque, Albuquerque, NM, 87185-1415, USA. E-mail: pjfeibe@sandia.gov; Fax: +1 (1)505 844 5470; Tel: +1 (1)505 844 6706

It was discovered, early on, that surfaces can reconstruct, with Si(100)–2 × 1[4] and Si(111)–7 × 7[5] among the first non-bulk-like crystal faces contemplated. Thought was devoted to their nature, particularly that of Si(100)–2 × 1, because of both its technological importance and its not-overly-large unit cell. The 7 × 7 structure, though a tantalizing objective, was viewed by many as beyond technical capabilities, and left for another day, when new methods would and did, in fact, allow its structure to be deciphered (first by transmission electron microscopy,[6] and later by scanning tunneling microscopy[7]).

Over the course of the first two decades of surface science, in summary, though far from loath to make promises on the important technological advances surface science would ultimately yield, surface scientists focused heavily on developing an "alphabet soup" of techniques, and on identifying benchmark systems where several techniques agreed on the surface atom arrangement.

The arrival of scanning tunneling microscopy (STM),[8] followed by another alphabet soup of complementary scanning probe techniques, changed everything. Once the STM existed it became possible to *image* the surface one was studying with atomic resolution, thus to appreciate how defective it might be and to see surface impurities. Scanning probe microscopies, largely, are what made "real world" surface studies timely.

The reason for these introductory remarks is not to set the stage for a review of the many phenomena surface scientists have successfully addressed since the mid-1980s, but to underline that in planning a research program, one needs to think not just about how to persuade granting officers and their referees that a particular project is "important," and that manpower and equipment to complete it are at hand, but also about *timeliness*. A project becomes "timely" when technical advances make it relatively easy to realize. If deriving new insight will be too hard, it is likely that the project in question is premature. In that case, stepping back to the study of more idealized systems, or focusing on technique development in whatever the area of research, might be the best way to make progress.

Again harking back to the first two decades of surface science, practitioners were virtually blindfolded then, working with reciprocal space data from diffraction experiments, sensitive only to the likely-as-not-unrepresentative, ordered regions of a surface, and using measures of surface composition of good but still limited sensitivity. The Nobel-winning invention of the STM amounted to taking the blindfold off, and made studies of real-world systems timely. Projects that might have absorbed several man-years of working in the dark of the pre-STM era suddenly became straightforward.

From the theorist's perspective, it should be added that the continuing advance of computing (Moore's law[9]) and algorithmic power was also very significant. It allowed for a parallel computational effort to decipher the behavior of more real-world surface systems.

With this broad-brush history in mind, and a charter to offer a critical discussion of the presentations at Faraday Discussion 141, my view of the biophysical science discussed is that the efforts devoted to developing new or improving older techniques, and to attempting to extract information from the most ideal systems were of the highest interest. Thus the application of terahertz spectroscopy to protein solvation dynamics,[10] NMR to water dynamics at proteins, peptides and in cells,[11] and the reported studies of periodic purple membranes[12] appealed most to this surface scientist.

The very first discussion, revolving around the population of H_3O^+ as against OH^- at the neat water surface was disturbing. Two experimental groups discussed plain evidence that the surface is negatively charged, by the presence of OH^- ad-ions.[13,14] In stark contrast, simulations conducted and discussed by plenary speaker P. Jungwirth[15] led to his conclusion that the same surface is positive, by virtue of the adsorption of H_3O^+. A contradiction like this is generally very valuable;

 This journal is © The Royal Society of Chemistry 2009

one cannot fail to learn something by resolving it. Thus it was disappointing that neither side addressed a possible resolution.

Such was offered in the course of discussion by surface experimentalist Salmeron, who pointed out the lessons from decades of surface science that very small impurity concentrations in bulk can be greatly magnified at a surface by a segregation energy, and that impurities from the ambient atmosphere, hydrocarbons for example, can rapidly change the nature of a surface under study. That this might be a relevant issue was discounted at the discussion by practitioners of ellipsometry. But dismissing the *only* suggested resolution is not helpful. The discrepancy must have a source, and if it is not contamination, one needs to ask if it can be a failure of the highly developed force-fields used in the simulations, or the quantum chemical calculations that confirm them. A positive outcome of the discussion, one hopes, is that a serious effort to resolve the discrepancy will now be undertaken.

Several studies were debated at our Discussion aimed at determining the range over which cellular water near a protein differs from bulk water. With the foregoing surface-science history in mind, one wonders if it might make sense at this juncture to step back from the intra-cellular studies in favor of measurements in the more controllable and characterizable environment of a surface science experiment. One such study was discussed in paper 21.[16] Another, allowing for examination of a broader range of surfaces, would be to use Houston's interfacial force microscope (IFM),[17] with which one can quantify forces between a tip of large curvature radius and a flat sample, both functionalized however the experimenter chooses, *i.e.* hydrophobic or hydrophilic, with or without self-assembled monolayers, *etc.*

The IFM is effectively a zero-compliance atomic force microscope (AFM). Instead of suspending a tip from a soft cantilever, whose deflection is a measure of the force between tip and sample, in the IFM the tip is mounted on one side of a stiff "seesaw" whose tendency to rotate in response to tip–sample interaction is balanced by delivering charge to capacitor pads on the seesaw's back surfaces. The amount of charge delivered measures the force.

The force balance approach has the advantage that force *versus* separation can be measured throughout the attractive region, without "snap to contact" which limits the range of separations over which AFM can generally be applied.[18] Force between tip and sample in a fluid, or connected by a meniscus, can be measured with the IFM *versus* approach or recession speed, or 90° out of phase with an imposed lateral oscillation, allowing measurements of viscosity *versus* separation. This capability has been used to quantify changes in viscosity attributable to confinement of water at the sub-nanometre level.

Fig. 1, for example, shows the force experienced by an OH-functionalized W tip as it approaches an OH-terminated silica sample, in water.[19] Note that resistance to drainage of the water between tip and sample is close to zero when the separation is ≥20 Å. But as the separation decreases toward 10 Å, the force rises, and it does so proportionately to the tip–sample approach speed, as it should if hydrodynamics is applicable to the observed behavior. From this data set, one can therefore extract a viscosity for the nano-confined water film. Remarkably (and controversially—compare *e.g.* ref. 16), it is roughly a million times the viscosity of bulk water. Refining this methodology will doubtless eliminate issues of surface characterization. Extending the results to surfaces functionalized to mimic the intra-cellular environment should then produce ideas transferable to the analysis of bio-systems.

It is worth expanding on the notion that what we should be about is producing *transferable* ideas (a.k.a. "take-home messages"). Several papers discussed succeeded in that, making them especially noteworthy and attractive for follow-up—whether or not one agrees with the conclusions. Here are some examples:

• Tobias *et al.*[12] – "The enhanced translational mobility of water above 200 K, seen in both the mean-squared amplitudes extracted from neutron data and the mean-squared displacements computed from the MD trajectories, can be attributed

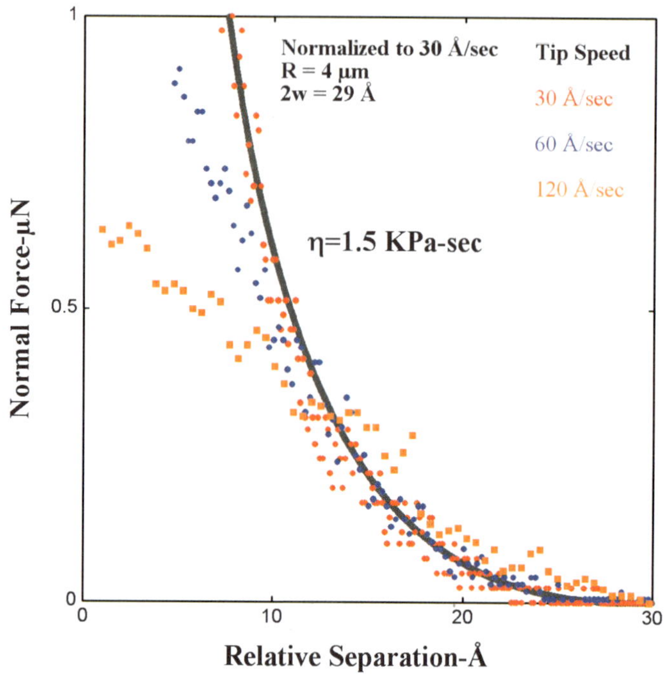

Fig. 1 The force × 30 Å s^{-1}/v experienced by an OH-functionalized W tip of curvature radius R = 4 μm, immersed in water, as it approaches an OH-coated, flat silica sample at three different approach speeds, v (see ref. 19). For separations between 10 and 30 Å, the force is proportional to the approach speed, as a hydrodynamic model would predict. Pursuing this interpretation, the solid line represents a solution to the Navier–Stokes equation, (see ref. 20) based on the assumption that tip and sample are each coated with a water "interphase" of thickness w = 14.5 Å and a viscosity η = 1.5 kPa s. The latter is about a million times the viscosity of bulk water.

to an increase in the population of water molecules that escape their cages with increasing temperature."

• Qvist *et al.*[11] – "The hydration layer can be regarded as a defect in the predominantly tetrahedrally coordinated H-bond network of bulk water, induced by a protein surface that provides fewer and less flexible H-bonding opportunities for the adjacent water molecules. These constraints slow down water rotation because they interfere with the cooperative mechanism that facilitates rotation in bulk water."

• Salmeron *et al.*[21] – "We propose that the hydrophilicity of mica above 0 °C is related to the fact that dangling H from the water molecules are formed easily above the monolayer upon adsorption of additional water."

Also: "All the examples discussed here support a mechanism for molecular-scale wetting by water that is strongly dependent on the ability of the surface to produce dangling H-bonds, from either molecular water (mica, Pd, Ru), or OH groups (oxides with defects)."

• Perkin *et al.*[16] – "This study has allowed us to formulate two conditions which must be satisfied in order for 'hydration lubrication' to take place: (i) ions must remain bound to the shearing surfaces under confinement; and (ii) the surface-bound ions must retain their hydration sheath under confinement and applied load."

Adding to the "toolkit" of big ideas for water–solid interactions has been the primary focus of my own efforts. To put the views I have expressed here in better perspective, I conclude with some brief examples.

 This journal is © The Royal Society of Chemistry 2009

The first involves my entrée into the science of water on surfaces, in which I aimed (a) to see whether density functional theory at the level of modern approximate functionals is up to the task of analyzing the structure of hydrogen-bonded systems, and (b) to explain what was, in 2001, the only water adsorption system for which atomic coordinates had been measured (in the LEED experiment and analysis of Held and Menzel[22]). Held and Menzel had expected that because of the small (~4%) mismatch between the lattice repeat of the basal plane of ice Ih and that of Ru(0001), the first water molecules deposited would aggregate into a "puckered hexagonal bilayer", similar to the water layers that stack to form the 3-dimensional ice-Ih crystal. To their surprise, the LEED data implied otherwise, specifically that in the wetting layer on Ru(0001), the O atoms are virtually coplanar.

Eight years later, a simple chemistry idea led to understanding.[23] It is that if the O atoms of the water layer are at roughly the same height above underlying Ru atoms, all of them must be forming bonds. In the ice-like bilayer, this is not possible for the half of the water molecules that have an H-atom "dangling" into the vacuum (see Fig. 2, left panel). The reason is that bonding between that H atom and its O neighbor uses up the power of the O(2p$_z$) orbital to bind, and this is just the orbital whose hybridization with the Ru d-band is needed to form an O–Ru bond. This argument immediately links coplanarity of the O atoms in the wetting layer to dissociation of the dangling H-atom, which then must (and easily does) find an alternative spot on the surface at which to adsorb.

Subsequent density functional calculations showed that (a) the half-dissociated water adlayer does indeed have coplanar O adatoms (see Fig. 2, right panel), and (b) the binding energy of the dissociated layer is substantially greater than one finds for a bilayer of intact water molecules. Thus, Held and Menzel's structural measurement ultimately yielded a simple and presumably transferable idea.

A follow-up study produced another important notion, this one more properly in the realm of physics. Although the half-dissociated layer was shown to be considerably better bound in the DFT calculation than an intact molecular layer, it was only *equally* well bound when compared to 3-dimensional ice Ih. Thus, there was still no explanation of why wetting should occur, thermodynamically. That emerged from a study of the zero-point corrections to water binding.[24] It made plain that zero-point energy favors dissociation, for a simple reason. The O–H bond is strong, short and stiff. Replacing it with an H–Ru bond that is weaker, longer and softer means substantially reducing $\frac{1}{2}\Sigma\hbar\omega$. Thus, it is the quantum nature of the proton that, ultimately, makes the dissociated adlayer the thermodynamically preferred state. The thought that zero-point corrections are not just a minor numerical "add-on", but

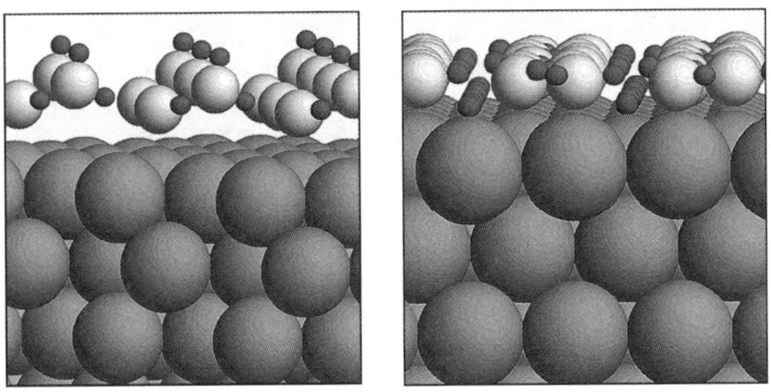

Fig. 2 Side views, after ref. 23, of an optimized, intact and a half-dissociated water adlayer on the Ru(0001) surface. Large, intermediate-sized and small balls represent Ru, O and H atoms respectively. Note, in the left panel, O–H bonds "dangling" into the vacuum above. In the right panel, formerly dangling H atoms are bound to the surface, and the O atoms to which they were attached have moved into chemical contact with underlying Ru atoms. The result is an adlayer with virtually coplanar O atoms.

can have important structural consequences, is surely an important and transferable one.

Research that helps us refine "conventional wisdom" is of great value. In water adlayers at low temperature, for example, one generally assumes that the best bound structures are characterized by a perfect H-bonding network, this despite the conflicting idea that wetting only occurs if the dominant energy involves water–substrate as against water–water bonding. Recent efforts on my part to explain the binding of water to Pt(111) suggest that relieving surface strain in a water adlayer may overcompensate the loss of binding associated with breaking H-bonds.[26] Consider, for example, the optimized, trial wetting structures illustrated in Fig. 3. The lower panel shows the experimentally observed $\sqrt{39} \times \sqrt{39}$–R16.1° supercell[27,28] filled by 32 water molecules, as assumed to be the right number by Glebov *et al.*,[27] because that allows for a perfectly H-bonded, if seemingly compressed[29] water layer. The upper panel shows a trial geometry in which one of the 32 water molecules has been removed, and the starting H-bond topology reorganized to allow the layer to remain fully H-bonded, albeit with two pentagonal water rings among the remaining hexagons.

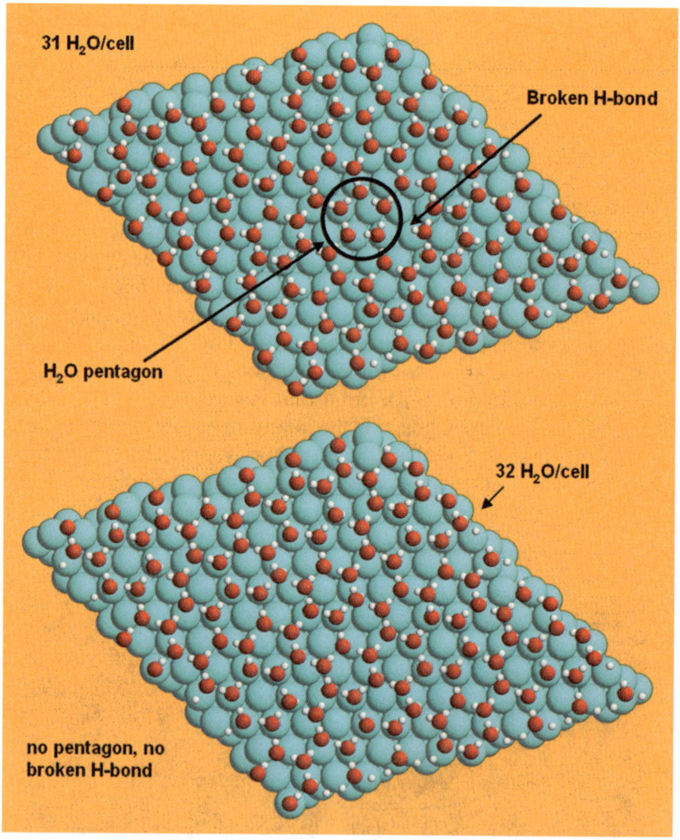

Fig. 3 Comparison of optimized $\sqrt{39} \times \sqrt{39}$–R16.1° H_2O/Pt(111) supercells containing 31 (upper panel) and 32 (lower panel) water molecules, with cyan, red and white balls representing Pt, O and H atoms. Note the broken H-bond and the H_2O pentagon in the upper panel. Despite these defects, the binding energy per H_2O is only 2 meV less than in the "perfect" wetting structure of the lower panel, an amount essentially "in the noise." For details of the optimization see ref. 25. Note that the water molecules whose O–H bonds are not roughly parallel to the surface are "H-down". That is, one of their H atoms lies between the O atom and the metal.

 This journal is © The Royal Society of Chemistry 2009

After optimization, as the panel shows, the 31-molecule layer has relaxed, such that one pentagon remains and one H-bond is broken. The binding of this structure is only 2 meV less than of the conventional 32 H_2O molecule arrangement, perhaps showing the way to a wetting motif quite different from what is anticipated based on our knowledge of the structure of ice Ih. Whether this is a valuable insight remains to be shown. In the high resolution images of Morgenstern *et al.*[30] one does not see pentagons or vacancies. On the other hand these images also do not manifest the $\sqrt{39} \times \sqrt{39}$–R16.1° supercell observed in the diffraction studies. Thus further experimental work may be the key to making pursuit of the strain-relief idea timely.

A last example concerning timeliness bears on whether the systematic error inherent in current approximate versions of density functional theory is confounding our efforts to understand water behavior near surfaces. Many efforts to find a first layer arrangement of water molecules on Pt(111), for example, including those represented in Fig. 3, have failed to reveal one whose binding exceeds that of ice Ih, and thus would be a thermodynamic wetting structure at 0 K. Pt(111) is, after all, observed to be a complete wetting substrate.[27,28] Is this a failure of the imagination, or of simulations to find the deepest well in a many dimensional hyperspace? Or is it a symptom of the inadequacy of today's approximate functionals? The answer might be found by continuing to spin the surface theorist's roulette wheel, *i.e.* by continuing to dream up new, potentially better structural motifs. What may be at stake, however, is an issue of timeliness, *i.e.* that we would spend our time more fruitfully doing something else till a functional is developed whose accuracy is an order of magnitude better than what is available today.

A related example is the subject of my contribution to the themed issue of *Physical Chemistry Chemical Physics* accompanying this Discussion,[31] on the relevance of lattice match to the effectiveness of a potential ice-nucleating nano-crystal. The use of AgI "smoke" to nucleate ice in clouds and thereby make it rain was suggested by Vonnegut, some 60 years ago, on the basis of his survey of lattice constants, which revealed that those of β-AgI and of ice Ih match to within ∼1%.[32] In ref. 31, I report that the match is much worse in a survey of DFT approximations, making them a poor way to explore what physical phenomena might make lattice match a relevant parameter. A technical advance is needed, in other words, before the lattice match problem becomes timely. Perhaps it will involve an improved account of van der Waals forces[33]—just the sort of advance that could convert many problems involving water–materials interactions to timely ones for the theorist.

Acknowledgements

Thanks to the organizers of Faraday Discussion 141 for the opportunity to present this overview. This work was supported by the DOE Office of Basic Energy Sciences, Division of Materials Science and Engineering. Sandia is operated by the Lockheed Martin Co. for the U. S. Department of Energy's National Nuclear Security Administration under contract DE-AC04-94AL85000. VASP was developed at T. U. Wien's Institut für Theoretische Physik.

References

1 See *e.g.* C. B. Duke, *Surf. Sci.*, 1994, **299/300**, 24; in the same volume, *The First Thirty Years*, ed. C. B. Duke [*Surf. Sci.*, 1994, **299/300**], one also finds historical perspectives from many other early contributors to the field.

2 See P. R. Watson, M. A. Van Hove and K. Hermann, *Atlas of Surface Structures*, American Institute of Physics, Woodbury, NY, 1994, for a compilation of structures determined in the first years of LEED analysis.

3 See *e.g.* P. M. Marcus, *Surf. Sci.*, 1994, **299/300**, 447, for one participant's perspective of the contentious development of LEED theory, and K. Heinz, *Rep. Prog. Phys.*, 1995, **58**, 637, for a review of more recent developments.

4 R. E. Schlier and H. E. Farnsworth, *J. Chem. Phys.*, 1959, **30**, 917.
5 J. J. Lander, G. W. Gobeli and J. Morrison, *J. Appl. Phys.*, 1963, **34**, 2298.
6 K. Takayanagi, Y. Tanishiro, M. Takahashi and S. Takahashi, *J. Vac. Sci. Technol., A*, 1985, **3**, 1502.
7 R. S. Becker, J. A. Golovchenko, D. R. Hamann and B. S. Swartzentruber, *Phys. Rev. Lett.*, 1985, **55**, 2032.
8 G. Binnig and H. Rohrer, *Helv. Phys. Acta*, 1982, **55**, 726; see also http://www.nobel.se/physics/laureates/1986/binnig-lecture.pdf.
9 G. E. Moore, *Electronics*, 1965, **38**, 114. For an official update, see http://www.intel.com/technology/mooreslaw/index.htm.
10 FD141/11: B. Born, S. J. Kim, S. Ebbinghaus, M. Gruebele and M. Havenith, The terahertz dance of water with the proteins: the effect of protein flexibility on the dynamical hydration shell of ubiquitin, *Faraday Discuss.*, 2009, DOI: 10.1039/b804734k.
11 FD141/9: J. Qvist, E. Persson, C. Mattea and B. Halle, Time scales of water dynamics at biological interfaces: peptides, proteins and cells, *Faraday Discuss.*, 2009, DOI: 10.1039/b806194g.
12 FD141/7: D. J. Tobias, N. Sengupta and M. Tarek, Hydration dynamics of purple membranes, *Faraday Discuss.*, 2009, DOI: 10.1039/b809371g.
13 FD141/2: J. K. Beattie, A. M. Djerdjev and G. G. Warr, The surface of neat water is basic, *Faraday Discuss.*, 2009, DOI: 10.1039/b805266b.
14 FD141/3: K. Hänni-Ciunel, N. Schelero and R. von Klitzing, Negative charges at the air/water interface and their consequences for aqueous wetting films containing surfactants, *Faraday Discuss.*, 2009, DOI: 10.1039/b809149h.
15 FD141/1: P. Jungwirth, Ions at aqueous interfaces, *Faraday Discuss.*, 2009, DOI: 10.1039/b816684f.
16 FD141/21: S. Perkin, R. Goldberg, L. Chai, N. Kampf and J. Klein, Dynamic properties of confined hydration layers, *Faraday Discuss.*, 2009, DOI: 10.1039/b805244a.
17 S. A. Joyce and J. E. Houston, *Rev. Sci. Instrum.*, 1991, **62**, 710.
18 Snap to contact occurs when the derivative of the tip–sample attraction with respect to separation exceeds the cantilever spring constant. The IFM seesaw is stiff enough that this essentially never occurs.
19 Fig. 1 was kindly provided by J. E. Houston (private communication). The data in the Figure are representative of work reported in detail by M. P. Goertz, J. E. Houston and X. Y. Zhu, *Langmuir*, 2007, **23**, 5491.
20 P. J. Feibelman, *Langmuir*, 2004, **20**, 1239.
21 FD141/13: M. Salmeron, H. Bluhm, M. Tatarkhanov, G. Ketteler, T. K. Shimizu, A. Mugarza, X. Deng, T. Herranz, S. Yamamoto and A. Nilsson, Water growth on metals and oxides: binding, dissociation and role of hydroxyl groups, *Faraday Discuss.*, 2009, DOI: 10.1039/b806516k.
22 G. Held and D. Menzel, *Surf. Sci.*, 1994, **316**, 92.
23 P. J. Feibelman, *Science*, 2002, **295**, 99.
24 P. J. Feibelman, *Phys. Rev. B*, 2007, **67**, 035420.
25 The structures illustrated in Fig. 3 were optimized with the VASP total energy computer code [G. Kresse and J. Hafner, *Phys. Rev. B*, 1993, **47**, 558; G. Kresse and J. Hafner, *Phys. Rev. B*, 1994, **49**, 14251; G. Kresse and J. Furthmüller, *Comput. Mater. Sci.*, 1996, **6**, 15; G. Kresse and J. Furthmüller, *Phys. Rev. B*, 1996, **54**, 11169] using the "PBE" generalized gradient approximation [J. P. Perdew, K. Burke and M. Ernzerhof, *Phys. Rev. Lett.*, 1996, **78**, 3865] and electron–nucleus interactions treated in the projector augmented wave (PAW) approximation [P. E. Blöchl, *Phys. Rev. B*, 1994, **50**, 17953; G. Kresse and D. Joubert, *Phys. Rev. B*, 1999, **59**, 1758]. The Pt(111) surface was modeled as a 3-layer slab, with the atoms of the lowest layer fixed at theoretical bulk separations, the water layer adsorbed on the top surface only, and the unphysical fields associated with this asymmetry compensated [J. Neugebauer and M. Scheffler, *Phys. Rev. B*, 1992, **46**, 16067]. The plane-wave basis was cut off at 700 eV, and the surface Brillouin zone sampled by an equally-spaced, 3×3 mesh of $\boldsymbol{k}$-vectors.
26 See also P. J. Feibelman, *Chem. Phys. Lett.*, 2005, **410**, 120.
27 A. Glebov, A. P. Graham, A. Menzel and J. P. Toennies, *J. Chem. Phys.*, 1997, **106**, 9382.
28 S. Haq, J. Harnett and A. Hodgson, *Surf. Sci.*, 2002, **505**, 171; G. Zimbita, S. Haq and A. Hodgson, *J. Chem. Phys.*, 2005, **123**, 174701.
29 Compression is defined here relative to the area density of water molecules in ice Ih. In reality, there is no way to define a "natural" standard for the density for the first water layer on a metal. For example, in the "H-down" structures of Fig. 3, bonding is not tetrahedral about the O atoms, unlike that in ice Ih. Standard or not, though, it is reasonable to assume that the constraint of pseudomorphism imposes a strain whose relief will benefit the binding of the wetting adlayer.

 This journal is © The Royal Society of Chemistry 2009

30 M. Morgenstern, J. Müller, T. Michely and G. Comsa, *Z. Phys. Chem.*, 1997, **198**, 43.
31 P. J. Feibelman, *Phys. Chem. Chem. Phys.*, 2008, **10**, 4688.
32 B. Vonnegut, *J. Appl. Phys.*, 1947, **18**, 593.
33 See *e.g.* S. D. Chakarova-Käck, E. Schröder, B. I. Lundqvist and D. C. Langreth, *Phys. Rev. Lett.*, 2006, **96**, 146107.

Poster titles

Water–methanol mixtures: topology of a hydrogen bonded network, **I. Bakó, T. Megyes, S. Bálint and V. Chihaia**, *Hungarian Academy of Sciences, Hungary*

Water adsorption isotherm on silicon oxide and its effects on nano-asperity adhesion and friction, **D. B. Asay, A. L. Barnette and S. H. Kim**, *The Pennsylvania State University, USA*

Is the water structure at oxide electrolyte interfaces causing interfacial charge? **J. Lützenkirchen, R. Zimmermann, T. Preočanin, D. Küttner, A. Filby, T. Rabung, D. Schild, M. Plaschke, H. Geckeis, C. Werner and N. Kallay**, *Forschungszentrum Karlsruhe, Germany*

Hydrogen bonding of alkanols in water identified from partial molar excess quantities, **S. Kato, J. Freitag and T. Hibino**, *Tokyo Metropolitan University, Japan*

Complementary organization in water and aqueous solutions of non-polar particles, **A. K. Lyashchenko, V. S. Dunyashev and L. V. Dunyashev**, *N. S. Kurnakov Institute of General and Inorganic Chemistry, Russia*

A surface reconstruction model for ice Ih, **D. M. Packwood and L. F. Phillips**, *University of Canterbury, New Zealand*

Chemical composition and reactivity of water on hexagonal Pt-group metal surfaces, **A. Shavorskiy, M. Gladys and G. Held**, *University of Reading, UK*

Adsorption of H_2O at the surface of atmospheric particles, **P. Cabrera-Sanfelix, D. Sanchez-Portal, A. Verdaguer, G. Darling, M. Salmeron, and A. Arnau**, *Donostia International Physics Center, Spain*

Low temperature spectroscopic and calorimetric investigation of water confined in cement pastes, **F. Ridi, P. Luciani, E. Fratini and P. Baglioni**, *University of Florence, Italy*

Growth and stability of ice multilayers studied with STM, **K. Thürmer and N. Bartelt**, *Sandia National Laboratories, USA*

Quantum dynamics study of Grotthuss diffusion in bulk liquid water, **G. Zechmann, D. E. Manolopoulos and U. W. Schmitt**, *University of Oxford, UK*

Estimation of the liquid–vapour spinodal for water and other liquids from interfacial properties, **A. R. Imre, G. Mayer, G. Házi, A. Márkus, R. Rozas, F. Römer and T. Kraska**, *KFKI Atomic Energy Research Institute, Hungary*

Thermodynamics and structural properties of supercooled water in ionic solutions, **D. Corradini, P. Gallo and M. Rovere**, *Università Roma Tre, Italy*

Cubic ice formation and its stability in highly concentrated aqueous solutions of polyhydroxy compounds, **K. Kajiwara, P. Thanatuksorn and F. Franks**, *Tokyo University of Technology, Japan*

Protein and cell water dynamics on a wide range of time scales, **E. Persson and B. Halle**, *Lund University, Sweden*

Ellipsometric study of depletion at the oil–water interface, **J. P. R. Day and C. D. Bain**, *Durham University, UK*

Down to atomic-scale intracellular water dynamics: a quasielastic neutron study, **M. Jasnin, M. Tehei, M. Moulin, M. Haertlein and G. Zaccai**, *Institut Laue Langevin, France*

Water hydrogen bond analysis on hydrophilic and hydrophobic bio-molecule sites, **D. Russo, J. Ollivier and J. Teixeira**, *Institut Laue Langevin, France*

Reconciling thermodynamic and kinetic views of bound water on protein surfaces, **M. Hamsa Priya, D. Asthagiri and M. E. Paulaitis**, *Ohio State University, USA*

SERS and quantum chemical studies of electrochemical interfacial water structures, **D.-Y. Wu, S. Duan, J.-F. Li, B. Ren and Z.-Q. Tian**, *Xiamen University, China*

Ab initio molecular dynamics study of the calcite–water interaction, **J. Lardge, D. Duffy and M. Gillan**, *University College London, UK*

Methanol and water layers on ZnO($000\bar{1}$): ordered mono- and bilayers and inter-diffusion of multilayers, **F. Traeger, N. Osterloh, D. Langenberg and Ch. Wöll**, *Ruhr-University Bochum, Germany*

Impact of HNO_3 and HCl on ice surface melting, **A. Křepelová, M. Ammann, J. T. Newberg, H. Bluhm and T. Huthwelker**, *Paul Scherrer Institut, Switzerland*

Thermally induced mixing of water-dominated interstellar ices, **D. J. Burke, J. L. Edridge and W. A. Brown**, *University College London, UK*

Partition functions of liquid water and dilute aqueous solutions, **R. H. Henchman and S. J. Irudayam**, *University of Manchester, UK*

Dual function of the hydration layer around an insect antifreeze protein, **D. R. Nutt and J. C. Smith**, *University of Reading, UK*

Adsorption and desorption of CO_2 from a water covered interstellar dust grain surface, **J. L. Edridge, K. Freimann, D. J. Burke and W. A. Brown**, *University College London, UK*

Water subnanoclusters on the surface of a rare gas solid: photodesorption and photochemistry, **I. Arakawa, T. Ide and T. Miura**, *Gakushuin University, Japan*

Structure of water at the solid/liquid interface by sum frequency generation spectroscopy, **K. Uosaki**, *Hokkaido University, Japan*

"Phase"-transitions in amorphous water: a thermal, volumetric and structural study, **K. Winkel, M. S. Elsaesser, E. Mayer and T. Loerting**, *University of Innsbruck, Austria*

Evidence for the transition from high-density to low-density hydrogen-bonded structure of confined supercooled water, **T. Yamaguchi, K. Yoshida, M. Taguchi, K. Ito, S. Kittaka, M.-C. Bellissent-Funel and P. Fouquet**, *Fukuoka University, Japan*

 This journal is © The Royal Society of Chemistry 2009

Microscopic wetting of mixed self-assembled monolayers: a molecular dynamics study, **M. Szöri and M. Roeselová**, *Academy of Sciences of the Czech Republic, Czech Republic*

A nuclear magnetic resonance study on the dynamics of water–methanol mixtures, **C. Corsaro, J. Spooren, C. Branca, N. Leone, M. Broccio, C. Kim, S.-H. Chen, H. E. Stanley and F. Mallamace**, *Università di Messina, Italy*

Surface potential at the ice/water interface, **T. Preočanin, A. Čop, J. Lützenkirchen and N. Kallay**, *University of Zagreb, Croatia*

The surface of water is denser than the bulk, **R. Greef and J. G. Frey**, *University of Southampton, UK*

Water interactions with crystalline polymers with large dipoles, **L. G. Rosa and P. A. Dowben**, *University of Puerto Rico – Humacao, Puerto Rico*

Water structure close to ordered and disordered hydrophobic surfaces studied by vibrational sum frequency spectroscopy, **E. C. Tyrode, T. Nishida and S. Ye**, *Royal Institute of Technology, Sweden*

The smallest droplet of acid—high resolution spectroscopy of $Cl^{-}(H_2O)_3H_3O^{+}$ @ He_n, **A. Gutberlet, Ö. Birer, G. Schwaab and M. Havenith**, *Ruhr-Universität Bochum, Germany*

Is NaCl in a frozen solution in a solid or liquid state? **V. Zelenay, T. Huthwelker, A. Krepelova, M. Janousch and M. Ammann**, *Paul Scherrer Institut, Switzerland*

Amorphous solid water probed by temperature programmed desorption: from compact to porous films, **J. H. Fillion, L. Amiaud, E. Congiu, F. Dulieu, A. Momeni and J. L. Lemaire**, *Université Pierre et Marie Curie, France*

The atom dipole interaction model revisited: TIP4P-like model of water with anisotropic polarizability, **J. Kolafa**, *Prague Institute of Chemical Technology, Czech Republic*

A new dye to probe the water–air and ice–air interfaces, **T. Bartels-Rausch and M. Ammann**, *Paul Scherrer Institut, Switzerland*

The thermodynamics and phase transitions in liquid water, **M. L. Rao and R. Roy**, *The Pennsylvania State University, USA*

Capillary condensation of water below zero—thermodynamic and kinetic supercooling in a wedge-pore, **H. K. Christenson and D. Nowak**, *University of Leeds, UK*

Ice nucleation in supercooled pure and salty water droplets—IR/VIS high-speed imaging and computer simulations, **S. Bauerecker, P. Ulbig, V. Buch, L. Vrbka and P. Jungwirth**, *Technische Universität Braunschweig, Germany*

Probing nanoscale water clusters by atomic force microscopy, **M. Lee, B. Sung and W. Jhe**, *Seoul National University, South Korea*

Hydrated electrons near liquid water interfaces studied with time-resolved photoelectron spectroscopy, **K. Siefermann, Y. Liu-Timmermann, E. Lugovoy and B. Abel**, *Universität Göttingen, Germany*

Reactivity of benzoyl chlorides in non-ionic microemulsions: potential application as indicators of system properties, **C. Cabaleiro-Lago and P. Hervés**, *University of Vigo, Spain*

Frequency- and temperature-dependent relaxation measurements, **J. Qvist, C. Mattea, M. Davidovic, D. Hamelberg and B. Halle**, *Lund University, Sweden*

The promotion of high electrical resistance in confined electrolytes between biological cell membranes and silicon oxide, **S. Pedersen, S. Friis, D. Nielsen and J. Kutchinsky**, *Sophion Bioscience A/S, Denmark*

Inhomogeneous behaviour near the Ice Ih surface, **M. Watkins and B. Slater**, *University College London, UK*

A new fluctuation force, **A. Gray-Weale and J. K. Beattie**, *University of Sydney, Australia*

Water and the second critical point: a calorimetric experiment proposal, **M. I. Marqués**, *Universidad Autónoma de Madrid, Spain*

Interactions between H_2O and preadsorbed D or ^{18}O on Pt(533), **M. J. T. C. van der Niet, I. Dominicus, O. T. Berg, L. B. F. Juurlink and M. T. M. Koper**, *Leiden University, The Netherlands*

Toward a flexible water model capable of polarisation and charge transfer, **J. Chen and T. J. Martínez**, *University of Illinois at Urbana-Champaign, USA*

Applications of a simple reaction force field for polarization, H-bonding and proton transfer in peptides and water, **S. Kale and J. Herzfeld**, *Brandeis University, USA*

Dynamic polarisable multipolar water potentials using machine learning trained on water cluster finite electron densities, **C. M. Handley and P. L. A. Popelier**, *University of Manchester, UK*

Computer simulations of the ultrafast phase transition of hydrogen-bonded liquids, **E. Vöhringer-Martinez, H. Grubmüller and B. Abel**, *MPI Göttingen, Germany*

Influence of wettability and surface charge on the interaction between an aqueous electrolyte solution and a solid surface, **S. Guriyanova, B. Semin and E. Bonaccurso**, *MPI for Polymer Research, Germany*

The formation of H_2O on interstellar dust grains, **V. L. Frankland and M. R. S. McCoustra**, *Heriot-Watt University, UK*

Excess electrons at water/air interfaces: from small to large water cluster anions, **Á. Madarász, P. J. Rossky and L. Túri**, *Institute of Chemistry, Budapest, Hungary*

The Skinner Prize for the best poster was awarded to Mr Gunther Zechmann of the University of Oxford, UK, for his poster on the quantum dynamics of Grotthuss diffusion in bulk liquid water.

 This journal is © The Royal Society of Chemistry 2009

List of participants

Professor B. Abel, *University of Leipzig, Germany*
Dr F. Abniki, *National Petrochemical Company, Iran*
Dr F. Al-Qalaf, *PAAET, College of Technological Studies, Kuwait*
Professor J. Anwar, *University of Bradford, United Kingdom*
Professor I. Arakawa, *Gakushuin University, Japan*
Professor C. Bain, *Durham University, United Kingdom*
Mr E. Bakhshi, *National Petrochemical Company, Iran*
Dr I. Bakó, *Hungarian Academy of Sciences, Hungary*
Dr T. Bartels-Rausch, *Paul Scherrer Institute, Switzerland*
Dr S. Bauerecker, *Technische Universität Braunschweig, Germany*
Dr J. Beattie, *University of Sydney, Australia*
Dr E. Bonaccurso, *Max Planck Institute for Polymer Research, Germany*
Professor M. Bonn, *FOM Institute for Atomic and Molecular Physics, The Netherlands*
Dr U. Bovensiepen, *Freie Universität Berlin, Germany*
Dr W. Brown, *University College London, United Kingdom*
Professor V. Buch, *Hebrew University, Israel*
Dr D. Burke, *University College London, United Kingdom*
Dr P. Cabrera-Sanfelix, *Donostia International Physics Center, Spain*
Dr M. Cacace, *CNR-ISMN, Italy*
Dr M. Chaplin, *London South Bank University, United Kingdom*
Dr M. Chapman, *Royal Society of Chemistry, United Kingdom*
Mr J. Chen, *University of Illinois, U.S.A.*
Dr H. Christenson, *University of Leeds, United Kingdom*
Professor D. Clary, *University of Oxford, United Kingdom*
Dr M. Collings, *Heriot Watt University, United Kingdom*
Mrs R. Cooper, *Royal Society of Chemistry, United Kingdom*
Dr F. Coudert, *Chimie Paris Tech, France*
Mr J. Day, *Durham University, United Kingdom*
Professor P. Debenedetti, *Princeton University, U.S.A.*
Mr F. Dulieu, *Universite de Cergy Pontoise, France*
Mr J. Edridge, *University College London, United Kingdom*
Dr P. Feibelman, *Sandia National Laboratories, U.S.A.*
Professor J. Fillion, *Université Pierre et Marie Curie, France*
Professor J. Finney, *University College London, United Kingdom*
Miss V. Frankland, *Heriot-Watt University, United Kingdom*
Dr J. Fraxedas, *CIN2, Spain*
Professor A. Fuchs, *Ecole Nationale Supérieure de Chimie de Paris, France*
Ms M. Gilbert, *Royal Society of Chemistry, United Kingdom*
Mrs S. Godfrey, *Royal Society of Chemistry, United Kingdom*
Dr R. Greef, *University of Southampton, United Kingdom*
Professor D. Greenhaigh, *Heriot-Watt University, United Kingdom*
Miss A. Gutberlet, *Ruhr University Bochum, Germany*
Mr D. Habig, *Ruhr University Bochum, Germany*
Professor B. Halle, *Lund University, Sweden*
Mr C. Handley, *University of Manchester, United Kingdom*
Professor M. Havenith, *Ruhr University Bochum, Germany*
Dr G. Held, *University of Reading, United Kingdom*
Dr R. Henchman, *University of Manchester, United Kingdom*
Dr P. Herves, *University of Vigo, Spain*
Professor A. Hodgson, *University of Liverpool, United Kingdom*

Mrs C. Hodkinson, *Royal Society of Chemistry, United Kingdom*
Mr F. Hooton, *University of Edinburgh, United Kingdom*
Mr X. Hu, *University College London, United Kingdom*
Dr T. Huthwelker, *Paul Scherrer Institute, Switzerland*
Dr A. Imre, *MTA KFKI Atomic Energy Research Institut, Hungary*
Miss M. Jasnin, *Institut Laue Langevin, France*
Professor W. Jhe, *Seoul National University, South Korea*
Dr S. Johnson, *Unilever R & D Port Sunlight, United Kingdom*
Professor P. Jungwirth, *Academy of Sciences of the Czech Republic, Czech Republic*
Professor K. Kajiwara, *Tokyo University of Technology, Japan*
Mr S. Kale, *Brandeis University, U.S.A.*
Dr S. Kato, *Tokyo Metropolitan University, Japan*
Professor S. Kim, *Pennsylvania State University, U.S.A.*
Professor J. Klein, *Weizmann Institute of Science, Israel*
Mr J. Klimes, *University College London, United Kingdom*
Dr J. Kolafa, *Institute of Chemical Technology, Czech Republic*
Dr A. Křepelová, *Paul Scherrer Institute, Switzerland*
Miss J. Lardge, *University College London, United Kingdom*
Dr M. Lee, *Seoul National University, South Korea*
Dr L. Liu, *University College London, United Kingdom*
Dr J. Lützenkirchen, *Forschungszentrum Karlsruhe, INE, Germany*
Professor A. Luzar, *Virginia Commonwealth University, U.S.A.*
Professor A. Lyashchenko, *Institute of General and Inorganic Chemistry, Russia*
Professor R. Lynden-Bell, *University of Cambridge, United Kingdom*
Mr A. Madarasz, *Eotvos University, Hungary*
Professor P. Madden, *University of Edinburgh, United Kingdom*
Dr M. Marques, *Universidad Autonoma de Madrid, Spain*
Professor N. Mason, *Open University, United Kingdom*
Professor M. McCoustra, *Heriot-Watt University, United Kingdom*
Dr F. McMillan, *Royal Society of Chemistry, United Kingdom*
Dr A. Michaelides, *University College London, United Kingdom*
Dr L. Milsom, *Royal Society of Chemistry, United Kingdom*
Ms H. Mohana Sundaram, *Ohio State University, U.S.A.*
Dr C. Morrison, *University of Edinburgh, United Kingdom*
Dr N. Nahler, *Durham University, United Kingdom*
Ms F. Nalden, *Royal Society of Chemistry, United Kingdom*
Dr D. Nutt, *University of Reading, United Kingdom*
Mr D. Packwood, *University of Canterbury, New Zealand*
Mr D. Pan, *University College London, United Kingdom*
Mr R. Pardoe, *University of Surrey, United Kingdom*
Dr A. Patel, *University of California, Berkeley, U.S.A.*
Dr M. Paulaitis, *John Hopkins University, U.S.A.*
Dr S. Pedersen, *Sophion Bioscience, Germany*
Dr S. Perkin, *University College London, United Kingdom*
Mr E. Persson, *Lund University, Sweden*
Mr T. Poerschke, *Ruhr University Bochum, Germany*
Dr T. Preocanin, *University of Zagreb, Croatia*
Mr J. Qvist, *Lund University, Sweden*
Dr M. Rao, *Pennsylvania State University, U.S.A.*
Professor M. Ricci, *Università degli Studi Roma Tre, Italy*
Dr F. Ridi, *University of Florence, Italy*
Dr M. Roeselova, *Institute of Organic Chemistry and Biochemistry, Czech Republic*
Professor L. Rosa, *University of Puerto Rico, Puerto Rico*

 This journal is © The Royal Society of Chemistry 2009

Professor M. Rovere, *University of Roma Tre, Italy*
Dr D. Russo, *CNR-INFM, C/O: Institut Laue Langevin, France*
Professor M. Salmeron, *University of California, Berkeley, U.S.A.*
Dr G. Schwaab, *Ruhr University Bochum, Germany*
Mr A. Shavorskiy, *University of Reading, United Kingdom*
Miss L. Shepherd, *University of Edinburgh, United Kingdom*
Ms K. Siefermann, *Universität Göttingen, Germany*
Dr B. Slater, *University College London, United Kingdom*
Mr J. Spooren, *Università di Messina, Italy*
Mr A. Stadler, *Institut Laue Langevin, France*
Miss C. Stockford, *University of Cambridge, United Kingdom*
Dr M. Szori, *Institute of Organic Chemistry and Biochemistry, Czech Republic*
Mr J. Thrower, *Heriot-Watt University, United Kingdom*
Dr K. Thürmer, *Sandia National Laboratories, U.S.A.*
Professor D. Tobias, *University Of California, Irvine, U.S.A.*
Dr F. Träger, *Ruhr University Bochum, Germany*
Dr E. Tyrode, *Royal Institute of Technology KTH, Sweden*
Professor K. Uosaki, *Hokkaido University, Japan*
Miss J. Van Der Niet, *Leiden University, The Netherlands*
Professor C. Vega, *Universidad Complutense de Madrid, Spain*
Mr A. Virshup, *University of Illinois, U.S.A.*
Mr E. Vöhringer-Martinez, *MPI Göttingen, Germany*
Prof Dr R. von Klitzing, *Technische Universität Berlin, Germany*
Dr M. Watkins, *University College London, United Kingdom*
Dr R. Wheatley, *University of Nottingham, United Kingdom*
Mr A. Willard, *University of California, Berkeley, U.S.A.*
Ms K. Winkel, *University of Innsbruck, Austria*
Professor D. Wu, *Xiamen University, China*
Professor T. Yamaguchi, *Fukuoka University, Japan*
Professor J. Yarwood, *Sheffield Hallam University, United Kingdom*
Dr J. Zaccai, *Institut Laue Langevin, France*
Mr G. Zechmann, *University of Oxford, United Kingdom*

Index of contributors*

*The page numbers in **bold** type indicate papers submitted for discussions.

 This journal is © The Royal Society of Chemistry 2009